Nanostructure Science and Technology

Series Editor

David J. Lockwood, FRSC
National Research Council of Canada
Ottawa, ON, Canada

More information about this series at http://www.springer.com/series/6331

Xiyun Yan
Editor

Nanozymology

Connecting Biology and Nanotechnology

Springer

Editor
Xiyun Yan
CAS Engineering Laboratory
for Nanozyme, Institute of Biophysics
Chinese Academy of Sciences
Beijing, China

ISSN 1571-5744 ISSN 2197-7976 (electronic)
Nanostructure Science and Technology
ISBN 978-981-15-1489-0 ISBN 978-981-15-1490-6 (eBook)
https://doi.org/10.1007/978-981-15-1490-6

This Springer imprint is published by the registered company Springer Nature Singapore Pte Ltd.
The registered company address is: 152 Beach Road, #21-01/04 Gateway East, Singapore 189721, Singapore

Foreword

Natural enzyme plays a critical role in the living system. Inspired by its remarkable catalytic efficiency, researchers have long sought to develop alternative material that would mimic the catalytic activity of the natural enzyme. This material has been defined as "artificial enzyme". Since the pioneering work of Ronald Breslow and coworkers, numerous types of materials have been used to mimic enzymes, including cyclodextrins, metal complexes, porphyrins, polymers, dendrimers, etc. Unexpectedly, Yan and coworkers discovered that Fe_3O_4 nanoparticles exhibited peroxidase mimicking activity in 2007. Since then, the functional nanomaterials with intrinsic enzyme-like properties have attracted enormous interest. These catalytic nanomaterials are collectively called as "nanozymes". Compared with conventional artificial enzymes, nanozymes are not only more stable and cost-effective but also exhibit unique properties emerged from their nanoscale sizes. For example, nanozymes have size-dependent catalytic activities, multi-functionalities as well as smart response to external stimuli. Over the past 10 years, various natural enzymes have been successfully mimicked by using different nanomaterials. These nanozymes have been explored for a wide range of applications, covering from molecular detection and tumor theranostics to antifouling and environment treatment. Though over thousands of literature (including original research articles, reviews, book chapters, and monograph) have been devoted to nanozymes, no comprehensive book has been edited yet.

The present book is intended to describe the concept, mechanism, characterization, and application of various nanozymes that have been developed over the past 10 years. It aims at serving as a textbook with coherent logic rather than a comprehensive review, which will not only bridge nanoscience and biology but also define the discipline of nanozymology. It provides a broad picture of nanozyme research, which covers three parts. The first part is about the basic concept, mechanism, and characterization; the second part is about nanomaterial-based nanozymes; and the third part is focused on their promising applications.

As the next generation of the artificial enzyme, it is important to have a clear core definition. On the other hand, nanozymology is a highly interdisciplinary field, covering biology, chemistry, physics, materials, nanoscience, and clinics. Therefore, a broad definition with fuzzy borders may help the growth of the field. These concerns have been highlighted in Chap. 1. In Chap. 2, the kinetics and mechanisms of nanozymes are discussed, which serve as basic evidence to determine if a nanomaterial with catalytic activity is a nanozyme or not. Chapter 3 describes the type of nanomaterials that have been explored to mimic various natural enzymes. The more detailed information is presented in the second part (Chaps. 5–11). To study and use nanozymes, it is important to prepare high-quality materials. Therefore, Chap. 4 is devoted to the preparation and characterization of nanozymes. Chapter 5 discusses iron oxide nanozyme, the first peroxidase mimicking nanozyme. Chapter 6 covers the Prussian blue based nanozymes. The multiple enzyme-like activities are addressed. Chapter 7 is devoted to carbon-based nanozymes, which covers fullerene, carbon nanotube, graphene, carbon dots, graphene quantum dots, etc. Chapters 8 and 9 are about vanadium oxide and cerium oxide nanozymes, two typical metal oxide based nanozymes. Metal-based nanozymes are discussed in Chap. 10 and hybrid nanozymes are covered in Chap. 11, respectively. Nanozymes have already found interesting applications in many fields. The third part is devoted to the applications of nanozymes. Chapter 12 describes the molecular detection in various scenarios by using nanozymes to replace their natural counterparts. The combination of diagnosis and theranostics may provide a promising way for future biomedicine. Chapter 13 summarizes the tumor theranostics with nanozymes and discusses the potential clinical practices. Besides tumor theranostics, nanozymes have also shown promising applications in other biomedical fields, such as anti-oxidation, anti-bacteria, anti-biofouling, etc. These topics are discussed in Chaps. 14 and 15. Chapter 16 covers another important application of nanozyme, i.e., environmental treatment. To show the emerging novel applications of nanozymes beyond the ones discussed in Chaps.12–17 discusses various innovative applications of nanozymes, covering from chemical synthesis and biomedical devices to logic gates. Chapter 18 is intended to discuss the future perspective and challenges of nanozymes.

I do hope this book will attract and inspire young researchers across various fields to study and explore the nanozyme research.

Erkang Wang
State Key Laboratory of Electroanalytical Chemistry
Changchun Institute of Applied Chemistry
Chinese Academy of Sciences
Changchun, P. R. China
ekwang@ciac.ac.cn

Preface

I am immensely grateful to Springer Nature, not only for inviting me to write this book in February 2016 and giving me this wonderful opportunity to look back on the past 12 years of great progress in the field of nanozyme, but also for giving this book its title. When the title *Nanozymology* was first brought up by her, I thought this was being a bit too overconfident; a field with only a decade of work was not nearly enough to be called something—ology. However, when we completed the manuscript in 2019, we all decided *Nanozymology* is the most accurate title for the book. As we think that nanozymology is a branch of artificial enzymes focusing on nanozymes, which is similar to the terms like enzyme and enzymology. We believe that nanozymology will not only provide a new concept of nanozymes with enzymatic features at nanoscale, but also boost new technologies for the applications of nanozymes in medicine, environment, and industry.

Nanozyme is a new concept that came from an unexpected discovery from my lab. My Ph.D. student in 2007, Lizeng Gao (now a professor and an expert in nanozyme), when verifying whether antibodies were labeled on nanobeads, was surprised to find that these nanomaterials exhibited unexpected enzyme-like activity with kinetics-like natural enzymes. To explore the reason behind this result, we worked with Prof. Taihong Wang's group (physics), Gu Ning's group (material science), and Sarah Perret (enzyme), and together confirmed that iron oxide nanoparticle possesses enzyme-like activity, which we later published in *Nature Nanotechnology*. At that time, however, we never expected it would turn out to be the foundation of nanozyme research and widely cited by everyone in this field. Up to now, there are more than 300 research groups around the world working on nanozyme research, and has been involved in the study of the design, mechanisms, and applications of nanozymes, which is sufficiently named as "Nanozymology". Importantly, it is worth mentioning that in 2018, the first nanozyme-based product has been approved.

During the preparation of the book, I was able to look back on the journey of nanozyme research, along the way I received support from countless people, and a lot of touching memories. From discovery, naming, and standardization, multiple students year after year have accompanied the growth of nanozyme field. I thank all my predecessors for their guidance and support, special thanks to Profs. Sishen Xie and Erkang Wang who gave me encouragement and support. Thanks to Gu Ning, Zhang Yu, and Xu Haiyan for their long-term friendship and cooperation; I thank Profs. Wolfgang Tremel, Heyon, and J. Manuel Perez for their continuous support as international peers.

The origination and rapid growth of nanozyme cannot be separated from the support of the Chinese government. The Chinese Ministry of Science and Technology, National Science Foundation of China have supported the study of nanozyme for the long term. In July 2019, Chinese Academy of Sciences has set up for the first time engineering laboratory for nanozymes; and Chinese Biophysical Society set up the first division of nanozymes in the world.

I am especially thankful to our writing team for this book, who are experts in their own fields and are very active at the forefront of nanozyme field. Without their dedication and time, we could not have completed the 18 chapters of the book. Dr. Lizeng Gao (Contributed to Chaps. 1, 2, 5, and 15), Dr. Hui Wei (Contributed to Chaps. 4, 14, 17), Dr. Kelong Fan (Contributed to Chaps. 9 and 13), Dr. Minmin Liang (Contributed to Chaps. 3 and 16), Dr. Lianbing Zhang (Contributed to Chap. 11), Dr. Yu Zhang (Contributed to Chap. 6), Dr. Rong Yang (Contributed to Chap. 10). Professor Xiaogang Qu contributed a major chapter (Chap. 7), in addition, Dr. Qu as the first president of Nanozyme Society with Gao and Wei contributed to addressing the challenges and perspectives of nanozymes in the last chapter (Chap. 18). I also thank Prof. Wolfgang Tremel from Johannes Gutenberg-Universität Mainz (Germany) and Juewen Liu from University of Waterloo (Canada) for kindly contributing to Chaps. 8 and 12, respectively. I appreciate all the authors in each chapter for their selfless dedication to this book. I am excited to see there are more young scientists involved in this field.

I dedicate this book to all those who have helped me along the way, including my husband and daughter, for their understanding, tolerance, and love. Finally, I would like to thank all of my readers, for your interest in nanozymology and for your active involvement in the research of nanozymes. Obviously, this book is not the last edition, research in nanozymology has just begun, there is going to be so much worth adding in the new editions to come.

Beijing, China Xiyun Yan

Contents

Part I Basic Concept, Mechanism and Characterization of Nanozymes

1 **Nanozymology: An Overview** 3
Xiyun Yan and Lizeng Gao

2 **Kinetics and Mechanisms for Nanozymes** 17
Lizeng Gao, Xingfa Gao and Xiyun Yan

3 **Types of Nanozymes: Materials and Activities** 41
Yongwei Wang, Minmin Liang and Taotao Wei

4 **Nanozymes: Preparation and Characterization** 79
Li Qin, Yihui Hu and Hui Wei

Part II Nanomaterial-Based Nanozymes

5 **Iron Oxide Nanozyme: A Multifunctional Enzyme Mimetics for Biomedical Application** 105
Lizeng Gao, Kelong Fan and Xiyun Yan

6 **Prussian Blue and Other Metal–Organic Framework-based Nanozymes** 141
Wei Zhang, Yang Wu, Zhuoxuan Li, Haijiao Dong, Yu Zhang and Ning Gu

7 **Carbon-based Nanozeymes** 171
Hanjun Sun, Jinsong Ren and Xiaogang Qu

8 **Functional Enzyme Mimics for Oxidative Halogenation Reactions that Combat Biofilm Formation** 195
Karoline Herget, Hajo Frerichs, Felix Pfitzner, Muhammad Nawaz Tahir and Wolfgang Tremel

9 Cerium Oxide Based Nanozymes 279
Ruofei Zhang, Kelong Fan and Xiyun Yan

10 Noble Metal-Based Nanozymes 331
Shuangfei Cai and Rong Yang

11 Hybrid Nanozyme: More Than One Plus One 367
Aipeng Li, Yao Chen and Lianbing Zhang

Part III Promising Applications of Nanozymes

12 Molecular Detection Using Nanozymes 395
Biwu Liu and Juewen Liu

13 Nanozyme-Based Tumor Theranostics 425
Xiangqin Meng, Lizeng Gao, Kelong Fan and Xiyun Yan

14 Nanozymes for Therapeutics 459
Wen Cao, Zhangping Lou, Wenjing Guo and Hui Wei

15 Nanozymes for Antimicrobes: Precision Biocide 489
Zhuobin Xu, Dandan Li, Zhiyue Qiu and Lizeng Gao

16 Nanozymes for Environmental Monitoring and Treatment 527
Jiuyang He and Minmin Liang

17 Beyond: Novel Applications of Nanozymes 545
Sheng Zhao, Sirong Li and Hui Wei

18 Nanozymology: Perspective and Challenges 557
Lizeng Gao, Hui Wei, Xiyun Yan and Xiaogang Qu

Correction to: Nanozymology: Perspective and Challenges C1
Lizeng Gao, Hui Wei, Xiyun Yan and Xiaogang Qu

Index 563

Part I
Basic Concept, Mechanism and Characterization of Nanozymes

Chapter 1
Nanozymology: An Overview

Xiyun Yan and Lizeng Gao

1.1 New Concept of Nanozyme

The concept of nanozyme in this book refers to a class of nanomaterials that possess intrinsic enzyme-like properties. Typically, nanozyme has the following features: First, it is made of a nanomaterial (inorganic or organic) with specific nanostructure which endows distinct physicochemical properties and catalyzes biochemical reactions of the substrates for natural enzymes under physiological conditions including mild temperature and physiological pH. It follows the same enzymatic kinetics (e.g., Michaelis–Menten equation) and catalytic mechanism (e.g., ping-pong, ordered, or random reaction) as a natural enzyme (Fig. 1.1). It also has inhibitors and activators to regulate its activity. The catalytic activity is from the nanozyme itself without modifying additional natural enzymes or chemical catalysts. In other words, the catalysis is an intrinsic enzyme-like activity. Second, typical nanoscale factors, such as size, morphology, and surface remarkably affect the activity of nanozymes, which provides a superior strategy for adjusting its activity. A nanozyme often has active centers or electron-transport structures which is similar to natural enzymes. The active site endows nanozyme with enzymatic or catalytic activity. For instance, carbon nanozymes doped with nitrogen mimic natural enzyme-like activity by having a similar structure as porphyrin in the active center of HPR. Third, nanozyme can be used as an enzyme substitute for human health and surpasses natural enzymes in many aspects such as stability, low cost, and ease of production.

The first evidence of nanozyme was reported in 2007, discovering that iron oxide nanoparticles possessed intrinsic peroxidase-like activity. Interestingly, this discovery comes from an unexpected result. While our team was working on precision cancer treatment through interdisciplinary collaboration by labeling our antibody

X. Yan (✉) · L. Gao
CAS Engineering Laboratory for Nanozyme, Key Laboratory of Protein and Peptide Pharmaceutical, Institute of Biophysics, Chinese Academy of Sciences, 15 Datun Road, Beijing 100101, China
e-mail: Yanxy@ibp.ac.cn

X. Yan (ed.), *Nanozymology*, Nanostructure Science and Technology,
https://doi.org/10.1007/978-981-15-1490-6_1

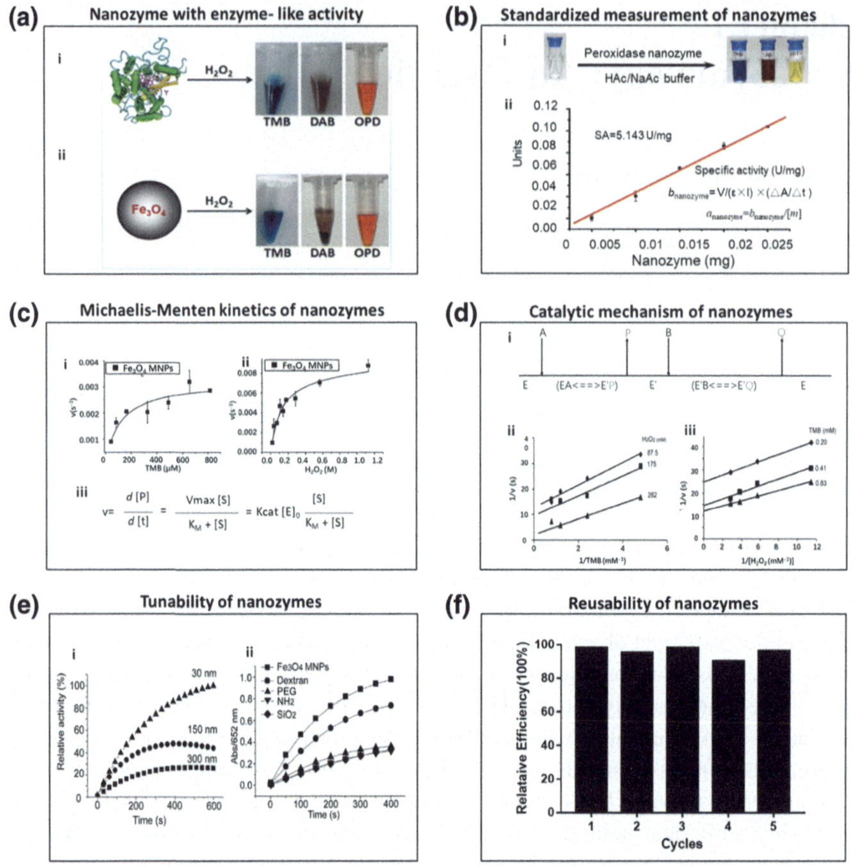

Fig. 1.1 The nanozyme has unique features. **a** Nanozyme with enzyme-like activity. (i) The colorimetric reaction catalyzed by horseradish peroxidase; (ii) Fe_3O_4 nanozymes catalyzed the same reactions in (i) by mimicking peroxidase activity. **b** Activity measurement for nanozymes. (i) Colorimetric reactions catalyzed by nanozymes with peroxidase-like activity. (ii) The calculation for the specific activity of nanozymes. **c** Enzymatic kinetics of nanozymes. (i) A curve of Michaelis–Menten kinetics for Fe_3O_4 nanozymes when TMB as the substrate; (ii) A curve of Michaelis–Menten kinetics for Fe_3O_4 nanozymes when H_2O_2 as the substrate; (iii) The equation for Michaelis–Menten kinetics. **d** Catalytic mechanism of nanozymes. (i) A scheme for the process of Ping-Pong mechanism; (ii) and (iii) Double-reciprocal plots of activity of Fe_3O_4 nanozymes at a fixed concentration of one substrate versus varying concentration of the second substrate for H_2O_2 and TMB. **e** Tunability of nanozymes. (i) Activity tuned by the size of nanozymes; (ii) Activity tuned by surface modification of nanozymes. **f** Reusability of nanozymes. Nanozymes can be used for many cycles due to high stability. Copyright 2007 Nature Publishing Group

onto nanomaterials with the hopes of creating new strategies for cancer diagnosis and treatment, we stumbled upon an amazing phenomenon of nanomaterial. Traditionally considered as an inert material, the magnetic iron oxide material, surprisingly, in nanoscale showed enzyme-like catalytic activity. After excluding all possible contamination factors, we, along with experts in physics, chemistry, materials, and enzymology, for the first time introduced the enzyme-studying methodology into nanotechnology, systematically studying its enzymatic properties, such as catalytic efficiency, enzyme-induced reaction dynamics and optimal reaction conditions, establishing methods for evaluating the catalytic activity of nanozymes, and using nanozyme as a substitute of natural enzymes for disease detection. Twelve years later, this paper has gone on to become the foundation of a new field, nanozyme [1, 2].

Since then, people began to accept that nanoparticles should no longer be considered as chemically inert, but are possible to be bioactive. Inspired by this, nanozyme study entered a rapid development period with more than 300 nanomaterials with enzymatic activity reported from 300 laboratories across 29 countries over the years (Fig. 1.2). As the enzymatic properties of nanomaterials received more and more attention, the term, nanozyme, was introduced to define these new materials. Professor Erkang Wang and Professor Wei Hui then published their review paper on Chemical Society Review, "Nanozymes: next-generation artificial enzymes", which had a large impact on promoting this new concept. At present, the new concept of nanozyme has been widely accepted by scientists at home and abroad, and has been included in the Encyclopedia of China and enzyme engineering textbooks [1, 2].

The reason why we have given the name "nanozyme" is because these nanomaterials have enzymatic features; nanozyme is defined as a nanomaterial with enzyme-like properties; it is both a nanomaterial and an artificial enzyme. The suffix "zyme" derives from the naming conventions for nonprotein enzymes. For instance, catalytic RNA is termed as a ribonucleic acid enzyme (Ribozyme), likewise, catalytic DNA is termed as Deoxyribozymes or DNAzymes, and catalytic antibody is termed as Abzyme. Thus, the term of "Nanozyme" strives for consistency to suggest artificial

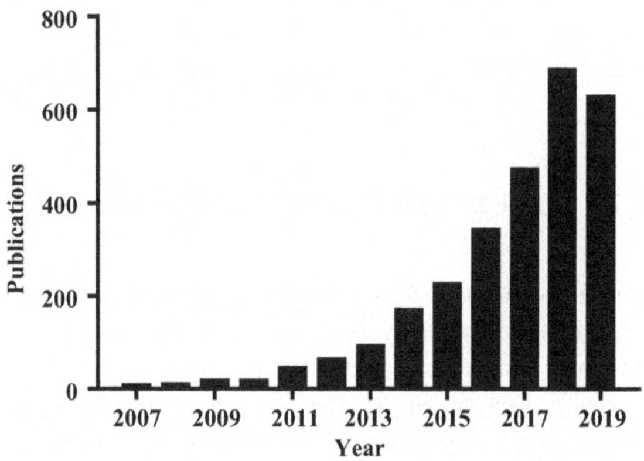

Fig. 1.2 The trend of publications in the field of nanozymes. Data was from the web of science until Aug 31, 2019. Searching keywords including "nanozymes", "nanozyme", "nano* peroxidase-like", "nano* catalase-like", "nano* oxidase-like", "nano* SOD-like", or "nano* artificial enzym"

enzymes based on nanomaterials, in which the nano prefix refers to both its size and its nano-effects.

Nanozyme is different from enzyme-integrated nanoparticles or nanomaterial hybrid enzymes, although it has also been called nanozyme in previous literature [3], in which the nanomaterials are often considered as catalytically inert and the catalysis is mostly perceived to come from its surface-modified enzymes. Here we emphasize that nanozymes have intrinsic catalytic activities from the nanomaterial itself. In addition, nanozyme as a nanomaterial is different from conventional artificial enzyme which are mostly organic supramolecular, such as cyclodextrin polymer. Because of its unique enzyme mimicking property, nanozyme is considered as the next generation of artificial enzymes, and can be used not only for human health but also for the environment, agriculture, energy, forensic science, and national security. Compared to traditional artificial enzymes, nanozymes have many superior features. First, nanozymes are made of inorganic or organic nanomaterials with the size ranging from several to hundreds of nanometers. They have unique nanostructural features, such as crystal form, lattice, plane, defect, vacancy, etc. These features provide plenty of variables to affect the catalytic activity. In contrast, traditional artificial enzymes are prepared with small organic chemicals, such as cyclodextrin, which only provide limited structural characteristics to mimic the active site of natural enzymes. Second, one nanozyme may be capable of performing several enzyme-like activities due to the above rich structural characteristics. For instance, iron oxide nanozymes perform peroxidase-like activity under acidic pH and catalase-like activity under neutral pH. The multiple activities allow nanozymes to be used for different requirements. Third, the activity, which is a limiting factor for traditional artificial enzymes or natural enzymes in terms of tunability, is easily adjustable for nanozymes by simply changing the component, size, morphology, chiral selectivity or surface properties. Fourth, nanozymes possess multifunctionality due to its distinct nanoscale properties. As nanomaterials, nanozymes may possess various other physicochemical properties, such as magnetism, fluorescence, conductivity, or photothermal effect, photodynamic effect (Fig. 1.3). These physicochemical features may interact or regulate their intrinsic enzyme-like activities, and also make it possible to design multifunctional materials or devices, demonstrating a significant advantage in practical applications. Now nanozyme has evolved from new concept to new materials, new technologies, and new applications, and has gradually developed a new branch in nanobiology.

1.2 Nanozymes is a New Generation of Artificial Enzymes

Enzymes are well known as the excellent catalysts in nature due to their high catalytic activity and selectivity. More than 5000 biochemical reaction types are catalyzed by enzymes in almost all metabolic processes. Since the first enzyme was determined by James B. Sumner in the first decades of the 1900s, people have been greatly enthusiastic to develop enzymes for industrial applications. However, the bottleneck for

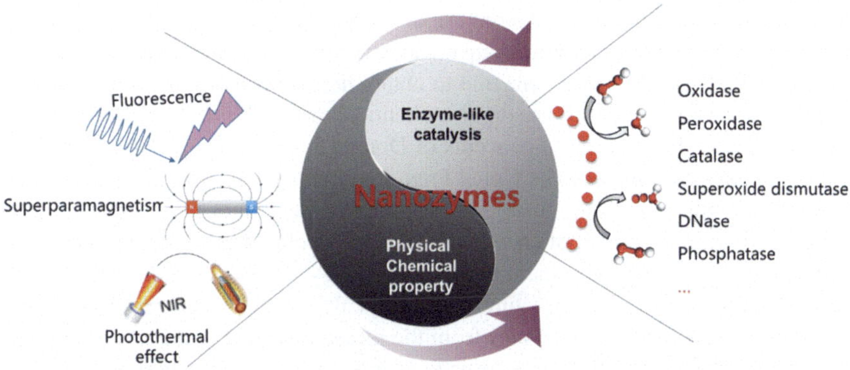

Fig. 1.3 Nanozymes possess enzyme-like catalysis and other physicochemical properties

enzyme application is that it has very low stability and high cost to produce. Protein-based enzyme is very sensitive and readily denatured, resulting in fast inactivation in a harsh environment. In addition, the enzyme is mainly synthesized biologically, difficult and costly to produce and purify at a large scale. These limitations heavily discourage the practical applications of natural enzymes.

Scientists explored many ways to overcome these difficulties, Frances H. Arnold proposed directed evolution of enzymes to improve the fitness of enzymes to tackle harsh conditions in the new and nonbiological environment, she won the Noble Prize in 2018 . While others developed artificial enzymes or enzyme mimics that substitute the function of natural enzymes. Inspired by the structure of the active site in enzyme, scientists have come up with a lot of strategies to design and prepare enzyme mimctics using chemical organic molecules (Fig. 1.4). The first mimetics were developed based on cyclodextrin in the 1960s. Molecular-imprinted polymers were also found with enzyme activity. Supramolecular chemistry was also introduced to design enzyme mimics based on the self-assembly of macromolecules. Amino acids or peptides are also used as characteristic molecular moieties to artificial enzymes. All the above

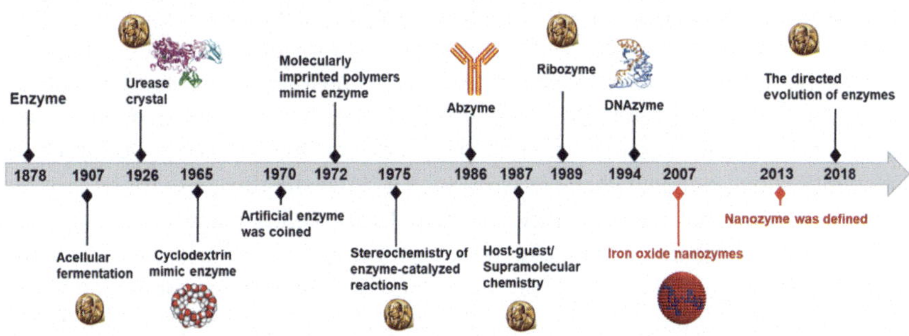

Fig. 1.4 The milestone in natural enzymes, artificial enzymes and nanozymes. The badges represent the work rewarded with Nobel prize

strategies focus on imitating the structure of the active domain in a protein enzyme and then embedding the catalytic group in the structure. The prerequisite is to prepare the structure similar to natural enzyme to obtain high selectivity or specificity. The majority of enzyme mimetics use organic chemical molecules as the designed backbone, following the host–guest concepts of Donald J. Cram, Jean-Marie Lehn, and Charle J. Pedersen who won the Nobel Prize for chemistry in 1987. However, such designed enzyme mimics are unable to match natural enzymes in terms of activity. Therefore, the development of enzyme mimics has declined in recent decades.

Recently, with the development of nanotechnology, many nanomaterials are synthesized with specific nanostructure and nanoscale effect, which provides a new source to develop artificial enzymes from the molecular level to the nanoscale. For instance, nanoceria and fullerene derivatives have been reported performing SOD mimicking activities [4, 5]. However, the concept of using nanomaterials to develop artificial enzymes has not been realized until 2007 when iron oxide nanoparticles were found to possess intrinsic peroxidase-like activity by Yan group [6]. Her group, not only made the discovery but more importantly introduced enzyme methodology to study nanomaterials, and was the first to systematically compare natural enzymes with nanozyme in catalytic efficiency and reaction dynamics and integrated the concepts and methods of biology and enzymology with nanoscience, and for the first time used nanozyme as an enzyme substitute in diagnosis and treatment of diseases, promoting nanomaterials and application in biomedical science.

Inspired by this work more than 300 types of nanomaterials, such as Fe_3O_4, Co_3O_4, CuO, FeS, V_2O_5, CeO_2, MnO_2, Au, Ag, Pt, Pd, carbon nanotube, graphene, metal–organic frameworks (MOFs) were found with intrinsic enzyme-like properties, indicating that may be a common nanoscale property for nanomaterials. Importantly, nanozyme has developed from discovery to rational design, up to chiral nanozymes, and single-atom nanozymes [7]. The catalytic activity of nanozymes also developed from mimicking natural enzymes to transcending its activity [8]. With applied study developing from in vitro detection to in vivo treatment, nanozyme as a new generation of artificial enzymes to attract multidisciplinary interest including chemical, material, physics, theoretical computing, biology, enzymatics, medicine, and others, once again set off a boom in artificial enzyme.

The discovery of nanozyme provides a new way to address the limitations of natural enzymes and conventional artificial enzyme. Nanozymes, like natural enzymes, can effectively catalyze the conversion of enzyme substrates under mild conditions and exhibit similar or even higher catalytic efficiency. With similar enzymatic reaction kinetics, nanozymes can be studied by enzyme methodology. Furthermore, nanozymes show excellent characteristics such as high stability, ease for large-scale preparation, low cost, and versatility. Compared with traditional artificial enzymes, nanozymes are more like natural enzymes, performing much higher catalytic activity under physical conditions and can be used as a natural enzyme for human health and environment. Moreover, the activity and selectivity/specificity of nanozymes are also adjustable by their special properties at the nanoscale, including nanometer size, abundant morphologies, large surface area, and distinct quantum effect. This sets nanozyme apart from traditional artificial enzymes and poses to be a promising next generation of enzyme mimics.

1.3 Nanozyme Application Evolving from Potential to Practice

The discovery of nanozyme broadly expands the application of nanomaterials. For example, iron oxide nanoparticles as an excellent magnetic nanomaterial have been long time used for bio-separation, biosensor, magnetic resonance imaging, and tumor hyperthermia therapy. Although it is known that iron oxide nanoparticles have a Fenton reaction, it was also considered to be inert in the biological system. Until 2007, Yan' group discovered that iron oxide nanoparticles have an intrinsic peroxidase-like activity and could be used as natural enzyme substitute, the application of iron oxide nanoparticles as peroxidase substitute have been expanded in vitro and in vivo for human health, including the diagnosis and therapy of diseases by regulating ROS balance under physiological conditions [9, 10]. Nowadays, the application of nanozymes has been extended from the molecular detection to environmental protection, and cancer diagnosis and treatment (Fig. 1.5a). Due to the superior properties of nanozymes with the advantages of both natural enzymes and artificial enzymes, including high activity and selectivity, tunable activity, high stability, and reusability, nanozymes have shown promising and broad applications in medicine, agriculture, environment, food, and pharmaceutical [11, 12]. The promising applications of nanozymes open up a new avenue for nanomaterials.

In bioanalysis, nanozymes could be used as a natural enzyme for clinical detection and environmental monitoring, but more stable, easy-make, and low cost. For instance, iron oxide nanoparticles like peroxidases can oxidize peroxidase substances, 3,3,5,5-tetramethylbenzidine (TMB) or 2,2′-azino-bis(3-ethylbenzothiazoline-6-sulfonic acid) (ABTS), to product color or enhance luminol chemiluminescence. As one type of the most widely studied nanozyme, peroxidase-mimic nanozymes, have been widely studied as a substitute for natural peroxidase for the detection of virus, tumor biomarkers, and other targets for diseases diagnosis, environmental monitoring.

In addition to catalysis, the nanozymes have their physicochemical properties, such as electricity, fluorescence, photo, sound, and magnetism. Importantly, the combination of intrinsic enzyme-like catalysis of nanozyme with their special nanoscale effects, such as magnetism and fluorescence, could achieve a multifunctional goal using one unique probe [13–17]. For example, in 2014 when Ebola had broken out in West Africa, a nanozyme-paper test was developed for the rapid diagnosis of Ebola. Compared with the traditional strip-test, the new nanozyme-paper test is simple, faster, cheaper, and ten times more sensitive, which solves the problem of the strip with low insensitivity. The principle for this new method is that the nanozyme conjugated with antibody showed three functions in one probe, binding to the Ebola virus,

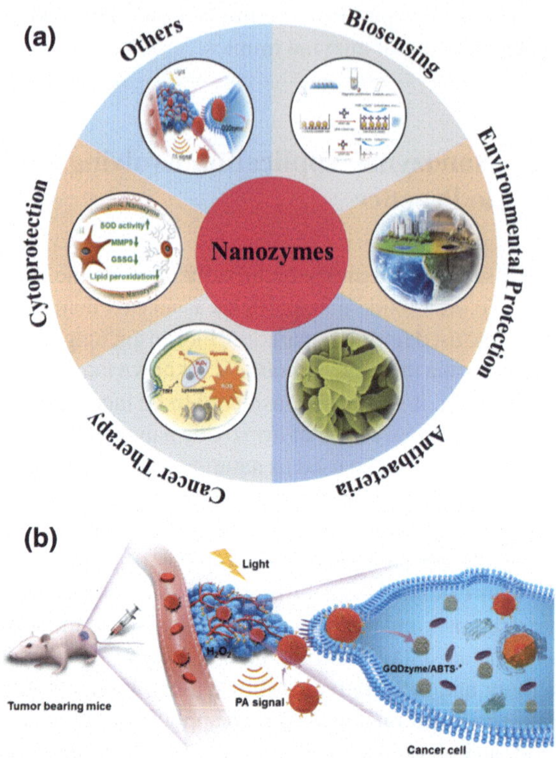

Fig. 1.5 The applications of nanozymes in vitro and in vivo. **a** The broad applications of nanozymes in different fields. **b** An example of nanozymes for in vivo imaging. Copyright © 2018 American Chemical Society

separating the virus from others in the sample by its magnetic property, and visualizing the virus by its catalyzing peroxidase substrate to enlarge signal. In 2018, the nanozyme-paper test, as the first nanozyme product in the world, was approved by CFDA. The nanozyme-paper test is used now not only for the diagnosis of infectious diseases but also for the detection of tumor biomarkers as well as food and agriculture. Furthermore, the nanozyme with peroxidase catalysis and magnetic function was also used for wastewater treatment, in which the peroxidase-like activity of nanozymes could degrade toxic phenol into CO_2, H_2O, and small organic acid, and their magnetic property could be easily collected and reusable. Therefore, the dual- or multifunction of nanozyme provides a smart new material that can be used for human health and living environment.

In in vivo study, nanozyme could be used as pharmaceuticals to regulate/balance metabolism in living cells. For instance, many nanozymes with peroxidase, oxidase, catalase, SOD-like activities have been used to regulate ROS in cells. Strong oxidizing active groups produced during the nanozyme catalysis, such as hydroxyl free radicals and single-line oxygen, can be used in inhibiting tumor cell growth [18]. In addition, using this ability, scientists developed versatile applications in anti-bacteria and anti-biofilm formation and other diseases caused by redox imbalance such as stroke, Parkinson, and Alzheimer's. For example, the intrinsic peroxidase-like activity of graphene quantum dot nanozyme effectively converts ABTS into its oxidized form in the presence of H_2O_2 to be an ideal contrast agent for photoacoustic imaging [19] (Fig. 1.5b).

Currently, nanozymes has been achieved great progress from random discovering enzyme-like activity to the rational design of nanozyme based on the demand, and their applications range from diagnosis in vitro to therapy in vivo. In the early stage, many in vitro applications have been developed, such as immunoassay and molecular detection, glucose detection, antibacterial. Later, in vivo study of nanozymes are now drawing more and more attention as it is possible to regulate ROS in cells and develop enzyme-alternative therapy using nanozymes to overcome the pathological status where natural enzymes are incompetent. Nanozyme-based catalytic therapy for tumor myocardial infarction, cerebral infarction, neurodegenerative disease is on the rise. We are pleased to see that the application of nanozyme is being applied from the potential to practical application.

1.4 Nanozyme is an Emerging Field Bridging Nanotechnology and Biology

As nanozyme is not only a new concept but also a new technology and application, nanozymes have drawn a lot of attention all over the world. More than 300 laboratories in the world have complemented nanozymes-related studies, including improving activity and disclosing catalytic mechanisms, modulating and functionalizing nanozymes for biomedical applications in vitro and in vivo. The new concept of nanozyme has been internationally accepted and published in the Encyclopedia of China and textbook [e.g.,《Enzyme Engineering》(Chinese version)]. More and more scientists, coming from chemistry, physics, material sciences, biology, enzymology, and medicine, have much interest in the research of nanozymes. The publications in nanozyme field are blooming, more than 2,200 papers are being published on scientific journals ranging from chemistry, material science to biology, and medicine. More and more patents have been transformed and the new nanozyme products have serve human health and environment. This tendency indicates that nanozyme is an emerging field bridging nanoscience and biology (Fig. 1.6).

Since 2015 the first nanozyme symposium hold in Hangzhou, China, the nanozyme symposium has been frequently introduced on international conferences in chemistry or life science (Table 1.1). We are very happy to see that many young scientists, like the authors for this book, are very active in this nanozyme field. For example, Dr. Hui Wei of Nanjing University has hosted the annual ACS-nanozyme session in Boston 2018 and San Diego 2019 in the USA, and he as the General-Secretary is preparing for the 2021 Gordon Research Conference for nanozyme. (For more information, please see the webpage constructed by Dr. Hui Wei: http://nanozymes.wixsite.com/nanozymes). Dr. Lizheng Gao, Dr. Kelong Fan, and Dr. Minmin Liang have served the 606th Xiangshan Science Conference for discussing the mechanism

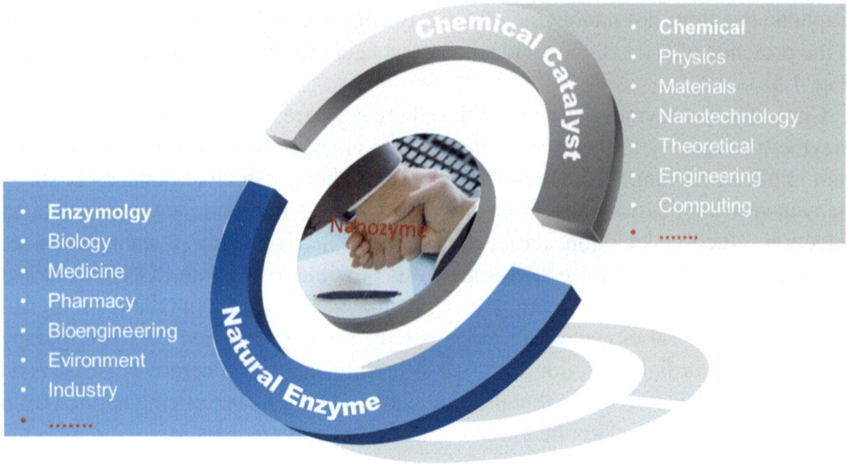

Fig. 1.6 Nanozymes is a field bridging chemistry, nanotechnology, and enzymology

Table 1.1 Selected nanozyme events

Timeline	Nanozyme events
2007/08/26	The first evidence of nanozyme was reported on Nature Nanotechnology
2013/06/05	Nanozyme was first clarified internationally
2015/05/09	The first Nanozyme workshop held in the 9th Asian Biophysics Association Symposium at Hangzhou, China
2015/05/11	Nanozyme-paper test for rapid local diagnosis of Ebola virus
2016/03/06	Nanozymes in analytical chemistry and beyond in Pittcon Conference and Expo in Atlanta, USA
2017/10/22	The 606th Xiangshan Science Conference for Nanozyme mechanism and applications
2018/03/16	The first Nanozyme product approved by the China Food and Drug Administration (CFDA)
2018/08/19	Nanozymes for bioanalysis was in the 256th ACS National Meeting and Exposition at Boston, USA
2018/11/02	International Nanozyme Symposium held at Kunshan, China
2019/08/02	The first Nanozyme Society established at the 17th Chinese Biophysics Congress at Tianjin, China
2019/07/01	The first Engineering Laboratory for Nanozyme was established Chinese Academy of Sciences
2019/08/27	Nanozymes for Bioanalysis and Beyond on 258th ACS National Meeting and Exposition at San Diego, USA

and applications of nanozyme. Importantly, the first Nanozyme Society has been established at the 17th Chinese Biophysics Congress at Tianjin, China. Dr. Xiaogang Qu was selected as President, Dr. Hui Wei and Dr. Kelong Fan were appointed as the General-Secretary and Vice General-Secretary, respectively. Dr. Junqiu Liu, Dr. Lizheng Gao, Dr. Lianbing Zhan, Dr. Xingfa Gao, and other main authors in this book as the members of the standing council have serviced the first nanozyme conference organized by Nanozyme Society in Tianjin, China. The Nanozyme Society gathered scientists from physics, chemistry, material science, biochemistry, biomedicine, and enzymology to discuss the key issues in nanozyme field, including the rational design of nanozymes, improving catalytic efficiency and specificity, exploring novel types of nanozymes, in vivo controlling, etc.

The study of nanozyme was initiated in china and has got strong financial support from Chinese government, including Key Research Program of Frontier Sciences, CAS (Grant No. QYZDBSSW-SM013, 2016–2020), CAS Engineering Laboratory for Nanozyme (2019–2021), National Key R&D Program of China (Grant No. 2017YFA0205501, 2017–2022), and the Key Project of Natural Science Foundation of China (Grant No. 81930050, 2020–2024). With these financial support, many research teams work on nanozyme, and new findings are reported up to 800 per year (Fig. 1.2), making us better understanding the mechanism of nanozymes and developing versatile applications in many important fields. We believe that nanozymes will contribute to developing cutting edge technologies to improve human life.

1.5 Nanozymology

As we understand, a term of nanozyme with "ology" as the suffix represents a study of nanozyme. For instance, enzymology is a branch of biochemistry dealing with enzymes, their nature, activity, and technologies for application. Correspondingly, nanozymology is a branch of artificial enzymes focusing on nanozymes. Nanozymology includes not only a new concept and new technology for better understanding the characterization of nanozymes at the nanoscale, but also systematically investigate the enzymatic features, kinetics, and mechanism, and combined with chemical catalysis and single-atom catalysis to rationally designing a new type of nanozymes for the applications in medicine, environment, and industry.

Since the first evidence of nanozyme reported in 2007, the knowledge about nanozymes is accumulated quickly. After a continuous study for the past 12 years, there are more than 300 types of nanozymes synthesized and their catalytic mechanisms are elucidated based on physicochemical principles, which makes it possible to understand the nature of nanozymes. Importantly, many new technologies are developed based on nanozymes, such as molecular detection and immunoassay, tumor catalytic therapy, catalytic antibacterial treatment. The applications based on the nanozymes range from biomedicine to environment treatment, from in vitro detection to in vivo enzymatic therapy, which significantly broadens the applications of enzymatic catalysis. These developments in the fundamentals and technologies of nanozymes spring out the field of nanozymology.

Although nanozyme has achieved great progress, it is quite a young field and there are more things unknown than that have been understood (Fig. 1.7). For instance, the catalytic mechanism and nature of nanozyme mimicking enzyme are still not been understood completely, since a significant difference exists between nanostructure and enzymatic protein structure. In addition, the de novo design of a nanozyme is still challenging both in nanosynthesis and mimicking the active center of the natural enzyme. Importantly, many studies have shown that nanozymes have great potential in the therapy of tumor, inflammation, and other diseases. However, controlling the performance of the nanozyme in a living system is another big challenge, because a nanozyme may have more than one intrinsic enzyme-like activity and probably perform the undesired activity to cause heavy side effects or off-target consequence.

Therefore, the main task of nanozymology will focus on solving these challenges and better understand the interface between inorganic and organic world. Also, nanozymology may open a new way for the study of the original life. For practical applications, we need to address the following questions. What does make a nanomaterial to perform enzyme-like activity? Is there any method to determine the active sites and their correlation between the structure and activity of nanozymes. Another challenge is what else are types of the natural enzymes that nanozyme can mimic, except for oxidoreductase-like activity, which is only one-sixth of the enzyme types regarding the catalytic activities. In particular, to improve the specificity of nanozymes close to the corresponding natural enzymes, a lot of studies are required. If we know more above questions, novel nanozyme-based technologies

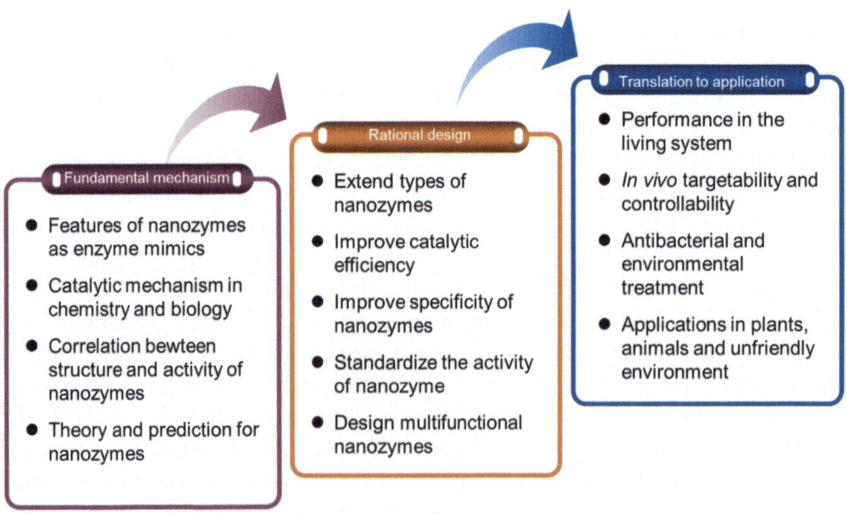

Fig. 1.7 Challenges in the field of nanozymology

and applications will be quickly developed to serve and improve the life of human beings.

In summary, this book will present a comprehensive illustration of the new concept of nanozyme and nanozymology, including developing history, mechanisms, classifications, preparations, and applications of nanozyme. We hope readers who are interested in nanozyme and find useful information in this book. We also provide several typical nanozymes as the representatives to learn the nanozymes regarding composition of nanomaterials. Taken together, this book will give state of the art of nanozymes and nanozymology from both aspects of nanotechnology and biology.

Acknowledgements This work was supported by the National Natural Science Foundation of China (Grant No. 81930050, 31871005, 31530026, 31900981), the Chinese Academy of Sciences under Grant No. YJKYYQ20180048, the Strategic Priority Research Program (No. XDB29040101), the Key Research Program of Frontier Sciences (No. QYZDY-SSW-SMC013), Chinese Academy of Sciences and National Key Research and Development Program of China (No. 2017YFA0205501), and Youth Innovation Promotion Association CAS (2019093).

References

1. Wei H, Wang EK (2013) Nanomaterials with enzyme-like characteristics (nanozymes): next-generation artificial enzymes. Chem Soc Rev 42(14):6060–6093
2. Wu J, Wang X, Wang Q, Lou Z, Li S, Zhu Y, Qin L, Wei H (2018) Nanomaterials with enzyme-like characteristics (nanozymes): next-generation artificial enzymes (II). Chem Soc Rev 48(4):1004–1076
3. Wang Z, Liu H, Yang SH, Wang T, Liu C, Cao YC (2012) Nanoparticle-based artificial RNA silencing machinery for antiviral therapy. Proc Natl Acad Sci USA 109(31):12387–12392
4. Kong L, Cai X, Zhou XH, Wong LL, Karakoti AS, Seal S, McGinnis JF (2011) Nanoceria extend photoreceptor cell lifespan in tubby mice by modulation of apoptosis/survival signaling pathways. Neurobiol Dis 42(3):514–523
5. Pagliari F, Mandoli C, Forte G, Magnani E, Pagliari S, Nardone G, Licoccia S, Minieri M, Di Nardo P, Traversa E (2012) Cerium oxide nanoparticles protect cardiac progenitor cells from oxidative stress. ACS Nano 6(5):3767–3775
6. Gao LZ, Zhuang J, Nie L, Zhang JB, Zhang Y, Gu N, Wang TH, Feng J, Yang DL, Perrett S, Yan X (2007) Intrinsic peroxidase-like activity of ferromagnetic nanoparticles. Nat Nanotechnol 2(9):577–583
7. Huang L, Chen J, Gan L, Wang J, Dong S (2019) Single-atom nanozymes. Sci Adv 5 (5):eaav5490
8. Komkova MA, Karyakina EE, Karyakin AA (2018) Catalytically synthesized Prussian blue nanoparticles defeating natural enzyme peroxidase. J Am Chem Soc 140(36):11302–11307
9. Gao LZ, Fan KL, Yan XY (2017) Iron oxide nanozyme: a multifunctional enzyme mimetic for biomedical applications. Theranostics 7(13):3207–3227
10. Wei H, Wang E (2008) Fe_3O_4 magnetic nanoparticles as peroxidase mimetics and their applications in H_2O_2 and glucose detection. Anal Chem 80(6):2250–2254
11. Gao LZ, Yan XY (2013) Discovery and current application of nanozyme. Prog Biochem Biophys 40(10):892–902
12. Wang XY, Hu YH, Wei H (2016) Nanozymes in bionanotechnology: from sensing to therapeutics and beyond. Inorg Chem Front 3(1):41–60

13. Duan DM, Fan KL, Zhang DX, Tan SG, Liang MF, Liu Y, Zhang JL, Zhang PH, Liu W, Qiu XG, Kobinger GP, Gao GF, Yan XY (2015) Nanozyme-strip for rapid local diagnosis of Ebola. Biosens Bioelectron 74:134–141
14. Gao N, Dong K, Zhao AD, Sun HJ, Wang Y, Ren JS, Qu XG (2016) Polyoxometalate-based nanozyme: design of a multifunctional enzyme for multi-faceted treatment of Alzheimer's disease. Nano Res 9(4):1079–1090
15. Cheng HJ, Zhang L, He J, Guo WJ, Zhou ZY, Zhang XJ, Nie SM, Wei H (2016) Integrated nanozymes with nanoscale proximity for in vivo neurochemical monitoring in living brains. Anal Chem 88(10):5489–5497
16. Huang Y, Liu Z, Liu C, Ju E, Zhang Y, Ren J, Qu X (2016) Self-assembly of multi-nanozymes to mimic an intracellular antioxidant defense system. Angew Chem Int Ed Engl 55(23):6646–6650
17. Wang ZZ, Dong K, Liu Z, Zhang Y, Chen ZW, Sun HJ, Ren JS, Qu XG (2017) Activation of biologically relevant levels of reactive oxygen species by Au/g-C3N4 hybrid nanozyme for bacteria killing and wound disinfection. Biomaterials 113:145–157
18. Fan K, Xi J, Fan L, Wang P, Zhu C, Tang Y, Xu X, Liang M, Jiang B, Yan X, Gao L (2018) In vivo guiding nitrogen-doped carbon nanozyme for tumor catalytic therapy. Nat Commun 9(1):1440
19. Ding H, Cai Y, Gao L, Liang M, Miao B, Wu H, Liu Y, Xie N, Tang A, Fan K, Yan X, Nie G (2019) Exosome-like nanozyme vesicles for H2O2-responsive catalytic photoacoustic imaging of xenograft nasopharyngeal carcinoma. Nano Lett 19(1):203–209

Chapter 2
Kinetics and Mechanisms for Nanozymes

Lizeng Gao, Xingfa Gao and Xiyun Yan

Abbreviations

ATP	Adenosine triphosphate
CAT	Catalase
DNA	Deoxyribonucleic acid
HRP	Horseradish peroxidase
RNA	Ribonucleic acid
SOD	Superoxide dismutase
TMB	3,3′,5,5′-tetramethylbenzidine

The kinetics and mechanisms of nanozymes determine if the nanomaterials are nanozymes and their catalytic properties. The understanding of mechanisms and active sites of nanozymes is critical to disclose the nature of nanozymes as enzyme mimetics and useful to design the desired nanozymes for practical applications.

L. Gao · X. Yan (✉)
CAS Engineering Laboratory for Nanozyme, Key Laboratory of Protein and Peptide Pharmaceutical, Institute of Biophysics, Chinese Academy of Sciences, 15 Datun Road, Beijing 100101, China
e-mail: Yanxy@ibp.ac.cn

X. Gao
College of Chemistry and Chemical Engineering, Jiangxi Normal University, Nanchang, Jiangxi 330022, China

X. Yan (ed.), *Nanozymology*, Nanostructure Science and Technology,
https://doi.org/10.1007/978-981-15-1490-6_2

2.1 Enzymatic Activities and Behavior of Nanozymes

As enzyme mimetics, nanozymes catalyze the biochemical reactions for the substrates of natural enzymes with under physiological conditions. First, the substrates for nanozymes are already known as those for natural enzymes in the standard enzyme assays or the metabolite derivatives. Second, the catalysis is performed under mild conditions, including ionic strength, pH, and temperature, which must be at physiological range.

1. **Peroxidase-like activity**

Currently, the peroxidase-like activity is one of the most popular activities for nanozymes. Peroxidase-like nanozymes can act on peroxide to generate transient reactive intermediates like free radicals, which further react with another substrate quickly. In this reaction, the first and second substrates are also called as hydrogen acceptor and donor, respectively. Thus, there are two substrates for peroxidase-like nanozymes (Fig. 2.1). According to the second substrates (i.e., hydrogen donors), peroxidase-like nanozymes consist of those mimicking peroxidase, haloperoxidase, lipid peroxidase, glutathione peroxidase, and so on.

(1) Substrates. The first substrate (i.e., hydrogen acceptor) for this activity is peroxide, primarily as hydrogen peroxide (H_2O_2), which is the substrate for horseradish peroxidase. It can be also peroxides in other forms such as lipid peroxide. Some peroxidase-like nanozymes may need the presence of halides to exhibit the activity (such as bromide for haloperoxidase-like nanozymes). The second substrates for this catalysis are general as long as they can provide electrons as hydrogen donors. Both small metabolites (or their derivatives) and large biomolecules including phenols, formic acid, formaldehyde, ethanol, nucleic acids (DNA/RNA), proteins, polysaccharides, and lipids can react with free radicals and are thus potential targets. Therefore, the specificity for peroxidase-like activity acts on the first peroxide substrate rather than the second hydrogen donor.

Peroxidase-like activity

$$H_2O_2 + AH_2 \rightarrow A + 2H_2O$$

Haloperoxidase-like activity

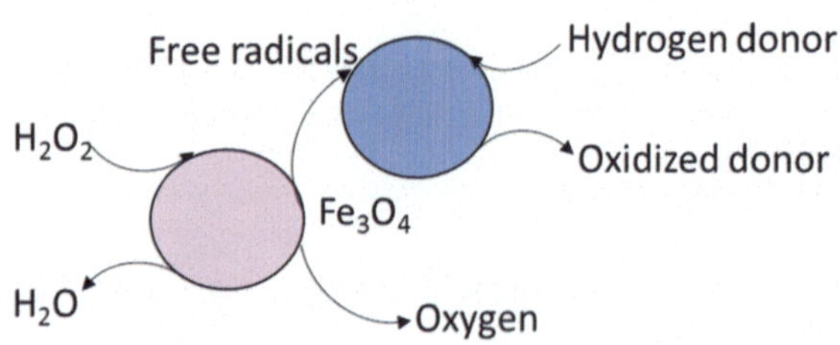

Fig. 2.1 Decomposing H_2O_2 into free radicals or oxygen by Fe_3O_4 nanozyme [1]. Copyright © 2017 Ivyspring International Publisher

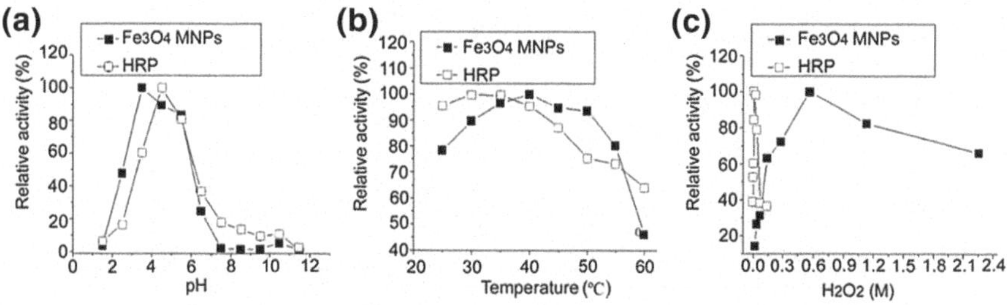

Fig. 2.2 Optimal reaction conditions for peroxidase-like nanozyme [2]: **a** pH, **b** temperature, and **c** substrate concentration. Copyright permission from 2007 Nature Publish Group

$$H_2O_2 + H^+ + X^- \rightarrow HOX + H_2O$$

$$HOX + AH \rightarrow AX + H_2O$$

The activity of peroxidase-like nanozyme is often assayed using the simple standard substrates as usually used for natural peroxidases. For instance, H_2O_2 and 3,3′,5,5′- tetramethylbenzidine (TMB) are often used for HRP colorimetric reaction, which is mostly used in enzyme-linked immunosorbent assay (ELISA). In this reaction, H_2O_2 generates free radicals and the radicals further oxidize TMB to form a blue product with characteristic absorbance at 652 nm. Similarly, H_2O_2 and TMB are used to assay the peroxidase-like activity of nanozymes.

(2) pH and buffer (Fig. 2.2a). This catalysis usually occurs under acidic condition. Therefore, the buffer is sodium acetate or sodium citrate with acid pH at 3–6. Usually, the optimal pH is around 4 and ionic strength is around 0.1–0.2 M.
(3) Temperature (Fig. 2.2b). The optimal temperature for this activity is around physiological temperature, e.g., 37 °C. The activity increases when the temperature rises to 37 °C from room temperature, but decreases once it is higher than 50 °C.
(4) Activator and inhibitor. The reaction can be inhibited by quenching free radicals. Antioxidant such as ascorbic acid, hypotaurine, sodium azide can inhibit the reaction. ATP and DNA are reported to enhance the activity.
(5) Stability. Nanozymes are much more stable than natural enzymes and traditional enzyme mimics. They are tolerant to extreme pH and temperature and can be separated for reuse.

2. **Catalase-like activity**. This activity of nanozymes decomposes H_2O_2 into molecular oxygen (O_2) and water (H_2O) as natural catalases do. Oxygen bubble is produced during the reaction. This activity prefers a neutral or basic pH condition (Fig. 2.3). Compared with natural catalases, the nanozymes often have broader optimal temperature ranges and thus higher stabilities.
3. **Oxidase-like activity**. Nanozymes catalyze the colorimetric oxidation reaction of TMB in the presence of O_2. H_2O_2 is not needed for this reaction. It shows similar pH and temperature dependence as those for peroxidase-like activity.

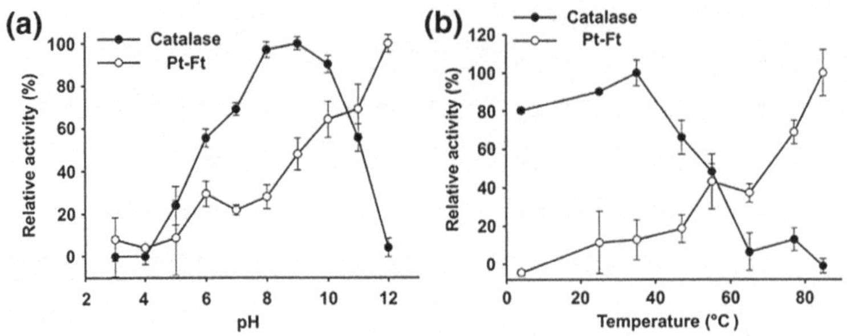

Fig. 2.3 Optimal reaction conditions for catalase-like nanozymes: **a** pH and **b** temperature [3]. Copyright © 2010 Elsevier Ltd

The activity is enhanced at a high O_2 atmosphere but inhibited under a high N_2 atmosphere.

4. **Superoxide dismutase-like activity**. Some nanozymes catalyze the dismutation reaction of superoxide ($O_2^{\bullet-}$) to produce O_2 and H_2O_2. This activity usually prefers a weak basic condition (pH = 8–9).
   SOD-like activity

$$O_2^{\bullet-} + M^{n+1} \rightarrow O_2 + M^n$$
$$O_2^{\bullet-} + M^n + 2H^+ \rightarrow H_2O_2 + M^{n+1}$$

5. **Sulfite oxidase-like activity**. This activity has been found for molybdenum trioxide (MoO_3), which catalyzes the oxidation of sulfite to sulfate.
   Sulfite oxidase-like activity

$$SO_3{}^{2-} + H_2O \rightarrow SO_4{}^{2-} + 2H^+ + 2e^-$$

6. **Protease-like activity**. Gold nanoparticles (AuNPs) and polyoxometalate (with Wells–Dawson structure, POMD) core–shell structures coated with N-Ac-Cys-heptapeptide complexes (AuNPs@POMD-8pep) possess protease-like activity with N-α-benzoyl-DL-arginine-4-nitroanilide (BAPNA) as the substrate [4]. The optimal pH and temperature of this activity were determined to be 8.0 and 55 °C, respectively. AuNPs@POMD-8pep has a higher protease activity than the commercial trypsin.
7. **Nuclease-like activity**. AuNPs that are passivated with multiple cerium (IV) complexes and supported on the surface of colloidal magnetic Fe_3O_4 /SiO_2 core/shell particles possess DNase-like activity, which can catalyze the hydrolysis of DNA or RNA [5]. This nanozyme shows the ability to cleave genomic DNA when treated with DMAE at pH 7.4 and 37 °C.

One outstanding feature of nanozymes is their multiple catalytic activities in especially biological redox reactions, which are in sharp contrast to traditional

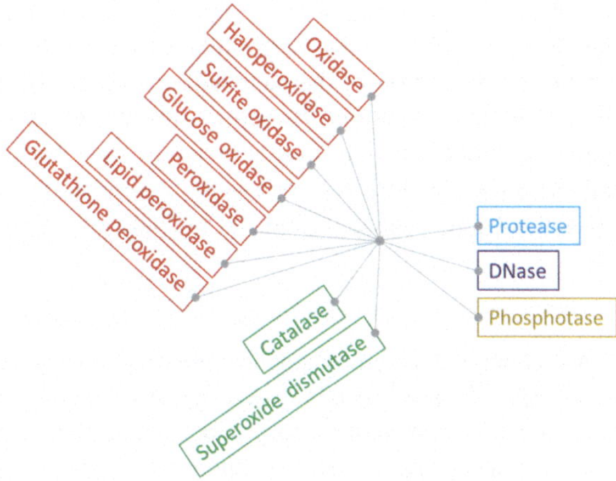

Fig. 2.4 Enzyme-like activities for nanozymes

enzyme mimics or natural enzymes that have specific catalytic activities. For example, iron oxide nanozyme possesses both peroxidase- and catalase-like activities. Gold nanoparticles possess glucose oxidase- and peroxidase-like activities. Nanoceria possesses oxidase-, peroxidase-, and SOD-like activities. These multiple activities provide nanozymes with environment-sensitive performances (Fig. 2.4).

Another feature is that different nanozymes may have the same catalytic activities. Although the catalytic behaviors of many nanozymes have been systematically studied, the comparison of the activities for different nanozymes is not straightforward because of their complicated compositions, structures, and surface modifications. Probably the specific activity can be used to assess the activity, namely, by determining the enzyme unit (U) (moles of substrate converted per unit time) with certain amount (e.g., mg) of nanozyme at the exactly same reaction conditions (substrate types and concentrations, pH, temperature and buffer, time, etc.), which may reflect the activity level of nanozymes [6].

2.2 Typical Kinetics and Catalytic Mechanisms for Nanozymes

Nanozymes typically follow very similar reaction kinetics and mechanisms as those of natural enzymes, which in turn characterize nanomaterials to be the nanozymes. Using kinetics assay, the reaction rate is measured to determine the catalytic efficiency of nanozymes as well as the dependence of the efficiency on varying reaction conditions. In terms of mechanisms, nanozymes first bind with the corresponding substrates and then convert them into products via surface catalytic reactions. Like natural enzymes, nanozymes just accelerate the reaction rates but do not alter the reaction equilibriums. The reaction catalyzed by a nanozyme also demonstrates a saturated kinetics, which means that the reaction rate (velocity) increases linearly with the substrate concentration at a relatively low substrate concentration but asymptotically approaches a plateau at a relatively high substrate concentration, provided that the concentration of the nanozyme is fixed (Fig. 2.5). When the reaction rate reaches the maximum, all the active sites are occupied by substrates. Nanozymes may have multiple active sites on their surface, which is different with that for natural enzymes. The kinetics assay is to investigate the relationship between reaction rate and substrate concentration and evaluate the nanozymes' activities.

$$E + S \rightleftharpoons ES \rightleftharpoons ES^* \rightleftharpoons EP \rightleftharpoons EP \rightleftharpoons E + P$$

A reaction catalyzed by a nanozyme may have more than one substrates as well as products. In the kinetics assay, the reaction rate can be determined based on either substrate consumption or product generation. Current methods for the kinetics assay use spectrophotometer or fluorescence spectrophotometer in most cases. If the reaction contains a single substrate, the kinetics assay is conducted by simply verifying the substrate concentration. However, if there are two or more substrates, the concentration of only one substrate is allowed to vary and those of the others need to be fixed when conducting the kinetics assay. The kinetics equations obtained from the assay are used to characterize the property of the nanozyme. Reportedly, the kinetics of most nanozymes follows the Michaelis–Menten kinetics equation with only few exceptions.

1. **Michaelis–Menten kinetics**

Michaelis–Menten kinetics is one of the best known models of enzyme kinetics. The model is described by the following equation, which relates the reaction rate (v) to substrate concentration $[S]$, V_{max}, and K_M. V_{max} represents the maximum rate achieved by the system, at saturating substrate concentration. K_M, also called the Michaelis constant, is the substrate concentration at which the reaction rate is half of V_{max}. At low substrate concentration $[S] \ll K_M$, the reaction rate (v) varies linearly with $[S]$ (first-order kinetics); at higher concentration $[S] \gg K_M$, the reaction rate

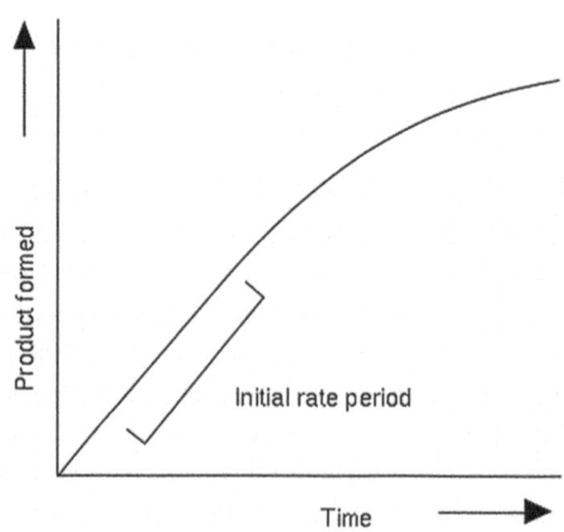

Fig. 2.5 Correlation of reaction rate with reaction time and product

reaches the maximum (V_{max}) and becomes independent on $[S]$ (zero-order kinetics). Under this condition, $V_{max} = K_{cat}[E_0]$, where $[E_0]$ is the initial concentration of nanozyme and K_{cat} is the maximum number of substrate molecules converted to product per enzyme molecule per second. According to this model of kinetics assay, the K_M value is inversely proportional with the affinity of the nanozyme to the substrate; a small K_M indicates a high affinity. K_M is used to determine the best substrate for a nanozyme. The ratio of K_{cat}/K_M is termed as catalytic efficiency, which reflects how efficiently a nanozyme converts a substrate into a product.

$$\mathrm{v} = \frac{\mathrm{V}_{max}[\mathrm{S}]}{K_{\mathrm{M}} + [\mathrm{S}]} \tag{2.1}$$

$$K_{cat} = V_{max}/[E] \tag{2.2}$$

(1) Kinetics for nanozymes with peroxidase-like activity. In the reaction catalyzed by peroxidase-like activity, there are two substrates which are peroxide and hydrogen donor. Therefore, the kinetics assay for one of them is often performed by fixing another's concentration. In H_2O_2-TMB colorimetric reaction, the kinetics assay for H_2O_2 is done with TMB fixed at saturated concentration and the assay for TMB is done with fixed H_2O_2 concentration. By this way, the reaction follows the typical Michaelis–Menten Kinetics as those for horseradish peroxidase. Usually, the apparent K_M value of the iron oxide nanozyme (IONzyme) for H_2O_2 was higher than that for HRP, indicating the lower affinity and efficiency to catalyze H_2O_2 (Fig. 2.6). In contrast, the apparent K_M value for the substrate of hydrogen donor (TMB) was much lower than that for HRP, suggesting the IONzyme has a higher affinity to TMB than natural enzyme. If calculated using nanoparticles' molar concentration, the IONzyme with the diameter at 300 nm showed a level of activity 40 times higher than that of HRP [2]. This may be due to the fact that a HRP molecule has only one iron ion, in contrast to an abundance of iron on the surface of an iron oxide nanoparticle. Besides H_2O_2-TMB reaction, the kinetics for peroxidase-like activity also can be measured by combining H_2O_2 with other substrates with chromogenic or fluorescence signal.

    IONzyme also showed a similar catalytic mechanism to that of HRP, following a ping-pong mechanism [2]. In the process of the reaction, the IONzyme binds and reacts with the first substrate (H_2O_2) to generate hydroxyl free radicals (•OH) as intermediate states and then the •OH captures a H from the hydrogen donor such as TMB. Electron spin resonance (ESR) was used to monitor the generation of •OH during the reaction and found that IONzyme could conduct the same •OH intermediate, which further confirmed its similarities to peroxidase activity.

(2) Kinetics for nanozymes with catalase-like activity. It is not easy to determine the kinetics for catalase-like activity as it usually generates O_2 in the form of bubbles by decomposing H_2O_2 in the neutral pH (Fig. 2.7a, b) [7]. The generated

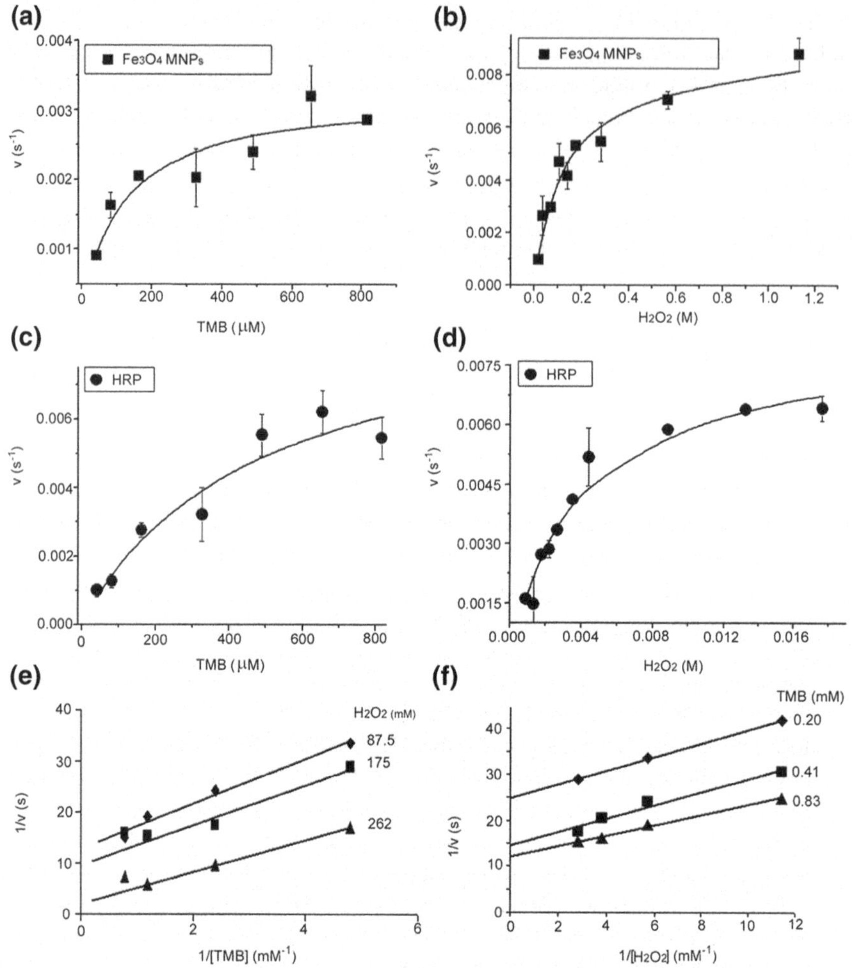

Fig. 2.6 Comparison of catalytic kinetics between iron oxide nanozyme and HRP [2]

bubble observed by naked eye was determined to be O_2 by a gas chromatography (unpublished data). For quantitative assay, the oximetry method can be used to detect the O_2 generation rate via oxygen electrode. The velocity of catalysis is proportional to the amount of generated molecular oxygen per second in the solution. As expected it also followed the typical Michaelis–Menten kinetics for catalase reaction [8] (Fig. 2.7c). However, the capacity of dissolved O_2 in aqueous buffer is affected by many exterior factors, such as temperature and pressure. O_2 from the air would enter the solution to interfere the measurement if H_2O_2 is in a low concentration and it may escape from the liquid if the reaction is too fast when H_2O_2 is at a too high concentration. The volumetric measurement for oxygen gas may be achieved via a volumetric bar-chart chip [9].

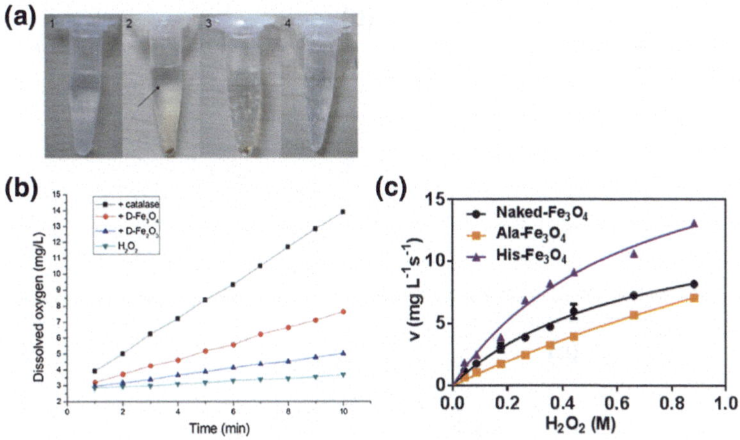

Fig. 2.7 Behaviors of catalase-like nanozyme [7, 8]. Copyright © 2012 American Chemical Society. Copyright © 2018 Royal Society of Chemistry

Other alternative methods may monitor H_2O_2 directly with spectrophotometer at 240 nm [10, 11] or measure hydroxyl radicals trapped by DEPMPO with ESR [2]. Therefore, more accurate strategy is needed for the characterization of catalase-like activity.

2. **Non-Michaelis–Menten kinetics**

Besides Michaelis–Menten kinetics, the catalysis of some nanozymes may follow other kinetics equations. For example, molybdenum trioxide (MoO_3-TPP) nanoparticles with intrinsic sulfide oxidase-like activity which catalyzes sulfite oxidation in the presence of ferricyanide, exhibited a sigmoidal behavior toward the substrate of sulfite while keeping the concentrations of MoO_3-TPP nanoparticles and ferricyanide constant [12] (Fig. 2.8). This behavior was ascribed to the cooperative binding of substrate to the active site, as sulfite anions are competing with the negatively charged ferricyanide on the surface of MoO_3 nanoparticles. Wolfgang Tremel group used Hill equation to analyze the kinetics parameters and found it followed the positive cooperative behavior, with a K_M value of 0.59 ± 0.02 mM for SO_3^{2-} and a Hill coefficient n (cooperativity constant) of 2.35 ± 0.15. Notably, the same sigmoidal behavior was observed for the natural enzyme and model complexes of SuO_x, indicating it is a general enzymatic kinetics.

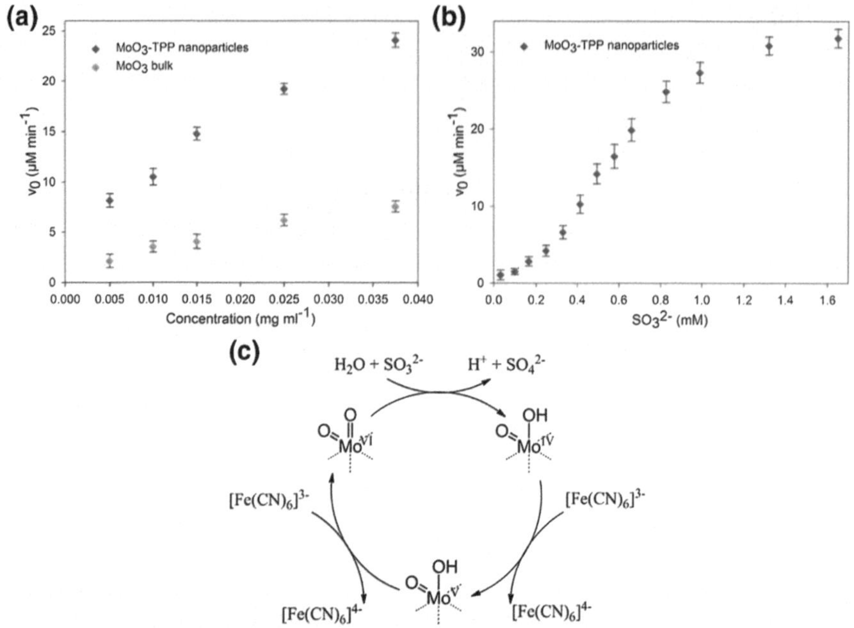

Fig. 2.8 Non-Michaelis–Menten kinetics of nanozyme [12]. Copyright © 2014 American Chemical Society

2.3 Characterization and Identification for Active Site in Nanozymes

Once a nanozyme is confirmed to mimic the activity of natural enzyme, it is critical to analyze the active site in order to understand the catalytic mechanism. There is a typical active site in the natural enzyme. The binding site of substrate is present in the protein structure of an enzyme, which may not be the same moiety in a nanozyme. As the flexibility of protein scaffold constituted by amino acids, substrate binding may cause structure conformation and trap the substrate into active site to complete the catalysis. However, nanozyme often does not have such structure flexibility as it is made by nanomaterial with very tough structure. Therefore, the active site and binding site share the same structural feature.

Natural enzymes are often constituted by amino acids, saccharide, metal atoms, and other cofactors. The amino acids form protein scaffold with specific structure which provides the stereoscopic space for cofactor and metal atom coordination and substrate binding. In addition, this scaffold is flexible and dynamic with response to the environmental factors including ionic strength, pH and temperature, etc. Such complicated structure endows natural enzyme with high activity, selectivity, and tunability. As most activities found in nanozymes belong to oxidoreductase, such as

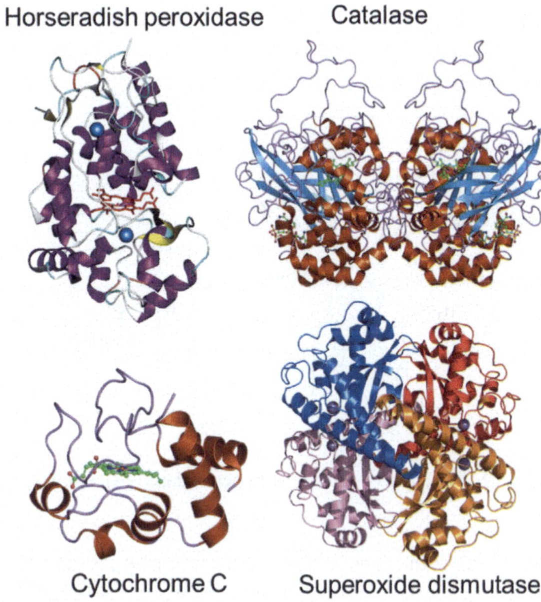

Fig. 2.9 Active sites in natural enzymes containing iron atoms. Protein structure information from PDB

peroxidase, catalase, oxidase, and SOD, here we focus on the structural features in these enzymes to probe the core information of catalysis.

In many peroxide-related enzymes, a common feature is that there is one or more heme molecules coordinated in the protein scaffold to serve as the active site, such as horseradish peroxidase (HRP), catalase (CAT), and cytochrome C (CytC) (Fig. 2.9). Although all have heme groups, the three enzymes perform different activities: HRP decomposes hydrogen peroxide into hydroxyl radical; CAT decomposes H_2O_2 into O_2 and water; CytC transfers O_2 into H_2O_2. As the center is occupied by an iron in heme, the metal atoms play key role in the catalysis in heme enzymes. In comparison, SOD containing metal atoms like Fe, Cu, Mn, or Zn can disproportionate superoxide into O_2 or H_2O_2, further indicating metal atoms are critical for the catalysis.

However, only metal atom cannot form the typical the structure of active site in natural enzyme. Here we are trying to elucidate our understanding using heme in HRP as the model. Heme is a coordination complex in which an iron ion is coordinated to a porphyrin with tetradentate structure. The iron ion is determined as the critical atom to perform the catalysis, responsible for O_2 or H_2O_2 adsorption to initiate the catalytic reaction. Besides the porphyrin ring, there are two carboxyl groups in the distal. The interaction between heme and protein scaffold in HRP. The protein crystal has provided high-resolution information to recognize the interaction between heme and residues of amino acids in HRP. The key residues include His170, Asp247, His142, Arg38, Asn70, and Phe41 [13] (Fig. 2.10). Among them, His170 directly coordinates with iron below the plane of heme while His142 and Arg38 are above the plane to facilitate the binding and stabilization of ligands and aromatic substrates. In addition heme, there are two calcium ions coordinated in the protein scaffold. The

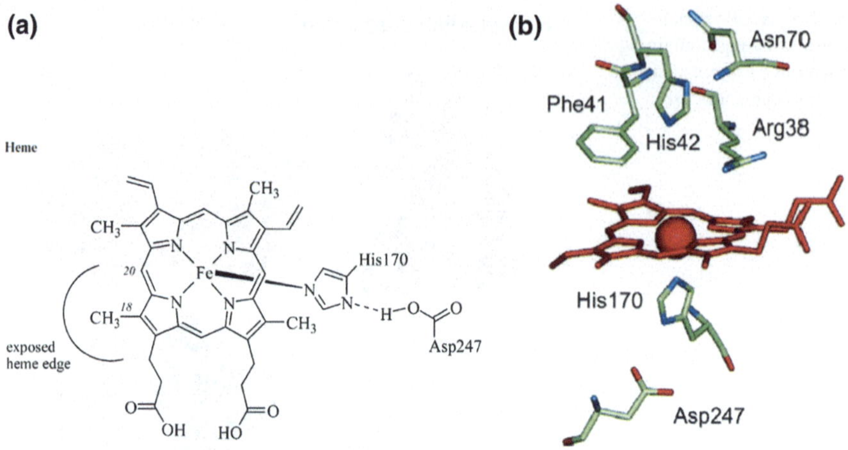

Fig. 2.10 Structure of active site in peroxidase [13]. Copyright © 2003 Elsevier Ltd

loss of calcium leads to 40% reduction of enzyme activity, indicating that calcium is important for the catalysis. Furthermore, glycosylated modification is present in HRP from different sources. The carbohydrate in HRP is up to 21.8% with eight N-linked glycans attached, which is related to substrate binding and protein heterogeneity. The above structure information indicates that the active site in HRP is constituted by heme with peripheral amino acid residues, and the distal calcium and glycosylation also affect the catalysis. The iron in the center of heme is responsible to adsorb oxygen or peroxide, the porphyrin provides a confinement for electron transfer, the carboxyl groups and proximal amino acid residues directly or indirectly participate in the catalytic process and endow substrate specificity, the presence of distal calcium ion and glycans maintains the high activity. These features are common in natural enzymes like CAT and CytC [14], providing insights for understanding and design the active site in nanozymes with enzyme-like activities.

1. **The active site for nanozyme mimicking natural enzyme**

Iron oxide nanozyme was reported with peroxidase-like activity by Yan group in 2007 and then with catalase-like activity by Gu group in 2012. As both peroxidase and catalase have heme as cofactor in the active site. The iron (Fe) play the key role in the catalysis by reversibly varying valency of iron. Since there are large amounts of iron on the surface of iron oxide nanoparticles, it is easy to understand that these nanoparticles could perform peroxidase or catalase-like activities (Fig. 2.11). If a nanostructure contains a heme-like structure, it may perform these activities in principle. However, the iron atom is not free but crystalized with oxygen atom (Fe_3O_4 or Fe_2O_3), which is different with that iron coordinated with nitrogen (N) in porphyrin. It is important to identify which valency of iron contributes to the catalysis. Although both iron oxides exert the activities, Fe_3O_4 often perform higher activity than Fe_2O_3, indicating that ferrous iron may have more contribution on the activity. It

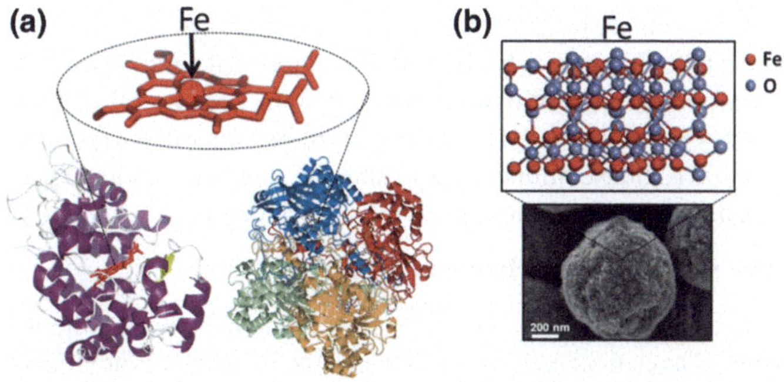

Fig. 2.11 The form of iron in natural enzyme and iron oxide nanoparticle [1]. Copyright © 2017 Ivyspring International Publisher

is also proved by reduction treatment with $NaBH_4$ for Fe_3O_4 to improve the catalytic efficiency. These experimental evidences suggest that iron in ferrous state is serving as the active site for the catalysis on the surface of iron oxide nanoparticles. The specific nanostructure of iron on the surface of iron oxide nanoparticles is still unknown in the active site. Simulation analysis may help to understand the mechanism, but the direct experimental evidences are required to verify it.

Besides iron role in the catalysis, other atomic structure may be involved into the catalytic process. Actually, there are also two carboxyl groups in heme structure. It is reported that the carboxyl group may benefit substrate binding. The improvement for the catalysis by carboxyl group in iron oxide nanoparticles was observed by Gao group [15]. Direct modification with porphyrin also improves the activity of Fe_3O_4 nanocomposites, which may be ascribed to the coordination between porphyrin and iron on the surface of nanoparticle [16]. In parallel, the residues from the amino acids may also take part in the catalytic process. Especially, imidazole of histidine in HRP plays an important role in iron coordination and H_2O_2 binding. Inspired by this feature, modifying the surface of iron oxide nanoparticles with histidine has been observed to significantly improve the peroxidase-like and catalase-like activities [8]. The single histidine modification improved the catalytic efficiency up to 20-fold compared with naked iron oxide nanoparticles. Simulation analysis indicates that the hydrogen bond formed between histidine and the hydrogen peroxide (initial state) not only weakens the O–H bond strength but also leads O to become more negatively charged. The former process is beneficial for splitting the O–O bond of hydrogen peroxide and the latter step enhances its adsorption onto the Fe_3O_4 nanozyme (final state). It thus serves in a similar role as His42 in the active site of HRP. The influence of iron and peripheral residues on catalytic activity indicates the certain structure is required to form active site of a nanozyme.

Even there is big difference between the active site of nanozyme and that of natural enzyme, it is still possible to design or improve substrate selectivity by surface modification on nanozyme. Juewen Liu group used molecular imprinting to create substrate-binding pockets on Fe_3O_4 nanozyme [17]. They designed the surface imprinting with selectivity to TMB and ABTS, respectively, as the former is positively charged and the latter is negatively charged. The imprinted nanozymes demonstrated nearly 100-fold increased specificity compared to the bare Fe_3O_4. Importantly, the imprinting strategy is also proved effective for gold nanoparticles (peroxidase mimics) and nanoceria (oxidase mimics), indicating the substrate-binding space could be artificially created to improve the selectivity of a nanozyme.

2. **The active structure in carbon nanozymes**

For above iron oxide nanozymes or other metal oxide nanozymes, the metal atoms are often assumed as critical active sites in which the same metals are found in natural enzyme. But there are still a lot nanozymes which do not contain metals, such as carbon nanozymes. They also perform similar peroxidase-like and/or catalase-like activity. Can similar feature of active site be found in these carbon nanozymes? Xiaogang Qu group deciphered the active site in carbon nanomaterials with peroxidase activity. They found that graphene oxide (GO), carbon dots, and carbon nanotubes (CNTs) have similar enzyme-like activities [18, 19]. The biggest feature for these carbon nanozymes is that they do not have metal atoms responsible for the catalysis. The main structural unit in these carbon nanocomposites is aromatic carbons arranged in hexagonal lattice at two dimensions. There are carbon atoms in dominant, plus minor oxygen-containing chemical groups. One interesting question is why these nonmetal carbons perform similar enzyme-like activities as those in metal nanomaterials. Coincidentally, there is also the aromatic structure in the natural enzymes. It is the heme that contains porphyrin ring chelating to iron as active center. As previous description for heme features, the porphyrin makes act as a good electron transferring system to facilitate catalysis and the distal carboxyl group affects the substrate binding. Correspondingly, it has been know that the aromatic unit in carbon nanomaterials, such as graphene, endows the material with excellent ability of electron transfer with high conductivity. Therefore, the electronic property may primarily contribute to the enzyme-like activities of graphene and other carbon nanomaterials. However, bare graphene is not very active for these catalysis. In comparison, graphene oxide performs increased peroxidase activity. Qu group analyzed the groups affecting catalytic activity in graphene quantum dots (GQDs) (Fig. 2.12). GQDs are a type of zero-dimensional material with the common features derived from graphene and carbon dots [20]. They identified that the –C=O groups were the catalytically active sites, whereas the O=C–O– groups acted as substrate-binding sites, and –C–OH groups decreased the activity of QCDs. Theoretical simulations indicate that H_2O_2 dissociation and hydroxyl radical formation on –C=O owe lower energy barrier than that on –COOH. Moreover, the binding energy of H_2O_2 on COOH (−0.67 eV) was lower than that for the –C=O (−0.30 eV) and –C–OH moieties (−0.17 eV). Chunying Shu group reported that carboxyfullerenes $C_{60}[C(COOH)_2]_2$ catalyzed the oxidation of TMB with H_2O_2, experimentally demonstrating its peroxidase-like activity [21]. Xingfa Gao group computationally verified that –COOH attached on a large aromatic carbon domain was able to catalyze the hemolysis of H_2O_2 to generate hydroxyl radicals [22]. These experimental evidence and simulation analysis demonstrate that heteroatomic groups play critical roles in catalysis and substrate binding.

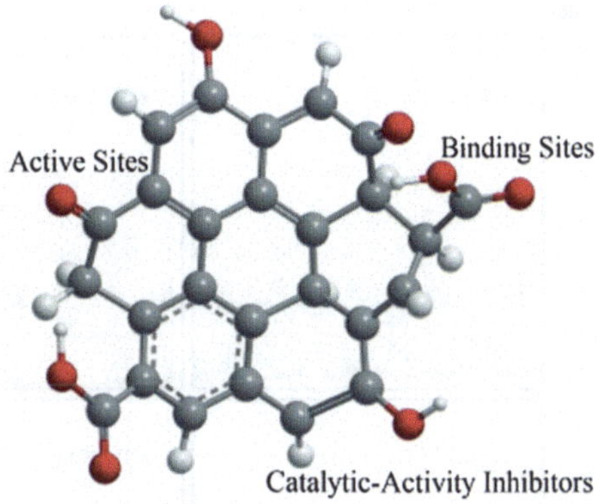

Fig. 2.12 The catalytic structure in carbon nanozyme [19]. Copyright © 2015 WILEY—VCH Verlag GmbH & Co. KGaA, Weinheim

Dopants in carbon nanozymes may be a general strategy to improve their catalytic activity. Inspired by the feature of heme which contains heteroatomic nitrogen and chelated iron, we doped nitrogen into carbon nanospheres and found the catalytic activity was not only dramatically increased but also diversified into four enzyme-like activities: peroxidase, oxidase, catalase, and SOD [23]. Recent work by Qu's group also proved that doping iron into carbon nanospheres can improve the peroxidase-like activity [24]. The nonmetal nanozymes provide an alternative concept to design and develop artificial enzymes by mimicking the feature from organic moieties and residues of amino acids in protein scaffold.

3. **Active site from crystal nanostructure**

With the same components, but the different surface arrangement of crystal planes may affect the formation of active site. The phenomena have been observed in metal and metal oxide nanozymes. For instance, Junjie Zhu reported that Fe_3O_4 nanostructures with different surface areas and exposed crystal planes exhibited different levels of peroxidase-like activity [25]. Among the crystal planes, the cluster spheres with (311) exposed planes performed higher catalytic efficiency than those for triangular plates with (220) and octahedral with (111) (Fig. 2.13). The cluster spheres showed higher affinity to the substrate of TMB, indicating there are more catalytically active irons exposed on the (311) plane. Actually, these three nanostructures with similar surface area, but the level of activity followed the order: (311) > (220) > (111), indicating different crystal planes formed different atomic structure as active sites for the catalysis.

More importantly, precious metals which do not exist in natural enzymes can form metal nanozymes with similar enzyme-like activities. The activities show the dependence on atomic structure at crystal planes. However, this type of nanozymes does not have the common features found in metal oxide or carbon nanozymes. They have neither transition metal atoms in oxygenated form, nor the organic moieties, such as porphyrin and carboxyl group. For example, gold nanozymes with oxidase activity follow the order: (211) > (110) > (111) [26, 27]. For the gold facets of (111) and (211), the former facet has a flat surface but the latter has a stepped structure which

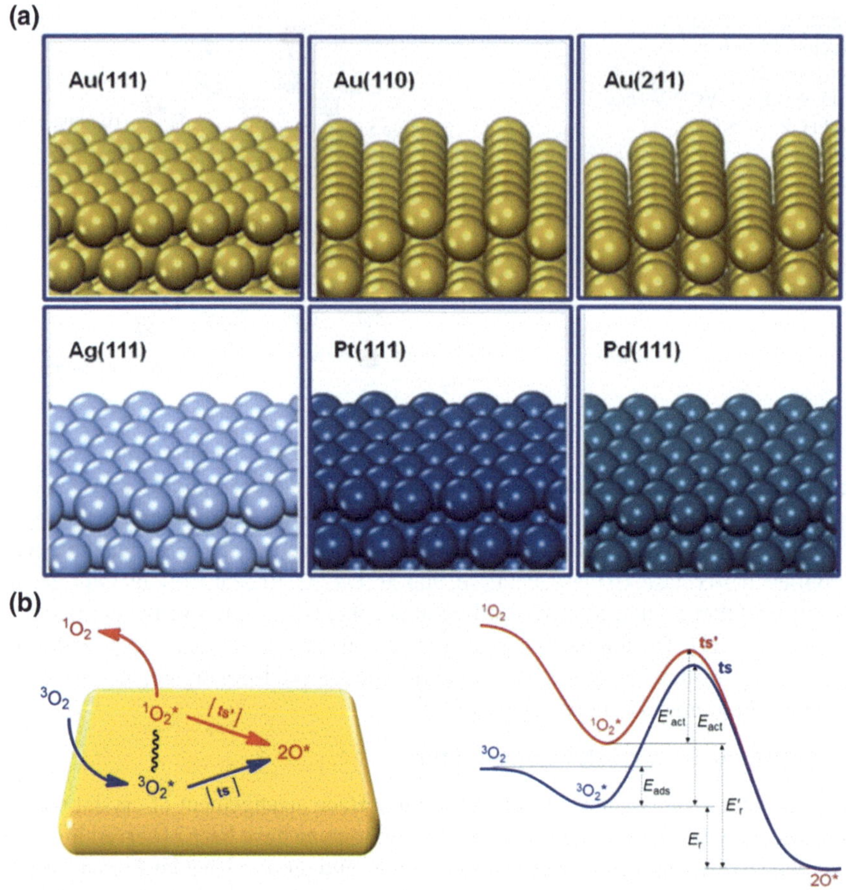

Fig. 2.13 Catalytic mechanism in noble metal nanozymes [26]. Copyright © 2015 American Chemical Society

exposes gold atoms with lower coordination numbers and thus larger affinities to oxygen. This phenomenon is consistent with the finding that gold nanoparticles only demonstrate catalytically active property when their sizes are at nanoscale (e.g., 3–5 nm) [28]. Similarly, palladium nanocrystals with (111) facets demonstrated higher catalase-like and SOD-like activities than those with (100) facets [29].

The catalytic performance of metal nanocrystals was affected by the atomic arrangement in the surface plane. To better understand the mechanism, simulation is used to analyze the activation energy on these crystal planes. Xingfa Gao group used first-principles density functional theory (DFT) calculations to analyze and predict the activities for series of metal and alloyed nanozymes, including Au, Ag, Pd, Pt, and alloys $Au_{4-x}M_x$ (x = 1, 2, 3; M = Ag, Pd, Pt). They proposed that the oxidase-like catalysis may follow the energy-based model in which the dissociative adsorption of

O_2 on the metal surfaces to form single-atomic O adatoms is the key step. Basically, the metals catalyze the dissociation of O_2 to yield O adatoms, which subsequently abstract hydrogens from substrates such as TMB and ascorbic acid, completing the oxidation of the substrates by O_2. In the O_2 dissociative adsorption process, the metal surface adsorbs $^3O_2$ and transfers electrons to the antibonding π^* orbitals of $^3O_2$, leading to the splitting of O–O bond. On the (111) facets of four different metals (Pd, Pt, Ag, and Au), the extent of O–O bond weakening caused by the metal adsorption follows the order of Pd(111) > Pt(111) > Ag(111) > Au(111). In addition, the activation energy barriers (E_{act}'s) of the dissociations and the corresponding reaction energies (E_r's) of between the dissociation and adsorption states are in the order of Pd(111) < Pt(111) < Ag(111) < Au(111). These energy analyses suggest that the order of oxidase-like activity is Pd(111) > Pt(111) $\gg$ Au(111), Ag(111). Further analysis indicated that high energy facet Au(211) can behave oxidase-like activity but Au (111) and Au(110) cannot. The peroxidase-like and catalase-like activities follow the similar order in Pd(111) > Pt(111) > Ag(111) > Au(111).

4. **Active site in SOD nanozymes**

Natural SOD-like activity is different from natural peroxidase- and catalase-like activities, as it has no typical heme cofactors. Natural SOD only requires metal atoms chelating into protein scaffold, such as Fe, Mn, Cu–Zn, and Ni (Fig. 2.14). These metal atoms are often coordinated with the residues of amino acid in protein scaffold, such as the imidazole moiety of histidine. Although without heme, such

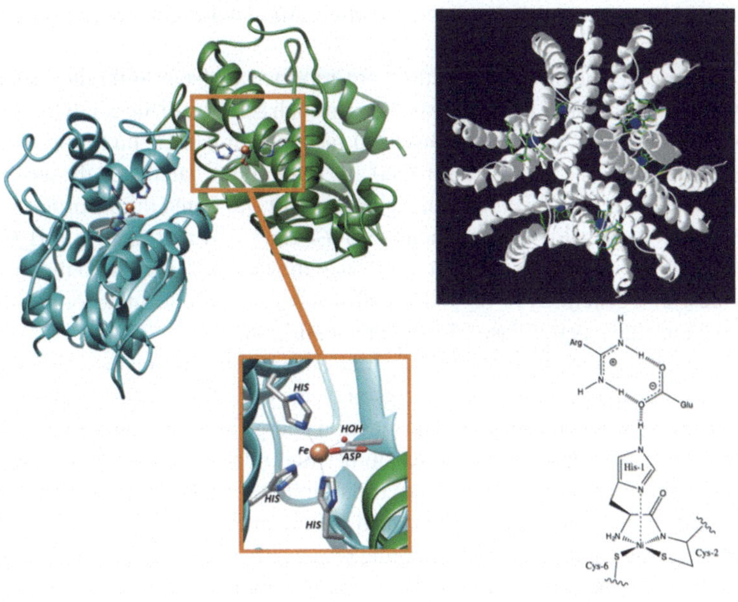

Fig. 2.14 Iron in the active site of natural SOD. Protein structure information from https://en.wikipedia.org/wiki/Superoxide_dismutase

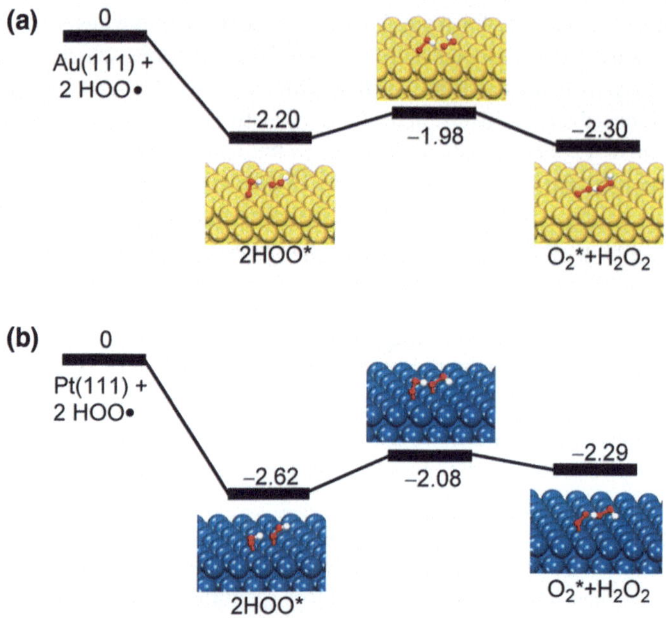

Fig. 2.15 Energy barriers in the catalysis of metal nanozymes [26]: **a** Au nanozyme and **b** Pt nanozymes. Copyright © 2015 American Chemical Society

coordination between metal atoms and protein scaffold determines the active site and affects the selectivity of catalysis.

Actually, some metal nanostructures are reported to have SOD-like activities. Other nanostructures such as ceria also perform such activity. In the catalytic process, superoxide radical ($O_2^{-\bullet}$) acts as a Brønsted base which can capture a proton from water to form $HO_2\bullet$ and OH^-. Xingfa Gao and co-workers calculated the adsorption energy of $HO_2\bullet$ on (111) facets of Au, Ag, Pd, and Pt and found the adsorptions were exothermic processes (Fig. 2.15). Then the $HO_2\bullet$ can easily rearrange into H_2O_2 and O_2 as the activation energies are quite low once they are adsorbed onto the facets of nanozymes. The generated H_2O_2 and O_2 may undergo subsequent decomposition if the nanozymes have peroxidase-like or catalase-like activity.

Taken together, there are typically substrate adsorption, intermediate dissociation, and electron transfer processes on the crystal facets of metal nanostructures, responsible for the enzyme-like catalysis. The reaction energy and activation energy for the reaction on the facet determines the feasibility and rate of the catalysis. Therefore, such nanozymes may be an ideal model for simulation analysis to reveal the mechanism based on the knowledge of classical chemical catalysis, such as adsorption energy, activation energy, etc. However, the practically as-prepared nanozymes may not have the uniform nanostructures. There may be abundant defects on the surfaces of the nanozymes, such as vacancy, edge, disorder, polymorph, dopant, or alloy. The defects may favor the generation of active site to improve the activity.

2.4 Nature of Nanozyme as Enzyme Mimetics

Currently most nanozymes demonstrate redox reactivities focusing on oxygen-related active substrates (e.g., H_2O_2 or free radicals). From chemistry, these nanozymes are made from inorganic elements with metal, metal oxides, or carbons. In comparison, natural enzymes are built with protein scaffolds with precise structures. Many enzymes reactive to oxygen have metal atoms in their active sites or proximal locations of the active centers. HRP is one of the most classical peroxidase enzymes which has been well studied from structure to catalytic mechanism. In the catalytic process, the Fe sitting in the center of heme accomplishes H_2O_2 decomposition by changing the state from Fe(III) to Fe(IV) with high-oxidation intermediates (Fig. 2.16) [30]. The valent-state change allows the enzyme to realize the catalytic cycle of peroxidase. In addition the change from Fe(II) to Fe(III) favors the oxidase cycle. Apparently, the valent-state change of Fe requires electron transfer from H_2O_2 or oxygen to Fe. Therefore, if a nanomaterial can fulfill the similar electron transfer, it can be theoretically assumed as a nanozyme to perform the corresponding enzyme-like activity.

Yu Zhang group tried to elucidate these electron transfer processes according to the theoretical redox potential of different redox couples to understand the mechanism of Prussian blue (PB) nanozymes [31]. They found that PB nanoparticles performed multienzyme-like activity including peroxidase, catalase, and SOD (Fig. 2.17). In general, PB can be reduced into Prussian white (PW) and oxidized into Berlin green (BG) or Prussian yellow (PY) via electron transfer. The different redox potentials of PB/PW, BG/PB, and PY/BG under different pH allow them to perform the above multiple enzyme-like activities. These compounds gradually and continuously changed along with the redox potential of the environment, which can explain why PB nanozymes demonstrated peroxidase, catalase or SOD activity under different pH (Fig. 2.17). For instance, H_2O_2 possesses strong oxidation ability

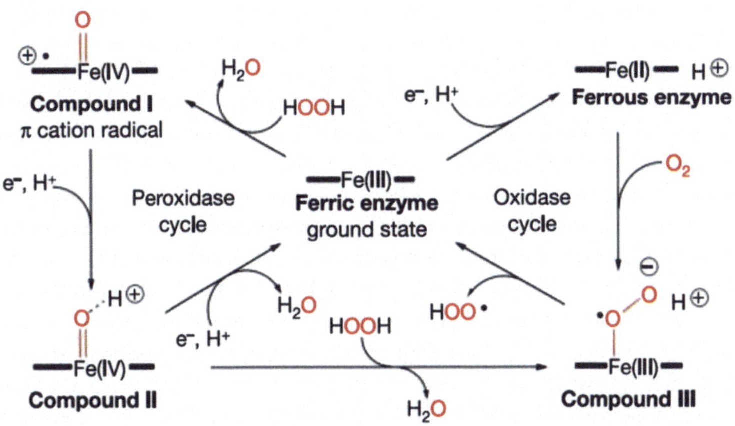

Fig. 2.16 Catalytic pathway in natural peroxidase [30]. Copyright © 2002, Springer Nature

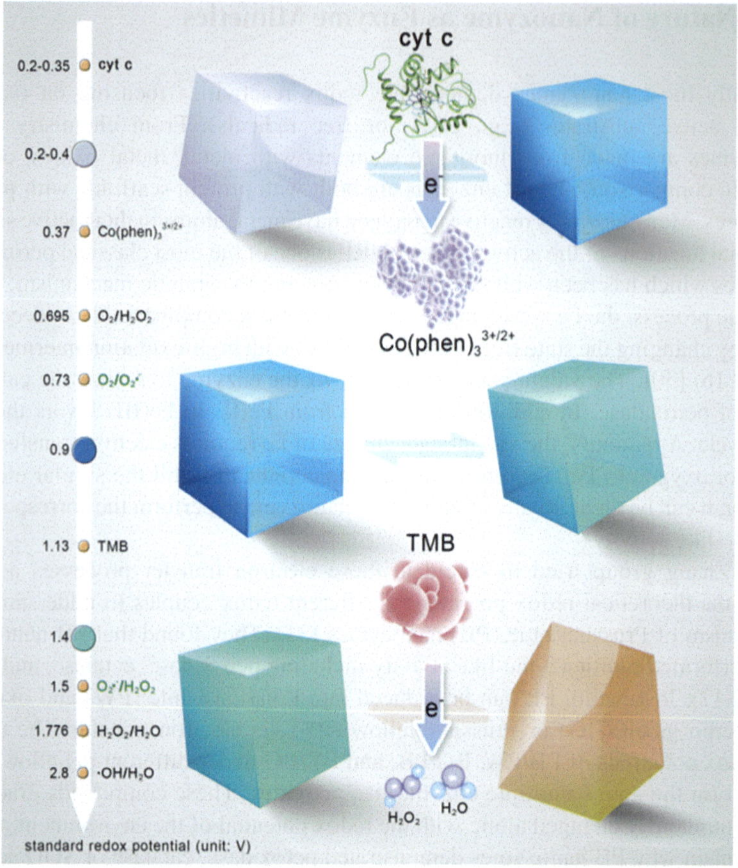

Fig. 2.17 The correlation between redox potential of substrates and Prussian blue nanozymes [31]. Copyright © 2016 American Chemical Society

in acidic environments, which prefers to oxidize PBNPs into BG or PY at pH 3.6. Thus PY/BG can transfer electrons from TMB (TMBOx/TMBRe 1.13 V) to H_2O_2 (H_2O_2/H_2O 1.776 V) at a suitable potential (1.4 V), and primarily contributes to the peroxidase-like activity of PBNPs.

This electron transfer model may be a general mechanism for metal or metal oxide nanomaterials to mediate biological oxidative reactions. Haiyuan Zhang et al. observed the difference in inducing oxygen radicals from 24 metal oxide (MOx) nanoparticles [32]. They used conduction band energy levels to delineate the phenomena. There is overlap between cellular redox potential (−4.12 to −4.84 eV) and conduction band energy (E_c) levels of TiO_2, Co_3O_4, Cr_2O_3, Ni_2O_3, Mn_2O_3, and CoO nanoparticles, which may allow electron transfers from biological redox couples to the conduction band of the semiconductor particles (Fig. 2.18). Therefore, the

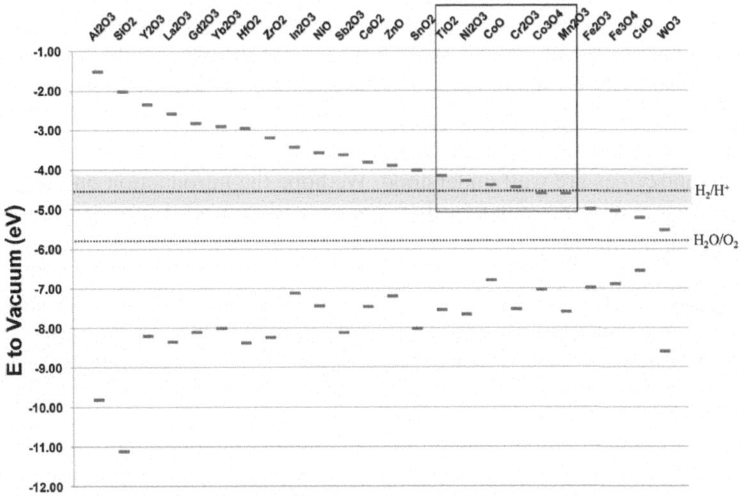

Fig. 2.18 Bandgap of conduction band energy (E_C) levels of metal oxide nanoparticles between cellular redox potential [32]. Copyright © 2012 American Chemical Society

electron transfers may be elucidated by the conduction band energy levels to understand the catalytic mechanism of nanozymes, even Zhang et al. did not investigate the enzyme-like property of the 24 metal oxide (MOx), some of which are already reported as nanozymes.

However, not all nanozymes have metals with such ability of valent-state changes or semiconductor levels, based on which straightforward electron transfer models can be established. There are no metals in carbon-based nanozymes. In addition, a lot of metals, which are demonstrated to be nanozymes, are not present in natural enzymes. Therefore, there must be other mechanisms to accomplish the electron transfer in the catalytic processes. Once a substrate is adsorbed, there must be certain nanoscale moiety with special physical or chemical feature responsible for the electron transfer. For instance, the heme-like structure in carbon nanomaterials may play the role for electron transfer.

2.5 Conclusion and Perspectives

Nanozymes perform intrinsic enzyme-like properties. They have not only similar catalytic behaviors including substrate, optimal pH, temperature, and activator/inhibitor, but also the similar enzymatic kinetics and mechanisms. Typical processes in nanozyme-mediated catalysis include substrate adsorption, electron transfer, and product release, which are the same as those in natural enzymes. The electron transfer processes for different types of nanomaterials have been described with different models, and the nature of many nanozymes still needs further investigation.

Typically, nanozymes have amorphous nanostructures in contrast to the protein scaffolds of enzymes with precise three-dimensional structures, which leads to low activity or selectivity for nanozymes. On the one hand, it is possible to improve the activity of nanozymes by mimicking the catalytic feature of natural enzymes. On the other hand, design and predict of nanozymes will be feasible if the mechanism and nature of nanozymes are well understood. We hope the introduction on the kinetics and mechanisms of nanozymes in this chapter will be useful for recognizing and understanding nanozymes.

Acknowledgements This work was supported in part by the Foundation of the Thousand Talents Plan for Young Professionals and Jiangsu Specially-Appointed Professor, the Interdisciplinary Funding at Yangzhou University, Strategic Priority Research Program of the Chinese Academy of Sciences (Grant No. XDA09030306), and National Natural Science Foundation of China (Grant No. 31530026 and 81671810).

References

1. Gao LZ, Fan KL, Yan XY (2017) Iron oxide nanozyme: a multifunctional enzyme mimetic for biomedical applications. Theranostics 7(13):3207–3227
2. Gao LZ, Zhuang J, Nie L, Zhang JB, Zhang Y, Gu N, Wang TH, Feng J, Yang DL, Perrett S, Yan X (2007) Intrinsic peroxidase-like activity of ferromagnetic nanoparticles. Nat Nanotechnol 2(9):577–583
3. Fan J, Yin JJ, Ning B, Wu XC, Hu Y, Ferrari M, Anderson GJ, Wei JY, Zhao YL, Nie GJ (2011) Direct evidence for catalase and peroxidase activities of ferritin-platinum nanoparticles. Biomaterials 32(6):1611–1618
4. Gao N, Dong K, Zhao AD, Sun HJ, Wang Y, Ren JS, Qu XG (2016) Polyoxometalate-based nanozyme: design of a multifunctional enzyme for multi-faceted treatment of Alzheimer's disease. Nano Res 9(4):1079–1090
5. Chen ZW, Ji HW, Liu CQ, Bing W, Wang ZZ, Qu XG (2016) A multinuclear metal complex based dnase-mimetic artificial enzyme: matrix cleavage for combating bacterial biofilms. Angew Chem Int Edit 55(36):10732–10736
6. Jiang B, Duan DM, Gao LZ, Zhou M, Fan KL, Tang Y, Xi JQ, Bi Y, Tong Z, Gao GF, Xie N, Tang A, Nie G, Liang MM, Yan XY (2018) Standardized assays for determining the catalytic activity and kinetics of peroxidase-like nanozymes. Nat Protoc 13:1506–1520
7. Chen ZW, Yin JJ, Zhou YT, Zhang Y, Song L, Song MJ, Hu SL, Gu N (2012) Dual enzyme-like activities of iron oxide nanoparticles and their implication for diminishing cytotoxicity. ACS Nano 6(5):4001–4012
8. Fan KL, Wang H, Xi JQ, Liu Q, Meng XQ, Duan DM, Gao LZ, Yan XY (2017) Optimization of Fe_3O_4 nanozyme activity via single amino acid modification mimicking an enzyme active site. Chem Commun 53(2):424–427
9. Song YJ, Xia XF, Wu XF, Wang P, Qin LD (2014) Integration of platinum nanoparticles with a volumetric bar-chart chip for biomarker assays. Angew Chem Int Edit 53(46):12451–12455
10. Mu JS, Zhang L, Zhao M, Wang Y (2014) Catalase mimic property of Co_3O_4 nanomaterials with different morphology and its application as a calcium sensor. ACS Appl Mater Inter 6(10):7090–7098
11. Pirmohamed T, Dowding JM, Singh S, Wasserman B, Heckert E, Karakoti AS, King JES, Seal S, Self WT (2010) Nanoceria exhibit redox state-dependent catalase mimetic activity. Chem Commun 46(16):2736–2738

12. Ragg R, Natalio F, Tahir MN, Janssen H, Kashyap A, Strand D, Strand S, Tremel W (2014) Molybdenum trioxide nanoparticles with intrinsic sulfite oxidase activity. ACS Nano 8(5):5182–5189
13. Veitch NC (2004) Horseradish peroxidase: a modern view of a classic enzyme. Phytochemistry 65(3):249–259
14. Campomanes P, Rothlisberger U, Alfonso-Prieto M, Rovira C (2015) The molecular mechanism of the catalase-like activity in horseradish peroxidase. J Am Chem Soc 137(34):11170–11178
15. Yang CH, Du JJ, Peng Q, Qiao RR, Chen W, Xu C, Shuai ZG, Gao MY (2009) Polyaniline/Fe_3O_4 nanoparticle composite: synthesis and reaction mechanism. J Phys Chem B 113(15):5052–5058
16. Liu QY, Li H, Zhao QR, Zhu RR, Yang YT, Jia QY, Bian B, Zhuo LH (2014) Glucose-sensitive colorimetric sensor based on peroxidase mimics activity of porphyrin-Fe_3O_4 nanocomposites. Mat Sci Eng C Mater 41:142–151
17. Zhang ZJ, Zhang XH, Liu BW, Liu JW (2017) Molecular imprinting on inorganic nanozymes for hundred-fold enzyme specificity. J Am Chem Soc 139(15):5412–5419
18. Lin YH, Ren JS, Qu XG (2014) Catalytically active nanomaterials: a promising candidate for artificial enzymes. Acc Chem Res 47(4):1097–1105
19. Song YJ, Qu KG, Zhao C, Ren JS, Qu XG (2010) Graphene oxide: intrinsic peroxidase catalytic activity and its application to glucose detection. Adv Mater 22(19):2206–2210
20. Sun HJ, Zhao AD, Gao N, Li K, Ren JS, Qu XG (2015) Deciphering a nanocarbon-based artificial peroxidase: chemical identification of the catalytically active and substrate-binding sites on graphene quantum dots. Angew Chem Int Edit 54(24):7176–7180
21. Li R, Zhen M, Guan M, Chen D, Zhang G, Ge J, Gong P, Wang C, Shu C (2013) A novel glucose colorimetric sensor based on intrinsic peroxidase-like activity of C60-carboxyfullerenes. Biosens Bioelectron 47:502–507
22. Zhao RS, Zhao X, Gao XF (2015) Molecular-level insights into intrinsic peroxidase-like activity of nanocarbon oxides. Chem-Eur J 21(3):960–964
23. Fan K, Xi J, Fan L, Wang P, Zhu C, Tang Y, Xu X, Liang M, Jiang B, Yan X, Gao L (2018) In vivo guiding nitrogen-doped carbon nanozyme for tumor catalytic therapy. Nat Commun 9(1):1440
24. Sang Y, Huang Y, Li W, Ren J, Qu X (2018) Bioinspired design of fe doped mesoporous carbon nanospheres for enhancement of nanozyme's activity. Chem-Eur J 24:7259–7263
25. Liu SH, Lu F, Xing RM, Zhu JJ (2011) Structural effects of Fe_3O_4 nanocrystals on peroxidase-like activity. Chem-Eur J 17(2):620–625
26. Shen XM, Liu WQ, Gao XJ, Lu ZH, Wu XC, Gao XF (2015) Mechanisms of oxidase and superoxide dismutation-like activities of Gold, Silver, Platinum, and Palladium, and their alloys: a general way to the activation of molecular oxygen. J Am Chem Soc 137(50):15882–15891
27. Li JN, Liu WQ, Wu XC, Gao XF (2015) Mechanism of pH-switchable peroxidase and catalase-like activities of Gold, Silver, Platinum and Palladium. Biomaterials 48:37–44
28. Hvolbaek B, Janssens TVW, Clausen BS, Falsig H, Christensen CH, Norskov JK (2007) Catalytic activity of Au nanoparticles. Nano Today 2(4):14–18
29. Ge CC, Fang G, Shen XM, Chong Y, Wamer WG, Gao XF, Chai ZF, Chen CY, Yin JJ (2016) Facet energy versus enzyme-like activities: the unexpected protection of palladium nanocrystals against oxidative damage. ACS Nano 10(11):10436–10445
30. Berglund GI, Carlsson GH, Smith AT, Szoke H, Henriksen A, Hajdu J (2002) The catalytic pathway of horseradish peroxidase at high resolution. Nature 417(6887):463–468
31. Zhang W, Hu SL, Yin JJ, He WW, Lu W, Ma M, Gu N, Zhang Y (2016) Prussian blue nanoparticles as multienzyme mimetics and reactive oxygen species scavengers. J Am Chem Soc 138(18):5860–5865
32. Zhang HY, Ji ZX, Xia T, Meng H, Low-Kam C, Liu R, Pokhrel S, Lin SJ, Wang X, Liao YP, Wang MY, Li LJ, Rallo R, Damoiseaux R, Telesca D, Madler L, Cohen Y, Zink JI, Nel AE (2012) Use of metal oxide nanoparticle band gap to develop a predictive paradigm for oxidative stress and acute pulmonary inflammation. Acs Nano 6 (5):4349–4368

Chapter 3
Types of Nanozymes: Materials and Activities

Yongwei Wang, Minmin Liang and Taotao Wei

Abbreviations

ABTS	2, 2′-Azino-bis (3-ethylbenzothiazoline-6-sulphonic acid)
BSA	Bovine serum albumin
C-Dots	Carbon nanodots
CNTs	Carbon nanotubes
CTAB	Cetyltrimethylammonium bromide
DBA	Diaminobenzidine
dsDNA	Double-strand DNA
ELISA	Enzyme-linked immunosorbent assay
EPR	Electron paramagnetic resonance
GOx	Glucose oxidase
GQDs	Graphene quantum dots
HRP	Horseradish peroxidase
LDH	Layered double hydroxide
MNPs	Magnetic nanoparticles
MRI	Magnetic resonance imaging
MWNTs	Multi-walled carbon nanotubes
NPs	Nanoparticles
OPD	O-phenylenediamine
PCR	Polymerase chain reaction
PLGA	Poly lactic-co-glycolic acid

Y. Wang · T. Wei (✉)
National Laboratory of Biomacromolecules, Institute of Biophysics, Chinese Academy of Sciences, 15 Datun Road, Beijing 100101, People's Republic of China
e-mail: weitt@ibp.ac.cn

M. Liang
Key Laboratory of Protein and Peptide Pharmaceutical, Institute of Biophysics, Chinese Academy of Sciences, 15 Datun Road, Beijing 100101, People's Republic of China
e-mail: mmliang@ibp.ac.cn

X. Yan (ed.), *Nanozymology*, Nanostructure Science and Technology,
https://doi.org/10.1007/978-981-15-1490-6_3

pNPP	p-Nitrophenol phosphate
ROS	Reactive oxygen species
SOD	Superoxide dismutase
SPIO	Superparamagnetic iron oxide
ssDNA	Single-strand DNA
SWNTs	Single-walled carbon nanotubes
TMB	Tetramethylbenzidine
UV	Under voltage
UV-Vis	Ultraviolet visible
V-HPO	Vanadium-dependent haloperoxidase

3.1 Nanomaterials as Nanozymes to Mimic Natural Enzymes

Enzymes are biological catalysts that play a crucial role in biochemical reactions by converting substrates into products under relatively mild conditions. Most natural enzymes are proteins with high catalytic activity and high specific activity. The catalytic activity and the specific activity of natural enzymes will be lost at extreme pH, high temperature, etc. Natural enzymes can also be degraded by proteases in the physiological environment and other appropriate conditions. Given these disadvantages of natural enzymes, it is very difficult to broadly realize their utility [1].

To overcome the drawbacks of the natural enzymes and produce more robust and economic enzyme analogues, researchers have mimicked the structural and functional aspects of natural enzymes over the past few decades. Artificial enzymes are a very important branch of biomimetic chemistry [2]. This term was coined by Ronald Breslow for enzyme mimics [3] that replicate the essential and general principles of natural enzymes with alternative materials [4]. Cyclodextrins, metal complexes, porphyrins, polymers, supramolecules, and biomolecules (such as nucleic acids, catalytic antibodies, and proteins) have similar structures and functions with natural enzymes via different approaches [5, 6].

As nanotechnology and nano-research has developed, scientists have found that some materials have unique properties at the nanoscale. Surprisingly, some nanomaterials such as fullerene derivatives, gold nanoparticles, rare earth nanoparticles, and ferromagnetic nanoparticles have unexpected enzyme-like activities [5, 6]. Now, more than 40 types of nanomaterial-based artificial enzymes (nanozymes) have subsequently been reported and were already used in biosensing, immunoassays, cancer diagnostics and therapy, neuroprotection, stem cell growth, pollutant removal, etc. Here, we describe various nanomaterials with peroxidase, oxidase, haloperoxidase, and superoxide dismutase activities.

3.2 Cerium Oxide-Based Nanomaterials

Cerium is the first element of the lanthanide group. It has 4f electrons making it an attractive element for material science, chemistry, biology, and engineering. Cerium oxide has a fluorite crystalline structure with possible applications in biomedical sciences [7, 8]. Cerium oxide (ceria) has a high catalytic activity in numerous applications due to the presence of mixed-valence states of Ce^{3+} and Ce^{4+} as well as the presence of oxygen vacancies [9, 10]. The redox couple can recycle Ce^{3+} and Ce^{4+}, which is the key to the catalytic activity [9, 10]. Oxygen vacancies can compensate for the reduction in the positive charge and thus stabilize the oxidation state of Ce^{3+}. The enzymatic activity of nanoceria (cerium oxide-based nanomaterials) has been explored and will be discussed below.

3.2.1 Nanoceria as Superoxide Dismutase Mimics

Superoxide dismutase (SOD) catalyzes the dismutation of superoxide anions into hydrogen peroxide and molecular oxygen. Superoxide anion is generated via the electron reduction of the oxygen molecules and is a reactive oxygen species (ROS) [11]. ROS can lead to tissue injury via protein, lipid, and nucleic acid damage. Previous studies revealed that the SOD can protect the tissue by eliminating superoxide anions. The first observation of a similar disproportionation reaction with nanoceria was reported by Self and colleagues [12, 13]. In 2007, Korsvik et al. showed that nanoceria had SOD-like activity. They demonstrated via X-ray photoelectron spectroscopy that the Ce^{3+}/Ce^{4+} in nanoceria has inherent valence switching ability making it a SOD mimic (Fig. 3.1). Their subsequent study clarified that the recycling of Ce^{3+} and Ce^{4+} oxidation states and regeneration of Ce^{3+} on the surface of the nanoceria allows the SOD mimic to have long-term enzyme activity. They also prepared two components of nanoceria with high and low ratios of Ce^{3+}/Ce^{4+} to show that the high ratio of Ce^{3+}/Ce^{4+} will create more high SOD-mimicking activity [12].

It is well known that the nanomaterials can interact with the biomolecules in tissue. They can combine with these biomolecules to modify the surface and thus more easily exhibit effective catalytic activity [14, 15]. Singh and collaborators reported the interactions of nanoceria with the biological microenvironment. They demonstrated that nanoceria could interact with phosphate buffer, which is a very common biological buffer in cell culture and animal models. The complex has a concentration-dependent shift in the UV-Vis absorbance spectrum [15]. Surprisingly, if we incubate the nanoceria with >50 μM phosphate, the nanoceria loses its SOD mimicry. The pre-interaction of nanoceria with phosphate increased catalase activity in a concentration-dependent manner. Both of these observations suggested that the presence of the biologically relevant buffer and components could clearly affect the intrinsic character of the nanoceria. For other anions, researchers showed that there is no change in UV-Vis spectra pattern, no drop in SOD activity, and no increase of the catalase activity, e.g.,

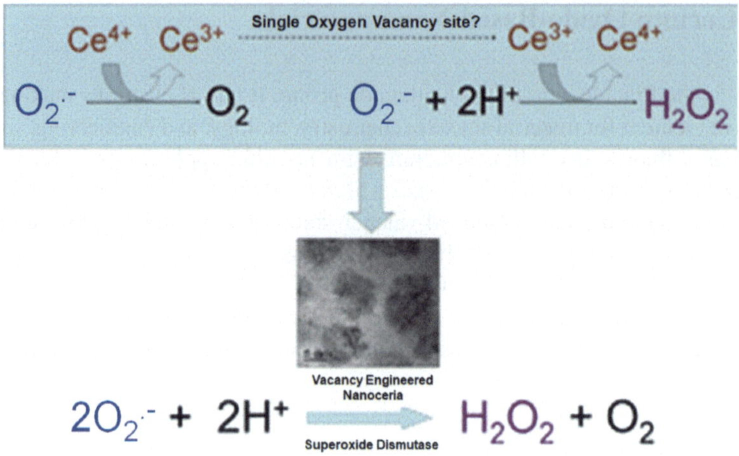

Fig. 3.1 The reaction scheme of Ce^{3+}/Ce^{4+} in nanoceria as superoxide dismutase (SOD) mimics. Reprinted with the permission from Ref. [12]. Copyright 2007 Royal Society of Chemistry

carbonate and sulfate anions. These results suggest that the interactions of nanoceria with different biological components might result in different properties [16]. Other work has demonstrated that the incubation of nanomaterials and protein complexes may lead to the formation of a protein corona on the surface of the particles. The protein molecules are strongly associated with the intrinsic properties of the nanomaterials and have a highly dynamic motion [17].

Similar to SOD, SOD mimics also play a very crucial role in redox-active process. The anti-inflammatory effects of nanoceria were reported in 2009 by Hirst and co-workers [18]. Nanoceria-based nanozymes show anti-inflammatory functions for the mixed valence and the oxygen defects making them highly effective catalysts. The cerium oxide-based nanomaterials also have been identified to have functions as antioxidants. Celardo examined the antioxidant effects and the biological antioxidant mechanisms of nanoceria via gradual doping of Sm^{3+} [19]. Another study revealed that if the nanoceria was encapsulated within a ferritin cage, then the activity of ROS scavenging of the 4.5 nm ceria will enhanced [20].

The nanoceria was also used to promote stem cell growth. Mandoli and co-workers showed that when nanoceria was fabricated together with PLGA scaffolds, the composite scaffold could promote murine-derived cardiac and mesenchymal stem cell growth [21]. Versus TiO_2 (known as directly induced cell growth), nanoceria-loaded PLGA offered higher cell proliferation activity perhaps because the Ce^{3+}/Ce^{4+} redox pair of nanoceria is recyclable, while the Ti^{3+}/Ti^{4+} redox pair of TiO_2 is not.

The nanoceria also has neuroprotective activity. The results of Chen and co-workers indicated that the accumulation of hydrogen peroxide-induced reactive oxygen intermediates could be eliminated when pretreated nanoceria was cultured with retinal neuron cells [22]. The animal study firmly suggested that the nanozymes could protect the rat retina photoreceptor cells from light-induced degeneration after intravitreal injection. The later studies showed that the neuroprotective activity of nanozymes could be realized in other systems such as adult rat spinal cord neuron and other diseases of the central nervous system [23].

3.2.2 Nanoceria as Catalase Mimic

Catalase catalyzes are found in all living organisms and are known to decompose hydrogen peroxide into molecular oxygen and water. Hydrogen peroxide is the stable product of the superoxide radicals' dismutation and plays an important role in biological systems. It is also the signaling molecule during signal transduction and a non-radical reactive oxygen species [9]. On the other hand, hydrogen peroxide can be converted into highly active and detrimental hydroxyl radicals via a Fenton chemistry reaction. This can damage the biomacromolecules. Fortunately, the natural catalase catalyzes are responsible for effective and efficient conversion of hydrogen peroxide to less active oxygen. Recently, researchers have demonstrated that many metal oxides including nanoceria have catalase activity.

Self and co-workers discovered that nanoceria exhibited a redox state-dependent biological catalase mimetic activity in 2010 and established the components of the catalase-like nanoceria at a high ratio of Ce^{4+}/Ce^{3+} (Fig. 3.2). They used several methods to monitor H_2O_2 at 240 nm. They then monitored release of oxygen via an oxygen electrode [24]. Subsequently, Celardo et al. proposed a mechanism for the nanoceria catalase mimetic: one molecule of H_2O_2 reacts with the Ce^{4+} of nanoceria

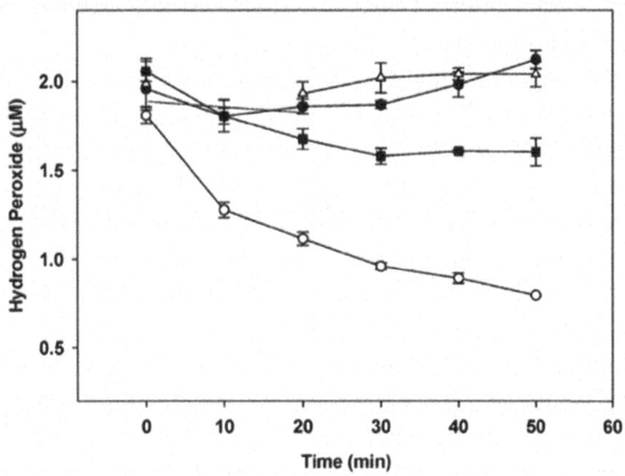

Fig. 3.2 Nanoceria exhibits catalase mimetic activity at physiologically relevant hydrogen peroxide concentrations in a redox state-dependent manner. Ceria preparation A (open triangle, 100 μM, filled circles, 300 μM), with a higher +3/+4 redox state ratio, exhibits no catalase mimetic activity at this concentration of hydrogen peroxide. Preparation B (filled squares, 100 μM and open circles, 300 μM), with relatively low +3/+4 redox state ratios, exhibits substantial catalase mimetic activity. Reprinted with the permission from Ref. [24]. Copyright 2010 Royal Society of Chemistry

to reduce it to Ce^{3+} with subsequent release of protons and oxygen; another H_2O_2 molecule reacts with Ce^{3+} of nanoceria to oxidize it back to Ce^{4+} and release water molecules [19]. The catalase-like activity of nanoceria is mainly at a low ratio of Ce^{3+}/Ce^{4+} while SOD enzymatic nanoceria was dominant at high Ce^{3+}/Ce^{4+} ratios.

Later studies showed that the catalase mimetic activity of nanoceria is very stable under exposure to different pH values (from 2 to 10) including both cell culture media and phosphate anions [14, 15]. Neither the UV-Vis spectra pattern nor the catalase-like activity changed under these conditions. The particle size and zeta potential also did not change under exposure to varying pH (up to pH 10). When the nanoceria was preincubated with phosphate (up to 100 μM concentration), there was a drastic increase in the particle size and the zeta potential was observed. The high affinity of cerium (Ce) to phosphate has been reported [25]. Based on the results obtained from in vitro studies, we know that the pH and the phosphate concentration do not have a significant effect on the free radical scavenging property of nanoceria; detailed in vivo studies were evaluated further.

3.2.3 Nanoceria as Oxidase Mimics

Oxidase enzyme catalyzes the oxidation–reduction reaction. This involves molecular oxygen, which can be converted into either water or hydrogen peroxide via donation of hydrogen atom and reduction of oxygen to O^{1-}/O^{2-}. The cytochrome C-oxidase, glucose oxidase (GOx), cytochrome-P450 oxidase, and xanthine oxidase are very important oxidases in living organisms. A color change can be observed during oxidase treatment for detection of analytes. Recent studies have demonstrated that some nanomaterials can imitate oxidase activity.

Nanoceria is an efficient antioxidant because it has SOD catalase mimicry based on the cycling of the Ce^{3+}/Ce^{4+}. Nanoceria also has oxidase-like activity, and surprisingly, the nanoceria could quickly oxidize a series of organic substrates under acidic pH and without any oxidizing agent (Fig. 3.3) [26]. This work was first reported by Asati in 2009. It showed that oxidase activity was determined by the particle size and the thickness of polymer coating on the surface of the nanoceria. In an immunoassay, nanoceria could oxidize the substrates even in the absence of the substrates. Later, Xu et al. demonstrated that the nanoceria recycles between Ce^{3+}/Ce^{4+} oxidation states under the physiological and alkaline pH conditions, and thus could maintain the catalytic property. Under acidic pH environment, the reactions of nanoceria with H^+ can produce dissoluble cerium and act as an oxidant. On the contrary, investigations made by Singh and co-workers indicated that both Ce^{3+} and Ce^{4+} do not compromise the antioxidant properties under acidic environments. This can also be studied in terms of synthesis method and temperature, impurities, lattice doping, surface modification, and particle size. These strongly influence nanoceria.

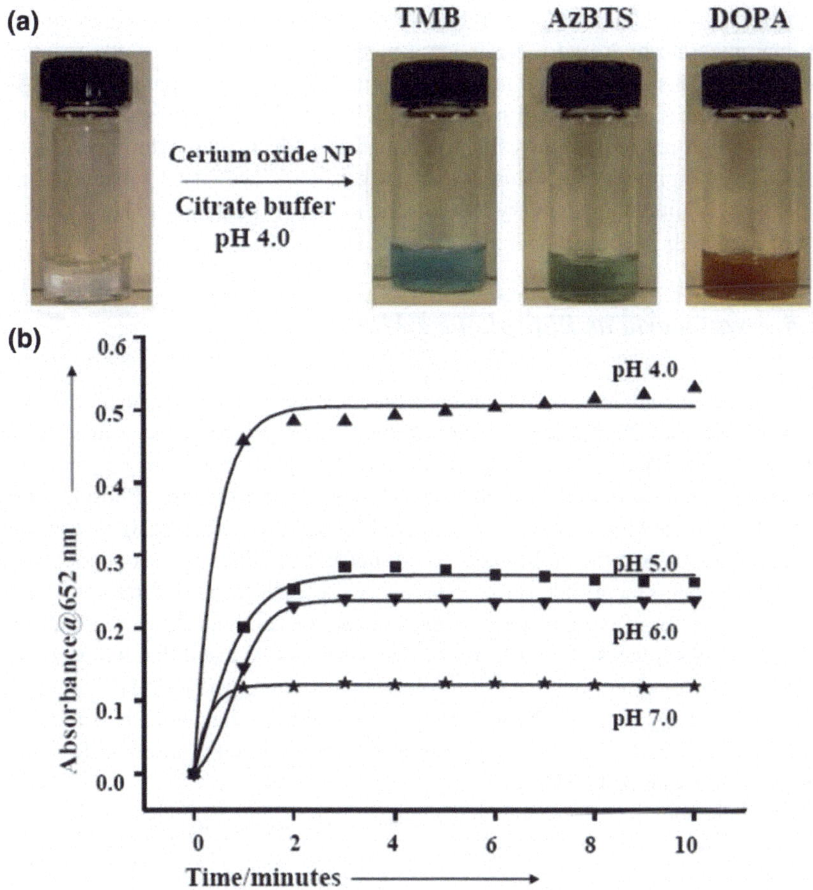

Fig. 3.3 Formation of color product due to substrate oxidation is pH-dependent. **a** Photographs show production of color product upon addition of nanoceria to TMB, AzBTS, and DOPA at pH 4.0. **b** Oxidation of TMB is pH-dependent with the optimum activity at pH 4.0. Reprinted with the permission from Ref. [26]. Copyright 2009 Angewandte Chemie-International Edition

3.2.4 Nanoceria as Peroxidase Mimetic

Peroxidase contains a heme as a cofactor in the active site or redox-active cysteine or selenium-containing cysteine residues (selenocysteine) [27]. Several types of peroxidases including glutathione peroxidase, myeloperoxidase, haloperoxidase, lactoperoxidase, peroxiredoxin, etc., are widely found in living organisms. The function of these enzymes is to convert the hydrogen peroxide to water and oxygen. Recently, some researchers have shown that some nanomaterials are stable and affordable to prepare and can also mimic peroxidase activity [28].

The transition metal ions (such as Fe) are very common in the peroxidase mimics, and they generate hydroxyl radicals via a mechanism similar to Fenton reactions. Heckert and colleagues showed that cerium ions may also represent the Fenton-like reaction and produce hydroxyl radicals [29]. Later, investigations carried out by Perez and other research groups suggested that nanoceria could exhibit peroxidase-like activity (Fig. 3.4) [30]. This colorimetric scheme was used for detection of glucose, environmental chemistry, biosensors, and nanomedicines [31].

3.2.5 Nanoceria as Phosphatase Mimic

Phosphorylation and dephosphorylation is the most common mechanism of switchable transduction. Phosphatase removes phosphate groups from substrates by hydrolyzing phosphoric acid monoesters into free phosphate ions. In the living organism, a very common phosphatase is alkaline phosphatase. Phosphatase is involved in several biological processes including cell proliferation, signal transduction, metabolism, and the cell-to-cell communications. Since the above mentioned activities had been identified, some researchers started to explore different types of enzymatic activity of the nanoceria. Among these researchers, Kuchma et al., first reported the phosphatase-like activity of nanoceria and showed that nanoceria can phosphorylate the ester bonds of p-nitrophenol phosphate (pNPP), ATP, o-phospho-L-tyrosine, and DNA [32]. The nanoceria (Ce^{3+}) did not cleave the plasmid DNA when incubated with it, while the phosphorus–oxygen bonds can be broken under treatment with nanoceria (Ce^{4+}).

3.3 Iron Oxide-Based Nanomaterials

Iron oxide nanoparticles have been well established and widely used in biological and biomedical areas. Examples include separation and capture of analytes, sensing, imaging, transfection, magnetic resonance imaging (MRI), hyperthermia therapy, and targeted drug delivery. They offer magnetism and superparamagnetism based on structure [6, 33]. They are chemically and biologically inert, and the iron is often conjugated into the metal catalysts, enzymes, or antibodies for further functionalization. It also serves as an ideal platform for diagnostic imaging and targeted therapy [34, 35], and the small peptide targets the particles for magnetic resonance imaging [36]. Several superparamagnetic iron oxide (SPIO) nanoparticles have already been approved by FDA with applications from the bench to the clinic [37, 38].

In 2007, Yan and co-workers discovered that ferromagnetic (Fe_3O_4) nanoparticles showed an intrinsic peroxidase-like activity similar to horseradish peroxidase (HRP) [39]. This was the first study to show that the inorganic nanoparticles are enzyme mimetics for biomedical applications. Later, various nanomaterials have been explored for their enzymatic activity. And today, they also presented a protocol

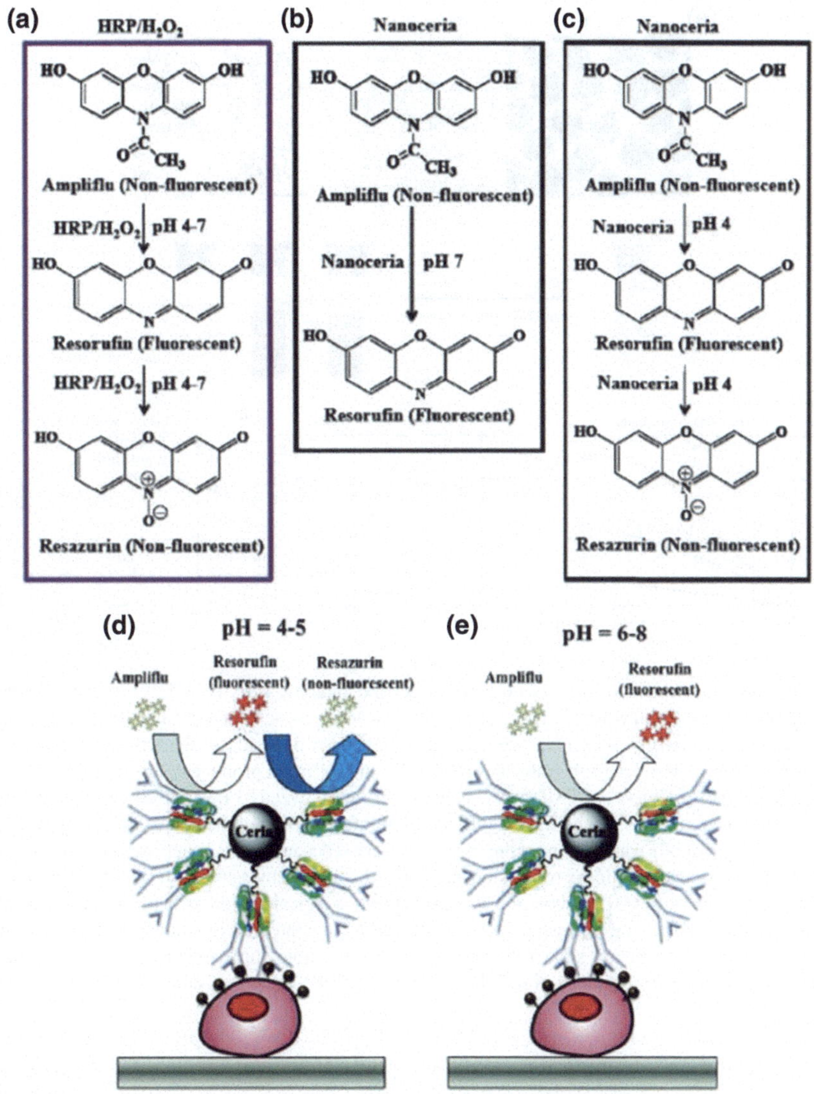

Fig. 3.4 Schematic showing the HRP/H_2O_2 and nanoceria mediated oxidation of Ampliflu. In the pH range 4–7, HRP/H_2O_2 oxidizes Ampliflu to a nonfluorescent final product (Resazurin) (**a**). In contrast, nanoceria oxidizes Ampliflu to the intermediate oxidation fluorescent product (Resorufin) at pH 7 (**b**), while at or below pH 5.0, nanoceria yields the terminal oxidized nonfluorescent product resazurin (**c**). The ability of nanoceria to oxidize Ampliflu to a stable fluorescent product in the pH range 6–8 will facilitate its use in ELISA (**d**, **e**) without the use of H_2O_2. Reprinted with the permission from Ref. [30]. Copyright 2011 American Chemical Society

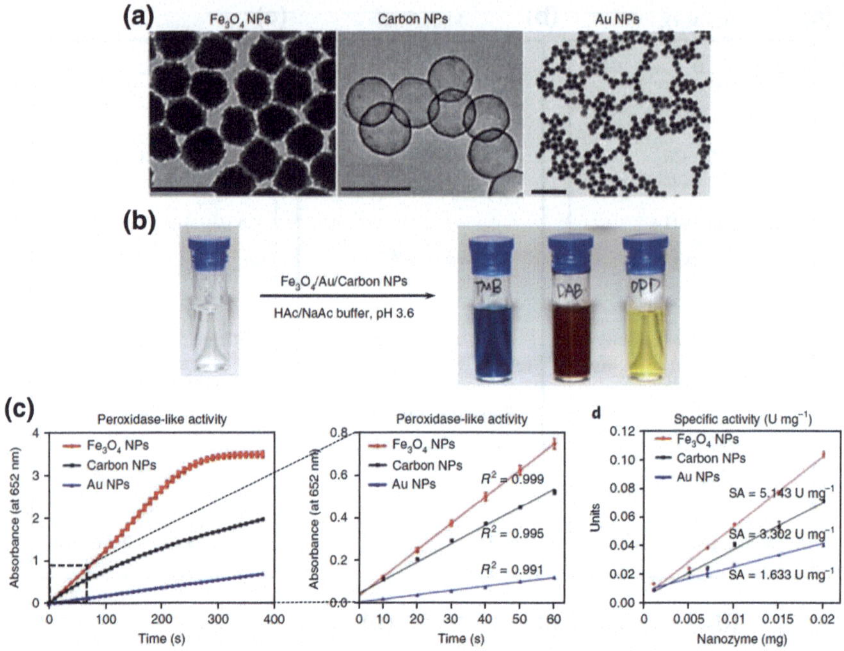

Fig. 3.5 Standardization of nanozyme peroxidase-like catalytic activity. **a** TEM images of three typical peroxidase nanozymes (Fe_3O_4, carbon, and Au NPs). Scale bars represent 200 nm, 200 nm, and 100 nm, respectively. **b** Fe_3O_4, carbon, and Au NPs show peroxidase-like activity and catalyze the oxidation of peroxidase substrates (TMB, DAB, and OPD) to produce colorimetric reactions. **c** Left, reaction–time curves of TMB colorimetric reaction catalyzed by Fe_3O_4 (red), carbon (black), and Au (blue) nanozymes. Right, the magnified initial linear portion of the nanozyme reaction–time curves. A length of 60 s was chosen for the initial rate period because the R2 coefficients were close to 1 during this period after a linear-regression analysis. Absorbance measured in arbitrary units. **d** The specific activities of Fe_3O_4, carbon and Au nanozymes were calculated as 5.143 U mg^{-1}, 3.302 U mg^{-1}, and 1.633 U mg^{-1}, respectively, using the nanozyme activity standardization method described herein. Error bars shown represent s.e. derived from three independent experiments. Reprinted with the permission from Ref. [40]. Copyright 2018 Springer Nature

for measuring and defining the catalytic activity units and kinetics for peroxidase nanozymes (Fig. 3.5) [40]. We will now explain the enzyme mimetic properties of iron oxide nanomaterials.

3.3.1 Iron Oxide as Peroxidase Mimics

Peroxidases are a large family of enzymes that catalyze the oxidation of the substrate with peroxide (hydrogen peroxide in most cases) ($AH_2 + ROOH \rightarrow A + ROH + H_2O$). The peroxidases play many crucial roles in biological systems via this catalysis: detoxifying reactive oxygen species (e.g., glutathione peroxidase) and defending against pathogens (e.g., myeloperoxidase). Peroxidase (especially HRP) is also widely used in bioanalytical and clinical chemistry. HRP is usually employed as a conjugate to an antibody for enzymatically catalyzing colorimetric substrates in signaling and imaging. Recent studies have shown that certain nanomaterials (such as iron oxide-based nanomaterials) have peroxidase catalytic activity.

Yan showed that Fe_3O_4 MNPs exhibited three different sizes (30, 50, and 300 nm) and all could oxidize TMB to the blue-colored product under treatment with H_2O_2 [39]. The other two substrates (DAB and OPD) were also oxidized to their corresponding products mimicking HRP (Fig. 3.6). The catalytic activity of the nanozyme was size-dependent—a smaller size has higher activity. Similar to native HRP, nanoparticle-based enzymes have a performance that also varies with pH and temperature. We note that the Fe_3O_4 MNPs are much more robust because they remain stable and retain their catalytic activity after incubation across a range of temperatures (4–90 °C) and pH (0–12). The robustness and the low cost of the nanozymes make them suitable for a wide range of applications. Later, Fe_3O_4 MNPs were developed as peroxidase mimics as a sensing platform for the detection of both H_2O_2 and glucose [41].

The results of these initial studies stimulated rapidly expanding interest in the use of iron oxide nanomaterial as a peroxidase mimic. To study the nanomaterials' peroxidase mimetic properties, it is common to use a substrate (such as ABTS or TMB) via the convenient colorimetric reaction in the presence of H_2O_2. In some reports, the activity of nanomaterials was compared to HRP [42, 43]. Wu et al. investigated the effects of aqueous-organic solvents on the activities of HRP and a

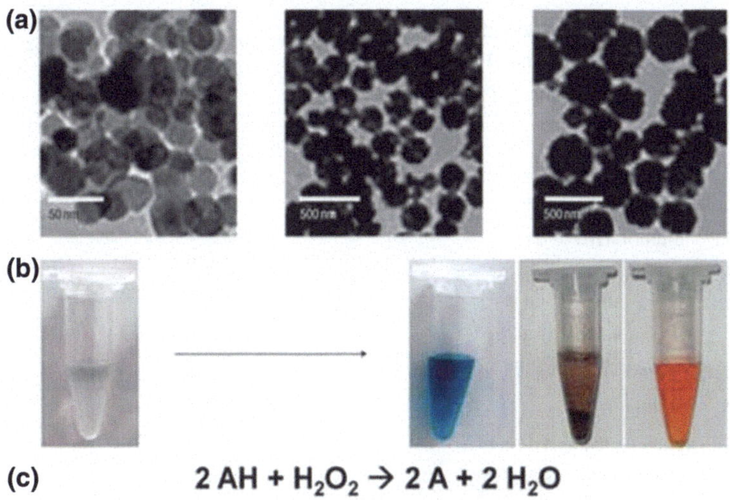

Fig. 3.6 Fe_3O_4 MNPs as nanozyme to mimic peroxidase. **a** TEM image of Fe_3O_4 MNPs with different sizes investigated. **b** The substrates, TMB, DAB, and OPD, were catalytically oxidized to form colorimetric products with Fe_3O_4 MNPs. **c** The reaction catalyzed by Fe_3O_4 MNPs. Reprinted with the permission from Ref. [39]. Copyright 2007 Springer Nature

Fe_3O_4-based nanozyme, and the results showed that the relative catalytic activities of the nanozyme under given conditions were generally better than that of HRP.

Fe_2O_3 (hematite) nanomaterials have also been studied. Magnetoferritin nanoparticles, also known as Fe_2O_3 in ferritin, are employed for targeting and visualizing tumor tissues [44]. The natural inorganic core of ferritin is hydrated iron oxide ferrihydrite, and it has very low peroxidase mimic activity. The magnetic core was post-constructed by in situ oxidation of iron ions inside the apo-ferritin. This results in magnetoferritin nanoparticles, which showed excellent peroxidase-like activity. A recent study also investigated that the Fe_3O_4 had higher peroxidase mimetic activity than Fe_2O_3 with similar size [45]. In addition, doped ferrites such as $MFeO_3$ (M = Bi, Eu) and MFe_2O_4 (M = Co, Mn, Zn) were also identified as peroxidase mimics [46–50] (Table 3.1).

Iron oxidase-based nanoparticles such as peroxidase have a wide range of applications in biological systems, clinical detection, and other areas. The concentration of H_2O_2 is important in many fields such as biology, medicine, environmental protection, and the food industry. H_2O_2 is involved in the catalytic reaction of peroxidase and its mimics, and it can measure the concentration of H_2O_2 via the color change of certain colorimetric substrates. Related work has been reported by Wei and Wang, and the results from the seminal report revealed that a colorimetric platform for H_2O_2

Table 3.1 Comparison of LOD of this work with some established methods using luminol-based CL for hydrogen peroxide and glucose. The summary data was referenced form the review of [46–50]

System	H_2O_2 (mM)	Glucose (mM)
$CoFe_2O_4$ MNPs–luminol–H_2O_2–GOx	0.01	0.024
β-CD/$CoFe_2O_4$ MNPs–luminol–H_2O_2	0.02	
Mg–Al-CO_3 LDHs–luminol–H_2O_2	0.02	
KIO_4–luminol–H_2O_2	0.03	0.44
Ferric oxide NPs–luminol–H_2O_2	1250	
Au nanoflower–luminol–H_2O_2	10	
Gold NPs–luminol–HRP–GOx		5
1-Ethyl-3-methylimidazolium ethylsulfate/Cu_{2+}–luminol–GOx	5	4
Co_3O_4 NPs–luminol–H_2O_2–GOx	0.0011	0.08
Cobalt(II)–luminol–H_2O_2	0.001	
HRP–MNP–luminol–H_2O_2	0.5	50
Luminol-capped Au NPs–H_2O_2 ECL	0.1	
Au NPs–Hb/PMMA–luminol–H_2O_2	0.2	
Pd NPs–MWCNTs–luminol–H_2O_2 ECL	0.0005	0.054
Gold NPs–GOx–HRP–luminol–H_2O_2		5
CF-$CoFe_2O_4$ NPs–luminol–H_2O_2–GOx	0.0005	0.01

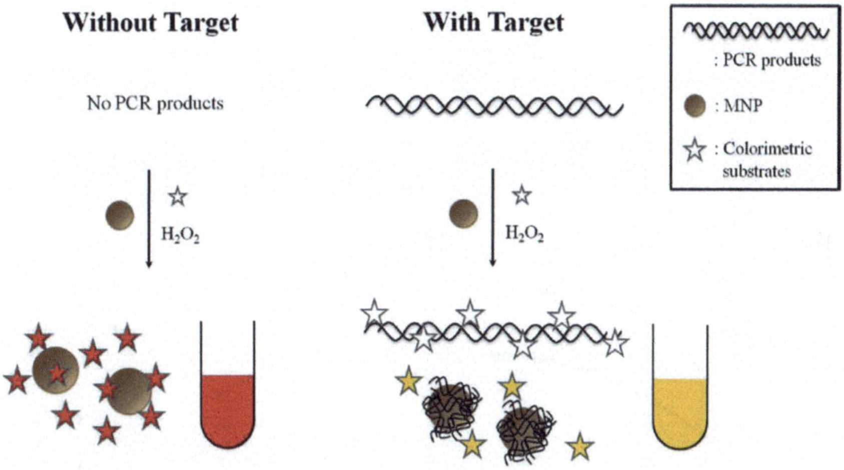

Fig. 3.7 Schematic illustration of the MNP-based, label-free, colorimetric detection of target DNA. Reprinted with the permission from Ref. [52]. Copyright 2011 John Wiley and Sons

determination was developed using ABTS as the substrate and Fe_3O_4 MNPs as the nanozyme [41].

Glucose detection is very important in clinical and food analysis [51]. Glucose oxidase (GOx) is involved in glucose detection because of its specificity and efficiency. After treatment with GOx, the products of the catalytic oxidation of glucose and molecular oxygen are H_2O_2. This can be measured with many analytical methods such as HRP-catalyzed colorimetric and electrochemical methods. Similar to H_2O_2 detection, numerous analytical approaches to glucose detection have been identified by varying the nanomaterials and substrates used. Several studies have reported novel methods for the determination of glucose concentration in real serum samples. When combined with PCR, double-stranded DNA can shield against MNPs' mimicking activity. This can detect DNA in a label-free colorimetric platform (Fig. 3.7) [52]. In addition, the peroxidase of iron-based oxidase nanoparticles can also be used in aptasensors, immunoassays, immunostaining, promotion of stem cell growth, pollutant removal, etc.

3.3.2 Iron Oxide as Both Peroxidase and Catalase Mimics

Gu et al. showed that iron oxide nanoparticles (both Fe_3O_4 and γ-Fe_2O_3) have dual enzyme mimetic properties including catalase and peroxidase [45]. The enzymatic activities were pH-dependent with catalase activity dominant at neutral conditions (pH = 7.4), and peroxidase dominant under acid conditions (pH = 4.8) (Fig. 3.8). The Fe_3O_4 MNPs and the γ-Fe_2O_3 have similar sizes and surface charges, and thus researchers have also made side-by-side comparisons of these two particles. Versus

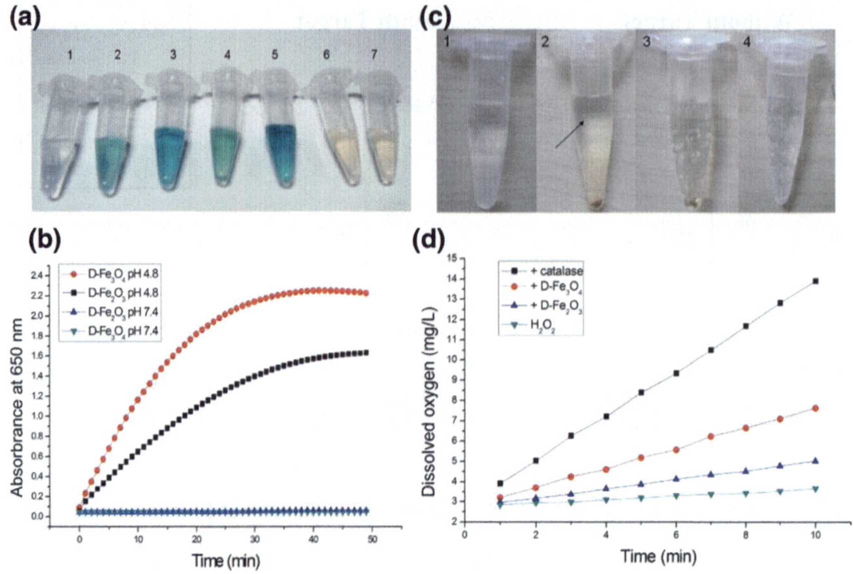

Fig. 3.8 Peroxidase-like activity and catalase-like activity of IONPs. **a**, **b** Peroxidase-like activity of IONPs. **a** Photograph of color reactions after 30 min incubation. Tubes 1–5: H_2O_2 +TMB in pH 4.8 buffer plus (1) none, (2) 10 μg/mL D-Fe_2O_3, (3) 10 μg/mL D-Fe_3O_4, (4) 20 μg/mL D-Fe_2O_3, and (5) 20 μg/mL D-Fe_3O_4. Tubes 6 and 7: H_2O_2 + TMB in pH 7.4 buffer plus (6) 20 μg/mL D-Fe_2O_3 and (7) 20 μg/mL D-Fe_3O_4. **b** UV-vis absorption–time course curves of the TMB–H_2O_2 reaction system catalyzed by 20 μg/mL D-Fe_2O_3 or D-Fe_3O_4 NPs in pH 4.8 or 7.4 buffer. **c**, **d** Catalase-like activity of IONPs. **c** Photograph of bubble reactions after 6 h incubation. Tubes 1–4: 100 mM H_2O_2 in pH 7.4 PBS buffer plus (1) none, (2) 20 μg/mL D-Fe_2O_3 (arrow indicating very small bubbles), (3) 20 μg/mL D-Fe_3O_4, and (4) 20 U/mL catalase. **d** Dissolved oxygen–time course curves of H_2O_2 in pH 7.4 buffer catalyzed by 20 μg/mL D-Fe_2O_3 or D-Fe_3O_4 NPs or 20 U/mL catalase. Reprinted with the permission from Ref. [45]. Copyright 2012 American Chemical Society

γ-Fe_2O_3, the Fe_3O_4 MNPs alone (or with H_2O_2 as peroxidase mimic) had higher cell toxicity. The γ-Fe_2O_3 was attributed to Fe_3O_4 MNPs' higher activity for both mimics. They also measured the formation of the hydroxyl radical at acidic conditions. This indicated the Fenton-like mechanism for peroxidase-mimicking activity.

Under neutral conditions, no hydroxyl radical was detected suggesting a different mechanism for the catalase-mimicking activity. Thus, they argued that at neutral or higher pH, the formation of HO_2 was faster, and it was then quickly ionized into O_2. This was further transformed into O_2 by reacting with hydroxyl radical. Thus, the cellular toxicity of the nanozymes was due to entrapment of the nanoparticles into acidic lysosomes to produce hydroxyl radical. They also suggested that the nanoparticles were trapped within the cytosol and can convert H_2O_2 into harmless products via catalase-mimicking activity.

3.3.3 *Iron Oxide as Oxidase Mimics*

Cao and Wang reported that Fe_2O_3 nanowires can also serve as a glucose sensor via an oxidase-like activity (Fig. 3.9) [53]. The nanozyme sensors were considered oxidase mimics rather than a peroxidase mimic because the Fe_2O_3 was fabricated into nanowires that mimicked the GOx behavior. This sensing system has several

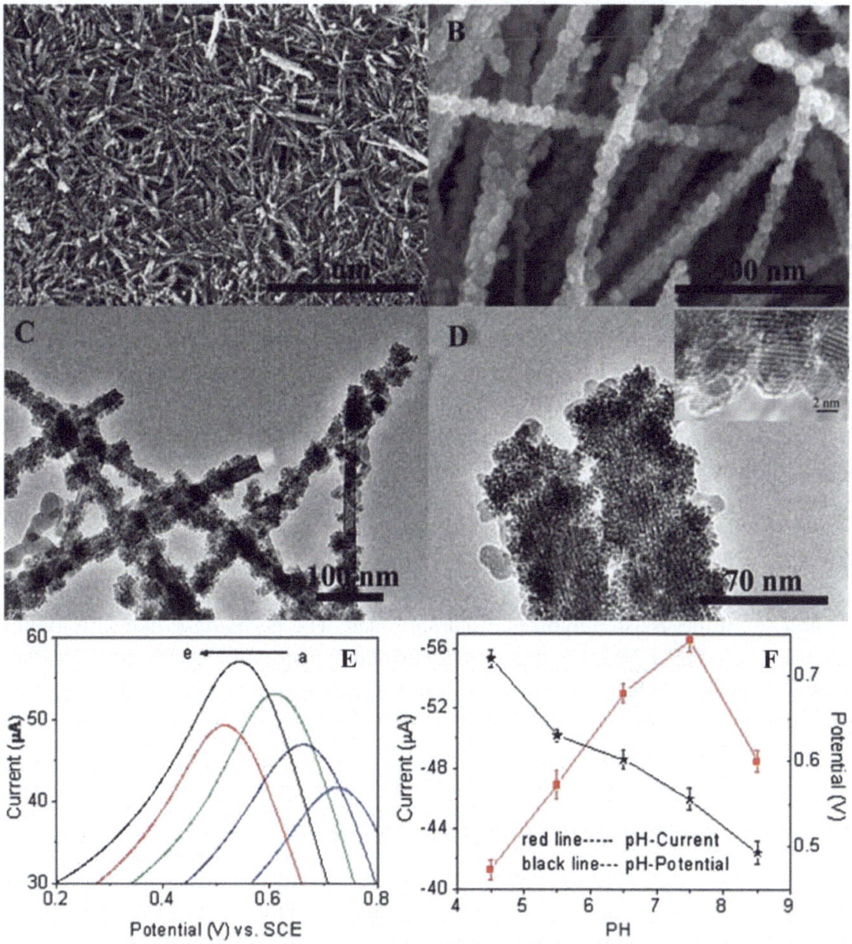

Fig. 3.9 SEM, TEM, and HRTEM (inset) images of the as-prepared Fe_2O_3 nanowire arrays (**a**, **b**, **c** and **d**). Cyclic voltammograms of 1 mM glucose at different pH values: (a) 4.5, (b) 5.5, (c) 6.5, (d) 7.5, and (e) 8.5 solutions at the n-FeGE electrode (from right to left) (**e**), and the dependencies of the glucose peak current and redox potential on the PBS solution pH (**f**) with a scanning rate of 0.1 V s^{-1}. Reprinted with the permission from Ref. [53]. Copyright 2011 Royal Society of Chemistry

properties: a linear range of 0.015–0 mM, a glucose detection limit of 6 μM, utility in serum, good reproducibility, and good storage stability.

3.4 Other Metal Oxide-Based Nanomaterials

Other metal oxide-based nanomaterials have also been explored to mimic enzymes. These include cobalt, copper, manganese, and vanadium oxide-based nanomaterials. We will describe these below.

3.4.1 Cobalt Oxide as Catalase and Peroxidase Mimics

Mu and co-workers observed that the cubic Co_3O_4 nanoparticles have dual intrinsic enzyme-mimicking activities, i.e., catalase and peroxidase-like activities [54]. They showed that the peroxidase-mimicking activity could identify the catalytic oxidation of TMB, and the results from the kinetics study indicated that when compared with HRP, this nanozyme had higher affinity toward TMB and lower affinity to H_2O_2. The potential mechanism was due to Co_3O_4 nanoparticles' ability for electron transfer between the substrate and H_2O_2 instead of the hydroxyl radical. The peroxidase-like activity was used in wastewater treatment via cobalt-doped graphitic carbon nitride ($Co–g–C_3N_4$) materials [55]. The catalase-like activity was also seen in the increased oxygen generation in the presence of H_2O_2 and the nanozyme. The catalase activity was then used as a fast sensor for H_2O_2 detection and calcium detection in milk [55]. In another work, they synthesized the gillyflower-shaped Co_3O_4 nanoparticles via an L-arginine-assisted hydrothermal approach. The peroxidase-like activity has also been tested [56]. Versus commercially available nanoparticles, the novel synthesized nanomaterials have better peroxidase-mimicking activity due to their unique gillyflower-like structures and larger specific surface area.

3.4.2 Copper Oxide as an Oxidase Mimic

Chen and co-workers reported that commercially available 30-nm CuO nanoparticles showed peroxidase-mimicking activity [57]. Versus HRP and other nanozymes, the CuO nanomaterials held the highest affinity to TMB. They then prepared the water-soluble CuO nanoparticles, and the smaller nanoparticles (6 nm) had higher affinity toward H_2O (Fig. 3.10). These had higher peroxidase-like activity than the commercial ones [58]. Another study showed that cupric oxide nanoparticle can be used as a chemiluminescent cholesterol sensor for detection via peroxidase mimicking [59, 60]. The workers also showed that the Zn–CuO material could be used as a peroxidase to detect tannic acid, tartaric acid, and ascorbic acid except for glucose

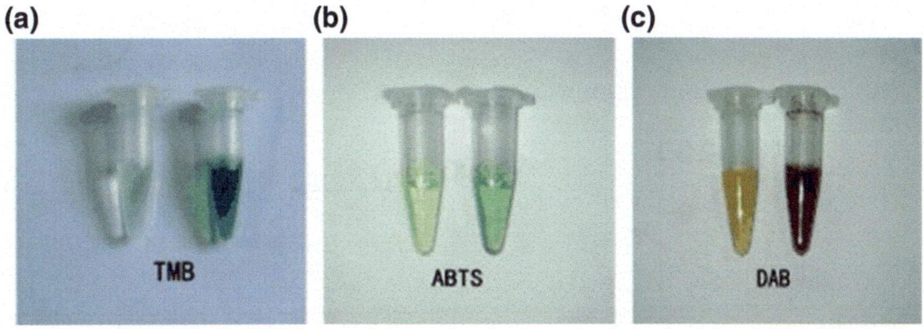

Fig. 3.10 Typical photographs of **a** TMB, **b** ABTS, and **c** DAB catalytically oxidized by the CuO NPs in the presence of H_2O_2. Reprinted with the permission from Ref. [57]. Copyright 2011 John Wiley and Sons

[61, 62]. The Zn–Cu NPs can also act like antibodies [61]. However, the Cu^{2+} is an oxidase mimic and is more efficient than Cu/Cuo nanoparticles for histidine detection even in serum. The NPs can also detect urea in the presence of O-phenylenediamine (OPD) as reported by Yan Xo and co-workers [63].

3.4.3 Manganese Dioxide as Oxidase Mimics

MnO_2 nanomaterials with different morphologies (i.e., nanosheets, nanospheres, nano-sticks, nano-complexes, and nanowires) have also been studied to determine their oxidase and peroxidase and catalase-mimicking activity [64, 65]. Of the different morphologies of the MnO_2, the MnO_2 nanowires have the highest and most stable activity and can be further used to label antibodies for immunoassay (Fig. 3.11). The workers carried out a side-by-side study to compare the performance of the new nanozyme-based immunoassay to traditional HRP-based ELISA. The result showed that both the MnO_2 nanowire-ELISA and HRP-ELISA exhibited good sensitivity and high selectivity. The MnO_2 nanowire-ELISA was more stable, more robust, and less expensive. The nanozyme oxidized the substrate without H_2O_2 eliminating the potential stability issue of H_2O_2. Another study also showed that the MnO_2 nanosheets could be used to detect glutathione via electrochemical luminescence [66].

3.4.4 Vanadium Pentoxide as Peroxidase Mimics

Tremel's group has demonstrated that the V_2O_5 nanowires had catalytic activity toward peroxidase substrates (such as ABTS and TMB) in the presence of H_2O_2 (Fig. 3.12) [67]. Versus natural vanadium-dependent haloperoxidase (V-HPO), the

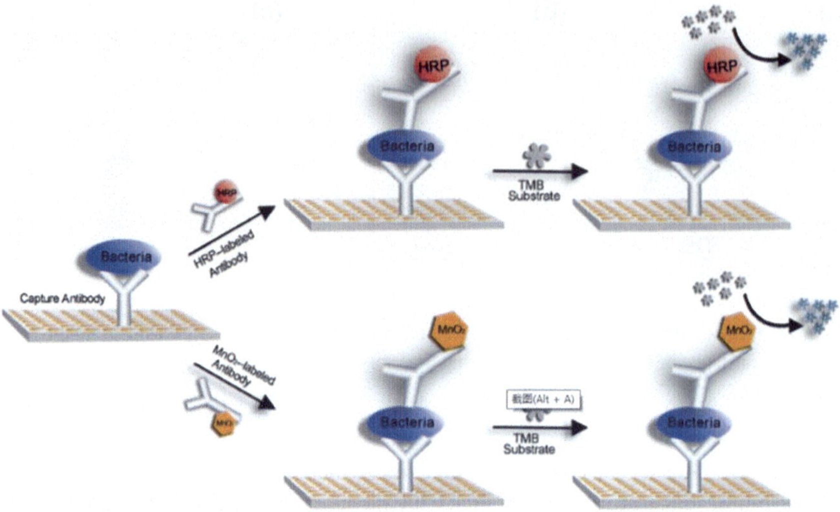

Fig. 3.11 Schematic representation of the detection method for bacteria based on the MnO_2-mediated immunoassay. HRP-mediated ELISA is presented for comparison. Reprinted with the permission from Ref. [65]. Copyright 2012 Elsevier

results showed that V_2O_5 nanowires exhibited similar kinetics but higher affinity toward both ABTS and H_2O_2. They then also declared that the V_2O_5-based nanozymes could be reused up to ten times while retaining their activity in several organic solvents tested.

In a subsequent study, Tremel and co-workers reported an excellent activity of the V_2O_5-based nanozymes in which the V-HPO-like V_2O_5 nanowires could prevent marine biofouling [68]. The excellent anti-biofouling capabilities of the V_2O_5 nanowire-based nanozymes have been demonstrated in a long-term (60 day) in situ study. The mechanism studies revealed that the HOBr and $^1O_2$ are key to anti-biofouling. In addition, the prepared functional materials also exerted a strong antibacterial activity against both Gram-negative and Gram-positive bacteria.

3.5 Metal-Based Nanomaterials

Similar to the metal oxidase-based nanomaterials, the metal-based nanomaterials can also mimic the natural enzymes. Various candidates have been reported in previous studies. In this section, we will discuss the enzymatic activity of the metal-based nanomaterials.

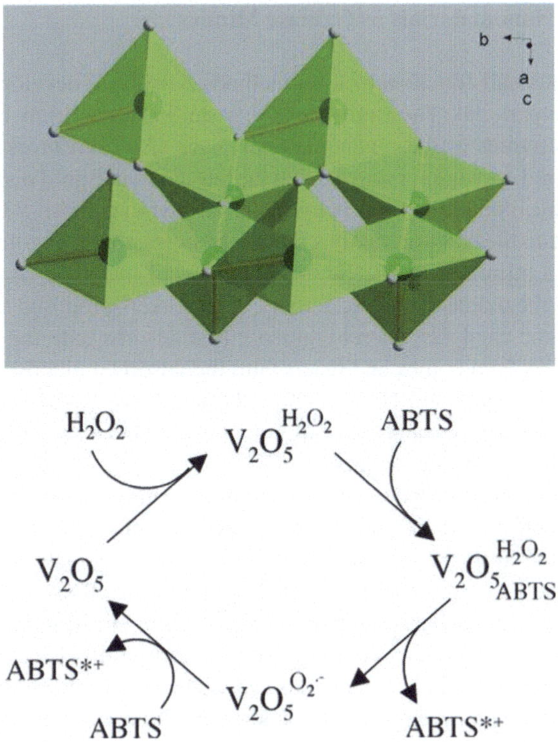

Fig. 3.12 The single-layer structure of V_2O_5 and the proposed mechanism for its peroxidase-mimicking activity. **a** Single layer from the V_2O_5 structure. In the V_2O_5 nanowires, these layers are stacked along [001], and wires extend along the [100] direction. A view of the exposed (110) planes shows a great similarity with the V–HPO active sites. **b** Proposed mechanism for the formation of ABTS with the formation of a vanadium peroxo complex intermediate and oxidation attack of ABTS and release of the product ($ABTS^{*+}$). As hydrogen peroxide is a two-electron oxidant, another molecule of ABTS is required for regeneration of the V_2O_5 nanowires leading to the formation of a new product ($ABTS^{*+}$). Reprinted with the permission from Ref. [67]. Copyright 2011 John Wiley and Sons

3.5.1 Gold Nanomaterials

Gold has been historically regarded as highly inert material with little or no reactivity. However, in recent years, nano-sized gold particles have been shown to serve as effective catalysts for a number of chemical reactions under a range of experimental conditions [69]. Later work demonstrated that gold nanoparticles can effectively catalyze the oxidation of CO at or even below room temperature [70]. Many efforts have been focused on the catalase of the gold-based nanomaterials. Thus, we will focus on the oxidase and peroxidase mimics of the gold nanomaterials.

3.5.1.1 Gold Nanomaterials as Oxidase Mimics

Routine synthesis and functionalization methods have been developed for citrate-coated gold nanoparticles. These are extensively studied for a variety of applications. The catalytic activities of these nanomaterials have also been investigated. To our surprise, Rossi and co-workers showed that the citrate-coated gold nanoparticles catalyzed the aerobic oxidation of glucose with dissolved oxygen [71–73]. The reaction was very similar to the reaction catalyzed by GOx suggesting that gold nanoparticles could serve as a mimic for GOx. Several other metal materials (Ag, Cu, Pd, and Pt) do not show the oxidase mimetic activity. The workers proposed an Eley–Rideal mechanism for the catalysis whereby glucose first adsorbs onto the gold nanoparticles. Later, an oxygen comes and reacts with the adsorbed glucose and forms the products (gluconic acids and H_2O_2). The gold nanozyme-based catalysis follows Michaelis–Menten kinetics, and the results demonstrated that the native enzyme was 55 times more active than the nanozyme.

Later these initial results led to an interesting self-catalyzed and self-limiting system for controllable growth of gold nanoparticles as reported by Fan et al. (Fig. 3.13) [74]. Based on the intrinsic oxidase activity of the gold nanomaterials, they further developed an innovative sensing strategy for DNA and micro-RNA detection [74, 75]. The growth of the nanoparticles can be tuned via the different affinities of ssDNA and dsDNA. The nanozyme activity could be used to successfully detect target DNA or microRNA because the gold nanomaterials can facilitate hybridization of the nucleic acids. We note the final output signals are adaptable. Furthermore, the colorimetric or chemiluminescent signals could be monitored under the coupled with HRP. On the other hand, without HRP, the plasmonic signals could be detected even at a single-nanoparticle level via dark field microscopy. In another work, the synergistic effects of graphene oxide–gold nanocluster (GO–AuNC) hybrid were used as an enzyme mimic to detect the cancer cells. This shows the high catalytic activity over a broad pH range—especially at neutral pH [76].

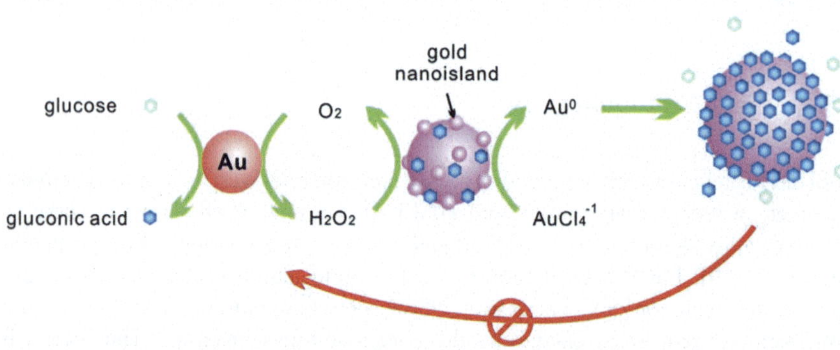

Fig. 3.13 Schematic demonstration of the AuNP-based self-limiting growth system. The size, shape, and catalytic activity of AuNPs are self-limited by the integrated influence from the catalytic reaction, seeded enlargement, and surface passivation of AuNPs. Reprinted with the permission from Ref. [74]. Copyright 2010 American Chemical Society

3.5.1.2 Gold Nanomaterials as Peroxidase Mimics

Gold nanoparticles with either positive or negative surface charges showed surprising peroxidase-mimicking activity [77, 78]. A regional study was performed by Chen and co-workers that focused on the origin of the peroxidase-like activity seen from gold nanoparticles. This explains that the enzymatic activity was indeed due to the gold nanomaterials [79]. They also examined the effect of surface modification on the activity. Here, the affinities change between the nanozymes, and the substrates could tune the enzymatic activity. Another study showed that BSA (bovine serum albumin) encapsulated in fluorescent gold nanoclusters could mimic peroxidase [80]. The gold nanomaterials were also used to fabricate Au–Pt nanorods and Au–M (M = Bi, Pd, Pt) nanostructures that represent enzymatic activity. To our surprise, versus other metals, the mercury ions could specifically enhance the peroxidase-like activity of gold nanoparticles, and a colorimetric sensor for mercury ions based on this phenomenon was developed (Fig. 3.14) [78].

3.5.2 Platinum Nanomaterials

Rather than using surface-bound ligands to drive the catalytic reactions, metallic nanoparticles can also be designed as the catalytic component of various enzyme mimics. As mentioned above, gold nanoparticles can mimic oxidase and peroxidase activity. Similar to nano-size gold particles, nano-platinum can also have the enzymatic activity.

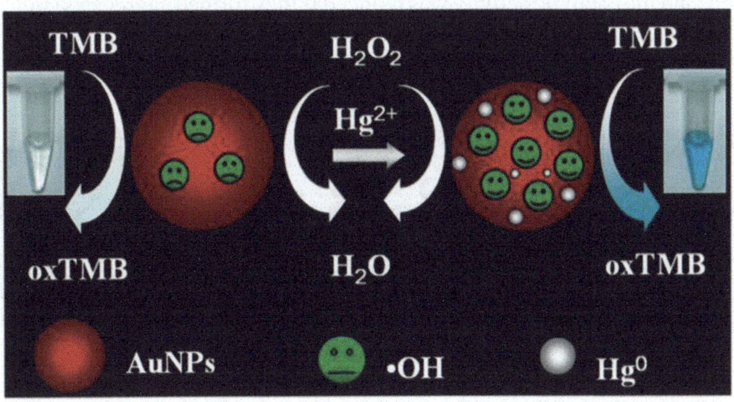

Fig. 3.14 The mercury enhanced peroxidase-like activity of citrate capped AuNPs. Hg^{2+} ions are reduced by citrate and form Hg^0 on the surface of AuNPs, changing the properties of the surface of AuNPs and stimulating the AuNPs-based catalytic oxidation of TMB with H_2O_2. Reprinted with the permission from Ref. [78]. Copyright 2011 Royal Society of Chemistry

3.5.2.1 Platinum Nanomaterials as SOD Mimics

In 2010, Zhang and co-workers synthesized platinum nanoparticles encapsulated by apo-ferritin (PtNP@apo-ferritin). They then tested their ability to detoxify reactive oxygen species [81]. The results showed that the PtNP@apo-ferritin particles offered long-term stability and good SOD-like activity in vitro. Similar to the uptake of ferritin, this mechanistic study also demonstrated that the PtNP@apo-ferritin were uptake by cells via the ferritin receptor-mediated process. Under externally induced stress, the uptake of the PtNP@apo-ferritin will lead to higher cell viability. Another study suggested that the SOD-like activity of PtNP@apo-ferritin was lower than ceria on a weight basis [82].

3.5.2.2 Platinum Nanomaterials as Catalase and/or Peroxidase Mimics

A study from Nie's group prepared 1–2 nm platinum nanoparticles within apo-ferritin (PtNP@apo-ferritin) with high stability (Fig. 3.15) [83]. Surprisingly, the PtNP@apo-ferritin exhibited dual enzyme mimic behaviors (catalase and peroxidase). Both activities are pH- and temperature-dependent. The results revealed that the catalase-like activity was enhanced by increasing the pH and temperature while the peroxidase-like activity had a maximum value at physiological temperature and slightly acidic conditions. The nanoparticles represented a higher catalytic activity under higher Pt content. Both reports suggested that PtNP@apo-ferritin should be investigated further for triple enzyme-mimicking [81, 83]. Another work synthesized 10-nm Pt nanocubes stabilized by cetyltrimethylammonium bromide (CTAB). These exhibited peroxidase activity [84]. An Eley–Rideal mechanism for the catalysis was proposed, and the catalytic activity was reduced if they formed aggregates (Fig. 3.16). This was likely due to the decreased surface area [84].

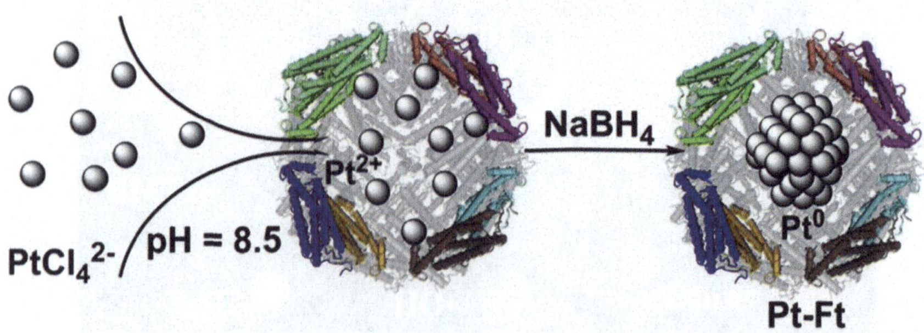

Fig. 3.15 Pt nanoparticles encapsulated within ferritin as dual enzyme mimics. Reprinted with the permission from Ref. [83]. Copyright 2011 Elsevier

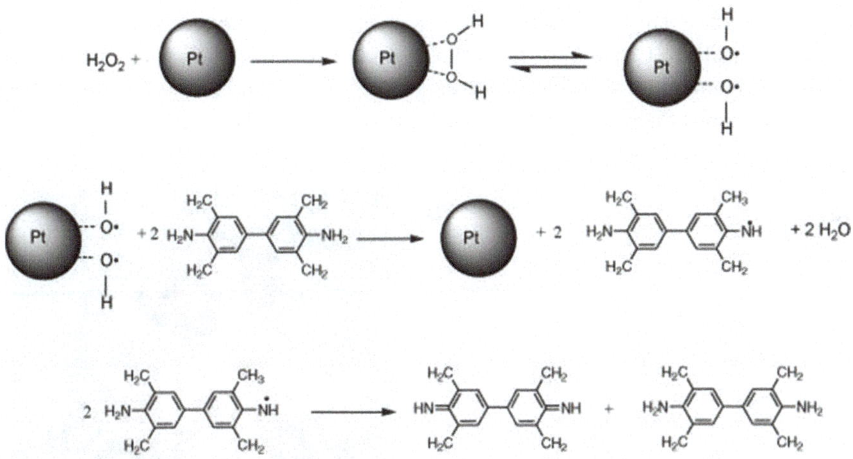

Fig. 3.16 Possible mechanism for the TMB–H_2O_2–Pt colloid system as the peroxidase mimics. Reprinted with the permission from Ref. [84]. Copyright 2011 Elsevier

3.5.3 Other Metal Nanomaterials

Au@M and AgM (Bi, Pd, and Pt) can also have enzymatic activity. Chang and co-workers demonstrated that bismuth–gold nanoparticles have peroxidase-like activity (Fig. 3.17) [85]. Wu, Yin, and co-workers extended the prior studies to bimetal nanoparticles, i.e., Au@Pt nanorods [86–89]. The Au@Pt nanorods have multiple enzyme mimetic capabilities. The oxidase and peroxidase mimetic properties were confirmed via OPD and TMB in the absence and presence of hydrogen peroxide,

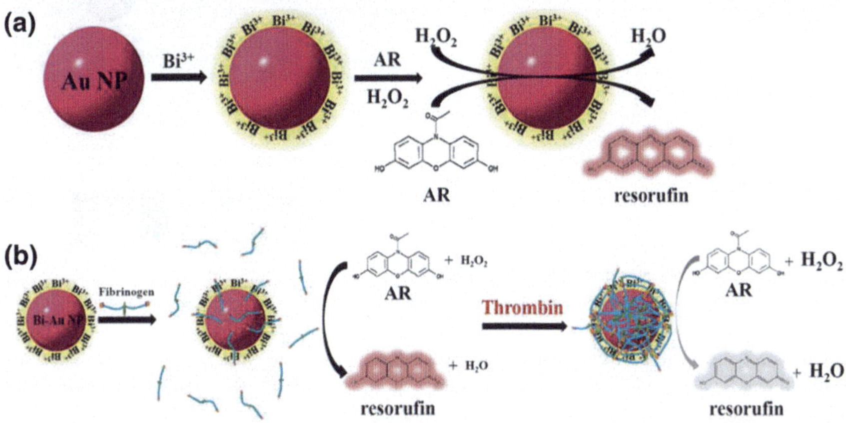

Fig. 3.17 Schematic drawing of **a** the preparation and catalyzing mechanism of the Bi–Au NPs and **b** preparation of the Fib-Bi–Au NP probe for detecting thrombin. Reprinted with the permission from Ref. [85]. Copyright 2012 Royal Society of Chemistry

respectively [86]. The catalase-like activity was further confirmed by EPR spectroscopic studies [86]. Other work from Wu, Yin, and co-workers showed that several silver-based bimetallic alloy nanostructures (AgM, M = Au, Pd, and Pt) can oxidize colorimetric substrates to the corresponding products with H_2O_2 [90].

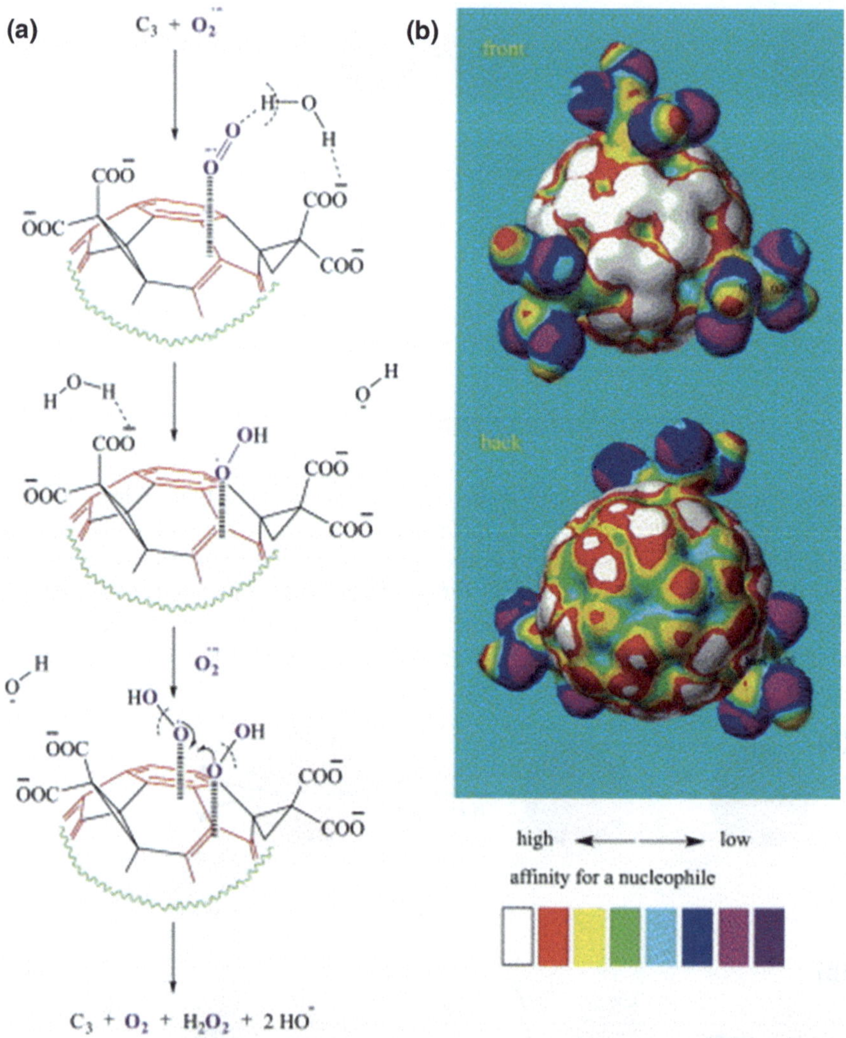

Fig. 3.18 **a** $C_{60}[C(COOH)_2]_3$ with C_3 symmetry (C_{60}–C_3) as nanozyme to mimic SOD and **b** the proposed catalytic mechanism. Reprinted with the permission from Ref. [103]. Copyright 2004 Elsevier

3.6 Carbon-Based Nanomaterials

Carbon-based nanomaterials including fullerene, carbon nanotubes (CNTs), and graphene and C-Dots (carbon nanodots, graphene quantum dots) are increasingly used for a diverse range of catalytic applications including as enzyme mimics [44, 91–99]. These carbon nanomaterials have also received considerable attention from the nanozyme community, and various studies have reported on their ability to mimic the activity of the natural enzymes.

3.6.1 *Fullerene and Derivatives as SOD Mimics*

Fullerene and its derivatives have attracted considerable attention in many research fields [100]. Based on its unique chemical reactivity toward radicals, early studies showed that [60] fullerene (i.e., C_{60}) could be used as a radical sponge [101]. Fullerenes without any modifications are insoluble in water, which makes them unlikely to interact with biomolecules in aqueous solution. Later work has developed water-soluble fullerene derivatives via chemical modification strategies. This offers several intriguing features such as interactions with biomolecules and enzyme-like activities.

In the first study, workers synthesized two polyhydroxylated fullerenes ($C_{60}(OH)_{12}$ and $C_{60}(OH)_nO_m$, n = 18–20, m = 3–7 hemiketal groups). These offer neuroprotective effects [102]. They also found that both of these derivatives retained the free radical scavenging capability of native C60 and could reduce the excitotoxic and apoptotic death of cultured cortical neurons. The mechanistic study revealed that the fullerenes could accept multiple free radicals and thus detoxify them and prevent adduct formation. The resulting adducts were diamagnetic (i.e., not radical species) and harmless. In this study, the researchers did not note SOD mimetic activity.

Later, several studies observed relatively consistent results, which showed that the fullerene derivatives (tris-malonic acid derivatives of C_{60} ($C_{60}[C(COOH)_2]_3$) could realize neuroprotective antioxidants both in vitro and in vivo [103–105]. Moreover, some detailed results confirmed the SOD mimetic property of C_{60}–C_3 via EPR and other techniques [103]. They also demonstrated that the mechanism of the C_{60}–C_3's SOD mimetic activity was the catalytic dismutation of superoxide rather than stoichiometric scavenging. This mechanism was experimentally validated by several observations such as the lack of structural modifications to C_{60}–C_3, the absence of detectable paramagnetic products, generation of hydrogen peroxide, and regeneration of oxygen. The semiempirical quantum mechanical calculations then revealed that the electron-deficient regions on C_{60}–C_3 electrostatically drew the substrate anions toward the C_{60}–C_3 surface and directed them for further dismutation. They later built a relationship between neuroprotection by carboxyl fullerenes (Fig. 3.19) and their affinity toward superoxide radicals by detailed structure-activity study [106]. Kinetic

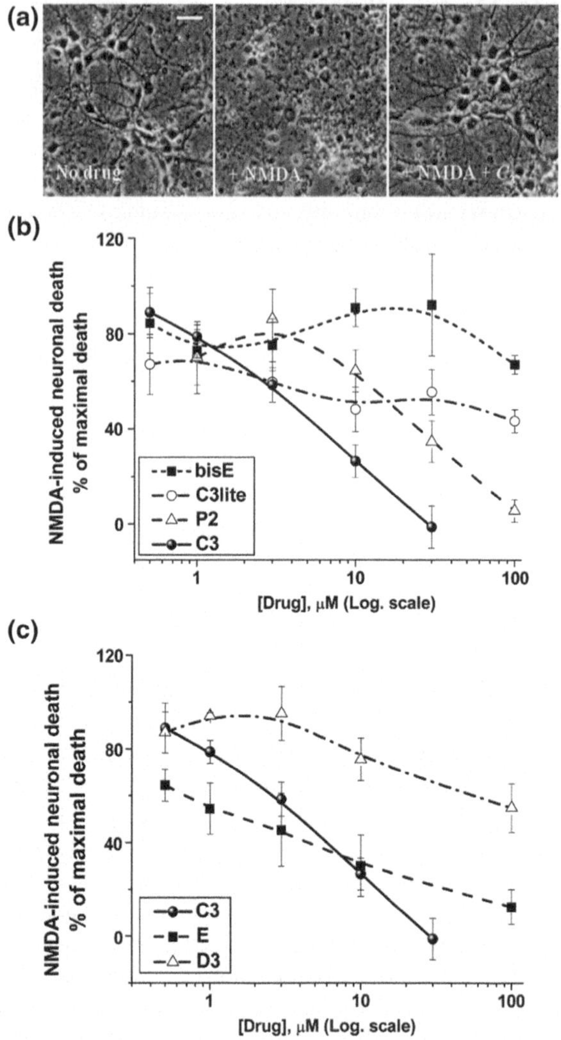

Fig. 3.19 Neuroprotection by carboxyfullerenes in NMDA receptor–mediated excitotoxic injury. Neurotoxicity produced by exposure to NMDA is reduced by co-application of the carboxyfullerenes. Data are presented as the percentage of cell death (LDH release) produced by NMDA alone, mean ± SEM, $n > 12$. **a** Phase-contrast images of cultured cortical neuronal 24 h after drug exposure; control (left), NMDA (middle), NMDA plus 30 μM C_3 (right). **b** Increasing the number of carboxylic group on the C_{60} surface increases neuronal survival, with the exception of the bisE compound (reviewed in more detail in the Discussion). **c** Neuronal protection provided by carboxyfullerenes that possess the same number of carboxylic groups but differ in their distribution over the C_{60} sphere. The white bar is 25 μm. Reprinted with the permission from Ref. [106]. Copyright 2008 Elsevier

studies were carried out by Ali and co-workers and compared the C_{60}–C_3's performance with native and other artificial enzymes (Fig. 3.18) [103]. In subsequent study, Gozin and co-workers showed that the C_{60}–C_3 nanozyme could form a non-covalent complex with human serum albumin with a binding constant of $1.2 * 10^7\ M^{-1}$ [107]. Fullerene and its derivatives based on nanoparticle SOD mimics offer protection in many redox-active biological processes including neuroprotection and antiaging [103–105, 108].

3.6.2 CNTs, Graphene, and Derivatives as Peroxidase Mimics

Recently, several groups have reported the peroxidase-like activity of CNTs (carbon nanotubes) [93, 109]. In 2010, Qu et al. first reported that the single-walled carbon nanotubes (SWCNTs) have peroxidase-like activity. The activity was pH-, temperature-, and H_2O_2-dependent [93]. If we remove the cobalt from the SWCNTs via sonication in concentrated sulfuric and nitric acids, then the SWNTs have similar peroxidase activity. This confirmed that the SWCNTs, rather than the trace amounts of metal, facilitate the catalytic activity [93]. Similar to Zhu's study, the Fe content did affect the activity of the helical CNTs [109]. The results showed that a higher Fe content offered better catalytic activity. They also showed that the helical CNTs had better catalytic activity than the multi-walled carbon nanotubes (MWCNTs). They did not compare the enzymatic activity between the helical CNTs and SWCNTs. The kinetic studies demonstrated that the helical CNTs have a higher affinity to TMB and a lower affinity to H_2O_2 than HRP [109]. The researchers also noted that the helical CNTs exhibited better affinity to H_2O_2 than Fe_3O_4 NPs.

Garg, Qu, Song, and co-workers revealed that the graphene also represented the peroxidase-mimicking activity [44, 94, 98]. Given that graphene oxide nanoparticles had a very high surface-to-volume ratio and a high affinity toward organic substrates, the graphene oxide had a higher affinity to TMB and a lower affinity to H_2O_2 versus HRP (Fig. 3.20).

In 2011, a functionalized graphene with hemin was first synthesized by Dong and co-workers and was reported to have peroxidase activity [110]. The functional hybrid nanosheets formed p-p interactions between graphene and hemin and represented the intrinsic peroxidase-like activity (Fig. 3.21). This was mainly due to hemin. These peroxidase-like activities were confirmed by the conversion of TMB, ABTS, and OPD to their colored product. The kinetics studies revealed that—consistent with the HRP and other carbon-based nanozymes—the graphene/hemin-based nanozymes have a ping-pong mechanism. In contrast to CNT and graphene oxide materials, the graphene/hemin-based nanozyme have a lower affinity to TMB but a higher affinity to H_2O_2 [111]. A later study from Quan and Wang demonstrated that the graphene could form a complex with other nanomaterials. This complex has a higher peroxidase activity with synergetic effects [112–114].

Other forms of carbon-based nanomaterials also have enzymatic activity. Carbon nanodots (C-Dots), graphene quantum dots (GQDs), doped GQDs/C-Dots, GQDs/C-dots conjugates and/or nanocomposites, and carbon nitride dots (CN-Dots) can mimic peroxidase activity [91, 115–117]. The peroxidase activity of the carbon-based nanomaterials have a wide range of applications for H_2O_2 and glucose detection, metal ion detection, DNA detection (Fig. 3.20), aptasensors, and immunoassays [52, 93, 112, 118].

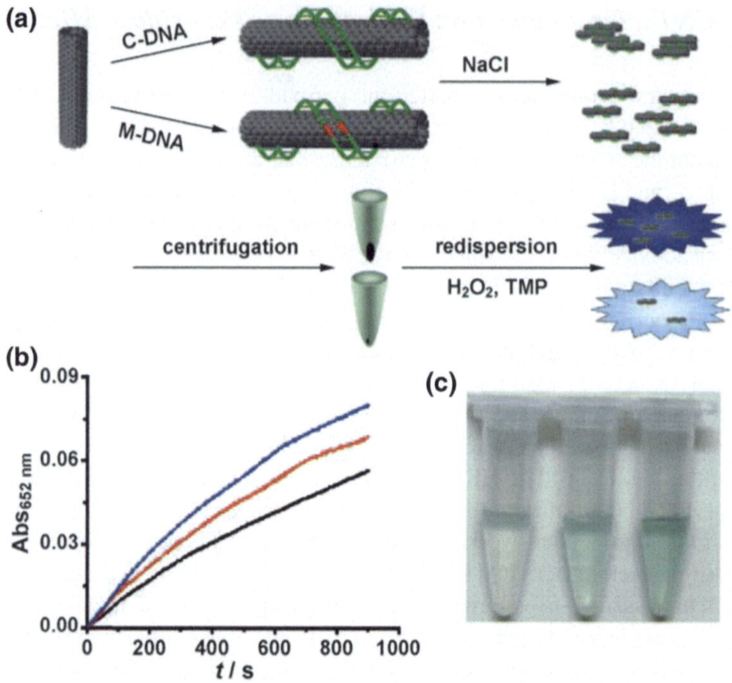

Fig. 3.20 **a** Protocol for SNP detection of complementary (C-DNA) and mismatched (M-DNA) duplex DNA. **b** The absorbance changes of the solution without (black line) or with 50 nm 15b (red line) and 50 nm 15B (blue line) over 15 min. The experimental errors were within 12%. **c** Distinguishable color difference of the corresponding solutions: without target DNA (left, colorless), with 200 nm 15b (middle, light blue), and 200 nm 15B (right, dark blue). Reprinted with the permission from Ref. [93]. Copyright 2010 John Wiley and Sons

3.7 Other Nanomaterials

Nanozymes can be divided into four types: metal-, metal oxide-, carbon-based, and others. Here, we describe some nontraditional tools.

3.7.1 Other Iron-Based Nanomaterials as Peroxidase Mimics

Other iron-based nanomaterials have also received considerable attention for their peroxidase-mimicking capability. In 2009, Ju et al. synthesized FeS nanosheets via a micelle-assisted approach and identified their peroxidase mimetic activities [119] (Fig. 3.22). Because of their large specific area, these FeS nanosheets had better peroxidase activity than spherical FeS nanomaterials. These FeS nanosheets were used for an amperometric sensor to detect H_2O_2. Later, two groups reported that

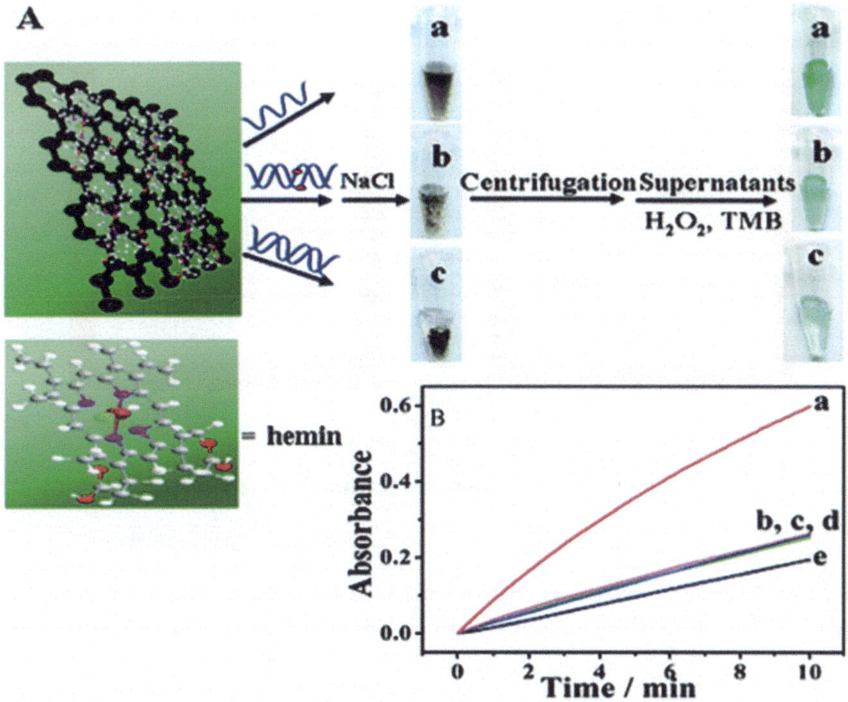

Fig. 3.21 The wet-chemical strategy for synthesizing hemin–graphene hybrid nanosheets (H-GNs) through the $\pi-\pi$ interactions to mimic the peroxidase activity. Reprinted with the permission from Ref. [110]. Copyright 2011 American Chemical Society

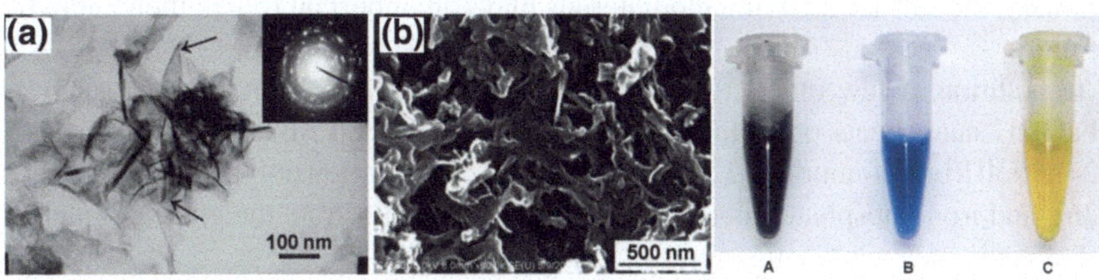

Fig. 3.22 FeS nanosheets as nanozyme to mimic peroxidase. The nanosheets were characterized by TEM and SEM imaging, and the peroxidase-like activity was confirmed through oxidation of TMB [images of the suspension of sheet-like FeS nanostructure (**a**), mixture of TMB and H_2O_2 after catalytic reaction by sheet-like FeS nanostructure (**b**), and mixture of TMB and H_2O_2 after adding H_2SO_4 to quench the catalytic reaction by sheet-like FeS nanostructure (**c**)]. Reprinted with the permission from Ref. [119]. Copyright 2009 John Wiley and Sons

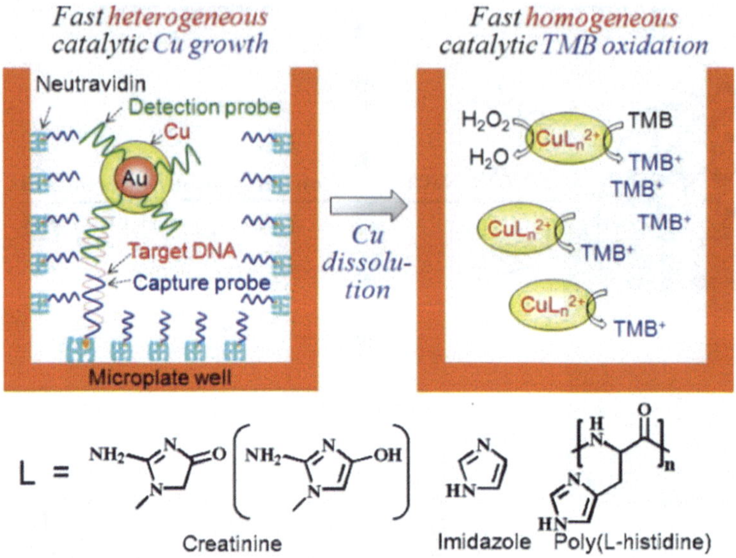

Fig. 3.23 Schematic representation of DNA detection. DNA detection using a peroxidase-like copper–creatinine complex. Reprinted with the permission from Ref. [134]. Copyright 2011 Elsevier

the FeS had peroxidase-like activity [120, 121]. Another study revealed that the FeS nano-needles had better activity than the FeS spherical nanoparticles [120]. A later study prepared FeTe nanorods and showed that they have higher activity than the Fe_3O_4 MNPs because the FeTe nanorods have a larger surface area [122]. In summary, the surface area of nanomaterials plays an important role in their catalytic activities.

In addition, a few other iron-based nanomaterials have been reported such as g-FeOOH nanosheets on graphene [123], iron-substituted SBA-15 microparticles [124], [FeIII(biuret-amide)] [125], Fe(III)-based coordination polymer nanoparticles [126], and iron phosphate microflowers [127]. These illustrate the growing interest and effort in new nanozyme mimics.

3.7.2 *Other Nanomaterials as Peroxidase Mimics*

Nanostructured layered double hydroxide (LDH) has peroxidase-mimicking activities. It acts as an electrochemical and colorimetric sensor [128–130]. Recently, both polyoxometalate and carboxyl-functionalized mesoporous polymers showed good peroxidase-mimicking activities [131, 132]. The CuS was synthesized using a solvothermal approach. The sample had a 2-fold higher peroxidase mimetic activity than CuS microspheres [133].

The enzymatic activities of most nanomaterials are confirmed via a colorimetric substrates in the presence of hydrogen peroxide. Usually, the potential applications of these nanozymes were hydrogen peroxide and/or glucose detection. DNA detection was proposed as a different approach for peroxidase-like copper–creatinine complexes (Fig. 3.23) [134]. The results showed that the nanozymes interacted with target DNA and formed a sandwich structure. This was then grown on a copper shell on the DNA–gold nanoparticle conjugate probes. The complex structure then released free copper ions, and the copper is complexed with creatinine to form a nanozyme that generates colorimetric signals.

Acknowledgements This work was supported by the National Key R&D Program of China (2017YFA0205501).

References

1. Gao L, Yan X (2016) Nanozymes: an emerging field bridging nanotechnology and biology. Sci China Life Sci 59(4):400–402
2. Walters DE (2006) Artificial enzymes. In: Breslow R (ed). Wiley-VCH Verlag, Weinheim, Germany, xii+181, pp 17.5 × 25 cm (2005). ISBN 3-527-31165-3. $99.95; J Med Chem 49(8):2668–2668
3. Breslow R, Overman LE (1970) An "artificial enzyme" combining a metal catalytic group and a hydrophobic binding cavity. J Am Chem Soc 92(4):1075–1077
4. Pollack RM (2009) From enzyme models to model enzymes. J Am Chem Soc 11(4):581–582
5. Liang M, Yan X (2019) Nanozymes: from new concepts, mechanisms, and standards to applications. Acc Chem Res. https://doi.org/10.1021/acs.accounts.9b00140
6. Wei H, Wang E (2013) Nanomaterials with enzyme-like characteristics (nanozymes): next-generation artificial enzymes. Chem Soc Rev 42(14):6060
7. Walkey C, Das S, Seal S, Erlichman J, Heckman K, Ghibelli L, Traversa E, McGinnis JF, Self WT (2015) Catalytic properties and biomedical applications of cerium oxide nanoparticles. Environ Sci Nano 2(1):33–53
8. Wason MS, Zhao J (2013) Cerium oxide nanoparticles: potential applications for cancer and other diseases. Am J Trans Res 5(2):126–131
9. Karakoti A, Singh S, Dowding JM, Seal S, Self WT (2010) Redox-active radical scavenging nanomaterials. Chem Soc Rev 39(11):4422–4432
10. Celardo I, Pedersen JZ, Traversa E, Ghibelli L (2011) Pharmacological potential of cerium oxide nanoparticles. Nanoscale 3(4):1411–1420
11. Kuciel R, Mazurkiewicz A (2004). Formation and detoxification of reactive oxygen species. Biochem Mol Biol Educ. A bimonthly publication of the International Union of Biochemistry and Molecular Biology 32(3):183–186
12. Korsvik C, Patil S, Seal S, Self WT (2007) Superoxide dismutase mimetic properties exhibited by vacancy engineered ceria nanoparticles. Chem Commun 10:1056–1058
13. Karakoti AS, Singh S, Kumar A, Malinska M, Kuchibhatla SVNT, Wozniak K, Self WT, Seal S (2009) PEGylated nanoceria as radical scavenger with tunable redox chemistry. J Am Chem Soc 131(40):14144–14145
14. Singh R, Singh S (2015) Role of phosphate on stability and catalase mimetic activity of cerium oxide nanoparticles. Colloid Surf B 132:78–84
15. Singh S, Dosani T, Karakoti AS, Kumar A, Seal S, Self WT (2011) A phosphate-dependent shift in redox state of cerium oxide nanoparticles and its effects on catalytic properties. Biomaterials 32(28):6745–6753

16. Kumar A, Das S, Munusamy P, Self W, Baer DR, Sayle DC, Seal S (2014) Behavior of nanoceria in biologically-relevant environments. Environ Sci Nano 1(6):516–532
17. Grafe C, Weidner A, von der Luhe M, Bergemann C, Schacher FH, Clement JH, Dutz S (2016) Intentional formation of a protein corona on nanoparticles: serum concentration affects protein corona mass, surface charge, and nanoparticle-cell interaction. Int J Biochem Cell B 75:196–202
18. Hirst SM, Karakoti AS, Tyler RD, Sriranganathan N, Seal S, Reilly CM (2009) Anti-inflammatory properties of cerium oxide nanoparticles. Small 5(24):2848–2856
19. Celardo I, De Nicola M, Mandoli C, Pedersen JZ, Traversa E, Ghibelli L (2011) Ce^{3+} ions determine redox-dependent anti-apoptotic effect of cerium oxide nanoparticles. ACS Nano 5(6):4537–4549
20. Liu XY, Wei W, Yuan Q, Zhang X, Li N, Du YG, Ma GH, Yan CH, Ma D (2012) Apoferritin-CeO2 nano-truffle that has excellent artificial redox enzyme activity. Chem Commun 48(26):3155–3157
21. Mandoli C, Pagliari F, Pagliari S, Forte G, Di Nardo P, Licoccia S, Traversa E (2010) Stem cell aligned growth induced by CeO_2 nanoparticles in PLGA scaffolds with improved bioactivity for regenerative medicine. Adv Func Mater 20(10):1617–1624
22. Chen JP, Patil S, Seal S, McGinnis JF (2006) Rare earth nanoparticles prevent retinal degeneration induced by intracellular peroxides. Nat Nanotechnol 1(2):142–150
23. Das M, Patil S, Bhargava N, Kang JF, Riedel LM, Seal S, Hickman JJ (2007) Auto-catalytic ceria nanoparticles offer neuroprotection to adult rat spinal cord neurons. Biomaterials 28(10):1918–1925
24. Pirmohamed T, Dowding JM, Singh S, Wasserman B, Heckert E, Karakoti AS, King JE, Seal S, Self WT (2010) Nanoceria exhibit redox state-dependent catalase mimetic activity. Chem Commun 46(16):2736–2738
25. Robinson JM, Karnovsky MJ (1983) Ultrastructural-localization of several phosphatases with cerium. J Histochem Cytochem 31(10):1197–1208
26. Asati A, Santra S, Kaittanis C, Nath S, Perez JM (2009) Oxidase-like activity of polymer-coated cerium oxide nanoparticles. Angew Chem Int Edit 48(13):2308–2312
27. van Bloois E, Pazmino DET, Winter RT, Fraaije MW (2010) A robust and extracellular heme-containing peroxidase from Thermobifida fusca as prototype of a bacterial peroxidase superfamily. Appl Microbiol Biot 86(5):1419–1430
28. He WW, Wamer W, Xia QS, Yin JJ, Fu PP (2014) Enzyme-like activity of nanomaterials. J Environ Sci Heal C 32(2):186–211
29. Heckert EG, Seal S, Self WT (2008) Fenton-like reaction catalyzed by the rare earth inner transition metal cerium. Environ Sci Technol 42(13):5014–5019
30. Asati A, Kaittanis C, Santra S, Perez JM (2011) pH-tunable oxidase-like activity of cerium oxide nanoparticles achieving sensitive fluorigenic detection of cancer biomarkers at neutral pH. Anal Chem 83(7):2547–2553
31. Celardo I, Traversa E, Ghibelli L (2011) Cerium oxide nanoparticles: a promise for applications in therapy. J Exp Ther Oncol 9(1):47–51
32. Kuchma MH, Komanski CB, Colon J, Teblum A, Masunov AE, Alvarado B, Babu S, Seal S, Summy J, Baker CH (2010) Phosphate ester hydrolysis of biologically relevant molecules by cerium oxide nanoparticles. Nanomed-Nanotechnol 6(6):738–744
33. Gao L, Fan K, Yan X (2017) Iron oxide nanozyme: a multifunctional enzyme mimetic for biomedical applications. Theranostics 7(13):3207–3227
34. Ho D, Sun XL, Sun SH (2011) Monodisperse magnetic nanoparticles for theranostic applications. Acc Chem Res 44(10):875–882
35. Lacroix LM, Ho D, Sun SH (2010) Magnetic nanoparticles as both imaging probes and therapeutic agents. Curr Top Med Chem 10(12):1184–1197
36. Xie J, Chen K, Lee HY, Xu CJ, Hsu AR, Peng S, Chen X Y, Sun SH (2008). Ultra-small c(RGDyK)-coated Fe(3)O(4) nanoparticles and their specific targeting to integrin alpha(v)beta(3)-rich tumor cells. J Am Chem Soc 130(24):7542–7543

37. Zhu L, Zhou ZY, Mao H, Yang LL (2017) Magnetic nanoparticles for precision oncology: theranostic magnetic iron oxide nanoparticles for image-guided and targeted cancer therapy. Nanomed UK 12(1):73–87
38. Anselmo AC, Mitragotri S (2015) A review of clinical translation of inorganic nanoparticles. AAPS J 17(5):1041–1054
39. Gao LZ, Zhuang J, Nie L, Zhang JB, Zhang Y, Gu N, Wang TH, Feng J, Yang DL, Perrett S, Yan X (2007) Intrinsic peroxidase-like activity of ferromagnetic nanoparticles. Nat Nanotechnol 2(9):577–583
40. Jiang B, Duan D, Gao L, Zhou M, Fan K, Tang Y, Xi J, Bi Y, Tong Z, Gao GF, Xie N, Tang A, Nie G, Liang M, Yan X (2018) Standardized assays for determining the catalytic activity and kinetics of peroxidase-like nanozymes. Nat Protoc 13(7):1506–1520
41. Wei H, Wang E (2008) Fe_3O_4 magnetic nanoparticles as peroxidase mimetics and their applications in H_2O_2 and glucose detection. Anal Chem 80(6):2250–2254
42. Wu Q, Rong J, Shan Z, Chen H, Yang W (2009) Effects of aqueous-organic solvents on peroxidase mimetic activity of Fe_3O_4 magnetic nanoparticles. Sheng wu gong cheng xue bao = Chin J Biotechnol 25(12):1976–1982
43. Zhang ZX, Wang ZJ, Wang XL, Yang XR (2010) Magnetic nanoparticle-linked colorimetric aptasensor for the detection of thrombin. Sensor Actuat B-Chem 147(2):428–433
44. Song YJ, Qu KG, Zhao C, Ren JS, Qu XG (2010) Graphene oxide: intrinsic peroxidase catalytic activity and its application to glucose detection. Adv Mater 22(19):2206–2210
45. Chen ZW, Yin JJ, Zhou YT, Zhang Y, Song L, Song MJ, Hu SL, Gu N (2012) Dual enzyme-like activities of iron oxide nanoparticles and their implication for diminishing cytotoxicity. ACS Nano 6(5):4001–4012
46. Fan YW, Huang YM (2012) The effective peroxidase-like activity of chitosan-functionalized $CoFe_2O_4$ nanoparticles for chemiluminescence sensing of hydrogen peroxide and glucose. Analyst 137(5):1225–1231
47. Shi WB, Wang H, Huang YM (2011) Luminol-silver nitrate chemiluminescence enhancement induced by cobalt ferrite nanoparticles. Luminescence 26(6):547–552
48. Luo W, Li YS, Yuan J, Zhu LH, Liu ZD, Tang HQ, Liu SS (2010) Ultrasensitive fluorometric determination of hydrogen peroxide and glucose by using multiferroic $BiFeO_3$ nanoparticles as a catalyst. Talanta 81(3):901–907
49. Bhattacharya D, Baksi A, Banerjee I, Ananthakrishnan R, Maiti TK, Pramanik P (2011) Development of phosphonate modified Fe_{1-x} $Mn_xFe_2O_4$ mixed ferrite nanoparticles: novel peroxidase mimetics in enzyme linked immunosorbent assay. Talanta 86:337–348
50. Ju LL, Chen ZY, Fang L, Dong W, Zheng FG, Shen MR (2011) Sol-Gel synthesis and photo-fenton-like catalytic activity of $EuFeO_3$ nanoparticles. J Am Ceram Soc 94(10):3418–3424
51. Wang K, Xu JJ, Sun DC, Wei H, Xia XH (2005) Selective glucose detection based on the concept of electrochemical depletion of electroactive species in diffusion layer. Biosens Bioelectron 20(7):1366–1372
52. Park KS, Kim MI, Cho DY, Park HG (2011) Label-free colorimetric detection of nucleic acids based on target-induced shielding against the peroxidase-mimicking activity of magnetic nanoparticles. Small 7(11):1521–1525
53. Cao X, Wang N (2011) A novel non-enzymatic glucose sensor modified with Fe_2O_3 nanowire arrays. Analyst 136(20):4241–4246
54. Mu JS, Wang Y, Zhao M, Zhang L (2012) Intrinsic peroxidase-like activity and catalase-like activity of Co_3O_4 nanoparticles. Chem Commun 48(19):2540–2542
55. Liang H, Lin FF, Zhang ZJ, Liu BW, Jiang SH, Yuan QP, Liu JW (2017) Multicopper laccase mimicking nanozymes with nucleotides as ligands. Acs Appl Mater Inter 9(2):1352–1360
56. Yin JF, Cao HQ, Lu YX (2012) Self-assembly into magnetic Co_3O_4 complex nanostructures as peroxidase. J Mater Chem 22(2):527–534
57. Chen W, Chen J, Liu AL, Wang LM, Li GW, Lin XH (2011) Peroxidase-like activity of cupric oxide nanoparticle. Chemcatchem 3(7):1151–1154
58. Chen W, Chen J, Feng YB, Hong L, Chen QY, Wu LF, Lin XH, Xia XH (2012) Peroxidase-like activity of water-soluble cupric oxide nanoparticles and its analytical application for detection of hydrogen peroxide and glucose. Analyst 137(7):1706–1712

59. Rahimi-Rad MH, Alizadeh E, Samarei R (2011) Aquatic leech as a rare cause of respiratory distress and hemoptysis. Pneumologia 60(2):85–86
60. Huang YY, Liu CQ, Pu F, Liu Z, Ren JS, Qu XG (2017) A GO-Se nanocomposite as an antioxidant nanozyme for cytoprotection. Chem Commun 53(21):3082–3085
61. Nagvenkar AP, Gedanken A (2016) $Cu_{0.89}Zn_{0.11}O$, a new peroxidase-mimicking nanozyme with high sensitivity for glucose and antioxidant detection. Acs Appl Mater Inter, 8(34):22301–22308
62. Gao LZ, Koo H (2017) Do catalytic nanoparticles offer an improved therapeutic strategy to combat dental biofilms? Nanomed UK 12(4):275–279
63. Xu Y, Wu XQ, Shen JS, Zhang HW (2015) Highly selective and sensitive recognition of histidine based on the oxidase-like activity of Cu^{2+} ions. Rsc Adv 5(112):92114–92120
64. Liu X, Wang Q, Zhao HH, Zhang LC, Su YY, Lv Y (2012) BSA-templated MnO_2 nanoparticles as both peroxidase and oxidase mimics. Analyst 137(19):4552–4558
65. Wan Y, Qi P, Zhang D, Wu JJ, Wang Y (2012) Manganese oxide nanowire-mediated enzyme-linked immunosorbent assay. Biosens Bioelectron 33(1):69–74
66. Gao WY, Liu ZY, Qi LM, Lai JP, Kitte SA, Xu GB (2016) Ultrasensitive glutathione detection based on lucigenin cathodic electrochemiluminescence in the presence of MnO_2 nanosheets. Anal Chem 88(15):7654–7659
67. Andre R, Natalio F, Humanes M, Leppin J, Heinze K, Wever R, Schroder HC, Muller WEG, Tremel W (2011) V_2O_5 nanowires with an intrinsic peroxidase-like activity. Adv Func Mater 21(3):501–509
68. Natalio F, Andre R, Hartog AF, Stoll B, Jochum KP, Wever R, Tremel W (2012) Vanadium pentoxide nanoparticles mimic vanadium haloperoxidases and thwart biofilm formation. Nat Nanotechnol 7(8):530–535
69. Hashmi ASK, Hutchings GJ (2006) Gold catalysis. Angew Chem Int Edit 45(47):7896–7936
70. Haruta M, Kobayashi T, Sano H, Yamada N (1987) Novel gold catalysts for the oxidation of carbon-monoxide at a temperature far below 0 °C. Chem Lett 2:405–408
71. Comotti M, Della Pina C, Matarrese R, Rossi M (2004) The catalytic activity of "Naked" gold particles. Angew Chem Int Edit 43(43):5812–5815
72. Beltrame P, Comotti M, Della Pina C, Rossi M (2006). Aerobic oxidation of glucose II. Catalysis by colloidal gold. Appl Catal A Gen 297(1):1–7
73. Beltrame P, Comotti M, Della Pina C, Rossi M (2004) Aerobic oxidation of glucose I. Enzymatic catalysis. J Catal 228(2):282–287
74. Luo WJ, Zhu CF, Su S, Li D, He Y, Huang Q, Fan CH (2010) Self-catalyzed, self-limiting growth of glucose oxidase-mimicking gold nanoparticles. ACS Nano 4(12):7451–7458
75. Zheng XX, Liu Q, Jing C, Li Y, Li D, Luo WJ, Wen YQ, He Y, Huang Q, Long YT, Fan CH (2011) Catalytic gold nanoparticles for nanoplasmonic detection of DNA hybridization. Angew Chem Int Edit 50(50):11994–11998
76. Zou HY, Yang T, Lan J, Huang CZ (2017) Use of the peroxidase mimetic activity of erythrocyte-like $Cu_{1.8}S$ nanoparticles in the colorimetric determination of glutathione. Anal Methods UK 9(5):841–846
77. Jv Y, Li BX, Cao R (2010) Positively-charged gold nanoparticles as peroxidiase mimic and their application in hydrogen peroxide and glucose detection. Chem Commun 46(42):8017–8019
78. Long YJ, Li YF, Liu Y, Zheng JJ, Tang J, Huang CZ (2011) Visual observation of the mercury-stimulated peroxidase mimetic activity of gold nanoparticles. Chem Commun 47(43):11939–11941
79. Chen YP, Xianyu YL, Jiang XY (2017) Surface modification of gold nanoparticles with small molecules for biochemical analysis. Acc Chem Res 50(2):310–319
80. Wang XX, Wu Q, Shan Z, Huang QM (2011) BSA-stabilized Au clusters as peroxidase mimetics for use in xanthine detection. Biosens Bioelectron 26(8):3614–3619
81. Zhang LB, Laug L, Munchgesang W, Pippel E, Gosele U, Brandsch M, Knez M (2010) Reducing stress on cells with apoferritin-encapsulated platinum nanoparticles. Nano Lett 10(1):219–223

82. Clark A, Zhu AP, Sun K, Petty HR (2011) Cerium oxide and platinum nanoparticles protect cells from oxidant-mediated apoptosis. J Nanopart Res 13(10):5547–5555
83. Fan J, Yin JJ, Ning B, Wu XC, Hu Y, Ferrari M, Anderson GJ, Wei JY, Zhao YL, Nie GJ (2011) Direct evidence for catalase and peroxidase activities of ferritin-platinum nanoparticles. Biomaterials 32(6):1611–1618
84. Ma M, Zhang Y, Gu N (2011) Peroxidase-like catalytic activity of cubic Pt nanocrystals. Colloid Surface A 373(1–3):6–10
85. Lien CW, Huang CC, Chang HT (2012) Peroxidase-mimic bismuth-gold nanoparticles for determining the activity of thrombin and drug screening. Chem Commun 48(64):7952–7954
86. He WW, Liu Y, Yuan JS, Yin JJ, Wu XC, Hu XN, Zhang K, Liu JB, Chen CY, Ji YL, Guo YT (2011) Au@Pt nanostructures as oxidase and peroxidase mimetics for use in immunoassays. Biomaterials 32(4):1139–1147
87. Liu JB, Hu XN, Hou S, Wen T, Liu WQ, Zhu X, Wu XC (2011) Screening of inhibitors for oxidase mimics of Au@Pt nanorods by catalytic oxidation of OPD. Chem Commun 47(39):10981–10983
88. Zhang K, Hu XN, Liu JB, Yin JJ, Hou SA, Wen T, He WW, Ji YL, Guo YT, Wang Q, Wu XC (2011) Formation of PdPt alloy nanodots on gold nanorods: tuning oxidase-like activities via composition. Langmuir 27(6):2796–2803
89. Liu JB, Hu XN, Hou S, Wen T, Liu WQ, Zhu X, Yin JJ, Wu XC (2012) Au@Pt core/shell nanorods with peroxidase- and ascorbate oxidase-like activities for improved detection of glucose. Sensor Actuat B-Chem 166:708–714
90. He WW, Wu XC, Liu JB, Hu XN, Zhang K, Hou SA, Zhou WY, Xie SS (2010) Design of AgM bimetallic alloy nanostructures (M = Au, Pd, Pt) with tunable morphology and peroxidase-like activity. Chem Mater 22(9):2988–2994
91. Garg B, Bisht T (2016) Carbon nanodots as peroxidase nanozymes for biosensing. Molecules 21(12)
92. Guo SJ, Dong SJ (2011) Graphene nanosheet: synthesis, molecular engineering, thin film, hybrids, and energy and analytical applications. Chem Soc Rev 40(5):2644–2672
93. Song YJ, Wang XH, Zhao C, Qu KG, Ren JS, Qu XG (2010) Label-free colorimetric detection of single nucleotide polymorphism by using single-walled carbon nanotube intrinsic peroxidase-like activity. Chem-Eur J 16(12):3617–3621
94. Garg B, Bisht T, Ling YC (2014) Graphene-based nanomaterials as heterogeneous acid catalysts: a comprehensive perspective. Molecules 19(9):14582–14614
95. Garg B, Ling YC (2013) Versatilities of graphene-based catalysts in organic transformations. Green Mater 1(1):47–61
96. Garg B, Bisht T, Ling YC (2014) Sulfonated graphene as highly efficient and reusable acid carbocatalyst for the synthesis of ester plasticizers. Rsc Adv 4(100):57297–57307
97. Garg B, Sung CH, Ling YC (2015) Graphene-based nanomaterials as molecular imaging agents. Wires Nanomed Nanobi 7(6):737–758
98. Garg B, Bisht T, Ling YC (2015) Graphene-based nanomaterials as efficient peroxidase mimetic catalysts for biosensing applications: an overview. Molecules 20(8):14155–14190
99. Garg B, Bisht T, Ling YC (2016) Graphene-based nanomaterials: versatile catalysts for carbon-carbon bond forming reactions. Curr Org Chem 20(15):1547–1566
100. Kroto HW, Heath JR, Obrien SC, Curl RF, Smalley RE (1985) C-60—Buckminsterfullerene. Nature 318(6042):162–163
101. Krusic PJ, Wasserman E, Keizer PN, Morton JR, Preston KF (1991) Radical reactions of C_{60}. Science 254(5035):1183–1185
102. Dugan LL, Gabrielsen JK, Yu SP, Lin TS, Choi DW (1996) Buckminsterfullerenol free radical scavengers reduce excitotoxic and apoptotic death of cultured cortical neurons. Neurobiol Dis 3(2):129–135
103. Ali SS, Hardt JI, Quick KL, Kim-Han JS, Erlanger BF, Huang TT, Epstein CJ, Dugan LL (2004) A biologically effective fullerene (C_{60}) derivative with superoxide dismutase mimetic properties. Free Radic Biol Med 37(8):1191–1202

104. Dugan LL, Turetsky DM, Du C, Lobner D, Wheeler M, Almli CR, Shen CK, Luh TY, Choi DW, Lin TS (1997) Carboxyfullerenes as neuroprotective agents. Proc Natl Acad Sci USA 94(17):9434–9439
105. Dugan LL, Lovett EG, Quick KL, Lotharius J, Lin TT, O'Malley KL (2001) Fullerene-based antioxidants and neurodegenerative disorders. Parkinsonism Relat Disord 7(3):243–246
106. Ali SS, Hardt JI, Dugan LL (2008) SOD Activity of carboxyfullerenes predicts their neuroprotective efficacy: a structure-activity study. Nanomed-Nanotechnol 4(4):283–294
107. Belgorodsky B, Fadeev L, Ittah V, Benyamini H, Zelner S, Huppert D, Kotlyar AB, Gozin M (2005) Formation and characterization of stable human serum albumin-tris-malonic acid [C_{60}]fullerene complex. Bioconjug Chem 16(5):1058–1062
108. Quick KL, Ali SS, Arch R, Xiong C, Wozniak D, Dugan LL (2008) A carboxyfullerene SOD mimetic improves cognition and extends the lifespan of mice. Neurobiol Aging 29(1):117–128
109. Cui RJ, Han ZD, Zhu JJ (2011) Helical carbon nanotubes: intrinsic peroxidase catalytic activity and its application for biocatalysis and biosensing. Chem-Eur J 17(34):9377–9384
110. Guo YJ, Deng L, Li J, Guo SJ, Wang EK, Dong SJ (2011) Hemin-graphene hybrid nanosheets with intrinsic peroxidase-like activity for label-free colorimetric detection of single-nucleotide polymorphism. ACS Nano 5(2):1282–1290
111. Guo YJ, Li J, Dong SJ (2011) Hemin functionalized graphene nanosheets-based dual biosensor platforms for hydrogen peroxide and glucose. Sensor Actuat B-Chem 160(1):295–300
112. Liu M, Zhao HM, Chen S, Yu HT, Quan X (2012) Interface engineering catalytic graphene for smart colorimetric biosensing. ACS Nano 6(4):3142–3151
113. Ye YP, Kong T, Yu XF, Wu YK, Zhang K, Wang XP (2012) Enhanced nonenzymatic hydrogen peroxide sensing with reduced graphene oxide/ferroferric oxide nanocomposites. Talanta 89:417–421
114. Liu M, Zhao HM, Chen S, Yu HT, Quan X (2012) Stimuli-responsive peroxidase mimicking at a smart graphene interface. Chem Commun 48(56):7055–7057
115. Liu S, Tian JQ, Wang L, Luo YL, Sun XP (2012) A general strategy for the production of photoluminescent carbon nitride dots from organic amines and their application as novel peroxidase-like catalysts for colorimetric detection of H2O2 and glucose. Rsc Adv 2(2):411–413
116. Shi WB, Wang QL, Long YJ, Cheng ZL, Chen SH, Zheng HZ, Huang YM (2011) Carbon nanodots as peroxidase mimetics and their applications to glucose detection. Chem Commun 47(23):6695–6697
117. Wang XH, Qu KG, Xu BL, Ren JS, Qu XG (2011) Multicolor luminescent carbon nanoparticles: synthesis, supramolecular assembly with porphyrin, intrinsic peroxidase-like catalytic activity and applications. Nano Res 4(9):908–920
118. Song YJ, Qu KG, Xu C, Ren JS, Qu XG (2010) Visual and quantitative detection of copper ions using magnetic silica nanoparticles clicked on multiwalled carbon nanotubes. Chem Commun 46(35):6572–6574
119. Dai ZH, Liu SH, Bao JC, Jui HX (2009) Nanostructured FeS as a mimic peroxidase for biocatalysis and biosensing. Chem-Eur J 15(17):4321–4326
120. Dutta AK, Maji SK, Srivastava DN, Mondal A, Biswas P, Paul P, Adhikary B (2012) Synthesis of FeS and FeSe nanoparticles from a single source precursor: a study of their photocatalytic activity, peroxidase-like behavior, and electrochemical sensing of H_2O_2. Acs Appl Mater Inter 4(4):1919–1927
121. Maji SK, Dutta AK, Biswas P, Srivastava DN, Paul P, Mondal A, Adhikary B (2012) Synthesis and characterization of FeS nanoparticles obtained from a dithiocarboxylate precursor complex and their photocatalytic, electrocatalytic and biomimic peroxidase behavior. Appl Catal A Gen 419:170–177
122. Roy P, Lin ZH, Liang CT, Chang HT (2012) Synthesis of enzyme mimics of iron telluride nanorods for the detection of glucose. Chem Commun 48(34):4079–4081
123. Peng C, Jiang BW, Liu Q, Guo Z, Xu ZJ, Huang Q, Xu HJ, Tai RZ, Fan CH (2011) Graphene-templated formation of two-dimensional lepidocrocite nanostructures for high-efficiency catalytic degradation of phenols. Energ Environ Sci 4(6):2035–2040

124. Liu S, Tian JQ, Wang L, Luo YL, Chang GH, Sun XP (2011) Iron-substituted SBA-15 microparticles: a peroxidase-like catalyst for H_2O_2 detection. Analyst 136(23):4894–4897
125. Malvi B, Panda C, Dhar BB, Sen Gupta S (2012) One pot glucose detection by [Fe-III(biuret-amide)] immobilized on mesoporous silica nanoparticles: an efficient HRP mimic. Chem Commun 48(43):5289–5291
126. Tian JQ, Liu S, Luo YL, Sun XP (2012) Fe(III)-based coordination polymer nanoparticles: peroxidase-like catalytic activity and their application to hydrogen peroxide and glucose detection. Catal Sci Technol 2(2):432–436
127. Wang W, Jiang XP, Chen KZ (2012) Iron phosphate microflowers as peroxidase mimic and superoxide dismutase mimic for biocatalysis and biosensing. Chem Commun 48(58):7289–7291
128. Wang YL, Chen SH, Ni F, Gao F, Li MG (2009) Peroxidase-like layered double hydroxide nanoflakes for electrocatalytic reduction of H_2O_2. Electroanal 21(19):2125–2132
129. Cui L, Yin H, Dong J, Fan H, Liu T, Ju P, Ai S (2011) A mimic peroxidase biosensor based on calcined layered double hydroxide for detection of H_2O_2. Biosens Bioelectron 26(7):3278–3283
130. Zhang YW, Tian JQ, Liu S, Wang L, Qin XY, Lu WB, Chang GH, Luo YL, Asiri AM, Al-Youbi AO, Sun XP (2012) Novel application of CoFe layered double hydroxide nanoplates for colorimetric detection of H_2O_2 and glucose. Analyst 137(6):1325–1328
131. Liu S, Wang L, Zhai JF, Luo YL, Sun XP (2011) Carboxyl functionalized mesoporous polymer: a novel peroxidase-like catalyst for H_2O_2 detection. Anal Methods UK 3(7):1475–1477
132. Liu S, Tian JQ, Wang L, Zhang YW, Luo YL, Li HY, Asiri AM, Al-Youbi AO, Sun XP (2012) Fast and sensitive colorimetric detection of H_2O_2 and glucose: a strategy based on polyoxometalate clusters. ChemPlusChem 77(7):541–544
133. He WW, Jia HM, Li XX, Lei Y, Li J, Zhao HX, Mi LW, Zhang LZ, Zheng Z (2012) Understanding the formation of CuS concave superstructures with peroxidase-like activity. Nanoscale 4(11):3501–3506
134. Singh A, Patra S, Lee JA, Park KH, Yang H (2011) An artificial enzyme-based assay: DNA detection using a peroxidase-like copper-creatinine complex. Biosens Bioelectron 26(12):4798–4803

Chapter 4
Nanozymes: Preparation and Characterization

Li Qin, Yihui Hu and Hui Wei

Abbreviations

ABTS	2,2′-Azinobis-(3-ethylbenzthiazoline-6-sulphonate)
AFM	Atomic force microscopy
BET	Brunauer–Emmett–Teller
BSA	Bovine serum albumin
HRP	Horseradish peroxidase
ROS	Reactive oxygen species
SOD	Superoxide dismutase
CLSM	Confocal laser scanning microscopy
DLS	Dynamic light scattering
DMAA	N, N-dimethylacrylamide
DTA	Differential thermal analysis
EDX	Energy-dispersive X-ray spectroscopy
EPR	Electron paramagnetic resonance
ESR	Electron spin resonance
FT-IR	Fourier transform infrared spectrometer
SEM	Scanning electron microscopy
OPD	O-Phenylenediamine
TEM	Transmission electron microscopy
SERS	Surface-enhanced Raman scattering
TGA	Thermogravimetric analysis

L. Qin · Y. Hu · H. Wei (✉)
Nanjing National Laboratory of Microstructures, Jiangsu Key Laboratory of Artificial Functional Materials, Department of Biomedical Engineering, College of Engineering and Applied Sciences, Nanjing University, Nanjing, Jiangsu 210093, China
e-mail: weihui@nju.edu.cn
URL: http://weilab.nju.edu.cn

X. Yan (ed.), *Nanozymology*, Nanostructure Science and Technology,
https://doi.org/10.1007/978-981-15-1490-6_4

TMB 3, 3′, 5, 5′-tetramethylbenzidine
XPS X-ray photoelectron spectroscopy
XRD X-ray diffraction

4.1 Nanozymes Preparation

Nanozymes are essentially nanomaterials with enzyme-like activities [1–3]. Therefore, the established nanomaterials preparation methods can be directly used to synthesize nanozymes. In order to explore novel physical properties and realize potential applications of nanozymes, researchers are making efforts to enhance the catalytic activities of these functional nanomaterials. Hence, developing effective methods to prepare nanozymes with good performance becomes an urgent demand. This section discusses conventional methods for nanozymes preparation, including hydrothermal method, solvothermal method, co-precipitation method, sol-gel method, and other methods.

4.1.1 Hydrothermal Method

Hydrothermal method refers to an aqueous reaction at high temperature and high vapor pressures to recrystallize materials in a closed system. Nanozymes with crystalline phases that are not stable at the melting point have been prepared by this method.

Metal oxides-based nanozymes have been successfully synthesized by hydrothermal methods [4–16]. For example, V_2O_5 nanowires with an intrinsic peroxidase-like activity were prepared by this method (Fig. 4.1) [4]. In brief, $VOSO_4·nH_2O$ and $KBrO_3$ were dissolved in water. Then, nitric acid was added dropwise under stirring until the solution pH reached 2. The solution was heated at 180 °C for 24 h in a Teflon-lined stainless steel autoclave. Peroxidase-like hexagonal tungsten oxide nanoflowers were prepared via a hydrothermal method by Park and coworkers [16].

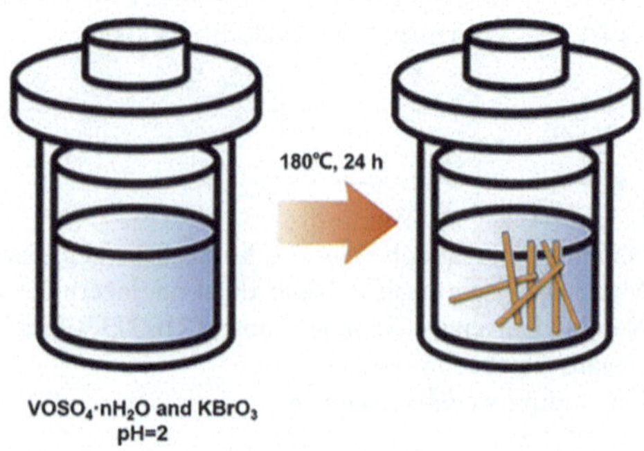

Fig. 4.1 Preparation of V_2O_5 nanowires via hydrothermal method Adapted with permission from [4]. Copyright (2017) Nature Publishing Group

Zhang et al. synthesized Co_3O_4 nanoparticles with intrinsic peroxidase-like activity and catalase-like activity by a hydrothermal method. They showed that the components and structures of Co_3O_4 nanoparticles could be modulated by varying the reaction conditions such as reaction time, temperature, and the amount of precursor. They demonstrated that the Co_3O_4 nanomaterials with different morphologies exhibited different catalytic activities in order of nanoplates > nanorods > nanocubes [12].

4.1.2 Solvothermal Method

Solvothermal synthesis involves the use of a solvent under high temperature (typically between 100 and 1000 °C) and high pressure (typically between 1 and 10,000 atm) to prepare a variety of materials such as metals, metal oxides, and semiconductors [17]. Notably, the morphology and crystalline of the materials can be modulated by the reaction temperature, reaction time, solvent, surfactant, precursor, etc.

Solvothermal method has been widely used in the preparation of nanozymes [18–21]. Au/CuS composites were prepared by a two-step method based on a facile solvothermal approach [21]. $Cu(NO_3)_2 \cdot 3H_2O$ and glycol were used as the precursor and solvent, respectively. It demonstrated that the Au/CuS composite exhibited excellent peroxidase-like catalytic activity. In another example, Zhu et al. used a solvothermal method to synthesize Fe_3O_4 nanocrystals with different structures and studied the structure effects of Fe_3O_4 nanocrystals on peroxidase-like activity [18]. They used $FeCl_3$, ethylene glycol, and anhydrous CH_3ONa as the precursor, solvent, and additive, respectively. They found that the structure of Fe_3O_4 nanocrystal could be controlled by the amount of anhydrous CH_3ONa. They successfully prepared three different structures of Fe_3O_4 nanocrystals and showed their peroxidase-like activities followed the order of cluster spheres > triangular plates > octahedra.

4.1.3 Co-precipitation Method

Co-precipitation method binds two solutes or more precipitate from the solution together, rather than remains dispersed in the solution. The main advantage of this method is the possibility of preparing pure and homogenous material. Nanozymes of more than one component can be prepared by a co-precipitation method [22–29].

As shown in Fig. 4.2, Das et al. reported that addition of NaOH solution to the metal precursor solution of Co^{2+} and Fe^{3+} ions at 80 °C resulted in the production of a brown colored metal hydroxide, which subsequently transformed into black colored $CoFe_2O_4$ nanoparticles. They exhibited significant peroxidase-like activity, which was evaluated by the chemiluminescent method [24, 29]. There are also lots of metal-based structured nanozymes synthesized by a co-precipitation method. Wei and Wang prepared magnetic Fe_3O_4 nanoparticles as peroxidase mimics by a co-precipitation method, which were used to detect H_2O_2 and glucose. In brief, ferric chloride aqueous

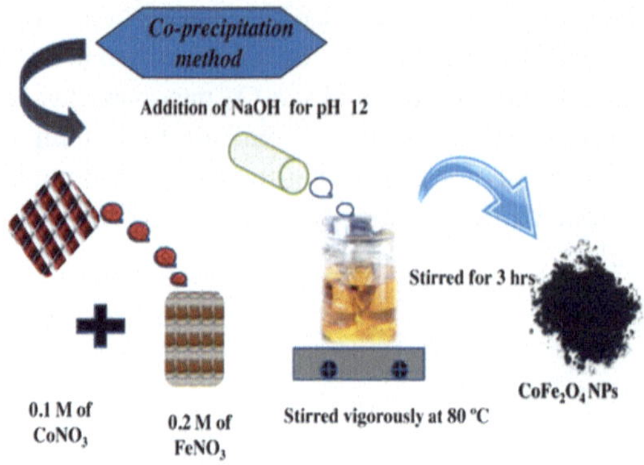

Fig. 4.2 The synthesis process of $CoFe_2O_4$ nanoparticles by co-precipitation method. Reprinted with permission from [29]. Copyright (2016) Elsevier

solution and ferrous chloride aqueous solution in HCl were mixed and deoxygenated, and then the mixture was added dropwise into oxygen-free ammonia solution. Black Fe_3O_4 colloidal particles were obtained by centrifugation and washing [23].

4.1.4 *Sol-Gel Method*

The sol-gel method is a wet chemical process to make solid-state materials. This synthesis technique involves the conversion of a colloidal liquid called sol into a semisolid gel phase. An overview of the sol-gel method is illustrated in Fig. 4.3. It can be used to prepare solid materials in a wide variety of morphologies, such as ultrafine or spherical powders, thin-film coatings, fibers, microporous inorganic membranes, or extremely porous aerogels. Moreover, it is possible to synthesize composite materials with high purity through the use of highly pure reagents [30, 31]. The sol-gel method can produce high purity and uniform structure at low temperature. But a sol-gel method still suffers from some limitations, such as substrate-dependence, high permeability, and difficult control of porosity.

Huang, Ye et al. synthesized $ZnFe_2O_4$-decorated ZnO nanofibers by a sol-gel process and co-electrospinning followed by calcination. The $ZnFe_2O_4$ with controllable particle sizes and uniform distribution was formed. The formed nanofibers showed good peroxidase-like activity, which were then used for colorimetric glucose detection in urine [31].

4.1.5 *Other Methods*

Nanozymes with smaller particle sizes usually have higher catalytic activities. To prepare small-sized nanozymes, a nanoemulsion strategy can be adopted, which formed water-in-oil (or oil-in-water) droplets [33, 34]. For example, around 3 nm

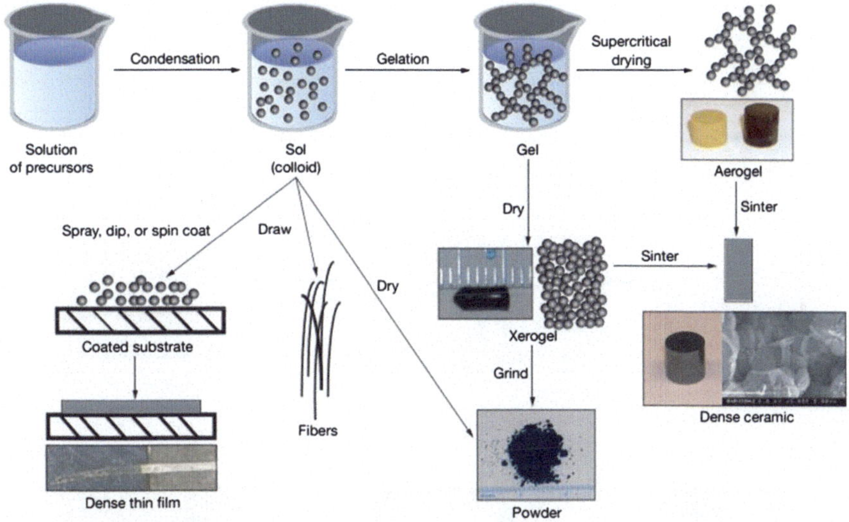

Fig. 4.3 The schematic diagram of sol-gel method. Reprinted with permission from [32]. Copyright (2010) Springer

ceria nanoparticles (NPs) were prepared by this method [33]. Due to the excellent reactive oxygen species (ROS) scavenging activity, the ceria NPs were used to protect brains from ischemic damages.

The seed-mediated growth strategy has been widely used to prepare metal nanozymes, such as gold, silver, platinum, and palladium [35–38]. The schematic diagram of seed-mediated method is shown in Fig. 4.4. For example, to understand the pH-dependent peroxidase- and catalase-like activities of noble metals (i.e., Au, Ag, Pt, and Pd), Wu, Gao, and co-workers prepared Au nanorods and then coated the Au nanorods with Ag, Pt, and Pd to form Au@Ag, Au@Pt, and Au@Pd. It should be noted that the Au nanorods themselves were also prepared by a classic seed-mediated method. By combining experiments with calculations, they revealed the enzyme-like activities were the intrinsic properties of the four noble metals [37]. At low pH, the base-like decomposition of hydrogen peroxide endowed them with peroxidase-like activities, while at high pH, the acid-like decomposition of hydrogen peroxide led to the catalase-like activities.

Microwave irradiation method takes advantages of microwave radiation to heat material containing mobile electric charges, such as polar molecules in a solvent or conducting ions in a solid [39]. Compared with conventional heating methods, it is much faster, cleaner, and more economical. A variety of materials like carbide, nitride, sulfides, and oxides have been synthesized by microwave irradiation methods [40–42].

Biosynthetic method is considered as a "green" and biocompatible approach to synthesize nanozymes. To date, some metal nanoclusters have been obtained by the biosynthetic method [43]. For example, Guo and his co-workers synthesized

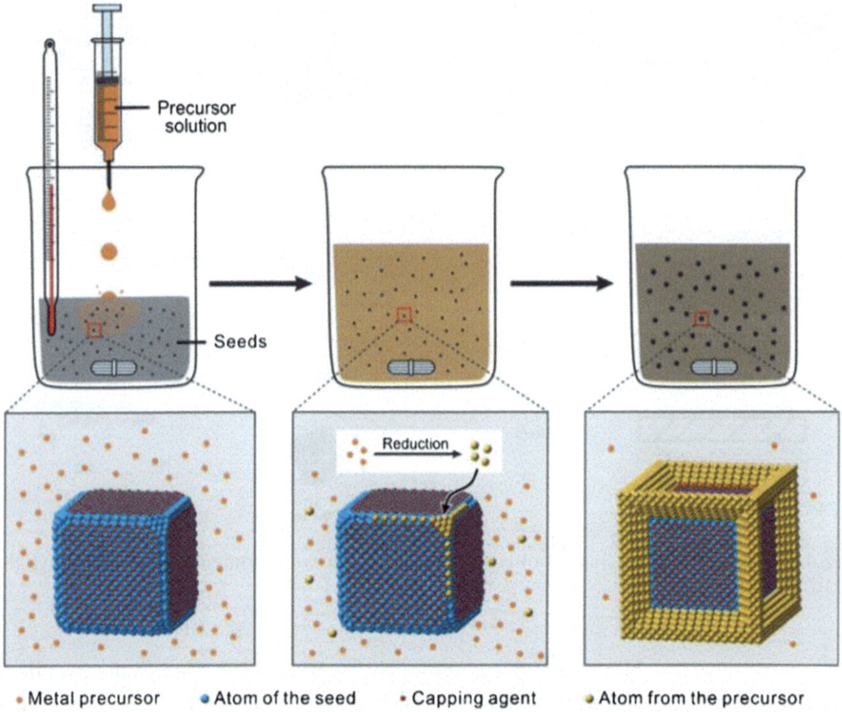

Fig. 4.4 General strategy for the seed-mediated growth of colloidal metal nanocrystals. Reprinted with permission from [38]. Copyright (2016) John Wiley and Sons

histidine (His) functionalized gold nanoclusters (His-AuNCs) with high peroxidase-like activity for sensitive and selective detection of Cu^{2+} and His [44] (Table 4.1).

4.2 Nanozymes Characterization

Generally speaking, there are four levels of nanozymes characterization: characterizing nanozymes as common nanomaterials, measuring the enzymatic kinetics of nanozymes, probing the catalytic process of nanozymes, and analyzing nanozymes in biological systems (such as cells).

4.2.1 Characterization as Common Nanomaterials

The general properties of nanozymes that need to be characterized include: size, shape, morphology, specific surface area, structure, composition, and so on. We will

Table 4.1 Preparation methods of various nanozymes

Methods	Nanozymes		Enzymatic activities	Refs.
Hydrothermal method	V_2O_5 nanowires		Peroxidase	[4]
	CeO_2 NPs		Peroxidase/oxidase/catalase/SOD	[6]
	Co_3O_4 NPs		Peroxidase/oxidase/catalase/SOD	[7, 12]
	$FePO_4$ microflowers		SOD	[8]
	Fe_3O_4 NPs		Peroxidase/catalase	[9]
	WC nanorods		Peroxidase	[10]
	CuO nanostructure		Peroxidase	[11]
	MnSe NPs		Peroxidase	[13]
	VO_2 nanoplates		Peroxidase	[14]
	Mn_3O_4 octahedrons		Oxidase	[15]
Solvothermal method	Au/CuS NC		Peroxidase	[21]
	Fe_3O_4		Peroxidase/catalase	[18]
	$CePO_4$:Tb,Gd NPs		Peroxidase	[19]
	$CoFe_2O_4$ NPs		Oxidase	[20]
Co-precipitation method	Fe_3O_4 NPs		Peroxidase/catalase	[23]
	$MnFe_2O_4$ NC		Oxidase	[26]
	CoFe LDHs		Peroxidase	[25]
Sol-gel method	$ZnFe_2O_4$/ZnO NC		Peroxidase	[31]
	Fe_2O_3 NPs		Peroxidase	[30]
Other methods	Nanoemulsion strategy	CeO_2 NPs	Peroxidase/catalase/catalase/SOD	[33]
	Seed-mediated growth method	Gold/silver/Pt NPs	Peroxidase/oxidase	[35–37]
	Microwave method	$CuInS_2$ NC	Peroxidase	[40]
	Biosynthetic method	His-AuNCs	Peroxidase	[44]

not include all the characterization methods but choose commonly used ones to describe in detail [45].

Dynamic light scattering (DLS) is the most conventional method to characterize the hydration kinetic particle size of nanozymes. The nanoparticles do irregular movements in liquid named Brownian motion. The smaller the nanoparticles and the viscosity of the medium are, the faster the Brownian motion is. When a light passes through nanozymes solution, the nanozyme particles will scatter the light and the instrument could detect the optical signals at an angle. Finally, the particle size and its distribution can be calculated by the light intensity fluctuation and the light intensity correlation function. Since the DLS measures the hydrodynamic size, its value is usually larger than that measured by transmission electron microscopy (TEM). Moreover, the DLS measurement can be used to study the surface modification

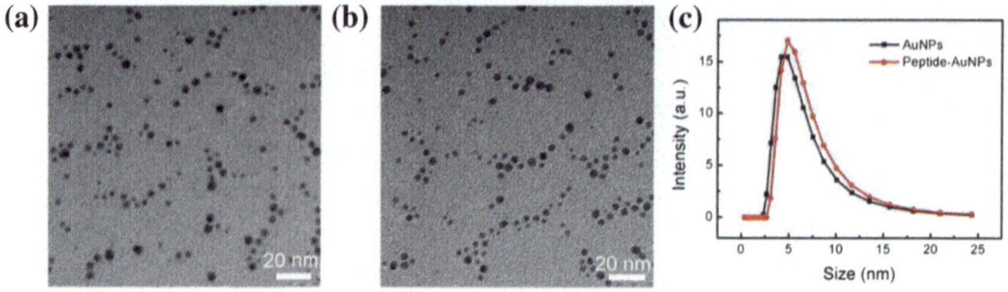

Fig. 4.5 TEM images of **a** AuNPs and **b** peptide–AuNPs. **c** Size distributions of AuNPs and peptide–AuNPs evaluated from DLS results. Reprinted with permission from [46]. Copyright (2015) American Chemical Society

of nanozymes. For example, the peptide-coated AuNPs in Fig. 4.5 showed a size of 3.8 nm under TEM imaging and a hydrodynamic size of 5.0 nm by DLS. For unmodified AuNPs, a hydrodynamic size of 4.0 nm was obtained by DLS. The increased hydrodynamic size of peptide–AuNPs compared with AuNPs alone confirmed the successful modification of peptide [46].

The zeta potential measurement function is often integrated with DLS. DLS can measure the mobility of charged particles by Doppler Effect, which can be converted to the corresponding zeta potential. Zeta potential, an indicative parameter of intensity of the repulsion or attraction between the particles, can be used to determine the stability of colloidal dispersions (including nanozyme solutions). The larger zeta potential (negative or positive) a nanozyme has, the more stable it is. For example, the peptide-coated AuNPs had a zeta potential of +33.0 mV, which showed good storage stability [46].

Transmission electron microscope (TEM) takes advantage of electron beam generated from an electronic gun to transmit through the samples. The transmitted electrons are focused by electron optic lenses to form images. Since the attenuation of electrons depends primarily on the density and thickness of the sample, the images of different shades will be displayed on an imaging device. TEM is of great importance to characterize the morphology, size, shape, and lattice fringes of nanomaterials. Figure 4.6 characterizes the different morphologies of the Pd nanocrystals. The typical TEM images clearly showed the Pd nanocrystals with an average edge length of 10 nm. In addition, the high-resolution TEM (HRTEM) images of Pd nanocrystals fully confirmed that these nanocubes and octahedrons are enclosed by {100} and {111} facets, respectively. Moreover, combined with XRD and EDX, TEM images can provide more precise information about element composition and crystal structure. With the development of aberration correctors for the objective lens, aberration-corrected TEM and STEM (scanning TEM) with resolution of 0.1 nm or better now available for characterizing nanomaterials. It should be noted that some vulnerable samples are susceptible to high energy electron beam damage. To avoid this issue, cryo-TEM could be used.

Scanning electron microscopy (SEM) is a convenient imaging technique to get high-resolution morphology information of nanomaterials. In principle, when high

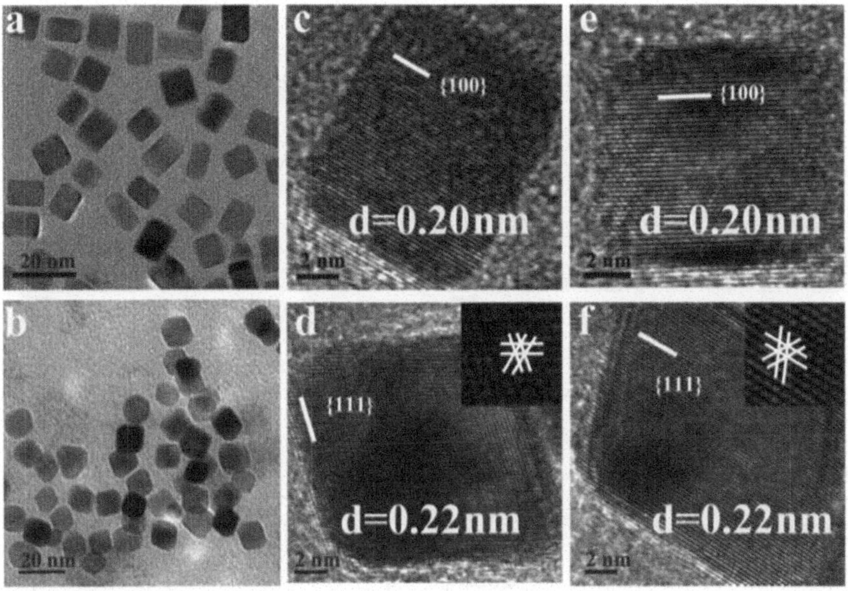

Fig. 4.6 TEM and HRTEM images of the Pd nanocubes (**a**, **c**, **e**) and octahedrons (**b**, **d**, **f**). Reprinted with permission from [47]. Copyright (2016) American Chemical Society

energy electron beams scan over a sample surface, a variety of physical information will be stimulated, especially the secondary electrons. The sample surface morphology can be observed by the acceptance and amplification of secondary electron signals. For conventional SEM imaging, the samples should be electrically conductive. Therefore, for nonconductive samples, a layer of electrically conducting materials (such as gold and carbon) is applied to cover the surface of the samples. Compared with TEM, SEM has a larger depth of field for imaging various uneven surfaces of fine structures. In addition, the preparation of samples for SEM is also easier than TEM. Oxidase-like $MnFe_2O_4$ nanozymes were synthesized and then characterized by SEM imaging. As shown in Fig. 4.7, $MnFe_2O_4$ nanozymes of 100–200 nm in size (average 150 nm) exhibited octahedron morphology [26]. Moreover, the morphology of the $MnFe_2O_4$ nanozymes was also checked after the catalytic reaction, revealing that the robustness of the nanozymes.

Energy-dispersive X-ray spectroscopy (EDX) is often equipped into a TEM or a SEM to carry out the elemental analysis of nanomaterials. An EDX spectrum is obtained by analyzing the elemental characteristic X-ray wavelength and intensity to measure elements contained in the sample. The content of the elements in a sample can be determined according to the intensity of different element lines. In Fig. 4.8, the EDX spectrum revealed that the Cu_2O nanowire mesocrystals contain the elements of Cu, O, N, and C. Their corresponding percentages were also determined [48]. Moreover, specific elements distribution of a sample could be obtained by EDX element mapping. For example, the Fe element of hemin was observed for

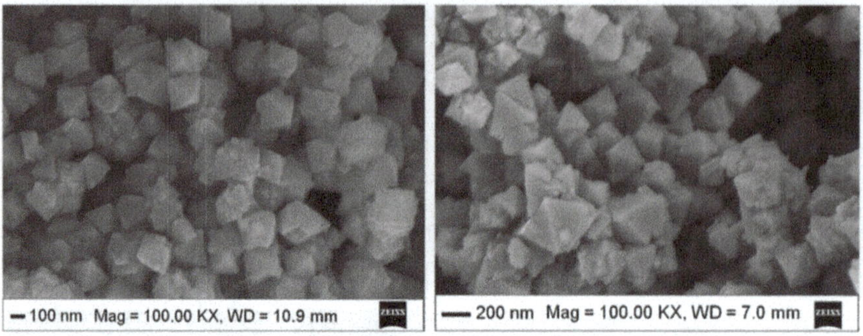

Fig. 4.7 Typical SEM images of oxidase-like $MnFe_2O_4$ nanozymes before (left) and after (right) the catalytic reaction. Reprinted with permission from [26]. Copyright (2016) John Wiley and Sons

Fig. 4.8 EDX spectrum of the Cu_2O nanowire mesocrystals. Reprinted with permission from [48]. Copyright (2016) Springer

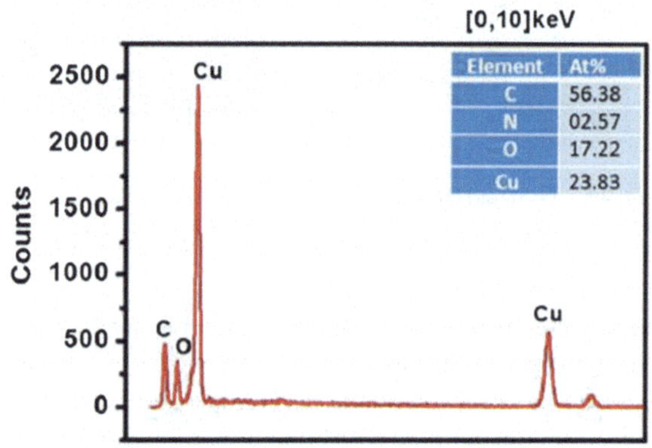

hemin@ZIF-8 nanozymes, confirmed the successful encapsulation of hemin within a ZIF-8 framework (Fig. 4.9) [49].

Atomic force microscopy (AFM) can study the surface structure and properties of a substance by detecting the extremely weak interatomic interactions between the surface and a micro-force sensitive probe, thereby one can get the surface topography structure and roughness information at nanometer resolution. Figure 4.10 shows the planar structures of the two-dimensional MoS_2 sample. Clearly, the multiple monolayers with a thickness of about 0.7–1.5 nm were observed, which was attributed to 1–2 monolayers of MoS_2 [50].

Compared with TEM and SEM, AFM provides real 3D images which can reveal surface information. In addition, AFM often works under normal pressure and even in liquid environment so that the images of biological macromolecules and tissues can be obtained. However, the micro-force sensitive probe is vulnerably affected by the samples and tough environments, interfering the image quality.

Fourier transform infrared spectroscopy (FT-IR Spectroscopy) is an infrared spectroscopy based on Fourier transform principle to analyze the interference infrared

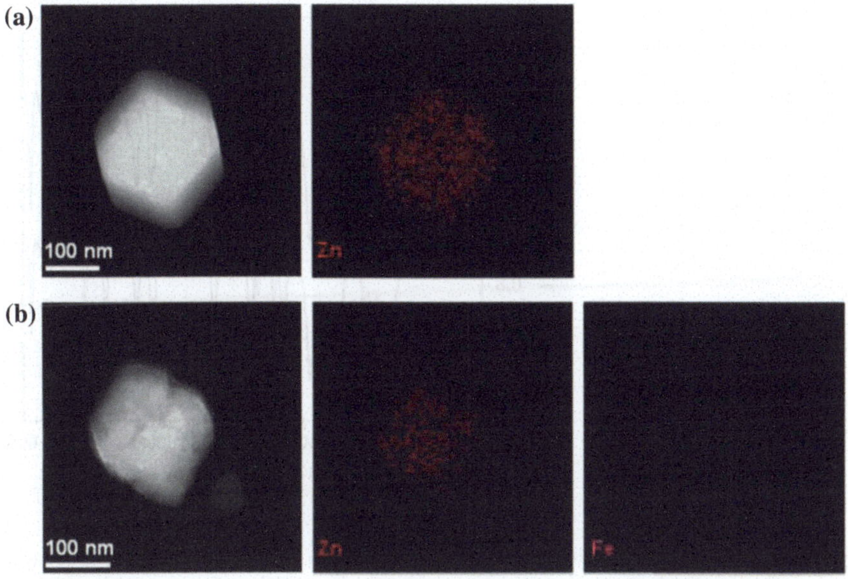

Fig. 4.9 TEM images and corresponding element mapping for ZIF-8 (**a**) and hemin@ZIF-8 (**b**). The Zn signals were from the ZIF-8 matrices while the Fe signals were from the encapsulated hemin. Reprinted with permission from [49]. Copyright (2016) American Chemical Society

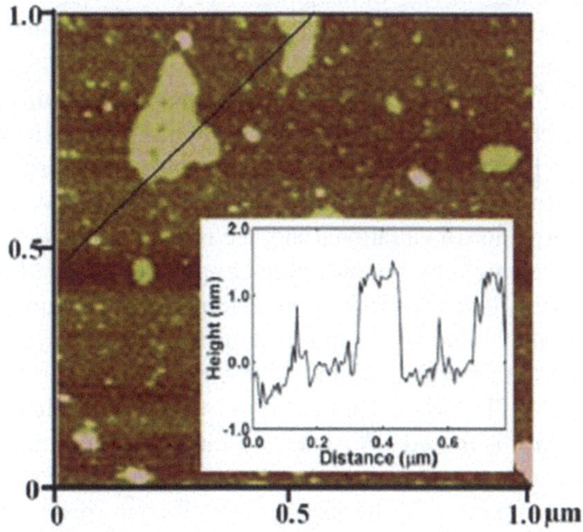

Fig. 4.10 AFM image of luminescent MoS_2 nanosheets with the inset representing the height profile along the black line overlaid on the image. Reprinted with permission from [50]. Copyright (2016) Elsevier

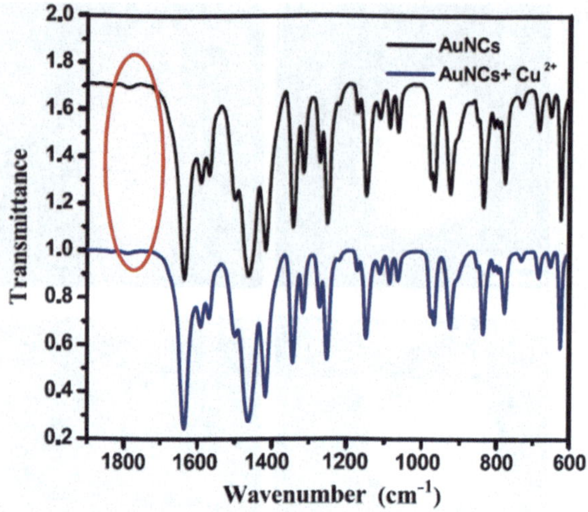

Fig. 4.11 FT-IR spectrum of His-Au nanoclusters with and without Cu^{2+}. Reprinted with permission from [44]. Copyright (2017) Elsevier

light and get the characteristic information. As shown in Fig. 4.11, FT-IR spectrum was used to confirm the binding sites of copper ions in histidine–Au nanoclusters (His-AuNCs) [44]. The stretching vibration at 1790 cm^{-1} of COO^- disappeared after the addition of copper ions. As speculated by the authors, the His-AuNCs were selectively coordinated with Cu^{2+} by the amino and carboxylate groups instead of the imidazole ring of histidine.

Generally speaking, FT-IR spectroscopy has the advantages of high signal-to-noise ratio, good reproducibility, and fast scanning rate. In the field of nanomaterials, FT-IR spectroscopy is often used to determine the kind of functional groups and characteristic surroundings. Accordingly, FT-IR spectroscopy requires relatively high purity samples to avoid interference of impurities.

Raman spectroscopy, complementary to IR spectroscopy, is a powerful tool to get information of vibrational and electronic structures based on the inelastic scattering of incident light by samples. Raman spectroscopy is of great value for both qualitative analysis (e.g., determination of molecular structures) and quantitative analysis. In Fig. 4.12, Raman spectra of both GO (a) and rGO (b) exhibited two remarkable peaks at around 1330 and 1598 cm^{-1}, which were assigned to the D band and G band in 2D carbon nanomaterials, respectively. Moreover, compared with GO, rGO had an increased D/G intensity ratio, and this index reflected that rGO had less in-plane sp^2 and more ordered crystal structure [51].

Compared with IR spectroscopy, Raman spectroscopy is an ideal tool for studying biological samples and chemical compounds in aqueous solutions because of the weak Raman scattering of water. However, Raman spectroscopy has a limitation of extremely low quantum efficiencies associated with the scattering process. Thus, it is essential to employ Surface-Enhanced Raman Scattering (SERS) to amplify the Raman signal. Wei group integrated the AuNPs into MIL-101 to provide a promising platform for real-time probing of AuNP-catalyzed reactions. In this system, the

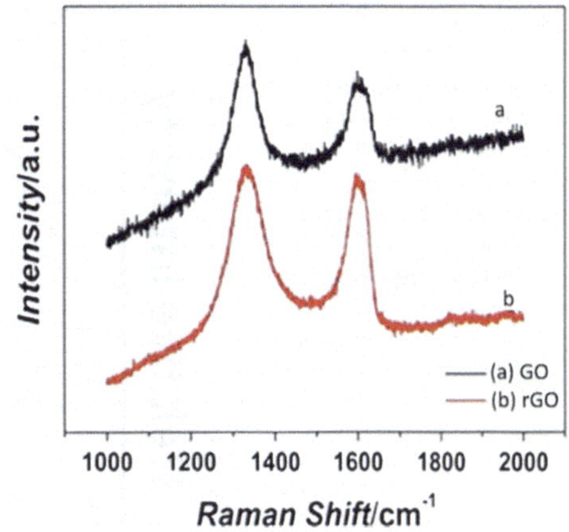

Fig. 4.12 Raman spectra of GO (**a**) and rGO (**b**). Reprinted with permission from [51]. Copyright (2016) Elsevier

AuNPs acted not only as the nanozymes but also as the SERS-active substrates to enhance the Raman signals [52].

X-ray diffraction (XRD) employs an X-ray to obtain the diffraction patterns of nanomaterials. One can get information about composition and crystalline phase of a nanomaterial by analyzing and comparing its characteristic diffraction patterns with the standard ones in the database (i.e., JCPDS cards). It can be seen from Fig. 4.13, diffraction peaks of $CeVO_4$-1.5, $CeVO_4$-2, $CeVO_4$-2.5, and $CeVO_4$-3 (synthesized

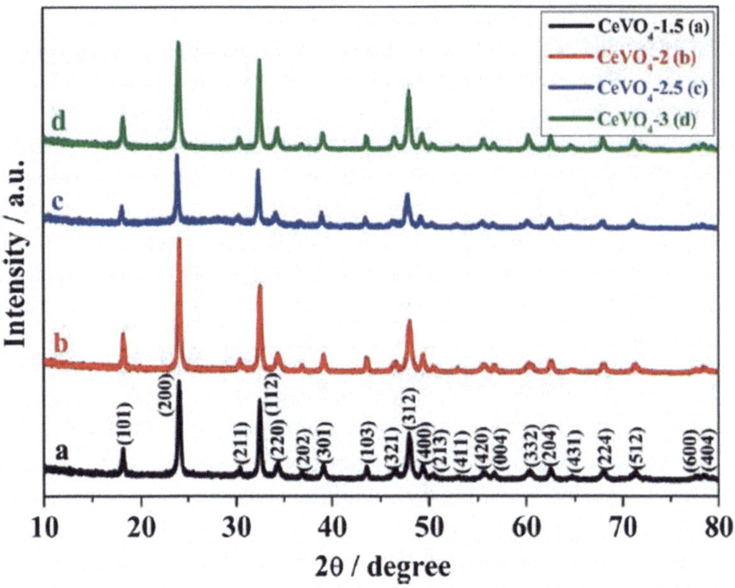

Fig. 4.13 XRD patterns of the as-prepared $CeVO_4$ samples. Reprinted with permission from [53]. Copyright (2016) Royal Society of Chemistry

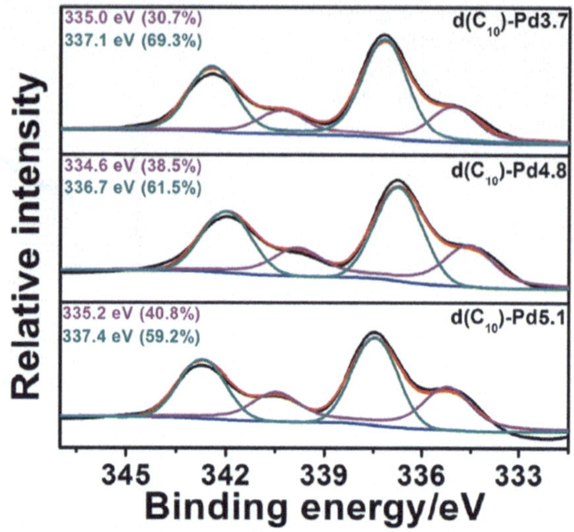

Fig. 4.14 Pd 3d XPS spectra of d (C_{10})–Pd. The black, red, cyan, and magenta lines represent the raw, fitted, Pd^{2+}, and Pd^0 component curves, respectively. Reprinted with permission from [54]. Copyright (2016) Royal Society of Chemistry

with different concentration of EDTA) matched well with the tetragonal phase zircon-type $CeVO_4$ [53]. Strong and sharp diffraction peaks indicated the high crystallinity of samples. Note, X-ray will do harm to the human body and it is essential to take necessary precautions if using XRD or other X-ray-assisted characterization instrumentations.

X-ray photoelectron spectroscopy (XPS) employs an X-ray radiation (200–1500 eV) to produce the valence electrons or inner electrons stimulated emission, so that the energy of excited photoelectrons can be measured. XPS is widely used in solid surface analysis, including the element composition, atomic valence state, surface energy distribution, and energy level structures. As shown in Fig. 4.14, the Pd $3d_{5/2}$ XPS curves for different $d(C_{10})$–Pd (with different average diameters of 3.7, 4.8, and 5.1 nm) could be deconvoluted into two typical bands, corresponding to Pd^0 (magenta) and Pd^{2+} (cyan) species, respectively. As the size of $d(C_{10})$–Pd increased, the percentage of Pd^0 increased while the Pd^{2+} decreased [54]. Besides XPS, ultraviolet photoelectron spectroscopy (UPS, 10–45 eV) and Auger electron spectroscopy (AES, 1000–10,000 eV) can be used to probe the valence and core levels, respectively.

Brunauer–Emmett–Teller method (BET) for materials' surface area analysis originates from the BET theory which is the theoretical basis of particle surface adsorption theory. For a BET test, the samples will be filled with an inert gas (mostly N_2 gas) and physically adsorbed by the inert gas. The specific surface area of the samples is calculated by measuring the equilibrium pressure and adsorption gas flow rate when the physical adsorption is in equilibrium. Multipoint BET method is the US standard surface test method, which measures the absolute adsorption of nitrogen under different partial pressures. The BET measurement can be further improved since the current test process is relatively complex and time-consuming. In Fig. 4.15, the specific surface areas of the CeO_2/NiO nanocomposites were determined by the

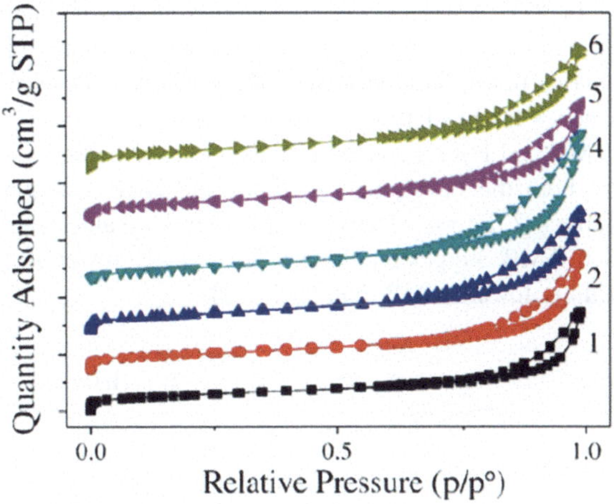

Fig. 4.15 N_2 adsorption–desorption isothermal curves of NiO and CeO_2–NiO nanocomposites. 1, NiO; 2, CeO_2/NiO-1; 3, CeO_2/NiO-2; 4, CeO_2/NiO-4; 5, CeO_2/NiO-6; and 6, CeO_2/NiO-8. Adapted with permission from [55]. Copyright (2017) Elsevier

N_2 adsorption and desorption isotherms and calculated by BET theory. The specific surface areas of samples 1–6 were calculated to be 192.3, 203.6, 212.1, 242.4, 217.5, and 229.8 m^2g^{-1}, respectively [55].

4.2.2 *Enzymatic Kinetics*

As emerging alternatives to natural enzymes, it is important to measure the enzymatic kinetics parameters of nanozymes, which can then be used to evaluate the catalytic efficiency of nanozymes and to make comparison between nanozymes and the corresponding natural enzymes.

The Michaelis–Menten equation represents the relationship between the initial velocity of the enzymatic reaction and the substrate concentration. In Michaelis–Menten equation, there are two basic parameters, K_m and $V_{\max}$. The Michaelis–Menten constant (K_m) is a characteristic physical quantity of an enzymatic catalyst and its corresponding substrate which is the concentration of the substrate when the enzymatic reaction reaches half of the maximum velocity ($V_{\max}$). K_m is the sign of affinity between an enzyme and its substrate. In general, the enzyme kinetics parameters are measured by monitoring the absorption (or fluorescent) signal of a colored (or fluorescent) product. For example, 3, 3′, 5, 5′-tetramethylbenzidine (TMB), o-phenylenediamine (OPD), and 2,2′-azinobis-(3-ethylbenzthiazoline-6-sulphonate) (ABTS) have been widely used as the substrates for peroxidase mimicking nanozymes. In addition to direct reading of the approximation of $V_{\max}$ and K_m from the plot of substrate concentration versus reaction velocity, it is more convenient to get the exact values of them from the corresponding Lineweaver–Burk graph.

In 2007, Yan and co-workers found that Fe_3O_4 nanoparticles could exhibit HRP mimicking activity [22]. To confirm the enzymatic mimicking activity of the Fe_3O_4 nanoparticles, they performed the kinetics analysis. Figure 4.16 is the steady-state kinetics assay of the Fe_3O_4 nanoparticles. Within the suitable range of substrate (TMB and H_2O_2) concentrations, typical Michaelis–Menten curves were observed for both the Fe_3O_4 nanoparticles and HRP. Compared with the data of HRP in Table 4.2, Fe_3O_4 MNPs had a higher K_m value with H_2O_2 and a lower K_m value with TMB, suggesting that the Fe_3O_4 had a lower affinity toward H_2O_2 but a higher affinity toward TMB than HRP [22].

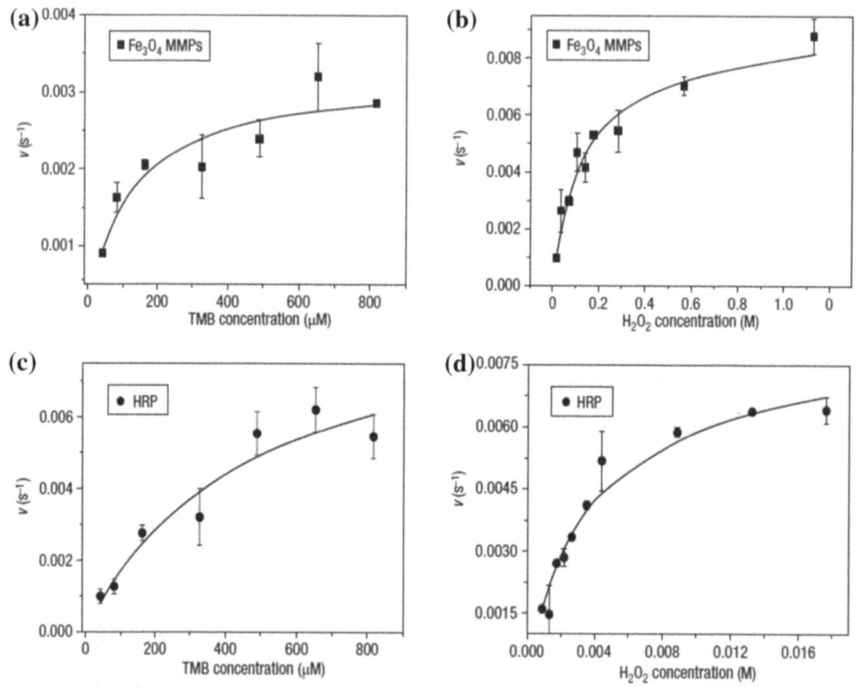

Fig. 4.16 Steady-state kinetics assay and catalytic mechanism of Fe_3O_4 MNPs. **a**–**d** The velocity (v) of the reactions was measured using 20 mg Fe_3O_4 MNPs (**a**, **b**) or 0.5 ng HRP (**c**, **d**) in 500 mL of 0.2 M NaAc (pH = 3.5) at 40°C. Error bars shown represent the standard error derived from three repeated measurements. **a**, **c** The concentration of H_2O_2 was 530 mM (Fe_3O_4 MNPs) or 8.8 mM (HRP) and the TMB concentration was varied. **b**, **d** The concentration of TMB was 816 mM and the H_2O_2 concentration was varied. Reprinted with permission from [22]. Copyright (2007) Nature Publishing Group

Table 4.2 Comparison of the kinetics parameters of Fe_3O_4 MNPs and HRP. [E] is the enzyme (or MNP) concentration, K_m is the Michaelis constant, V_{max} is the maximal reaction velocity, and K_{cat} is the catalytic constant, where $K_{cat} = V_{max}/[E]$. Reprinted with permission from [22]. Copyright (2007) Nature Publishing Group

	[E] (M)	Substrate	K_m (mM)	V_{max} (M s^{-1})	K_{cat} (s^{-1})
Fe_3O_4 MNPs	11.4×10^{-13}	TMB	0.098	3.44×10^{-8}	3.02×10^4
Fe_3O_4 MNPs	11.4×10^{-13}	H_2O_2	154	9.78×10^{-8}	8.58×10^4
HRP	2.5×10^{-11}	TMB	0.434	10.00×10^{-8}	4.00×10^3
HRP	2.5×10^{-11}	H_2O_2	3.70	8.71×10^{-8}	3.48×10^3

4.2.3 Probing the Catalytic Process

The characterization of a catalytic process mainly refers to the detection of various products (especially intermediates), possible reaction pathways and changes of parameters in the catalytic process.

Electron paramagnetic resonance (EPR) or electron spin resonance (ESR) originates from the magnetic moments of unpaired electrons. It can be used to detect the unpaired electrons contained in atoms or molecules from qualitative and quantitative aspects and explore the surroundings. To free radicals, the orbital magnetic moments are almost inactive, and the majority of the total magnetic moments (more than 99%) contributed from electron spin. In Fig. 4.17a, the EPR spectra of the solution containing CuO NPs, GOx, and glucose showed characteristic peaks of hydroxyl radicals originated from reactions between CuO and H_2O_2. Since the hydroxyl radical could then attack reductive substrates to form carbon radical intermediates, the typical carbon radical signals were detected and confirmed after the addition of DMAA monomer (Fig. 4.18b) [56].

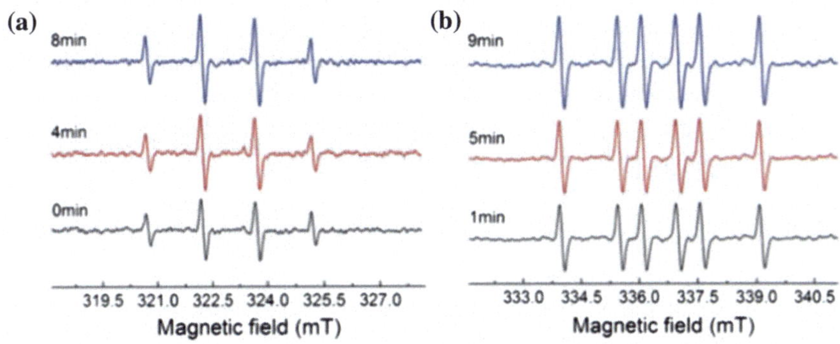

Fig. 4.17 **a** EPR spectra of the solution of CuO/GOx/glucose. **b** EPR spectra of the solution of CuO/GOx/glucose/DMAA. Reprinted with permission from [56]. Copyright (2017) Royal Society of Chemistry

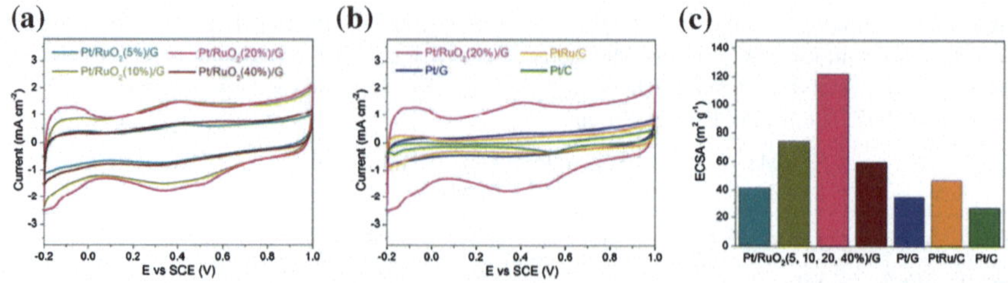

Fig. 4.18 Electrocatalytic activities of 3D $Pt/RuO_2/G$ architectures toward methanol electrooxidation. CVs of **a** the $Pt/RuO_2/G$ architectures with varying RuO_2 contents, and **b** Pt/RuO_2 (20%)/G, Pt/G, PtRu/C, and Pt/C in 1 M H_2SO_4 solution at 20 mVs^{-1}. **c** Specific ECSA values for the catalysts, showing the highest Pt utilization in the Pt/RuO_2 (20%)/G architecture. Reprinted with permission from [57]. Copyright (2017) Royal Society of Chemistry

Electrochemistry has an ability to convert the chemical signals related to redox reactions into electron signals. It can provide deep insights into a reaction from the thermodynamics, kinetics, absorption and desorption, diffusion, and other aspects. It is also an effective method to study multiple complex areas. Its precision, high sensitivity, and strong stability make it popular in catalytic field. As shown in Fig. 4.18a, b, the electrocatalytic properties of the Pt/RuO_2/graphene (different RuO_2 contents) architectures were evaluated by cyclic voltammetry. The electrochemically active surface areas (ECSAs) of electrode catalysts in Fig. 4.18c were calculated from the prominent peaks which were related to the hydrogen adsorption/desorption process. The Pt/RuO_2/graphene architectures with 20% Ru have the highest ECSA according to the results [57].

Differential thermal analysis (DTA) and Thermogravimetric analysis (TGA) are two commonly used thermal analysis methods in nanomaterial field. DTA refers to the measurement of the temperature differences between an inert control sample and a sample of interest under the same analysis environment (e.g., the same thermal cycles and the same quality of samples). TGA refers to a thermal analysis technique which determines the relationship between the mass and temperature of samples to study the thermal stability and composition. Figure 4.19 is a typical TGA graph conducted for bovine serum albumin (BSA) as control and biocompatible nanoceria encapsulated albumin nanoparticles (BCNPs). The results demonstrated that BCNPs had a slower rate of degradation, suggesting the enhanced stability of BCNPs compared with pristine BSA [58].

4.2.4 *Analyzing the Nanozyme Properties in Biological Systems*

Fluorescence microscopy (such as confocal laser scanning microscopy, CLSM) and **flow cytometry** are widely used to observe the survival state, morphology, and

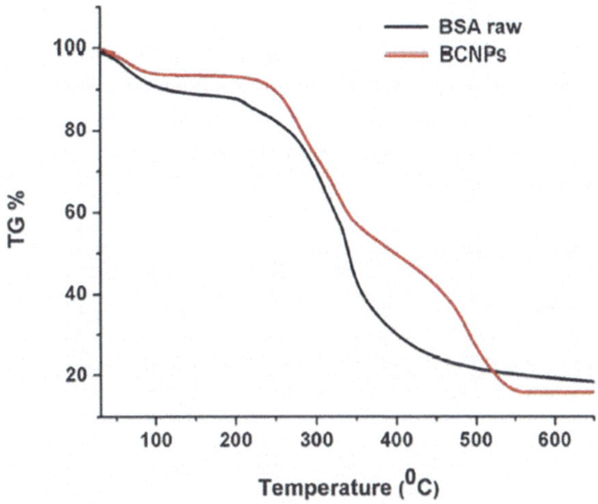

Fig. 4.19 TGA data curves of BSA (control) and biocompatible nanoceria encapsulated albumin nanoparticles. Reprinted with permission from [58]. Copyright (2015) Royal Society of Chemistry

distribution of cells. They have been used to assess the performance of nanozymes by detection of various physiological indicators of cells. **Small animal in vivo imaging system/multifunction in vivo imaging system** is a micro-optics system that helps us detect weak fluorescence and light-emitting signals in small animals preprocessed. Figure 4.20 shows the in vivo fluorescence images of mice after treatment with PMA. PMA will induce the inflammation and the release of ROS, which can oxidize the DCFH-DA to produce the fluorescence signals under the imaging system. After

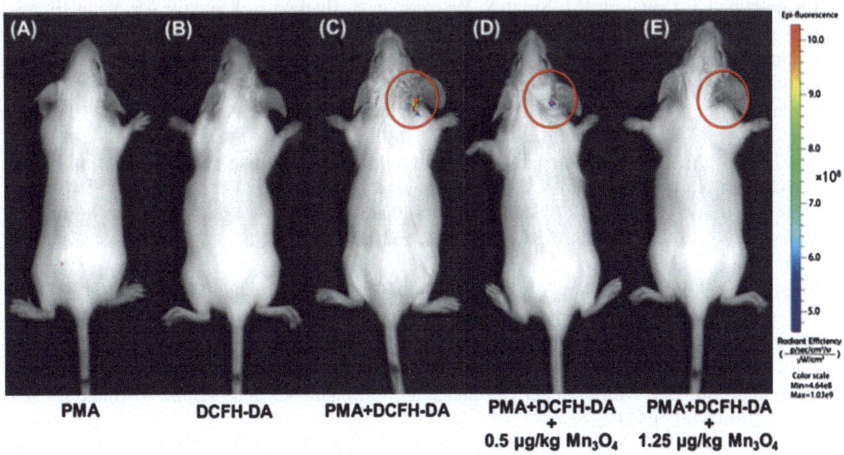

Fig. 4.20 In vivo fluorescence imaging of mice with PMA-induced ear inflammation after treatment with **a** PMA, **b** DCFH-DA, **c** PMA and DCFH-DA, **d** PMA and DCFH-DA with 0.5 μg kg^{-1} Mn_3O_4 NPs, and **e** PMA and DCFH-DA with 1.25 μg kg^{-1} Mn_3O_4 NPS. Reprinted with permission from [59]. Copyright (2018) Royal Society of Chemistry

treatment with Mn_3O_4 NPs, the signals weakened, which proved that Mn_3O_4 NPs have excellent SOD mimicking activity for scavenging ROS [59].

In summary, we can easily obtain the information of nanozymes about interactions with cells and biodistribution in tumor tissues by these noninvasive imaging systems. In this way, more nanozymes based therapies can be developed to conquer diseases including cancers.

Acknowledgements We thank National Natural Science Foundation of China (21722503 and 21874067), 973 Program (2015CB659400), PAPD program, Shuangchuang Program of Jiangsu Province, Open Funds of the State Key Laboratory of Analytical Chemistry for Life Science (SKLA-CLS1704), Open Funds of the State Key Laboratory of Coordination Chemistry (SKLCC1819), Fundamental Research Funds for the Central Universities (021314380103), and Thousand Talents Program for Young Researchers for financial support. We thank Jia Yao for help with the writing.

References

1. Wei H, Wang E (2013) Nanomaterials with enzyme-like characteristics (nanozymes): next-generation artificial enzymes. Chem Soc Rev 42(14):6060–6093
2. Wang X, Guo W, Hu Y, Wu J, Wei H (2016) Nanozymes: next wave of artificial enzymes. Springer, Berlin, Heidelberg
3. Wang X, Hu Y, Wei H (2016) Nanozymes in bionanotechnology: from sensing to therapeutics and beyond. Inorg Chem Front 3(1):41–60
4. André R, Natálio F, Humanes M, Leppin J, Heinze K, Wever R, Schröder H, Müller WE, Tremel W (2011) V2O5 nanowires with an intrinsic peroxidase-like activity. Adv Funct Mater 21(3):501–509
5. Pirmohamed T, Dowding JM, Singh S, Wasserman B, Heckert E, Karakoti AS, King JE, Seal S, Self WT (2010) Nanoceria exhibit redox state-dependent catalase mimetic activity. Chem Commun 46(16):2736–2738
6. Jiao X, Song H, Zhao H, Bai W, Zhang L, Lv Y (2012) Well-redispersed ceria nanoparticles: promising peroxidase mimetics for H_2O_2 and glucose detection. Anal Methods 4(10):3261–3267
7. Mu J, Wang Y, Zhao M, Zhang L (2012) Intrinsic peroxidase-like activity and catalase-like activity of Co_3O_4 nanoparticles. Chem Commun 48(19):2540–2542
8. Wang W, Jiang X, Chen K (2012) Iron phosphate microflowers as peroxidase mimic and superoxide dismutase mimic for biocatalysis and biosensing. Chem Commun 48(58):7289–7291
9. Shen L, Bao J, Wang D, Wang Y, Chen Z, Ren L, Zhou X, Ke X, Chen M, Yang A (2013) One-step synthesis of monodisperse, water-soluble ultra-small Fe_3O_4 nanoparticles for potential bio-application. Nanoscale 5(5):2133–2141
10. Li N, Yan Y, Xia BY, Wang JY, Wang X (2014) Novel tungsten carbide nanorods: an intrinsic peroxidase mimetic with high activity and stability in aqueous and organic solvents. Biosens Bioelectron 54(12):521–527
11. Liu Y, Zhu G, Bao C, Yuan A, Shen X (2014) Intrinsic peroxidase-like activity of porous CuO micro-/nanostructures with clean surface. Chin J Chem 32(2):151–156
12. Mu J, Zhang L, Zhao M, Wang Y (2014) Catalase mimic property of Co_3O_4 nanomaterials with different morphology and its application as a calcium sensor. ACS Appl Mater Interfaces 6(10):7090–7098
13. Qiao F, Chen L, Li X, Li L, Ai S (2014) Peroxidase-like activity of manganese selenide nanoparticles and its analytical application for visual detection of hydrogen peroxide and glucose. Sensor Actuat B-Chem 193(3):255–262

14. Zhang L, Xia F, Song Z, Webster NA, Luo H, Gao Y (2015) Synthesis and formation mechanism of VO_2(A) nanoplates with intrinsic peroxidase-like activity. RSC Adv 5(75):61371–61379
15. Zhang X, Huang Y (2015) Evaluation of the antioxidant activity of phenols and tannic acid determination with Mn_3O_4 nano-octahedrons as an oxidase mimic. Anal Methods 7(20):8640–8646
16. Park CY, Seo JM, Jo H, Park J, Ok KM, Park TJ (2017) Hexagonal tungsten oxide nanoflowers as enzymatic mimetics and electrocatalysts. Sci Rep 7:40928–40938
17. Liu J, Qiao S, Hu Q, Lu G (2011) Magnetic nanocomposites with mesoporous structures: synthesis and applications. Small 7(4):425–443
18. Liu S, Lu F, Xing R, Zhu J (2011) Structural effects of Fe_3O_4 nanocrystals on peroxidase-like activity. Chem Eur J 17(2):620–625
19. Wang W, Jiang X, Chen K (2012) $CePO_4$: Tb, Gd hollow nanospheres as peroxidase mimic and magnetic-fluorescent imaging agent. Chem Commun 48(54):6839–6841
20. Zhang X, He S, Chen Z, Huang Y (2013) $CoFe_2O_4$ nanoparticles as oxidase mimic-mediated chemiluminescence of aqueous luminol for sulfite in white wines. J Agric Food Chem 61(4):840–847
21. Cai Q, Lu S, Liao F, Li Y, Ma S, Shao M (2014) Catalytic degradation of dye molecules and in situ SERS monitoring by peroxidase-like Au/CuS composite. Nanoscale 6(14):8117–8123
22. Gao L, Zhuang J, Nie L, Zhang J, Zhang Y, Gu N, Wang T, Feng J, Yang D, Perrett S, Yan X (2007) Intrinsic peroxidase-like activity of ferromagnetic nanoparticles. Nat Nanotechnol 2(9):577–583
23. Wei H, Wang E (2008) Fe_3O_4 magnetic nanoparticles as peroxidase mimetics and their applications in H_2O_2 and glucose detection. Anal Chem 80(6):2250–2254
24. Shi W, Zhang X, He S, Huang Y (2011) $CoFe_2O_4$ magnetic nanoparticles as a peroxidase mimic mediated chemiluminescence for hydrogen peroxide and glucose. Chem Commun 47(38):10785–10787
25. Zhang Y, Tian J, Liu S, Wang L, Qin X, Lu W, Chang G, Luo Y, Asiri AM, AlYoubi AO, Sun X (2012) Novel application of CoFe layered double hydroxide nanoplates for colorimetric detection of H_2O_2 and glucose. Analyst 137(6):1325–1328
26. Vernekar AA, Das T, Ghosh S, Mugesh G (2016) A remarkably efficient $MnFe_2O_4$-based oxidase nanozyme. Chem Asian J 11(1):72–76
27. Tourinho FA, Franck R, Massart R (1990) Aqueous ferrofluids based on manganese and cobalt ferrites. J Mater Sci 25(7):3249–3254
28. Schwarz JA, Contescu C, Contescu A (1995) Methods for preparation of catalytic materials. Chem Rev 95(3):477–510
29. Vinosha A, Raja K, Christina Fernandez A, Krishnan S, Das J (2016) Investigation of optical, electrical and magnetic properties of cobalt ferrite nanoparticles by naive co-precipitation technique. Optik 127(20):9917–9925
30. Liu X, Tao S, Shen Y (1997) Preparation and characterization of nanocrystalline α-Fe_2O_3 by a sol-gel process. Sensor Actuat B-Chem 40(2):161–165
31. Zhao M, Huang J, Zhou Y, Pan X, He H, Ye Z, Pan X (2013) Controlled synthesis of spinel $ZnFe_2O_4$ decorated ZnO heterostructures as peroxidase mimetics for enhanced colorimetric biosensing. Chem Commun 49(69):7656–7658
32. Lee M, Kim T, Bae C, Shin H, Kim J (2010) Fabrication and applications of metal-oxide nanotubes. JOM 62(4):44
33. Kim CK, Kim T, Choi I, Soh M, Kim D, Kim Y, Jang H, Yang H, Kim JY, Park H (2012) Ceria nanoparticles that can protect against ischemic stroke. Angew Chem Int Ed 51(44):11039–11043
34. Gupta A, Eral HB, Hatton TA, Doyle PS (2016) Nanoemulsions: formation, properties and applications. Soft Matter 12(11):2826–2841
35. He W, Liu Y, Yuan J, Yin J, Wu X, Hu X, Zhang K, Liu J, Chen C, Ji Y (2011) Au@ Pt nanostructures as oxidase and peroxidase mimetics for use in immunoassays. Biomaterials 32(4):1139–1147

36. Nangia Y, Kumar B, Kaushal J, Suri CR (2012) Palladium@gold bimetallic nanostructures as peroxidase mimic for development of sensitive fluoroimmunoassay. Anal Chim Acta 751(21):140–145
37. Li J, Liu W, Wu X, Gao X (2015) Mechanism of pH-switchable peroxidase and catalase-like activities of gold, silver, platinum and palladium. Biomaterials 48:37–44
38. Xia Y, Gilroy KD, Peng HC, Xia X (2016) Seed-mediated growth of colloidal metal nanocrystals. Angew Chem Int Ed 56(1):60
39. Zhu Y, Chen F (2014) Microwave-assisted preparation of inorganic nanostructures in liquid phase. Chem Rev 114(12):6462–6555
40. Liu H, Gu C, Xiong W, Zhang M (2015) A sensitive hydrogen peroxide biosensor using ultra-small $CuInS_2$ nanocrystals as peroxidase mimics. Sensor Actuat B-Chem 209:670–676
41. Garg B, Bisht T (2016) Carbon nanodots as peroxidase nanozymes for biosensing. Molecules 21(12):1653
42. Dong W, Liu X, Shi W, Huang Y (2015) Metal–organic framework MIL-53(Fe): facile microwave-assisted synthesis and use as a highly active peroxidase mimetic for glucose biosensing. RSC Adv 5(23):17451–17457
43. Hu Y, Guo W, Wei H (2015) Protein- and peptide-directed approaches to fluorescent metal nanoclusters. Isr J Chem 55(6–7):682–697
44. Liu Y, Ding D, Zhen Y, Guo R (2017) Amino acid-mediated 'turn-off/turn-on' nanozyme activity of gold nanoclusters for sensitive and selective detection of copper ions and histidine. Biosens Bioelectron 92:140–146
45. Sadik OA, Du N, Kariuki V, Okello V, Bushlyar V (2014) Current and emerging technologies for the characterization of nanomaterials. ACS Sustain Chem Eng 2(7):1707–1716
46. Gao L, Liu M, Ma G, Wang Y, Zhao L, Yuan Q, Gao F, Liu R, Zhai J, Chai Z, Zhao Y, Gao X (2015) Peptide-conjugated gold nanoprobe: intrinsic nanozyme-linked immunsorbant assay of integrin expression level on cell membrane. ACS Nano 9(11):10979–10990
47. Ge C, Fang G, Shen X, Chong Y, Wamer W, Gao X, Chai Z, Chen C, Yin J (2016) Facet energy versus enzyme-like activities: the unexpected protection of palladium nanocrystals against oxidative damage. ACS Nano 10(11):10436–10445
48. Li G, Ma P, Zhang Y, Liu X, Zhang H, Xue W, Mi Y, Luo Y, Fan H (2016) Synthesis of Cu_2O nanowire mesocrystals using PTCDA as a modifier and their superior peroxidase-like activity. J Mater Sci 51(8):3979–3988
49. Cheng H, Zhang L, He J, Guo W, Zhou Z, Zhang X, Nie S, Wei H (2016) Integratednanozymes with nanoscale proximity for in vivo neurochemical monitoring in living brains. Anal Chem 88(10):5489–5497
50. Wang Y, Hu J, Zhuang Q, Ni Y (2016) Enhancing sensitivity and selectivity in a label-free colorimetric sensor for detection of iron(II) ions with luminescent molybdenum disulfide nanosheet-based peroxidase mimetics. Biosens Bioelectron 80:111–117
51. Yan X, Gu Y, Li C, Tang L, Zheng B, Li Y, Zhang Z, Yang M (2016) Synergetic catalysis based on the proline tailed metalloporphyrin with graphene sheet as efficient mimetic enzyme for ultrasensitive electrochemical detection of dopamine. Biosens Bioelectron 77:1032–1038
52. Hu Y, Cheng H, Zhao X, Wu J, Muhammad F, Lin S, He J, Zhou L, Zhang C, Deng Y, Wang P, Zhou Z, Nie S, Wei H (2017) Surface-Enhanced Raman Scattering active gold nanoparticles with enzyme-mimicking activities for measuring glucose and lactate in living tissues. ACS Nano 11(6):5558–5566
53. Ju P, Yu Y, Wang M, Zhao Y, Zhang D, Sun C, Han X (2016) Synthesis of EDTA-assisted $CeVO_4$ nanorods as robust peroxidase mimics towards colorimetric detection of H_2O_2. J Mater Chem B 4(38):6316–6325
54. Li W, Zhi X, Yang J, Zhang J, Fu Y (2016) Colorimetric detection of cysteine and homocysteine based on an oligonucleotide-stabilized Pd nanozyme. Anal Methods 8(25):5111–5116
55. Mu J, Zhao X, Li J, Yang E, Zhao X (2017) Coral-like CeO_2/NiO nanocomposites with efficient enzyme-mimetic activity for biosensing application. Mater Sci Eng C 74:434–442
56. Ye Y, Xiao L, He B, Zhang Q, Nie T, Yang X, Wu D, Cheng H, Li P, Wang Q (2017) Oxygen-tuned nanozyme polymerization for the preparation of hydrogels with printable and antibacterial properties. J Mater Chem B 5(7):1518–1524

57. Huang H, Zhu J, Li D, Shen C, Li M, Zhang X, Jiang Q, Zhang J, Wu Y (2017) Pt nanoparticles grown on 3D RuO_2-modified graphene architectures for highly efficient methanol oxidation. J Mater Chem A 5(9):4560–4567
58. Bhushan B, Gopinath P (2015) Antioxidant nanozyme: a facile synthesis and evaluation of the reactive oxygen species scavenging potential of nanoceria encapsulated albumin nanoparticles. J Mater Chem B 3(24):4843–4852
59. Yao J, Cheng Y, Zhou M, Zhao S, Lin S, Wang X, Wu J, Li S, Wei H (2018) ROS scavenging Mn_3O_4 nanozymes for in vivo anti-inflammation. Chem Sci 9(11):2927–2933

Part II
Nanomaterial-Based Nanozymes

Chapter 5
Iron Oxide Nanozyme: A Multifunctional Enzyme Mimetics for Biomedical Application

Lizeng Gao, Kelong Fan and Xiyun Yan

Abbreviations

ABTS	2, 2′-azino-bis (3-ethylbenzothiazoline-6-sulfonic acid)
DAB	3, 3′-Diaminobenzidine
DMSA	Dimercaptosuccinic acid
DNA	Deoxyribonucleic acid
ELISA	Enzyme-linked immunosorbent assay
ESR	Electron spin resonance
GO	Graphene oxide
GOx	Glucose oxidase
HRP	Horseradish peroxidase
ION	Zyme iron oxide nanozyme
MRI	Magnetic resonance imaging
MNPs	Magnetic nanoparticles
OPD	o-phenylenediamine
ROS	Reactive oxygen species
SPIO	Superparamagnetic iron oxide
TA	Terephthalic acid
TMB	3, 3′, 5, 5′-Tetramethylbenzidine

The content in this chapter has been published on Theranostics and reformatted in this book.

L. Gao (✉) · K. Fan · X. Yan
CAS Engineering Laboratory for Nanozyme. Key Laboratory of Protein and Peptide Pharmaceutical, Institute of Biophysics, Chinese Academy of Sciences, 15 Datun Road, Beijing 100101, China
e-mail: gaolizeng@ibp.ac.cn

X. Yan (ed.), *Nanozymology*, Nanostructure Science and Technology,
https://doi.org/10.1007/978-981-15-1490-6_5

5.1 Introduction

Iron oxide nanoparticle is a well-known nanomaterial in biomedical applications due to its special magnetic property at nanoscale. Typical iron oxide particles show coercivity and remanence (retentivity) in bulk format. However, as the particle size decreases into nanoscale, it tends to display paramagnetic or superparamagnetic magnetization behavior, which means iron oxide nanoparticles can be easily aggregated in the presence of external magnetic field and readily re-dispersed with the removal of magnetic field. Given excellent magnetic property (superparamagnetism), iron oxide nanoparticles has greatly inspired a lot of applications in various fields, especially for biomedicine, including bio-separation and purification, biosensor, transfection, magnetic resonance imaging (MRI), hyperthermia therapy, targeted drug delivery, etc. [1–5] (Fig. 5.1). It also serves as an ideal platform to customize diagnostic imaging and targeted therapy together for theranostic treatment [6, 7]. Several types of superparamagnetic iron oxide (SPIO) nanoparticles have been already approved by FDA to facilitate the translation of them for clinical applications [8, 9]. In all these applications, iron oxide nanoparticles are generally assumed biologically inert or coated to avoid any unforeseen activities.

In 2007, a surprising discovery was reported that ferromagnetic (Fe_3O_4) nanoparticles showed intrinsic peroxidase-like activity which had a similar catalytic behavior as horseradish peroxidase (HRP) [10]. This was the first time an inorganic nanoparticle was considered as an enzyme mimetics for biomedical application. This pioneering work soon inspired the finding of a variety of nanomaterials with enzyme-like activities, including metal, metal oxides, and metal–carbon compound nanomaterials with similar catalysis as peroxidase, haloperoxidase, NADH peroxidase, catalase, oxidase, glucose oxidase, sulfite oxidase, superoxide dismutase, etc. [11–17]. More importantly, these catalytic nanomaterials may become a new generation of artificial enzyme or enzyme mimic [18, 19]. A specified term, “Nanozyme” [20], was introduced to define the nanomaterials with intrinsic enzyme-like activities, in order to

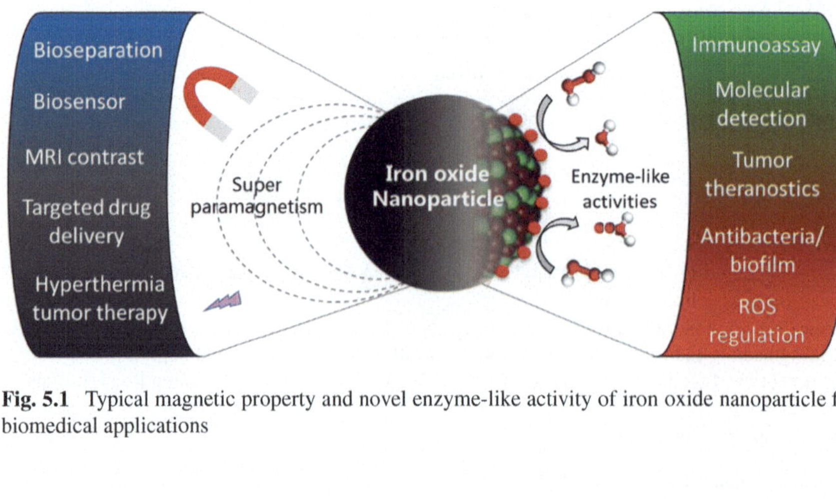

Fig. 5.1 Typical magnetic property and novel enzyme-like activity of iron oxide nanoparticle for biomedical applications

distinguish those nanocomplexes with external immobilized enzymes. A series of novel applications have widely been developed based on the featured mimetic activities to improve human health from diagnosis to antibacteria and cancer therapy, engineer the environment for pollutant monitor and removal and refine the efficiency in chemical industry [21–25]. Now, nanozyme is becoming a novel interdisciplinary field bridging nanotechnology and biology [12, 26–31].

Among the reports in this field, iron oxide nanoparticle (iron oxide nanozyme (IONzyme) including Fe_3O_4 and Fe_2O_3) is one of the most classical types of nanozymes. The analytical methods developed for IONzyme characterization have been referred to exploring other nanozymes, especially for kinetics and mechanism assay to determine the mimicking property. In addition, more and more researchers are intending to use iron oxide nanomaterials as enzyme mimetics and have developed many novel biomedical applications based on their enzyme-like activities. Although many review articles about iron oxide nanoparticles have been published [32–34], none of them summarized the intrinsic enzyme-like properties for biomedical application. To advance the concept of nanozyme and highlight the importance of IONzyme, we will systematically introduce enzymatic properties of IONzyme and summarize the latest biomedical applications based on its biomimetic activities in this review article (Fig. 5.1).

5.2 IONzyme: A Novel Enzyme Mimetics

IONzyme exhibits peroxidase-like activity or catalase-like activity under physiological reaction conditions with pH dependence. Both activities show typical catalytic features that are similar to natural enzymes, including substrates, optimal pH, and temperature. More importantly, all the IONzyme catalyzes follow Michaelis–Menten kinetics and possess same mechanism as natural enzyme, affirming that IONzyme is a new type of enzyme mimetics. All these enzyme-like activities are from the IONzyme themselves synthesized with typical methods for nanomaterials preparation rather than modification with catalytic groups or natural enzymes on the surface. Therefore, IONzyme has intrinsic enzyme-like activities, representing a typical type of nanozymes.

5.2.1 Enzymatic Activities of IONzyme

Currently, IONzyme is reported to mimic two enzymatic activities: peroxidase (EC 1.11.1.7) and catalase (EC 1.11.1.6), which both have porphyrin heme as the cofactor in the active site and catalyze the substrate of hydrogen peroxide but the former generates free radicals to react with hydrogen donor (AH_2) and the latter generates oxygen (Eqs. 5.1 and 5.2). Both peroxidase and catalase belong to the oxidoreductase family and play a critical role in preventing cellular oxidative damage in aerobically respiring organism.

$$H_2O_2 + AH_2 \overset{\text{peroxidase}}{\longrightarrow} A + 2H_2O \tag{5.1}$$

$$2H_2O_2 \overset{\text{Catalase}}{\longrightarrow} O_2 + 2H_2O \tag{5.2}$$

(1) **Peroxidase-like activity**: Peroxidase-like activity, which catalyzes the typical colorimetric reaction containing hydrogen peroxide (H_2O_2) and chromogenic reagents (Fig. 5.2a), was the first enzyme-like activity found in IONzyme. Both Fe_2O_3 and Fe_3O_4 nanomaterials have this activity but usually the latter has better activity than the former [35]. In this reaction process, H_2O_2 is catalyzed to generate free radicals as the intermediate and then free radicals react with hydrogen donor (usually chromogenic molecules) to form water and oxidized donor. IONzyme catalysis requires the same optimal conditions as those for HRP, including the optimal temperature at 37–40 °C under the optimal pH (pH 3–6.5) in acidic buffer [10] (Table 5.1). The H_2O_2 concentration must be in a proper range for both IONzyme catalysis and HRP to maintain the activity, because excessive amount of H_2O_2 could inhibit the colorimetric reaction when

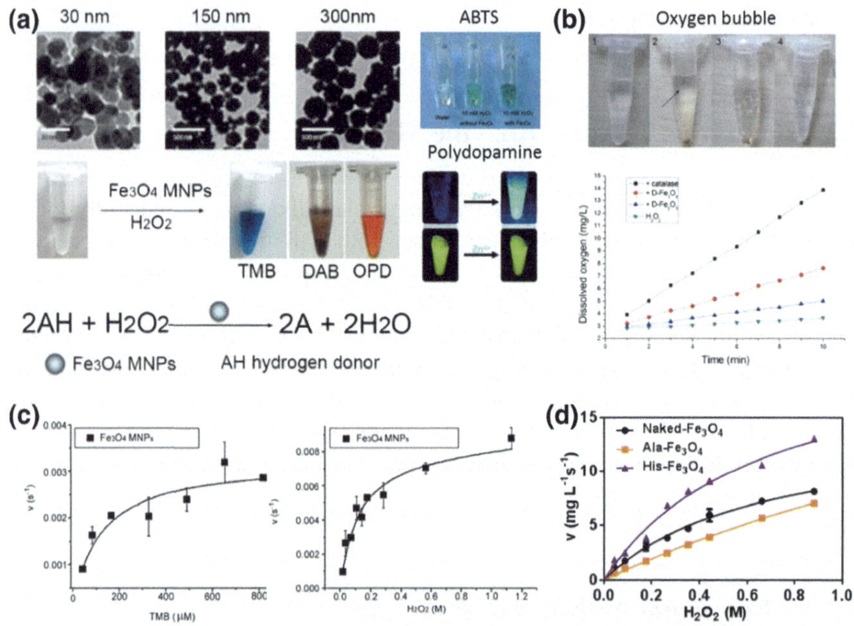

Fig. 5.2 Enzyme-like activities of IONzyme. **a** Colorimetric reaction catalyzed by peroxidase-like activity of IONzyme with substrates, such as TMB, DAB, OPD [10], ABTS [36], and fluorescent polydopamine [37]. **b** H_2O_2 decomposition by catalase-like activity of IONzyme with oxygen (bubble) generation and identification with dissolved oxygen measurement [35]. **c** Michaelis–Menten kinetics for peroxidase-like activity [10]. **d** Michaelis–Menten kinetics for catalase-like activity [48]. Reproduced with the permission from references [10, 35–37, 48]

Table 5.1 Substrates, optimal conditions, and kinetic parameters for IONzyme activities

IONzyme and size (diameter)	Substrate	K_M (mM)	V_{max} (M s^{-1})	K_{cat} (s^{-1})	pH	Temperature (°C)	Mimicking activity
Fe_3O_4, 300 nm [10]	H_2O_2	154	9.78×10^{-8}	8.58×10^4	3.5	40	Peroxidase
	TMB	0.098	3.44×10^{-8}	3.02×10^4			
Fe_3O_4, 13 ± 3.5 nm [44]	H_2O_2	54.6	1.8×10^{-8}		7.4	30	Peroxidase
	TMB	0.374	2.6×10^{-8}				
GO-Fe_3O_4 [49]	H_2O_2	0.71	5.31×10^{-8}		4	40	Peroxidase
	TMB	0.43	13.08×10^{-8}				
Fe3O4@Pt [50]	H_2O_2	702.6	7.136×10^{-7}	84.420	4.4	RT	Peroxidase
	TMB	0.147	0.711×10^{-7}	84.112			
Fe_3O_4@Carbon, 120 nm [51]	H_2O_2	0.38	73.99×10^{-8}		3	45	Peroxidase
	TMB	0.072	17.99×10^{-8}				
Magnetosome [52]	H_2O_2	170.65	9.33×10^{-9}		4	28, light	Peroxidase
	TMB	0.90	4.45×10^{-9}				
Fe_3O_4@Cu@Cu_2O, 50 nm [53]	H_2O_2	2.3	11.9×10^{-8}		8	25	Peroxidase
	OPDA	0.85	13.2×10^{-8}				
$Mn_{0.5}Fe_{0.5}Fe_2O_4$, 10–11 nm [54]	H_2O_2	310	3.63×10^{-6}	1.134×10^4	4	RT	Peroxidase
	TMB	0.139	4.5×10^{-6}	1.4062×10^4			
Fe_3O_4, 34–46 nm [55]	TMB	0.24–0.71	$0.42–2.4 \times 10^{-7}$	$0.2–1.14 \times 10^{-4}$		RT	Peroxidase
Fe_3O_4, 34–46 nm [55]	ABTS	0.12–0.96	$0.52–6.10 \times 10^{-7}$	$0.25–2.9 \times 10^{-4}$		RT	Peroxidase
Fe_3O_4, 13 nm [43]	Amplex ultrared	3.6×10^{-5}	4.15×10^{-6}		7.0	37	Peroxidase

(continued)

Table 5.1 (continued)

IONzyme and size (diameter)	Substrate	K_M (mM)	V_{max} (M s^{-1})	K_{cat} (s^{-1})	pH	Temperature (°C)	Mimicking activity
PB-γ-Fe_2O_3, 9.8 nm [56]	H_2O_2	323.6	1.17×0^{-6}	3.79×0^{3}	4.6	RT	Peroxidase
	TMB	0.307	1.06×10^{-6}	3.43×0^{3}			
PB-Fe_2O_3, 46 nm [57]	H_2O_2	0.015×10^{-3}	2.28×10^{-7}		4		Peroxidase
	TMB	9.95×10^{-3}	1.23×10^{-7}				
PB–Fe_2O_3 [58]	H_2O_2	91.54	8.308×0^{-8}		7	40	Peroxidase
	3,5-DTBC	1.22	4.431×0^{-8}				
γ-Fe_2O_3, 122.4 nm [59]	H_2O_2	21.14	1.319×10^{-9}		5	30	Peroxidase
	TMB	0.1709	2.647×10^{-9}				
γ-Fe_2O_3, 20–50 nm [60]	H_2O_2	157.19	1.284×10^{-8}		4	RT	Peroxidase
	TMB	0.0887	0.97×10^{-8}				
GO-Fe_2O_3 [61]	H_2O_2	305	1.01×10^{-7}		3.6	25	Peroxidase
	TMB	0.118	5.38×10^{-8}				
Pd@γ-Fe_2O_3 [62]	H_2O_2	0.254	1.28×10^{-7}	0.094	3.6	37	Peroxidase
	ABTS	0.049	1.02×10^{-8}	0.075	2		
α-Fe_2O_3 [63]	TMB	0.257	1.65×10^{-8}	5.5	4	37	Peroxidase
Fe_3O_4, 200 nm [48]	H_2O_2	571.6	20.45[a]		7	RT	Catalase
GO-Fe_2O_3 [61]	H_2O_2	–	–		7.4		Catalase
Horseradish peroxidase [10]	H_2O_2	3.7	8.71×10^{-8}	3.48×10^{3}	3.5	40	Peroxidase
	TMB	0.434	10.00×10^{-8}	4.00×10^{3}			
Catalase from *bovine liver* [64]	H_2O_2	54.3	1.6510^{-5}	6.09×10^{4}	8		Catalase

[a]The unit is mg L^{-1} s^{-1} of O_2

3, 3′, 5, 5′-Tetramethylbenzidine (TMB) was the hydrogen donor. Many chromogenic substrates were able to be catalyzed by IONzyme, for example, TMB, o-phenylenediamine (OPD), 3, 3′-Diaminobenzidine (DAB), and 2, 2′-azino-bis (3-ethylbenzothiazoline-6-sulfonic acid) (ABTS) [10, 35, 36] and even fluorescent polydopamine [37] (Fig. 5.2a), terephthalic acid (TA) [38], luminol, benzoic acid [39], etc. Apart from the chromogenic substrates, biomolecules also can be the target including proteins, nucleic acids, polysaccharides [40], and lipids [41]. Especially, IONzyme can induce lipid oxidation by catalyzing lipid peroxides (LOOH) generated from unsaturated lipids via chain reactions–propagation [41], indicating that IONzyme has broad substrates.

Like natural enzymes, some chemicals may stimulate or inhibit the catalysis, which are usually termed as activator or inhibitor. Theoretically, all the chemicals having effect on HRP should display similar role in IONzyme catalysis. Currently, the reported activators include ATP, ADP, AMP [42, 43], and DNA. Especially, ATP can enhance the peroxidase-like activity at neutral pH by complexation with Fe_3O_4 nanoparticles to participate in single electron transfer reaction [44]. Liu Juewen group found that ssDNA can enhance nearly 10-fold activity by DNA adsorption on Fe_3O_4 nanoparticles facilitating TMB binding [45]. And the inhibitors usually can bleach the free radicals, like sodium azide, ascorbic acid, hypotaurine, and catecholamines [46]. Controversially, it was found that ssDNA and dsDNA can reduce the catalytic activity by the screening effect of adsorbed DNA on the surface of iron oxide nanoparticles to OPD [47], indicating the inhibition is from blocking the affinity of substrate to IONzyme rather than bleaching the free radicals. Although the catalytic behaviors have been systematically studied, the comparison of the activity between IONzymes is still challenging due to their complicated nanostructures and surface modifications. Probably, specific activity can be used to assess the peroxidase-like activity by determining the enzyme unit (U) (moles of substrate converted per unit time) with certain amount (e.g., mg) of IONzyme at the exactly same reaction conditions (substrate types and concentrations, pH, temperature and buffer, time, etc.), which may reflect the activity level of IONzymes.

(2) **Catalase-like activity**: In addition to peroxidase-like activity, IONzyme also shows catalase-like activity to decompose H_2O_2 into oxygen and water. Gu group first reported that both dimercaptosuccinic acid (DMSA)-coated γ-Fe_2O_3 and Fe_3O_4 nanoparticles decomposed H_2O_2 to oxygen under neutral and basic pH (pH 7–10) [35] (Fig. 5.2b). Similar to peroxidase-like catalysis, Fe_3O_4 has higher catalase-like activity than Fe_2O_3 [35]. The oxygen bubble was observed by naked eye after the catalytic reaction, and the generation of oxygen thus could be measured by the oximetry with oxygen electrode. Up to now, IONzyme can perform the aforementioned two activities with proper pH control.

5.2.2 *Kinetics and Mechanism*

The assay based on Michaelis–Menten kinetics is the standard method for enzymatic characterization in the field of nanozyme. IONzyme follows the typical Michaelis–Menten kinetics similar to both HRP and catalase (Eq. 5.3). Most work for determining the apparent steady-state kinetic parameters is based on the peroxidase reaction (Table 5.1). In H_2O_2–TMB system, both substrates of H_2O_2 and TMB followed Michaelis–Menten behavior [10] (Fig. 5.2c). Usually, the apparent K_M value of the IONzyme for H_2O_2 was higher than that for HRP, indicating the lower affinity and efficiency to catalyze H_2O_2 (Table 5.1). In contrast, the apparent K_M value for the substrate of hydrogen donor (TMB) was much lower than that for HRP, suggesting the IONzyme has a higher affinity to TMB than native enzyme. If calculated using nanoparticles' molar concentration, the IONzyme with the diameter at 300 nm showed a level of activity 40 times higher than that of HRP [10]. This may be due to the fact that a HRP molecule has only one iron ion, in contrast to an abundance of iron on the surface of an iron oxide nanoparticle. IONzyme also showed similar catalytic mechanism as HRP, following a ping-pong mechanism [10]. In the process of the reaction, the IONzyme binds and reacts with the first substrate (H_2O_2) to generate hydroxyl free radicals (•OH) as an intermediate state and then the •OH captures a H^+ from the hydrogen donor such as TMB. Electron spin resonance (ESR) was used to monitor the generation of •OH during the reaction and found that IONzyme could conduct the same •OH intermediate, which further confirmed its similarities to peroxidase activity. As free radicals attack doesn't have specificity, the substrates are not only limited to chromophores. Theoretically, any molecules that can serve as hydrogen donor can be the potential substrates in this catalytic process. This gives IONzyme a broad scope of application, especially acting on various biological molecules, such as protein, nucleic acid, polysaccharide [40, 48], and lipid [41].

$$\nu = (V_{\max}[S])/(K_M + [S]) \tag{5.3}$$

where $[S]$ is the substrate concentration, v and $V_{\max}$ represent the velocity and maximum velocity, respectively, K_M is the Michaelis constant representing the substrate concentration at which the velocity is half of $V_{\max}$.

In contrast, it is not easy to determine the kinetics for catalase-like activity as it usually generates oxygen in the form of bubbles by decomposing H_2O_2 in the neutral pH (Fig. 5.2b) [35]. The generated bubble observed by naked eye was determined to be oxygen by a gas chromatography (unpublished data). For quantitative assay, the oximetry method can be used to detect the O_2 generation rate via oxygen electrode. The velocity of catalysis is proportional to the amount of generated molecular oxygen in the solution. As expected it also followed the typical Michaelis–Menten kinetics for catalase reaction [48] (Fig. 5.2d). However, the capacity of dissolved oxygen in aqueous buffer is affected by many exterior factors, such as temperature and diffusion. Oxygen from the air would enter the solution to interfere the measurement if H_2O_2 is in low concentration and it may escape from the liquid if the reaction is too fast when H_2O_2 is at too high concentration. The volumetric measurement for oxygen gas may be achieved via a volumetric bar-chart chip [65]. Other alternative methods may monitor H_2O_2 directly with spectrophotometer at 240 nm [66, 67] or measure hydroxyl radicals trapped by DEPMPO with ESR [68]. Therefore, more accurate strategy is needed for the characterization of catalase-like activity.

In general, IONzyme exhibits dual enzyme-like activities catalyzing H_2O_2 into oxygen or free radicals (Fig. 5.3a). However, the mechanism for IONzyme to mimic HRP/catalase catalysis is still under investigation. Fundamentally, the catalysis belongs to the Fenton chemistry as the iron plays the key role to engage with

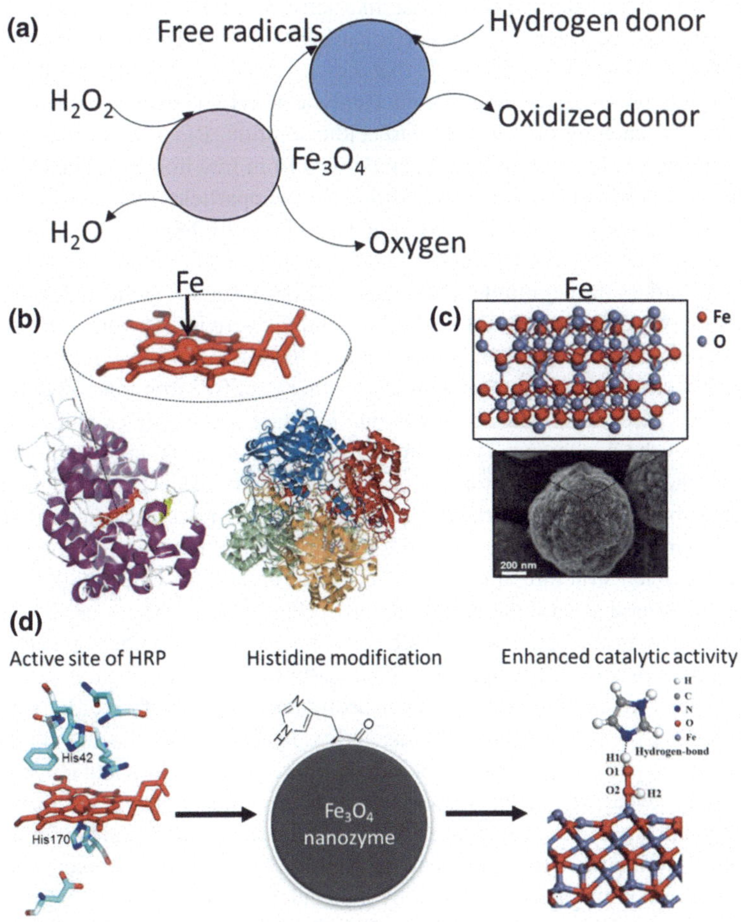

Fig. 5.3 Mechanism for IONzyme mimicking enzyme activity. **a** Reaction pathway for IONzyme. **b** Protein structures for HRP (RCSB PDB-1HCH) and catalase (RSCB PDB-1A4E) and iron in the center of active site [71]. **c** Plenty of iron on iron oxide nanoparticles. **d** Mimicking the active site of natural enzyme to improve the activity of IONzyme [48]. Reproduced with the permission from references [48, 71]

H_2O_2 for advanced oxidation, including iron ion, iron chelates, nano iron oxide, and peroxidase/catalase. In this process, three types of radical species are generated: hydroxyl radicals (OH•), hydroperoxyl radical (HO_2•), and ferryl ion (FeO^{2+}). The former two radicals have been observed in IONzyme catalysis, indicating that IONzyme precedes Fenton reaction [69, 70]. However, there is still no direct detection for ferryl ion which is typically formed in peroxidase catalysis (compound I) [71]. In addition, compared to free ferrous (Fe^{2+}) dominated in the traditional Fenton reagent, there are two types of irons (Fe^{2+}/Fe^{3+}) confined and coordinated tetrahedrally or octahedrally in the crystallized nanostructure of IONzyme [72]. Indeed, the valence state of iron could affect the catalytic activity in which Fe^{2+} ions may play a dominant role in the catalytic peroxidase-like activity of IONzyme, demonstrated by adjusting the ratio between two types of iron using $NaIO_4$ oxidation (increasing Fe^{3+}) or $NaBH_4$ reduction (increasing Fe^{2+}) that increasing Fe^{2+} led to better activity than Fe^{3+} [10]. It is important to note that the catalytic activity is from the nanoparticles rather than by leaching of iron ions into acidic solution. Reviewers often question whether the peroxidase activity might be derived from free iron ions via the Fenton reaction, as opposed to catalytic activity from the nanoparticles themselves. There is indeed a trace amount of iron released from the surface of IONzyme due to the acidic dissolution in NaAc buffer. However, the released Fe content in the supernatant was around two orders of magnitude lower than the concentration required for the Fenton reaction, which only showed negligible catalytic activity. To verify the activity is from nanozymes, Gao [10] and other studies [36, 41, 73, 74] have proved that the catalytic effect arose from the nanoparticles rather than free iron ions, showing that the observed peroxidase-like activity is due to intact nanoparticles and occurs on the surface of the nanozyme. Therefore, IONzyme catalysis is more accurate to be ascribed to heterogeneous Fenton system, which usually behaves kinetic processes including substrate binding, surface reaction, and product releasing, showing similar enzymatic kinetics [75]. For instance, the iron in IONzyme has similar confinement with the restricted structure at nanoscale as that in natural enzymes (both HRP and catalase are around several nanometers in size) (Fig. 5.3b, c). There is a heme containing one iron in the active site of HRP or catalase, in which the iron is chelated with porphyrin to achieve efficient electron transfer for catalyzing the redox reaction. Similarly, the iron is confined within the nanostructure on the surface of IONzyme to form active site with electron-transferring ability. In this respect, it is not unordinary for IONzyme to perform peroxidase-like and catalase-like activities, but it is still hard to accurately reveal why IONzyme demonstrates such many enzymatic features in the catalytic process. Some evidences implicate that the complex surface or defects may provide versatile active sites, endowing more features as enzymatic kinetics. For example, changing surface charge by molecular or imprinting can improve the affinity and selectivity to the substrates [55, 76]. Furthermore, partially grafting functional group (imidazole) in the active site of HRP could improve the affinity of IONzyme to H_2O_2 [48], enhancing the mimicking ability of IONzyme to natural enzyme (Fig. 5.3d). Taken together, though the mechanism is still not well disclosed, the catalytic behaviors, kinetics, and confined iron at nanoscale have proven that IONzyme is a novel enzyme mimetics.

5.2.3 IONzyme Synthesis

IONzyme can be synthesized using typical chemical methods for the synthesis of iron oxide nanomaterials, which are simple to be scaled up with low cost [30]. Different synthetic strategies produce IONzymes with variable size, morphology, and structure (Table 5.2). For instance, IONzymes made by co-precipitation usually have a relatively small size of below 30 nm in diameter [10, 35, 36, 70, 77–86]. In comparison, solvothermal preparation forms nanoparticles with larger size, from 100 to 500 nm [10, 39, 40, 48, 73, 87–93]. Other methods, such as sol–gel [77, 94], oxidative hydrolysis [41], thermal decomposition [95–99], and Massart hydrolysis [100, 101], can also be used to synthesize iron oxide nanomaterials with varying size and morphologies. IONzymes made by these methods therefore may exhibit different activities. In addition, surface modification can be introduced on IONzyme either during the synthetic process or after preparation, facilitating further applications in biological system.

Besides pure iron oxide, nanozymes can be integrated into other nanomaterials to form multifunctional hybrid nanocomplex. For instance, iron oxide NPs with the diameter from 2 to 3 nm [130] can be deposited onto graphene oxide (GO) surfaces [131]. It can also be integrated into hydrogel by in situ precipitation [132] and loaded onto silver nanowire to form Ag@Fe_3O_4 nanocomposite [133]. The above chemical syntheses allow to make the nanozyme at large scale with low cost, which is a notable advantage in practical application compared to natural enzymes or traditional enzyme mimetics.

Biogenic methods provide another route to make IONzymes with biomimetic mineralization [134] (Table 5.2). For example, bacterial magnetosomes are biomineralized inorganic ferromagnetic nanoparticles within the single-domain size range

Table 5.2 Some typical synthesis methods for IONzymes

Methods	Structure/morphology	Range of size (nm, diameter)
Co-precipitation [10, 35, 36, 54, 70, 77–86, 102–115]	Spheres	8–30
Sol–gel [77, 94]	Spheres	10, 150
Solvothermal [10, 39, 40, 51, 73, 87–93, 116–123]	Spheres, octahedra, and triangular plates	100–500
Oxidative hydrolysis [41]	Spheres	30–40
Thermal decomposition [95–99]	Cubic, starlike	5–13
Massart hydrolysis [100, 101]	Spheres	6–30
Ferritin [124–126]	Spheres	2.7–16
Mineralization [127]	Spherical	30–300
Magnetotactic bacteria [128, 129]	Cubic, particle	20–80

of 35–120 nm [127–129]. However, these nanoparticles are usually covered by bacterial membrane which may reduce the activity. Magnetoferritin is another ideal candidate of IONzyme [124–126]. Ferritin is an iron storage protein composed of 24 subunits made up of the heavy-chain ferritin (HFn) and the light-chain ferritin. Ferritin is spherical, with an outer diameter of 12 nm and interior cavity diameter of 8 nm. The cavity has been used as a reaction chamber to synthesize highly crystalline and monodisperse nanoparticles through biomimetic mineralization within the protein shell. These biogenic approaches would be expected to show fine and uniform particle size, good dispersity, and biocompatibility without extra surface modification.

Although abovementioned varied methods for nanomaterials synthesis have been used for IONzyme preparation, there is still no serious report to carefully compare the activities of IONzyme made by different methods. It is challenging because IONzymes have diverse size, morphology, shape, nanostructure (facets), and surface modification. Here, we suggest to use the concept of specific activity for enzyme purity determination to assess the activity of IONzymes from different sources by normalizing the activity as unit/mass.

5.3 Extraordinary Property of IONzyme

Despite similarities with enzymes in catalysis, IONzyme has many advantages in stability, tunability of activity, and multifunctionality, compared to traditional enzyme mimetics and natural enzymes because of its nanoscale effects. These features further ensure its versatile applications in biomedicine.

5.3.1 Stability

IONzyme shows enhanced stability under extreme conditions such as acidic or basic environment and high temperature and is much more robust than natural enzymes. For instance, Fe_3O_4 nanozyme remained stable over a wide range of pH (1–12) and temperatures (4–90 °C) [84]. In contrast, the natural enzyme HRP did not show any activity after treatment at pH lower than 5 and lost activity rapidly when the temperature was greater than 40 °C. The super high stability endows IONzyme long-term storage of up to 40 days in sealing vessel under ambient conditions to keep good activity. In addition, it showed excellent reusability, maintaining activity after multiple times of recycled use [79, 106]. More impressively, the hybrid nanozyme of Fe_3O_4 nanospheres/reduced graphene oxide (Fe_3O_4 NSs/rGO NCs) stored at 4 °C was found to be stable for more than 3 months [119]. However, biosensor based on IONzyme showed shorter stability which was around 2–3 weeks, but still are acceptable for the practical application [123, 135]. The high stability allows IONzyme to be widely used in biomedicine.

5.3.2 Tunability of Activity

One of the main features for nanozyme, compared to natural enzymes and other mimetics, is that its activity can be tuned by modulating the size, structure/morphology, dopant, and surface modification [70, 136], which means that it is possible to design the nanozyme with proper activity according to the need. First, the activity of IONzyme can be modulated by controlling the size and morphology of nanoparticle. Usually, the smaller the size, the higher the catalytic activity. This phenomenon can be explained by the fact that the smaller nanoparticles have a greater surface area to interact with substrates. For instance, the IONzymes showed different activities toward TMB in the order 30 nm > 150 nm > 300 nm when the nanoparticles were at the same mass amount [10] (Fig. 5.4a). The same phenomenon was reproduced on the Fe_3O_4 nanoparticles with average diameters of 11, 20, and 150 nm, that is, the catalytic activity increased when the nanoparticle size reduced [77]. Aside from the size, the structures/morphologies also affect the activity of the IONzyme. Liu et al. reported that three Fe_3O_4 nanostructures, cluster spheres, octahedra, and triangular plates (Fig. 5.4b), showed different peroxidase-like activities following the order of cluster spheres > triangular plates > octahedral. This phenomenon was closely related to their preferential exposure of catalytically active iron atoms or crystal planes [73]. High-energy facet like (110) may also contribute to enhancing the catalytic activity [137]. Similarly, magnetic cobalt ferrite ($CoFe_2O_4$) nanozyme had peroxidase-like activity following the order: spherical > near corner-grown cubic > starlike > near cubic > polyhedron [97]. Taken together, these results indicate that it is possible to adjust the activity of IONzyme by controlling the size and morphology at nanoscale.

Second, the activity could be further improved by doping other elements in the IONzyme or integrated with other nanomaterials (Fig. 5.4c). Several metal dopants are able to increase the activity of IONzyme, such as Au, Ag, and Pt. The peroxidase-like activity of Au@Fe_3O_4 nanoparticles (NPs) was effectively enhanced due to the synergistic effect between the Fe_3O_4 NPs and Au NPs [93]. Lee et al. reported that Au-Fe_3O_4 nanoparticles are catalytically more active compared to Au or Fe_3O_4 nanoparticles, respectively, which is attributed to polarization effects from Au to Fe_3O_4 [98] and the special electronic structure at the interfaces between the Fe_3O_4 aggregates and the gold nanoparticles [117]. Ag also facilitates Fe_3O_4 nanowire to enhance peroxidase-like activity with good stability and high absorbance [133]. Pt-modified Fe_3O_4 magnetic nanoparticles (Fe_3O_4@Pt NPs) showed strong affinity with substrates and enhanced catalytic activity compared to Fe_3O_4 nanoparticles [50]. Strikingly, dumbbell-like $Pt_{48}Pd_{52}$–Fe_3O_4 NPs even showed more activity than the natural enzyme [99]. Other metals who can form metal oxide complex with iron oxide also showed the capability to improve the catalytic activity of nanozyme. For example, MFe_2O_4 (M = Co, Mn, Ni, Cu) hollow nanostructures performed tunable activity by adjusting size, shape, and composition [138]. In particular, Mn^{2+}-doped $Fe_{1-x}Mn_xFe_2O_4$ nanoparticles had remarkably enhanced the peroxidase activity and

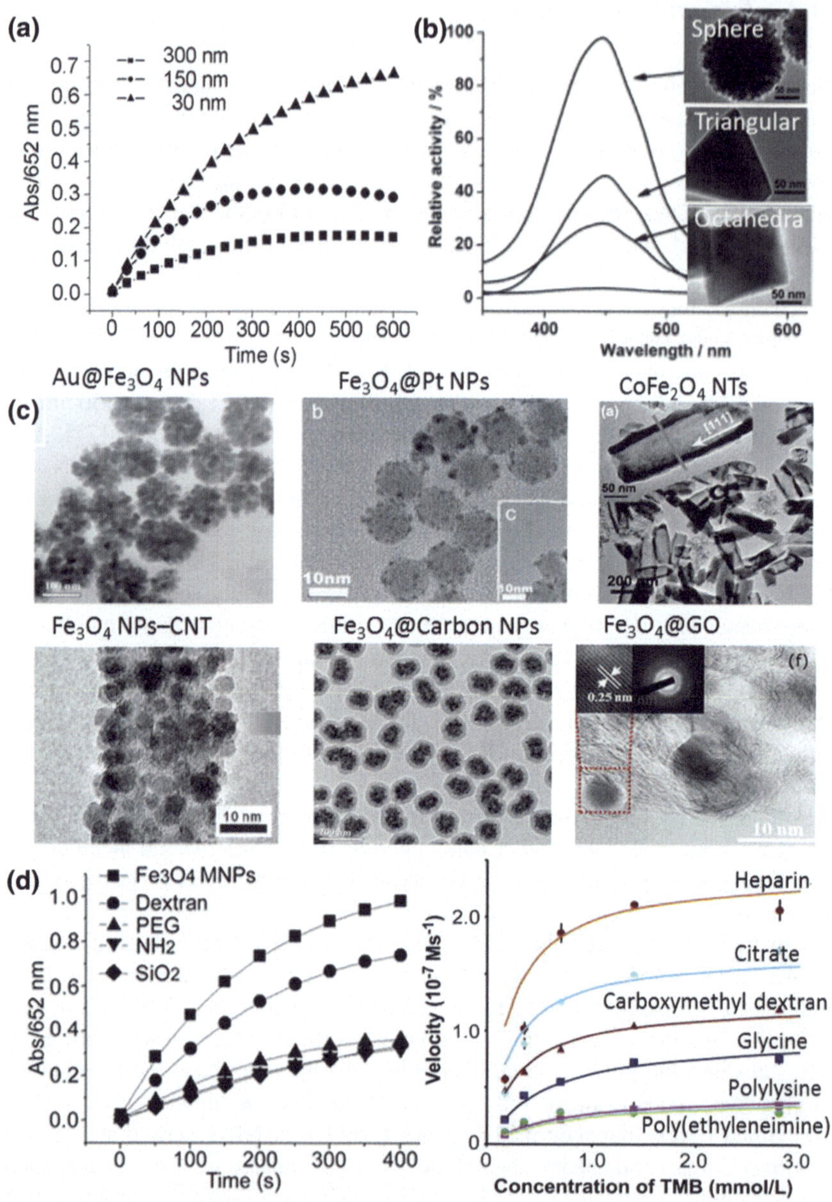

Fig. 5.4 Activity tunability based on **a** size [10], **b** morphologies [73], **c** dopants/integration with nanocomplex, and **d** surface modification with different molecules [10, 55]. Au@Fe_3O_4 NPs [93], Fe_3O_4@Pt NPs [50], $CoFe_2O_4$ NTs [138], Fe_3O_4 NPs@CNT [139], Fe_3O_4@Carbon NPs [51], and Fe_3O_4@GO [130] (NPs: nanoparticles; NWs: nanowires; NTs: nanotubes; CNT: carbon nanotube; GO: graphene oxide). Reproduced with the permission from references [10, 50, 51, 55, 73, 93, 130, 138, 139]

magnetism with the increase of Mn^{2+} proportion [54]. Cobalt-doped magnetic composite nanoparticles ($Co_xFe_{3-x}O_4$ MNPs) possessed higher activities mimicking both peroxidase and catalase comparing with Fe_3O_4 nanoparticle, although they were similar in crystal structure, size distribution, and morphology [109]. More impressively, the nanoreactor of Fe_3O_4@SiO_2 core–shell structure with nanochannels showed outstanding activity for H_2O_2 reduction compared with bare Fe_3O_4 or Fe_3O_4@SiO_2 core–shell nanoparticles, because the outer silica shells with nanochannels were permeable and the Fe_3O_4 cores were accessible to the reactants [140]. Most of dopants are achieved in the synthetic process of iron oxide nanoparticles, therefore providing a simple way to improve the activity.

Another route to tune activity is to integrate iron oxide nanoparticles with carbon nanomaterials to get synergistic effect (Fig. 5.4c). The integrated nanocomplex exhibited ultrahigh peroxidase mimetic activity and better water solubility compared to those of pure nanoparticles [139]. For instance, Fe_3O_4-multi-walled carbon nanotube (Fe_3O_4-MWCNT) magnetic hybrids can be used as an efficient peroxidase mimic catalyst that could overcome such pH limitations in a Fenton-like reaction [74]. Fe_3O_4@Carbon nanoparticles also have enhanced peroxidase-like activity due to partially graphitized carbon which facilitates electron transfer in the catalytic decomposition of H_2O_2, leading to the production of highly reactive hydroxyl radicals [51, 122]. Synergistic structural and functional effect of the combined graphene oxide (GO) and Fe_3O_4 nanoparticles was discovered by Zubir et al. according to the presence of strong interfacial interactions (Fe–O–C bonds) between both components [130]. The complex had enhanced affinity toward H_2O_2 [49]. Actually, the synergistic effect may be ascribed to the peroxidase-like activity of GO itself based on the work from Qu's group [11]. Furthermore, more complicated composites, graphene quantum dots (GQDs/Fe_3O_4), showed superb peroxidase-like activities, which were much higher than composites of GO and Fe_3O_4 NPs (GO/Fe_3O_4), individual GQDs, and individual Fe_3O_4 NPs. These enhanced peroxidase activities of the GQDs/Fe_3O_4 composites can be attributed to the unique properties of GQDs and the multiple synergistic interactions between the GQDs and Fe_3O_4 NPs [106]. Reduced graphene oxide (RGO) also can improve the activity of Iron oxide nanoparticles [107, 119]. The above dopants or integrations provide a variety of ways to improve the activity of IONzyme.

Third, the activity could be controlled by surface modifications introduced in the process of nanoparticle synthesis or directly on the as-prepared nanoparticles (Fig. 5.4d). In the early stage of the investigation, our group observed that the enzyme activity of the modified Fe_3O_4 MNPs decreased after modification in general. But the degree of decrease is different according to the molecules and methods. Dextran modified the Fe_3O_4 NPs (during synthesis) showed comparable activity with naked samples. However, polyethylene glycol (PEG) modification (during synthesis) dramatically decreased the activity which is similar as SiO_2, 3-aminopropyltriethoxysilane (APTES) coating (after synthesis) [10]. Furthermore, the activity could be affected by surface charge from coated molecules (Fig. 5.4d). It was found that iron oxide nanoparticle with negative surface charge (heparin) exhibited a 5.9-fold higher peroxidase activity than that with the positive charge (ethyleneimine) when TMB was the substrate. However, the conversed result displayed an 11.5-fold increase in catalytic activity when ABTS was the substrate. In addition, under the similar charges, less thickness would increase the catalytic efficiency, indicating there is competition between thickness and charge effects on activities [55]. These evidences indicate that the influence of surface modification on nanozyme activity is complicated with multiple factors.

The activity could be improved by modifying IONzyme with the molecules having similar structure from the active site of natural enzyme. Prussian blue modification could effectively improve the activity as the molecule of Prussian blue itself performs high peroxidase activity [56, 141]. Porphyrin-derivative functionalized Fe_3O_4 nanocomposites exhibited ultrahigh peroxidase-like activity with enhanced affinity toward H_2O_2 compared with pure Fe_3O_4 nanoparticles [142]. The catalytic activity was further enhanced by the attachment of β-cyclodextrin polymer (Pβ-CD) on the surfaces of the Fe_3O_4 [91]. Histidine modification effectively enhanced the affinity to H_2O_2 for IONzyme because the imidazole group from histidine can provide a similar microenvironment as that in the active site of HRP [48]. The biomimetic modification may be a new way to improve the selectivity of IONzyme.

Finally, modification or integration with biomolecules also effectively enhances the activity of IONzyme. DNA-capped iron oxide nanoparticles are nearly 10-fold more active as a peroxidase mimic for TMB oxidation than naked nanoparticles and the enhancement is related to the length of DNA [45]. Natural biomacromolecule, like magnetoferritin, in which apoferritin is surrounding Fe_3O_4 nanoparticles exhibits a tunable peroxidase-like activity via loading various iron contents [125, 143]. Interestingly, compared with bare magnetic nanoparticles (MNPs), peroxidase-like casein MNPs exhibit good catalytic properties as casein incorporated on MNPs notably improved the affinity toward both H_2O_2 and TMB [84]. In summary, the activity of IONzyme could be regulated by size, structure/morphology, dopants/integration, and surface modification, which is one of the biggest advantages compared to other enzyme mimetics.

5.3.3 *Multifunctionality*

In general, IONzymes are all magnetic nanomaterials with superparamagnetism, while enzyme-like catalysis is another novel common nanoscale feature for iron oxide nanomaterials. Therefore, IONzyme possesses two basic functions: enzyme-like activity and superparamagnetism, endowing multipurpose performances especially when combined them together. Regarding the enzyme-like activities, IONzyme can mimic peroxidase and catalase activities at acidic and neutral pH, respectively [35]. Therefore, the activity can be easily controlled by pH under special circumstances like tumors or biofilms. More importantly, IONzyme can be used as a universal vehicle to load other functional molecules or reagents in attempt to construct cascade reactions. For example, glucose oxidase (GOx) can be conjugated onto the surface of IONzymes to form a new nanocomplex. GOx catalyzes glucose to generate hydrogen peroxide which then can be catalyzed by nanozymes to induce colorimetric reaction. By two-step catalysis, the nanocomplex can perform sequential reaction using glucose as initial substrate [36]. All these features indicate that IONzyme is a novel and multifunctional enzyme mimetics, which has given IONzyme a lot of advantages in practical applications, especially in the field of biomedicine. It no longer seems to be confined to the magnetic property when using iron oxide nanomaterials.

5.4 Extensive Applications of IONzyme in Biomedicine

The discovery of enzymatic activity from IONzyme provides an opportunity to use it as an active nanomaterial in biomedical filed. Currently, most of applications are developed based on the peroxidase-like activity which is able to amplify detection signal by colorimetric reaction and generate free radicals to kill bacteria and cells or interfere ROS level.

5.4.1 Enzyme Alternative for Immunoassay and Pathogen Detection

Regarding to the peroxidase-like activity, one of the first applications taking advantage of this characteristic was the use of IONzyme to replace HRP in enzyme-linked immunosorbent assay (ELISA) and other HRP-related molecular detection [144, 145]. IONzyme can be directly used as HRP alternative by conjugating with antibody to amplify signal via colorimetric reaction (Fig. 5.5a). In addition, it can be applied to capture or enrich trace amount antigen by its superparamagnetism, which helps to improve the sensitivity and efficiency [54]. For example, the capture-detection immunoassay based on chitosan modified magnetic nanoparticles (CS-MNPs) has a detection limit of 1 ng/ml for carcinoembryonic antigen (CEA) [144]. It has been successfully demonstrated in sandwich ELISA and direct ELISA (Fig. 5.5a–c). Right now many antigens or pathogens have been detected by these novel immunoassays, including IgG, hepatocellular carcinoma biomarker Golgi protein 73 (GP73) [146], human chorionic gonadotropin (HCG) [147], mycoplasma pneumonia [148], Vibrio cholera, rotavirus [105], and cancer cells with human epidermal growth factor receptor 2 (HER2) [105, 149] and epidermal growth factor receptor (EGFR) [80].

Recently, a lateral flow test was developed using iron oxide nanozyme strip to detect Ebola virus (EBOV) [92] (Fig. 5.5b). This nanozyme strip can detect the glycoprotein of EBOV as low as 1 ng/ml, which is 100-fold more sensitive than the standard strip method (colloidal gold strip). In addition, the sensitivity of the nanozyme strip for EBOV detection and diagnostic accuracy for New Bunyavirus clinical samples is comparable with ELISA but is much faster (within 30 min) and simpler (without

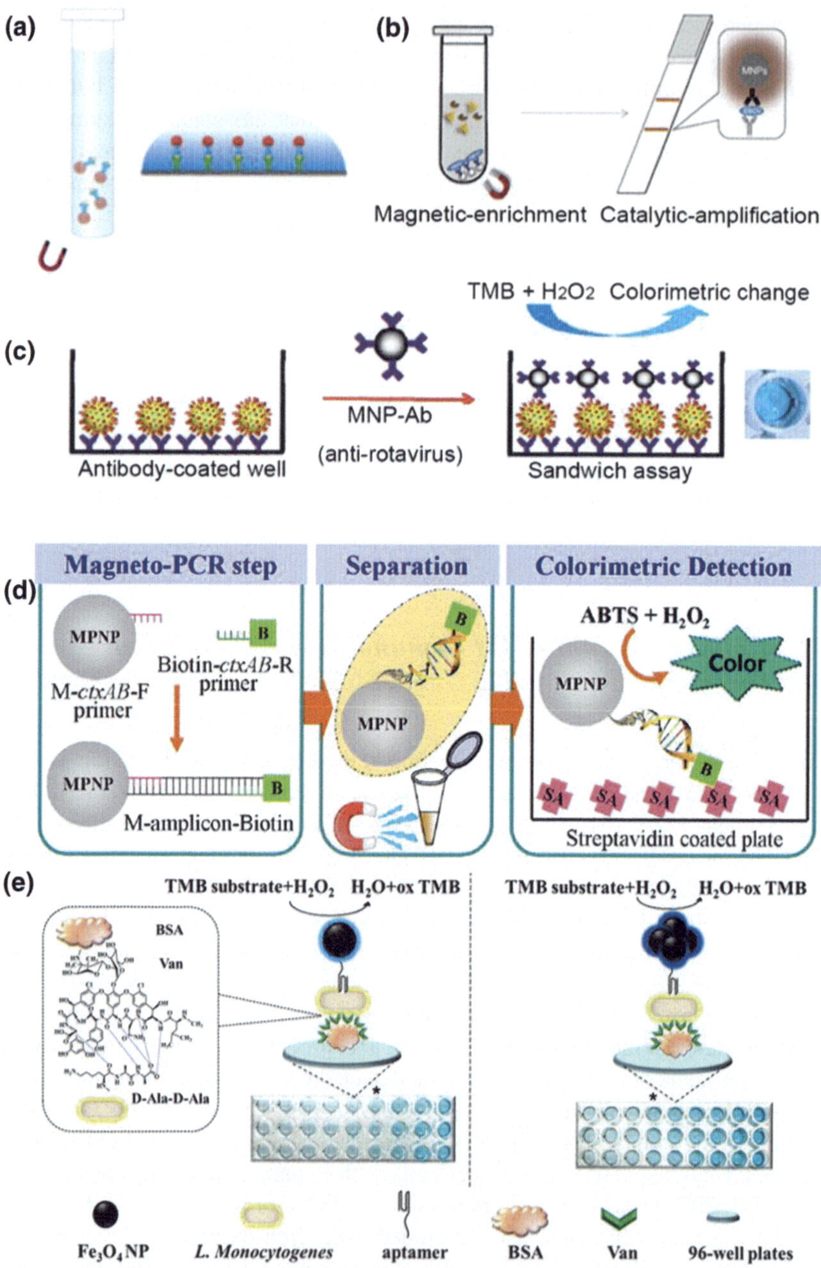

Fig. 5.5 Novel immunoassays and pathogen detections based on IONzyme. **a** Regular immunoassay [150]. **b** Nanozyme strip for Ebola detection [92]. **c** Virus detection with sandwich format [105]. **d** DNA detection via hybridization [151]. **e** Bacteria detection via aptamer recognition [152]. Reproduced with the permission from references [92, 105, 150–152]

need of specialist facilities). The results demonstrated that the nanozyme strip test can rapidly and sensitively detect EBOV, providing a valuable simple screening tool for diagnosis of infection in Ebola-stricken areas.

Besides immunoassay via antibody–antigen recognition, other detection strategies have been developed based on the molecular recognition using DNA or aptamer. DNA hybridization could be achieved on the surface of IONzyme coated with proper primer or probe. Then the primer can be used to amplify target DNA by PCR in combination with another biotinylated primer. IONzyme still kept high activity after many thermal cycles in PCR reaction. Therefore, a sandwich assay can be developed on a surface coated with streptavidin (Fig. 5.5d) and achieve signal amplification via IONzyme catalysis. Thiramanas Raweewan et al. [151] have used this system to detect *Vibrio cholerae* (*V. cholerae*) with a sensitivity of 10^3 CFU/ml, displaying high specificity and efficiency for the detection of various bacterial DNAs in drinking and tap water. However, DNA may have inhibition effect on the activity of IONzyme. The presence of large amount of DNA was found to effectively reduce the colorimetric reaction in the IONzyme–H_2O_2–OPD system [153]. Therefore, the inhibition of color signal can be used to evaluate the target DNA after PCR amplification, by which *Chlamydia trachomatis* in human urine was successfully detected. The aptamers also can be used in similar way as they specifically recognize the target molecules with high affinity (Fig. 5.5e), working like antibodies to form the typical sandwich structure with antigens. Therefore, it is rational to obtain sensitive detection for many targets, such as thrombin [154] and *Listeria monocytogenes* (*L. monocytogenes*) in food bacteria detection [152]. Altogether, many immunoassays and molecular detections have been successfully developed using IONzyme as HRP alternative, showing very promising applications in biomedicine. However, until now there is still no real product or kit developed based on IONzyme for immunoassay.

5.4.2 Cascade Enzymatic Reaction for Substrate-based Detection

Enzymes can be assembled onto IONzyme to perform cascade catalytic reactions from enzymes to iron oxide nanoparticle if the enzyme can generate intermediate H_2O_2. The integration may improve the catalytic efficiency by proximity effect which helps to overcome the diffusion-limited kinetics and intermediate instability [22]. Wei and Wang group first reported to combine glucose oxidase (GOx) on to Fe_3O_4 nanoparticles for glucose detection (Fig. 5.6). In this system, GOx catalyzes glucose to generate H_2O_2 which in turn is catalyzed by iron oxide nanoparticle through peroxidase activity. If a chromogenic substrate is present, the color signal can be produced in proportional to glucose concentration [36]. By this way, a typical glucose concentration—response curve where as low as 3 μM glucose was achieved with a linear range from 50 μM to 1 mM. Since then, many groups have used this method to combine GOx with or integrate GOx on IONzyme for glucose detection [39, 49,

59, 81, 84, 91, 95, 121, 122, 128, 131, 142, 155–160]. In particular, other nanomaterials with intrinsic GOx activity can be integrated with IONzyme to form specific artificial enzymatic system. For example, Au nanoparticles with GOx activity can be assembled on the surface of iron oxide nanoparticles with peroxidase activity to get sequential reactions [88]. This functional units can be used a robust nanoreactor to perform self-organized cascade reaction which can detect glucose concentration in a linear range from 10 to 130 μM with a detection limit of 0.5 μM. All these systems have the potential to be used to measure blood glucose level in diabetes diagnosis.

Alternatively, other oxidase which generates intermediate H_2O_2 can be integrated with IONzyme to detect the corresponding substrates besides glucose, including cholesterol oxidase (ChOx) for cholesterol [155, 161], galactose oxidase (Gal Ox) for galactose [161], and alcohol oxidase (Al Ox) for alcohol [162] (Fig. 5.6). More complicated system could be used to detect acetylcholine (ACh) by combing IONzyme with acetylcholine esterase (AChE) and choline oxidase (CHO) which can sequentially catalyze ACh to form H_2O_2 in turn catalyzed by IONzyme for colorimetric reaction [119] (Fig. 5.6). This triplicate cascade system also can be used to detect organophosphorous pesticides and toxic nerve agents which inhibit the enzymatic

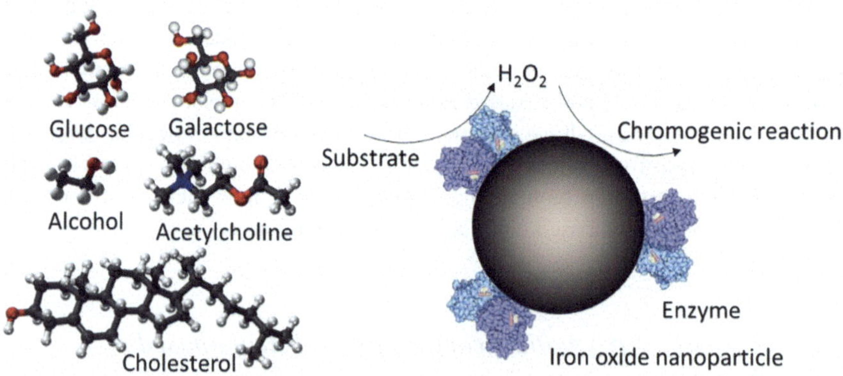

Glucose + O_2 $\xrightarrow{\text{Glucose oxidase}}$ Gluconic acid + H_2O_2

Galactose + O_2 $\xrightarrow{\text{Galactose oxidase}}$ 1,6-Dialdehyde + H_2O_2

Alcohol + O_2 $\xrightarrow{\text{Alcohol oxidase}}$ Aldehyde + H_2O_2

Cholesterol + O_2 $\xrightarrow{\text{Cholesterol oxidase}}$ Cholest-5-en-3 + H_2O_2

Acetylcholine + H_2O $\xrightarrow{\text{Acetylcholinesterase}}$ Acetic acid + Choline $\xrightarrow{\text{Choline oxidase}}$ Betaine + H_2O_2

Fig. 5.6 Cascade enzymatic reactions for substrate-based detection activity of AChE to produce H_2O_2 [89]. Therefore, the absorbance signal from colorimetric reaction is in inverse proportion to the concentration of organophosphorus compounds including acephate, methylparaoxon, and Sarin (neurotoxic). Taken together, the concept of integrating other enzymes with IONzyme to construct the cascade system provides a new strategy for the rapid and sensitive detection of small molecules taking part in one of the sequential steps.

5.4.3 Tumor Diagnosis and Therapy

Besides immunoassay for the detection of tumor biomarker or cells, IONzyme shows potential application for tumor diagnosis and therapy. First, tumor pathological analysis can be achieved with magnetoferritin nanoparticles (M-HFn) [124], a specific type of IONzyme. Iron oxide nanoparticles are encapsulated inside the recombinant human heavy-chain ferritin (HFn) protein shell, which binds to tumor cells that overexpress transferrin receptor 1 (TfR1) (Fig. 5.7a). The Fe_3O_4 core catalyzes the oxidation of peroxidase substrates in the presence of hydrogen peroxide to produce a color reaction that is used to visualize tumor tissues. Almost 474 clinical specimens from patients with nine types of cancer were examined and verified that these nanoparticles can distinguish cancerous cells from normal cells with a sensitivity of 98% and specificity of 95%. This result suggested that ferritin-based IONzyme has the potential to become a diagnostic tool for rapid, low-cost, and universal assessment of malignant tumors.

IONzyme also shows the potential for direct tumor elimination. Usually, iron oxide nanoparticles are used as the contrast for cancer imaging or the carrier for targeted drug delivery. The enzyme-like activity of iron oxide nanoparticles is neglected in tumor therapy for a long time. Theoretically, it can affect tumor viability by catalyzing H_2O_2 to generate toxic radicals, but the intracellular amount of H_2O_2 may not be enough to sustain an antitumor effect. To address this problem, two options can be considered: direct injection of H_2O_2 to the body or combine an enzyme to generate H_2O_2 using in vivo substance as substrate. The first way has been proved using magnetite Fe_3O_4 nanozyme and H_2O_2 in a mice model bearing subcutaneous Hela tumor [163] (Fig. 5.7b). Significant inhibition efficacy on tumor growth was shown by the combination of Fe_3O_4 NPs and H_2O_2 after treatment. Tumor imaging was also achieved by combining the enhanced T_2-weighted signal from Fe_3O_4 in MR imaging, indicating that IONzyme could be used for cancer theranostics, especially in epidermal diseases. However, H_2O_2 injection in this model may not be a practical option due to the potential high toxicity for normal tissue. In comparison, integrating enzyme which can consume cellular substance to generate H_2O_2 may be a better way because the system can make H_2O_2 locally therefore with low toxicity to normal tissue. More strikingly, recent studies showed that iron itself within nanomaterials can cause enough ROS to lead tumor cells into apoptosis (also called ferroptosis) [165] or induce pro-inflammatory macrophage polarization in tumor tissues to inhibit tumor growth [166]. No external H_2O_2 administration is needed in these tumor therapies.

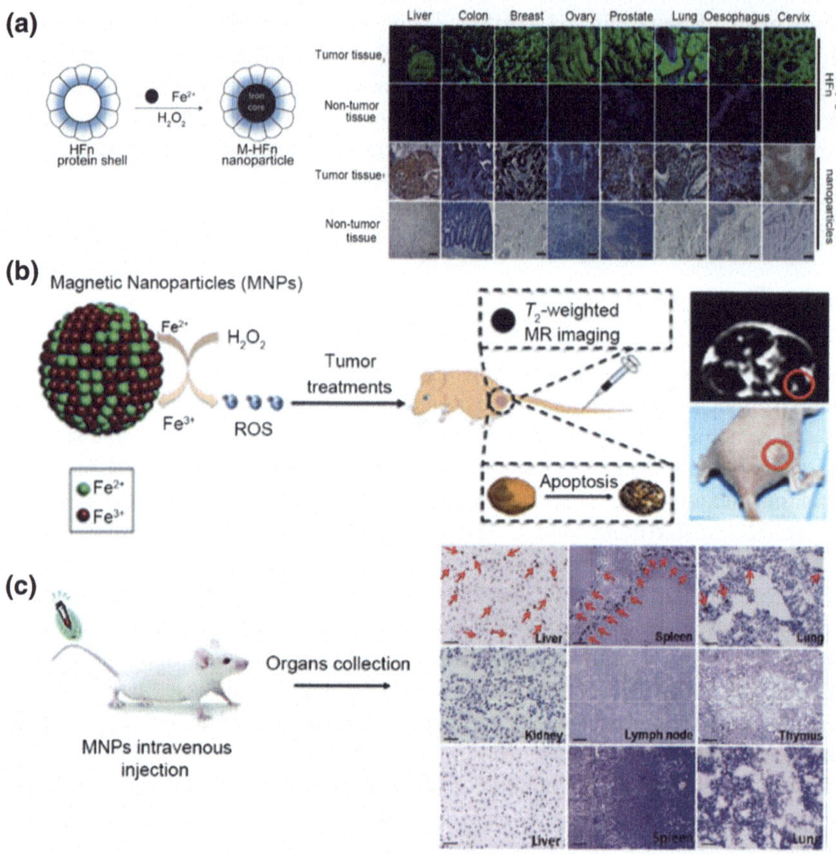

Fig. 5.7 IONzyme activity for tumor diagnosis and therapy. **a** Tumor assessment with magnetoferritin [124]. **b** Tumor therapy and imaging with IONzyme [163]. **c** Nanoparticle distribution via enzyme-like activity [164]. Reproduced with the permission from references [124, 163, 164]

Such antitumor effects may be ascribed to the catalytic activity of iron in nanomaterial which is same as IONzyme.

It is also possible to track the distribution of IONzyme in a tumor therapy. Quantitative analysis of the bio-distribution, pharmacokinetics, and organ clearance in animal model is important to understand the in vivo behavior and evaluate the biosafety of IONzyme. Histochemical visualization of IONzyme in mouse tissues would be demonstrated by employing its intrinsic peroxidase activity which could catalyze the oxidation of peroxidase substrates to produce a color reaction at the onset of nanoparticle (Fig. 5.7c). Based on this principle, dextran-coated Fe_3O_4 nanoparticles was found mainly localized in liver, spleen, and lung rather than kidney, lymph node, and thymus [164]. In combination with H and E staining, cellular location was further examined to show that the nanoparticles were uptaken mainly by reticuloendothelial system (RES) in these organs, which is Kuppfer macrophage cells in liver, alveolar macrophages in lung, and macrophage perifollicular areas in spleen. Organ clearance also could be evaluated with the same way, showing the uptake and then rapid clearance with different time windows. This approach provides more detailed information about distribution and location compared to in vivo imaging, which can be used to match up with MRI analysis to better understand the in vivo behavior of iron oxide nanomaterial. Presumably, other nanoparticles having intrinsic peroxidase activity could also be considered in similar way to track their in vivo behavior. Altogether, IONzyme shows great potentials in tumor diagnosis and therapy and ex vivo tracking for bio-distribution evaluation.

5.4.4 Antibacteria and Biofilm Elimination

In the catalytic process, IONzyme reacts with H_2O_2 to generate free radicals (intermediate product) which is highly toxic to bacteria because it attacks cell membrane, proteins, and nucleic acids and finally induces bacterial malfunction. Hydrogen peroxide is a common biocidal chemical that has various cleaning and disinfectant uses, including as an antibacterial agent for hygienic and medical treatments. The mechanism is also through the reaction of H_2O_2 with cell components but the efficiency is usually low to those resistant bacteria. Regarding to the ping-pong mechanism, IONzyme with peroxidase activity may help to enhance the antibacterial effects of H_2O_2. The enhancement effect has been demonstrated on *Escherichia coli* (*E. coli*) [163] and methicillin-resistant *Staphylococcus aureus* (MRSA) [167]. This antibacterial property showed beneficial effect for wound healing associated with bacteria-infected abscesses and other treatment modalities against multiple-drug-resistant bacteria.

The enhancement on H_2O_2 catalysis to generate radicals also provide opportunity to eliminate biofilm, a special bacteria community which helps bacteria to develop drug resistance by limiting the penetration of antibiotics or other biocides into the protective, organic matrix (Fig. 5.8). Fe_3O_4 nanozyme with peroxidase-like activity could potentiate the efficacy of H_2O_2 in biofilm degradation and prevention via enhanced oxidative cleavage of biofilm components (model nucleic acids, proteins, and oligosaccharides) in the presence of H_2O_2 [40]. The combination of Fe_3O_4 nanozyme with H_2O_2 cleaved nucleic acids (DNA), proteins (BSA, antibody, cell lysis), and polysaccharide into small pieces (Fig. 5.8a). In comparison, IONzyme or H_2O_2 only failed to complement the oxidative degradation [40]. The capability of cleavage of biomolecules allows the Fe_3O_4 NP-H_2O_2 system efficiently to break down the existing biofilm matrix and prevent new biofilms from forming, killing both planktonic bacteria and those within the biofilm, providing a novel strategy for biofilm elimination, and other applications. This strategy has been successfully applied on dental biofilm elimination and caries prevention [168] (Fig. 5.8b–d). IONzyme with peroxidase-like activity not only effectively degrade glucans from oral biofilm matrix into glucose, but also dramatically kill the embedded *Streptococcus mutans* (*S. mutans*) from 10^8 CFU/ml to 10^2 CFU/ml (up to 6-log reduction).

Moreover, it displays an additional property of reducing apatite demineralization in acidic conditions. In the topical daily treatments akin to a clinical situation, IONzyme in combination with H_2O_2 effectively suppress the onset and severity of dental caries while sparing normal tissues in vivo (Fig. 5.8d, e). These results revealed the potential to exploit nanozymes as a potent alternative approach for treatment of a prevalent biofilm-associated disease.

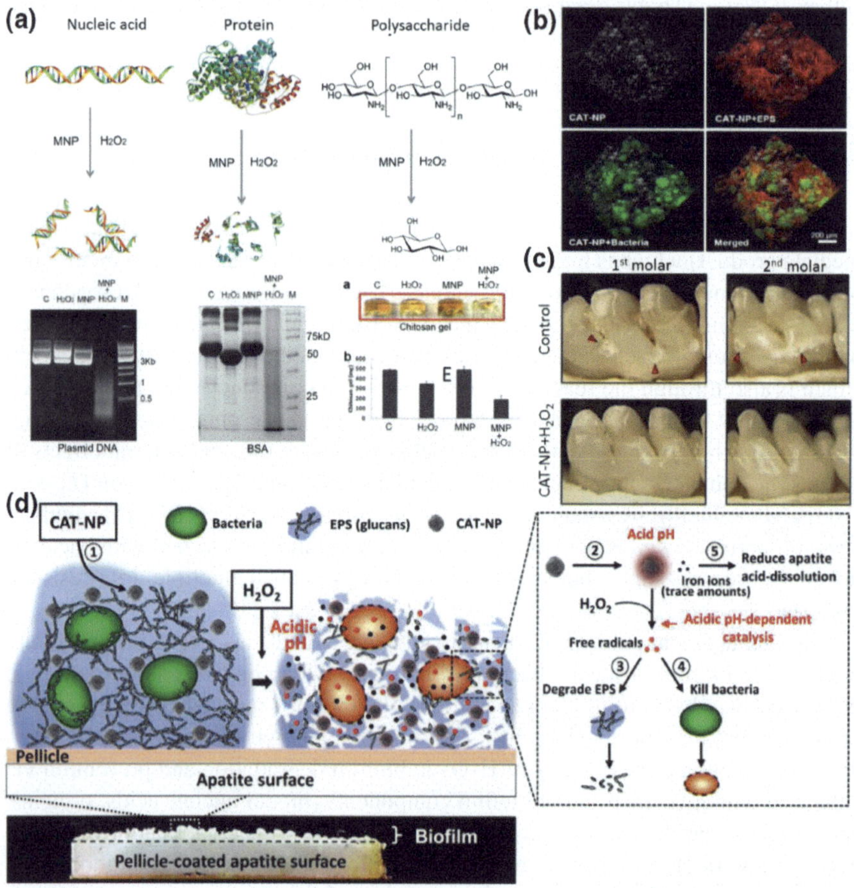

Fig. 5.8 IONzyme for biofilm elimination and dental caries prevention [40, 168]. **a** Degradation on nucleic acid, protein, and polysaccharide with IONzyme. **b** 3D distribution of IONzyme in oral biofilm. **c** Dental caries prevention with IONzyme. **d** Multiple functions of IONzyme in dental application. Reproduced with the permission from references [40, 168]

5.4.5 Modulation of Cellular Oxidative Stress

Since iron oxide nanoparticles were first found with enzyme-like activity, a question has arisen: does it maintain enzymatic function in a living cell or in vivo? Indeed, iron oxide nanoparticles are often used as a contrast for in vivo cancer imaging or a carrier for drug delivery. Therefore, the influence on cell viability needs to be thoroughly considered. It may change the level of reactive oxygen species (ROS) by diminishing H_2O_2 or generating free radicals. As introduced previously, IONzyme demonstrated dual enzyme-like activities, peroxidase, and catalase, under acidic and neutral pH, respectively [35] (Fig. 5.9a). Therefore, contradictive outcomes may be obtained in various organelles and cytosomes having different intracellular microenvironments.

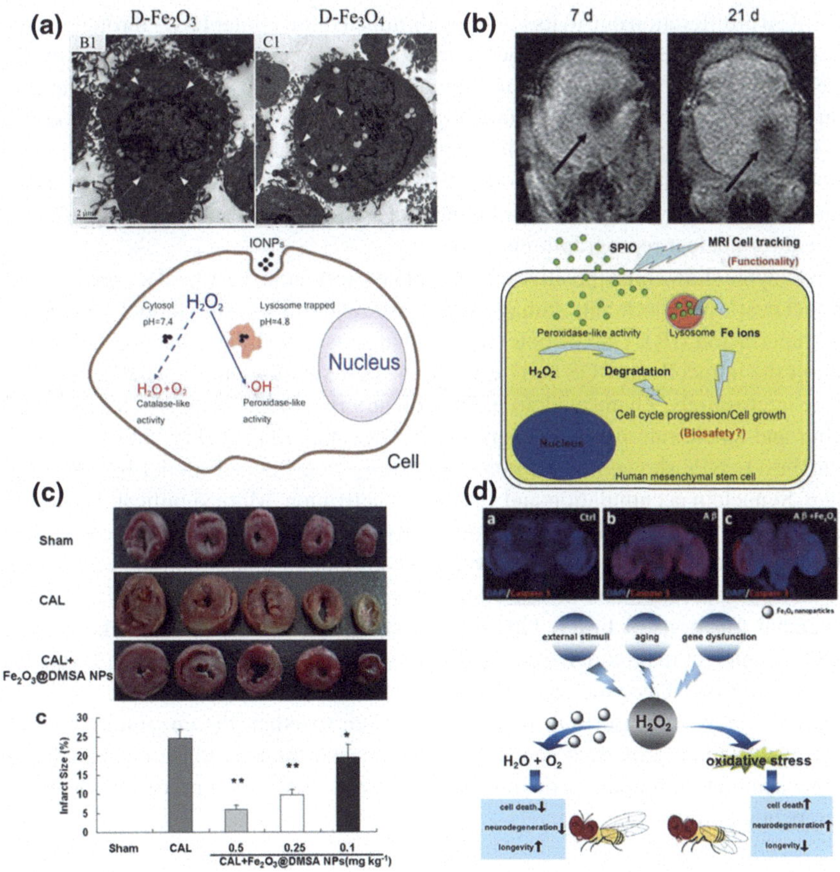

Fig. 5.9 ROS modulation with IONzyme. **a** Dual enzyme-like activities in the cell [35]. **b** Stimulation on stem cell proliferation [169]. **c** Cardioprotective activity [122]. **d** Aging delay and neurodegeneration amelioration in *Drosophila* [171]. Reproduced with the permission from references [35, 122, 169, 171]

For example, in lysosome mimic condition, the nanozyme could catalyze H_2O_2 to produce hydroxyl radicals which induce cell damage dramatically. However, it may diminish H_2O_2 in neutral cytosol mimic condition because of the decomposition of H_2O_2 into H_2O and O_2 directly. The dual activities should be taken into consideration when evaluating the cytotoxicity based on the intracellular location of IONzyme. Besides potential ROS impact, IONzyme may also cause liposome membrane damage according to lipid oxidation [41], which can catalyze pre-existing lipid peroxides (LOOH) or H_2O_2 as a substrate to initiate the chain reaction process. Until now, many types of iron oxide nanoparticles were found with high biocompatibility and biosafety to mammalian cells, indicating the adverse influence from intrinsic activity of IONzyme may not be very significant to the cells.

In contrast, beneficial effects of IONzyme are observed on stem cells. Ferucarbotran, an ionic SPIO, was found to increase human mesenchymal stem cells (hMSCs) proliferation due to its ability to diminish intracellular H_2O_2 through intrinsic peroxidase-like activity and accelerate cell cycle progression mediated by the free iron released from lysosomal degradation to alter the expression of the protein regulators of the cell cycle [169] (Fig. 5.9b). Similarly, poly(L-lysine)-modified Fe_3O_4 nanoparticles could promote the proliferation of cancer stem cells from U251 glioblastoma multiform by reducing intracellular H_2O_2 [82]. These results indicate that IONzyme may cause more influence on stem cells because ROS signaling is critical for nuclear reprogramming [170].

IONzyme shows more positive effect in in vivo test. First, Fe_2O_3 NPs were found to protect hearts from ischemic damage at the tissue and cell level in a model of ischemia and reperfusion (IR). One of potential mechanisms is that these nanoparticles may inhibit the intracellular ROS and decrease the peroxidation injury [122] (Fig. 5.9c). More interestingly, Zhang et al. found that dietary iron oxide nanoparticles can delay aging and ameliorate neurodegeneration in Drosophila [171] (Fig. 5.9d). Intracellular Fe_3O_4 NPs showed neuroprotective ability in a PD cell model by diminishing α-Synuclein accumulation and Caspase-3 activation. More significantly, dietary Fe_3O_4 NPs could enhance the climbing ability of aged *Drosophila* and prolong their life span by reducing in vivo ROS levels. In addition, these nanoparticles can alleviate neurodegeneration and increase longevity in a *Drosophila* AD model. All these beneficial functions of Fe_3O_4 NPs may be conscribed to the catalase-like activity reducing intracellular oxidative stress. Although optimistic results are displayed, the cellular roles of IONzyme still need to be evaluated carefully, as some work has reported it may increase ROS level [114]. Taken together, the enzyme-like activities of IONzyme provide a new way to understand the potential function of iron oxide nanomaterials uptaken by the cell and may be used to control cellular ROS for therapeutic purpose.

5.5 Conclusion and Perspective

It has been a decade since intrinsic enzyme-like activity was first discovered in iron oxide nanoparticles. As the new generation of enzyme mimetics, the catalytic behaviors and kinetics of IONzyme have been systematically investigated to understand the mechanism and further improve its activity by controlling certain parameters, such as the size, morphology, nanostructure, dopants, and surface modification or integration with other nanomaterials, which enables us to rationally design the proper nanozymes according to the practical applications. Compared to the traditional enzyme mimetics or natural enzymes, IONzyme has much better stability and is a multifunctional and versatile platform able to be functionalized with additional molecules or labels. These advantages have opened a new way to use iron oxide nanomaterials independent of magnetism. It is already known that magnetic iron oxide materials have been widely used in biomedical and clinics, such as DNA extraction, gene delivery, cell sorting, and tumor imaging. For example, it can be used to sort T-cell for the chimeric antigen receptor (CAR) T-cell therapy which is one of the most promising immunotherapies for cancer treatment. All these applications are based on the magnetism of these materials. The enzyme-like activities will bring extra benefits in medical applications, such as immunoassay and pathogen detection, tumor diagnosis and therapy, biofilm elimination, and ROS modulations at multilevels for cell differentiation, cardioprotection, and neuroprotection. More inspiring findings are that by in vivo ROS regulation iron oxide nanomaterials not only polarize the macrophage cells and wake up immune attack in tumor tissue [166], but also induce ferroptosis directly to inhibit tumor growth [165]. In addition, the most recent work reported that iron oxide nanozymes can suppress the viability of intracellular bacteria through peroxidase-like activity [172]. These nanozymes also have the ability to destruct influenza virus by inducing lipid peroxidation in viral envelope [173]. Altogether, the enzyme-like activities of IONzyme may inspire more cutting-edge technologies in many important fields to improve human health.

Although many types of IONzymes have been reported, some fundamental challenges are need to be addressed. (1) There is still no standard way to evaluate and compare the activity level between IONzymes from different preparations and modifications and nanozymes from different nanomaterials. We suggest to use the specific activity termed to evaluate the enzyme activity which reflects the unit value at certain amount of enzyme proteins. (2) The activities of IONzyme are still lower than natural enzymes and the way to improve its selectivity is still limited. A potential approach is to refer to the catalytic mechanism and molecular structure of active site in natural enzymes. The feature for the interaction between substrate and enzyme will help to design certain nanostructure to improve the selectivity. Liu Juewen group just reported a new way using molecular imprinting that may improve the selectivity of IONzyme [76]. (3) Current activities of IONzyme are limited just mimicking peroxidase and catalase. Actually, there are hundreds of natural enzymes using iron as the cofactor for their catalysis besides the redox reaction. It is still challenging to design an IONzyme with a needed activity. The potential way may still lie on learning the structure and mechanism in natural enzymes. (4) Although some beneficial effects have been demonstrated, the intracellular role of IONzyme remains unclear, especially for the correlation among the catalytic activity, therapeutic effect, and biocompatibility in vivo. For instance, iron oxide nanoparticles have already been used for tumor in vivo MRI with clinical permission, but the biocompatibility needs to be carefully evaluated with respect to their intrinsic peroxidase/catalase-like activities. The ultimate influence on ROS-sensitive biological system needs to be carefully investigated, including immune activation, neural development, heart stress, stem cell proliferation, and differentiation and tumor growth. Therefore, great efforts are needed to tackle the fundamental challenges and further improve the activity of IONzyme for both in vitro and in vivo biomedical applications.

Acknowledgements This work was supported by Young Elite Scientist Sponsorship Program by the National Key R&D Program of China (Grant No. 2018YFC1003500), CAST, Beijing Natural Science Foundation (Grant No. 5164037), China Postdoctoral Science Foundation (Grant No. 2015M570158) and the China Postdoctoral Science Special Foundation (Grant No. 2016T90143), Strategic Priority Research Program of the Chinese Academy of Sciences (Grant No. XDA09030306), National Natural Science Foundation of China (Grant No. 81930050, 31530026 and 81671810), Key Research Program of Frontier Sciences, CAS (Grant No. QYZDB-SSW-SMC013), and Natural Science Foundation of Jiangsu, China (BK20161333).

References

1. Unsoy G, Gunduz U, Oprea O et al (2015) Magnetite: from synthesis to applications. Curr Top Med Chem 15(16):1622–1640
2. Xie J, Huang J, Li X et al (2009) Iron oxide nanoparticle platform for biomedical applications. Curr Med Chem 16(10):1278–1294
3. Pan Y, Du XW, Zhao F et al (2012) Magnetic nanoparticles for the manipulation of proteins and cells. Chem Soc Rev 41(7):2912–2942
4. Frimpong RA, Hilt JZ (2010) Magnetic nanoparticles in biomedicine: synthesis, functionalization and applications. Nanomed UK 5(9):1401–1414
5. Colombo M, Carregal-Romero S, Casula MF et al (2012) Biological applications of magnetic nanoparticles. Chem Soc Rev 41(11):4306–4334
6. Ho D, Sun XL, Sun SH (2011) Monodisperse magnetic nanoparticles for theranostic applications. Acc Chem Res 44(10):875–882
7. Lacroix LM, Ho D, Sun SH (2010) Magnetic nanoparticles as both imaging probes and therapeutic agents. Curr Top Med Chem 10(12):1184–1197
8. Zhu L, Zhou Z, Mao H et al (2017) Magnetic nanoparticles for precision oncology: theranostic magnetic iron oxide nanoparticles for image-guided and targeted cancer therapy. Nanomedicine (Lond) 12(1):73–87
9. Anselmo AC, Mitragotri S (2015) A review of clinical translation of inorganic nanoparticles. AAPS J 17(5):1041–1054
10. Gao L, Zhuang J, Nie L et al (2007) Intrinsic peroxidase-like activity of ferromagnetic nanoparticles. Nat Nanotechnol 2(9):577–583
11. Song Y, Qu K, Zhao C et al (2010) Graphene oxide: intrinsic peroxidase catalytic activity and its application to glucose detection. Adv Mater 22(19):2206–2210
12. Ragg R, Tahir MN, Tremel W (2016) Solids go bio: Inorganic nanoparticles as enzyme mimics. Eur J Inorg Chem 13–14:1906–1915

13. Tao Y, Ju EG, Ren JS et al (2015) Bifunctionalized mesoporous silica-supported gold nanoparticles: Intrinsic oxidase and peroxidase catalytic activities for antibacterial applications. Adv Mater 27(6):1097–1104
14. Cheng HJ, Lin SC, Muhammad F et al (2016) Rationally modulate the oxidase-like activity of nanoceria for self regulated bioassays. Acs Sensors 1(11):1336–1343
15. Wang S, Cazelles R, Liao WC et al (2017) Mimicking horseradish peroxidase and NADH peroxidase by heterogeneous Cu^{2+}-modified graphene oxide nanoparticles. Nano Lett 17(3):2043–2048
16. Singh S (2016) Cerium oxide based nanozymes: Redox phenomenon at biointerfaces. Biointerphases 11(4)
17. Vázquez-González M, Liao W-C, Cazelles R et al (2017) Mimicking horseradish peroxidase functions using Cu^{2+}-modified carbon nitride nanoparticles or Cu^{2+}-modified carbon dots as heterogeneous catalysts. ACS Nano 11(3):3247–3253
18. Dong ZY, Luo Q, Liu JQ (2012) Artificial enzymes based on supramolecular scaffolds. Chem Soc Rev 41(23):7890–7908
19. He WW, Wamer W, Xia QS et al (2014) Enzyme-like activity of nanomaterials. J Environ Sci Heal C 32(2):186–211
20. Wei H, Wang E (2013) Nanomaterials with enzyme-like characteristics (nanozymes): next-generation artificial enzymes. Chem Soc Rev 42(14):6060–6093
21. Chen ZW, Ji HW, Liu CQ et al (2016) A multinuclear metal complex based DNase-mimetic artificial enzyme: matrix cleavage for combating bacterial biofilms. Angew Chem Int Edit 55(36):10732–10736
22. Cheng HJ, Zhang L, He J et al (2016) Integrated nanozymes with nanoscale proximity for in vivo neurochemical monitoring in living brains. Anal Chem 88(10):5489–5497
23. Korschelt K, Ragg R, Metzger CS et al (2017) Glycine-functionalized copper(II) hydroxide nanoparticles with high intrinsic superoxide dismutase activity. Nanoscale 9(11):3952–3960
24. Gao N, Dong K, Zhao AD et al (2016) Polyoxometalate-based nanozyme: design of a multifunctional enzyme for multi-faceted treatment of Alzheimer's disease. Nano Res 9(4):1079–1090
25. Sun A, Mu L, Hu X (2017) Graphene oxide quantum dots as novel nanozymes for alcohol intoxication. ACS Appl Mater Interfaces 9(14):12241–12252
26. Gao L, Yan X (2016) Nanozymes: an emerging field bridging nanotechnology and biology. Science China, Life sciences
27. Gao LZ, Yan XY (2013) Discovery and current application of nanozyme. Prog Biochem Biophys 40(10):892–902
28. Shin HY, Park TJ, Kim MI (2015) Recent research trends and future prospects in nanozymes. J Nanomater
29. Lin Y, Ren J, Qu X (2014) Catalytically active nanomaterials: a promising candidate for artificial enzymes. Acc Chem Res 47(4):1097–1105
30. Wang XY, Hu YH, Wei H (2016) Nanozymes in bionanotechnology: from sensing to therapeutics and beyond. Inorg Chem Front 3(1):41–60
31. Wang X, Guo W, Hu Y et al (2016) Nanozymes: next wave of artificial enzymes. Springer, Berlin, Heidelberg
32. Shen Z, Wu A, Chen X (2016) Iron oxide nanoparticle based contrast agents for magnetic resonance imaging. Mol Pharm
33. Lee N, Yoo D, Ling D et al (2015) Iron oxide based nanoparticles for multimodal imaging and magnetoresponsive therapy. Chem Rev 115(19):10637–10689
34. Liu G, Gao J, Ai H et al (2013) Applications and potential toxicity of magnetic iron oxide nanoparticles. Small 9(9–10):1533–1545
35. Chen ZW, Yin JJ, Zhou YT et al (2012) Dual enzyme-like activities of iron oxide nanoparticles and their implication for diminishing cytotoxicity. ACS Nano 6(5):4001–4012
36. Wei H, Wang E (2008) Fe_3O_4 magnetic nanoparticles as peroxidase mimetics and their applications in H_2O_2 and glucose detection. Anal Chem 80(6):2250–2254

37. Liu BW, Han X, Liu JW (2016) Iron oxide nanozyme catalyzed synthesis of fluorescent polydopamine for light-up Zn^{2+} detection. Nanoscale 8(28):13620–13626
38. Zhang RZ, He SJ, Zhang CM et al (2015) Three-dimensional Fe- and N-incorporated carbon structures as peroxidase mimics for fluorescence detection of hydrogen peroxide and glucose. J Mater Chem B 3(20):4146–4154
39. Shi Y, Su P, Wang YY et al (2014) Fe_3O_4 peroxidase mimetics as a general strategy for the fluorescent detection of H_2O_2-involved systems. Talanta 130:259–264
40. Gao L, Giglio KM, Nelson JL et al (2014) Ferromagnetic nanoparticles with peroxidase-like activity enhance the cleavage of biological macromolecules for biofilm elimination. Nanoscale 6(5):2588–2593
41. Wang LJ, Min Y, Xu DD et al (2014) Membrane lipid peroxidation by the peroxidase-like activity of magnetite nanoparticles. Chem Commun 50(76):11147–11150
42. Lin YH, Huang YY, Ren JS et al (2014) Incorporating ATP into biomimetic catalysts for realizing exceptional enzymatic performance over a broad temperature range. Npg Asia Mater 6
43. Yang YC, Wang YT, Tseng WL (2017) Amplified peroxidase-like activity in iron oxide nanoparticles using adenosine Monophosphate: application to urinary protein sensing. ACS Appl Mater Interfaces 9:10069–10077
44. Vallabani NV, Karakoti AS, Singh S (2017) ATP-mediated intrinsic peroxidase-like activity of Fe_3O_4-based nanozyme: one step detection of blood glucose at physiological pH. Colloids Surf B Biointerfaces 153:52–60
45. Liu BW, Liu JW (2015) Accelerating peroxidase mimicking nanozymes using DNA. Nanoscale 7(33):13831–13835
46. Liu CH, Yu CJ, Tseng WL (2012) Fluorescence assay of catecholamines based on the inhibition of peroxidase-like activity of magnetite nanoparticles. Anal Chim Acta 745:143–148
47. Kim YS, Jurng J (2013) A simple colorimetric assay for the detection of metal ions based on the peroxidase-like activity of magnetic nanoparticles. Sensor Actuat B-Chem 176:253–257
48. Fan KL, Wang H, Xi JQ et al (2017) Optimization of Fe_3O_4 nanozyme activity via single amino acid modification mimicking an enzyme active site. Chem Commun 53(2):424–427
49. Dong YL, Zhang HG, Rahman ZU et al (2012) Graphene oxide-Fe_3O_4 magnetic nanocomposites with peroxidase-like activity for colorimetric detection of glucose. Nanoscale 4(13):3969–3976
50. Ma M, Xie J, Zhang Y et al (2013) Fe_3O_4@Pt nanoparticles with enhanced peroxidase-like catalytic activity. Mater Lett 105:36–39
51. An Q, Sun C, Li D et al (2013) Peroxidase-like activity of Fe_3O_4@carbon nanoparticles enhances ascorbic acid-induced oxidative stress and selective damage to PC-3 prostate cancer cells. ACS Appl Mater Interfaces 5(24):13248–13257
52. Li KF, Chen CF, Chen CY et al (2015) Magnetosomes extracted from *Magnetospirillum magneticum* strain AMB-1 showed enhanced peroxidase-like activity under visible-light irradiation. Enzyme Microb Tech 72:72–78
53. Wang ZH, Chen M, Shu JX et al (2016) One-step solvothermal synthesis of Fe_3O_4@Cu@Cu_2O nanocomposite as magnetically recyclable mimetic peroxidase. J Alloy Compd 682:432–440
54. Bhattacharya D, Baksi A, Banerjee I et al (2011) Development of phosphonate modified $Fe_{1-x}Mn_xFe_2O_4$ mixed ferrite nanoparticles: novel peroxidase mimetics in enzyme linked immunosorbent assay. Talanta 86:337–348
55. Yu FQ, Huang YZ, Cole AJ et al (2009) The artificial peroxidase activity of magnetic iron oxide nanoparticles and its application to glucose detection. Biomaterials 30(27):4716–4722
56. Zhang XQ, Gong SW, Zhang Y et al (2010) Prussian blue modified iron oxide magnetic nanoparticles and their high peroxidase-like activity. J Mater Chem 20(24):5110–5116
57. Dutta AK, Maji SK, Srivastava DN et al (2012) Peroxidase-like activity and amperometric sensing of hydrogen peroxide by Fe_2O_3 and Prussian Blue-modified Fe_2O_3 nanoparticles. J Mol Catal A-Chem 360:71–77

58. Dutta AK, Maji SK, Biswas P et al (2013) New peroxidase-substrate 3,5-di-tert-butylcatechol for colorimetric determination of blood glucose in presence of Prussian Blue-modified iron oxide nanoparticles. Sensor Actuat B-Chem 177:676–683
59. Liu QY, Zhang LY, Li H et al (2015) One-pot synthesis of porphyrin functionalized gamma-Fe_2O_3 nanocomposites as peroxidase mimics for H_2O_2 and glucose detection. Mat Sci Eng C-Mater 55:193–200
60. Roy A, Sahoo R, Ray C et al (2016) Soft template induced phase selective synthesis of Fe_2O_3 nanomagnets: one step towards peroxidase-mimic activity allowing colorimetric sensing of thioglycolic acid. Rsc Adv 6(38):32308–32318
61. Song LN, Huang C, Zhang W et al (2016) Graphene oxide-based Fe_2O_3 hybrid enzyme mimetic with enhanced peroxidase and catalase-like activities. Colloid Surface A 506:747–755
62. Kluenker M, Tahir MN, Ragg R et al (2017) Pd@Fe_2O_3 superparticles with enhanced peroxidase activity by solution phase epitaxial growth. Chem Mater 29(3):1134–1146
63. Chaudhari KN, Chaudhari NK, Yu JS (2012) Peroxidase mimic activity of hematite iron oxides (alpha-Fe_2O_3) with different nanostructres. Catal Sci Technol 2(1):119–124
64. Mu JS, Zhang L, Zhao M et al (2013) Co_3O_4 nanoparticles as an efficient catalase mimic: properties, mechanism and its electrocatalytic sensing application for hydrogen peroxide. J Mol Catal A-Chem 378:30–37
65. Song YJ, Xia XF, Wu XF et al (2014) Integration of platinum nanoparticles with a volumetric bar-chart chip for biomarker assays. Angew Chem Int Edit 53(46):12451–12455
66. Mu JS, Zhang L, Zhao M et al (2014) Catalase mimic property of Co_3O_4 nanomaterials with different morphology and its application as a calcium sensor. ACS Appl Mater Interfaces 6(10):7090–7098
67. Pirmohamed T, Dowding JM, Singh S et al (2010) Nanoceria exhibit redox state-dependent catalase mimetic activity. Chem Commun 46(16):2736–2738
68. Fan J, Yin JJ, Ning B et al (2011) Direct evidence for catalase and peroxidase activities of ferritin-platinum nanoparticles. Biomaterials 32(6):1611–1618
69. Zhang JB, Zhuang J, Gao LZ et al (2008) Decomposing phenol by the hidden talent of ferromagnetic nanoparticles. Chemosphere 73(9):1524–1528
70. Wang N, Zhu LH, Wang DL et al (2010) Sono-assisted preparation of highly-efficient peroxidase-like Fe_3O_4 magnetic nanoparticles for catalytic removal of organic pollutants with H_2O_2. Ultrason Sonochem 17(3):526–533
71. Veitch NC (2004) Horseradish peroxidase: a modern view of a classic enzyme. Phytochemistry 65(3):249–259
72. Li XK, Paier J (2016) Adsorption of water on the Fe_3O_4(111) surface: structures, stabilities, and vibrational properties studied by density functional theory. J Phys Chem C 120(2):1056–1065
73. Liu S, Lu F, Xing R et al (2011) Structural effects of Fe_3O_4 nanocrystals on peroxidase-like activity. Chemistry 17(2):620–625
74. Wang H, Jiang H, Wang S et al (2014) Fe_3O_4-MWCNT magnetic nanocomposites as efficient peroxidase mimic catalysts in a Fenton-like reaction for water purification without pH limitation. Rsc Adv 4(86):45809–45815
75. He J, Yang XF, Men B et al (2016) Interfacial mechanisms of heterogeneous Fenton reactions catalyzed by iron-based materials: a review. J Environ Sci China 39:97–109
76. Zhang ZJ, Zhang XH, Liu BW, Liu JW (2017) Molecular imprinting on inorganic nanozymes for hundred-fold enzyme specificity. J Am Chem Soc. https://doi.org/10.1021/jacs.7b00601
77. Peng FF, Zhang Y, Gu N (2008) Size-dependent peroxidase-like catalytic activity of Fe_3O_4 nanoparticles. Chinese Chem Lett 19(6):730–733
78. Chang Q, Deng KJ, Zhu LH et al (2009) Determination of hydrogen peroxide with the aid of peroxidase-like Fe_3O_4 magnetic nanoparticles as the catalyst. Microchim Acta 165(3–4):299–305
79. Zhang SX, Zhao XL, Niu HY et al (2009) Superparamagnetic Fe_3O_4 nanoparticles as catalysts for the catalytic oxidation of phenolic and aniline compounds. J Hazard Mater 167(1–3):560–566

80. Wu YH, Song MJ, Xin ZA et al (2011) Ultra-small particles of iron oxide as peroxidase for immunohistochemical detection. Nanotechnology 22(22)
81. Gao Y, Wang GN, Huang H et al (2011) Fluorometric method for the determination of hydrogen peroxide and glucose with Fe_3O_4 as catalyst. Talanta 85(2):1075–1080
82. Wang XQ, Tu Q, Zhao B et al (2013) Effects of poly(L-lysine)-modified Fe_3O_4 nanoparticles on endogenous reactive oxygen species in cancer stem cells. Biomaterials 34(4):1155–1169
83. Guan GJ, Yang L, Mei QS et al (2012) Chemiluminescence switching on peroxidase-like Fe_3O_4 nanoparticles for selective detection and simultaneous determination of various pesticides. Anal Chem 84(21):9492–9497
84. Liu Y, Yuan M, Qiao LJ et al (2014) An efficient colorimetric biosensor for glucose based on peroxidase-like protein-Fe_3O_4 and glucose oxidase nanocomposites. Biosens Bioelectron 52:391–396
85. Wei SL, Li JW, Liu Y (2015) Colourimetric assay for beta-estradiol based on the peroxidase-like activity of Fe_3O_4@mSiO(2)@HP-beta-CD nanoparticles. Rsc Adv 5(130):107670–107679
86. Liu SJ, Fu JW, Wang MH et al (2016) Magnetically separable and recyclable Fe_3O_4-polydopamine hybrid hollow microsphere for highly efficient peroxidase mimetic catalysts. J Colloid Interf Sci 469:69–77
87. Zhang ZX, Wang XL, Yang XR (2011) A sensitive choline biosensor using Fe_3O_4 magnetic nanoparticles as peroxidase mimics. Analyst 136(23):4960–4965
88. He X, Tan L, Chen D et al (2013) Fe_3O_4-Au@mesoporous SiO_2 microspheres: an ideal artificial enzymatic cascade system. Chem Commun 49(41):4643–4645
89. Liang MM, Fan KL, Pan Y et al (2013) Fe_3O_4 magnetic nanoparticle peroxidase mimetic-based colorimetric assay for the rapid detection of organophosphorus pesticide and nerve agent. Anal Chem 85(1):308–312
90. Zeng T, Zhang XL, Wang SH et al (2014) Assembly of a nanoreactor system with confined magnetite core and shell for enhanced Fenton-like catalysis. Chem-Eur J 20(21):6474–6481
91. Shi Y, Huang J, Wang JN et al (2015) A magnetic nanoscale Fe_3O_4/P beta-CD composite as an efficient peroxidase mimetic for glucose detection. Talanta 143:457–463
92. Duan DM, Fan KL, Zhang DX et al (2015) Nanozyme-strip for rapid local diagnosis of Ebola. Biosens Bioelectron 74:134–141
93. Wang CQ, Qian J, Wang K et al (2016) Colorimetric aptasensing of ochratoxin A using Au@Fe_3O_4 nanoparticles as signal indicator and magnetic separator. Biosens Bioelectron 77:1183–1191
94. Shi JH, Tong LZ, Liu DM et al (2012) Fabrication, structure, and properties of Fe_3O_4@C encapsulated with $YVO_4:Eu^{3+}$ composites. J Nanopart Res 14(4)
95. Wang H, Li S, Si YM et al (2014) Recyclable enzyme mimic of cubic Fe_3O_4 nanoparticles loaded on graphene oxide-dispersed carbon nanotubes with enhanced peroxidase-like catalysis and electrocatalysis. J Mater Chem B 2(28):4442–4448
96. JaR Guivar, Fernandes EGR, Zucolotto V (2015) A peroxidase biomimetic system based on Fe_3O_4 nanoparticles in non-enzymatic sensors. Talanta 141:307–314
97. Zhang K, Zuo W, Wang ZY et al (2015) A simple route to $CoFe_2O_4$ nanoparticles with shape and size control and their tunable peroxidase-like activity. Rsc Adv 5(14):10632–10640
98. Lee Y, Garcia MA, Frey Huls NA et al (2010) Synthetic tuning of the catalytic properties of Au–Fe_3O_4 nanoparticles. Angew Chem 49(7):1271–1274
99. Sun X, Guo S, Chung CS et al (2013) A sensitive H_2O_2 assay based on dumbbell-like PtPd–Fe_3O_4 nanoparticles. Adv Mater 25(1):132–136
100. Cheng R, Li GQ, Cheng C et al (2015) Catalytic oxidation of 4-chlorophenol with magnetic Fe_3O_4 nanoparticles: mechanisms and particle transformation. Rsc Adv 5(82):66927–66933
101. Chen CX, Lu LX, Zheng Y et al (2015) A new colorimetric protocol for selective detection of phosphate based on the inhibition of peroxidase-like activity of magnetite nanoparticles. Anal Methods UK 7(1):161–167
102. Gao YY, Li HX, Ou ZZ et al (2011) Enhancing the catalytic activity of peroxidase by adsorption onto Fe_3O_4 magnetic nanoparticle/multiwalled carbon nanotube composite surfaces. Acta Phys-Chim Sin 27(10):2469–2477

103. Niu HY, Dizhang Meng ZF et al (2012) Fast defluorination and removal of norfloxacin by alginate/Fe@Fe_3O_4 core/shell structured nanoparticles. J Hazard Mater 227:195–203
104. Ai X, Wang Y, Hou XD et al (2013) Advanced oxidation using Fe_3O_4 magnetic nanoparticles and its application in mercury speciation analysis by high performance liquid chromatography-cold vapor generation atomic fluorescence spectrometry. Analyst 138(12):3494–3501
105. Woo MA, Kim MI, Jung JH et al (2013) A novel colorimetric immunoassay utilizing the peroxidase mimicking activity of magnetic nanoparticles. Int J Mol Sci 14(5):9999–10014
106. Wu XC, Zhang Y, Han T et al (2014) Composite of graphene quantum dots and Fe_3O_4 nanoparticles: peroxidase activity and application in phenolic compound removal. Rsc Adv 4(7):3299–3305
107. Xiong LY, Zheng LZ, Xu JP et al (2014) A non-enzyme hydrogen peroxide biosensor based on Fe_3O_4/RGO nanocomposite material. Ecs Electrochem Lett 3(12):B26–B29
108. Wang W, Liu Y, Li TL et al (2014) Heterogeneous Fenton catalytic degradation of phenol based on controlled release of magnetic nanoparticles. Chem Eng J 242:1–9
109. Niu XY, Xu YY, Dong YL et al (2014) Visual and quantitative determination of dopamine based on $Co_xFe_{3-x}O_4$ magnetic nanoparticles as peroxidase mimetics. J Alloy Compd 587:74–81
110. Ai X, Wu L, Zhang MN et al (2014) Analytical method for the determination of trace toxic elements in milk based on combining Fe_3O_4 nanoparticles accelerated UV Fenton-like digestion and solid phase extraction. J Agr Food Chem 62(34):8586–8593
111. Zhang XL, He ML, Liu JH et al (2014) Fe_3O_4@C nanoparticles as high-performance Fenton-like catalyst for dye decoloration. Chin Sci Bull 59(27):3406–3412
112. Bhalkikar A, Gernhart ZC, Cheung CL (2015) Recyclable magnetite nanoparticle catalyst for one-pot conversion of cellobiose to 5-hydroxymethylfurfural in water. J Nanomater 2015:264037
113. Jia Y, Yu HM, Wu L et al (2015) Three birds with one Fe_3O_4 nanoparticle: integration of microwave digestion, solid phase extraction, and magnetic separation for sensitive determination of arsenic and antimony in fish. Anal Chem 87(12):5866–5871
114. Sang JL, Wu RL, Guo PP et al (2016) Affinity-tuned peroxidase-like activity of hydrogel-supported Fe_3O_4 nanozyme through alteration of crosslinking concentration. J Appl Polym Sci 133(8)
115. Nie DX, Shi GY, Yu YY (2016) Fe_3O_4 magnetic nanoparticles as peroxidase mimetics used in colorimetric determination of 2,4-dinitrotoluene. Chin J Anal Chem 44(2):179–184
116. Zhang ZX, Zhu H, Wang XL et al (2011) Sensitive electrochemical sensor for hydrogen peroxide using Fe_3O_4 magnetic nanoparticles as a mimic for peroxidase. Microchim Acta 174(1–2):183–189
117. Sun HY, Jiao XL, Han YY et al (2013) Synthesis of Fe3O4-Au nanocomposites with enhanced peroxidase-like activity. Eur J Inorg Chem 1:109–114
118. Wang XS, Huang H, Li GQ et al (2014) Hydrothermal synthesis of 3D hollow porous Fe_3O_4 microspheres towards catalytic removal of organic pollutants. Nanoscale Res Lett 9
119. Qian J, Yang XW, Jiang L et al (2014) Facile preparation of Fe_3O_4 nanospheres/reduced graphene oxide nanocomposites with high peroxidase-like activity for sensitive and selective colorimetric detection of acetylcholine. Sensor Actuat B-Chem 201:160–166
120. Xiao SW, Zhang CT, Chen R et al (2015) Selective oxidation of benzyl alcohol to benzaldehyde with H_2O_2 in water on epichlorohydrin-modified Fe_3O_4 microspheres. New J Chem 39(6):4924–4932
121. Wang YH, Zhou B, Wu S et al (2015) Colorimetric detection of hydrogen peroxide and glucose using the magnetic mesoporous silica nanoparticles. Talanta 134:712–717
122. Xiong F, Wang H, Feng YD et al (2015) Cardioprotective activity of iron oxide nanoparticles. Scientific Reports 5:8579
123. Qi CC, Zheng JB (2015) Novel nonenzymatic hydrogen peroxide sensor based on Fe_3O_4/PPy/Ag nanocomposites. J Electroanal Chem 747:53–58
124. Fan KL, Cao CQ, Pan YX et al (2012) Magnetoferritin nanoparticles for targeting and visualizing tumour tissues. Nat Nanotechnol 7(7):459–464

125. Melnikova L, Pospiskova K, Mitroova Z et al (2014) Peroxidase-like activity of magnetoferritin. Microchim Acta 181(3–4):295–301
126. Cai Y, Cao CQ, He XQ et al (2015) Enhanced magnetic resonance imaging and staining of cancer cells using ferrimagnetic H-ferritin nanoparticles with increasing core size. Int J Nanomed 10:2619–2634
127. Yin GF, Huang ZB, Deng M et al (2011) Preparation and cell response of bio-mineralized Fe_3O_4 nanoparticles. J Colloid Interf Sci 363(1):393–402
128. Pan Y, Li N, Mu JS et al (2015) Biogenic magnetic nanoparticles from Burkholderia sp. YN01 exhibiting intrinsic peroxidase-like activity and their applications. Appl Microbiol Biot 99(2):703–715
129. Guo FF, Yang W, Jiang W et al (2012) Magnetosomes eliminate intracellular reactive oxygen species in Magnetospirillum gryphiswaldense MSR-1. Environ Microbiol 14(7):1722–1729
130. Zubir NA, Yacou C, Motuzas J et al (2014) Structural and functional investigation of graphene oxide-Fe_3O_4 nanocomposites for the heterogeneous Fenton-like reaction. Sci Rep 4:4594
131. Chang Q, Tang HQ (2014) Optical determination of glucose and hydrogen peroxide using a nanocomposite prepared from glucose oxidase and magnetite nanoparticles immobilized on graphene oxide. Microchim Acta 181(5–6):527–534
132. Gao Y, Wei Z, Li F et al (2014) Synthesis of a morphology controllable Fe_3O_4 nanoparticle/hydrogel magnetic nanocomposite inspired by magnetotactic bacteria and its application in H_2O_2 detection. Green Chem 16(3):1255–1261
133. Chen JZ, Liu YJ, Zhu GX et al (2014) Ag@Fe_3O_4 nanowire: fabrication, characterization and peroxidase-like activity. Cryst Res Technol 49(5):309–314
134. Mirabello G, Lenders JJM, NaJM Sommerdijk (2016) Bioinspired synthesis of magnetite nanoparticles. Chem Soc Rev 45(18):5085–5106
135. Yang X, Wang LN, Zhou GZ et al (2015) Electrochemical detection of H_2O_2 based on Fe_3O_4 nanoparticles with graphene oxide and polyamidoamine dendrimer. J Clust Sci 26(3):789–798
136. Liu B, Liu J (2017) Surface modification of nanozymes. Nano Research 10(4):1125–1148
137. Cheng XL, Jiang JS, Jiang DM et al (2014) Synthesis of rhombic dodecahedral Fe_3O_4 nanocrystals with exposed high-energy 110 facets and their peroxidase-like activity and lithium storage properties. J Phys Chem C 118(24):12588–12598
138. Fan HM, Yi JB, Yang Y et al (2009) Single-crystalline MFe_2O_4 nanotubes/nanorings synthesized by thermal transformation process for biological applications. ACS Nano 3(9):2798–2808
139. Lee JW, Jeon HJ, Shin HJ et al (2012) Superparamagnetic Fe_3O_4 nanoparticles-carbon nitride nanotube hybrids for highly efficient peroxidase mimetic catalysts. Chem Commun 48(3):422–424
140. Zhang LH, Guo SJ, Dong SJ (2009) Nanoreactor of Fe_3O_4@SiO_2 core-shell structure with nanochannels for efficient catalysis. J Biomed Nanotechnol 5(5):586–590
141. Hu SL, Zhang XQ, Zang FC et al (2016) Surface modified iron oxide nanoparticles as Fe source precursor to induce the formation of Prussian Blue nanocubes. J Nanosci Nanotechno 16(2):1967–1974
142. Liu QY, Li H, Zhao QR et al (2014) Glucose-sensitive colorimetric sensor based on peroxidase mimics activity of porphyrin-Fe(3)o(4) nanocomposites. Mat Sci Eng C-Mater 41:142–151
143. Tang ZW, Wu H, Zhang YY et al (2011) Enzyme-mimic activity of ferric nano-core residing in ferritin and its biosensing applications. Anal Chem 83(22):8611–8616
144. Gao LZ, Wu JM, Lyle S et al (2008) Magnetite nanoparticle-linked immunosorbent assay. J Phys Chem C 112(44):17357–17361
145. Wang X, Niessner R, Tang DP et al (2016) Nanoparticle-based immunosensors and immunoassays for aflatoxins. Anal Chim Acta 912:10–23
146. Yang MZ, Guan YP, Yang Y et al (2014) Immunological detection of hepatocellular carcinoma biomarker GP73 based on dissolved magnetic nanoparticles. Colloid Surface A 443:280–285
147. Yang MZ, Guan YP, Yang Y et al (2013) Peroxidase-like activity of amino-functionalized magnetic nanoparticles and their applications in immunoassay. J Colloid Interf Sci 405:291–295

148. Yang MZ, Guan YP, Yang Y et al (2014) A sensitive and rapid immunoassay for mycoplasma pneumonia based on Fe_3O_4 nanoparticles. Mater Lett 137:113–116
149. Il Kim M, Kim MS, Woo MA et al (2014) Highly efficient colorimetric detection of target cancer cells utilizing superior catalytic activity of graphene oxide-magnetic-platinum nanohybrids. Nanoscale 6(3):1529–1536
150. Perez JM (2007) Iron oxide nanoparticles—hidden talent. Nat Nanotechnol 2(9):535–536
151. Thiramanas R, Jangpatarapongsa K, Tangboriboonrat P et al (2013) Detection of *Vibrio cholerae* using the intrinsic catalytic activity of a magnetic polymeric nanoparticle. Anal Chem 85(12):5996–6002
152. Zhang LS, Huang R, Liu WP et al (2016) Rapid and visual detection of Listeria monocytogenes based on nanoparticle cluster catalyzed signal amplification. Biosens Bioelectron 86:1–7
153. Park KS, Kim MI, Cho DY et al (2011) Label-free colorimetric detection of nucleic acids based on target-induced shielding against the peroxidase-mimicking activity of magnetic nanoparticles. Small 7(11):1521–1525
154. Zhang ZX, Wang ZJ, Wang XL et al (2010) Magnetic nanoparticle-linked colorimetric aptasensor for the detection of thrombin. Sensor Actuat B-Chem 147(2):428–433
155. Kim MI, Shim J, Li T et al (2011) Fabrication of nanoporous nanocomposites entrapping Fe_3O_4 magnetic nanoparticles and oxidases for colorimetric biosensing. Chem-Eur J 17(38):10700–10707
156. Liu CH, Tseng WL (2011) Oxidase-functionalized Fe_3O_4 nanoparticles for fluorescence sensing of specific substrate. Anal Chim Acta 703(1):87–93
157. Ma YH, Zhang ZY, Ren CL et al (2012) A novel colorimetric determination of reduced glutathione in A549 cells based on Fe_3O_4 magnetic nanoparticles as peroxidase mimetics. Analyst 137(2):485–489
158. Yang ZH, Chai YQ, Yuan R et al (2014) Hollow platinum decorated Fe_3O_4 nanoparticles as peroxidase mimetic couple with glucose oxidase for pseudobienzyme electrochemical immunosensor. Sensor Actuat B-Chem 193:461–466
159. Kim MI, Cho D, Park HG (2015) Colorimetric quantification of glucose and cholesterol in human blood using a nanocomposite entrapping magnetic nanoparticles and oxidases. J Nanosci Nanotechno 15(10):7955–7961
160. Zhang J, Yang C, Chen CX et al (2013) Determination of nitrite and glucose in water and human urine with light-up chromogenic response based on the expeditious oxidation of 3,3′,5,5′-tetramethylbenzidine by peroxynitrous acid. Analyst 138(8):2398–2404
161. Kim MI, Shim J, Li T et al (2012) Colorimetric quantification of galactose using a nanostructured multi-catalyst system entrapping galactose oxidase and magnetic nanoparticles as peroxidase mimetics. Analyst 137(5):1137–1143
162. Kim MI, Shim J, Parab HJ et al (2012) A convenient alcohol sensor using one-pot nanocomposite entrapping alcohol oxidase and magnetic nanoparticles as peroxidase mimetics. J Nanosci Nanotechno 12(7):5914–5919
163. Zhang D, Zhao YX, Gao YJ et al (2013) Anti-bacterial and in vivo tumor treatment by reactive oxygen species generated by magnetic nanoparticles. J Mater Chem B 1(38):5100–5107
164. Zhuang J, Fan KL, Gao LZ et al (2012) Ex vivo detection of iron oxide magnetic nanoparticles in mice using their intrinsic peroxidase-mimicking activity. Mol Pharmaceut 9(7):1983–1989
165. Kim SE, Zhang L, Ma K et al (2016) Ultrasmall nanoparticles induce ferroptosis in nutrient-deprived cancer cells and suppress tumour growth. Nat Nanotechnol 11(11):977–985
166. Zanganeh S, Hutter G, Spitler R et al (2016) Iron oxide nanoparticles inhibit tumour growth by inducing pro-inflammatory macrophage polarization in tumour tissues. Nat Nanotechnol 11(11):986–994
167. Pan WY, Huang CC, Lin TT et al (2016) Synergistic antibacterial effects of localized heat and oxidative stress caused by hydroxyl radicals mediated by graphene/iron oxide-based nanocomposites. Nanomed-Nanotechnol 12(2):431–438
168. Gao LZ, Liu Y, Kim D et al (2016) Nanocatalysts promote *Streptococcus mutans* biofilm matrix degradation and enhance bacterial killing to suppress dental caries in vivo. Biomaterials 101:272–284

169. Huang DM, Hsiao JK, Chen YC et al (2009) The promotion of human mesenchymal stem cell proliferation by superparamagnetic iron oxide nanoparticles. Biomaterials 30(22):3645–3651
170. Zhou G, Meng S, Li YH et al (2016) Optimal ROS signaling is critical for nuclear reprogramming. Cell Rep 15(5):919–925
171. Zhang Y, Wang ZY, Li XJ et al (2016) Dietary iron oxide nanoparticles delay aging and ameliorate neurodegeneration in *Drosophila*. Adv Mater 28(7):1387–1393
172. Shi SR, Wu S, Shen YR et al (2018) Iron oxide nanozyme suppress intracellular *Salmonella Enteritidis* growth and alleviates infection in vivo. Thernostics 8(22):6149–6162
173. Qin T, Ma RN, Yin YY et al (2019) Catalytic inactivation of influenza virus by iron oxide nanozyme. Theranostics (in press). https://doi.org/10.7150/thno.37380

Chapter 6
Prussian Blue and Other Metal–Organic Framework-based Nanozymes

Wei Zhang, Yang Wu, Zhuoxuan Li, Haijiao Dong, Yu Zhang and Ning Gu

Abbreviations

AA	Ascorbic acid
AAO	Ascorbic acid oxidase
ABTS	2,2′-azinobis(3-ethylbenzothiazoline)-6-sulfonic acid
ALP	Alkaline phosphatase
ApoA1	Apolipoprotein A1
BDC	1,4-benzenedicarboxylate
BG	Berlin green
BTB	1,3,5-benzenetribenzoate
BTC	1,3,5-benzentricarboxylate
CD	Cyclodextrin
CV	Coefficients of variation
Cys	Cysteine

Z. Li · H. Dong · Y. Zhang (✉) · N. Gu (✉)
State Key Laboratory of Bioelectronics, Jiangsu Key Laboratory for Biomaterials and Devices, School of Biological Science and Medical Engineering, Southeast University, Nanjing 210096, People's Republic of China
e-mail: zhangyu@seu.edu.cn

N. Gu
e-mail: guning@seu.edu.cn

Collaborative Innovation Center of Suzhou Nano Science and Technology, Southeast University, Nanjing 210096, People's Republic of China

W. Zhang
Core Facility, The First Affiliated Hospital of Nanjing Medical University, 300 Guangzhou Road, Nanjing, Jiangsu 210029, People's Republic of China

Y. Wu
Research Center of Clinical Oncology, The Affiliated Cancer Hospital of Nanjing Medical University, Nanjing 210009, People's Republic of China

X. Yan (ed.), *Nanozymology*, Nanostructure Science and Technology,
https://doi.org/10.1007/978-981-15-1490-6_6

FDA	Food and Drug Administration
GOx	Glucose oxidase
GSH	Gluthione
GMP	Guanosine monophosphate
HAP	Hydroxyapatite
Hcy	Homocysteine
HDS	Hydroxy double salts
Hep	Heparin
HKUST	Hong Kong University of Science and Technology
HRP	Horseradish peroxidase
HTA	2-hydroxy terephthalic acid
IBD	Inflammatory bowel disease
IONP	Iron oxide nanoparticle
IRMOF	Isoreticular metal–organic framework
LDA	Linear discriminant analysis
LMG	Leucomalachite green
LOx	Lactate oxidase
MB	Methylene blue
MEKP	Methyl ethyl ketone peroxide
MG	Malachite green
MIP	Molecularly imprinted polymer
MIL	Materials of Institute Lavoisier
MOF	Metal–organic framework
MTV-MOF-5	Multivariate MOF-5
NBT	Nitro blue tetrazolium
NIR	Near-infrared
PB	Prussian blue
PBA	Prussian blue analogs
PBNPs	Prussian blue nanoparticles
PCN	Porous coordination network
PDT	Photodynamic therapy
PEC	Photoelectrochemical
PMGO	Prussian blue-incorporated magnetic graphene oxide
Pt NPs	Platinum nanoparticles
PW	Prussian white
PY	Prussian yellow
ROS	Reactive oxygen species
SAN	Single-atom nanozyme
SDM	Sulfadimethoxine
SERS	Surface-enhanced Raman scattering
TA	Terephthalic acid
TBHP	Tert-butyl hydroperoxide
TCPP(Fe)	Fe-bound tetrakis(4-carboxyphenyl) porphyrin
TH	Thiamine
TLR	Toll-like receptor

TMB	3,3′,5,5′-tetramethylbenzidine
UCNP	Upconversion nanoparticle
ZIF	Zeolitic imidazolate frameworks

Coordination polymers, colloquially known as metal–organic frameworks (MOFs), are a series of polymers rapidly developed in nearly two decades [1]. MOFs are porous materials possess three-dimensionally ordered structure taking metal ions as the junction points and linked by ligands. MOFs have showed great prospect in the construction of nanozymes. On one hand, MOFs are suitable for enzyme immobilization due to their porosity and multiplex structures. On the other hand, transition metal nodes containing MOFs themselves can play as biomimetic catalysts. In this chapter, we will focus on the discussion about the properties of MOF-based nanozymes and summarize the current research status and the remained problems. We not only introduce the structure of MOF nanozymes but also summarize their applications in biodetection and disease therapy.

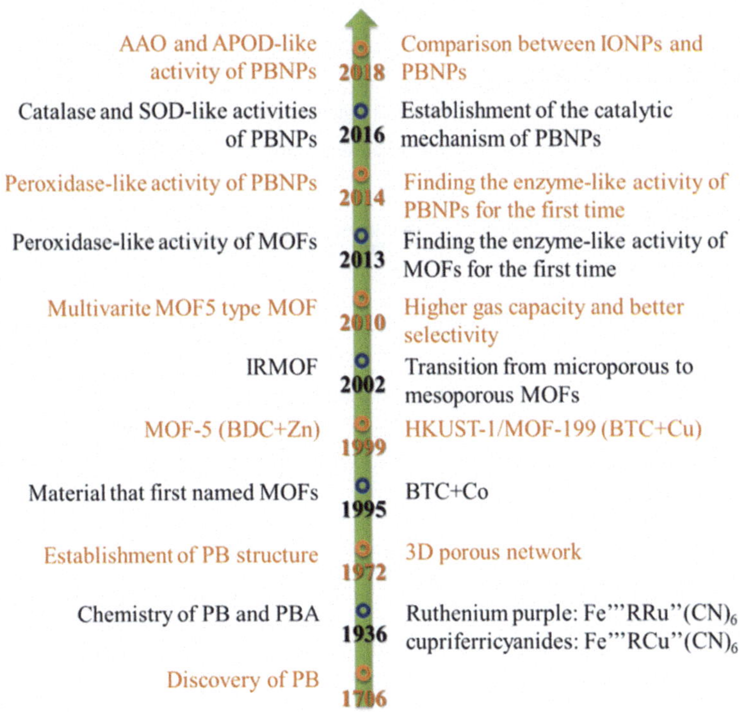

Fig. 6.1 A brief timeline for the development MOFs

6.1 History of Metal–Organic Frameworks (MOFs)

As shown in Fig. 6.1, a paint-maker named Diesbach discovered a blue segment Prussian blue (PB) in 1706, which is a well-known coordination polymer. The chemical component of PB is potassium ferrous ferricyanide (equivalent to potassium ferric ferrocyanide). After the discovery of PB, various Prussian blue analogs (PBAs) were synthesized, characterized, and applied in different fields [2, 3]. PBAs are a series materials prepared via replacing the ferrous ions in PB with other transition metal M (M = Mn, Co, Ni, Cu, Zn, Cd, Ru, etc.) ions. The formula of PBA is usually $A_xMFe(CN)_6$ and their crystallographic structure is face-centered cubic. The Fe(III) and M(II) in PBA adopt the six-coordination with C and N in the cyanides, respectively. In 1972, Ludi et al. established the structure of PB, namely, PB is a kind of coordination polymer possessing 3D porous network [4]. Not strictly, PB was the first MOF material in history [5].

In 1995, Yaghi's group prepared the first named MOFs using 1,3,5-benzentricarboxylate (BTC) as the building block and transition metal Co. This work is a milestone in the history of MOFs [6]. Since then, many MOFs composed of different metal nodes and ligands (Table 6.1) were synthesized and studied intensively.

Table 6.1 Structure of ligands

Structure	Name of ligands
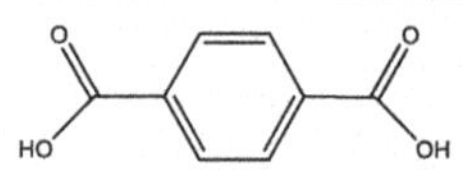	1,4-benzenedicarboxylate (1,4-BDC)
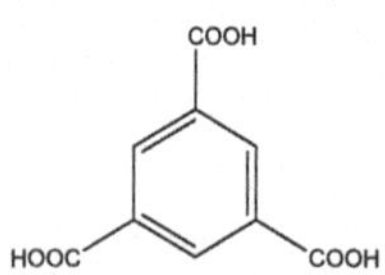	1,3,5-benzentricarboxylate (BTC)
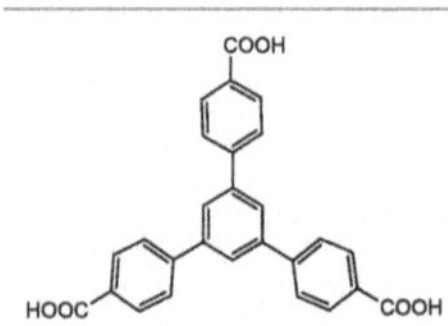	1,3,5-benzenetribenzoate (BTB)
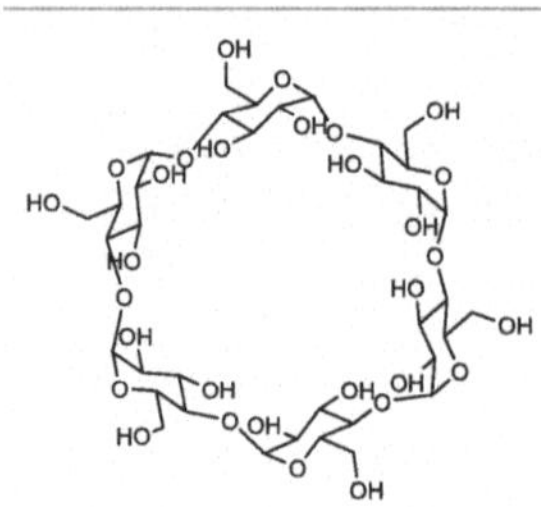	Cyclodextrin (CD)

(continued)

Table 6.1 (continued)

Structure	Name of ligands
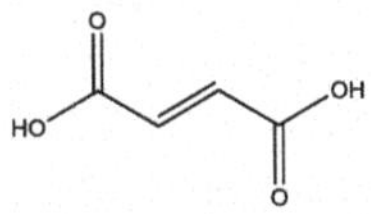	Fumaric acid
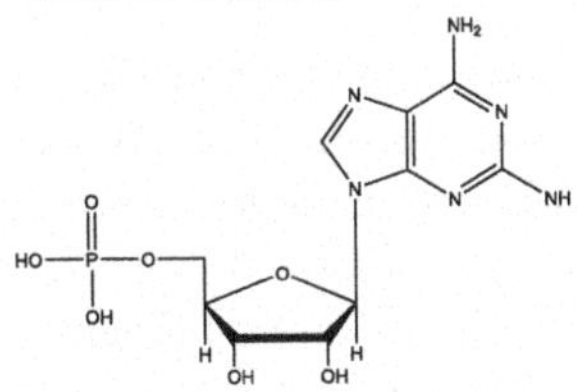	Nucleotide
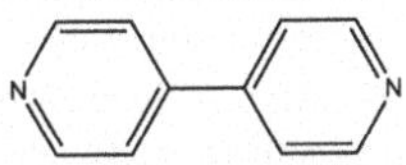	4,4′-Bipyridine (BPY)
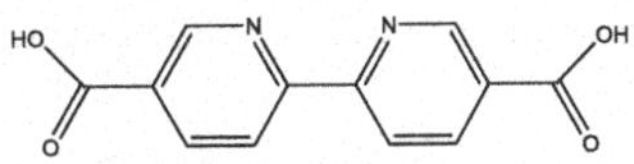	2,2′-bipyridine-5,5-dicarboxylic acid
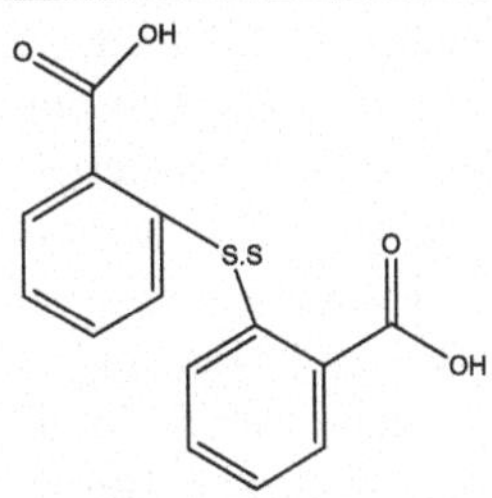	2,2′-Dithiosalicylic acid (DTDBA)
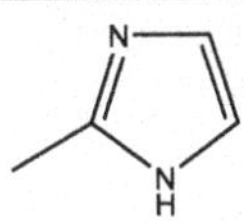	2-methylimidazole
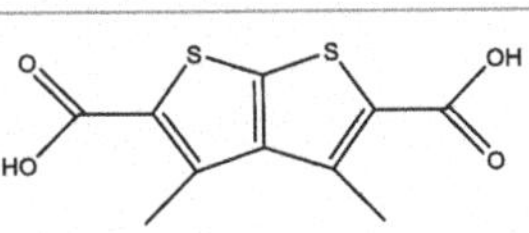	3,4-dimethylthieno[2,3-b] thiophene-2,5-dicarboxylic acid (H_2DMTDC)
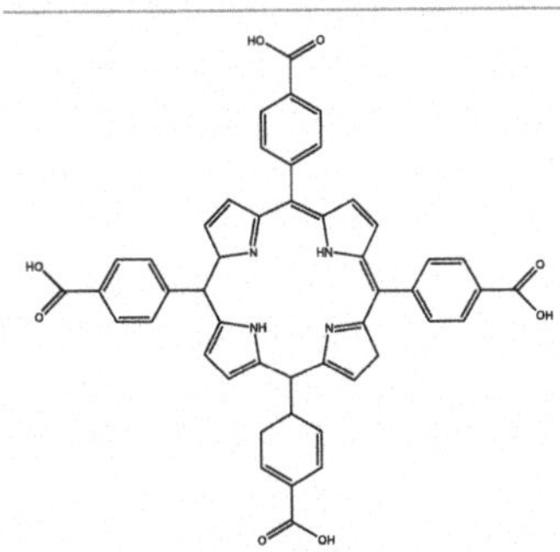	TCPP

In 1999, Yaghi's group improved this kind of MOFs to synthesize a cubic 3D MOF material called MOF-5 using organic ligand 1,4-benzenedicarboxylate (1,4-BDC) and Zn^{2+} [7]. In the same year, MOF-199 was obtained by Hong Kong University of Science and Technology (HKUST) and named HKUST-1, which is a Cu-BTC MOF. In 2002, Yaghi's group designed an isoreticular metal–organic framework (IRMOF) series with homogeneous periodic pores varied from 3.8 to 28.8 angstroms, which realized the transition from microporous MOFs to mesoporous MOFs [8, 9]. In 2004, using a triangular unit carboxylate derivative 1,3,5-benzenetribenzoate (BTB), Yaghi's group established MOF-177 with high surface area [10]. In 2005, Materials of Institute Lavoisier (MIL)-101 was synthesized. This kind of MOFs has large pore sizes and surface area [11]. Until 2006, Yaghi's group synthesized more than 100 kinds of zeolitic imidazolate frameworks (ZIFs) [12]. In 2010, Yaghi's group prepared 18 multivariate MOF-5 (MTV-MOF-5)-type MOFs. Comparing to normal MOF-5, the obtained MTV-MOF-5 has higher gas uptake capacity and better selectivity [13]. In 2011, macromolecular ligands such as cyclodextrin (CD) were used to synthesize a series of MOFs composed of CD, namely, CD-MOFs [14, 15]. Since then, the catalytic properties including enzyme-like activity of MOFs were intensively studied by experts from various areas.

The catalytic properties of MOFs strongly depend on the crystallinity. Attributing to their plentiful cavity construction and high surface area, all kinds of catalytic active sites can be incorporated. MOFs can be chemically modified easily and their pore size can be regulated, which make them excellent biomimetic catalytic materials. Firstly, MOFs can integrate more than one kind of catalytic active site to facilitate multifunctional catalysis. Secondly, because of the supramolecular interactions (such Van der Waals force, hydrogen bond, hydrophilicity, etc.) between metal ions and the reactant molecules as well as that between ligands and the reactant molecules, MOFs have the advantages of high selectivity, high sensitivity, and expedience. Thirdly, the catalytic efficiency of MOFs can be optimized via regulating the hydrophilic–lipophilic balance. Fourthly, MOFs enjoy the advantage of recoverability.

6.2 Enzyme Immobilization in MOFs

The well-defined pore structure of MOFs can control the distribution of guests incorporated inside them, and subsequently enhance their stability and prevent their aggregation. Some research suggests that incorporating enzymes in MOFs can improve the efficiency of the catalysis. Controlling the particle morphology is the key point for enzyme immobilization using MOFs.

MOF UiO-66 was synthesized by Xie's group taking 1,4-BDC as organic ligand and Zr^{4+} as metal node. In order to increase the protein loading efficiency, UiO-66-NH_2 was obtained by functionalizing UiO-66 with NH_2. They then immobilized cellulase to UiO-66-NH_2 and studied the catalytic activity of cellulose@UiO-66-NH_2. The data indicate the positive role of UiO-66-NH_2 in increasing the activity, strength, and stability of cellulase. The high cost of cellulase impedes its application in biorefinery industry, while cellulose@UiO-66-NH_2 paves the road for a new form of innovations with stable and reusable cellulose hydrolysis method [16]. Jia's group [17] successfully synthesized a novel mesoporous ZIF-8 with high protein loading capacity and embedded catalase molecules into uniformly sized ZIF crystals. The obtained catalase@ZIF composites with cruciate flow-like morphology possess 400% higher catalytic activity than the ones with standard rhombic dodecahedral morphology. Moreover, when embedded in ZIF nanocrystals via a de novo method, catalase maintains its biological function under unfolding conditions [18].

Surface modification with enzymes has also been reported. Tang's group grafted catalase to the surface of UCNPs/MB@ZIF-8 (UCNP = upconversion nanoparticle; MB = methylene blue). The coordination interaction between UCNPs and MB with Zn^{2+} during the on-pot supramolecular assembling process raised the loading amount of UCNPs and MB in ZIF-8. The obtained UCNPs/MB@ZIF-8@catalase integrates the near-infrared (NIR) imaging ability of UCNPs, $^1O_2$ generation ability of MB and the catalytic ability of catalase. The grafted catalase can catalyze the breakdown of H_2O_2 to O_2, and the aim of $^1O_2$ generation in hypoxic tumors is achieved. Their work opens new opportunities for creating MOF-based tumor photodynamic therapy (PDT) [19].

Hemin, an iron-containing porphyrin with chlorine that can be formed from a haem group, is the active center of catalase, peroxidase, hemoglobin, and other hemeproteins, which has catalytic activity originated from the reversible transformation of Fe(III)/Fe(II). However, hemin has the shortcoming of low stability. Hence, some work immobilized hemin to MOFs to improve its enzyme-mimicking performance and evaluated the biodetection ability of hemin–MOFs. Moreover, hemin–MOFs were also used to dispose organic pollutants such as phenol [20]. Wei's group synthesized integrated nanozymes GOx/hemin@ZIF-8 by embedding two catalysts GOx and hemin into ZIF-8. With the aid of in vivo microdialysis, GOx/hemin@ZIF-8 was successfully applied to the colorimetric visualization of cerebral glucose in the brain of living rats, with a detection limit of 1.7 μM [21]. Similar metalloligands like hemoglobin and Fe-bound tetrakis(4-carboxyphenyl) porphyrin (TCPP(Fe)) were also embedded in MOFs to achieve high peroxidase activity [22, 23].

Despite natural enzymes, nanozymes were also immobilized to MOFs to implement different functions [24]. Platinum nanoparticles (PtNPs) have ultrasmall size and high peroxidase-like activity, and thus they have been immobilized to MOFs. Ultrathin Cu-TCPP(Fe) nanosheets with thickness less than 10 nm were synthesized for the growth of PtNPs and the obtained PtNPs/Cu-TCPP(Fe) hybrid nanosheets were used to detect H_2O_2 and the glucose concentration, with a detection limit of 0.357 μM, 0.994 μM, respectively [25]. Qu's group prepared Pt decorated porous coordination network-224 (PCN-224-Pt) via in situ formation of Pt NPs on the surface of PCN-224. In this work, PCN-224 is a kind of porphyrinic Zr-MOF nanoparticle, which is a photosensitizer-integrated MOF. The obtained PCN-224-Pt possesses high catalase-like activity attributed to the catalytic property of Pt nanozymes. At the hypoxic tumor site, PCN-224-Pt can catalyze the decomposition of H_2O_2 to produce O_2 and facilitate the formation of cytotoxic $^1O_2$, which can enhance the PDT efficiency [26]. Introducing bimetallic nanozymes Fe_3O_4 and AuNPs to the 2D MOF, nanosheets Cu(HBTC)-1 obviously improved their peroxidase-like activity. Taking use of Cu(HBTC)-1/Fe_3O_4–AuNPs, the authors achieved ultra-sensitive detection of H_2O_2 and glucose [27]. PB was found to possess peroxidase-like activity in 2014 (Fig. 6.1), and then it was also composed of MOFs to obtain high catalytic activity for glucose detection [28]. AuNPs have been reported to possess glucose oxidase (GOx)-like activity, based on this, Zhang's group synthesized AuNPs/Cu-TCPP(Co) hybrid nanosheets for the detection of glucose and got a detection limit of 8.5 μM [29]. Besides GOx-like activity, AuNPs possess intrinsic peroxidase-like activity and have surface-enhanced Raman scattering (SERS) property. AuNPs@MIL-101 was designed by in situ growing AuNPs into MOF MIL-101 and can act as peroxidase mimics and SERS substrates. AuNPs@MIL-101 can catalyze H_2O_2 to oxidize caged Raman reporter leucomalachite green (LMG) into Raman-active malachite green (MG); simultaneously, AuNPs@MIL-101 play as the SERS substrates to enhance the Raman signals of the as-produced MG. The authors then assembled GOx and lactate oxidase (LOx) onto AuNPs@MIL-101 to form integrative nanozymes AuNPs@MIL-101@GOx and AuNPs@MIL-101@LOx, respectively. The obtained integrative nanozymes have cascade catalytic activities that can be applied to detect glucose and lactate in living rats' brains. The assembled GOx and LOx oxidize glucose and lactate to produce H_2O_2, which continue to oxidize LMG into MG via the catalysis of AuNPs. The Raman signals generated by MG can be measured and were verified to have positive correlation with glucose and lactate concentration [30]. Many other studies also immobilized oxidases to MOF to detect the given compounds, making use of the intrinsic peroxidase-like activity of MOFs, indicating MOFs are both proper enzyme immobilization matrix and effective enzyme mimics [31]. MIL-101(Cr)@PB was fabricated to study the difference with the peroxidase-like activity of bare PB. Although MIL-101(Cr) could not prevent PB from alkali attack, it protected the iron-hydroxo species produced from PB decomposition, thus MIL-101(Cr)@PB displays excellent catalytic activity [32].

MOF-Au–Ce, Au-doped MOF MIL-88B(Fe) assembled with Ce, was verified to possess DNase- and peroxidase-like activities. The Ce complexes grafted on MOF showed admirable DNase-like activity that can catalyze hydrolysis of extracellular DNA (eDNA) and can be used to eliminate bacterial biofilms. Due to the synergistic effect between Au and MOF, when doped with 2.5 μmol Au, the peroxidase-like activity of the obtained MOF-2.5Au grew 2.7 times than the pristine MOF. Less Au (1.6 μmol) couldn't provide enough active centers, while superfluous Au (4.9 μmol) would block pores of MOF and prevent substrates from contacting active sites. Via its peroxidase-like activity, MOF-2.5Au–Ce can catalyze the transformation of H_2O_2 into •OH, which can kill the bacteria on site. MOF-2.5Au–Ce and H_2O_2 can significantly reduce inflammatory cells and promote wound healing in vivo; moreover, they can also avoid persistent inflammation via killing bacteria in the wound issues [33].

As a series of typical photosensitizers, porphyrin and its derivatives can be used in cancer PDT. Porphyrinic MOFs are used as photosensitizers for cancer PDT and can be combined with other therapies. The difficulty of the spreading of PDT is that its efficiency is severely limited by tumor hypoxia. Usually, to kill cancer cells more efficiently, the researchers establish the bioreactors taking porphyrinic MOFs as carriers of other molecules that can facilitate the generation of O_2. Zhang's group synthesized porphyrin-based Zr-MOF PCN-224 and loaded GOx and catalase, after that they modified them with a cancer cell membrane. The obtained bioreactor mem@catalase@GOx@PCN-224 (mCGP) led a cascade catalytic reaction in tumor. GOx accelerated the decomposition of glucose. Also, the endogenous H_2O_2 was transferred to O_2 and H_2O with the existence of catalase. When irradiated with 660 nm laser, O_2 can be transformed into $^1O_2$, and thus the high O_2 content can enhance the production of $^1O_2$. Benefit from the cancer cell membrane functionalization, mCGP has immune escape and homotypic targeting behaviors, endowing them cancer targeting and retention abilities. Consequently, the cascade bioreactor mCGP efficiently inhibited the cancer growth after a single administration [34]. As mentioned before, Qu's group decorated platinum nanozymes on photosensitizer-integrated MOFs to promote the PDT efficiency, because platinum nanoparticles have catalase-like activity. The nanoplatform PCN-224-Pt can facilitate the production of $^1O_2$ in hypoxic tumor site, which highlights the composites of nanozymes, and MOFs are potential agents for cancer therapy [26]. A new UiO-66 type of NMOF-based organic molecular photosensitizer (UiO-66-TPP-SH) was synthesized from UiO-66 NMOF and S-ethylthiol ester monosubstituted metal-free porphyrin (TPP-SH) via a facile post-synthetic approach. The surface-decorated porphyrinic NMOF UiO-66-TPP-SH demonstrates obviously higher photodynamic activity than the interior-located porphyrinic NMOF and higher photodynamic treatment efficiency [35].

6.3 MOF-Derived Nanozymes

Generally, reducing the sizes of nanozymes can provide a larger active surface area and thus increase the catalytic activity. However, the enhanced surface free energy will very likely lead the aggregation of these nanozymes and further deprive the catalytic ability. To overcome this, many studies prepared uniformly dispersed nanozymes using MOFs as precursors. Using MOF-5-L (leaf-like metal–organic framework-5) as a precursor, Wang and colleagues synthesized Cu hydroxy double salts (HDSs) nanoflowers, which possess peroxidase-like activity and can be used in the colorimetric detection of H_2O_2. Based on this, they developed a facile glucose detection method by using Cu HDSs nanoflowers and GOx, with a linear range between 0.5–10 μM and a low detection limit of 0.5 μM [36]. Cu NPs@C was fabricated using Cu-MOF as a precursor and applied to detect ascorbic acid (AA), with a linear range between 10 μM–1 mM and a detection limit of 1.21 μM [37]. FeP nanozyme was synthesized taking PB as precursors and applied to detect H_2O_2, with a linear range between 2–130 μM and a detection limit of 0.62 μM [38]. CoNPs was embedded in NH_2-MIL-88(Fe)-derived magnetic carbon via in situ reduction method. The obtained CoNPs/MC was successfully applied to detect glucose content in real biological samples [39]. Pyrolyze Fe-doped ZIF-8 under N_2 and NH_3 atmospheres at 900 °C, Lin's group obtained Fe–N–C single-atom nanozyme (SAN) maintaining the dodecahedral structure of the precursor ZIF-8 [40]. Fe–N–C SAN has the features of consisting of outstanding single-atom active Fe dispersion and large surface area, providing a specific activity of 57.76 U/mg^{-1}, which was even comparable to horseradish peroxidase (HRP).

For the first time, Qu and colleagues obtained uniformly dispersed ultrasmall CeO_2-based nanozymes isolated within porous carbonaceous frameworks taking cerium-contained MOFs (Ce-MOFs) as sacrificed precursors [41]. These monodispersed Ce-MOFs have high oxidase-like activity, efficient ATP-deprivation ability, and good drug loading capacity. Ce-MOFs can transform oxygen into reactive oxygen species (ROS) and cause the oxidative damage of tumor cells. Because of their strong coordination ability toward phosphates, Ce-MOFs can hydrolyze ATP to release phosphate groups and adenosine, suggesting their ATP-deprivation capacity that can also damage cancer cells. Moreover, the porous structure made Ce-MOFs excellent drug delivery vectors for loading of anticancer drugs. Altogether, doxorubicin loading Ce-MOFs were applied to synergistic cancer therapy combing enhanced oxidative damage, reduced energy supply, and chemotherapy.

6.4 Intrinsic Enzyme-like Activities of MOFs

As described before, MOFs have been used as an excellent immobilization matrix to carry natural enzymes. With the deeper study of MOFs, their intrinsic enzyme-like activities have been found and explored. Iron-contained MOFs were widely studied for their peroxidase-like activity [42, 43]. Because of their ability to catalyze redox-active reactions involving molecular oxygen and other reactive oxygen metabolites, ligands porphyrin and its derivatives were most widely used to compose MOFs with intrinsic enzyme-like activities. Among all kinds of applications of MOFs, chemical catalysis caused great attention. However, less attention has been paid to the biological catalysis using MOFs. Biological catalysis applications of iron-contained MOFs have been extensively studied and reveal good prospects. Wei's group established a kind of 2D MOF nanosheets and successfully used them as a diagnostic platform for in vivo heparin (Hep) activity monitoring (Table 6.2) [23]. As shown in Fig. 6.2a, they chose iron-porphyrin derivative TCPP(Fe) ligands to synthesize 2D Zn-TCPP(Fe) nanosheets with intrinsic peroxidase-like activity. They also studied the effect of porphyrin-coordinated metal; the results show Zn-TCPP(Zn) and Zn-TCPP(Co) exhibited similar nanosheet structure to Zn-TCPP(Fe). However, Zn-TCPP(Mn) exhibited nanobelt morphology. The peroxidase-like activity of Zn-TCPP(Fe) is ~7–20 times higher than MOFs with non-Fe porphyrin ligands. Modifying 2D Zn-TCPP(Fe) nanosheets with anti-Hep peptides AG73 shields their peroxidase-like activity (Fig. 6.2b). Using AG73-modified 2D Zn-TCPP(Fe) nanosheets with quenched peroxidase-like activity, the authors demonstrated the bioassay for Hep detection (Fig. 6.2c). AG73 peptides on the surface of Zn-TCPP(Fe) nanosheets could especially recognize Hep and trigger the release of AG73, leading to the recovery of peroxidase-like activity and produce fluorescence signal, which

Table 6.2 Application of MOFs in biodetection

Ligand	Transition metal nodes	Name of MOFs	Detection method	Test sample	Linear detection range	Detection limit	References
2-methylimidazole	Zn^{2+}	ZIF-8@bovine hemoglobin (ZIF8@BHb)	Colorimetric method (TMB)	H_2O_2	0–800 μM	1 μM (naked-eye)	[22]
				Phenol	0–200 μM	1 μM (naked-eye)	
TCPP(M)	Cu^{2+}	Au NPs/Cu-TCPP(Fe) hybrid nanosheets	Colorimetric method (TMB)	Glucose	10–300 μM	8.5 μM	[29]
TCPP(Fe)	Zn^{2+}	2D Zn-TCPP(Fe) nanosheets	Fluorescence method (Amplex Red)	Hep	0.1–10 μg/mL	15 ng/mL	[23]
TCPP(Fe)	Cu^{2+}	Pt NPs/Cu-TCPP(Fe)	Colorimetric method (TMB)	H_2O_2	2–100 μM	0.357 μM	[25]
				Glucose	2–200 μM	0.994 μM	
1,4-BDC	Cr^{3+}	AuNPs@MIL-101	SERS (LMG) in vivo	Glucose	10–200 μM	4.2 μM	[30]
				Lactate	10–200 μM	5.0 μM	
1,4-BDC	Zr^{4+}	Hierarchically porous MOF HP-PCN-224(Fe) GOx@HP-PCN-224(Fe) Uricase@HP-PCN-224(Fe)	Colorimetric method (ABTS)	Glucose	5–300 μM	0.87 μM	[31]
			Fluorescence method (4-aminophenazone/2,4-dichlorophenol sulfonate, DCPS)	Uricase	5–100 μM	1.8 μM	
Hemin	Cu^{2+}	Cu-Hemin MOFs/CS-rGO	Electrode method	H_2O_2	0.065–410 μM	0.019 μM	[44]
1,4-BDC	Al^{3+}	Hemin@MIL-101(Al)-NH_2	Colorimetric method (TMB)	H_2O_2	5–200 μM	N/A	[45]
				Glucose	10–300 μM	N/A	
1,4-BDC	Fe^{3+}	PB/MIL-101(Fe)	Colorimetric method (TMB)	H_2O_2	2.40–100 μM	0.15 μM	[28]
				Glucose	0.1–10 mM	0.4 μM	
BTC	Cu^{2+}	Cu(HBTC)-1/Fe_3O_4-AuNPs	Colorimetric method (TMB)	H_2O_2	2.86–71.43 nM	1.10 nM	[27]
				Glucose	12.86–257.14 μM	12.20 μM	

(continued)

Table 6.2 (continued)

Ligand	Transition metal nodes	Name of MOFs	Detection method	Test sample	Linear detection range	Detection limit	References
				SDM	0.01–1.15 μM	0.005 μM	
BTC	Fe^{3+}	Fe_3O_4@MIL-100(Fe)	Colorimetric method (TMB)	cholesterol	2–50 μM	0.8 μM	[24]
Fumaric acid	Fe^{3+}	MIL-88	Colorimetric method (TMB)	H_2O_2	2.0–20.3 μM	0.562 μM	[46]
				Glucose	2.57–10.1 μM	1.03 μM	
BDC	Fe^{3+}	Fe-MIL-88NH_2	Chemiluminescence method	H_2O_2	0.1–10.0 μM	0.025 μM	[47]
BDC	Fe^{3+}	Fe-MIL-88NH_2	Colorimetric method (TMB)	Glucose	2–300 μM	0.48 μM	[48]
BDC	Fe^{3+}	Fe-MIL-88B-NH_2-GOx	Colorimetric method (TMB)	Glucose	1–500 μM	0.478 μM	[49]
BDC	Fe^{3+}	Fe-MIL-88NH_2	Colorimetric method (TMB)	GSH	1.0–100.0 μM	0.45 μM	[50]
				Hcy	1.0–80.0 μM	0.40 μM	
				Cys	1.0–80.0 μM	0.39 μM	
H_2DMTDC	Ce^{3+}	Ce-based MOF	Colorimetric method (TMB)	GSH	N/A	0.125 μM	[51]
				Hcy	N/A	0.132 μM	
				Cys	N/A	0.150 μM	
BDC	Fe^{3+}	Fe-MIL-88	Fluorescence method	GSH	50 nM–10 μM	30 nM	[52]
				Hcy	50 nM–10 μM	40 nM	
				Cys	50 nM–10 μM	40 nM	
TCPP	Zr^{4+}	PCN-222	Photoelectrochemical (PEC) measurements and cyclic voltammetry	α-casein	N/A	0.13 μg/mL	[53]
BDC	Fe^{3+}	HAP@MIL-100(Fe)	Colorimetric method (TMB)	H_2O_2	0.95–28.57 μM	N/A	[42]
				Glucose	2–50 μM	N/A	
			Colorimetric method (OPD)	AA	9.52–142.86 μM	N/A	

(continued)

Table 6.2 (continued)

Ligand	Transition metal nodes	Name of MOFs	Detection method	Test sample	Linear detection range	Detection limit	References
BTC	Fe^{3+}	Fe-MIL-101 MOF	Colorimetric method (TMB)	Glucose	10.6–150 μM	2.5 μM	[54]
1,4-BDC	Fe^{3+}	MIL-53(Fe)	Colorimetric method (TMB)	AA	28.6–190.5 μM	15 μM	[55]
1,4-BDC	Fe^{3+}	MIL-53(Fe)	Chemiluminescence method (luminol)	Glucose	0.1–10 μM	0.05 μM	[56]
1,4-BDC	Fe^{3+}	MIL-68 or MIL-100	Colorimetric method (TMB)	AA	30–485 μM	6 μM	[57]
		MIL-68		H_2O_2	3.0–40 μM	0.256 μM	
		MIL-100		H_2O_2	3.0–40 μM	0.155 μM	
BDC	Fe^{3+}	MIL-53(Fe)	Colorimetric method (TMB)	Glucose	0.25–20 μM	0.25 μM	[43]
1,4-BDC	Fe^{3+}	Glycine-MIL-53(Fe)	Colorimetric method (TMB)	H_2O_2	0.10–10 μM	49 nM	[58]
				Glucose	0.25–10 μM	0.13 μM	
TCPP	Zr^{4+}	Fe@PCN-224 NPs	Colorimetric method (TMB)	H_2O_2	N/A	1.60 μM	[59]
				Glucose	N/A	22 μM	
BPY	Cu^{2+}	Cu-MOFs	Colorimetric method (ABTS)	ALP	1–34 U/L	0.19 U/L	[60]
1,4-BDC	Ni^{2+}	Ni-MOF Nanosheets	Colorimetric method (TMB)	H_2O_2	0.04–160 μM	8 nM	[61]
1,4-BDC	Ni^{2+}	Ni-MOF/Ni/NiO	Electrode method	Glucose	0.4–900 μM	0.1 μM	[62]
BTC	Cu^{2+}	HKUST-1	Fluorescence method	TH	4–700 μM	1 μM	[63]
BTC	Fe^{3+}	MOF(Co/2Fe)	Colorimetric method (TMB)	H_2O_2	10–100 μM	5 μM	[64]
BTC	Cu^{2+}	CuNPs@C	Colorimetric method (TMB)	AA	10–1000 μM	1.41 μM	[37]
BDC	Zn^{2+}	AgNPs@ZnMOF nanocomposite	Fluorescence method (terephthalic acid, TA)	Patulin	0.1–10 μM	0.06 μM	[65].
BDC	Fe^{3+}	CoNPs/NH_2-MIL-88(Fe), namely, CoNPs/MC	Colorimetric method (TMB)	Glucose	0.25–30 μM	156 nM	[66]

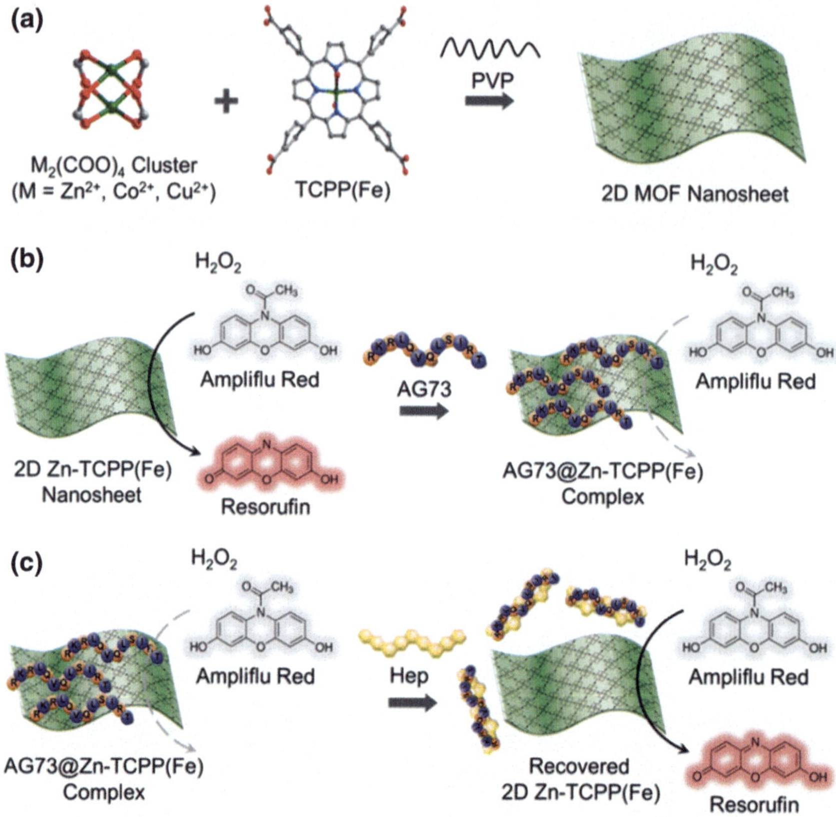

Fig. 6.2 **a** Scheme of the synthesis of 2D MOF nanosheets. **b** AG73 inhibits the peroxidase-like activity of 2D Zn-TCPP(Fe) nanosheets taking Ampliflu Red as the fluorogenic substrates. **c** Hep detection using AG73-modified 2D MOF-based bioassay. Copyright 2017 ACS Publications [23]

could be collected to quantify Hep concentration. This study provides a novel idea for MOF-based platform design. In another study [67], they applied this series of MOFs to discriminate phosphates and probe their enzymatic hydrolysis processes (Fig. 6.3). The obtained Zn-TCPP(Fe), Co-TCPP(Fe), and Cu-TCPP(Fe) 2D nanozymes show peroxidase-like activity taking TCPP(Fe) as the main active site and the divalent metal ions Zn^{2+}, Co^{2+}, and Cu^{2+} as the junction sites. As the binding force of the junction metal ions with phosphates is stronger than that with the carboxyl groups of TCPP(Fe), phosphates can peel off the junction ions from the 2D-MOF nanosheets, leading to the inhibition the peroxidase mimicking activity. Zn-TCPP(Fe), Co-TCPP(Fe), and Cu-TCPP(Fe) 2D nanozymes can serve as three signal channels of sensor array, because different phosphates have different affinities to the same junction metal ions. As shown in Fig. 6.2b, a training data matrix was set up by the data collected from the three kinds of 2D-MOF nanozymes, namely, a matrix of 5 phosphates × 3 arrays × 6 replicates. Since the sensor arrays were also applied

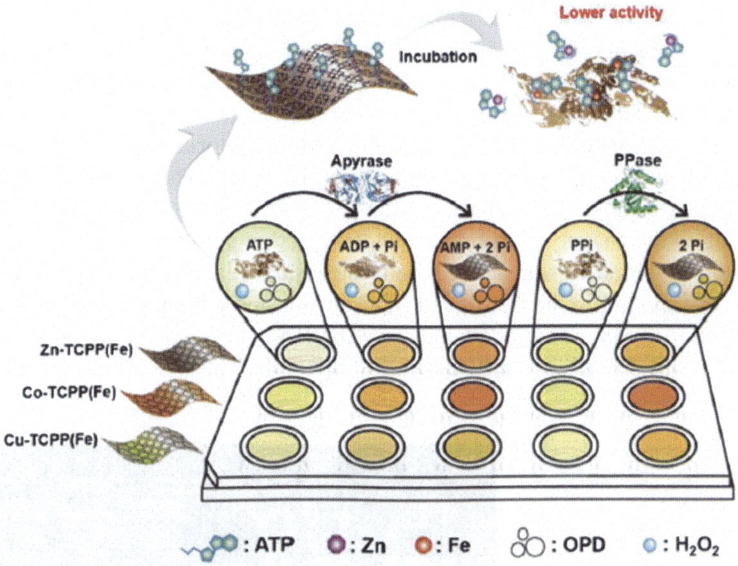

Fig. 6.3 Schematic illustration of probing phosphates and their enzymatic hydrolytic process with 2D-MOF nanozyme sensor arrays. Copyright 2017 ACS Publications [67]

to monitor the hydrolytic reactions of ATP and pyrophosphate (PPi) catalyzed by apyrase and PPase, respectively. This hydrolytic process monitoring was conducted by the detection of ATP, AMP, ADP, PPi, and phosphate (Pi) concentration using the molybdenum blue colorimetric method. Finally, linear discriminant analysis (LDA) was used to convert the training matrix into three canonical scores, and the first two most significant discrimination factors were used to generate a 2D canonical score plot, displaying the whole hydrolytic process in the fingerprint pattern.

Similarly, a zirconium (Zr)-based porphyrinic MOF PCN-222 with a 3D network formed by Zr–O clusters connected by TCPP ligands was synthesized and studied. The obtained PCN-222 possesses high O_2 storage capacity and has large binding sites for reaction substrates. In the O_2-saturated aqueous media, PCN-222 was found to enhance the photocurrent response toward dopamine (DA). After the phosphoprotein α-casein absorbed, the current signal was significantly decreased, indicating the steric hindrance effect of α-casein prevented the interface electron transfer among DA and O_2. Using the photoelectrochemical (PEC) measurements and cyclic voltammetry, this study established a PCN-222 based biosensor for α-casein detection [53].

Li's group first proposed a nanosized MOF, Fe-MIL-88NH$_2$ with a uniform octahedral shape possesses intrinsic peroxidase-like activity [48]. The catalytic behavior of Fe-MIL-88NH$_2$ fits well with typical Michaelis–Menten kinetics. The peroxidase-like activity can efficiently catalyze the decomposition of H_2O_2 into strong oxidizing hydroxyl radical (•OH). Hence, Fe-MIL-88NH$_2$ can significantly enhance the chemiluminescence of luminol-H_2O_2 reaction and used to detect the H_2O_2 content in milk sample (Fig. 6.4a) [47]. Similarly, with the assistance of GOx, Fe-MIL-

88NH$_2$ was used to detect glucose concentration through measuring the concentration of by-product H_2O_2 (Fig. 6.4b) [48]. The same method was used for colorimetric determination of thiol compounds including glutathione (GSH), cysteine (Cys), and homocysteine (Hcy) (Fig. 6.4c). Fe-MIL-88NH$_2$ has peroxidase-like activity that can catalyze the oxidation of TMB by H_2O_2, and this colorimetric procedure can be inhibited by thiol compounds due to the competitive reaction between TMB and thiol compounds with H_2O_2. The concentrations of GSH, Cys, and Hcy were calculated according to the change of absorbance in the reaction systems [50]. Another study shows that •OH radicals, generated from the Fenton reaction between Fe-MIL-88 and H_2O_2, can oxidize the free ligand BDC outside and within the MOF into highly fluorescent hydroxylated terephthalic acid (OHBDC). Noteworthy, the •OH radicals generation was promoted in the existence of biothiols. Thus, biothiol content can be

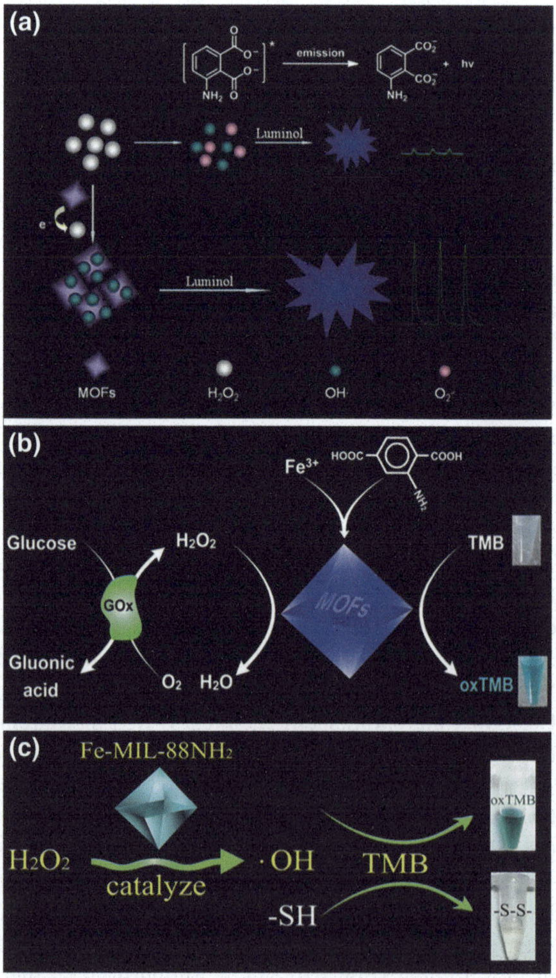

Fig. 6.4 **a** Reaction mechanism of catalyzing luminol–H_2O_2 reaction using Fe-MIL-88NH$_2$. Copyright 2016 Springer Nature [47]. **b** Scheme illustration of glucose detection using a TMB-H_2O_2 system taking Fe-MIL-88NH$_2$ as the peroxidase mimics. Copyright 2013 The Royal Society of Chemistry [48]. **c** Detection of thiol compounds such as glutathione (GSH), cysteine (Cys), and homocysteine (Hcy) using Fe-MIL-88NH$_2$ as the catalyst. Copyright 2014 The Royal Society of Chemistry [50]

detected by the Fe-MIL-88/H_2O_2 system without adding any external chemical tracers [52]. The inherent oxidase-mimicking activity of Ce-MOF allows it to catalyze the oxidation of TMB into blue ox-TMB without the need of any external oxidizing agents such as H_2O_2. The authors established a colorimetric sensing platform for biothiol detection in pH 4 NaAc buffer using Ce-MOF, both as a catalyst and a colorimetric sensor, by recording the decolorization of ox-TMB [51].

Comparing with separately using GOx and Fe-MIL-88NH_2, immobilization of GOx onto Fe-MIL-88B-NH_2 simplified the glucose detecting procedure and achieved good results [49]. Other iron-contained MOFs such as Fe@PCN-224 NPs, fabricated by incorporating free iron ions into the center of porphyrin unit of PCN-224, showed high peroxidase-like activity and were applied to the detection of H_2O_2 and glucose concentration [59]. If modified with antibodies (Ab), Fe-MOF NPs can especially detect *Salmonella* Enteritidis with a detection limit of 34 CFU/mL. Following the basic principle shown in Fig. 6.5, a sandwich construction of "Ab1-modified nanozymes—*Salmonella* Enteritidis-Ab2 modified magnetic beads" will form after binding to bacterial cells. This sandwich construction can be removed from the supernatant using a magnet, and the amount of the residual nanozyme can be quantified taking TMB as chromogenic substrate; finally, *Salmonella* Enteritidis can be successfully detected by calculating the removed nanozyme. The coefficients of variation (CV) were less than 7.0% for 30 days' storage at 4 °C, which indicates favorable stability that is much more higher than HRP-based immunoassays [68]. The biological detection efficiency of MOFs can also be enhanced by surface modification. When functionalized with glycine, MIL-53(Fe) MOF has higher affinity with the substrates, and therefore shows higher peroxidase-like activity. Also, the obtained

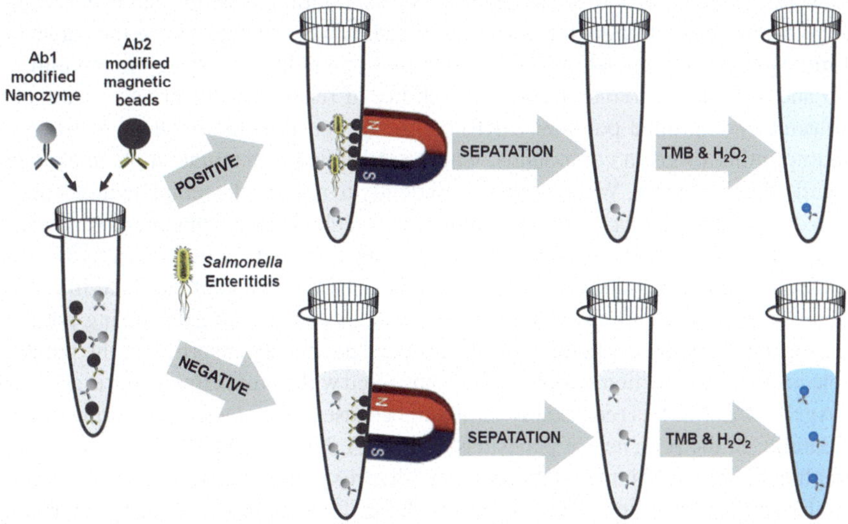

Fig. 6.5 Schematic illustration of the immunodetection of *Salmonella* Enteritidis using nanozyme Fe-MOF NPs Copyright 2018 The Nonferrous Metals Society of China [68]

glycine-MIL-53(Fe) is also more stable under alkaline and acidic conditions. Comparing to MIL-53(Fe), glycine-MIL-53(Fe) can detect glucose concentration with a lower detection limit of 0.13 μM [58].

Despite Fe-containing MOFs, other transition metals (Cu^{2+}, Ni^{2+}, Rh^{3+}, Zn^{2+}, Ru^{2+}) containing MOFs were also found to possess peroxidase-like activity [61–64, 69] and used in biodetection of hydrogen peroxide, glucose, thiamine (TH) and alkaline phosphatase (ALP), and so on.

Based on guanosine monophosphate (GMP) coordinated copper, a kind of multicopper laccase mimicking nanozyme, denoted as Cu/GMP MOF, was developed by Liu's group [70]. Laccase is a copper-containing oxidase that can catalyze the oxidation of phenolic compounds by oxygen. The laccase-like activity of Cu/GMP MOF was verified by the chromogenic reaction of phenolic compounds (hydroquinone, naphthol, catechol, and epinephrine) with 4-aminoantipyrine (4-AP). By a series of contrast experiments, they found that the catalytic activity of Cu/GMP MOF is originated from the interaction between Cu^{2+} and guanosine instead of phosphate binding in GMP. Also, Cu^{2+} is required and cannot be substituted by other metal ions such as Tb^{3+}, Eu^{3+}, Sm^{3+}, La^{3+}, Gd^{3+}, Zn^{2+}, and Fe^{3+}. The as-prepared Cu/GMP MOF is approximately 2400-fold more cost-effective than the commercial laccase, offering us new perspectives regarding the lower cost and good reproducibility. Qu's group fabricated another Cu MOF nanozyme Cu-TCPP MOF nanodots and denoted them as CTMDs [71], which have both glutathione peroxidase (GPx)- and superoxide dismutase (SOD)-like activities. Through a modified nitro blue tetrazolium (NBT) assay, they verified the SOD-like activity, which is because CTMDs consist of a similar Cu active site coordination environment to that of natural SOD. Depart from SOD-like activity, they found the GPx-like activity via a glutathione reductase (GR) coupled assay; however, this article did not give a deep exploration to the mechanism. Owing to the dual-enzyme-mimicking activities, CTMDs were found to have the capacity of highly efficient removal of ROS, so that they can reduce the systemic inflammation and alleviate endotoxemia with the ability of renal enrichment and clearance. Molecularly imprinted polymer (MIP) can be designed as the polymer matrix for the imprinting of given compounds, which are applied as template molecules with specific shape and size. When wash up efficiently to remove the template molecules, a key-lock connection between the template molecules (given compounds) and MIP sites can be developed to adsorb the same molecules with a high selectivity and specificity. Khataee and the colleagues synthesized Ag nanoparticle/flake-like Zn-based MOF (AgNPs@ZnMOF) composite and capped it with MIP, taking patulin as template. The obtained empty MIP sites can specifically absorb with the patulin molecules in the detection samples. When absorbed with patulin, the catalytic activity of AgNPs@ZnMOF is remarkably inhibited because of the electron capturing effect of patulin. In the presence of peroxidase mimic AgNPs@ZnMOF, H_2O_2 can oxide terephthalic acid (TA) into high fluorescent 2-hydroxy terephthalic acid (HTA), while the absorption of patulin will lead to a quenching effect on the fluorescence intensity. Via detecting the fluorescence signal, they successfully detected the concentration of patulin, with a linear range of 0.1–10 μM and a detection limit of 0.06 μM [65].

6.5 Structure and Enzyme-like Activities of Prussian Blue

In the days that followed the growing understanding of PB and PBAs, they were widely used in the fields of electrochemistry, magnetism, photology, and biomedicine [72, 73]. Specifically, Food and Drug Administration (FDA) authorized PB to be applied as an antidote for certain kinds of heavy metal (thallium and cesium) poisoning [72]. PB nanoparticles (PBNPs) were synthesized and explored for mimicking natural enzymes, which led to the long-lasting research interests in iron-based nanozymes. Till now, PBNPs have been reported for their peroxidase-, catalase-, SOD-, and ascorbic acid oxidase (AAO)-like activities.

PB is prepared by Fe^{3+} and $K_4Fe^{II}(CN)_6$ solutions, and its molecular formula is $KFe^{III}[Fe^{II}(CN)_6]$. In the crystalline structure of PB (Fig. 6.6), the distances of Fe–C, Fe–N, and C–N are 1.92 Å, 2.03 Å, and 1.13 Å, respectively [74, 75]. PB was classified into "insoluble" $Fe_4(Fe(CN)_6)_3$ and "soluble" $KFeFe(CN)_6$. In fact, PB is insoluble, potassium in $KFe^{III}[Fe^{II}(CN)_6]$ will dissociate in aqueous medium and form reticular negative ionization layer on the surface of crystal, making the PB solution a stable dispersoid [76]. The "insoluble" PB contains interstitial water molecules and can be expressed as $Fe_4(Fe(CN)_6)_3 \cdot nH_2O$, n = 14–16 [2, 77–79]. Half of irons (low spin) in PB are diamagnetic, who interact with C and form covalent bond, while other irons (high spin) link with N through ionic bonds and they are paramagnetic. Fe(II) and Fe(III) interact with each other via cyano group (–CN–), and thus PB is a kind of molecular magnet, which fits Curie's law in the range of 200 and 300 °K [74, 75, 80–82].

PB has good redox activity due to the bridging function of –CN–, which making it an excellent electron transporter applying to electrochemical field [83, 84]. As shown in Eqs. 6.1–6.7, PB has different redox statuses including Prussian white (PW), Berlin green (BG), and Prussian yellow (PY) [85–87].

$$[Fe(III)Fe(II)(CN)_6]^{-}(PB) \tag{6.1}$$

$$[Fe(II)Fe(II)(CN)_6]^{2-}(PW) \tag{6.2}$$

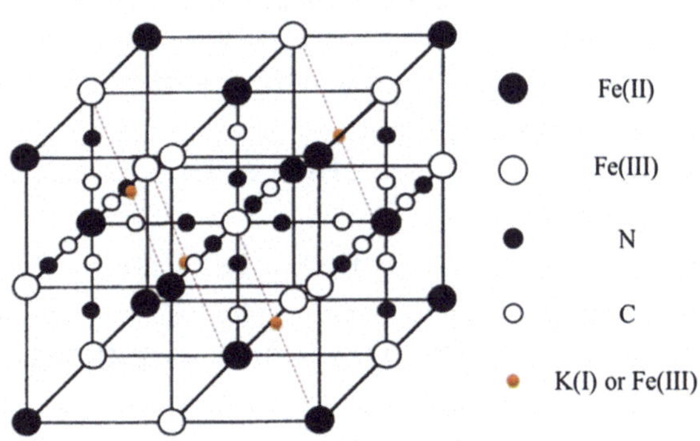

Fig. 6.6 Crystalline structure of PB

$$\{Fe(III)_3[Fe(III)(CN)_6]_2[Fe(II)(CN)_6]\}^-(BG) \tag{6.3}$$

$$[Fe(III)Fe(III)(CN)_6](PY) \tag{6.4}$$

$$PB + e^- \rightarrow PW\,(6.1) \tag{6.5}$$

$$3PB \rightarrow BG + 2e^-\,(6.2) \tag{6.6}$$

$$PB \rightarrow PY + e^-\,(6.3) \tag{6.7}$$

Prussian blue nanoparticle (PBNP) is the earliest and mostly researched polymer nanomaterials [88]. Along with the development of nanotechnology, the researchers have proposed many methods to synthesize PBNPs, such as hydrothermal process [89–91], polymer protection method [79, 92, 93], hydrolysis method [94, 95], micro-emulsion method [96, 97], and template method [78, 98]. Based on these synthetic methods, PBNPs have been intensively studied and applied. Our initial study reported that PB coating could enhance the peroxidase-like activity of γ-Fe_2O_3 nanoparticles [99]. Later, we found that PBNPs exhibited catalase-like activity at pH 7.4. Taking use of the generated O_2 gas bubbles, we applied them to in vivo ultrasound imaging of H_2O_2 overproduced tissues experiencing oxidative stress [73]. Recently, we further confirmed that PBNPs can mimic multienzymes including peroxidase-, catalase-, SOD-, and AAO-like activities [100, 101].

PB was widely used to modify electrodes to detect many biological markers due to its good redox activity [102–110]. Hence, we inferred that PBNPs possess multienzyme-like activities seemed to be particularly associated with their redox property. Our previous study showed that PBNPs did not follow the Fenton reaction mechanism and conversely, they could scavenge·OH (Eq. 6.8) [100]. From the redox potential aspect (Fig. 6.7), we illustrated the reaction mechanism of the peroxidase-like activity of PBNPs. More specifically, H_2O_2 tends to exhibit reducibility at neutral and alkaline pH (Eq. 6.9) and oxidizability at acidic pH (Eq. 6.10). Hence, PB was first oxidized into BG and PY by H_2O_2 at acidic pH, and then, PB/BG oxidized TMB; in other words, PB/BG transferred electrons from TMB to H_2O_2 (Eq. 6.11).

$$PB + H^+ + \bullet OH \rightarrow PY + H_2O \tag{6.8}$$

$$H_2O_2 \rightarrow O_2 + 2e^- + 2H^+ \tag{6.9}$$

$$H_2O_2 + 2e^- + 2H^+ \rightarrow 2H_2O \tag{6.10}$$

$$TMB + H_2O_2 + 2H^+ \xrightarrow{PY} TMB(oxidized) + 2H_2O \tag{6.11}$$

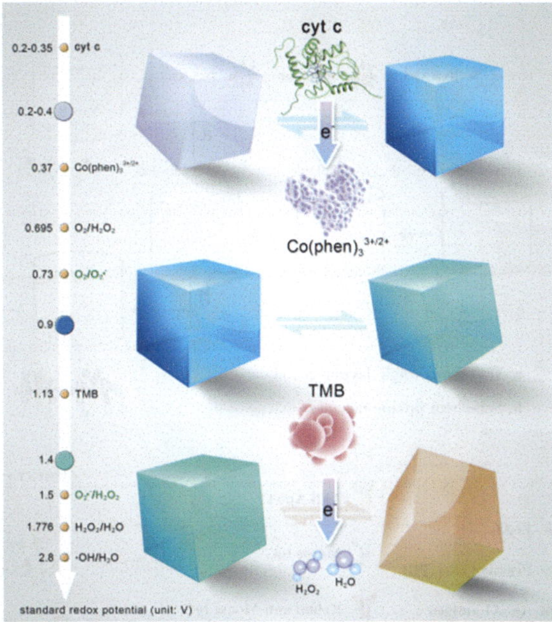

Fig. 6.7 The standard redox potentials of different compounds in the reaction systems taking PBNPs as nanozymes

Taking use of their peroxidase-like activity, PBNPs were used to detect H_2O_2 and glucose via a sensitive and selective colorimetric method [111, 112]. Besides H_2O_2 and glucose, PBNPs were also used to detect other biological sample. For instance, self-linkable PB-incorporated magnetic graphene oxide (PMGO) was constructed to detect apolipoprotein A1 (ApoA1) in the urine of bladder cancer patients (Fig. 6.8), with a linear detection range of 0.05 to 100 ng/mL [113]. The as-prepared PMGO presents high peroxidase-like activity, strong thermal stability, and low cost. The immunosensing of ApoA1 in the urine samples shows a reasonable variation with a standard reference from the hospital. The PMGO nanozyme-based immunosensing is simple, convenient, sensitive, highly selective, and of low cost, providing a promising tool for ApoA1 detection, which is important for the understanding of pathophysiological mechanisms of bladder cancer.

In neutral and alkaline conditions, the presence of PBNPs would catalyze H_2O_2 into H_2O and O_2 via Eqs. 6.12–6.15, which illustrated the reaction mechanisms of their catalase-like activity [100] (Fig. 6.9).

$$3PB + H_2O_2 \rightarrow BG + 2OH^- \tag{6.12}$$

$$2BG + H_2O_2 \rightarrow 6PY + 2OH^- \tag{6.13}$$

$$6PY + H_2O_2 + 2OH^- \rightarrow 2BG + O_2 + 2H_2O \tag{6.14}$$

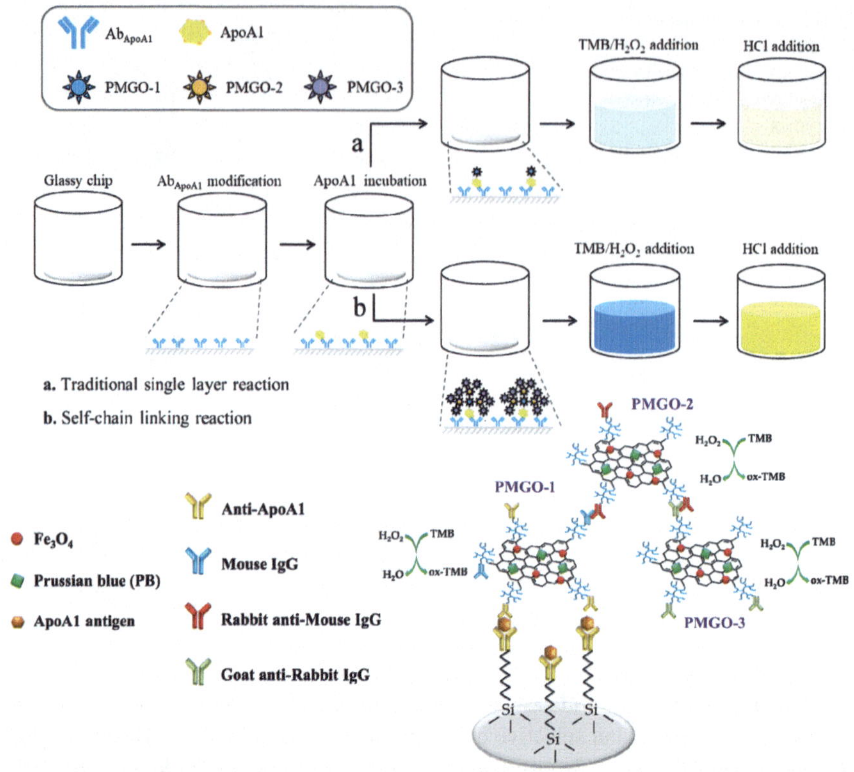

Fig. 6.8 Illustration of the ApoA1 detection procedures with **a** traditional single layer reaction and **b** self-chain linking reaction for signal amplification in TMB/H_2O_2 systems. Copyright 2018 Elsevier B.V. [113]

$$BG + H_2O_2 + 2OH^- \rightarrow 3PB + O_2 + 2H_2O \tag{6.15}$$

When playing catalase-like activity, PBNPs catalyze breakdown of H_2O_2 into oxygen molecules at neutral pH, and the generated O_2 gas bubbles can be used to enhance ultrasound imaging. In tissue experiencing oxidative stress, PBNPs were applied to detect the overproduce H_2O_2 by ultrasound imaging [73]. Since H_2O_2 could act both as oxidizing and reducing agents, superoxide anion could be converted into H_2O_2 and O_2 via the mechanisms shown in Eqs. 6.16–6.20. These imply that PBNPs possess SOD-like activity (Eq. 6.16) [100].

$$2{\bullet}OOH \overset{SOD/PBNPs}{\longrightarrow} H_2O_2 + O_2 \tag{6.16}$$

$$2{\bullet}OOH + 2H^+ + 3PB \rightarrow 2H_2O_2 + BG \tag{6.17}$$

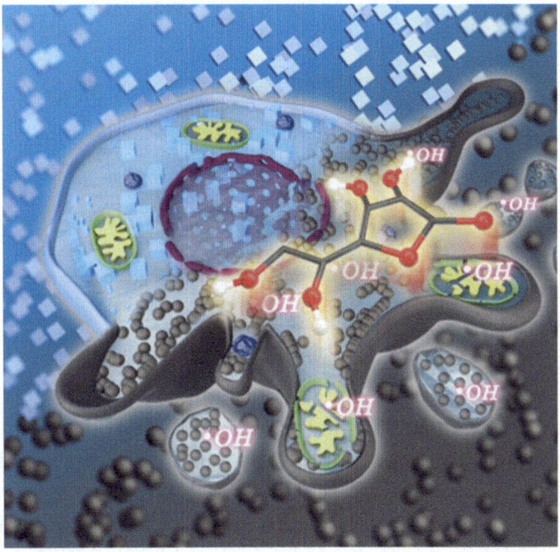

Fig. 6.9 Schematic illustration of the influence of PBNPs (blue ones) and IONPs (black ones) on the anticancer ability of AA

$$2\bullet OOH + BG \rightarrow 2O_2 + 2H^+ + 3PB \tag{6.18}$$

$$\bullet OOH + H^+ + BG \rightarrow H_2O_2 + 3PY \tag{6.19}$$

$$\bullet OOH + 3PY \rightarrow O_2 + H^+ + BG \tag{6.20}$$

PBNPs were peroxidase, catalase, SOD mimetics, and •OH scavengers; hence, they could scavenge (ROS). Based on these properties, we further demonstrated that PBNPs could alleviate liver inflammation in vivo in lipopolysaccharide-treated mice model [100]. Manganese Prussian blue nanozymes (MPBZs) with multienzyme activity were also fabricated and were approved to be a new generation of ROS nanoscavengers [114]. A dextran sulfate sodium (DSS)-induced mouse colitis model was built to evaluate the ROS scavenging and anti-inflammatory effects of the obtained MPBZs. Owing to the increased intestinal permeability and the positively charged surfaces of the inflamed mucosa in the colitis mice, the oral administrated negatively charged MPBZs of approximately 120 nm prefer to accumulate at the inflamed mucosa, making the MPBZs inflammation-targeting ROS nanoscavengers. MPBZs successfully treated inflammatory bowel disease (IBD) via a primary effect on the toll-like receptor (TLR) signaling pathway, which is closely related to ROS.

Using ESR detection, we revealed PBNPs possessed AAO-like activity via Eqs. 6.21–6.24. Taking iron oxide nanoparticles (IONPs) as control, PBNPs did not enhance the anticancer ability of AA; conversely, they showed suppression effects due to the AAO and ascorbic acid peroxidase-like activities (Fig. 6.9) [101].

$$2AA + H_2O_2 \xrightarrow{\text{PBNPs/APOD}} 2AA^{\bullet -} + 2H_2O \tag{6.21}$$

$$2AA^{\bullet -} + H_2O_2 \overset{PBNPs/APOD}{\longrightarrow} 2DHA + 2H_2O \tag{6.22}$$

$$2AA + O_2 \overset{PBNPs/AAO}{\longrightarrow} 2AA^{\bullet -} + H_2O_2 \tag{6.23}$$

$$2AA^{\bullet -} + H_2O_2 \overset{PBNPs/APOD}{\longrightarrow} 2DHA + 2H_2O \tag{6.24}$$

6.6 Perspective and Challenges

MOF-based nanozymes were commonly used in biomedical fields including biodetection, bioimaging, and treatment of some diseases. However, many more breakthroughs are expected in the research of MOF-based nanozymes. First, the uniformity, catalytic activity, and selectivity of MOF-based nanozymes are in need of optimization. Second, new catalytic properties of MOFs are needed to be developed and MOF-based nanozymes should be designed beyond the catalytic properties of the natural enzymes. Third, the kinetics and biosafety of MOF-based nanozymes should be systematically evaluated. Finally, more potential applications of MOF-based nanozymes are waiting for a breakthrough, especially in translational medicine. Future work on MOF-based nanozymes may focus on the controllable and standard synthesis, the explanation of their catalytic mechanisms, and the exploitation of new applications. In this chapter, we summarized the development of MOF-based nanozymes. Despite the remaining problems and challenges, as a young research field, the enzymatic properties of MOFs will encourage the researchers to continue to explore and innovate in the day to come.

Acknowledgements This work was supported in part by the National Key Research and Development Program of China (No. 2017YFA0205502), National Natural Science Foundation of China (No. 81801827, 81901833), and the Basic Research Program of Jiangsu Province (Natural Science Foundation, No. BK20181086, BK20191080).

References

1. Batten SR, Champness NR, Chen XM, Garciamartinez J, Kitagawa S, Öhrström L, O'Keeffe M, Paik Suh M, Reedijk J (2013) Terminology of metal–organic frameworks and coordination polymers (IUPAC Recommendations 2013). Pure Appl Chem 85(8):1715–1724
2. Keggin JF, Miles FD (1936) Structures and formulae of the Prussian blues and related compounds. Nature 137:577–578
3. Sato O, Iyoda T, Fujishima A, Hashimoto K (1996) Photoinduced magnetization of a cobalt–iron cyanide. Science 272(5262):704
4. Buser HJ, Ludi A, Petter W, Schwarzenbach D (1972) Single-crystal study of Prussian Blue: $Fe_4[Fe(CN)_6]_2$, $14H_2O$. J Chem Soc, Chem Commun 23(23):1299–1299

5. Kitagawa S, Kitaura R, Noro SI (2010) Functional porous coordination polymers. Angew Chem Int Ed 35(29):2334–2375
6. Yaghi OM, Li G, Li H (1995) Selective binding and removal of guests in a microporous metal–organic framework. Nature 378(6558):703–706
7. Li H, Eddaoudi M, O'Keeffe M, Yaghi OM (1999) Design and synthesis of an exceptionally stable and highly porous metal-organic framework. Nature 402(6759):276–279
8. Eddaoudi M, Kim J, Rosi N, Vodak D, Wachter J, O'Keeffe M, Yaghi OM (2002) Systematic design of pore size and functionality in isoreticular MOFs and their application in methane storage. Science 295(5554):469–472
9. Braga D (2003) Hydrogen storage in microporous metal-organic frameworks. Science 300(5622):1127–1129
10. Chae HK, Siberio-Pérez DY, Kim J, Go YB, Eddaoudi M, Matzger AJ, O'Keeffe M, Yaghi OM (2004) A route to high surface area, porosity and inclusion of large molecules in crystals. Nature 427(6974):523–527
11. Férey G, Mellot-Draznieks C, Serre C, Millange F, Dutour J, Surblé S, Margiolaki I (2005) A chromium terephthalate-based solid with unusually large pore volumes and surface area. Science 309(5743):2040–2042
12. Park KS, Ni Z, Côté AP, Choi JY, Huang R, Uribe-Romo FJ, Chae HK, O'Keeffe M, Yaghi OM (2006) Exceptional chemical and thermal stability of zeolitic imidazolate frameworks. Proc Natl Acad Sci USA 103(27):10186–10191
13. Deng H, Doonan CJ, Furukawa H, Ferreira RB, Towne J, Knobler CB, Wang B, Yaghi OM (2010) Multiple functional groups of varying ratios in metal-organic frameworks. Science 327(5967):846–850
14. Gassensmith JJ, Furukawa H, Smaldone RA, Forgan RS, Botros YY, Yaghi OM, Stoddart JF (2011) Strong and reversible binding of carbon dioxide in a green metal-organic framework. J Am Chem Soc 133(39):15312–15315
15. Wu D, Gassensmith JJ, Ushakov S, Stoddart JF, Navrotsky A (2013) Direct calorimetric measurement of enthalpy of adsorption of carbon dioxide on CD-MOF-2, a green metal-organic framework. J Am Chem Soc 135(18):6790–6793
16. Nasser I, Yang XL, Dubale AA, Li RF, Xie MH (2018) Hydrolysis of cellulose using cellulase physically immobilized on highly stable zirconium based metal-organic frameworks. Bioresour Technol 270:377–382
17. Cui J, Feng Y, Lin T, Tan Z, Zhong C, Jia S (2017) Mesoporous metal-organic framework with well-defined cruciate flower-like morphology for enzyme immobilization. ACS Appl Mater Interfaces 9(12):10587–10594
18. Liao FS, Lo WS, Hsu YS, Wu CC, Wang SC, Shieh FK, Morabito JV, Chou LY, Wu KCW, Tsung CKF (2017) Shielding against unfolding by embedding enzymes in metal-organic frameworks via a de novo approach. J Am Chem Soc 139(19):6530–6533
19. Cai H, Shen T, Zhang J, Shan C, Jia J, Li X, Liu W, Tang Y (2017) A core-shell metal-organic-framework (MOF)-based smart nanocomposite for efficient NIR/H_2O_2-responsive photodynamic therapy against hypoxic tumor cells. J Mater Chem B 5(13):2390–2394
20. Jiang W, Yang J, Wang X, Han H, Yang Y, Tang J, Li Q (2017) Phenol degradation catalyzed by a peroxidase mimic constructed through the grafting of heme onto metal-organic frameworks. Bioresource Technol 247:1246–1248
21. Cheng H, Zhang L, He J, Guo W, Zhou Z, Zhang X, Nie S, Wei H (2016) Integrated nanozymes with nanoscale proximity for in vivo neurochemical monitoring in living brains. Anal Chem 88(10):5489–5497
22. Yin Y, Chen LG, Qi X, Guo L, Lin Z, Cai Z, Yang HH (2016) Protein-metal organic framework hybrid composites with intrinsic peroxidase-like activity as a colorimetric biosensing platform. ACS Appl Mater Interfaces 8(42):29052–29061
23. Cheng H, Liu Y, Hu Y, Ding Y, Lin S, Cao W, Wang Q, Wu J, Muhammad F, Zhao X (2017) Monitoring of heparin activity in live rats using metal-organic framework nanosheets as peroxidase mimics. Anal Chem 89(21):11552–11559

24. Wu Y, Ma Y, Xu G, Wei F, Ma Y, Song Q, Wang X, Tang T, Song Y, Shi M (2017) Metal-organic framework coated Fe_3O_4 magnetic nanoparticles with peroxidase-like activity for colorimetric sensing of cholesterol. Sensors Actuat B-Chem 249:195–202
25. Chen H, Qiu Q, Sharif S, Ying S, Ying Y, Wang Y (2018) Solution-phase synthesis of platinum nanoparticles decorated metal-organic framework hybrid nanomaterials as biomimetic nanoenzyme for biosensing applications. ACS Appl Mater Interfaces 10(28):24108–24115
26. Zhang Y, Wang F, Liu C, Wang Z, Kang L, Huang Y, Dong K, Ren J, Qu X (2018) Nanozyme decorated metal-organic frameworks for enhanced photodynamic therapy. ACS Nano 12(1):651–661
27. Tan B, Zhao H, Wu W, Liu X, Zhang Y, Quan X (2017) Fe_3O_4-AuNPs anchored 2D metal-organic framework nanosheets with DNA regulated switchable peroxidase-like activity. Nanoscale 9(47):18699–18710
28. Cui F, Deng Q, Sun L (2015) Prussian blue modified metal–organic framework MIL-101(Fe) with intrinsic peroxidase-like catalytic activity as a colorimetric biosensing platform. RSC Adv 5(119):98215–98221
29. Huang Y, Zhao M, Han S, Lai Z, Yang J, Tan C, Ma Q, Lu Q, Chen J, Zhang X (2017) Growth of Au nanoparticles on 2D metalloporphyrinic metal-organic framework nanosheets used as biomimetic catalysts for cascade reactions. Adv Mater 29(32):1700102
30. Hu Y, Cheng H, Zhao X, Wu J, Muhammad F, Lin S, He J, Zhou L, Zhang C, Deng Y (2017) Surface-enhanced Raman scattering-active gold nanoparticles with enzyme mimicking activities for measuring glucose and lactate in living tissues. ACS Nano 11(6):5558–5566
31. Liu X, Qi W, Wang YF, Lin DW, Yang XJ, Su RX, He ZM (2018) Rational design of mimic multienzyme systems in hierarchically porous biomimetic metal-organic frameworks. ACS Appl Mater Interfaces 10(39):33407–33415
32. Su L, Xiong Y, Yang H, Zhang P, Ye F (2015) Prussian blue nanoparticles encapsulated inside a metal–organic framework via in situ growth as promising peroxidase mimetics for enzyme inhibitor screening. J Mater Chem B 4(1):128–134
33. Liu Z, Wang F, Ren J, Qu X (2019) A series of MOF/Ce-based nanozymes with dual enzyme-like activity disrupting biofilms and hindering recolonization of bacteria. Biomaterials 208:21–31
34. Li SY, Cheng H, Xie BR, Qiu WX, Zeng JY, Li CX, Wan SS, Zhang L, Liu WL, Zhang XZ (2017) Cancer cell membrane camouflaged cascade bioreactor for cancer targeted starvation and photodynamic therapy. ACS Nano 11(7):7006–7018
35. Kan JL, Jiang Y, Xue A, Yu YH, Dong YB (2018) Surface decorated porphyrinic nanoscale metal-organic framework for photodynamic therapy. Inorg Chem 57(9):5420–5428
36. Li A, Mu X, Li TR, Wen H, Li W, Li Y, Wang B (2018) Formation of porous Cu hydroxy double salts nanoflowers derived from metal-organic frameworks with efficient peroxidase-like activity for label-free detection of glucose. Nanoscale 10:11948–11954
37. Tan H, Ma C, Gao L, Li Q, Song Y, Xu F, Wang T, Wang L (2015) Metal-organic framework-derived copper nanoparticle@carbon nanocomposites as peroxidase mimics for colorimetric sensing of ascorbic acid. Chemistry 20(49):16377–16383
38. Yang W, Hao J, Zhang Z, Zhang B (2015) Metal-organic frameworks-derived synthesis of porous FeP nanocubes: an effective peroxidase mimetic. J Colloid Interface Sci 460:55–60
39. Dong W, Zhuang Y, Li S, Zhang X, Chai H, Huang Y (2017) High peroxidase-like activity of metallic cobalt nanoparticles encapsulated in metal–organic frameworks derived carbon for biosensing. Sensors Actuat B-Chem 255:2050–2057
40. Niu X, Shi Q, Zhu W, Liu D, Tian H, Fu S, Cheng N, Li S, Smith JN, Du D (2019) Unprecedented peroxidase-mimicking activity of single-atom nanozyme with atomically dispersed Fe–Nx moieties hosted by MOF derived porous carbon. Biosens Bioelectron 142:111495
41. Cao F, Zhang Y, Sun Y, Wang Z, Zhang L, Huang Y, Liu C, Liu Z, Ren J, Qu X (2018) Ultrasmall nanozymes isolated within porous carbonaceous frameworks for synergistic cancer therapy: enhanced oxidative damage and reduced energy supply. Chem Mater 30(21):7831–7839
42. Chen FF, Zhu YJ, Xiong ZC, Sun TW (2016) Hydroxyapatite nanowires@metal-organic framework core/shell nanofibers: templated synthesis, peroxidase-like activity, and derived flexible recyclable test paper. Chemistry 23(14):3328–3337

43. Dong W, Liu X, Shi W, Huang Y (2015) Metal–organic framework MIL-53(Fe): facile microwave-assisted synthesis and use as a highly active peroxidase mimetic for glucose biosensing. RSC Adv 5(23):17451–17457
44. Wang L, Yang H, He J, Zhang Y, Yu J, Song Y (2016) Cu-hemin metal-organic-frameworks/chitosan-reduced graphene oxide nanocomposites with peroxidase-like bioactivity for electrochemical sensing. Electrochim Acta 213:691–697
45. Qin FX, Jia SY, Wang FF, Wu SH, Song J, Liu Y (2013) Hemin@metal–organic framework with peroxidase-like activity and its application to glucose detection. Catal Sci Technol 3(10):2761–2768
46. Gao C, Zhu H, Chen J, Qiu H (2017) Facile synthesis of enzyme functional metal-organic framework for colorimetric detecting H_2O_2 and ascorbic acid. Chin Chem Lett 28(5):1006–1012
47. Li Y, You X, Shi X (2016) Enhanced chemiluminescence determination of hydrogen peroxide in milk sample using metal–organic framework Fe–MIL–88NH_2 as peroxidase mimetic. Food Anal Meth 10(3):1–8
48. Liu YL, Zhao XJ, Yang XX, Li YF (2013) A nanosized metal-organic framework of Fe-MIL-88NH_2 as a novel peroxidase mimic used for colorimetric detection of glucose. Analyst 138(16):4526–4531
49. Xu WQ, Jiao L, Yan HY, Wu Y, Chen LJ, Gu WL, Du D, Lin YH, Zhu CZ (2019) Glucose oxidase-integrated metal-organic framework hybrids as biomimetic cascade nanozymes for ultrasensitive glucose biosensing. ACS Appl Mater Interfaces 11(25):22096–22101
50. Jiang Z, Liu Y, Hu X, Li Y (2014) Colorimetric determination of thiol compounds in serum based on Fe-MIL-88NH_2 metal–organic framework as peroxidase mimetics. Anal Methods 6(15):5647–5651
51. Dalapati R, Sakthivel B, Ghosalya MK, Dhakshinamoorthy A, Biswas S (2017) A cerium-based metal-organic framework having inherent oxidase-like activity applicable for colorimetric sensing of biothiols and aerobic oxidation of thiols. CrystEngComm 19(39):5915–5925
52. Juan SZ, Jun Ze J, Fang LY (2015) A sensitive and selective sensor for biothiols based on the turn-on fluorescence of the Fe-MIL-88 metal-organic frameworks-hydrogen peroxide system. Analyst 140(24):8201–8208
53. Zhang GY, Zhuang YH, Shan D, Su GF, Cosnier S, Zhang X (2016) A zirconium based porphyrinic metal-organic framework (PCN-222): enhanced photoelectrochemical response and its application for label-free phosphoprotein detection. Anal Chem 88(22):11207–11212
54. Ortiz-Gómez I, Salinas-Castillo A, García AG, Álvarez-Bermejo JA, Orbe-Payá ID, Rodríguez-Diéguez A, Capitán-Vallvey LF (2018) Microfluidic paper-based device for colorimetric determination of glucose based on a metal-organic framework acting as peroxidase mimetic. Microchim Acta 185(1):47
55. Lunhong A, Lili L, Caihong Z, Jian F, Jing J (2013) MIL-53(Fe): a metal-organic framework with intrinsic peroxidase-like catalytic activity for colorimetric biosensing. Chemistry 19(45):15105–15108
56. Yi X, Dong W, Zhang X, Xie J, Huang Y (2016) MIL-53(Fe) MOF-mediated catalytic chemiluminescence for sensitive detection of glucose. Anal Bioanal Chem 408(30):1–8
57. Jian-Wei Z, Hao-Tian Z, Zi-Yi D, Xueqing W, Shu-Hong Y, Hai-Long J (2013) Water-stable metal-organic frameworks with intrinsic peroxidase-like catalytic activity as a colorimetric biosensing platform. Chem Commun 50(9):1092–1094
58. Dong W, Yang L, Huang Y (2017) Glycine post-synthetic modification of MIL-53(Fe) metal-organic framework with enhanced and stable peroxidase-like activity for sensitive glucose biosensing. Talanta 167:359–366
59. Li T, Hu P, Li J, Huang P, Tong W, Gao C (2019) Enhanced peroxidase-like activity of Fe@ PCN-224 nanoparticles and their applications for detection of H_2O_2 and glucose. Colloids Surf Physicochem Eng Aspects 577:456–463
60. Wang C, Gao J, Cao Y, Tan H (2018) Colorimetric logic gate for alkaline phosphatase based on copper (II)-based metal-organic frameworks with peroxidase-like activity. Anal Chim Acta 1004:74–81

61. Chen J, Yun S, Li H, Qin X, Hu X (2018) Nickel metal-organic framework 2D nanosheets with enhanced peroxidase nanozyme activity for colorimetric detection of H_2O_2. Talanta 189:254–261
62. Shu Y, Chen J, Xu Q, Hu X (2018) Synthesis of a novel Au nanoparticles decorated Ni-MOF/Ni/NiO nanocomposite and electrocatalytic performance for the detection of glucose in human serum. Talanta 184:136–142
63. Tan H, Li Q, Zhou Z, Ma C, Song Y, Xu F, Wang L (2015) A sensitive fluorescent assay for thiamine based on metal-organic frameworks with intrinsic peroxidase-like activity. Anal Chim Acta 856:90–95
64. Yang H, Yang R, Zhang P, Qin Y, Chen T, Ye F (2017) A bimetallic (Co/2Fe) metal-organic framework with oxidase and peroxidase mimicking activity for colorimetric detection of hydrogen peroxide. Microchim Acta 184(12):4629–4635
65. Bagheri N, Khataee A, Habibi B, Hassanzadeh J (2018) Mimetic Ag nanoparticle/Zn-based MOF nanocomposite (AgNPs@ZnMOF) capped with molecularly imprinted polymer for the selective detection of patulin. Talanta 179:710–718
66. Dong W, Zhuang Y, Li S, Zhang X, Chai H, Huang Y (2017) High peroxidase-like activity of metallic cobalt nanoparticles encapsulated in metal–organic frameworks derived carbon for biosensing. Sensors Actuat B Chem
67. Qin L, Xiaoyu W, Yufeng L, Hui W (2018) 2D-MOF nanozyme sensor arrays for probing phosphates and their enzymatic hydrolysis. Anal Chem 90(16):9983–9989
68. Cheng N, Zhu C, Wang Y, Du D, Zhu M-J, Luo Y, Xu W, Lin Y (2019) Nanozyme enhanced colorimetric immunoassay for naked-eye detection of salmonella enteritidis. J Anal Testing 3(1):99–106
69. Chen WH, Vázquez-González M, Kozell A, Cecconello A, Willner I (2017) Cu_{2+}-modified metal-organic framework nanoparticles: a peroxidase-mimicking nanoenzyme. Small 23(62):1703149
70. Liang H, Lin F, Zhang Z, Liu B, Jiang S, Yuan Q, Liu J (2016) Multicopper laccase mimicking nanozymes with nucleotides as ligands. ACS Appl Mater Interfaces 9(2):1352–1360
71. Zhang L, Zhang Y, Wang Z, Cao F, Sang Y, Dong K, Pu F, Ren J, Qu X (2019) Constructing metal–organic framework nanodots as bio-inspired artificial superoxide dismutase for alleviating endotoxemia. Mater Horiz
72. Jing L, Liang X, Deng Z, Feng S, Li X, Huang M, Li C, Dai Z (2014) Prussian blue coated gold nanoparticles for simultaneous photoacoustic/CT bimodal imaging and photothermal ablation of cancer. Biomaterials 35(22):5814–5821
73. Yang F, Hu S, Zhang Y, Cai X, Huang Y, Wang F, Wen S, Teng G, Gu N (2012) A hydrogen peroxide-responsive O_2 nanogenerator for ultrasound and magnetic-resonance dual modality imaging. Adv Mater 24(38):5205–5211
74. Davidson D, Welo LA (1928) The nature of Prussian blue. J Phys Chem 32(8):1191–1196
75. Robin MB (1961) The color and electronic configurations of Prussian blue. Spectrochim Acta 17(9–10):1095–1095
76. DeLongchamp DM, Hammond PT (2004) High-contrast electrochromism and controllable dissolution of assembled Prussian blue/polymer nanocomposites. Adv Funct Mater 14(3):224–232
77. Buser H, Schwarzenbach D, Petter W, Ludi A (1977) The crystal structure of Prussian blue: $Fe_4[Fe(CN)_6]_3.xH_2O$. Inorg Chem 16 (11):2704–2710
78. Johansson A, Widenkvist E, Lu J, Boman M, Jansson U (2005) Fabrication of high-aspect-ratio Prussian blue nanotubes using a porous alumina template. Nano Lett 5(8):1603–1606
79. Shokouhimehr M, Soehnlen ES, Hao JH, Griswold M, Flask C, Fan XD, Basilion JP, Basu S, Huang SPD (2010) Dual purpose Prussian blue nanoparticles for cellular imaging and drug delivery: A new generation of T-1-weighted MRI contrast and small molecule delivery agents. J Mater Chem 20(25):5251–5259
80. McConnell H, Davidson N (1950) Spectrophotometric investigation of the interaction between iron(II) and iron(III) in hydrochloric acid solutions. J Am Chem Soc 72(12):5557–5560

81. Thompson RC (1948) Some exchange experiments involving hexacynoferrate(II) and hexacyanoferrate(III) ions. J Am Chem Soc 70(3):1045–1046
82. Ibers JA, Davidson N (1951) On the interaction between hexacyanatoferrate(III) ions and (a) hexacyanatoferrate(II) or (b) iron(III) ions. J Am Chem Soc 73(1):476–478
83. Wood KC, Zacharia NS, Schmidt DJ, Wrightman SN, Andaya BJ, Hammond PT (2008) Electroactive controlled release thin films. Proc Natl Acad Sci USA 105(7):2280–2285
84. Neff VD (1978) Electrochemical oxidation and reduction of thin films of Prussian blue. J Electroanal Chem 125(6):886–887
85. Qiu J-D, Peng H-Z, Liang R-P, Li J, Xia X-H (2007) Synthesis, characterization, and immobilization of Prussian blue-modified Au nanoparticles: application to electrocatalytic reduction of H_2O_2. Langmuir 23(4):2133–2137
86. Zhao G, Feng J-J, Zhang Q-L, Li S-P, Chen H-Y (2005) Synthesis and characterization of Prussian blue modified magnetite nanoparticles and its application to the electrocatalytic reduction of H_2O_2. Chem Mater 17(12):3154–3159
87. Itaya K, Ataka T, Toshima S (1982) Spectroelectrochemistry and electrochemical preparation method of Prussian blue modified electrodes. J Am Chem Soc 104(18):4767–4772
88. Larionova J, Guari Y, Sangregorio C, Guérin C (2009) Cyano-bridged coordination polymer nanoparticles. New J Chem 33(6):1177–1190
89. Zheng X-J, Kuang Q, Xu T, Jiang Z-Y, Zhang S-H, Xie Z-X, Huang R-B, Zheng L-S (2007) Growth of Prussian blue microcubes under a hydrothermal condition: possible nonclassical crystallization by a mesoscale self-assembly. J Phys Chem C 111(12):4499–4502
90. Niederberger M, Cölfen H (2006) Oriented attachment and mesocrystals: Non-classical crystallization mechanisms based on nanoparticle assembly. Phys Chem Chem Phys 8(28):3271–3287
91. Hu M, Jiang J-S (2010) Non-classical crystallization controlled by centrifugation. Cryst Eng Comm 12(11):3391–3393
92. Uemura T, Ohba M, Kitagawa S (2004) Size and surface effects of Prussian blue nanoparticles protected by organic polymers. Inorg Chem 43(23):7339–7345
93. Shokouhimehr M, Soehnlen ES, Khitrin A, Basu S, Huang SD (2010) Biocompatible Prussian blue nanoparticles: preparation, stability, cytotoxicity, and potential use as an MRI contrast agent. Inorg Chem Commun 13(1):58–61
94. Wu X, Cao M, Hu C, He X (2006) Sonochemical synthesis of Prussian blue nanocubes from a single-source precursor. Cryst Growth Des 6(1):26–28
95. Ding Y, Hu YL, Gu G, Xia XH (2009) Controllable synthesis and formation mechanism investigation of Prussian blue nanocrystals by using the polysaccharide hydrolysis method. J Phys Chem C 113(33):14838–14843
96. Liang GD, Xu JT, Wang XS (2009) Synthesis and characterization of organometallic coordination polymer nanoshells of prussian blue using miniemulsion periphery polymerization (MEPP). J Am Chem Soc 131(15):5378–5379
97. McHale R, Ghasdian N, Liu YB, Ward MB, Hondow NS, Wang HH, Miao YQ, Brydson R, Wang XS (2010) Prussian blue coordination polymer nanobox synthesis using miniemulsion periphery polymerization (MEPP). Chem Commun 46(25):4574–4576
98. Domínguez-Vera JM, Colacio E (2003) Nanoparticles of Prussian blue ferritin: a new route for obtaining nanomaterials. Inorg Chem 42(22):6983–6985
99. Zhang X-Q, Gong S-W, Zhang Y, Yang T, Wang C-Y, Gu N (2010) Prussian blue modified iron oxide magnetic nanoparticles and their high peroxidase-like activity. J Mater Chem 20(24):5110–5116
100. Zhang W, Hu S, Yin J-J, He W, Lu W, Ma M, Gu N, Zhang Y (2016) Prussian blue nanoparticles as multienzyme mimetics and reactive oxygen species scavengers. J Am Chem Soc 138(18):5860–5865
101. Zhang W, Wu Y, Dong H-J, Yin J-J, Zhang H, Wu H-A, Song L-N, Chong Y, Li Z-X, Gu N, Zhang Y (2018) Sparks fly between ascorbic acid and iron-based nanozymes: a study on Prussian blue nanoparticles. Colloid Surf B 163(1):379–384

102. Li R, Guo D, Ye J, Zhang M (2015) Stabilization of Prussian blue with polyaniline and carbon nanotubes in neutral media for in vivo determination of glucose in rat brains. Analyst 140(11):3746–3752
103. Peng KF, Zhao HW, Wu XF (2015) Signal-enhanced electrochemical immunosensor for CD36 based on cascade catalysis of a GOx labeled Prussian blue functionalized Ceria nanohybrid. RSC Adv 5(3):1812–1817
104. Shen GY, Hu X, Zhang SB (2014) A signal-enhanced electrochemical immunosensor based on dendrimer functionalized-graphene as a label for the detection of alpha-l-fetoprotein. J Electroanal Chem 717:172–176
105. Liu Z-H, Zhang G-F, Chen Z, Qiu B, Tang D (2014) Prussian blue-doped nanogold microspheres for enzyme-free electrocatalytic immunoassay of p53 protein. Microchim Acta 181(5–6):581–588
106. Wang G, Chen L, Zhu Y, He X, Xu G, Zhang X (2014) Prussian blue-Au nanocomposites actuated hemin/G-quadruplexes catalysis for amplified detection of DNA, Hg^{2+} and adenosine triphosphate. Analyst 139(20):5297–5303
107. Wang T, Fu YC, Chai LY, Chao L, Bu LJ, Meng Y, Chen C, Ma M, Xie QJ, Yao SZ (2014) Filling carbon nanotubes with Prussian blue nanoparticles of high peroxidase-like catalytic activity for colorimetric chemo- and biosensing. Chem-Eur J 20(9):2623–2630
108. Xu TS, Zhang HY, Li XG, Xie ZH, Li XY (2015) Enzyme-triggered tyramine-enzyme repeats on prussian blue-gold hybrid nanostructures for highly sensitive electrochemical immunoassay of tissue polypeptide antigen. Biosens Bioelectron 73:167–173
109. Gao ZD, Qu YF, Li TT, Shrestha NK, Song YY (2014) Development of amperometric glucose biosensor based on prussian blue functionlized TiO_2 nanotube arrays. Sci Rep 4:6891
110. Pandey PC, Singh R, Pandey Y (2015) Controlled synthesis of functional Ag, Ag–Au/Au–Ag nanoparticles and its nanocomposite with Prussian blue for bioanalytical applications. RSC Adv 5:49671–49679
111. Zhang WM, Ma D, Du JX (2014) Prussian blue nanoparticles as peroxidase mimetics for sensitive colorimetric detection of hydrogen peroxide and glucose. Talanta 120:362–367
112. Zhang W, Zhang Y, Chen YH, Li SY, Gu N, Hu SL, Sun Y, Chen X, Li Q (2013) Prussian blue modified ferritin as peroxidase mimetics and its applications in biological detection. J Nanosci Nanotechnol 13(1):60–67
113. Peng C, Hua M-Y, Li N-S, Hsu Y-P, Chen Y-T, Chuang C-K, Pang S-T, Yang H-W (2019) A colorimetric immunosensor based on self-linkable dual-nanozyme for ultrasensitive bladder cancer diagnosis and prognosis monitoring. Biosens Bioelectron 126:581–589
114. Zhao JL, Gao W, Cai XJ, Xu JJ, Zou DW, Li ZS, Hu B, Zheng YY (2019) Nanozyme-mediated catalytic nanotherapy for inflammatory bowel disease. Theranostics 9(10):2843–2855

Chapter 7
Carbon-based Nanozeymes

Hanjun Sun, Jinsong Ren and Xiaogang Qu

Abbreviations

ABTS	2,2′-azino-bis(3-ethylbenzothiazoline-6-sulphonic acid)
CNMs	Carbon nanomaterials
CNTs	Carbon nanotubes
CQDs	Carbon quantum dots
DNase	Deoxyribonuclease
EPR	Electron paramagnetic resonance
g-C_3N_4	Graphitic carbon nitride
GO	Graphene oxide
GO-COOH	Carboxyl-modified graphene oxide
GOx	Glucose oxidase
GQDs	Graphene quantum dots
HCC	Hydrophilic carbon clusters
HRP	Horseradish peroxidase
K_m	Michaelis–Menten constant
MWCNTs	Multi-walled carbon nanotubes
NHE	Normal hydrogen electrode
NIR	Near-infrared
NR	Nanoribbon
PDI	Perylene diimide
PEG	Polyethylene glycol
SOD	Superoxide dismutase
SWCNTs	Single-walled carbon nanotubes
TA	Terephthalic acid

H. Sun · J. Ren · X. Qu (✉)
Laboratory of Chemical Biology and State Key Laboratory of Rare Earth Resource Utilization, Chinese Academy of Sciences, 5625 Renmin Street, Changchun, Jilin 130022, China
e-mail: xqu@ciac.ac.cn

X. Yan (ed.), *Nanozymology*, Nanostructure Science and Technology,
https://doi.org/10.1007/978-981-15-1490-6_7

TAOH	2-hydroxy terephthalic acid
TMB	3,3,5,5-tetramethylbenzidine
V_{max}	Maximum initial velocity

After decades of development, CNMs, including fullerene, CNTs, graphene, GQDs, and CQDs, have been regarded as rising stars in many active research fields, such as electronics, energy, catalysis, imaging, sensing, and biomedicine, based on their excellent physical and chemical properties [1–15]. Among numerous applications, CNMs have attracted growing interest as metal-free catalysts for some chemical [5, 15] and biochemical reactions [3, 6–8, 13, 14]. As a class of nanozymes for biochemical reactions, CNMs have a higher operational stability and a lower cost than natural enzymes [3, 6–8, 13, 14]. Meanwhile, they are prepared more easily and have great robustness against stringent conditions [3, 6–8, 13, 14]. Because of these above advantages of CNMs, they have been paid great attention and widely applied as nanozymes in sensors, catalysis, and therapy [3, 6–8, 13, 14]. In this chapter, we will focus on the discussion about the nature of CNMs-based nanozymes and summarize the current status and give the future perspectives, hoping it will bring some new breakthroughs in the near future.

7.1 Introduction of CNMs

Carbon is a class of materials which the human have known since antiquity [16]. The three forms of natural carbon, namely, amorphous carbon, graphite, and diamond have been widely utilized by human from their early civilization [16]. The diverse hybridization states (sp, sp^2, sp^3) of carbon result in their different structural conformations and unique physical, chemical, and electronic properties [1, 9, 17]. With the development of nanotechnology, some novel allotropes of carbon at nanoscale have been synthesized (Fig. 7.1) [18–20]. The discovery of fullerene C_{60} created an entirely new branch of carbon chemistry [18]. Subsequently, the discovery of CNTs opened up a new area in materials science [19]. After 13 years, the coming of graphene brought the Noble prize to carbon materials again in 2010 after fullerene [20]. Due to their excellent physical and chemical properties, the zero-dimensional fullerene, one-dimensional CNTs, two-dimensional graphene as well as their derivatives, CQDs [21] and GQDs [22] have been regarded as a new class of promising materials for a wide range of potential applications, such as electronics, sensing, imaging, catalysis, medicine, energy conversion, and storage [1–15]. In this chapter, we will focus on the enzymatic activity of CNMs [3, 6–8, 13, 14].

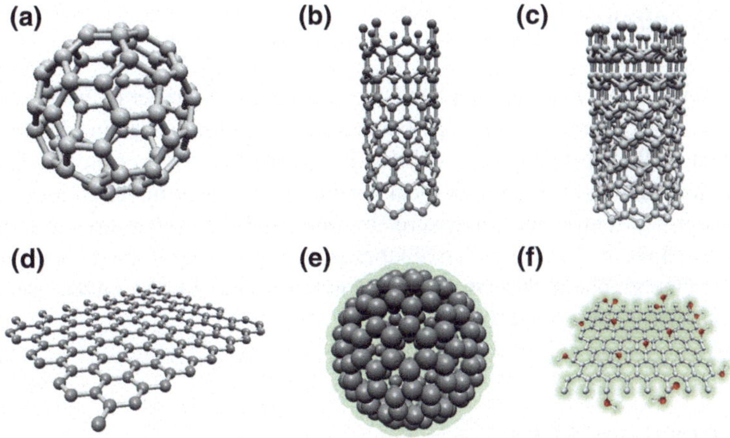

Fig. 7.1 Carbon nanomaterials. **a** fullerene (C_{60}). **b** single-walled carbon nanotube. **c** multi-walled carbon nanotube. **d** graphene. **e** carbon quantum dots. **f** graphene quantum dots

7.1.1 Fullerene

The first fullerene molecule C_{60} with a football-like structure was found in 1985 by Kroto et al. [18]. A C_{60} molecule consists of 12 pentagons and 20 hexagons facing symmetrically with a diameter of 0.71 nm [18]. Later a simple synthesis method was put forward by Kratschmer and Huffman [23], the fullerene chemistry started being developed fast. With the increasing amount and molecule species of fullerene, more and more knowledge about their physical and chemical properties had been obtained; meanwhile, their applications in photonic, electronic, superconducting, magnetic, and biomedical fields had been expanded as well [1, 2, 4, 9–12]. The fullerene and their derivatives can work as superoxide dismutase mimics [24–28] and peroxidase mimics [29].

7.1.2 CNTs

The one-dimensional materials CNTs, which were unearthed by Iijima in 1991, are well-ordered and hollow graphitic nanomaterials with high aspect ratios [19]. According to their diameter (φ) and the number of graphene layers, CNTs can be divided into SWNTs (0.4–2 nm in diameter) and MWNTs (2–100 nm in diameter) [9, 12]. CNTs have high mechanical strength, high surface areas, excellent chemical and thermal stability, and rich electronic, optical and enzymatic properties [1, 2, 4, 9–12]. For their enzymatic activity like HRP, the SWCNTs were often chosen as research object [30, 31].

7.1.3 Graphene

The two-dimensional material graphene, the new coming member of carbon family, is one-atom-thick planar sheets of sp^2-bonded carbon atoms packed in a honeycomb crystal lattice [9, 12]. Compared with CNTs, graphene have even higher fracture strength, thermal conductivity, electrical conductivity, charge mobility, and surface areas [2, 4, 9–12]. Thus, since discovered by Geim and Novoselov through mechanical exfoliated method using cohesive tape, graphene has been always a star material [20]. Some graphene derivatives, for example, GO and GO-COOH, also show outstanding peroxidase like activity [32].

7.1.4 CQDs and GQDs

CQDs were accidentally discovered by Xu's group in 2004 when they separated and purified SWCNTs [21]. Then the studies of fluorescence properties of CQDs make them become a new class of viable fluorescent nanomaterials, which are referred to as carbon nanolights [33]. CQDs are usually quasispherical nanoparticles with sizes below 10 nm [9, 33, 34]. Typically, CQDs contain abundant oxygen-containing groups, especially carboxylic groups on their surface, which endows them with excellent water solubility and suitability for further functionalization [9, 33, 34]. Like most of CNMs, the CQDs have the peroxidase like activity [35–39].

GQDs, as defined, are a kind of zero-dimensional graphene derivative with characteristics derived from both graphene and CQDs, which can be considered as incredibly small pieces of graphene [9, 40, 41]. The sizes of most GQDs range from 3 to 20 nm and they usually consist of no more than five layers of graphene sheets [9, 40, 41]. As a derivative of graphene, GQDs have similar composition and functional groups with GO; their crystalline nature also resembles that of graphene [9, 40, 41]. Therefore, GQDs process peroxidase like activity which is identical with GO [42–48].

7.2 Carbon-Based Nanozeymes

Even pristine fullerene, CNTs and graphene are insoluble in water, which is unfavorable to their applications as nanozymes, the multiform binding modes of carbon atoms made the modification on CNMs easy, thus leading to improved water solubility [24–32, 35–39, 42–52]. Since 1996, the superoxide dismutase like activity of fullerene and its derivatives had been discovered [24], various kinds of carbon nanomaterials, for instant, CNTs, GO, GQDs, CQDs and so on, have been regarded as nanozymes. According to the enzymatic activity they mimic, the carbon-based

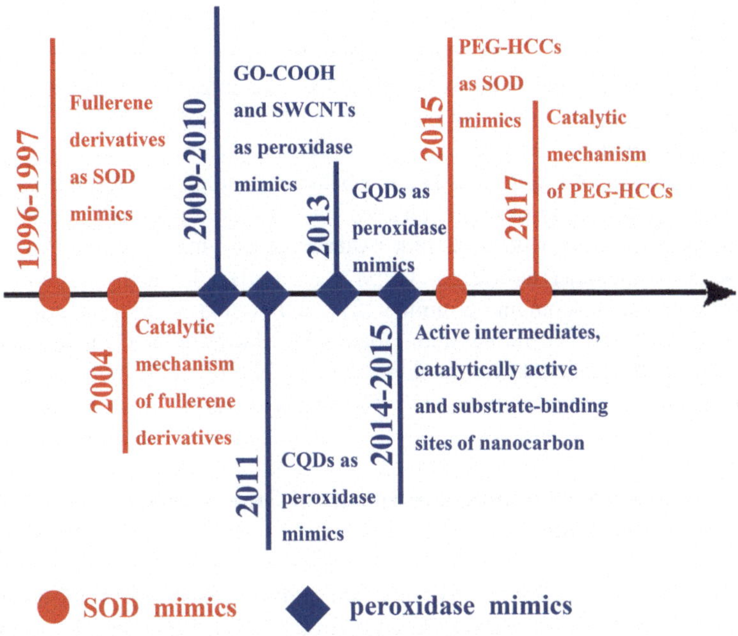

Fig. 7.2 A brief timeline for the development of carbon-based nanozymes

nanozymes can be mainly divided into two categories, superoxide dismutase mimics and peroxidase mimics (Fig. 7.2).

7.2.1 Carbon-Based Superoxide Dismutase Mimics

7.2.1.1 Superoxide Dismutase-like Activity of Fullerene and Its Derivatives

Fullerene and their derivatives as SOD mimics were the first carbon-based nanozeymes which were found in 1996 [24]. Even though the fullerene was considered as a class of radical sponge due to their unique chemical reactivity toward radicals [53], since the pristine fullerene were water-insoluble, above all, for working as a nanozyme, the fullerene was chemically modified to improve their solubility in water [24]. Two polyhydroxylated fullerenes ($C_{60}(OH)_{12}$ and $C_{60}(OH)_nO_m$, $n = 18–20$, $m = 3–7$ hemiketal groups) were prepared by Choi's group [24]. Then they found both derivatives retained the free radical scavenging capability of native C_{60} and significantly reduced excitotoxic and apoptotic death of cultured cortical neurons [24]. Even there was no concept about nanozyme at that time, in fact, this was the first example of fullerene derivatives as superoxide dismutase mimics.

Then, tris-malonic acid derivatives of C_{60} ($C_{60}[C(COOH)_2]_3$) showed higher neuroprotective activity than polyhydroxylated fullerenes both in vitro and in vivo [25–27]. Interestingly, the superoxide dismutase like activity of $C_{60}[C(COOH)_2]_3$ was depended on the symmetry of molecules. $C_{60}[C(COOH)_2]_3$ with C_3 symmetry (C_{60}–C_3) showed higher activity than $C_{60}[C(COOH)_2]_3$ with D_3 symmetry (C_{60}-D_3) because the polarity of C_{60}–C_3 resulted in a better cell membrane penetrating ability [27]. Taken C_{60}–C_3 as an example, kinetics studies were performed to evaluate its performance as well. By using a xanthine oxidase/cytochrome c assay, the V_{max} value of C_{60}–C_3 was calculated to be $2.2 \times 10^6\ M^{-1}\ s^{-1}$ at pH 7.4, which was comparable to several manganese-containing mimics [27]. But compared with natural SOD, it was 100-fold lower [27]. In addition, Gozin's lab showed C_{60}–C_3 nanozyme and human serum albumin could form a non-covalent complex with a K_m of 1.2×10^7 M^{-1} and a similar activity with free C_{60}–C_3 nanozyme [28].

7.2.1.2 Superoxide Dismutase-like Activity of Other Carbon Nanomaterials

Besides fullerene and its derivatives, some other carbon-based superoxide dismutase mimics have been found as well very recently (Fig. 7.3) [51, 52]. Tour et al. developed a novel metal-free carbon nanomaterials, PEG-HCCs, by oxidizing SWCNTs using a mixture of oleum and nitric acid (Fig. 7.3a) [51]. From the EPR measurement, they found nanomolar concentrations of PEG-HCCs could scavenge micromolar to millimolar concentrations of $O_2^{\bullet-}$ in a short time. Excitedly, the turnover number of PEG-HCCs was compared with bovine Cu/Zn SOD, the most efficient native SOD enzyme at physiological pH [51].

7.2.1.3 Catalytic Mechanism

Basically, similar to the natural superoxide dismutase, the C_{60}–C_3 catalyzed reactions followed the ping-pong mechanism [27]. For further study, EPR and some other techniques were used to confirm the superoxide dismutase like activity and study the catalytic mechanism of C_{60}–C_3. Ali and his colleagues considered the C_{60}–C_3's superoxide dismutase like activity was caused by the catalytic dismutation of superoxide rather than stoichiometric scavenging. Then from the observed experimental results about the lack of structural modifications to C_{60}–C_3, absence of detectable paramagnetic products, generation of hydrogen peroxide and regeneration of oxygen, this catalytic mechanism was successfully proved [27]. The reaction Eq. (7.1) was as followed.

$$O_2^{\bullet-} + 2H_2O \rightarrow H_2O_2 + O_2 + 2OH^- \tag{7.1}$$

Furthermore, semiempirical quantum mechanical calculations to simulate the interactions between the superoxide radical substrate and the C_{60}–C_3 nanozyme had

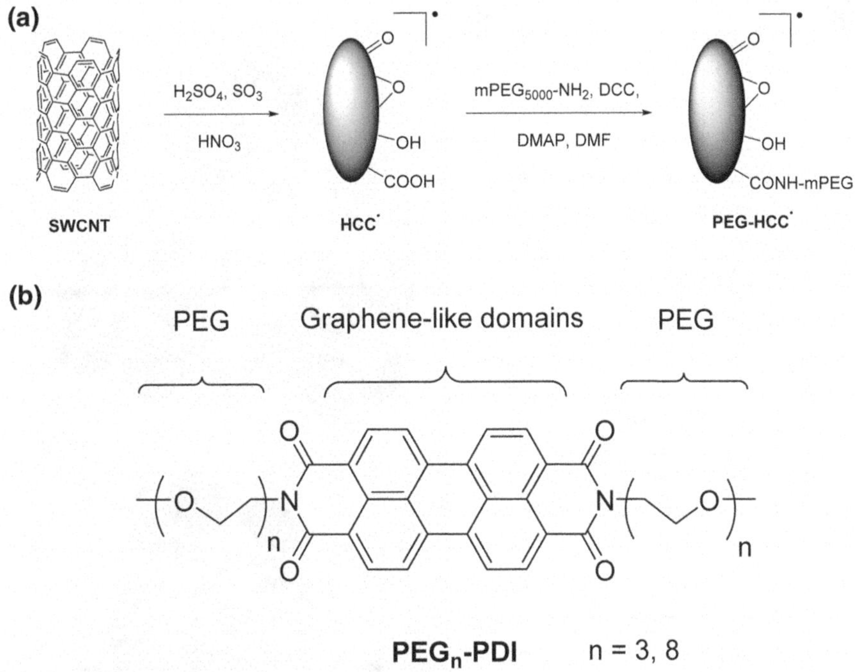

Fig. 7.3 **a** Synthesis of PEG-HCCs from SWCNTs as the starting material. **b** Structural formula of PEG_n−PDI. Reprinted with permission from [52]. Copyright (2017) American Chemical Society

been also implemented [27]. From the electron distribution map, Ali et al. concluded that the electron-deficient regions on C_{60}–C_3 electrostatically drew the substrate anions toward the C_{60}–C_3 surface and directed them for further dismutation with the help of protons supplied from the carboxyl groups together with water molecules around C_{60}–C_3 nanozyme (Fig. 7.4). After that, a more detailed structure–activity study for carboxyfullerenes indicated that the activity was closely related to both the number and distribution of carboxyl groups on the fullerene ball [54].

Apart from studying the catalytic mechanism of fullerene and their derivatives as SOD mimics, the catalytic mechanism of PEG-HCCs and their analogues have been well investigated [51, 52]. Through EPR measurement, Tour et al. had proposed the catalytic mechanism of PEG-HCC as superoxide dismutase mimic. $O_2^{\bullet -}$ could be first oxidized into O_2 catalyzed by PEG-HCC, one of the two half-reactions processes (7.2) of SOD, and this might also be the rate-determining step of the SOD catalytic cycle. Then it is followed by the reduction of $O_2^{\bullet -}$ in the presence of water to form H_2O_2 (7.3) [51].

$$\mathrm{PEG-HCC^{\bullet}} + \mathrm{O_2^{\bullet -}} \rightarrow \mathrm{PEG-HCC^{-}} + \mathrm{O_2} \quad (7.2)$$

$$\mathrm{PEG-HCC^{-}} + \mathrm{O_2^{\bullet -}} + 2\mathrm{H_2O} \rightarrow \mathrm{PEG-HCC} \bullet + \mathrm{H_2O_2} + 2\mathrm{OH^{-}} \quad (7.3)$$

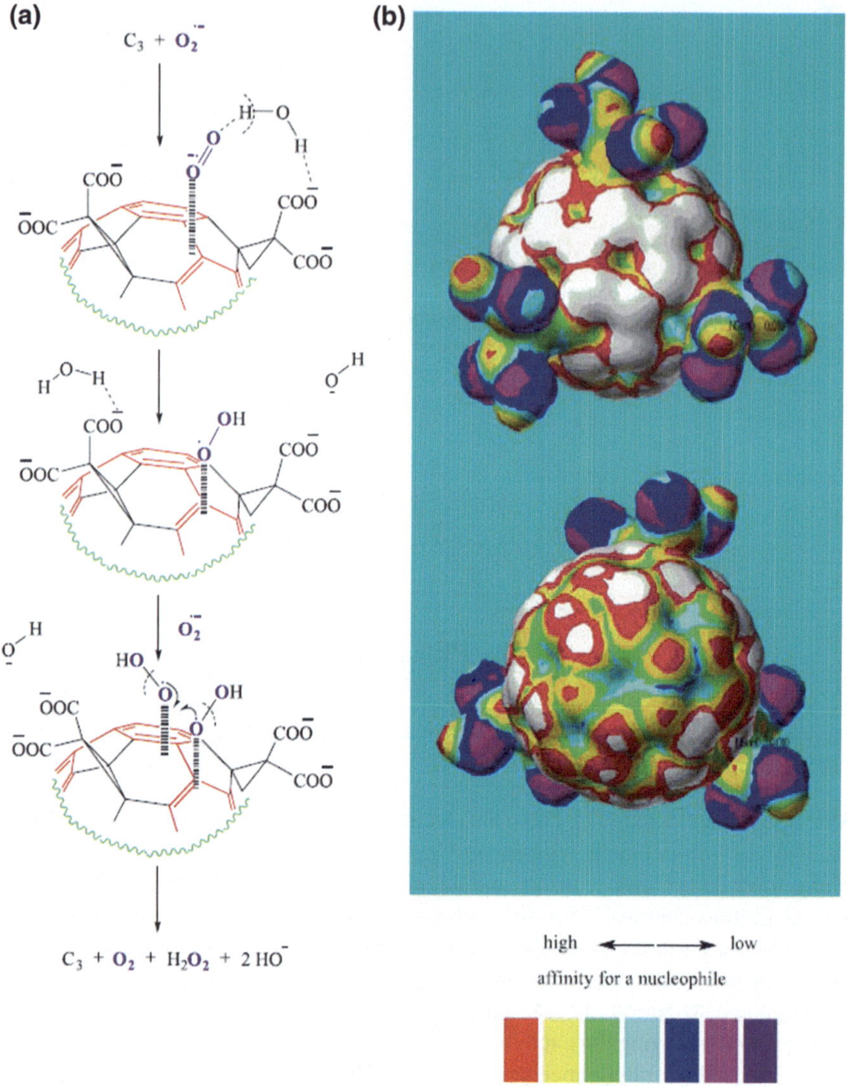

Fig. 7.4 **a** The proposed catalytic mechanism of C_{60}–C_3. **b** The electron distribution map of C_{60}–C_3. Reprinted with permission from [27]. Copyright (2004) Elsevier

To better understand the catalytic mechanisms of PEG-HCCs as superoxide dismutase mimic, Tour and his colleagues synthesized the small well-defined conjugated molecular analogues, PEG_n-PDIs (n = 3, 8), as a model (Fig. 7.3b) [52]. Similar to PEG-HCC, PEG_n-PDI was able to oxidize $O_2^{\bullet-}$ (7.3), and then the forming PEG_n-$PDI^{\bullet-}$ intermediate react with $O_2^{\bullet-}$ in the presence of protons to form H_2O_2 (7.4) [52].

$$PEG_n - PDI^{\bullet} + O_2^{\bullet -} \rightarrow PEG_n - PDI^- + O_2 \quad (7.4)$$

$$PEG_n - PDI^- + O_2^{\bullet -} + 2H_2O \rightarrow PEG_n - PDI^- + H_2O_2 + 2OH^- \quad (7.5)$$

From the electrochemical measurements, they discovered the redox potential of PEG_n-PDI located between the following two half-reactions (7.6 and 7.7),

$$O_2^{\bullet -} \rightarrow O_2 + e^- \quad \varphi = -0.16 \text{ V versus NHE} \quad (7.6)$$

$$O_2^{\bullet -} + 2H^+ + 2e^- \rightarrow H_2O_2 \quad \varphi = +0.94 \text{ versus NHE} \quad (7.7)$$

which drove that the PEG-HCC or PEG_n-PDI played the role of superoxide dismutase mimics [52].

7.2.2 *Carbon-Based Peroxidase Mimics*

7.2.2.1 Peroxidase-like Activity of Carbon Nanomaterials

CNMs could catalyze H_2O_2 to oxidize TMB or ABTS and exhibit peroxidase-like activity, which was the most reported enzymatic property of CNMs [29–32, 35–39, 42–44, 47–50]. Our group contributed much effort to CNMs as peroxidase mimics [30, 32, 36, 45, 46]. In 2010, we first surprisingly discovered that GO-COOH could catalyze the colorimetric reaction of peroxidase substrate 3,3,5,5-tetramethylbenzidine (TMB) in thc presence of H_2O_2 (Fig. 7.5) [32]. Similar to HRP, the catalytic activity of the GO-COOH was dependent on pH, temperature, and H_2O_2 concentration [32]. For another kind of CNMs, the peroxidase like activity

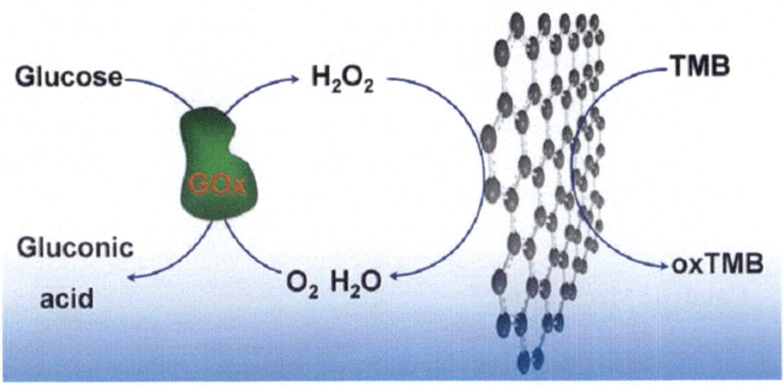

Fig. 7.5 Colorimetric detection of glucose by using GOx and the peroxidase-like activity of GO-COOH. Reprinted with permission from [32]. Copyright (2010) Wiley-VCH

Table 7.1 Typical parameters for Michaelis–Menton kinetics of some typical carbon-based peroxidase mimics

Materials	Substrate	K_m (mM)	V_{max} (10^{-8} M s^{-1})
GO-COOH [32]	H_2O_2	3.99	3.85
CQDs [35]	H_2O_2	26.77	30.61
GQDs [42]	H_2O_2	0.49	2.62
$C_{60}[C(COOH)_2]_2$ [29]	H_2O_2	24.58	4.01
Carbon nanohorn [49]	H_2O_2	49.8	2.07

of SWCNTs was first found by our lab as well [30]. The cobalt catalyst in SWNTs was carefully removed by sonication in concentrated sulfuric and nitric acids, confirmed the activity originated from the SWNTs itself [30]. Zhu et al. also showed that the helical CNTs had better catalytic performance than MWNTs [31]. For photoluminescent CQDs or carbon nanoparticles, Huang's [35] and our groups [36] found that they exhibited high peroxidase like activity at almost the same time. With iron and sliver doping, the peroxidase like activity of CQDs would be further improved [55, 56]. In addition, GQDs, which can be regarded as incredibly small pieces of graphene, displayed even higher peroxidase like activity than GO [42–44]. Even there were few reports, despite the abovementioned CNMs, some other CNMs, for example, carbon nanohorn and C_{60}-carboxyfullerenes ($C_{60}[C(COOH)_2]_2$) had a high peroxidase-like activity as well. Moreover, CNMs-based hybrid displayed the same enzymatic activity [50]. Wang et al. prepared MWCNT@rGONR by the unzipped oxidation of MWCNTs and the subsequent chemical reduction with hydrazine. The obtained MWCNT@rGONR heterostructure has a higher peroxidase-like activity than that of MWCNTs and its unreduced form because the heterostructures benefitted the acceleration of their electron transfer process and the consequent facilitation of ·OH radical generation [50]. Through Michaelis–Menten model, the V_{max} and K_m for H_2O_2 of different CNMs were calculated and listed in Table 7.1. It was worth mentioning that even though the CNMs exhibited relative high peroxidase like activity, there was still a big gap compared with natural peroxidases, such as HRP.

Apart from CNMs, some analogues of CNMs, such as g-C_3N_4, also exhibited certain peroxidase like activity [57]. And similar to CQDs, the iron ion doping could further enhance their catalytic activity [57]. Recently, our group synthesized Au/g-C_3N_4 hybrid nanozyme, and found the hybrid nanozyme showed an enhanced catalytic activity due to the synergetic effect [58].

7.2.2.2 Catalytic Mechanism

Although the abovementioned nanocarbon-based peroxidase mimics have exhibited enzymatic activity in diagnostics and therapeutic applications, compared with natural peroxidases, for example, HRP, the low efficiency of the CNMs has unfortunately restricted their further applications. Understanding their catalytic mechanism should facilitate the design and synthesis of more effective nanocarbon-based peroxidase mimic. Since our group first found the GO exhibited intrinsic peroxidase catalytic activity in 2010, there have been a few reports discussing the catalytic mechanism of reactions catalyzed by nanocarbon-based peroxidase mimics [32, 45, 46, 59, 60].

Our group first studied the dynamic process of the reaction which H_2O_2 oxidized TMB catalyzed by GO-COOH [32]. Through Michaelis–Menten model and analyzing and comparing the kinetic parameters $V_{\max}$ and K_m of GO-COOH by Lineweaver–Burk plot, we found similar to HRP catalyzed reaction, the GO-COOH-catalyzed reaction is inhibited at high H_2O_2 concentration and followed the ping-pong mechanism [32]. A bathochromic shift (ca. 19 nm) from the UV spectrum of GO-COOH after the addition of 88.2 mM H_2O_2 could be observed, suggested electron transfer occurred from the top of the valence band of graphene to the lowest unoccupied molecular orbital (LUMO) of H_2O_2 [32].

For a long time, there were few reports discussing the detailed catalytic mechanism of carbon-based nanocarbon-based peroxidase mimic [59, 60]. Even some literature pointed out that the peroxidase-like activity of nanocarbon originated from the iron impurities in nanocarbon materials instead of themselves [60]. In order to deepen the understanding and clarify, the dispute taken GQDs which have a higher activity than GO as an example, we further studied the catalytic mechanism of nanocarbon-based peroxidase mimics [45, 46]. The active intermediates of the catalytic reaction were identified first (Fig. 7.6) [45]. By using TA, a fluorescence probe could capture •OH and generate TAOH, we have found that the peroxidase-like activity of GQDs stems from their ability to catalyze the decomposition of H_2O_2, generating OH [45]. Through selectively deactivating these specific oxygen-containing functional groups and investigating the catalytic activity of different GQD derivatives as peroxidase mimics, we further studied the functions of different oxygen-containing functional groups in this peroxidase mimic [46]. From calculating and comparing the kinetic parameters $V_{\max}$ and K_m by Lineweaver–Burk plots of GQDs and their derivatives, the –C=O groups act as the catalytically active sites for converting H_2O_2 to OH and the O=C–O– groups serve as the substrate binding sites for H_2O_2 (Fig. 7.7) [46]. The existence of –C–OH groups decreases the catalytic activity of GQDs and a theoretical study further proved the above conclusion [46]. Our works on catalytic mechanism of nanocarbon-based peroxidase mimic have not only provided new insights into deciphering carbon-based nanozyme enhanced our understanding of nanocarbon-based artificial enzymes, but will also facilitate the design and construction of high catalytically active and other target-specific artificial enzymes [45, 46].

7.3 Carbon Nanomaterials as Modulators for Nanozymes

Apart from their intrinsic enzymatic activity, CNMs, especially for CNTs and graphene, also play an important role in nanozymes hybrid. Considering the large specific surface areas, excellent physical properties and rich surface chemistry of

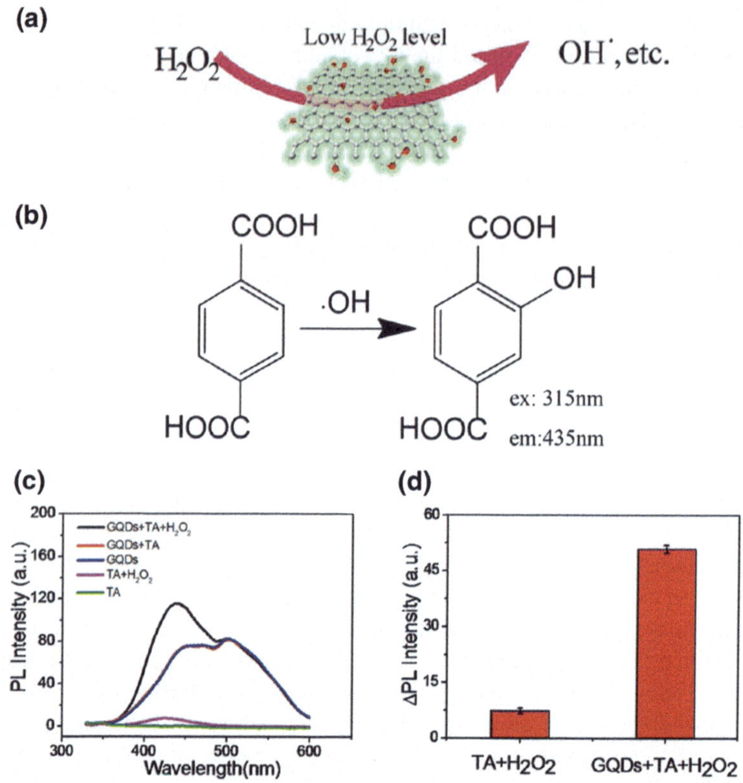

Fig. 7.6 **a** Active intermediates of the GQDs-catalyzed reactions. **b** Reaction between •OH and TA. **c** Fluorescence spectra of the PBS solution (pH = 7.4, 25 mM) include only TA; TA and H_2O_2; only GQDs; TA and GQDs; and TA, GQDs and H_2O_2 after 12 h reaction. The concentrations of TA, H_2O_2 and GQDs were 0.5 mM, 1 mM and 100 μg/mL, respectively. **d** Histograms of ΔPL intensity showed the catalysis effect of GQDs; error bars were taken from three parallel experiments. Reprinted with permission from [45]. Copyright (2014) American Chemical Society

CNMs and their derivatives, they can also be utilized as modulators to further promote the catalytic activity and stability of nanozymes. Some classic nanozymes, for instance, metal oxide (Fe_3O_4 [61–76], FeO_xH [77], Fe_2O_3 [78, 79], Co_3O_4 [80, 81], ZnO [82], CuO [83–86], Cu_2O [87], CeO_2 [76], etc.), noble metal nanoparticles (Pt [76, 88–92], Pd [72, 93], Au [70, 72, 76, 94–104], Ag [105], etc.), metal chalcogenides ($CoSe_2$ [106], CuS [107, 108], etc.), noble metal alloy (PtPd [109], PdNi [110], AuPd [111, 112], CuAg [113], etc.), metal ion (Cu^{2+} [114–116], etc.), metal complexes (hemin [117–127], chiral metallo-supramolecular complex ($[Fe_2L_3]^{4+}$ [128], Fe-phenanthroline [129], prussian blue [130], etc.) have been made use of constructing nanozyme-CNMs hybrids. The progress in this field has been recently detailed summarized [131, 132].

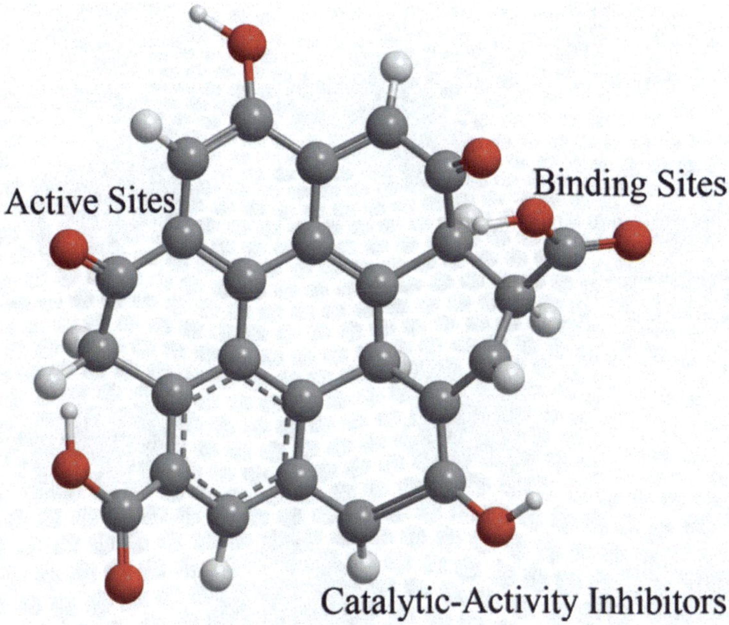

Fig. 7.7 Deciphering a nanocarbon-based artificial peroxidase: –C=O groups act as the catalytically active sites for converting H_2O_2 to ·OH and the O=C–O– groups serve as the substrate binding sites for H_2O_2. The existence of –C–OH groups decreases the catalytic activity of GQDs. Reprinted with permission from [46]. Copyright (2015) Wiley-VCH

7.3.1 Dispersing and Stabilizing Nanozymes by CNMs

Among all of the nanozyme-CNMs hybrids, the most used function of CNMs is that they can disperse and stabilize nanoparticles and small molecules effectively [61, 118]. Fan's group prepared the Fe_3O_4/MWCNTs nanocomplex [61]. The stabilized and uniformed dispersion of Fe_3O_4 nanoparticles on MWCNTs endue their higher peroxidase-like activity than Fe_3O_4 alone [61]. To solve the problem that molecular aggregation of hemin in aqueous solution causes passivation of its catalytic activity as peroxidase mimic, Duan and Huang's group reported the synthesis of a hemin–graphene conjugate (Fig. 7.8) through π–π stacking interactions [118]. The GO in this hemin–graphene conjugate prevented iron-porphyrin from self-dimerization, thus keeping the high catalytic activity of hemin [118]. Besides that, the GO support could block one side of the porphyrin molecule and prevent hydrogen peroxide attack from both sides, lowering the possibility of oxidative destruction of the hemin catalyst [118]. Moreover, the two-dimensional graphene sheets not only provided the large surface area for the diffusion of substrate and product but also worked as π donor to iron center of hemin through cation-π interaction, which was also crucial for the catalytic activity [118]. Taking the above advantages, the obtained hemin–graphene conjugate exhibited an amazing activity more than two orders of magnitude

Fig. 7.8 Formation of hemin–graphene conjugates through π–π stacking interactions. Reprinted with permission from [118]. Copyright (2012) Wiley-VCH

better than free hemin, and more than one order of magnitude better than any other supported system [118]. This kind of hemin–graphene nanocomposites had been also prepared by our lab and applied to realize ultrasensitive sensing for biomolecules [121]. In addition, at the same time that CNMs dispersed and stabilized nanozymes, for someone with poor conductivity such as Fe_3O_4, the high conductivity of CNMs was another important property to improve their activities [66].

7.3.2 *Modulating the Substrate Adsorption by CNMs*

Besides dispersing and stabilizing the nanozymes, CNMs can improve the catalytic activity of hybrid nanozymes by modulating the absorbing of substrate on the surface of nanozymes as well [97]. Our group found that the peroxidase-like activity of lysozyme-stabilized AuNCs could be regulated by GO [97]. Compared with AuNCs and GO alone, the GO-AuNCs hybrid displayed high catalytic activity over a broad pH range, even at neutral pH (Fig. 7.9) [97]. The significant activity enhancement of GO-AuNCs hybrid could be assigned to that GO have high surface to volume ratios as well as high affinity for hydrophobic molecules are able to absorb the TMB

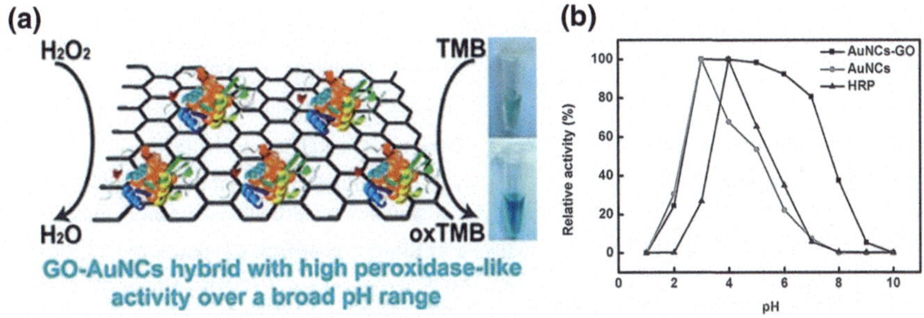

Fig. 7.9 **a** Schematic illustration of the high catalytic activity of synergistic GO-AuNCs hybrid. **b** GO-AuNCs hybrids showed the high peroxidase-like activity over a board pH range. Reprinted with permission from [97]. Copyright (2013) Wiley-VCH

efficiently, thus reducing the distance between the active site on AuNCs and TMB, further resulting in the great enhancement of activity [97].

7.3.3 Influence from Physical and Chemical Properties of CNMs

Some excellent physical and chemical properties of CNMs have also been utilized to modulate the activity of nanozeymes [76, 114, 115, 128]. Taking the advantages of the NIR photothermal effect of graphene, our group has realized the modulating for enzymatic activity of chiral metallo-supramolecular complex, Fe_3O_4, Au, CeO_2, and Pt by NIR heating the GO nanosheets [76, 128]. Under NIR irradiation, the temperature of solution could reach the optimal temperature of the above nanozymes in the presence of GO, thus resulting in the activation of catalytic activity [76, 128]. In view of the fact that GO and GQDs can intercalate into dsDNA at the major groove, Guo and Zhang's lab used GO and GQDs to improve the DNase-like activity of Cu^{2+} [114, 115]. From the result of agarose gel electrophoresis experiment, the activity of Cu^{2+}/CNMs hybrid for DNA cleavage was strongly depended on the binding strength between dsDNA and CNMs, which demonstrated the important role the GO or GQDs played in the system [114, 115].

7.4 Perspective and Challenges

Undoubtedly, the carbon-based nanozymes have all the advantages of nanozymes with the comparison to natural enzyme, such as low costs, facile production on large scale, the possibility of long-term storage, and high stability in harsh environments [3, 6]. In addition to the above advantages shared with other nanozymes, the large

Table 7.2 Main advantages and disadvantages of carbon-based nanozymes

Main advantages	Main disadvantages or challenges
(i) Low costs	(i) Lower catalytic activity
(ii) Facile production on large scale	(ii) Unclear catalytic mechanism
(iii) Long-term storage	(iii) Difficulties in rational design and construction of carbon-based nanozymes
(iv) High stability against harsh environments	(iv) Limited nanozymes types
(v) Large surface area	
(vi) Rich surface chemistry	

surface area and rich surface chemistry of nanocarbon also benefit their enzymatic activity. Aside from these advantages, there are still four exciting challenges remained for carbon-based nanozymes. (i) Nanocarbon usually exhibits relative lower catalytic activity than natural enzymes, especially for the carbon-based peroxidase mimics. (ii) Even devoted much efforts, some catalytic mechanism of carbon-based nanozyme has not been fully understood. Taking the carbon-based peroxidase mimics as examples, although the active intermediates, catalytically active and substrate-binding sites are identified now [45, 46], the reaction progresses are not clear. Therefore, further investigations are indeed essential to study the catalytic mechanism, which will be helpful to understand the working principle of carbon-based nanozymes. (iii) Rational design and construction of high catalytic active and other target-specific carbon-based nanozymes is still a significant challenge. The solution to this issue may be supplied by the fast developing of nanotechnology and better understanding of catalytic mechanism [6, 46]. (iv) Up to now, the nanocarbon only have two kinds of enzymatic activity, superoxide dismutase-like activity and peroxidase-like activity. The exploration of new enzymatic reactions catalyzed nanocarbon is on desire for expanding their applications as nanozymes (Table 7.2).

In spite of facing the abovementioned challenges, considering their extremely low toxicity in vivo, the future work on carbon-based nanozymes may focus on the exploitation of their potential applications in vivo like natural enzymes.

7.5 Conclusion

In this chapter, we summarized the development of carbon-based nanozymes. Both of the enzymatic properties and the corresponding catalytic mechanisms were detaily discussed. These researches featured here attempted to better understand what carbon-based nanozymes are and how carbon-based naonozyems work. What's more, the CNMs-based modulators in nanozymes hybrids were investigated. Despite that there are still some challenges and unsolved problems, the unique enzymatic properties and advantages of carbon-based nanozymes promised them to continue to enliven in the field of nanozymes for the years to come.

Acknowledgements This work was supported in part by the National Natural Science Foundation of China (Grants 21431007, 21533008, 21871249, 91856205, and 21820102009), and Key Program of Frontier of Sciences, CAS QYZDJ-SSW-SLH052.

References

1. Mauter MS, Elimelech M (2008) Environmental applications of carbon-based nanomaterials. Environ Sci Technol 42(16):5843–5859
2. Dai LM, Chang DW, Baek JB et al (2012) Carbon nanomaterials for advanced energy conversion and storage. Small 8(8):1130–1166
3. Wei H, Wang EK (2013) Nanomaterials with enzyme-like characteristics (nanozymes): next-generation artificial enzymes. Chem Soc Rev 42(14):6060–6093
4. Jariwala D, Sangwan VK, Lauhon LJ et al (2013) Carbon nanomaterials for electronics, optoelectronics, photovoltaics, and sensing. Chem Soc Rev 42(7):2824–2860
5. Su DS, Perathoner S, Centi G (2013) Nanocarbons for the development of advanced catalysts. Chem Rev 113(8):5782–5816
6. Lin YH, Ren JS, Qu XG (2014) Catalytically active nanomaterials: a promising candidate for artificial enzymes. Accounts Chem Res 47(4):1097–1105
7. Garg B, Bisht T, Ling YC (2015) Graphene-based nanomaterials as efficient peroxidase mimetic catalysts for biosensing applications: an overview. Molecules 20(8):14155–14190
8. Shin HY, Park TJ, Kim MI (2015) Recent research trends and future prospects in nanozymes. J Nanomater 2015:756278
9. Georgakilas V, Perman JA, Tucek J et al (2015) Broad family of carbon nanoallotropes: classification, chemistry, and applications of fullerenes, carbon dots, nanotubes, graphene, nanodiamonds, and combined superstructures. Chem Rev 115(11):4744–4822
10. Hong GS, Diao SO, Antaris AL et al (2015) Carbon nanomaterials for biological imaging and nanomedicinal therapy. Chem Rev 115(19):10816–10906
11. Bartelmess J, Quinn SJ, Giordani S (2015) Carbon nanomaterials: multi-functional agents for biomedical fluorescence and Raman imaging. Chem Soc Rev 44(14):4672–4698
12. Sun HJ, Ren JS, Qu XG (2016) Carbon nanomaterials and DNA: from molecular recognition to applications. Accounts Chem Res 49(3):461–470
13. Wang XY, Hu YH, Wei H (2016) Nanozymes in bionanotechnology: from sensing to therapeutics and beyond. Inorg Chem Front 3(1):41–60
14. Liu BW, Liu JW (2017) Surface modification of nanozymes. Nano Res 10(4):1125–1148
15. Navalon S, Dhakshinamoorthy A, Alvaro M, Antonietti M, Garcia H (2017) Active sites on graphene-based materials as metal-free catalysts. Chem Soc Rev 46(15):4501–4529
16. Hirsch A (2010) The era of carbon allotropes. Nat Mater 9(11):868–871
17. Ajayan PM (1999) Nanotubes from carbon. Chem Rev 99(7):1787–1800
18. Kroto HW, Heath JR, O'Brien SC et al (1985) C_{60}: buckminsterfullerene. Nature 318(6042):162–163
19. Iijima S (1991) Helical microtubules of graphitic carbon. Nature 354(6348):56–58
20. Novoselov KS, Geim AK, Morozov SV et al (2004) Electric field effect in atomically thin carbon films. Science 306(5696):666–669
21. Xu XY, Ray R, Gu YL et al (2004) Electrophoretic analysis and purification of fluorescent single-walled carbon nanotube fragments. J Am Chem Soc 126(40):12736–12737
22. Ponomarenko LA, Schedin F, Katsnelson MI et al (2008) Chaotic dirac billiard in graphene quantum dots. Science 320(5874):356–358
23. Kratschmer W, Lamb LD, Fostiropoulos K et al (1990) Solid C60: a new form of carbon. Nature 347(6291):354–358

24. Dugan LL, Gabrielsen JK, Yu SP et al (1996) Buckminsterfullerenol free radical scavengers reduce excitotoxic and apoptotic death of cultured cortical neurons. Neurobiol Dis 3(2):129–135
25. Dugan LL, Turetsky DM, Du C et al (1997) Carboxyfullerenes as neuroprotective agents. Pro Natl Acad Sci USA 94(17):9434–9439
26. Dugan LL, Lovett EG, Quick KL et al (2001) Fullerene-based antioxidants and neurodegenerative disorders. Park Relat D 7(3):243–246
27. Ali SS, Hardt JI, Quick KL et al (2004) A biologically effective fullerene (C_{60}) derivative with superoxide dismutase mimetic properties. Free Radic Bio Med 37(8):1191–1202
28. Belgorodsky B, Fadeev L, Ittah V et al (2005) Formation and characterization of stable human serum albumin-tris-malonic acid [C_{60}]fullerene complex. Bioconjugate Chem 16(5):1058–1062
29. Li R, Zhen M, Guan M et al (2013) A novel glucose colorimetric sensor based on intrinsic peroxidase-like activity of C_{60}-carboxyfullerenes. Biosens Bioelectron 47:502–507
30. Song Y, Wang X, Zhao C et al (2010) Label-free colorimetric detection of single nucleotide polymorphism by using single-walled carbon nanotube intrinsic peroxidase-like activity. Chem Eur J 16(12):3617–3621
31. Cui R, Han Z, Zhu JJ (2011) Helical carbon nanotubes: intrinsic peroxidase catalytic activity and its application for biocatalysis and biosensing. Chem Eur J 17(34):9377–9384
32. Song Y, Qu K, Zhao C et al (2010) Graphene oxide: intrinsic peroxidase catalytic activity and its application to glucose detection. Adv Mater 22(19):2206–2210
33. Baker SN, Baker GA (2010) Luminescent carbon nanodots: emergent nanolights. Angew Chem Int Ed 49(38):6726–6744
34. Zhao AD, Chen ZW, Zhao CQ et al (2015) Recent advances in bioapplications of C-dots. Carbon 85:309–327
35. Shi W, Wang Q, Long Y et al (2011) Carbon nanodots as peroxidase mimetics and their applications to glucose detection. Chem Commun 47(23):6695–6697
36. Wang X, Qu K, Xu B et al (2011) Multicolor luminescent carbon nanoparticles: synthesis, supramolecular assembly with porphyrin, intrinsic peroxidase-like catalytic activity and applications. Nano Res 4(9):908–920
37. Liu S, Tian J, Wang L et al (2012) A general strategy for the production of photoluminescent carbon nitride dots from organic amines and their application as novel peroxidase-like catalysts for colorimetric detection of H_2O_2 and glucose. RSC Adv 2(2):411–413
38. Zhu W, Zhang J, Jiang Z et al (2014) High-quality carbon dots: synthesis, peroxidase-like activity and their application in the detection of H_2O_2, Ag^+ and Fe^{3+}. RSC Adv 4(33):17387–17392
39. Long Y, Wang X, Shen D et al (2016) Detection of glucose based on the peroxidase-like activity of reduced state carbon dots. Talanta 159:122–126
40. Li LL, Wu GH, Yang GH et al (2013) Focusing on luminescent graphene quantum dots: current status and future perspectives. Nanoscale 5(10):4015–4039
41. Sun HJ, Wu L, Wei WL et al (2013) Recent advances in graphene quantum dots for sensing. Mater Today 16(11):433–442
42. Zhang Y, Wu C, Zhou X et al (2013) Graphene quantum dots/gold electrode and its application in living cell H_2O_2 detection. Nanoscale 5(5):1816–1819
43. Zheng AX, Cong ZX, Wang JR et al (2013) Highly-efficient peroxidase-like catalytic activity of graphene dots for biosensing. Biosens Bioelectron 49:519–524
44. Wu X, Tian F, Wang W et al (2013) Fabrication of highly fluorescent graphene quantum dots using l-glutamic acid for in vitro/in vivo imaging and sensing. J Mater Chem C 1(31):4676–4684
45. Sun H, Gao N, Dong K et al (2014) Graphene quantum dots-band-aids used for wound disinfection. ACS Nano 8(6):6202–6210
46. Sun H, Zhao A, Gao N et al (2015) Deciphering a nanocarbon-based artificial peroxidase: chemical identification of the catalytically active and substrate-binding sites on graphene quantum dots. Angew Chem Int Ed 54(24):7176–7180

47. Tang D, Liu J, Yan X et al (2016) Graphene oxide derived graphene quantum dots with different photoluminescence properties and peroxidase-like catalytic activity. RSC Adv 6(56):50609–50617
48. Nirala NR, Khandelwal G, Kumar B et al (2017) One step electro-oxidative preparation of graphene quantum dots from wood charcoal as a peroxidase mimetic. Talanta 173:36–43
49. Samuel ELG, Marcano DC, Berka V et al (2015) Highly efficient conversion of superoxide to oxygen using hydrophilic carbon clusters. Pro Natl Acad Sci USA 112(8):2343–2348
50. Jalilov AS, Nilewski LG, Berka V et al (2017) Perylene diimide as a precise graphene-like superoxide dismutase mimetic. ACS Nano 11(2):2024–2032
51. Zhu S, Zhao XE, You J et al (2015) Carboxylic-group-functionalized single-walled carbon nanohorns as peroxidase mimetics and their application to glucose detection. Analyst 140(18):6398–6403
52. Qian J, Yang X, Yang Z et al (2015) Multiwalled carbon nanotube@reduced graphene oxide nanoribbon heterostructure: synthesis, intrinsic peroxidase-like catalytic activity, and its application in colorimetric biosensing. J Mater Chem B 3(8):1624–1632
53. Krusic PJ, Wasserman E, Keizer PN er al (1991) Radical reactions of C_{60}. Science 254(5035):1183–1185
54. Ali SS, Hardt JI, Dugan LL (2008) SOD Activity of carboxyfullerenes predicts their neuroprotective efficacy: a structure-activity study. Nanomed Nanotechnol 4(4):283–294
55. Yang W, Huang T, Zhao M et al (2017) High peroxidase-like activity of iron and nitrogen co-doped carbon dots and its application in immunosorbent assay. Talanta 164:1–6
56. Jin J, Zhu S, Song Y et al (2016) Precisely controllable core–shell Ag@carbon dots nanoparticles: application to in situ super-sensitive monitoring of catalytic reactions. ACS Appl Mater Interfaces 8(41):27956–27965
57. Tian J, Liu Q, Asiri AM et al (2013) Ultrathin graphitic carbon nitride nanosheets: a novel peroxidase mimetic, Fe doping-mediated catalytic performance enhancement and application to rapid, highly sensitive optical detection of glucose. Nanoscale 5(23):11604–11609
58. Wang Z, Dong K, Liu Z et al (2017) Activation of biologically relevant levels of reactive oxygen species by Au/g-C3N4 hybrid nanozyme for bacteria killing and wound disinfection. Biomaterials 113:145–157
59. Zhao R, Zhao X, Gao X (2015) Molecular-level insights into intrinsic peroxidase-like activity of nanocarbon oxides. Chem Eur J 21(3):960–964
60. Dong Y, Li J, Shi L et al (2015) Iron impurities as the active sites for peroxidase-like catalytic reaction on graphene and its derivatives. ACS Appl Mater Interfaces 7(28):15403–15413
61. Zuo X, Peng C, Huang Q et al (2009) Design of a carbon nanotube/magnetic nanoparticle-based peroxidase-like nanocomplex and its application for highly efficient catalytic oxidation of phenols. Nano Res 2(8):617–623
62. Ye Y, Kong T, Yu X et al (2012) Enhanced nonenzymatic hydrogen peroxide sensing with reduced graphene oxide/ferroferric oxide nanocomposites. Talanta 89:417–421
63. Dong YL, Zhang HG, Rahman ZU et al (2012) Graphene oxide-Fe_3O_4 magnetic nanocomposites with peroxidase-like activity for colorimetric detection of glucose. Nanoscale 4(13):3969–3976
64. Wu X, Zhang Y, Han T et al (2014) Composite of graphene quantum dots and Fe_3O_4 nanoparticles: peroxidase activity and application in phenolic compound removal. RSC Adv 4(7):3299–3305
65. Qian J, Yang X, Jiang L et al (2014) Facile preparation of Fe_3O_4 nanospheres/reduced graphene oxide nanocomposites with high peroxidase-like activity for sensitive and selective colorimetric detection of acetylcholine. Sensor Actuat B-Chem 201:160–166
66. Wang H, Li S, Si Y et al (2014) Recyclable enzyme mimic of cubic Fe_3O_4 nanoparticles loaded on graphene oxide-dispersed carbon nanotubes with enhanced peroxidase-like catalysis and electrocatalysis. J Mater Chem B 2(28):4442–4448
67. Li Q, Tang G, Xiong X et al (2015) Carbon coated magnetite nanoparticles with improved water-dispersion and peroxidase-like activity for colorimetric sensing of glucose. Sens Actuat B-Chem 215:86–92

68. Zhang S, Li H, Wang Z et al (2015) A strongly coupled Au/Fe_3O_4/GO hybrid material with enhanced nanozyme activity for highly sensitive colorimetric detection, and rapid and efficient removal of Hg^{2+} in aqueous solutions. Nanoscale 7(18):8495–8502
69. Xu HY, Shi TN, Zhao H et al (2016) Heterogeneous Fenton-like discoloration of methyl orange using Fe_3O_4/MWCNTs as catalyst: process optimization by response surface methodology. Front Mater Sci 10(1):45–55
70. Yuan F, Zhao H, Zang H et al (2016) Three-dimensional graphene supported bimetallic nanocomposites with DNA regulated-flexibly switchable peroxidase-like activity. ACS Appl Mater Interfaces 8(15):9855–9864
71. Shi B, Su Y, Zhang L et al (2016) Facilely prepared Fe_3O_4/nitrogen-doped graphene quantum dot hybrids as a robust nonenzymatic catalyst for visual discrimination of phenylenediamine isomers. Nanoscale 8(20):10814–10822
72. Zhang J, Ma J, Fan X et al (2017) Graphene supported Au-Pd-Fe_3O_4 alloy trimetallic nanoparticles with peroxidase-like activities as mimic enzyme. Catal Commun 89:148–151
73. Wang Q, Zhang X, Huang L et al (2017) One-pot synthesis of Fe_3O_4 nanoparticle loaded 3D porous graphene nanocomposites with enhanced nanozyme activity for glucose detection. ACS Appl Mater Interfaces 9(8):7465–7471
74. Yousefinejad S, Rasti H, Hajebi M et al (2017) Design of C-dots/Fe_3O_4 magnetic nanocomposite as an efficient new nanozyme and its application for determination of H_2O_2 in nanomolar level. Sensor Actuat B-Chem 247:691–696
75. Lu N, Zhang M, Ding L et al (2017) Yolk-shell nanostructured Fe_3O_4@C magnetic nanoparticles with enhanced peroxidase-like activity for label-free colorimetric detection of H_2O_2 and glucose. Nanoscale 9(13):4508–4515
76. Xu C, Bing W, Wang F et al (2017) Versatile dual photoresponsive system for precise control of chemical reactions. ACS Nano ACS Nano 11(8):7770–7780
77. Hsu KI, Lien CW, Lin CH et al (2014) Immobilization of iron hydroxide/oxide on reduced graphene oxide: peroxidase-like activity and selective detection of sulfide ions. RSC Adv 4(71):37705–37713
78. Li L, Zeng C, Ai L et al (2015) Synthesis of reduced graphene oxide-iron nanoparticles with superior enzyme-mimetic activity for biosensing application. J Alloy Compd 639:470–477
79. Song L, Huang C, Zhang W et al (2016) Graphene oxide-based Fe_2O_3 hybrid enzyme mimetic with enhanced peroxidase and catalase-like activities. Colloid Surf A 506:747–755
80. Zhang Z, Hao J, Yang W et al (2013) Porous Co_3O_4 nanorods–reduced graphene oxide with intrinsic peroxidase-like activity and catalysis in the degradation of methylene blue. ACS Appl Mater Interfaces 5(9):3809–3815
81. Fan S, Zhao M, Ding L et al (2017) Preparation of Co_3O_4/crumpled graphene microsphere as peroxidase mimetic for colorimetric assay of ascorbic acid. Biosens Bioelectron 89:846–852
82. Hayat A, Haider W, Raza Y et al (2015) Colorimetric cholesterol sensor based on peroxidase like activity of zinc oxide nanoparticles incorporated carbon nanotubes. Talanta 143:157–161
83. Ragavan KV, Rastogi NK (2016) Graphene–copper oxide nanocomposite with intrinsic peroxidase activity for enhancement of chemiluminescence signals and its application for detection of Bisphenol-A. Sens Actuat B-Chem 229:570–580
84. Singh P, Nath P, Arun RK et al (2016) Novel synthesis of a mixed Cu/CuO-reduced graphene oxide nanocomposite with enhanced peroxidase-like catalytic activity for easy detection of glutathione in solution and using a paper strip. RSC Adv 6(95):92729–92738
85. Zhang L, Hai X, Xia C et al (2017) Growth of CuO nanoneedles on graphene quantum dots as peroxidase mimics for sensitive colorimetric detection of hydrogen peroxide and glucose. Sens Actuat B-Chem 248:374–384
86. Chen M, Ding Y, Gao Y et al (2017) N, N [prime or minute]-di-caboxy methyl perylene diimide (PDI) functionalized CuO nanocomposites with enhanced peroxidase-like activity and their application in visual biosensing of H_2O_2 and glucose. RSC Adv 7(41):25220–25228
87. Guo Y, Wang H, Ma X et al (2017) Fabrication of Ag–Cu_2O/reduced graphene oxide nanocomposites as surface-enhanced raman scattering substrates for in situ monitoring of peroxidase-like catalytic reaction and biosensing. ACS Appl Mater Interfaces 9(22):19074–19081

88. Wang H, Li S, Si Y et al (2014) Platinum nanocatalysts loaded on graphene oxide-dispersed carbon nanotubes with greatly enhanced peroxidase-like catalysis and electrocatalysis activities. Nanoscale 6(14):8107–8116
89. Lin XQ, Deng HH, Wu GW et al (2015) Platinum nanoparticles/graphene-oxide hybrid with excellent peroxidase-like activity and its application for cysteine detection. Analyst 140(15):5251–5256
90. Chau LY, He Q, Qin A et al (2016) Platinum nanoparticles on reduced graphene oxide as peroxidase mimetics for the colorimetric detection of specific DNA sequence. J Mater Chem B 4(23):4076–4083
91. Chen J, Ge J, Zhang L et al (2016) Poly(styrene sulfonate) and Pt bifunctionalized graphene nanosheets as an artificial enzyme to construct a colorimetric chemosensor for highly sensitive glucose detection. Sens Actuat B-Chem 233:438–444
92. Wang Y, Qi W, Song Y (2016) Antibody-free detection of protein phosphorylation using intrinsic peroxidase-like activity of platinum/carbon dot hybrid nanoparticles. Chem Commun 52(51):7994–7997
93. Li S, Li H, Chen F et al (2016) Strong coupled palladium nanoparticles decorated on magnetic graphene nanosheets as enhanced peroxidase mimetics for colorimetric detection of H_2O_2. Dyes Pigments 125:64–71
94. Liu M, Zhao H, Chen S et al (2012) Interface engineering catalytic graphene for smart colorimetric biosensing. ACS Nano 6(4):3142–3151
95. Liu M, Zhao H, Chen S et al (2012) Stimuli-responsive peroxidase mimicking at a smart graphene interface. Chem Commun 48(56):7055–7057
96. Zhang Y, Xu C, Li B et al (2013) In situ growth of positively-charged gold nanoparticles on single-walled carbon nanotubes as a highly active peroxidase mimetic and its application in biosensing. Biosens Bioelectron 43:205–210
97. Tao Y, Lin Y, Huang Z et al (2013) Incorporating graphene oxide and gold nanoclusters: a synergistic catalyst with surprisingly high peroxidase-like activity over a broad pH range and its application for cancer cell detection. Adv Mater 25(18):2594–2599
98. Chen X, Tian X, Su B et al (2014) Au nanoparticles on citrate-functionalized graphene nanosheets with a high peroxidase-like performance. Dalton Trans 43(20):7449–7454
99. Zhan L, Li CM, Wu WB et al (2014) A colorimetric immunoassay for respiratory syncytial virus detection based on gold nanoparticles-graphene oxide hybrids with mercury-enhanced peroxidase-like activity. Chem Commun 50(78):11526–11528
100. Chen X, Zhai N, Snyder JH et al (2015) Colorimetric detection of Hg^{2+} and Pb^{2+} based on peroxidase-like activity of graphene oxide-gold nanohybrids. Anal Methods-UK 7(5):1951–1957
101. Haider W, Hayat A, Raza Y et al (2015) Gold nanoparticle decorated single walled carbon nanotube nanocomposite with synergistic peroxidase like activity for d-alanine detection. RSC Adv 5(32):24853–24858
102. Zheng C, Ke W, Yin T et al (2016) Intrinsic peroxidase-like activity and the catalytic mechanism of gold@carbon dots nanocomposites. RSC Adv 6(42):35280–35286
103. Kumar S, Bhushan P, Bhattacharya S (2016) Development of a paper-based analytical device for colorimetric detection of uric acid using gold nanoparticles-graphene oxide (AuNPs-GO) conjugates. Anal Methods-UK 8(38):6965–6973
104. Ahmed SR, Takemeura K, Li TC et al (2017) Size-controlled preparation of peroxidase-like graphene-gold nanoparticle hybrids for the visible detection of norovirus-like particles. Biosens Bioelectron 87:558–565
105. Ju J, Zhang R, Chen W (2016) Photochemical deposition of surface-clean silver nanoparticles on nitrogen-doped graphene quantum dots for sensitive colorimetric detection of glutathione. Sens Actuat B-Chem 228:66–73
106. Tian X, Wang X, Dai C (2017) Visual and quantitative detection of glucose based on the intrinsic peroxidase-like activity of $CoSe_2$/rGO nanohybrids. Sens Actuat B-Chem 245:221–229

107. Nie G, Zhang L, Lu X et al (2013) A one-pot and in situ synthesis of CuS-graphene nanosheet composites with enhanced peroxidase-like catalytic activity. Dalton Trans 42(38):14006–14013
108. Dutta S, Ray C, Mallick S et al (2015) A gel-based approach to design hierarchical CuS decorated reduced graphene oxide nanosheets for enhanced peroxidase-like activity leading to colorimetric detection of dopamine. J Phys Chem C 119(41):23790–23800
109. Chen X, Su B, Cai Z et al (2014) PtPd nanodendrites supported on graphene nanosheets: a peroxidase-like catalyst for colorimetric detection of H_2O_2. Sens Actuat B-Chem 201:286–292
110. Wang A, Zhao H, Chen X et al (2017) A colorimetric aptasensor for sulfadimethoxine detection based on peroxidase-like activity of graphene/nickel@palladium hybrids. Anal Biochem 525:92–99
111. Chen H, Li Y, Zhang F et al (2011) Graphene supported Au-Pd bimetallic nanoparticles with core-shell structures and superior peroxidase-like activities. J Mater Chem 21(44):17658–17661
112. Yang L, Liu X, Lu Q et al (2016) Catalytic and peroxidase-like activity of carbon based-AuPd bimetallic nanocomposite produced using carbon dots as the reductant. Anal Chim Acta 930:23–30
113. Darabdhara G, Sharma B, Das MR et al (2017) Cu-Ag bimetallic nanoparticles on reduced graphene oxide nanosheets as peroxidase mimic for glucose and ascorbic acid detection. Sens Actuat B-Chem 238:842–851
114. Ren H, Wang C, Zhang J et al (2010) DNA cleavage system of nanosized graphene oxide sheets and copper ions. ACS Nano 4(12):7169–7174
115. Zhou X, Zhang Y, Wang C et al (2012) Photo-fenton reaction of graphene oxide: a new strategy to prepare graphene quantum dots for DNA Cleavage. ACS Nano 6(8):6592–6599
116. Wang S, Cazelles R, Liao WC et al (2017) Mimicking horseradish peroxidase and NADH peroxidase by heterogeneous Cu^{2+}-Modified graphene oxide nanoparticles. Nano Lett 17(3):2043–2048
117. Guo Y, Deng L, Li J et al (2011) Hemin-graphene hybrid nanosheets with intrinsic peroxidase-like activity for label-free colorimetric detection of single-nucleotide polymorphism. ACS Nano 5(2):1282–1290
118. Xue T, Jiang S, Qu Y et al (2012) Graphene-supported hemin as a highly active biomimetic oxidation catalyst. Angew Chem Int Ed 51(16):3822–3825
119. Zhang Y, Xu C, Li B (2013) Self-assembly of hemin on carbon nanotube as highly active peroxidase mimetic and its application for biosensing. RSC Adv 3(17):6044–6050
120. Zhang Y, Xia Z, Liu H et al (2013) Hemin-graphene oxide-pristine carbon nanotubes complexes with intrinsic peroxidase-like activity for the detection of H_2O_2 and simultaneous determination for Trp, AA, DA, and UA. Sens Actuat B-Chem 188:496–501
121. Tao Y, Lin Y, Ren J et al (2013) Self-assembled, functionalized graphene and DNA as a universal platform for colorimetric assays. Biomaterials 34(20):4810–4817
122. Sun R, Wang Y, Ni Y et al (2014) Spectrophotometric analysis of phenols, which involves a hemin-graphene hybrid nanoparticles with peroxidase-like activity. J Hazard Mater 266:60–67
123. Bi S, Zhao T, Jia X et al (2014) Magnetic graphene oxide-supported hemin as peroxidase probe for sensitive detection of thiols in extracts of cancer cells. Biosens Bioelectron 57:110–116
124. Fang M, Wang L, Wang P et al (2016) Selective and sensitive determination of copper ions in soft drink based on high catalysis of hemin–graphene hybrid nanosheets coupled with enzyme inhibitions. J Iran Chem Soc 13(10):1937–1944
125. Magerusan L, Socaci C, Pogacean F et al (2016) Enhancement of peroxidase-like activity of N-doped graphene assembled with iron-tetrapyridylporphyrin. RSC Adv 6(83):79497–79506
126. Xu X, Wei M, Liu Y et al (2017) A simple, fast, label-free colorimetric method for detection of telomerase activity in urine by using hemin-graphene conjugates. Biosens Bioelectron 87:600–606
127. Wang X, Hou C, Qiu W et al (2017) Protein-directed synthesis of bifunctional adsorbent-catalytic hemin-graphene nanosheets for highly efficient removal of dye pollutants via synergistic adsorption and degradation. ACS Appl Mater Interfaces 9(1):684–692

128. Xu C, Zhao C, Li M et al (2014) Artificial evolution of graphene oxide chemzyme with enantioselectivity and near-infrared photothermal effect for cascade biocatalysis reactions. Small 10(9):1841–1847
129. Zhang R, He S, Zhang C, Chen W (2015) Three-dimensional Fe- and N-incorporated carbon structures as peroxidase mimics for fluorescence detection of hydrogen peroxide and glucose. J Mater Chem B 3(20):4146–4154
130. He Y, Niu X, Shi L et al (2017) Photometric determination of free cholesterol via cholesterol oxidase and carbon nanotube supported Prussian blue as a peroxidase mimic. Microchim Acta 184(7):2181–2189
131. Chen C, Wang Z, Ren J, Qu X (2018) Enzyme mimicry for combating bacteria and biofilms. Acc Chem Res 51:789–799
132. Huang Y, Ren J, Qu X (2019) Nanozymes: classification, catalytic mechanisms, activity regulation, and applications. Chem Rev 119:4357–4412

Chapter 8
Functional Enzyme Mimics for Oxidative Halogenation Reactions that Combat Biofilm Formation

Karoline Herget, Hajo Frerichs, Felix Pfitzner, Muhammad Nawaz Tahir and Wolfgang Tremel

Abbreviations (Alphabetic)

AHL	Acylated homoserine lactones
AI	Autoinducers
AIP	Autoinducing peptide
CB	Celestine blue
CSP	Competence-stimulating peptide
DAHL	Dibromoacetyl homoserine lactone
DBHL	Dibromo-homoserine lactone
DHP	2,3-Dihydro-4*H*-pyran
DMP	3,5-Dimethylphenol
FAAH	Fatty acid amide hydrolase
FAD	Flavin adenine dinucleotide
FDA	5′-fluoro-5′-deoxyadenosine (5′-FDA)
GFP	Green fluorescent protein
HPA	Heteropolyacid
HPLC-MS	High-performance liquid chromatography–mass spectrometry
HPO	Haloperoxidase
HO	Halogenase
IPO	Iodoperoxidase
LDH	Layered double hydroxides
MBHL	Monobromo–homoserine lactone

K. Herget · H. Frerichs · F. Pfitzner · W. Tremel (✉)
Institut für Anorganische Chemie und Analytische Chemie, Johannes Gutenberg-Universität, Duesbergweg 10-14, 55099 Mainz, Germany
e-mail: tremel@uni-mainz.de

M. N. Tahir
Department of Chemistry, King Fahd University of Petroleum and Minerals, P.O. Box 5048, Dhahran 31261, Kingdom of Saudi Arabia

X. Yan (ed.), *Nanozymology*, Nanostructure Science and Technology,
https://doi.org/10.1007/978-981-15-1490-6_8

MCD	Monochlorodimedone
MCM	Mobil Composition of Matter, mesoporous silica
MMO	Methane monooxygenases
NAD(P)H	Nicotinamide adenine dinucleotide phosphate
NMR	Nuclear magnetic resonance
OHHL	N-3-oxohexanoyl homoserine lactone
PET	Positron emission tomography
POM	Polyoxometalate
PR	Phenol red
PS	Polystyrene
QS	Quorum sensing
RNA	Ribonucleic acid
REACH	Registration, Evaluation, Authorisation of Chemicals
RNS	Reactive nitrogen species
ROS	Reactive oxygen species
SaA	Salicyl aldehyde
SAM	S-adenosyl-L-methionine (SAM)
SAH	*S*-adenosylhomocysteine (SAH)
SEM	Scanning electron microscopy
TB	Thymol blue
TEM	Transmission electron microscopy
TMB	1,3,5-Trimethoxybenzene
TNB	2-Nitro-5-thiobenzoate
TOF	Turnover frequency
TON	Turnover number
VCPO	Vanadium-dependent haloperoxidase
VHPO	Vanadium-dependent haloperoxidase
VBPO	Vanadium-dependent bromoperoxidase
XPS	X-ray photoelectron spectroscopy
ZSM-5	Zeolite Socony Mobil-5, an aluminosilicate zeolite

8.1 Introduction

Catalysis is at the heart of chemistry as it provides tools for efficiently and selectively making and breaking chemical bonds. It is the basis of large-scale processes in bulk chemistry and petrochemistry. Future environmental requirements will necessitate new catalytic solutions. Enzymes are among the most effective catalysts known in terms of their efficiency and selectivity. They can achieve rate accelerations by several orders of magnitude and sometimes react exclusively with only one single stereoisomer of a substrate. The number of industrial applications has increased, mainly because of advances in protein engineering and environmental and economic necessities. A number of enzyme-based processes have been commercialized for producing valuable products [1–4]. Their industrial applications are still restricted because of stability, catalytic efficiency, and specificity. Enzyme engineering allowed to overcome such shortcomings and to extend their substrate ranges [5, 6].

Chemists have tried to mimic enzymes with enzyme models. Breslow originally coined the terms "biomimetic chemistry" and "artificial enzyme" to describe this field [7]. Enzymes bind their substrates, and catalysis is achieved with the aid of two or more functional groups. This allows for substrate selectivity, reaction selectivity, and stereoselectivity. Binding occurs by metal coordination, ion pairing, metal or Lewis acid–base coordination, and hydrophobic interactions with roughly half the known enzymes using metal ions. The other half relies on organic functional groups. Stability is still a key problem.

To engineer the stability of a protein, modifications at the active site are a good starting point. This may appear paradoxical, as substitutions at the active site typically reduce the activity of a mutant protein. Still, functional artificial enzyme models for different types and classes of enzymes have been developed, some mimicking the mode of action of their natural counterparts while others just adopt some general features. Attempts to develop genuine enzyme-mimetic systems have been successful in achieving substrate binding, but less successful in accomplishing high turnover rates. Large and ingeniously designed molecular binding structures have been synthesized for the purpose of catalysis [8–10].

Based on reports that nanomaterials can exhibit enzyme-like activities and show in vitro biocompatibility, new nanomaterial-based artificial enzymes have been investigated that might have the potential to be functional even inside cells or living organisms [11–14]. An important question is, whether inorganic nanomaterials can mimic the high catalytic efficiency of their natural analogues which speed up a wide variety of different reactions by binding the substrates with exceptional specificity [15–25]. The activity and specificity of natural enzymes depend on the specific reaction conditions like temperature, pH, and the chemical structure of the substrate, but enzymes suffer typically from low stability, short shelf life, and high production costs which limits their versatility compared to nanoparticle mimics. In contrast, nanoparticles can be surface functionalized for specific targeting and tolerate changes in reaction conditions more easily [26, 27]. Furthermore, nanoparticle mimics can operate in nonaqueous systems, and they can be manufactured cost-efficiently up to industrial scale.

One of the key features related to nanoparticles is their enhanced chemical activity due to their large surface area which leads to high catalytic activities. Great efforts have been made in recent years to identify new materials with enzyme-like activities. An exceeding number of reports on peroxidase nanoparticle mimics have been reported [15, 28], but very few other enzymatic systems have been explored so far. A number of enzymes have been particularly intriguing because they catalyze oxidation and/or halogenation reactions that hold great promise for synthetic applications.

Nature oxidizes methane to methanol enzymatically in a single step under aerobic conditions with methane monooxygenases (MMO), a group of enzymes with active sites containing iron– or copper–oxygen centers [29–31]. Industrial processes that convert alkanes, in particular, methane, to more valuable products operate at high temperature and pressure and lead to a variety of products. The activity of MMOs has been mimicked at oxygen-activated iron sites [32], and oxygen-activated copper sites in zeolite Cu-ZSM-5 selectively convert methane to methanol at low temperatures (100 °C) [33, 34].

Another striking examples are cytochrome P450 enzymes which catalyze the insertion of one atom of an oxygen molecule into the aliphatic position of an organic substrate (RH), while the other one is reduced to water. Cytochromes P-450 are a superfamily of heme-type mixed-function oxidases of animal or microbial origin. The natural use of P450 requires cofactors such as NAD(P)H and electron transfer proteins such as P450-reductase, which is one of the main obstacles to commercial implementation. Regeneration systems are available to solve the cofactor problem [35, 36], but these electron transport chains are more difficult to optimize and do not have the robustness required for a chemical transformation at large scale. This is a reason why cofactor-dependent oxidoreductases are lagging behind "simple" cofactor-independent enzymes (e.g., hydrolases, oxidases, peroxidases) considering their implementation in synthetic organic chemistry. The catalytic activities of P450 for the substrate conversion are maintained when a cofactor is replaced by an electrode or metal nanoparticles as an electron donor to P450 dissolved in solution, thereby shortcutting the complex natural electron transport chain [37–39]. A redox mediator between the electrode and P450 may or may not be required depending on systems [40–47].

Cytochrome P450 enzymes have promise in the synthesis and discovery of drugs and in drug development. The enzymes have been used already in the 1950s to introduce functional groups into drugs that would be difficult to introduce chemically [48]. The success of P450 enzymes as selective biocatalysts for oxygenation reactions of complex substrates has stimulated the study of metalloporphyrin model compounds that mimic P450 reactivity. Key intermediates of metal porphyrin-catalyzed oxygenation reactions include oxo- and dioxo-metal(V) species that transfer an oxygen atom to the substrate through a hydrogen abstraction/oxygen recombination pathway known as the oxygen rebound mechanism [49].

Important in the context of this review is that carbon–halogen bond formations, including fluorination reactions, are also catalyzed by manganese porphyrins and related salen complexes. Biphasic sodium hypochlorite/manganese porphyrin systems convert even nonactivated aliphatic C–H to C–Cl bonds efficiently and selectively [50, 51]. The oxygen rebound rate in Mn-mediated hydroxylation is correlated with the nature of the trans-axial ligands L bound to the manganese center ($L{-}Mn^{V}{-}O$). In particular, Mn-catalyzed aliphatic C–H fluorination reactions could be carried out with simple fluorides [52, 53]. Very few enzymes have been described that can convert fluoride to organic fluorine [54, 55]. *Streptomyces cattleya* have evolved a fluorinase that overcomes the chemical challenges of using aqueous fluoride to form carbon–fluorine bonds [56]. It catalyzes an S_N2-type nucleophilic substitution at the C-5′ position of S-adenosyl-L-methionine (SAM) and fluoride ions to form 5′-fluoro-5′-deoxyadenosine (5′-FDA) and L-methionine as neutral leaving group [57]. The fluorinase-catalyzed reaction shows a significant rate enhancement (acceleration by a factor of 10^6 to 10^{15}).

Halogenation reactions belong to the standard toolbox of synthetic organic chemistry because the polarity makes C–X bonds reactive enough to replace the halogen (X) by other substituents (Y). Halogenations are, however, as free-radical reactions unselective. In industry, halogenation reactions are typically performed with hazardous, toxic, and corrosive halogenating reagents, often in chlorinated solvents [58–60]. Vinyl chloride, for example, is synthesized by oxidative halogenation in a gas-phase reaction from hydrochloric acid, oxygen, ethylene, and a heterogeneous copper catalyst [61]. N-bromo- and N-chlorosuccinimide are convenient sources of chlorine and bromine radicals for radical substitution and electrophilic addition reactions. Late-stage fluorination is a popular and growing technology in medicinal chemistry with applications such as blocking site-specific metabolism and development of potential positron emission tomography (PET) ligands [62, 63].

Nature relies on the oxidative halogenation approach by making the halogenating agent in situ from halide anions under acidic conditions. This leads to highly efficient halogenations with a 100% halogen atom economy. Suitable oxidants are either hydrogen peroxide or oxygen. Therefore, biochemical halogenation reactions catalyzed by HPOs and halogenases (HGs) have attracted growing interest in academic and industrial research [58, 59, 64–71]. These enzymes are considered "green" catalysts for the synthesis of antiviral and anticancer drugs, but also with respect to their antimicrobial activity. Fluorinated compounds [62, 63] and halogenated [72–75] natural products are described elsewhere [76–78].

In recent years, HPO mimics and the corresponding halogenated products have been investigated because of better cost economy, upscaling, and stability under operating conditions (pH, salt concentrations, organic solvents) [79] compared to native enzymes. We review halogenating enzymes and relevant assays to monitor oxidative halogenation reactions and highlight new functional HPO mimics for the halogenation of organic compounds and their potential to combat biofouling.

8.2 Halogenating Enzymes

Natural halogenating enzymes encountered in eukaryotes and prokaryotes [80] have versatile applications [81]. Many halogenating enzymes in bacteria, archaea, fungi, and algae are vanadium-dependent, and many hosts are marine organisms. These enzymes catalyze the electrophilic oxidative halogenation of organic compounds via reactive oxygen species (ROS) [82, 83]. (Pseudo-)halides, especially Cl^-, Br^-, I^-, and SCN^- are oxidized in two-electron transitions to the corresponding hypohalous acids or hypohalites [84]. Depending on the required co-oxidants, the enzymes are classified into haloperoxidases (HPOs) [85, 86] with H_2O_2 and halogenases (HGs) using O_2 as co-substrate.

With regard to the nature of their active centers, halogenating enzymes can be subdivided into (i) flavin-, non-heme iron(II)- and S-adenosyl-L-methionine (SAM)-dependent HGs (F-, NI-, S-HG) and (ii) cofactor-free-, heme iron(II)- or vanadium-dependent HPOs (HPO, V-, HI-HPO) [87]. These six families of halogenating enzymes are suggested to have evolutionary and functional relationships to the α/β hydrolases, phosphatases, peroxidases, oxidoreductases, and the SAM hydroxide adenosyltransferases [87]. According to the most electronegative halide that the enzymes can oxidize they are labeled chloroperoxidase (CPO, substrate Cl^-, Br^-, I^-, e.g., myeloperoxidase in neutrophils [88]), bromoperoxidase (BPO, substrate: Br^-, I^-, e.g., eosinophil peroxidase in phagocytes [89], and lactoperoxidase in human exocrine secretions [90]) and iodoperoxidase (IPO, substrate: I^-, e.g., thyroid peroxidase [91]). The highly reactive intermediates, e.g., hypohalous acids (HOX, X = Cl, Br, I, CN, SCN), react subsequently with different nucleophilic acceptors (R-H) to form a variety of halogenated metabolic compounds (R-X, X = (pseudo-) halogen, Eqs. 8.1 and 8.2). When the nucleophilic acceptors contain amino groups, halo-amines are formed (Eq. 8.3) [66]. This reaction occurs mainly in the biological inflammatory response involving myeloperoxidase as halogenating enzyme [92–94]. The resulting chloramines are uncharged, reactive halogenating, and oxidizing agents [93, 95], and can diffuse through the membranes of bacteria like *Escherichia coli* [96].

$$R-H+H_2O_2+X^-+H^+ \xrightarrow[\text{“HOX”}]{\text{HPO/HPO-mimetics}} X-R+2H_2O \qquad (8.1)$$

$$R-H+O_2+X^-+H^+ \xrightarrow[\text{“HOX”}]{\text{HG/HG-mimetics}} X-R+2H_2O \qquad (8.2)$$

$$R_2NH+HOX \longrightarrow R_2NX+H_2O \qquad (8.3)$$

In the absence of an organic acceptor, at alkaline pH values and at high H_2O_2 concentrations, singlet oxygen is formed ($^1O_2$) by HPOs via proxobromide intermediates (Eq. 8.4). Singlet oxygen formation can be detected by its chemiluminescence at 1268 nm ($^1O_2(^1\Delta_g) \rightarrow {}^3O_2\left(^3\sum_{g^-}\right)$ transition) [85, 97–104].

$$HOX+H_2O_2 \xrightarrow{\text{HPO/HPO-mimetics}} {}^1O_2+H^++X^-+H_2O \qquad (8.4)$$

The highly reactive intermediates, in particular, hypohalite and hypohalous compounds (e.g., HOX), halo-amines (R_2NX), and singlet oxygen ($^1O_2$), participate in oxidative halogenation reactions at both electron-rich and non-activated compounds as well as in “simple” oxidation reactions [105]. Halogenating enzymes have been identified as important biocatalysts during the formation of reactive precursors for natural products [106] or metabolic reactions [107, 108], e.g., via epoxidation [105, 109, 110] or cyclization reactions, enantioselective sulfoxidations [105, 107], in the oxidation [111, 112] of ascorbate, urate, pyridine nucleotides, tryptophan, in halohydrin formation as well as in oxidative stress response and signal transduction [113]. The respective enzymes are involved in reactions leading to the formation of (volatile) halogenated compounds in the atmosphere [114–117] (e.g., halomethane [82, 118–121]) and in specific host defense and antimicrobial systems [88]. HPOs and HGs participate in the biosynthesis of natural products such as halogenated metabolites

[76, 83, 102, 106–108, 122–126] like terpenes, indoles, phenols, and others derived from amino acids. In mammals, they are responsible for the biosynthesis of the essential tyrosine-based thyroid hormones thyroxine (T4, **11**) and triiodothyronine (T3, **12**) [127, 128]. Mammalian laccases synthesize iodovanillin with antifungal activity [129]. Halogenating enzymes in marine organisms are more popular because the bioavailability of the reagents and cofactors (e.g., vanadium, sources of bromine and oxidants, H_2O_2, and O_2) are omnipresent. Although chlorinated [130] and iodinated [114, 128] products have great application potential, this review is confined to oxidative bromination reactions because the resulting products are easier to synthesize in industrial processes.

An example, isolated from marine algae, catalyzes the oxidative bromination of the monoterpenes geraniol and nerolidol. In a subsequent step, these compounds react to bromocyclic structures encountered in many marine metabolites [122] (Fig. 8.1).

The enzymes are moderately substrate- and stereo(regio)-selective [82, 131], as indicated by the regiospecific oxidation of 1,3-di-tert-butylindole [82], the competition kinetics for the oxidative bromination of phenol red in combination with geraniol or nerolidol [124], asymmetric halogenation reactions yielding α, β, or γ-synderol [124], and enantioselective sulfoxidation reactions [107]. This is accomplished with the aid of custom-made substrate pockets and tunnels buffed with hydrophobic amino acids at the catalytic center. In the X-ray structure of VBPO from *Corallina officinalis,* hydrophilic residues are located close to the entrance of the active site [81, 132]. This protein scaffold pushes the organic substrate to the active center, where it promotes the reaction by providing an electrophilic reactive bromine atom in a non-nucleophilic environment. In addition, it protects the local environment from an uncontrolled release of hypohalous species [107]. A lack of selectivity

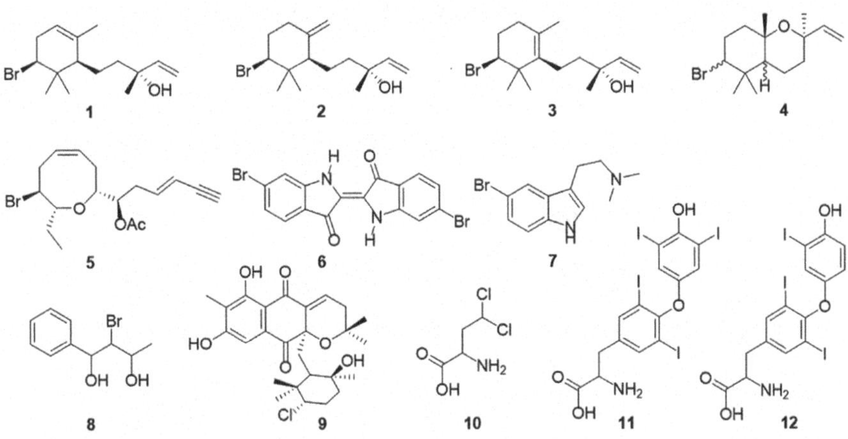

Fig. 8.1 Examples of natural halogenated compounds with biological activity [83, 102, 106, 108, 123, 126, 536]. **1** α-, **2** β-, **3** γ-synderol [122, 124]; **4** 3-β-bromo-8- epicarrapioxide [124]; **5** laurencin [122]; **6** tyrion purple [102]; **7** 5-bromo-*N,N*-dimethyltryptamine [102]; **8** bromohydrin [107]; **9** napyradiomycin (derivative) [537]; **10** armentomycin [190, 537]; **11** thyredoxine (T4) [83, 128]; **12** triiodothyronine (T3) [83, 128]

occurred (i) when the substrates are present only in substoichiometric amounts compared to the hypohalous species [133], and (ii) when oxidized bromine species (e.g., HOBr, OBr^-, Br_2, Br_3^-) [124] were present outside of the active enzyme site [134, 135]. Martinez et al. [82] suggested HPOs to have different selectivities for different substrates: Some compounds bind at the active site and block the release of oxidized intermediates [82, 102, 107, 136]. As a highly reactive and diffusible molecule [135], unblocked HOX is even more beneficial: hypohalite ions play a critical role in the natural protection and host defense systems of organisms, especially in marine plants such as red and brown seaweed (Fig. 8.2).

Although heme iron chloroperoxidase (CPO) was the first halogenating enzyme to be discovered from the terrestrial fungus *Caldariomyces fumago* [118], vanadium-dependent HPOs are the most prominent halogenating enzymes. They occur in a variety of marine organisms and are suggested to be involved in the synthesis of a broad variety of halogenated marine natural products. Most chemical enzyme mimics are based on VHPOs. The similarity of the HPO reaction mechanism to that proposed for many homogeneous catalysts prompted the study of structural and functional model compounds, and VHPOs are described in more detail below together with activity assays, enzyme mimics, and applications. The cofactor and catalytic center of VHPO, particularly the trigonal-bipyramidally coordinated orthovanadate unit, is stabilized through a highly complex hydrogen-bonded network from amino acids such as histidine, lysine, arginine, and serine (Fig. 8.3, **I**) [86]. The active sites

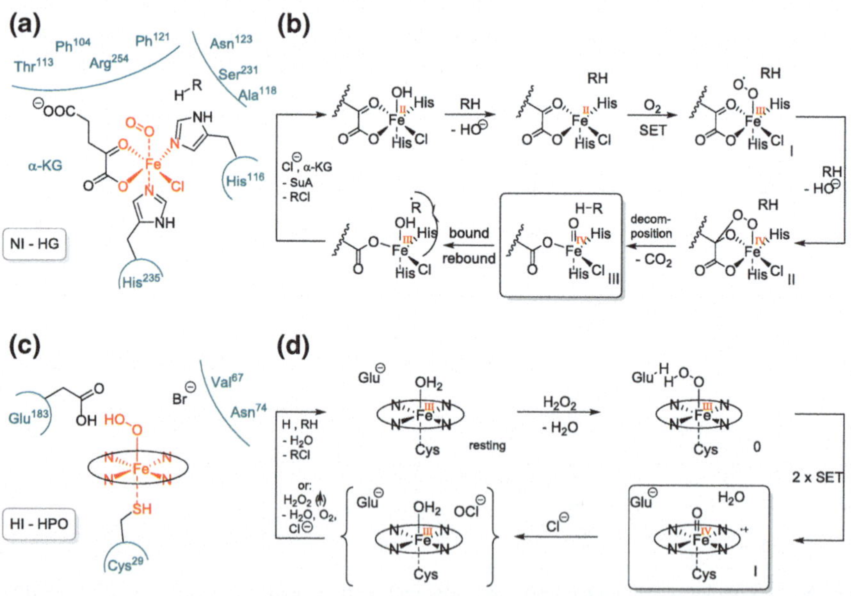

Fig. 8.2 Active centers of halogenating enzymes and proposed mechanisms of HOX formation for **a, b** iron non-heme HG (NI-HG) [70, 139, 190, 378, 538] and **c, d** iron-hemechloroperoxidase (HI-CPO, in general: HI-HPO) [70, 131, 539, 540]

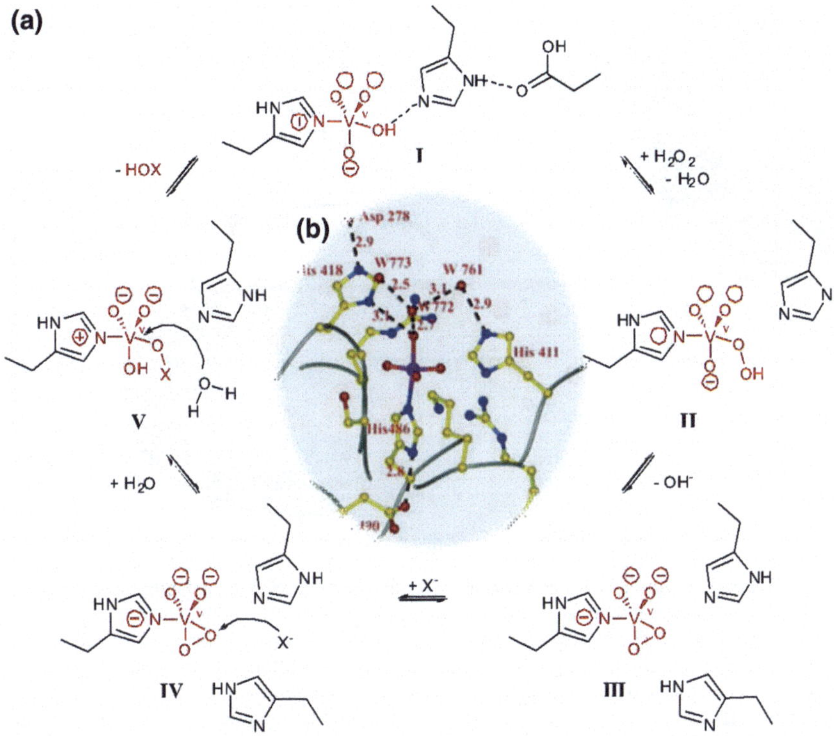

Fig. 8.3 (a) Proposed HOX formation mechanism and (b) active center of vanadium-dependent bromoperoxidase from *A. nodosum* [79, 103, 107, 134]. The active site reproduced with permission from Ref. [103] Copyright (1999) Elsevier. The orthovanadate unit is highlighted in red, and the different postulated steps of H_2O_2 binding (**I–III**), halide ($X^- = Cl^-, Br^-, I^-$) attack (**IV**), and HOX (**V–I**) release are shown. In each step, vanadium(V) retains its oxidation state. For the sake of simplicity, the orthovanadate group is shown in its highest deprotonated state and the surrounding protein and hydrogen-bonded network is only indicated

of different HPOs are structurally almost superimposable, especially the conserved distal histidine appears in every HPO structure [103, 132, 137]. Interestingly, the conserved active sites are virtually identical to those found in acid phosphatases, which indicates an evolutionary relationship [137, 141]. The activity differences between VBPO and VCPO are believed to result from slight structural differences of the active centers. As an example, replacing a histidine group in VBPO by a phenylalanine unit in VCPO [142, 143] leads to a change in hydrogen bonding and thus reactivity [143, 144].

A kinetic study of the VHPO-catalyzed halide oxidation to the corresponding hypohalous species under steady-state conditions revealed a high degree of similarity between different bromoperoxidases [79, 81, 134, 145]. Figure 8.4 shows the catalytic center of VBPO of *Ascophyllum nodosum* and the proposed reaction mechanism for the formation of HOX [146]. Although the biochemical pathways are not

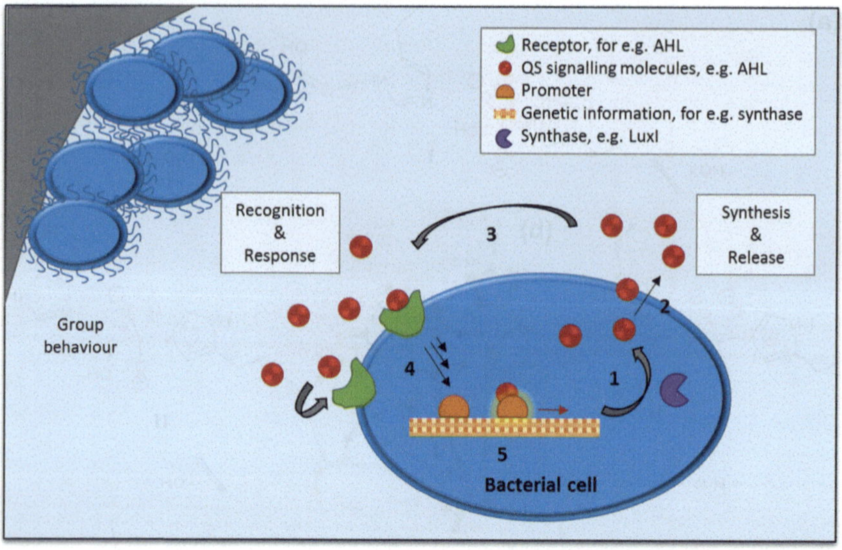

Fig. 8.4 Natural quorum-sensing mechanism of marine bacteria via QS signaling molecules (e.g., AHL) that regulate and coordinate group behavior (e.g., biofilm formation, bacterial attachment, swarming mobility, virulence, antibiotic resistance, synthesis of organic compounds) for the marine bacterial symbiont *Vibrio fischeri* [160]. The numbers represent different steps in the QS process of bacteria. AI molecules **1** pass the membrane **2** via diffusion (short-chain AHLs) or active transport (long-chain AHLs). **3** Receptor binding of AHLs leads to multiple responses which can affect gene expression **4**, e.g., activation of promoter complexes **5,** and stimulate the synthesis of signaling molecules **1** or bioluminescent phenotypes [159, 160]

fully understood, DFT investigations [147, 148] of the vanadium cofactor of VHPO suggest protonation to be an important step. The vanadium(V) site binds a peroxo species (η^2-peroxo) in a side-on fashion without changing its oxidation state (Fig. 8.3, **I–III**), i.e., it acts as a Lewis acid by expanding its coordination sphere. The intermediate complex can be attacked by halides (e.g., Br^-) (**IV**), and HOX is released after hydrolysis (**V**). In summary, an "ionic mechanism" was postulated involving an electrophilic "Br^+" rather than radical "Br·"species as for the Fe-dependent HPOs or HOs [79, 149, 150].

This is associated with vanadium-dependent bromo- or chloroperoxidases (BPO, CPO) [83], prevalent in all classes of marine algae. Their structure contains dodecamers (six homodimers, occurring mainly in red seaweed like *Carex pilulifera, Corallina officinalis*) or single, "homodimeric" forms (fungus *Curvularia inaequalis*, brown seaweed *Ascophyllum nodosum*) [137]. Figure 8.2 compares the "active states" of the catalytic centers of these three halogenating metal enzymes, where "active state" refers to the proposed redox state and environment of the metal center just before the halide attack. The structural differences between non-heme iron HG, heme iron, or vanadium HPOs give rise to an enhanced stability of the heme-independent enzymes and different HOX formation mechanisms.

Vanadium haloperoxidases (VHPO) are chemically more stable than heme iron HPOs, because deactivation of the active site by oxidation is more difficult [79, 98, 138]. The structure of the non-heme iron HG requires a stepwise radicalic HOX formation. Similar as for the iron-dependent peroxidases, a high-valent non-heme iron(IV)-oxo intermediate is formed (Fig. 8.2a, b), and a substrate is halogenated in a bound–rebound mechanism by another co-substrate (e.g., α-ketoglurate) [139, 140]. In contrast, heme iron- or vanadium-dependent HPOs catalyze electrophilic halogenations, where the porphyrin complex is redox active (Fig. 8.1c, d), while the vanadium metal center does not change its oxidation state (Fig. 8.2) [139].

8.3 Antimicrobial Activity of HPOs

Oxidative bleaching and the antimicrobial activity of singlet oxygen and HOX generated with the aid of halogenating enzymes have been reported and patented years ago [66, 88, 151–153]. HPOs of marine organisms have been studied with regard to their "antifouling activity" to combat bacterial attack, adhesion, and colonization that is controlled by quorum sensing (QS), i.e., cell-to-cell communication in bacteria.

Many bacteria benefit from social interactions. Intercellular signaling provides an opportunity to interfere with their ability to coordinate efforts to invade human, animal, or plant hosts. Communication interference in the microbial world is a ploy to gain an advantage over competitors. QS relies on small, secreted signaling molecules known as autoinducers (AIs) to initiate coordinated responses across a population. In many cases, the responses elicited by QS contribute to pathogenesis through the synchronized production of virulence determinants, like toxins, proteases, and other immune-evasive factors. Furthermore, QS can enable bacteria to resist antimicrobial compounds or drugs, such as biofilm development. If these efforts to coordinate are blocked, bacteria lose their ability to carry out an organized assault on the host's defense or immune system or would be less able to form organized community structures promoting pathogenesis, such as biofilms [154–158].

The most common AIs belong to one of the following three categories: acylated homoserine lactones (AHLs), used by Gram-negative bacteria (also referred to as autoinducer-1 [AI-1]), peptide signals, used by Gram-positive bacteria, and autoinducer-2 (AI-2 furanosyl compounds, e.g., furanosyl borate) [140, 159] used by Gram-negative and Gram-positive bacteria. There are other quorum-sensing signals that go beyond these classes. Figure 8.4 summarizes the mode of action of AI compounds for the AHL regulatory system of the marine symbiotic bacterium *Vibrio fischeri* [159–161].

AHL-Based Quorum Sensing. AHLs are synthesized and released by bacterial cells via different pathways (Fig. 8.4, **1+2**), and their structure determines their signaling function and their modes of interaction with environmental factors during cell-to-cell transit [162]. Typically, AHLs contain a five-membered ring with different amide-linked side chains and a range of 4–18 carbons in length, but many structural variants of the basic AHL molecule have been described. The acyl side chain may be saturated or unsaturated and contain a substituted hydroxyl- or oxo-functional group at the C3 position [163, 164]. QS depends on signaling molecules which are stable on the timescale of diffusion through the extracellular matrix, i.e., quorum-sensing regulated gene expression depends on factors like the rate of cue degradation as a function of environmental conditions, calling distance, and mobility of the cue [165].

Extracellular AI compounds have cognate receptors (Fig. 8.4, **3**), thereby regulating specific response mechanisms via modification of gene expressions (Fig. 8.4, **4+5**). The autoinducer synthase gene is a target for Lux proteins and their homologues, common in Gram-negative bacteria [161]. Activation of the quorum-sensing cascade results in increased expression of autoinducer synthase, leading to the production of more AHLs. This acts as a positive feedback loop and amplifies the quorum-sensing effect [160, 161, 166]. The synthesis of new AHL autoinducers is triggered by the translation of LuxI enzymes. Receptor binding and thereby stimulating the AI-dependent transcriptional activator LuxR leads to an enhanced synthesis of signaling molecules via LuxI synthase or bioluminescent phenotypes [167]. This model roughly describes the QS mechanism in bacteria (e.g., possible crosstalk). LuxR and LuxI homologues are encountered in different species [160].

Peptide-Based Quorum Sensing. Gram-positive bacteria do not have LuxI or LuxR homologues. They rely on modified oligopeptides as autoinducer molecules. These are genetically encoded and generated ribosomally within the cell. As these peptides cannot permeate the cell membrane, specialized transporters are needed for active transport. At points between translation, export, and detection, peptides are subject to various modifications, including processing and/or cyclization. Detection of these peptide signals can occur either at the cell surface or inside the cells. Many peptide autoinducers are detected by a membrane-bound sensor kinase, which switches its kinase/phosphatase activity in response to the interaction with a peptide, which changes the phosphorylation state of the cognate response regulator and results eventually in the activation or repression of QS target genes. Systems with extracellular detection include the *agr* system of *Staphylococcus aureus* and the *fsr* system of *Enterococcus faecalis*, both of which control virulence factor production. The *agr* system of *S. aureus* is based on cyclic autoinducing peptides (AIPs) that belong to four distinct groups that interact with cognate AgrC sensor kinases of the same group to regulate exotoxin production and biofilm dispersal [168]. The *fsr* system uses a different cyclic peptide, gelatinase biosynthesis-activating pheromone (GBAP), which is detected by the sensor kinase FsrC and induces the formation of gelatinase [169]. Other bacteria rely on linear peptide autoinducers that are detected extracellularly, including the competence-inducing QS system of *Streptococcus pneumoniae,* which is mediated by the competence-stimulating peptide (CSP) [170]. Alternatively, some linear-peptide-based QS systems actively transport the autoinducers back into the cell where the peptide signal can interact directly with a cognate regulator to modify target gene expression. A typical example is the PrgX system of *E. Facaelis* [171].

AI-2-Based Quorum Sensing. AI-2 is formed from the *S*-adenosylhomocysteine (SAH) precursor by the sequential enzymatic activities of 5-methylthioadenosine/*S*-adenosylhomocysteine nucleosidase and the metalloenzyme LuxS [172]. 4,5-Dihydroxy-2,3-pentanedione (DPD), the product of these reactions, is unstable in aqueous solution and undergoes spontaneous rearrangement into multiple interconvertible cyclic furanones of the AI-2 group. AI-2 can freely diffuse out of both Gram-negative and Gram-positive bacteria and, as for all other QS signals, accumulates extracellularly. Many species produce and respond to AI-2, and the corresponding AI-2 receptors have been identified in *Vibrio harveyi* and *Salmonella enterica serovar Typhimurium. Vibrio harveyi* detects a borate-complexed form of AI-2 by virtue of the LuxP/LuxQ receptor/sensor kinase complex [173, 174]. In contrast, the *S. Typhimurium* transporter LsrB interacts with a nonborated form of AI-2, followed by internalization, phosphorylation, and interaction with the cytoplasmic transcriptional regulator LsrR [175]. Thus, bacterial species can detect different forms of AI-2, and detection can take place either extra- or intracellularly, depending on the bacterium.

QS depends on signaling molecules which are stable on the timescale of diffusion through the extracellular matrix. Modification (e.g., halogenation) or displacing of the AI molecules ("biomimicry") disrupts the *quorum sensing* through destructive interference with receptors and/or inactivating enzymes [159, 160, 176, 177]. Some marine algae, e.g., *Corallina officinalis* or *Delisea pulchra*, excrete iron- and vanadium-dependent HPOs/HOs that participate in the disruption of QS as well as in the synthesis of halogenated organic antagonist [159, 176]. Figure 8.5 illustrates selected examples of autoinducers and natural compounds being active against them.

VBPOs localized in the exterior region of *D. pulchra* disrupt the bacterial quorum sensing in different ways: (i) By the bactericidal effect of HOX itself, (ii) by the synthesis of bromofuranone compounds (**17**) (antagonists both against AI-1 and AI-2 systems) [178–180], and (iii) by the oxidative bromination of the bacterial signaling

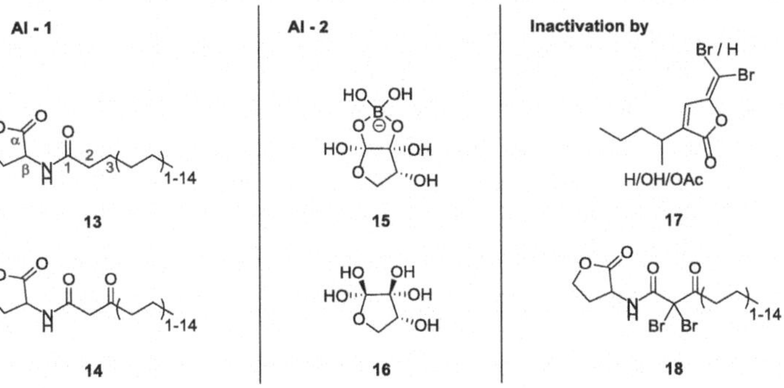

Fig. 8.5 Selected natural QS molecules (autoinducers) AI-1 (**13, 14**), AI-2 (**15, 16**) and their inactivation by brominated furanones (**17**) or lactones (**18**), e.g., isolated from *D. pulchra*. [140] **13** shows the typical numbering and denotation of AHLs

compound 3-oxo-acylhomoserine lactone (3-oxo-AHL, **14**) to the corresponding bromoacyl-AHL (inactivation, **18**) [140, 177, 181].

Furanones are ubiquitous (e.g., as pheromones [182]), and as structural AHL analogues they represent a tool to combat bacterial infection and biofilm formation [183]. Natural halogenated furanones are supposed to disrupt the AI-2 biosynthetic pathway, and brominated furanones target the AI-2 producing recombinant *Bacillus subtilis* enzyme LuxS (S-ribosylhomocysteine lyase) [180, 184]. An addition–elimination mechanism for the covalent modification of LuxS was proposed to yield an inactive enzyme and to disrupt further quorum signaling [180]. Kjelleberg and coworkers [160, 178, 185] described that halogenated furanones occurring in the marine red alga *D. pulchra* inhibit QS by downregulating the LuxR protein of *V. fischeri* overexpressed in *E. coli* (AHL-regulated bioluminescence assay) [178, 185]. This leads to the conclusion that furanones, at concentrations produced by the alga, control bacterial colonization of surfaces by specifically interfering with AHL-mediated gene expression at the level of the LuxR protein. An evidence for the inhibition of QS by the modulation of gene expression was provided by Ren et al. [179] who demonstrated using DNA/RNA microarrays that brominated furanones are able to inactivate LuxI while gene expression of LuxS remains constant.

Several test studies have been carried out with libraries of natural or synthetic furanones and AHLs. Martinelli et al. [183] tested a variety of furanones regarding their functional role on QS in *Chromobacterium violaceum* mutants. The biosynthesis of the violacein, a bisindole formed by the condensation of two tryptophan molecules, depends on the addition of AHLs for this mutant. As a compound, violacein has several biological activities, including being an anticancer agent and an antibiotic against *Staphylococcus aureus* and other Gram-positive pathogens. In most of violacein-producing bacterial strains isolated from nature, this bisindole is a secondary metabolite that is associated with biofilm production [186]. Its production within *C. violaceum* and other strains is regulated by QS [187]. Violacein production by *C. violaceum* has become a useful indicator of quorum-sensing molecules and their inhibitors [188].

The length of the 3-alkyl chain and the bromination pattern of the ring structure have a major effect on the biological activity of the 1′-unsubstituted furanones. The introduction of a bromine atom on the 1′ position of the 3-alkyl chain enhanced the activity of the furanones in both biological test systems. The potential of the (bromo)alkylmaleic anhydrides as a new and chemically accessible class of biofilm and quorum-sensing inhibitors has been demonstrated.

A library of different substituted furanones and (bromo)alkylmaleic anhydrides was tested for the antagonistic effect against biofilm formation by *S. Typhimurium* and the QS-regulated bioluminescence of *V. harveyi* [189]. The length of the 3-alkyl chain and the bromination pattern of the ring structure had a major effect on the biological activity of the 1′-unsubstituted furanones, whereas a bromine atom on the 1′ position of the 3-alkyl chain enhanced the activity of the furanones and decreased the QS-regulated bioluminescence and virulence of *V. harveyi*. This suggested a negative competition for the activation of LuxR, as the transcriptional activator protein is involved in the regulation of the expression of phenotypes such as bioluminescence, siderophore production, biofilm formation, and virulence [189].

The inactivation of native bacterial AHLs has been targeted by a variety of marine organisms. AHLs can easily be halogenated through electrophilic halogenation at the alpha position, and the receptor recognition is disturbed by the corresponding mono- and di-halogenated AHLs [140, 190]. Syrpas et al. [191] studied the HPO-mediated defense system of the benthic diatom *Nitzschia cf pellucida* system against bacteria (violacein assay). The halogenation–degradation pathway of QS molecules like*N*-β-ketoacylated homoserine lactones (AHLs) was analyzed with HPLC-MS in combination with synthetic halogenated AHL reference compounds. The degradation pathway of β-keto-AHLs triggered by the halogenating enzymes is shown in Fig. 8.6 [181, 191]. *N*-3-oxohexanoyl homoserine lactone (OHHL, **19**) is attacked by electrophiles (e.g., hypobromite HOBr) at the α-position. The resulting mono- and dibromo-homoserine lactones (MBHL, **20**; DBHL, **21**) were hydrolyzed with hydrolases or by basic hydrolysis. After cleaving off butyric acid **22**, *N*-(α,α-dibromoacetyl) homoserine lactone (DAHL, **23**) and the hydrolyzed product **24** were formed. This shows that the decrease of the QS activity in *Escherichia coli* JB523 microplate assays correlates with halogenation and the cleavage of the halogenated *N*-acyl chain of the lactone. Similarly, the production of green fluorescent protein (GFP) by *E. coli* JB523 depends on the addition of AI, the effects of AHLs, and their halogenated counterparts. The ability to induce GFP formation decreases in the order Br > I > Cl and DAHL (no GFP formation).

The ubiquitous occurrence and the involvement in communication and regulatory systems demonstrate the importance of halogenating enzymes in Nature and their high potential for a broad range of application, e.g., in fouling control and in drug design. Quorum sensing in marine organisms and potential applications have been reviewed by several authors [159, 162, 192–202]. QS inhibitors (e.g., furanone derivates) do not only affect bacterial biofilm formation in a destructive and concentration-dependent manner (violacein test) [195]; they also prevent the larval settling of the polychaete *Hydroides elegans* and the bryozoan *Bugula neritina*. The inhibition of biofilm formation of *Pseudomonas aeruginosa* [203] by sesquiterpene lactones from the seeds of the tropical tree *Annona cherimola* is a prominent example.

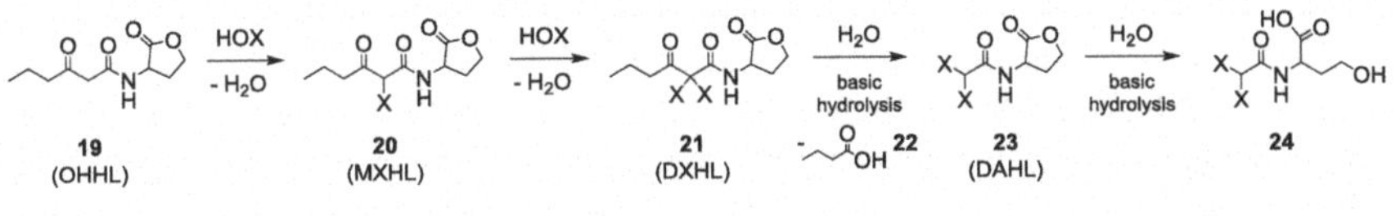

Fig. 8.6 Degradation pathway of β-keto-AHLs for *N*-3-oxohexanoyl homoserine lactone (OHHL **19**) (Scheme adapted from Refs. [181, 191]). The AHL molecule suffers electrophilic attack at the α-position by HOX (X = Cl, Br, I) formed by enzyme catalysis. The resulting mono- and di-halogenated homoserine lactones (MXHL **20**, DXHL **21**) are hydrolyzed by hydrolases or basic hydrolysis. After cleavage of butyric acid **22**, *N*-(α,α-dihaloacetyl) homoserine lactone (DAHL **23**) and the follow-on product **24** are formed

Native AHLs of a variety of bacteria modulate the response of the model plant hosts *Arabidopsis thaliana* and *Medicago truncatula*. Here, AHL amidolysis is catalyzed by a fatty acid amide hydrolase (FAAH) yielding L-homoserine [204], which enhances transpiration or inhibits plant growth in a concentration-dependent manner. As FAAHs exist in animals and also in humans (often in the brain and/or liver), AHL-degradation products also affect the response in their metabolism [204] via inter-kingdom signaling [205]. Another study on root-associated fungi highlights their ability to degrade AHLs (violacein assay) by lactonases. Thus, endophytic fungi might function as biocontrol agents against pathogenic bacteria [206].

In summary, bacteria can communicate with each other in a very complex manner. They play dual roles in maintaining their host's health and contributing to host's illness. Chemical communication between bacteria is based on different types of interactions: pathogenic, symbiotic, and transient. Many systems point to a chemical interplay, and yet many molecules involved are not characterized. In instances where the chemistry is known or elucidated, important implications for the host's fate can be proposed and tested.

8.4 Analytical Assays for Oxidative Halogenation

The development and validation of reliable enzyme assays is critical for an understanding of enzyme biochemistry and kinetics. This section provides a brief overview of assays developed for HPOs and HPO enzyme mimics.

I_3^- Assay. Historically, the I_3^- assay was established to detect the halogenating activity of enzymes. It is based on the oxidation of iodine by hypoiodite or bromine by hypobromite species (e.g., HOI, HOBr). The triiodide anion can be detected spectrophotometrically by its specific absorption at 350 nm ($\varepsilon_{350} = 25\ M^{-1}\ cm^{-1}$) [124, 207]. As high iodine concentrations and acidic conditions (pH < 6.5) are needed to overwhelm the chemical instability of I_3^- and possible side reactions (e.g., the oxidation of H_2O_2 by HOI) can occur, this is a more qualitative test system (Table 8.1, entry 1) [207].

MCD Assay. Based on the natural product caldariomycin, Hager et al. [118] developed a qualitative test method for the oxidative bromo- and chlorination in 1966. In the standard procedure, monochlorodimedone (MCD, $\varepsilon_{290} \approx 20\,000\ M^{-1}cm^{-1}$) [118] is oxidatively halogenated to dichlorodimedone or monobromo–monochlorodimedone in aqueous solution containing hydrogen peroxide, a bromide source (KBr or NH_4Br, in general) and the catalyst [118, 124]. As the halo-MCD has a much lower extinction coefficient ($\varepsilon_{290} = 100\ M^{-1}cm^{-1}$), the reaction can be monitored by the absorption changes at 290 nm. The rate of the decrease of the absorbance is related directly to the formation of hypohalite (HOX). As MCD is not reactive toward iodination, only CPO- and BPO-like activity can be determined (Table 8.1, entry 2) [207]. Due to the straightforward detection and the reaction kinetics, the MCD assay has been established as a standard test method to explore halogenating enzymes [134]. The nature of the mechanism and the underlying enzyme kinetics have been

Table 8.1 Assays to monitor HPO activity. II. Entry 1: Sulfonephthalein dyes. Entry 2: Side reaction of sulfonephthalein dyes (X = I, Br, Cl). Phenol red (R_1, R_2 = H) or thymol blue (R_1 = iPr, R_2 = Me) are oxidatively halogenated stepwise [e.g., bromophenol blue (X_4PR, X, R_1 = Br and R_2 = H) or dibromo-thymol blue (X_2TB, X = Br, R_1 = iPr, R_2 = Me)]

Entry	Dyes/substrates
1	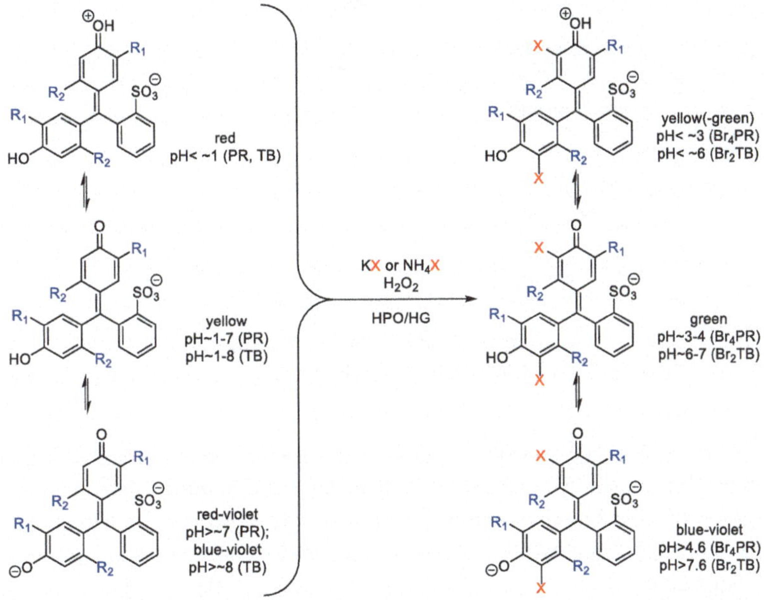
2	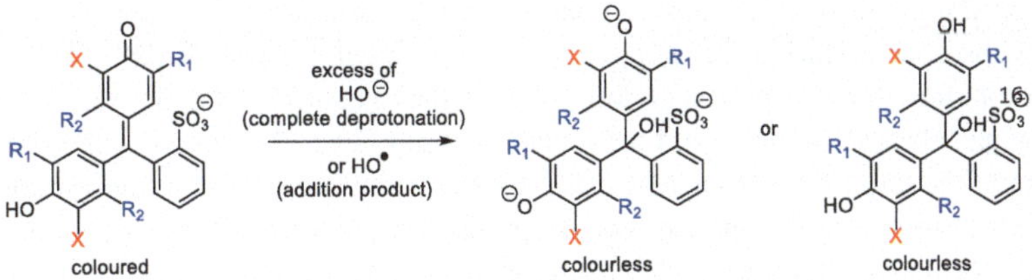

studied in detail [104, 134, 208, 209]. The drawback of the MCD assay is that it focuses on substrate consumption rather than on product formation. This may lead to erroneous results, as a decrease in the MCD concentration does not automatically indicate the formation of halogenated compounds [210]. MCD may be inactivated by O_2 depending on the experimental conditions and radical forms of MCD [211–213]. Monitoring the reaction based on UV–Vis absorption and TLC [118] does not give any clues, which products are formed. NMR and mass spectrometry provide information concerning the halogenated species but are difficult to carry out under operating conditions. As nanoparticles and enzyme mimics are less specific, a “blind activity” in the MCD assay is possible. If HPO/HG-like activity is established, this assay is a valuable tool to study enzyme and catalyst kinetics.

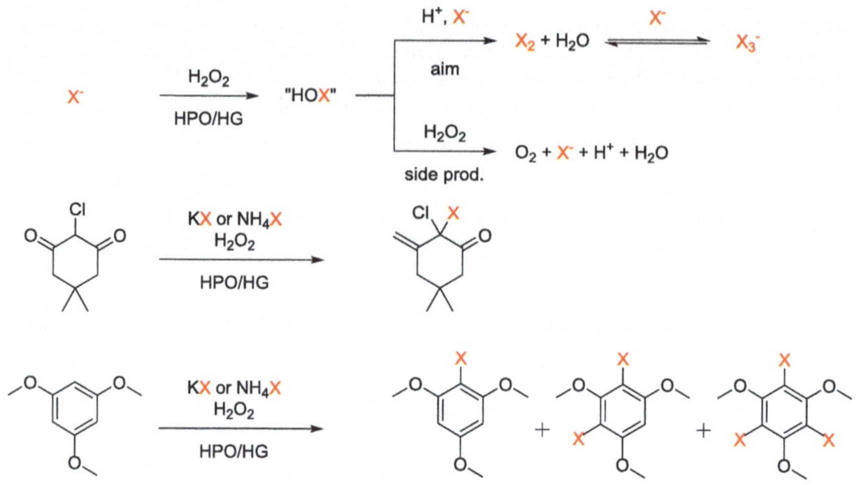

Fig. 8.7 Assays to monitor HPO activity. I. Entry 1: I_3^- assay (X = I, Br), entry 2: MCD assay (X = Br, Cl), entry 3: TMB/DMP assay (X = I, Br, Cl)

TMB and DMP Assays. Electron-rich aromatic systems are accessible to an electrophile attack by hypohalites. This is the basis for the 1,3,5-trimethoxybenzene (TMB) or 3,5-dimethylphenol (DMP) assays, where 1,3,5-trimethoxybenzene (TMB) or 3,5-dimethylphenol (DMP) are used as substrates to explore oxidative halogenation activity (Fig. 8.7). A typical reaction mixture contains halide, H_2O_2, and the catalyst under acidic conditions (e.g., citrate buffer, pH 3). The change in absorbance during the halogenation reaction can be monitored by UV–Vis spectrophotometry. The products (mono-, di-, trihalogenated molecules) can be identified by gas chromatography and mass spectrometry. The TMB assay was used to determine the chlorine isotope effect for a CPO-catalyzed oxidative chlorination [130]. Since highly acidic conditions are required and TMB is sensitive to oxidation due to the formation of strongly absorbing or fluorescent products [214], this assay is not well established.

Sulfonephthalein Dyes. Sulfonephthalein dyes overcome the problems of side reactions. They combine an optical color change (qualitative "quick screening" assay) with quantitative UV–Vis, NMR, and mass detection for bromination, chlorination, and even iodination. Phenol red (PR, phenolsulfonephthalein) or thymol blue (TB, thymolsulfonephthalein) are used as indicators. The reaction scheme is shown in Table 8.1, entry 1. A subsequent oxidative halogenation reaction at the free ortho positions leads to tetra-halophenol blue (X_4PR, X = Br, Cl, I) or di-halothymol blue (X_2TB, X = Br, Cl, I). Kinetic studies are evaluated with respect to the brominated products: Turnover numbers (TON, K_{cat}), reaction constants, reaction orders, and kinetic parameters of enzymes (e.g., Michaelis–Menten constants) are easily accessible, although the reaction mechanism is more complicated than that of the MCD assay. The stability of the end products and their distinct spectral properties facilitate the optical analysis. Thymol blue and phenol red are common pH indicators, and their color changes in the neutral regime (i) from yellow to red-violet (PR, between pH 6.8 and 8.2) [215] and (ii) from yellow to blue (TB, between pH 8.0 and 9.2) [216]. Oxidative halogenation of the chromophore changes the pKa values by affecting the delocalization of the π system [207]. The fully halogenated products can be assigned spectrophotometrically and even by eye inspection. Br_4PR shows a color change between pH 3.0 (yellow) and 4–6 (blue) [217], Br_2TB between pH 6.0 (yellow) and 7.6 (blue) [218]. Enzymatic bromination of TB can be used as qualitative and quantitative assay of the oxidative halogenation activity [207].

Phenol red is used to detect HPO- or HG-like activity. It is water-soluble in the absence of organic solvents. Although the turnover of phenol red to the fully halogenated end product is more complex regarding the kinetics, the validation for the fourfold halogenated phenol red is simple because it has a distinct absorption maximum at 590 nm at pH > 4.6 (typical operating conditions), and its absorbance differs from that of the educts in any pH range. However, it should be noted that at high peroxide concentrations phenol red can be halogenated even in the absence of HPO [219]. Kinetic studies have shown that bleaching in basic media might be due to reaction with hydroxyl radicals (Table 8.1, entry 2) [220, 221], which destroy the quinoid system of PR and yield colorless products with a large hypsochromic shift. Monitoring the halogenation by NMR and mass spectrometry, etc. are alternatives to avoid these intricacies.

Luminescence Assays. Assays (chemiluminescence or fluorescence) based on rhodamine, coumarin, and fluorescein derivatives [222, 223] are popular for detecting reactive oxygen or nitrogen species (ROS, RNS) in biological systems [224]. The detection of HPO-like reactions is based on the oxidative activity of HOX (X = Cl or Br). Luminescent dyes are useful for accurate sensing even in multicomponent systems (e.g., living cells) [223]. Nonfluorescent aminophenyl fluorescein (APF) or hydroxyphenyl fluorescein (HPF) is converted to fluorescein by oxidative deacylation (Table 8.2, entry 1) via HOX intermediates or by adding HOX [225, 226]. The fluorescent product (525 nm) can be monitored and localized in laboratory test systems with immobilized HPOs [226], but also in native human neutrophils and eosinophils [225]. Hypochlorite-promoted ring-opening mechanisms have been proposed for fluorescein [227] (Table 8.2, entry 2) or rhodamine derivatives [228] (Table 8.2, entry 3). Rhodamine has been used to demonstrate the existence of phagosomal and mucosal HOCl formation in the intestinal epithelia of *Drosophila melanogaster* [228, 229]. These dyes extend the range of applications to detect HOCl in immune response system and in the pathogenesis of human diseases in cells and animal tissues [228, 229]. Xanthene derivatives and oximes of phenanthroimidazole derivatives [230] have been used for HOX sensing (deoximination via HOCl leads to a redshift relative to the starting compound due to intramolecular charge transfer (ICT), Table 8.3 entry 4). Chemiluminigenic substrates such as cyclic hydrazides, dioxetane precursors [231], or ferrocene-modified pyridylthiazole derivatives [232], which are oxidized by HOCl selectively, have been used.

Despite recent advantages in the development of luminescence assays, limitations arise as the luminescent products depend on the oxidizing power of HOX rather than on oxidative halogenation [233]. Thus, a distinction between HPO- and PO-like

Table 8.2 Assays to monitor HPO activity. III. Entry 1: Oxidative deacylation of nonfluorescent aminophenyl fluorescein (APF, R = N) or hydroxyphenyl fluorescein (HPF, R = O) to fluorescein. Proposed hypochlorite-promoted ring-opening mechanism for nonfluorescent xanthene dyes (entry 2: fluorescein-like, entry 3: rhodamine-like) to their fluorescent counterparts. Entry 4: Deoximination reaction of oxime derivatives of phenanthroimidazole-based compounds (X = Br or Cl)

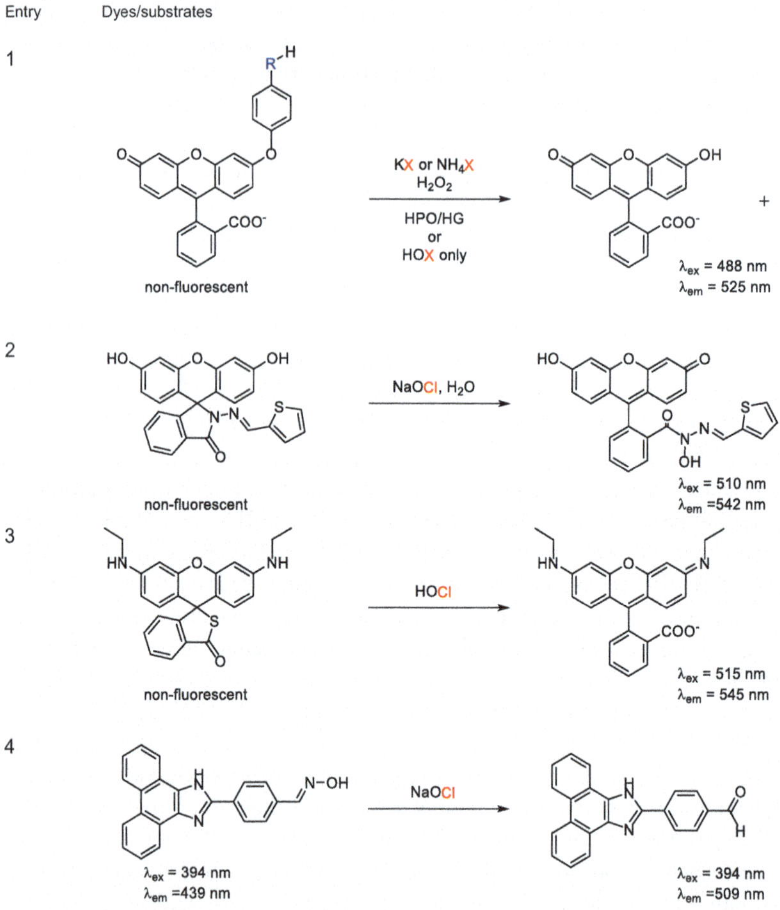

activities is difficult. Luminescence methods are useful tools for a spatially resolved analysis even in multicomponent systems if HPO-like activity can be demonstrated with alternative assays.

Taurine–Haloamine Assay. Haloamine is formed by the reaction of amines with HOX, and reactive chloramines [234] and bromamines [235] are natural disinfectants in biological systems [92, 112]. Thiol oxidation by hypochlorous acid and chloramines is a favorable reaction and may be responsible for alterations in regulatory or signaling pathways in cells exposed to neutrophil oxidants. As a consequence, detection methods for HOX are based on the haloamine reaction (related to the amine

Table 8.3 Assays to monitor HPO activity. IV. Entry 1: HOX-catalyzed reaction of taurine-to-taurine–haloamine (X = I, Br, Cl), entry 2–4 trapping reagents for haloamines. Entry 2: TNB method, entry 3: TMB method, entry 4: CB method

Entry	Dyes
1	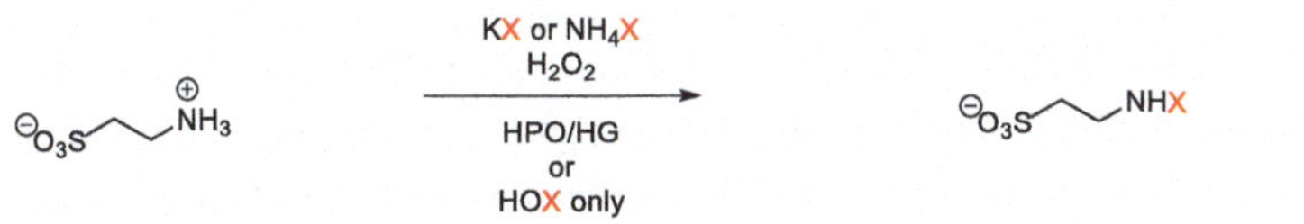
2	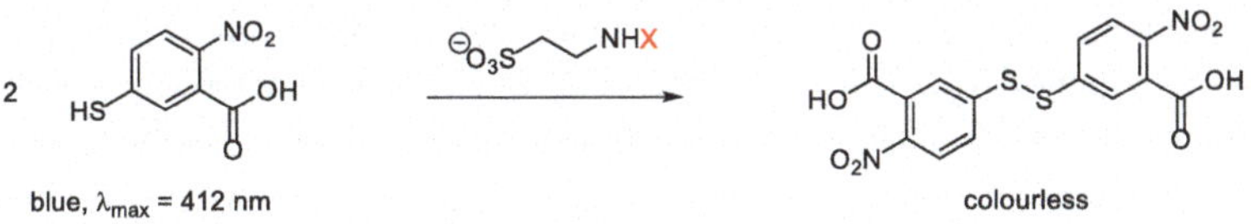
3	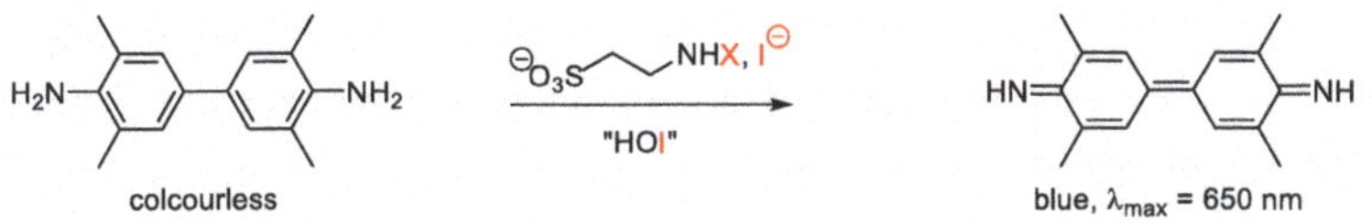
4	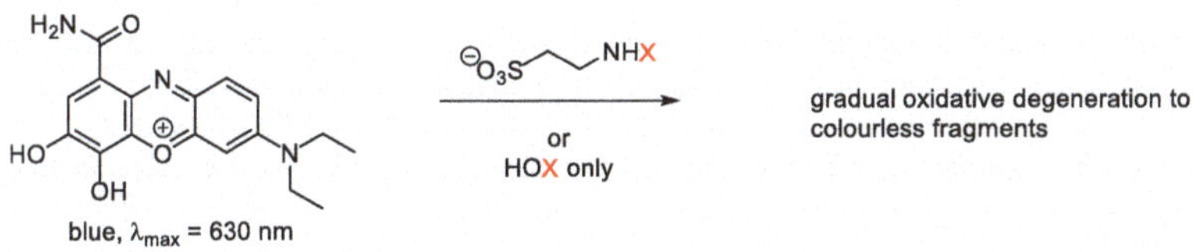

taurine reaction, Table 8.3, entry 1) [92, 95, 112]. The resulting taurine haloamines react with trapping agents. They can be oxidized or halogenated with haloamine intermediates, for example, through disulfide formation with 2-nitro-5-thiobenzoate (TNB, TNB method, Table 8.3, entry 2) [236, 237] or by oxidation of 3,3′,5,5′-tetramethylbenzidine (TMB) with additional iodide as co-substrate (TMB method, Table 8.3, entry 3) [214]. Kettle reviewed methods for the detection of HOCl in biological systems [93]. TNB-taurine/chloramine is suitable for detecting purified peroxidases and isolated neutrophils. HOCl chlorinates the tyrosyl residues in small peptides. This leads to an increase of the 3-chlorotyrosine and 3,5-dichlorotyrosine levels in proteins after exposure to (low concentrations) of hypochlorous acid. Analysis by gas chromatography and mass spectrometry can probe the oxidation by myeloperoxidase in the pathology of several diseases. Fluorescent probes or the derivatization of tyrosine or methionine [238] are suitable markers for the halogenation of bacterial proteins in neutrophil phagosomes [239]. HOCl, HOBr, and their corresponding taurine haloamines can be detected by oxidative degradation of celestine blue (CB method, Table 8.3, entry 4) [240]. This allows HOX detection with increased selectivity and specificity over a broad pH range to monitor the reaction kinetics colorimetrically.

Natural Compounds and Their Analogues as Test Systems. Hypochlorous acid is the major oxidant generated by neutrophils. Myeloperoxidase catalyzes the biosynthesis of hypochlorous acid from hydrogen peroxide and chloride and is involved in tissue damage associated with many diseases that rely on inflammatory cells. Difficulties to demonstrate its crucial role in pathology could be overcome with the advent of sensitive biomarkers for hypochlorous acid. The use of other biomarkers provides complementary information [93]. Table 8.4 compiles test reactions and marker molecules modeled after natural compounds and reactions [122, 124]. The monoterpene nerol (Table 8.2, entry 1) is formed by ring closure similarly as laurencins. Geraniol and its derivates are converted to synderol and bromo-8-epicaparrapi oxide sesquiterpenes [122, 190] (Table 8.2, entry 2–4), and the bacterial signaling molecule 3-acylhomoserine lactone (AHL) is transformed to its inactive analogue via oxidative bromination (Table 8.2, entry 5). The bactericidal antagonist bromofuranone is synthesized with VBPO (Table 8.2, entry 6) and utilized to disrupt the bacterial QS [140]. Since the yields of the halogenated products are lower and (different from pure model substrates) the formation of side products may be expected, kinetic studies should not rely on these assays only.

Table 8.4 Assays to monitor HPO activity. V. Entry 1: Oxidative bromination reaction of nerol to brominated laurencin analogue. Entry 2–4: Oxidative bromination reaction of geraniol (and derivatives) to brominated synderol (and derivatives) and sesquiterpenes (entry 4). Entry 5: Oxidative bromination reaction of AHL to brominated AHL and subsequent degradation by basic or enzymatic hydrolysis. Entry 6: Oxidative bromination cyclization reaction of 4-pentynoic acid to bromofuranone

Entry Substrates

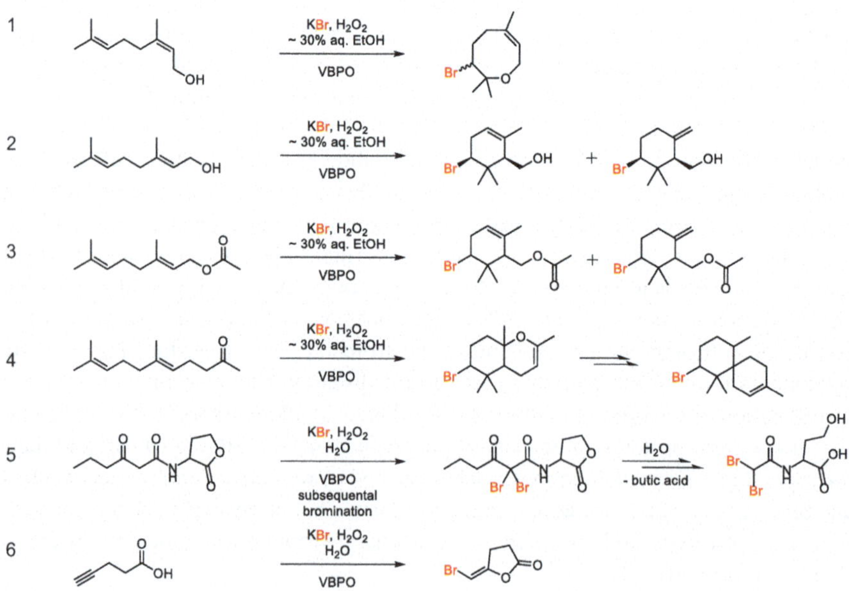

8.5 Homogeneous Biomimetic HPO/HG-Like Catalysts

Based on the structure of the active centers of halogenating metalloenzymes and the mechanistic hypothesis outlined in Figs. 8.3 and 8.4, the requirements for catalysts are as follows: (i) They should have a tunable redox system with metal atoms in a high oxidation state whose coordination sphere can easily expand, (ii) good binding and release capabilities of oxygen compounds with low energy barriers, (iii) low toxicity, (iv) eco-friendly extraction of the raw materials, (v) large-scale applicability to improve the enzyme-like activity, (vi) high operational stability (especially with regard to temperature, pH, solvents), (vii) high enzyme-like selectivity, and (viii) regioselectivity (enantioselective synthesis). Transition metal ions/complexes may fulfill many of these demands. Current research is focused on a fundamental understanding of structural and functional HPO/HG-like mimics [241].

Organometallic vanadium compounds [97, 242] modeling the structure of the active site of VHPO and peroxo-compounds [97, 243, 244] have shown good efficiency and selectivity for a variety of oxidation reactions [97, 245, 246]. Examples are halogenation [247], decarbonylative halogenation [248], (enantioselective) sulfoxidation in the absence of halides [249–251], epoxidation [252], hydroxylation of alkenes [253], and oxidation of primary and secondary alcohols [254] analogous to the native enzyme [105]. Homogenous catalysts are Schiff-base complexes of vanadium [247, 255], N-heterocycle-vanadium pincer compounds [256], imidazole, scorperate [257, 258], and amino acid complexes [259, 260]. Vanadium acetylacetonate [261, 262] and ammonium metavanadate [263–265] have found applications as "green" catalysts [266]. Oxo-peroxo-vanadium (V) complexes are the active species [267, 268]. Many of the active catalysts require pretreatment in strong acidic media to reduce a catalase-type degradation of H_2O_2 by halides and halogens [266], and their application in biological systems at neutral or moderate pH is limited.

Although vanadium catalysts show promising activity, their industrial application is problematic according to European Union regulations (REACH) [269, 270]. Hetero-polyoxometalates of Mo(VI) [267, 268, 271–275], W(VI) [243, 266, 267, 276–284], Re(VII) [211, 285], Mn(II-VII) [50, 51, 286, 287], Cu(II) [288–298], Ag(I)/Cu(II) [299, 300], Ru(III) [301–303], Ce(IV) [304–312], Fe(III) [313–316], Pd(II) [317–323], Ti(II,IV) [324, 325], and Nb(V) [326] are useful, but often more expensive and less active alternatives [281, 327–330]. The HPO-like activity of main group compounds of B(III) [331], Ba(II) [332], Bi(III) [333], Sb(V) [334], Se [335–338], and Te [266, 335, 337, 339–342] has been reviewed [303, 329, 330, 343–345] with a focus on the synthesis of (hetero) aryl halides, vinyl halides, and alkyl halides by C–H activation under homogeneous conditions.

8.6 Heterogeneous Versus Enzyme HPO/HG-Like Catalysts

New directions are explored to meet the demand for sustainable and environmentally friendly solutions that avoid the formation of hazardous waste while preserving high selectivity, stability, low energy consumption at an industrial scale [346]. Reactions without (metal) catalysts [266] appear to be a good and cost-saving option, but they require harsh reaction conditions that are incompatible with the thermal stability of most organic compounds. Biocatalysis is a viable alternative, where enzymes or microorganisms catalyze various biochemical reactions. Advantages of enzymes have advantages over conventional catalysts are chemo-, regio-, and stereospecificity and mild reaction conditions [347, 348]. Still, an industrial-scale application of biocatalysts is modest, as enzyme-catalyzed reactions depend on temperature, pH, or the chemical structure of the substrates, and enzymes suffer from low stability in nonaqueous solvents, short shelf life, and high production costs. They can be immobilized (on porous glass, SiO_2, and organic polymers) [349–352]. Prominent examples of biochemical reactions are the isomerization of glucose to fructose, important in the production of soft drinks by glucoamylase immobilized on SiO_2 and the conversion of acrylonitrile to acrylamide by corynebacteria entrapped in polyacrylamide gel. Enzymes and microorganisms are used in sewage treatment plants, biosensors, organic synthesis, the food and textile industry, and pharmaceutics [16, 353–356]. HGs have a high potential for bioengineering and the biosynthesis of new halogenated metabolites in organic synthesis or as potent drugs [70, 357, 358]. Enzymes are used for oxidation reactions and hydrolysis [359] by companies like BASF, DSM, and Lonza. The shortcomings of enzymatic reactions are their complexity, their stability under operating conditions (solvent, pH, temperature, etc.), and their lifetime. Strategies have been developed to produce enzyme-organic/inorganic hybrid materials in combination with recombinant enzymes to enhance stability, activity, and lifetime of the proteins [360, 361].

The goal of biomimetic chemistry is to emulate the structural and functional aspects of natural enzymes [10, 362, 363]. Nature has served as a source of inspiration for "designing" functional (supra)molecular enzyme models, "artificial enzymes" [9]. Much effort has been invested in the synthesis of (supra)molecular models that are more stable and cost efficient compared to their natural counterparts [8]. Most supracompounds contain a receptor or cavitand connected to an active site, with the aim to mimic reactions that are carried out by enzymes [364, 365]. This approach has led to catalytically active model compounds that display enhanced selectivity and/ or activity. Still, model catalysts, especially in the presence of other competing reactions inside living cells or even organisms, remain a challenge [366].

Although biochemical processes are triggered by enzymes, biocatalysis is mechanistically a special case of heterogeneous catalysis [367, 368]. In analogy to enzymes, the catalyst may be stabilized by a matrix or support [369, 370]. A surface reaction occurs according to $A + S \rightleftharpoons AS \rightarrow$ products, where A is the reactant, S is an adsorption site of the enzyme/surface, AS is the substrate complex, and k_1, k_{-1}, and k_2 are the rate constants for the adsorption, desorption, and reaction to the final product.

The global reaction rate r is given by $k_2\theta C_S$ (where θ is surface coverage of the substrate molecules S, C_S is the total number of—occupied and unoccupied—sites, provided that k_2, the rate constant of the enzyme/substrate complex, is $\gg k_1/k_{-1}$ (= K_m, the Michaelis constant). The functions of enzymes are governed by enzyme dynamics [371], and enzymes are typically much larger than their substrates. The reaction environment it exposes is so different from the surrounding solvent that it may be considered a solid catalyst [367, 372]. The radii of enzymes and nanoparticles are comparable to nanometer-sized cellular features, and the interactions of enzymes and nanoparticles with soluble molecules have comparable strength [373].

Nanoparticles are highly reactive, as demonstrated by transition metal or metal oxide nanoparticles in catalysis [4, 15, 368]. They are stabilized in solution with ligands, surfactants, or polymers, and their activity and selectivity are determined by size and the atomic-level structure of the planes they expose. The most active particles in catalysis have diameters of several nanometers, i.e., they contain only *several hundred* atoms. Their reactivity may even be higher than that of enzymes, because *any* surface site can be catalytically active, whereas enzymes have only a *single* binding site, although with exceptional specificity and complexity. Therefore, enzyme-mimetic catalysis with nanoparticles has been successful in those cases, where small species (peroxides [374, 375], superoxides [376], or sulfites [13]) are involved and steric demands play only a minor role for the specificity of the reaction [15, 368, 377].

The reaction at the active site of vanadium-dependent HPOs proceeds without a change in the oxidation state by addition of a η^2-peroxo group (Fig. 8.3). Functional HPO mimics have been devised based on that mechanism. Similarly, transition metals like tungsten, molybdenum, and titanium that are known to form related peroxo-complexes are active in halogenation reactions. In contrast, non-heme iron HGs generate HOX via a radical mechanism HOX (Fig. 8.3) [139, 378], and other one-electron-transfer redox catalysts might also exhibit HG-like activity. However, most of them also show high peroxidase rather than HPO activity. Their HOX selectivity can be increased by reducing their activity.

Enzymes and biomimetic catalysts enhance the reaction rate by providing an alternative reaction pathway for breaking and making of bonds. The activation energy for the new pathway is less than the activation energy of the conventional route, because transition states and intermediates are stabilized by the catalyst (Fig. 8.8a). Enzymes are classified according to their catalytic performance and rate constants under steady-state conditions, where the concentration of the enzyme–substrate complex remains constant (rate constants $k_1 = k_{-1} > k_2$, Fig. 8.8a). The quantitative connection between the decrease of the substrate concentration ($[S]_0$ = initial substrate concentration) and the reaction rate (v; v_{max} = maximal reaction rate) is given by the Michaelis–Menten equation (Eq. 8.5). The formation of the enzyme/catalyst complex is the rate-determining step. The Michaelis constant, $K_m = k_1/k_{-1}$, indicates the substrate affinity of the catalysts/enzyme (Eq. 8.6). Additionally, K_m corresponds to the half-maximal reaction rate (Eq. 8.6, via Eq. 8.5), i.e., it represents the substrate concentration where half of the enzyme exists as an enzyme–substrate ([ES]) complex. The reaction shows a Langmuir-like behavior (Fig. 8.8b).

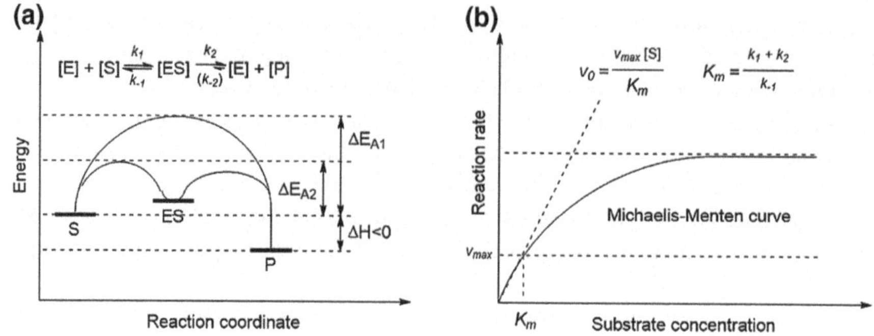

Fig. 8.8 **a** Energy profile of an enzyme-catalyzed reaction. The substrate (S) and the enzyme (E) reversibly form an enzyme-substrate (ES) complex that dissociates to the final product (P). The enzyme offers an alternate reaction pathway (shown in red) where the rate-determining step has a smaller activation barrier ($\Delta E_{A1} > \Delta E_{A2}$). **b** Michaelis–Menten plot, v_0: initial rate, v_{max}: maximum rate, K_m: Michaelis–Menten constant

$$v = \frac{v_{max} \cdot [S]_0}{K_m + [S]_0} \tag{8.5}$$

$$K_m = \frac{k_{-1} + k_2}{k_1} = \frac{1}{2} v_{max} \tag{8.6}$$

$$\frac{1}{v} = \frac{K_m}{v_{max}} \cdot \frac{1}{[S]} + \frac{1}{v_{max}} \tag{8.7}$$

As an accurate determination of the kinetic parameters from the Michaelis–Menten plot is difficult, several linearization methods have been established, the most popular being the double reciprocal representation (Lineweaver Burk plot, Eq. 8.7).

The activity of HPOs is evaluated with the model substrates monochlorodimedone (MCD) or phenol red (PR). Jittam et al. [379] reported the oxidative bromination of sulfonephthalein dyes with bromoperoxidases (BPO) extracted from native red algae. The K_m values with respect to the co-substrates H_2O_2, bromide, and myeloperoxidase (MPO) in comparison are compiled in Table 8.5.

Table 8.5 K_m values of phenol red (PR) and dibromo/dichlorophenol red (Br_2PR/Cl_2PR) for extracted bromoperoxidase (BPO) of the marine red alga *Gracilaria tangi* and commercial myeloperoxidase (MPO). Reprinted with permission from Ref. [366]

Enzyme	K_m (Dye) (mM)	K_m (H_2O_2) (mM)	K_m (KBr) (mM)
BPO (crude)	PR, $1.5 \cdot 10^{-2}$ (pH 5.8)	$3.7 \cdot 10^{-2}$	$3.7 \cdot 10^{-1}$
BPO (crude)	Br_2PR, $3.1 \cdot 10^{-2}$ (pH 6.5)	$1.5 \cdot 10^{-1}$	$2.9 \cdot 10^{-1}$
MPO (commercial)	PR, $8.0 \cdot 10^{-3}$ (pH 5.8)	$3.5 \cdot 10^{-1}$	$4.5 \cdot 10^{-0}$
MPO (commercial)	Cl_2PR, $2.3 \cdot 10^{-2}$ (pH 5.8)	$3.5 \cdot 10^{-2}$	$2.7 \cdot 10^{-0}$

The K_m values are in the range of ca. 10^{-5} M for both, the organic substrate and H_2O_2, indicating a strong affinity of the substrate to the enzyme. As bromide is always present in sufficient concentration, the affinity of the enzyme toward bromide is less pronounced in order to provide HOBr in the required amounts. The catalytic efficiency as a measure of substrate specificity of an enzyme is given by the turnover rate (k_{cat}) over K_m. For surface reactions, a Michaelis–Menten analysis is less common. Kinetic studies for HPO or HG in wastewater treatment support the formation of hypobromite [84, 380] and bromamine [381] as active species. Hypobromite is formed according to Eq. 8.4. Besides HOBr, the oxidants, and halogenating agents, mono- and dibromamines are formed from the reaction of NH_4^+ with HOBr [381] or through the formation of bromochloramines [382, 383].

$$[\mathrm{HOBr}] \equiv \left[\mathrm{Br_2} + \mathrm{Br^-}\right] \equiv \left[\mathrm{Br_3}^-\right] \quad (8.8)$$

The oxidation of bromine (compared to that chlorine and iodine) prevails in most cases, because bromine is more reactive than chlorine (c.f. drinking water treatment) by several orders of magnitude [103]. The specific rate constants for a homolytic cleavage of H_2O_2 suggest that a slow radical transfer from a hydroxyl radical (HO•) to bromide (Br–) and hypobromite species (HOBr) is dominant [380]. The kinetics of bromide oxidation for wastewater treatment with hydroxyl radicals or ozone was analyzed experimentally and theoretically [84, 149, 380, 384–386]. An overview about possible oxidation reactions and their equilibrium constants via hydroxyl radicals/H_2O_2 is given in Refs. [149, 380]. Under laboratory conditions, a constant supply of hydroxyl radicals is ensured by γ-radiolysis of H_2O_2. The initial rate of HOBr formation with respect to oxidation and reduction processes is calculated as $6.5 \cdot 10^{-8}$ M s^{-1} [149].

The rate constants, the reaction conditions, and organic "trapping agents" depend on the oxidized bromine species—in the range between 10^7 and 10^{10} M^{-1} s^{-1} [149]. A study of the rate constants for the oxidative bromination of phenolic compounds showed maximum values for the reaction of HOBr with phenolate ions and approx. 2×10^8 M^{-1} s^{-1} with respect to the ionic forms of *p*-cresol or phenol (pH ~ 7–8) [84, 381, 383]. The reaction rates for the oxidative bromination reaction are ~10^3 times higher than for the oxidative chlorination [381]. Analogous results were obtained from a kinetic study for HOBr in comparison to HOCl [383]. HOCl reacts with bromide according to $\mathrm{HOCl} + \mathrm{Br^-} \rightarrow \mathrm{HOBr} + \mathrm{Cl^-}$, and thus the bromination reaction prevails as long as bromide ions are present in the reaction mixture. The ortho positions of the phenolic compounds are preferred over the para position [383]. The kinetic studies show that the oxidative halogenation proceeds via complex multistage reactions and includes radical, electrophilic, or combined mechanisms with respect to the formation of HOBr and the reactive centers of the substrate.

In enzymology, the turnover number (TON) is defined as the maximum number of chemical conversions of substrate molecules at a single catalytic site per second for a given enzyme concentration. It is calculated from the maximum reaction rate and catalyst site concentration. The catalysts can be assessed with respect to their specific catalytic activity through the TON, turn over frequency or rate (TOF, k_{cat}), and the specific activity (SA) [387]. TOF and SA are the most significant indicators for catalytic efficiency. In organometallic catalysis, the turnover number (TON) is defined by the number of moles of product (P) that a mole of catalyst (C) can convert before becoming inactivated (Eq. 8.9). The TOF is defined here TON/per time unit (per hour or second, dimension: 1/t, Eq. 8.10). This corresponds to the rate constant of the rate-determining step in enzyme kinetics (k_{cat}) and is typically determined by Michaelis–Menten-type analysis via the maximum reaction rate (v_{max}) per of catalytically active site (Eq. 8.10). The catalytically active sites may be the enzyme (i.e., its active site) or the metal sites on a nanoparticle surface.

Different from enzymes, nanoparticles have many independent catalytic surface sites. As turnover numbers (TON) and catalytic rate constants are given *per* active subunit (as in enzymology) [346], the activity of nanoparticle enzyme mimics has to be normalized to one active site [376, 388]. The specific activity (SA) is given by the formation rate of the product (in mole) versus time and the supplied amount of catalytic materials (in gram, Eq. 8.11). The SA value allows estimating efficiency, costs, and even the design of production units.

$$TON = \frac{n(P)_{\text{total}}}{n(C)_0} = \frac{[P]_{\text{total}}}{[C]_0} \tag{8.9}$$

$$TOF_{[t^{-1}]} = k_{\text{cat}} = \frac{d[P]_t/dt}{[C]_0} = \frac{v_{\text{max}}}{[C]_0} \tag{8.10}$$

$$SA_{[mole\ g^{-1}t^{-1}]} = \frac{n(P)_t/dt}{m(C)_0} \tag{8.11}$$

8.7 Supported Biomimetic Catalysts for Halogenation Reactions

Current research focuses on bromoperoxidase-like reactions. A variety of catalytically active transition metal complexes in combination with different support materials have been described. Table 8.6 compiles bromoperoxidase-like catalysts sorted by active metal, support, and reaction conditions. Environmentally friendly (in terms of solvent, starting compounds, temperature, and pressure) catalysts are highlighted, and suitable assays (e.g., phenol red (PR), phenol (Ph), salicylaldehyde (SaA), thymol blue (TB), and monochlorodimedone (MCD)) for validation are given. In addition, oxidative halogenation reactions beyond HOBr catalysis are considered. Major examples are oxidation, hydrohalogenation, or epoxidation. Multibromination (Eq. 8.12) and conversion (Eq. 8.13) are considered as well.

Table 8.6 Heterogeneous catalysts for oxidative halogenation reactions sorted by their active elements and supports (entry 8–59). (i) Organic/polymer (PS = polystyrole); (ii) inorganic (LDH = layered double hydroxide, ZrO_2 = zirconia, SiO_2 = silica, MCM = mesoporous silica, zeolite), (iii) without support (entry 60–65). Reaction conditions (educts, educt concentrations and assays, solvent, pH, temperature). Bromine sources: (a) NH_4Br, (b) KBr, (c) NaBr; substrate/dye: MCD = monochlorodimedone, PR = phenol red, Ph = phenol, SaA = salicylaldehyde, TB = thymol blue; DHP = 2,3-dihydro-4*H*-pyran, St = *trans*-stilbene, CA = cinnamic acid, PA = 4-pentenoic acid, An = anisole, Ch = cyclohexanone. Solvents: (A) water/methanol/tetrahydrofuran (4:3:2), (B) acetonitrile/water (1:4), (C) acetic acid, (D) methanol, (E) HEPES buffer, (F) phosphate buffer, (G) diethylether, (H) carbon tetrachloride, (I) acetonitrile/diethylether (1:3), (K) dichloromethane/water (1:1), (L) phosphate-buffered saline, PBS, (M) Tris-SO4 buffer. Comparison of catalytic activities by TOF = turnover frequency, SA = specific activity, listed or determined (*) based on the oxidized Br with the given specifications (TOF [1/h] = $mol(Br^{ox})/(mol(cat) \cdot h)$; SA [mmol/g h] = $mmol(Br^{ox})/(g(cat) \cdot h)$, Table S1 and S2). Reference materials (native enzyme, supports) are listed in entries 1–7

Entry	Catalyst	Support	Total cat.	Br^- (mM)	H_2O_2 (mM)	Dye (in mM)	Conditions	TOF (1/h)	SA (mmol/g h)	Refs.
1	VHPO	None		100[(a)]	2.5–5	MCD (0.5)	(A) 20 mL, pH 6.5	n.a.	600	[389]
2	VHPO	Immobilized		100[(a)]	2.5–5	MCD (0.5)	(A) 20 mL,pH 6.5	n.a.	$1.5 \cdot 10^{-1}$ – 6.0	[389]
3	MgAl	(MgAl)LDH	2.7 mg	100[(a)]	2.5	PR (0.05)	(A) 10 mL	n.a.	$0.3 \cdot 10^{-1}$	[390]
4	Zr	$mZrO_2$ (m-700; m-630)	20 mg	100[(a)]	10.3	PR (0.05)	H_2O, 32 mL	n.a.	$4.2 \cdot 10^{-1}$; $8.4 \cdot 10^{-1}$	[424]
5	Zr	ZrO_2	200 mg	550[(b)]	550	PR (0.5)	(C) 4 mL, 5 h	$1.2 \cdot 10^{-1}$*	0.989 . $8 \cdot 10^{-1}$ *	[412]
6	Al, Si	HZSM-5(30) zeolite	200 mg	550[(b)]	550	Ph (500)	(C) 4 mL, 5 h	n.a.	1.8*	[396]
7	Al, Si	HY zeolite	200 mg	550[(b)]	550	Ph (500)	(C) 4 mL, 5ht	n.a.	1.5*	[396]
8	V(IV)O complex	(modified) salen-polymer	20 mg	1000[(b)]	2500	SaA (500)[(13)]	4 mL H_2O, 1 M $HCClO_4$, 2 h	31.5–33.8	141.0–158.0*	[397]

(continued)

Table 8.6 (continued)

Entry	Catalyst	Support	Total cat.	Br^- (mM)	H_2O_2 (mM)	Dye (in mM)	Conditions	TOF (1/h)	SA (mmol/g h)	Refs.
9	$V(V)O_2$ complex	PS	30 mg	2000[(b)]	2000	SaA (1000)	10 mL H_2O, 2M H_2SO_4, 2 h	100.0–104.3	137.5*	[398]
10	$V(V)O_2$ complex	PS	5 mg	500[(b)]	500	SaA (250)	20 mL H_2O, 1M $HCLO_4$, 2 h	2254	634*	[399]
11	$V(V)O_2$ complex	PS	20 mg, ~1 mM V	12,500[(b)]	3750	SaA (2500)[(11)]	4 mL H_2O, rt, 2 h, 10M $HClO_4$	775; 800	192.6*; 183.0*	[400]
12	$V(V)O_3^-$	(MgAl)LDH	5.4 mg, 200 μM V	100[(a)]	2.5	PR (0.05)	(A) 10 mL	0.1	$0.4 \cdot 10^{-1}$	[390]
13	$V_{10}O_{28}^{6-}$	(MgAl)LDH	6 mg, 2 mM V	100[(a)]	2.5	PR (0.05)	(A) 10 mL, ~2 h	$1.3 \cdot 10^{-2}$	$0.4 \cdot 10^{-2}$	[390]
14	V(IV)O complex, 0.65 wt%V	MCM-41	25 mg, 640 μM V	1000[(b)]	6000	SaA (1000)[(5)]	5 mL H_2O/ACN (4:1), 0.4M $HClO_4$ 3.5 h	448	56.6*	[402]
15	$V(V)O_2$ complex	Zeolite-Y	20 mg, 0.8; 0.6 wt% V	1000[(b)]	3750	SaA (500)	4 mL H_2O, 0.2 mL 70% $HClO_4$, 4 h	52; 67	8.5; 6.7*	[403]
16	$WO_2(O_2)(CN_2)$	PAN	~ 200 mg, 3 mM W	2000[(b)]	80	PR (1.6)[(9)]	(F) 25 mL, 50 mM, pH 5.5, 30 °C, 1.33 h	274.0*	0.6	[541]
17	WO_4^{2-}	(MgAl)LDH	2.7 mg, 50 μM W	100[(a)]	2.5	PR (0.05)[(2)]	(A) 10 mL	21.6	4.0	[390]
18	WO_4^{2-}	(MgAl)LDH	50 μM W	100[(a)]	2.50	PR (0.05)[(1)]	(A) 20 mL, pH 6–8	33	6.2	[389]

(continued)

Table 8.6 (continued)

Entry	Catalyst	Support	Total cat.	Br^- (mM)	H_2O_2 (mM)	Dye (in mM)	Conditions	TOF (1/h)	SA (mmol/g h)	Refs.
19	WO_4^{2-}	(MgAl)LDH	50 μM W	100[(a)]	5.00	MCD (0.5)[(1)]	(A) 20 mL, pH 6–8	71	13.3	[389]
20	WO_4^{2-}	(NiAl)LDH	50 μM W	100[(a)]	5.00	MCD (0.5)[(1)]	(A) 20 mL, pH 6-8	48	9	[389]
21	WO_4^{2-}	(NiAl)LDH (Taktovite)	80 mg, 4 mM W	800[(a)]	440	DHP (360)[(3)]	(B), 10 mL, 25 min	192.8 (r.t)*; 497.2 (35 °C)*	96.4 (r.t)*; 248.6 (35 °C)*	[404]
22	15 wt% $H_3PW_{12}O_4$	(ZnAl)LDH (hydrotalcite-like)	200 mg, 31 mM W	550[(b)]	550	Ph (500)	(C) 4 mL, 5 h, 110 °C	2.3*	1.4*	[542]
23	WO_x	$mZrO_2$	0.3–1.6 wt% W	100[(a)]	27.7	Ph (0.05)	H_2O	57–84	n.a.	[413]
24	15 wt% $H_4SiW_{12}O_{40}$	ZrO_2	200 mg, 31 mM W	550[(b)]	550	Ph (500)	(C) 4 mL, 5 h	2.9*	1.8*	[394]
25	15 wt% $H_3PW_{12}O_{40}$	ZrO_2	200 mg, 31 mM W	550[(b)]	550	PR (500)	(C) 4 mL, 5 h	3.0*	1.8*	[394]
26	15 wt% $H_3PW_{12}O_{40}$	ZrO_2	200 mg, 31 mM W	550[(b)]	550	PR (500)	(C) 4 mL, 5 h	3.0*	1.9*	[412]
27	12 wt% $H_3PW_{12}O_{40}$	ZrO_2	500 mg, 25 mM W	1800[(d)]		An (1800)[(5)]	(D) 10 mL, 1.5 h	46.6*	23.3**	[410]
28	15 wt% $H_4SiW_{12}O_{40}$	TiP	200 mg, 31 mM W	550[(b)]	550	Ph (500)	(C) 4 mL, 5 h	3.1*	1.9*	[414]
29	15 wt% $H_3PW_{12}O_{40}$	TiP	200 mg, 31 mM W	550[(b)]	550	Ph (500)	(C) 4 mL 5 h	3.1*	2.0*	[414]

(continued)

Table 8.6 (continued)

Entry	Catalyst	Support	Total cat.	Br^- (mM)	H_2O_2 (mM)	Dye (in mM)	Conditions	TOF (1/h)	SA (mmol/g h)	Refs.
30	$Mo(VI)O_2$ complex	PS	30 mg	2000$^{(b)}$	2000	SaA (1000)	10 mL H_2O, 2M H_2SO_4, 2 h	213.3	170.8*	[398]
31	$Mo(VI)O_2$ complex	PS	15 mg, 169 μM Mo	500$^{(b)}$	500	St	(K) 40 mL, 0.5 M $HClO_4$, 2 h, 40 °C	512.0*	230.4*	[543]
32	$Mo(VI)O_2(O_2)(CN)_2$	PAN	~100 mg, 3 mM Mo	2000$^{(b)}$	80	PR (1.6)$^{(9)}$	(F) 25 mL, 50 mM, pH 5.5, 30 °C, 1.33 h	240.0*	1.2*	[541]
33	$Mo(VI)O_4{}^{2-}$ 9.9% Mo	(MgAl) LDH	250 mg, ~ , 26 mM Mo	200$^{(b)}$	650	CA (100)$^{(4)}$	10 mL H_2O, 20 min	9.3*	9.6*	[405]
34	$Mo(VI)O_4{}^{2-}$	(MgAl) LDH	2.7 mg, 50 μM Mo	100$^{(a)}$	2.5	PR (0.05)	(A) 10 mL	20.1	3.7	[390]
35	15 wt% $H_3PMo_{12}O_{40}$	(ZnAl) LDH	200 mg, 49 mM Mo	550$^{(b)}$	550	Ph (500)	(C) 4 mL, 5 h, 110°C	1.7*	1.7*	[542]
36	15 wt% $H_4SiMo_{12}O_{40}$	ZrO_2	200 mg, 49 mM Mo	550$^{(b)}$	550	Ph (500)	(C) 4 mL, 5 h	1.7*	1.6*	[394]
37	15 wt% $H_3PMo_{12}O_{40}$	ZrO_2	200 mg, 49 mM Mo	550$^{(b)}$	550	Ph (500)	(C) 4 mL, 5 h	1.8*	1.7*	[394]
38	5 wt% Mo	ZSM-5(30) Zeolite	200 mg, 26 mM Mo	550$^{(b)}$	550	Ph (500)	(C) 4 mL, 5 h	4.3*	2.2*	[396]
39	15 wt% $H_4SiMo_{12}O_{40}$	TiP	200 mg 49 mM Mo	550$^{(b)}$	550	Ph (500)	(C) 4 mL, 5 h	1.7*	1.7*	[414]

(continued)

Table 8.6 (continued)

Entry	Catalyst	Support	Total cat.	Br^- (mM)	H_2O_2 (mM)	Dye (in mM)	Conditions	TOF (1/h)	SA (mmol/g h)	Refs.
40	15 wt% $H_3PMo_{12}O_{40}$	TiP	200 mg, 49 mM Mo	550[(b)]	550	Ph (500)	(C) 4 mL, 5 h	2.0*	2.0*	[414]
41	5–18 wt% Ti (IV)	MCM-48	7–30 mg	100[(b)]	10	PR (0.2)	(E) 3 mL, 0.1 M pH 6.5	$1.6 \cdot 10^{-1}$ [377]	$1.6 \cdot 10^{-1}$ [377]	[390,416]
42	Ti 4 wt%	MCM-41	0.5 mM	100[(b)]	2.5	PR (0.05)	(A) 10 mL	$5.6 \cdot 10^{-2}$	$4.7 \cdot 10^{-2}$	[390]
43	Ti (IV)	TS1 zeolite	200 mg	550[(b)]	550	Ph (500)	(C) 4 mL, 5 h	n.a.	1.8*	[396]
44	Ti (IV)	TS1 zeolite	0.5 mM Ti	100[(a)]	2.5	PR (0.05)	(A) 10 mL	$1.0 \cdot 10^{-3}$	$2.5 \cdot 10^{-4}$	[390]
45	$Ti(IV)O_2$ 4 wt%	ZrP	200 mg, 41.8 mM Ti	550[(b)]	550	Ph (500)	(C) 4 mL 5 h	2.0*	1.6*	[325]
46	$Ti(IV)O_2$ 4 wt%	TiP	200 mg, 41.8 mM Ti	550[(b)]	550	Ph (500)	C) 4 mL 5 h	2.7*	2.3*	[325]
47	$Zr(IV)/SO_4^{2-}$	SBA-15-620	50 mg	370[(a)]	310	Ph (74)[(7)]	(B) (7:3) 6.75 mL, 4 h	n.a.	2.3*	[417]
48	Cr 5 wt%	ZSM-5(30) Zeolite	200 mg, 48 mM Cr	550[(b)]	550	Ph (500)	(C) 4 mL, 5 h	1.7*	1.7*	[396]
49	Fe(III) complex	PS	50 mg, 3.66% Fe	2000[(b)]	2000	SaA (1000)[(10)]	H_2O, 10 mL, 2 M conc. H_2SO_4	99.0*	65.3*	[420]
50	$Fe(bipyridine)_2$	NaY zeolite	50 μM Fe	100[(a)]	2.50	PR (0.05)	(A) 20 mL	$1.3 \cdot 10^{-3}$	$7.8 \cdot 10^{-4}$	[389]

(continued)

Table 8.6 (continued)

Entry	Catalyst	Support	Total cat.	Br^- (mM)	H_2O_2 (mM)	Dye (in mM)	Conditions	TOF (1/h)	SA (mmol/g h)	Refs.
51	Fe 5 wt%	ZSM-5(30) Zeolite	200 mg, 4.5 mM Fe	550$^{(b)}$	550	Ph (500)	(C) 4 mL, 5 h	2.2*	2.0*	[396]
52	Cu(II)	PS	15 mg, 1.1 mM Cu	440$^{(b)}$	1000	Ph (400)$^{(11)}$	(C) 5 mL, 2 h	174.8*	64.7*	[421]
53	Cu(II) complex	SiO_2	50 mg, 563 μM Cu	550$^{(b)}$	550	Ph (500)$^{(5)}$	(C) 4 mL, 60 °C, 1.5 h	586.7*	26.4*	[422]
54	Cu 5 wt%	ZSM-5(30) Zeolite	200 mg, 3.9 mM Cu	550$^{(b)}$	550	Ph (500)	(C) 4 mL, 5 h	2.2*	1.7*	[396]
55	Re 0.04–0.67 wt%	$mZrO_2$-630/700	20 mg	100$^{(a)}$	10.3	PR (0.05)$^{(5)}$	H_2O	n.a.	0.1–1.3	[424]
56	Se 2.5 mol%	Xerogel		1400$^{(c)}$	350	PA (140)	(F) pH 6.2, 5 mL	$9.1 \cdot 10^{-3}$*	n.a.	
57	Te 2.5 mol%	Xerogel		1400$^{(c)}$	350	PA (140)	(F) pH 6.2, 5 mL	$4.1 \cdot 10^{-2}$*	n.a.	[426]
58	$NaHSO_4$	SiO_2	100 mg	240$^{(d)}$		Ch (200)	(G or H) 5 mL, 0.5 h	n.a.	18.6*	[427]
59	H_2SO_4	SiO_2	15 mg	210$^{(d)}$		Ph (200)$^{(5)}$	(I) 5 mL, 0.5 h	n.a.	110.7*	[428]
60	V	V_2O_5	91 mg, 40 mM V	120$^{(e)}$	636	*o*-cresol (40)	(B) (1:1), 25 mL, pH ~ 2.1, 5°C, 1.5 h	0.6*	6.7*	[433]
61	V	V_2O_5 nw	0.02 mg, 220 μM V	1$^{(b)}$	0.01	MCD (0.05)	(M) 100 mM, 1 mL, pH 8.3	$1.4 \cdot 10^{-2}$*	$1.5 \cdot 10^{-1}$*	[430]

(continued)

Table 8.6 (continued)

Entry	Catalyst	Support	Total cat.	Br^- (mM)	H_2O_2 (mM)	Dye (in mM)	Conditions	TOF (1/h)	SA (mmol/g h)	Refs.
62	V	V_2O_5 nw	1 mg	40 (KI)	1	TB (25)*(16)*	(L) pH 8.0, 1 mL, 1 min	136.4*	1500*	[432]
63	Ce	CeO_2 nr	50 mg, 291 mM	500 (I_2)	2 mL 35%	DMB (1000)	H_2O, 1 mL, 100 °C, pH 3.0	6.7*	38.8*	[442]
64	Ce	CeO_2 np	80 mg, 613 mMl	250(b)	6 bar O_2	TMB (83)(14)	0.75 mL, H_2O, 100 °C, 20 h	$6.1 \cdot 10^{-3}$*	$3.5 \cdot 10^{-2}$*	[437]
65	Ce	CeO_2 nr	25 µg, 145 µM	25(a)	0.3	PR (0.01)(15)	1 mL, H_2O, (v_{max} = 0.4µM_{HOBr}/min)	0.2*	1.0*	[221]

(1) DMT, $^1O_2$, dimedone, methoxystyrene, methylcyclohexene; (2) DMT, $^1O_2$, MCD, (3) oxidative hydrohalogenation, alkoxyhalo-adducts, anhydro-sugars, (4) halodecarboxylation of aromatic derivates, (5) diverse (hetero)aromatics, (6) oxybromination as side reaction, (7) other aromatics, dimethylaniline, methylaniline, aniline, Ph, SaA, (8) (a)cyclic ketone, amide, β-ketoester, (9) aniline, *m/p/o*-aminophenol, *m/p/o*-nitroaniline, quinol, pyrogallol, resorcinol, acetamide, salicylaldehyde, o-methoxytoluene, catechol, (10) oxidation of styrenes, benzyl alcohols, various alkenes, sulfides, aromatic alcohols, ethylbenzene, (11) sulfoxidation, aromatic (aliphatic) compounds, (12) hydroxylation of Ph, (13) photocatalytic activity, anisole, olefinic substrates, (14) halogen exchange, (15) halogenation of lactones, (16) *L*-dopamine, (17) anisole, toluene, ethylbenzene, Ph

$$nBr^{ox}_{(\mathrm{mmol})} = x \cdot nBr_x dye_{(\mathrm{mmol})} \tag{8.12}$$

$$nBr^{ox}_{(\mathrm{mmol})} = \sum \mathrm{conversion}_{(\%)} \cdot x \cdot nBr_x dye_{(\mathrm{mmol})} \tag{8.13}$$

Sels et al. [389, 390] reported a specific activity of 600 mmol of oxidized bromine per gram of enzyme per hour for V-bromoperoxidase isolated from *C. officinalis* (Table 8.6, entry 1). The activity decreases by two orders of magnitude after immobilization on surfaces (Table 8.6, entry 2). TOF and SA values of synthetic catalysts are comparable with those of native enzymes. It is low for magnesium/aluminum layered double hydroxides (Mg, Al-LDH (Table 8.6, entry 3), but sizeable for zirconia and zeolites (Table 8.6, entry 4–7).

Many structural and functional enzyme mimics modeling the active site of marine bromoperoxidases have been devised. Besides vanadium complexes, molybdenum and tungsten compounds containing the metal atoms in the highest oxidation states (ionic radii of V^{5+}: 46 pm, Mo^{6+}: 59 pm and W^{6+}: 60 pm) [391] are the most efficient HPO mimics with metal-η^2-peroxo-intermediates. Oxo- and peroxo-ligands are preferred according to Pearso's HSAB model [392]. Many compounds suffer from the formation of catalytically less or inactive polyoxometalates (POM), heteropolyacids (HPA), or gel-like structures (e.g., vanadium pentoxide gels [393]) depending on the solution pH.

Attempts have been undertaken to overcome this problem by "heterogenization". Several homogenous organo-transition metal complexes have been immobilized on polymer supports [268], layered double hydroxides, LDHs) [389, 390], zirconia (ZrO_2) [394], silica [395], or in zeolites [396]. A series of oxovanadium(V) and oxomolybdenum(VI) complexes have been attached to a polymer backbone through the anchor groups of a modified salen-like structure (Table 8.6, entry 8) [397]. Likewise, vanadium(V) Schiff-base and related complexes were grafted to a polystyrene (PS) backbone without [398] or with spacer units [399–401] (Table 8.6, entry 9–11). The structure and synthesis of a PS-bound oxovanadium(V)-complex are shown in Fig. 8.9a [399]. The different activities of the polymer-bound catalysts (Table 8.6, entry 8–11) are related to differences in the degree of loading and the different coordination environments of the vanadium centers. Ligands with electron donor atoms (N- and O-functions) have an impact on the accessibility and the stability of the metal oxidation state, thereby affecting the oxygen storage (η^2-peroxo groups) and release (HOBr formation) behavior. The PS-grafted compounds have high SA and TOF values and high selectivity with respect to para-bromo salicylic aldehyde [397, 398]. Their catalytic performance can be enhanced by decreasing the pH (pH $\approx$ 2–4), increasing the temperature (e.g., 80 °C) and the amount of hydrogen peroxide [397, 398].

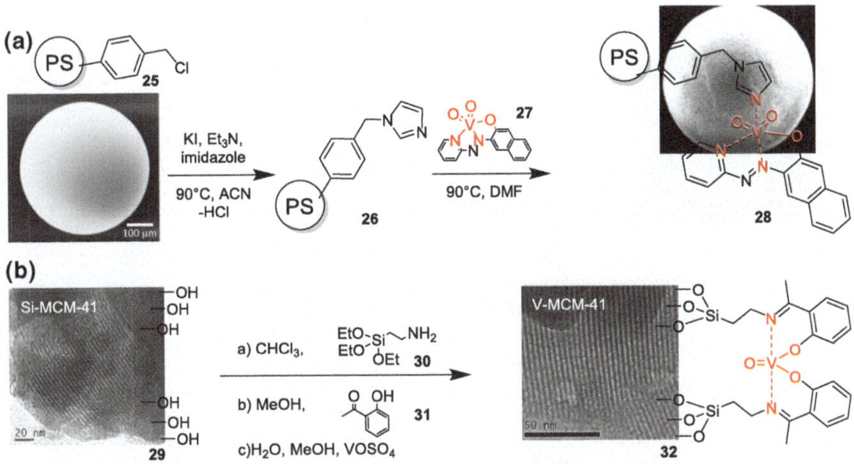

Fig. 8.9 Synthesis of **a** the polystyrene (PS)-grafted dioxovanadium(V) complex **28** [399] and **b** a vanadium complex incorporated in mesoporous silica (MCM) **32** [402]. The Fe-SEM and the TEM images (left) show the undoped support and (right) the incorporation of the transition metal. The surface of the spherical PS matrix is roughened for V-PS **28**. Fe-SEM and TEM images reprinted with permission from Refs. [399, 402]. Copyright (2014) Elsevier. Copyright (2011) American Chemical Society

LDH-supported catalysts [390] are up to four orders of magnitude less reactive (Table 8.6, entry 12) than their PS-grafted counterparts (Table 8.6, entry 3). Likewise, polyoxometalates showed lower oxidative bromination activities than the corresponding metallates, as demonstrated by the decavanadate/vanadate pair (Table 8.6, entry 13) [390]. Vanadium(IV) complexes grafted on mesoporous silica (MCM) [402] or dioxovanadium(V) Schiff-base complexes [403] in zeolite-Y (Fig. 8.10b) showed enhanced TOF and SA values (Table 8.6, entry 14, 15). The larger active surface areas of MCMs lead to higher activity combined with good selectivity toward monobrominated salicylaldehyde.

Molybdenum and tungsten complexes supported on organic matrices are established as well (Table 8.6, entry 16, 30–32). The metal coordination is comparable to that of the V-complexes (Figs. 8.10, 8.11), and the overall catalytic activities are similar (TOF for V 108, W 274, Mo 240 h^{-1}; Table 8.6, entry 8, 16, 34) despite the use of different assays, reaction conditions, and polymer supports [PS or polyacrylonitrile (PAN)]. The highest SA (634 mmol $Br^{ox} g_{cat}^{-1} h^{-1}$) and TOF (2254 h^{-1}) values have been reported for a vanadium-PS complex (Table 8.6, entry 10). Still, vanadium and molybdenum on polymer supports require acidic conditions (H_2SO_4, $HClO_4$), while tungsten catalysts are still active (TOF ~ 274 h^{-1}) in phosphate buffer (pH 5.5, Table 8.6, entry 16).

There are distinct trends for inorganic supports. LDHs exchanged with tungstate and molybdates retain their catalytic activity (SA) in different assays (PR, Ph, MCD;

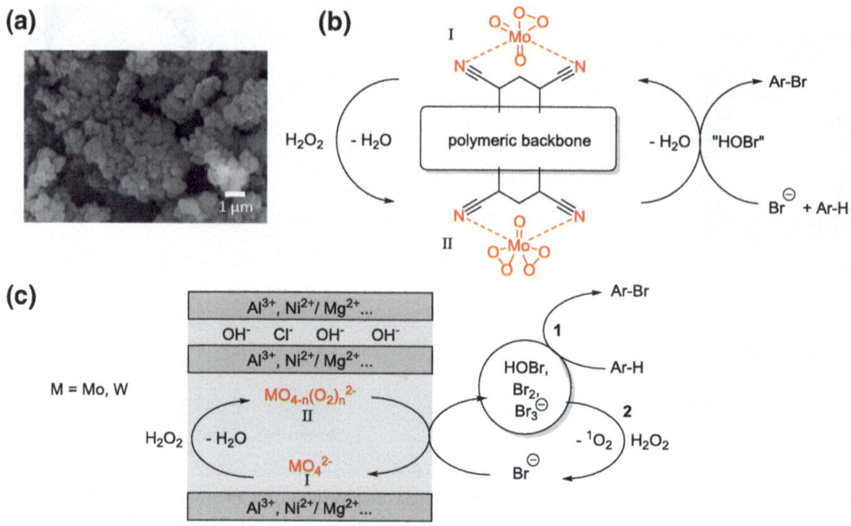

Fig. 8.10 Proposed mechanism for the oxidative bromination with immobilized Mo-PAN complexes. **a** Fe-SEM image of the Mo-PAN-complex (reprinted with permission from Ref. [541]), Copyright (2013) Elsevier. **b** Mo-PAN catalyzed oxidative bromination [541], **c** oxidative bromination catalyzed by $WO_4{}^{2-}$ or $MoO_4{}^{2-}$ exchanged layered double hydroxides (LDHs). Both reactions involve η^2peroxo species (I to II). A two-electron oxidation of bromide to bromite (HOBr, ^-OBr, Br_2, $Br_3{}^-$) is compatible with the electrophilic bromination of an aromatic substrate (**1**). With excess H_2O_2 (**2**) $^1O_2$ is generated via a peroxo-bromide species [389]

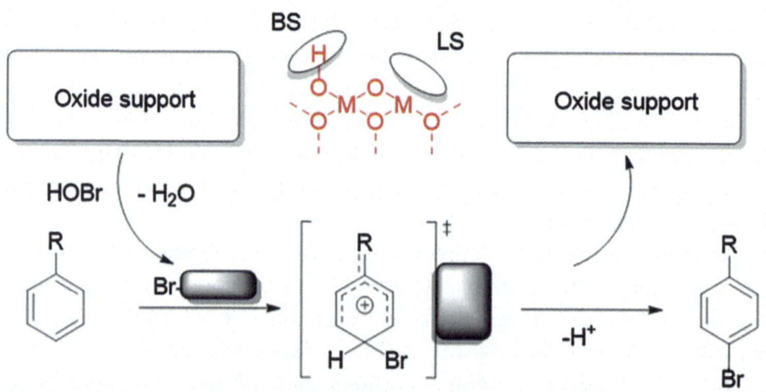

0 I II

Fig. 8.11 Halogenation of reactions with metal (M = Zr, Si, Al or Ti, Mo, W, etc.) substituted oxide supports (zirconia, zeolites, silica). Brønsted (BS) and Lewis (LS) acidic sites are involved in the oxidative halogenation. BS: HOBr formation, LS: peroxo group binding and stabilization of σ-complexes 0-II (Wheland intermediates) [417]

Table 8.6, entries 17–22, 33–35) at neutral pH (pH ≈ 6–8). The increased hydrobromination rate (MCD reaction) of 2,3-dihydro-4*H*-pyran (DHP) with tungstate-exchanged takovite (Ni, Al-LDH) shows that the bromination rates can be enhanced by adjusting the reaction conditions and substrate dyes (Table 8.6, entry 20, 21). The activity can be further enhanced by increasing the reaction temperature [404]. DHP and other cyclic enol ethers are susceptible for hydro- or methoxybromination, iodination, and Br-assisted epoxidation reactions [404]. These intermediates are useful for the synthesis of glycoconjugates or oligosaccharides under mild conditions in water [404]. WO_4^{2-}/LDH-based HPO mimics are more effective than their MoO_4^{2-} or VO_4^{3-} analogues [266, 404]. The metal ions of the LDHs affect the reactivity of the catalyst in the order (Mg, Al) ≥ (Ni, Al; takovite-like) ≥ (Zn, Al; hydrotalcite-like)-LDH [389, 390].

Replacement of LDH-intercalated WO_4^{2-} or MoO_4^{2-} by polyoxometalates (e.g., heteropolyacids, HPA) decreases the activity to the level of self-activity. This can be rationalized from the structure of the highly active LDH/WO_4^{2-} (Table 8.6, entry 19) where isolated WO_4^{2-} anions are mainly located at the edge positions [389]. Peroxotungsten intermediates identified by UV–Vis spectroscopy appear to be the key species for the formation of hypohalous acid (HOBr). The reaction pathway shown in Fig. 8.10 is assumed to be identical for tungstate and molybdate [389, 405, 406]. The mechanism via Br^+ species has been demonstrated with 2,3-dimethoxytoluene (DMT), as the methyl group would be attacked in a radical mechanism as well. The formation of ring-substituted Br-DMT is compatible with an electrophilic bromination mechanism [390]. The stability of the exchanged catalyst in LDHs is critical in aqueous media because of deactivation due to leaching and polycondensation. Decatungstate ($W_{10}O_{32}^{4-}$) encapsulated in the organic resin Amberlite IRA-900 shows reasonable activity for photocatalytic bromination, bromohydrin formation, or epoxidation [407].

An alternative approach has been pursued by impregnating mesoporous oxides supports like ($mZrO_2$) [408], silica (MCM) [402, 409], or zeolites with transition metal ions. The exposed hydroxyl groups and the large surface area of the support materials allow the attachment of organometallic complexes or W and Mo heteropolyacids (HPA) in combination with Si and P (Table 8.6, entry 22, 24–29, 35–37, 39, 40). Mononuclear metal complexes are typically more active than polyoxometalates. The activity increases when halogenating co-agents like *N*-bromosuccinimide (NBS) [410] or Br_2 [411] are used instead of bromides. The calcination temperature of the catalyst after doping and the degree of doping are crucial for the activity [394, 412]. An optimum doping of 15% may be rationalized by the presence of highly dispersed WO_x species that are required for efficient oxidation of bromine and enough free surface area for dye adsorption without blocking catalytic sites [413]. HPA(Mo)-impregnated titanium phosphate (TiP) showed increased SA values compared to molybdenum Keggin clusters [394, 414].

The activity and regioselectivity (e.g., increased yield of *para*-bromophenol) [394] of oxidative bromination reactions depend on the number of exposed Lewis acidic metal sites on the catalyst surface. This leads to an enhanced yield of halogenated compounds through the formation of tungsten or molybdenum peroxo species, because Brønsted acidic sides are assumed to be involved in the electrophilic bromination process [412, 415]. Figure 8.11 illustrates the proposed reaction mechanism on oxide supports (zirconia, silica, zeolites). Transition metal catalysts on oxide supports can be recycled. They show reduced activity compared their polymer-grafted counterparts, although they are stable in acidic media. Regiospecific bromination is possible, and catalysts leaching can be suppressed.

Ti/MCM materials have been the first heterogeneous oxidative halogenation catalysts in neutral aqueous and organic–aqueous media in the absence of acids [416]. Titanium oxide is a photocatalysts that can carry surface-bound peroxo groups as possible intermediates for electrophilic halogenation reactions. Still, no HPO- or HO-like activity has been reported for the pure TiO_2. Grafting Ti(IV) compounds on mesoporous silica or zeolites (Table 8.6, entry 41–44) results in very low specific activities (Table 8.6, entry 4–7). The situation is improved for mesoporous silica [409], cubic (MCM-48) being preferred compared to the hexagonal pore array (MCM-41) [416]. TiO_x cluster formation has been assumed as a possible reason for the activity decrease of different catalysts with increasing amounts of Ti [416]. The activity is slightly increased when titanium phosphate (ZrP, Table 8.6, entry 45) was used as a support instead of zirconium phosphate (TiP, Table 8.6, entry 46). The proposed Ti (IV) peroxo complex on a titanium phosphate surface shows values comparable to those of HPA-impregnated TiP supports (Table 8.6, entry 28, 29, 39, 40). Ti pillared on ZrP or TiP in combination with the calcination temperature control the regioselectivity of oxidative bromination reactions. Materials calcined at 110 °C exhibit *ortho*-selectivity in the bromination of phenol, whereas *para*-selectivity was observed for catalysts calcined at 500 °C [325]. Zirconium(IV) oxide showed moderate oxidative bromination activity (Table 8.6, entry 4, 5), but the reactivity was increased for sulfated zirconia on mesoporous silica (SBA-15-620, Table 8.6, entry 47). *para*-selectivity was observed for the bromination of phenol. The good catalytic performance of the sulfated zirconia on SBA-15 obtained by calcination at higher temperature was attributed to the good dispersion of sulfate groups to give sulfated silica-zirconia with efficient redox behavior and a high density of acidic sites (Fig. 8.11) [417].

Chromium is expected to catalyze oxidative halogenation reactions through metal-bound peroxo groups, as shown by Narender et al. [396] for Cr-zeolites (CrZSM-5 or 30, Table 8.6, entry 48). Although the selectivity for *para*-bromophenol is high, the activities are only moderate (Table 8.6, entry 6, 7). In addition, chromate can be formed in side reactions with subsequent Cr leaching (~3%) [396]. No HPO-like activity has been reported for manganese compounds. Pillared zeolites with Mn–bipyridine complexes showed no bromination activity [389, 390].

Only a few Fe and Cu compounds show oxidative halogenation activity. Plausible reasons may be the redox potential for higher oxidation states and the inability to bind peroxo units. Enzymes overcome this problem with specially designed protein scaffolds having active side pockets that allow the stabilization of organic radicals by iron and copper complexes rather than the formation of hypohalites. The peroxidase-, catalase-, and SOD-like activities of iron and copper compounds have been reviewed before [418, 419]. Iron and copper compounds have been used as HG mimics. When complexed by chloromethylated polystyrene iron(III) is an active and reusable catalyst for oxidation and oxidative bromination reactions (Table 8.6, entry 49) [420] with an activity close to that of polymer-bound vanadium and molybdenum compounds (Table 8.6, entry 9, 30). Figure 8.14a illustrates the synthesis of a resin-supported iron(II) salen complex which can be recycled up to five times. An analogous copper complex on polystyrene has TOF values comparable to those of Mo-, W-, and V-containing polymers (Fig. 8.12b, Table 8.6, entry 50). The immobilized compounds also catalyze the oxidation of thioethers to sulfoxides or sulfones in a selective manner [421]. Similarly, copper(II) perfluorophthalocyanine immobilized on silica gel (Fig. 8.12c) showed promising activity and recyclability (up to six times) without loss of activity (Table 8.6, entry 53) [422]. Thomas and Raja [409], Raja and Ratnasamy [423] reported the oxyhalogenation of aromatic compounds with hexadecachloro copper phthalocyanine complexes incorporated in zeolites ($CuCl_{16}Pc$–Na–X, 0.27 wt% Cu) with TOF values ~ 160–380 h^{-1}. These values, however, describe the total substrate conversion including simple oxidation reactions. Halogenated products (chlorination and bromination) contribute only 10–30 wt% in assays with benzene, toluene, phenol, aniline, anisole, or resorcinol [423]. With cobalt or iron instead of copper the TOF value decreases (TOF(Co) = 107 h^{-1}, TOF(Fe) = 210 h^{-1}), but the relative amount of brominated products increased (up to 50% for Co and 40% for Fe) [423].

Zeolites showed activity with H_2O_2 and O_2 as oxidizing agents [423]. The impregnation of zeolites with iron bipyridine complexes suppressed this activity (Table 8.6, entry 50). The deposition of iron or copper salts (instead of coordination compounds)

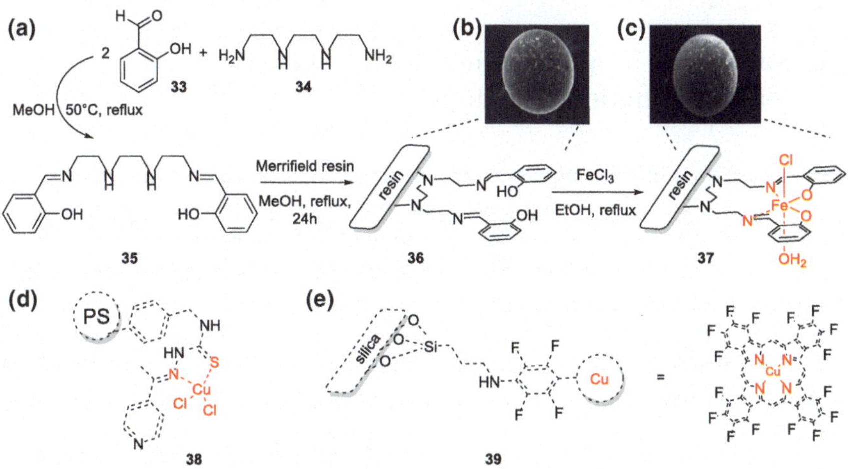

Fig. 8.12 **a** Synthesis of the polystyrene (PS) anchored Fe(III) complex **37** and **b**, **c** of the corresponding Fe-SEM images (left: PS-anchored ligand; right: PS-anchored Fe(III) complex (reprinted with permission from Ref. [420]). Copyright (2013) Springer Nature) **d** PS-anchored Cu(II) complex **38** [421], **e** silica-supported Cu(II) complex **39** [422]

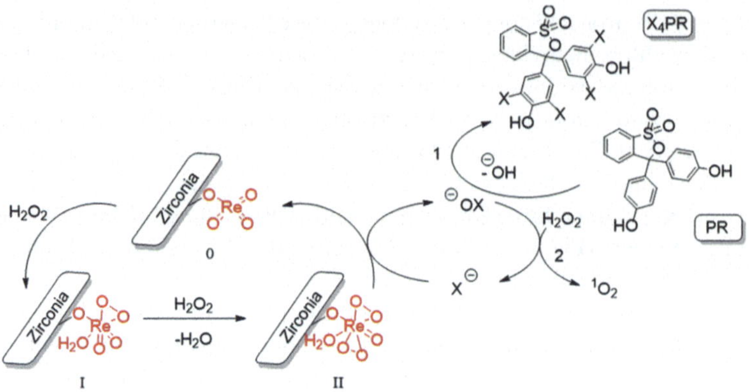

Fig. 8.13 Hypothetic mechanism of the **1** oxidative halogenation (X = Br) reaction of PR to X_4PR over Re/m-ZrO_2 catalyst (0-II) and **2** side reaction yielding O_2 (scheme in analogy to [424])

on inorganic supports yielded similar activities as for surface grafted transition metal oxides (Table 8.6, entry 51, 54).

Although rhenium and rhodium compounds showed oxidative bromination activity in homogenous reactions rhenium-promoted mesoporous zirconia [424] showed this activity only under heterogeneous conditions (Table 8.6, entry 55). Re-assisted chlorination of phenol and iodination of phenol red has been reported [424]. The recyclability of the catalyst (four cycles) was explained by Re–O–Zr condensation. Figure 8.13 illustrates the HPO-like mechanism involving Re-peroxo groups.

8.8 Supported Non-transition Metal Catalysts for Halogenation Reactions

Besides transition metal compounds, metal-free systems have been reported as heterogeneous oxidative bromination catalysts. Surprisingly, the HPO-like activity of selenium and tellurium compounds is not limited to homogenous reactions [266]. Organo-Se or organo-Te complexes stabilized in silica gel ("xerogel") display oxidative bromination activity for the substrate 4-pentenoic acid [425, 426] (Table 8.6, entry 56, 57). The product, 4,5-dibromopentanoic acid, can be cyclized subsequently to the corresponding bromolactone (Fig. 8.14a). This bromolactone bears structural analogies to the natural AI-antagonist bromofuranone (Fig. 8.5). The covalently modified xerogel was synthesized by nucleophilic substitution of a chlorine-substituted silica precursor by diphenyl ditelluride or diselenide, respectively (Fig. 8.14a).

Different oxidic forms of organotellurides and organoselenides may be involved in the proposed mechanism of the oxidative halogenation (Fig. 8.14b) suggested [425]. Depending on the reaction with hydrogen peroxide, telluride is oxidized to

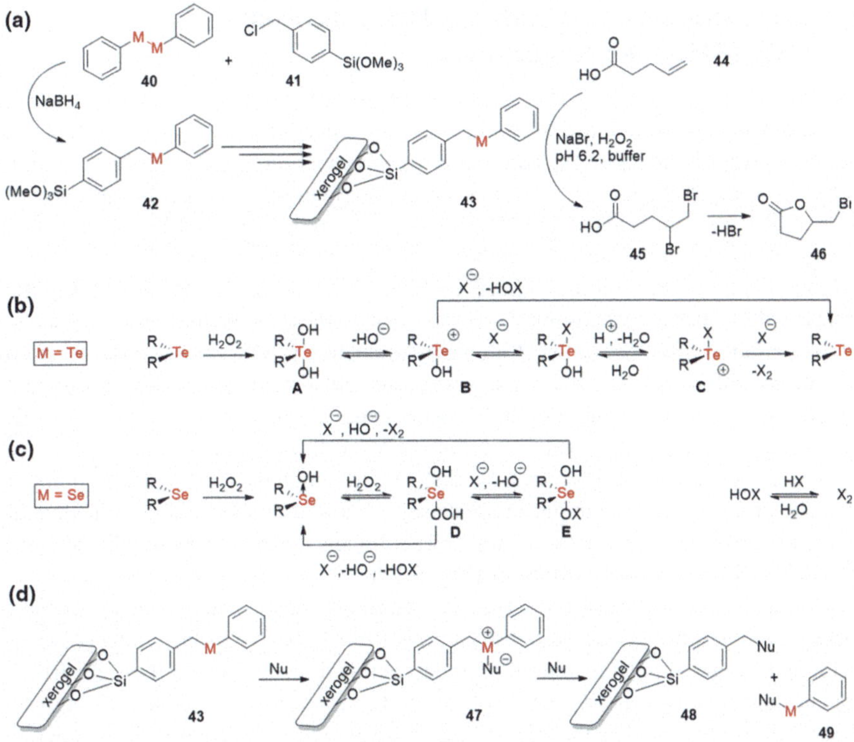

Fig. 8.14 **a** Synthesis of chalcogenide xerogel (M = Se or Te) **43** and the oxidative bromination of 4-pentoic acid **44** to bromofuranone **46** with modified xerogel **43**, **b**, **c** proposed Se and Te intermediates, and **d** proposed inactivation mechanism of the xerogel **43** (Nu = e.g., H_2O_2, OH, Cl). Illustration adapted from Refs. [425, 426]

tellurium oxide with subsequent hydration to dihydroxy tellurane (A). Organoselenides showed a different mechanism and formed hydroxy perhydroxy selenanes (D) with H_2O_2. Oxidized halide species were formed subsequently [or the halide exchanged form (E)] [425]. A loss in activity was observed after recycling (up to four times) due to a subsequent nucleophilic (Nu) attack at the Se/Te center [425]. This leads to the formation of phenyl chalcogenic acids (Nu = H_2O_2/OH^-) and phenyl chalcogenic chloride (Nu = Cl^-) (Fig. 8.14c). Cyclic ketones and lactams [427] or (hetero)aromatics [428] were brominated with sodium bisulfate or sulfonated silica (Table 8.6, entry 57, 58) as Brønstedt acids in the presence of *N*-bromosuccinimide (NBS) instead of bromide and H_2O_2 as oxidant. Zeolite (HZSM-5) bound NBS as "hidden" bromine source allowed bromination of aniline in CCl_4 [429].

8.9 Transition Metal Oxides as Heterogeneous HPO/HG-Like Nanozymes

In analogy to vanadium enzymes and complexes, vanadium pentoxide (V_2O_5) is an active catalyst for the regioselective oxidative halogenation, specifically bromination [430, 431] and iodination [432] of aromatic compounds even in the absence of acids (Table 8.6, entry 59–62) [433]. Regioselectivity is promoted by the substrate (charge-donating and/or steric hindrance) and depends on the source of bromide, e.g., tetrabutylammonium bromide (TBAB), which can stabilize Br_3^- species during the reaction [433]. Peroxovanadium(V) species were believed to be the active species in the oxidation of bromide. Later, Rothenberg and Clark [431] optimized the reaction conditions with respect to total costs, waste, risk, and hazard factors and showed that HBr or Br_2 could be replaced by halide salts under acidic conditions. Strong acids as co-catalyst enhance the catalytic activity of V_2O_5 through the formation of active Br^+ species and the disproportionation of H_2O_2. Although the total costs, the amount of waste products, and the required halide salts and acids are favorable for technical applications, increasing environmental and ecological awareness eventually demand nontoxic and sustainability solutions (Fig. 8.15).

Nanomaterials are much more reactive than bulk metal oxides because of their surface/volume ratio. Their properties are controlled also by surface defects [434]. Point and extended defects determine their chemical reactivity and their electrical, optical, and mechanical properties [435], and are key to heterogeneous catalysis [436]. This is illustrated by the HPO-like activity of highly defective V_2O_5 nanowires even in seawater at ambient temperature (Table 8.6, entry 61) [430] and ceria nanorods [221]. Based on the kinetic parameters and the crystal structure, a surface-bound η^2-peroxo group was proposed as transient active species in harmony with the proposed reaction mechanism for HPO (Fig. 8.16) [430].

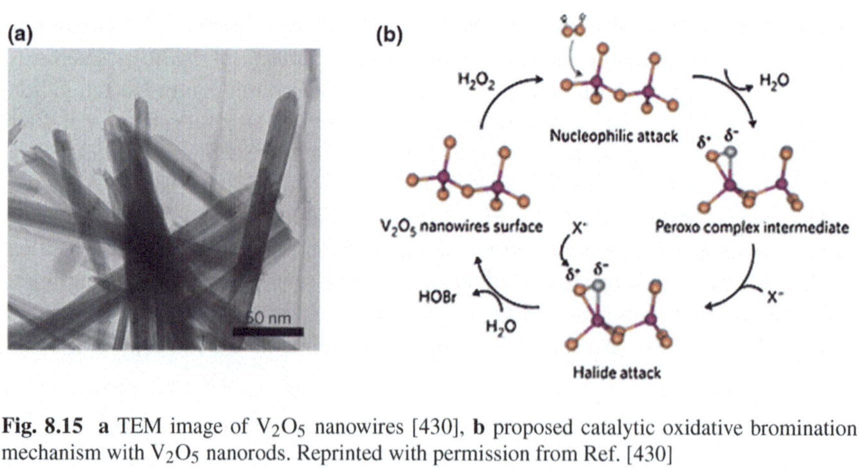

Fig. 8.15 **a** TEM image of V_2O_5 nanowires [430], **b** proposed catalytic oxidative bromination mechanism with V_2O_5 nanorods. Reprinted with permission from Ref. [430]

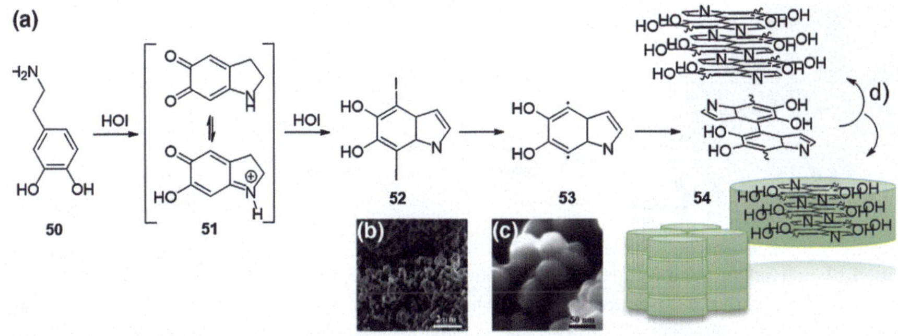

Fig. 8.16 **a** Proposed formation mechanism of melanin-like biopolymer **54** catalyzed by V_2O_5 nanowires, KI, H_2O_2, and dopamine [432]. **b**, **c** SEM image displaying the fibrillary and porous structure of the spherical domains (diameter of ~50 nm). Reprinted with permission from Ref. [432]. **d** Proposed structure of the resulting melanin-like biopolymer **54**

Besides being active in oxidative bromination, V_2O_5 nanowires catalyze iodination reactions. Iodination of dopamine leads to the formation of melanin with SA and TOF values that are two times higher than those of the native enzyme. The structure of the product and the proposed reaction mechanism are shown in Fig. 8.16. The synthetic biopolymer is analogous to the natural fibrillar and porous structure with spherical domains and diameters of ~50 nm. The mechanism is based on the catalytic formation of OI^-/HOI catalyzed by V_2O_5 nanowires. Dopamine is oxidized by the hypoiodite intermediate and undergoes subsequent cyclization, iodination, and radical polymerization. The planar rings assemble to nanoplates via π-stacking and finally aggregate to clusters and filamental biopolymers. Although V_2O_5 is a potent haloperoxidation catalysts, serious shortcomings for everyday applications are its toxicity and possible carcinogenicity under REACH safety criteria.

Nanocrystalline cerium dioxide (nanoceria) is a viable alternative. In oscillating reactions, the Ce^{4+}/Ce^{3+} redox couple is known to catalyze the halogenation of malonic acid in the Belousov–Zhabotinsky reaction [304]. CeO_2 is active in halogen exchange reactions and the O_2-dependent halogenation (oxyhalogenation) of methoxybenzenes (Table 8.6, entry 64) [437]. These reactions were carried out between 100 and 140 °C in organohalide solvents that simultaneously served as halide source. A plausible mechanism was proposed based on the redox potentials of the Ce^{4+}/Ce^{3+} (1.44 V), $Cl_2/2Cl^-$ (1.36 V), $Br_2/2Br^-$ (1.07 V), and $I_2/2I^-$ (0.58 V) redox couples. Fluorination was not possible ($F_2/2F^-$: 2.87 eV). The halogenation proceeds slowly under anaerobic conditions, is accelerated in the presence of O_2 (6 bar), and is stopped by radical scavengers [437]. Ce^{3+} is re-oxidized under aerobic conditions (Fig. 8.17b).

Reactions with acetylated arenes and substitution reactions with deuterated starting compounds showed an inverse kinetic isotopic effect for chlorine and iodine which is typical for electrophilic aromatic substitutions. The basic sites of nanoceria facilitate a heterolytic cleavage of the organic halide, which are oxidized via single electron transfer reactions to the corresponding element. Thus, the reaction can be

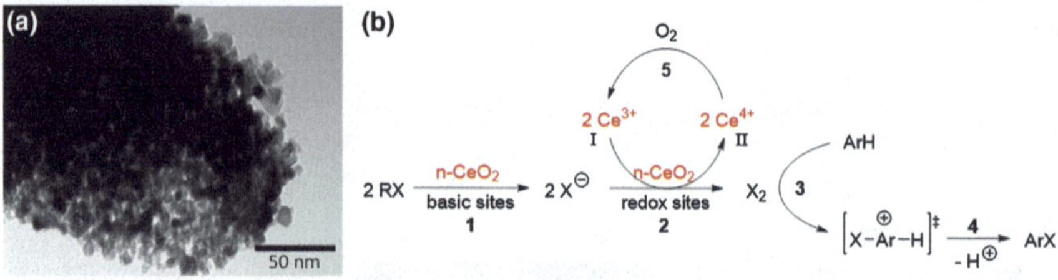

Fig. 8.17 **a** Transmission electron micrograph of ceria nanoparticles (diameter $\leq$ 10 nm). **b** Proposed mechanism of the O_2-dependent halogenation reaction of electron-rich aromatic compounds (ArH) catalyzed by nanoceria (n-CeO_2) and organic halides (RX, X = Cl, Br, I). The basic sites of n-CeO_2 aid in the decomposition of the organic halide (**1**). The redox sites allow the oxidation by one-electron transfer reactions (**2**). The X^+ species oxidize the arenes via σ complexes Wheland intermediates (**3, 4**). The catalyst is reactivated by O_2 (**5**, I-II). TEM image reprinted with permission from Ref. [437]. Copyright (2013) American Chemical Society

described as a combination of an O_2-dependent HG and HPO-like mechanism. Particle size, surface area, and morphology are very important, as bulk ceria or iron(III) oxide did not show any activity. The exposed crystal planes determine the reactivity in the order cube < octahedra < nanorods [437, 438]. The catalyst was inactivated by substitution of oxide by halide in the CeO_2 structure (indicated by a color change from pale yellow to orange). Another shortcoming of using molecular oxygen is the formation of volatile side products such as methanol and methyl bromide.

The activity of nanoceria could be extended from O_2-dependent HG-like to more controllable H_2O_2-dependent HPO-like reactions. CeO_2 nanoparticles are functional mimics of halogenating enzymes in aqueous media containing halide anions, hydrogen peroxide (rather than O_2) as co-oxidant, and phenol red as substrate [221]. The oxidative bromination reaction of phenol red to bromophenol blue allowed a comparison with native enzymes and an analysis of the reaction kinetics in order to derive a reaction mechanism. The reaction follows a pseudo-third order kinetic (in total, Eq. 8.14) with k_{gen} as general reaction constant, and $[S]$ and $[S]_0$ as substrate and initial substrate concentrations. Other specific kinetic parameters were determined by Lineweaver–Burk analysis [221] according to

$$v = \frac{d[\text{Br}_4\text{PR}]}{dt} = k_{\text{gen}}\left[\text{CeO}_{2-\text{x}}\right]_0[\text{NH}_4\text{Br}]_0[\text{H}_2\text{O}_2] \tag{8.14}$$

Even for very low catalyst and substrate concentrations (25 μg mL^{-1} catalyst, 300 μM H_2O_2 and 25 mM halide salt, Table 8.6, entry 65), similar to those of marine HPOs in natural environments, CeO_{2-x} nanorods showed a catalytic activity comparable to that of immobilized enzymes (Table 8.6, entry 4 and 65). The activity and the reaction rate can be increased by adjusting the reaction conditions, e.g., increasing the temperature, the concentrations of H_2O_2 and bromide [221], or by increasing the defect density of the CeO_{2-x} nanorods [439].

A radical mechanism was proposed for the formation of halogenated ROS based on experimental data and calculations with high-level methods [221], which highlights the key role of Ce^{3+} and the Ce^{3+}/Ce^{4+} ratio. By virtue of the Ce^{4+}/Ce^{3+} redox potential cerium can switch between its tetra- and trivalent states. As a result, ceria is nonstoichiometric with oxygen vacancies in the fluorite lattice. Trivalent Ce^{3+} is associated with reactive defect sites in the CeO_{2-x} structure, which are located preferentially at the particle surface (as anticipated from the ionic radii and demonstrated by XPS measurements) [221]. A Ce^{3+} defect density of ~0.5 wt% was determined from magnetic susceptibility measurements. The proposed radical process contrasts with the mechanism suggested for vanadium, tungsten, or rhenium compounds involving metal-η^2-peroxo-intermediates, but it is in accordance with the mechanism suggested for some natural heme-containing HPOs [139, 378, 440]. The difference in the active halogen species (X_2 [437] or HOX [221]) may be related to the co-oxidant. The degradation of celestine blue (CB) shows HOX to be the dominant species [221].

Ceria nanorods preferentially expose their (110) facets. Quantum chemical calculations at various levels of theory (periodic calculations at the generalized gradient approximation–density functional theory (GGA-DFT) level and calculations of small finite model compounds at the metahybrid-DFT level) including hydration (Ce^{4+}–OH^- and Ce^{3+}–H_2O coverage, Fig. 8.18a, b) support the experimental findings with HOBr generation at Ce^{3+} defect sites in a radical reaction, the release of HOBr being the rate-limiting step. Antioxidants like cysteine and ascorbic acid slow down the reaction [221].

Marine HPOs catalyze the halogenation of organic substrates via reactive HOBr intermediates. This leads to an inactivation of the bacterial signaling compounds. CeO_{2-x} nanorods emulate this reaction by catalyzing the conversion of bromide ions to reactive HOBr intermediates that target natural signaling compounds associated with bacterial quorum sensing and biofouling. This could be demonstrated for the bacterial signaling molecule 3-oxo-acylhomoserine lactone (3-oxo-HL, Fig. 8.19) which was halogenated to the mono- and dibrominated lactones within 30–150 min under ambient conditions with H_2O_2 and KBr as bromine source. The formation of the brominated lactone was demonstrated by HPLC combined with mass spectrometry [221].

Sulfated ceria–zirconia (especially $SO_4^{2-}/Ce_{0.07}Zr_{0.93}O_2$) is also a potent iodination catalyst. Phenol, aniline, and their derivatives were oxidatively halogenated with I_2 in PEG-200 with high product yields (up to 97% iodoaniline after 24 h) [441]. The activity was attributed to the large surface area, the number of acidic sites, and their acidity. Similarly, benzylic compounds such as dimethoxybenzene (DMB) could be iodinated with I_2, H_2O_2 in the presence of ceria nanorods in high yields (up to 97%, 0.5 h, mono-iodination) [442], but temperatures of 100 °C and an excess of H_2O_2 were needed. A Fenton-like decomposition of H_2O_2 at Ce^{3+} centers was proposed (Fig. 8.20), as the presence of hydroxyl radicals could be demonstrated by electron paramagnetic resonance (EPR) spectroscopy with the spin trap agent 5,5-dimethyl-pyrroline *N*-oxide (DMPO). Four characteristic signals (ratio of 1:2:2:1) of the DMPO-OH spin adduct were observed at 334.6−335.0 B_0/mT and *g*-value of 1.992. Further EPR analysis revealed that a decrease of the pH lead to an increase of

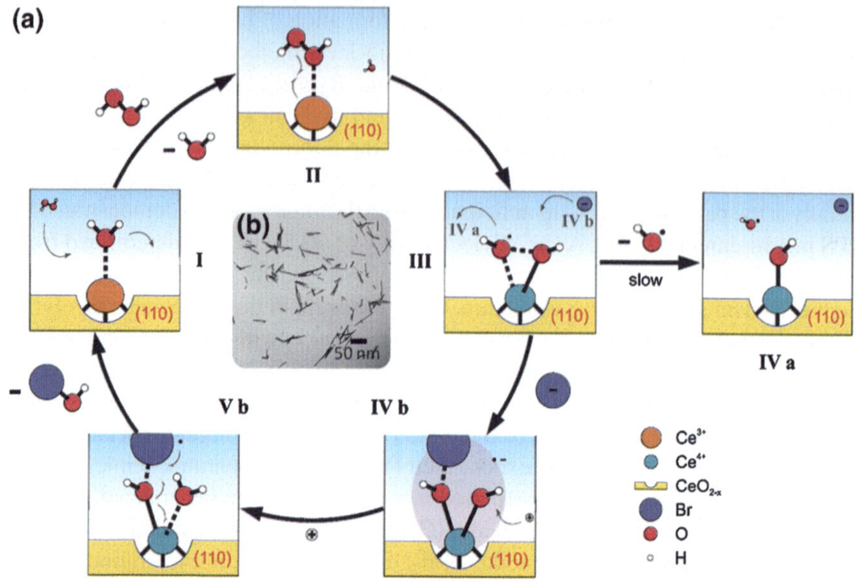

Fig. 8.18 **a** Proposed catalytic bromination mechanism of the CeO_{2-x} nanorods. On a (110) model surface (I), the H_2O ligand can be exchanged against H_2O_2 (II). With a $\Delta G°$ close to zero this exchange is unlikely for low H_2O_2 concentrations. However, the exchange itself is favorable ($\Delta G°$ < 0) and becomes fully irreversible (more negative $\Delta G°$) for a (partial) dissociation of H_2O_2, which leads to an oxidation of the Ce^{3+} site with OH^- anion and OH• radical ligands (III). These two groups are bound to each other through a weak two-center three-electron bond. As both Ce–O distances in this species are equal, the description involves two mesomeric boundary structures to account for the chemical bonding. A release of an OH• radical from this species into solution may occur, but it represents a step "uphill" in Gibbs free energy and is therefore slow (IVa). A Br^- anion can add to one of the O atoms to form species which is best described as an anionic surface site (with two OH^- ligands) where one of the OH^- anions interacts with a Br• radical (IVb). The other noninteracting OH^- anion is protonated to restore a neutral surface site (V). Dissociation of HOBr finally regenerates the initial Ce^{3+} site (I). Although the final step is associated with a positive $\Delta G°$, it occurs because (i) the back reaction is energetically uphill as well and (ii) the concentration dependence of ΔG (= $\Delta G° + RT\ln K$) makes ΔG favorable for low HOBr concentrations. (b) Transmission electron microscopy (TEM) of ceria nanoparticles. Images and text reprinted with permission from Ref. [221]

the EPR response which correlates with the amount of OH radicals. The pH required for oxidative iodination is lower than for bromination ((pH $\leq$ 3.0) [442] vs. pH $\leq$ 7.0 [221]). Increasing the pH to values $\geq$ 6.0 resulted only hydroxylated rather than iodinated products (e.g., for benzene). Computational studies suggested a stepwise reaction mechanism with four different iodine compounds [442] and iodites (I^+) as key species. A stepwise oxidation of one iodine molecule (I_2) to two HOI molecules that are involved in the actual oxidative iodination step (Fig. 8.20) was proposed. The electrophilicity of HOI is not sufficient to promote iodination, but protonation to H_2OI^+ under acidic conditions increases the reactivity and drives the reaction [442].

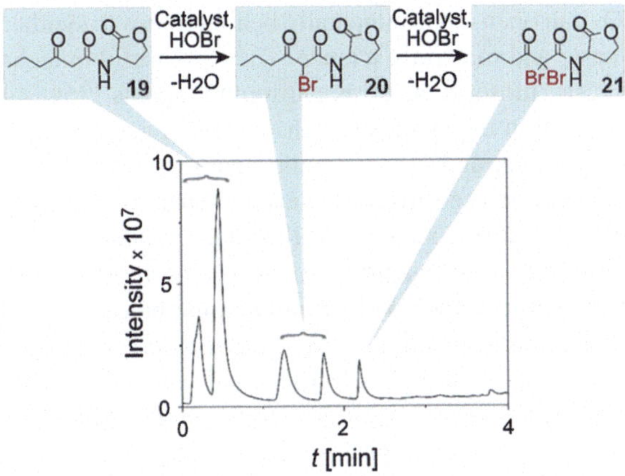

Fig. 8.19 Oxidative bromination of signaling molecules. CeO_{2-x} nanorods catalyze the oxidative bromination of 3-oxo-hexanoylhomoserine lactone **19** (3-oxo-HL) to 2,2′ dibromo-3-oxo-hexanoylhomoserine lactone **21** identified by HPLC/ESI-MS. The halogenated products are assigned to their specific signals in the total ion chromatogram (reprinted with permission from Ref. [221])

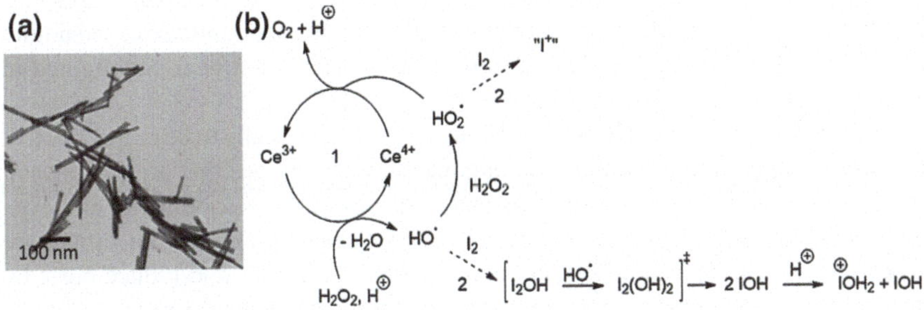

Fig. 8.20 **a** TEM image of ceria nanorods for oxidative iodination (reprinted with permission from Ref. [442], Copyright (2012) John Wiley and Sons). **b** Proposed formation mechanism of oxidized iodine species ("I^+") by a Fenton-like activity of $Ce^{4+/3+}$ (reaction 1). Hydroxyl radicals are formed by one-electron transfer reaction and oxidize I_2 via hydroxyl-iodite transition states (I_2OH, $I_2(OH)_2$), IOH). IOH is protonated in strong acidic media to the reactive species H_2OI^+ (reaction path 2). Singlet oxygen can be formed in side reactions with excess H_2O_2 [442]

8.10 Antimicrobial and Antifouling Agents

Enzymes are used in industrial applications and processes, including the food and beverage, textile, detergent, or medical industries. Enzymes screened from natural origin are often engineered because their native forms do not meet the requirements for industrial application. Protein engineering is performed to design and construct new enzymes with tailored functional properties, including stability, catalytic activity, reaction product inhibition, and substrate specificity [356].

HPOs and HOs can open environmentally benign routes to synthetic halogenated compounds for basic and materials science. They are useful in technical processes for organic synthesis, biosensing, or as antifouling agents [354, 443, 444]. HOX formation is a key step in the synthesis of halogenated natural products [357, 359, 445] and precursor compounds for coupling reactions [446–448]. The oxidizing properties of HOX can be used for the regio- and enantioselective oxygenation of substrates [70, 105, 449, 450], oxidative cyclization, epoxidation [124], or sulfoxidation reactions [250]. Heme iron enzymes [70] are used currently in depolymerization processes to obtain synthetic fuel [451]. Similarly, the biocatalysis of sulfoxidation might find potential applications in the desulfurization process of diesel fuel [452], and enzymatic biosensors with high selectivity and sensitivity are being developed [453]. Oxidative halogenation can be used for the chromogenic detection of halide [65, 454–456] or peroxide contaminations [457, 458]. With isotope-labeled substrates [130], HPOs allowed identifying the unknown origin of environmental or samples using isotope analysis [130]. In addition, HPOs, H_2O_2, and halide constitute potent antimicrobial systems whose cytotoxic effect and antitumor properties have been demonstrated [459–461]. HPOs are involved in the synthesis and post-modification of natural products and their analogues with anthelmintic, anti-inflammatory, or antimicrobial activity [74, 77, 359, 462–472]. Brominated cyclic sesquiterpenes [473], acetogenin, and antiviral compounds could be made [124, 445]. Bacteria and viruses could even be inactivated with a resistant mutant of vanadium chloroperoxidase [474]. The pharmaceutical use of HPOs has been highlighted in several reviews [73, 74, 76–78, 123, 201, 462, 475–478].

By protein engineering, HPOs have been modified genetically to be more effective at desired temperatures, pH, or other manufacturing conditions that are detrimental to the enzyme activity otherwise. This makes them suitable for industrial and even home applications. Still, as heterogeneous catalysts nanoparticle enzyme mimics are easy to recover, to recycle [479] and may facilitate processes and reduce costs by replacing enzymes in mass applications, where these factors are essential.

Halogenating enzymes are useful in generating HOX and ROS species. The inactivation of spores, viruses, and bacteria [67] by HPOs and HGs is a prime goal for their use as surface disinfectants [480], sterilizing agents in laundry [481] with antimicrobial and bleaching properties, in industrial wastewater treatment [482], and on mammalian skin [66]. A patent for using recombinant HPO as antimicrobial agents was filed in 1995 [483], and HPOs were established disinfection agents in organic matrices [474, 484].

The colonization of microorganisms that serve as food supply of higher organisms on surfaces exposed to water causes massive problems in the domestic, industrial, medical, and marine sectors. The development of antimicrobial coating systems is a challenging example for the application of HPO mimics. Microbial biofilms are ubiquitous in flora and fauna, and all structures and interfaces in contact with natural waters suffer from biofouling due to the colonization of bacteria and aquatic organisms [471]. Figure 8.21 illustrates the process of marine biofouling. First, organic molecules (proteins and polysaccharides) form a "conditioning film" as basis for the settlement and colonization of microorganisms (I), bacteria (II), and plants (IV).

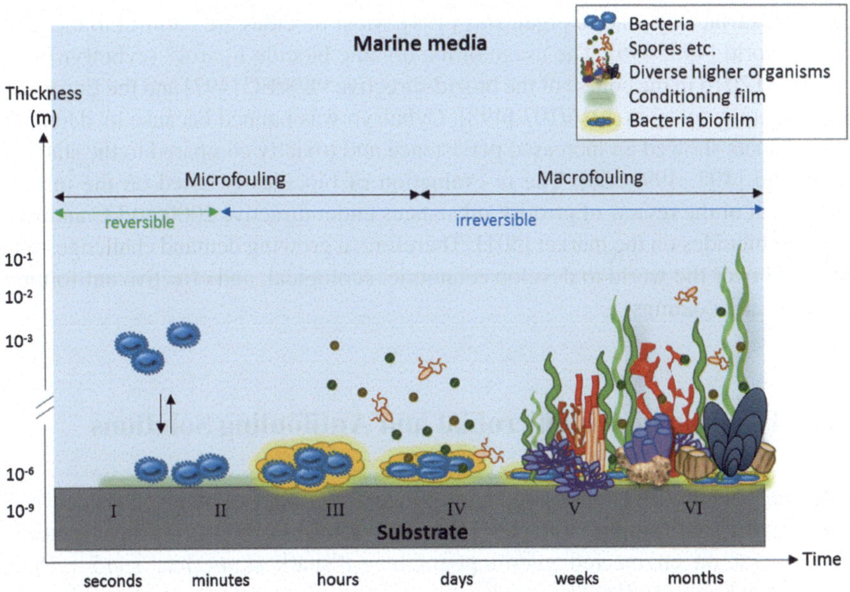

Fig. 8.21 The fouling process: (I) Conditioning film of adsorbed small organic/inorganic particles; (II) reversible adhesion of bacteria; (III) bacterial biofilm formation by covalent linkage due to expressed, extracellular adhesion molecules and polymers; (IV) microfouling: slime formation via the incorporation of spores, germs, and larvae of higher organisms (diatoms, barnacles, algae); (V, VI) macrofouling: settlement and growth of soft foulers (green/brown/red macroalgae, tunicates, sponges, anemones) and hart foulers (barnacles, tubeworms, bryozoans, mussels). Adapted from Refs. [486–488]

Within minutes the process becomes irreversible when the attachment of the bacteria to the surface is enhanced by extracellular adhesion molecules and polymers (III). The term “biofilm” describes the covalent adhesion of unicellular microorganisms through interactions with the surrounding, e.g., via nutrient supply cycles. The formation of the biofilm improves the accessibility and attractiveness for the deposition of spores, germs, and larvae of higher organisms yielding a slimy layer (IV), and higher multicellular organisms like macroalgae or tunicates (soft foulers) grow out of the slime. Macrofouling (V) proceeds and eventually animals with exoskeletons (hard foulants) settle (VI) [486–488].

Biofilm formation is a major concern for applications ranging from biomedical implants, devices, and food packaging to industrial and marine equipment [489]. Marine facilities, bridge piers, naval vessels, offshore wind parks, and ship hulls are particularly affected, which leads to economic losses [137, 432, 443, 444, 485, 489–491]. Modern ship coatings want to minimize hull roughness by biofouling, because higher roughness leads to increased drag, higher fuel consumption, and increased greenhouse gas emissions [489, 492, 493]. Current antifouling coatings release copper and other co-biocides [488] including organic booster biocides like copper pyrithione [493] to impede the settlement, colonization, and destruction of the metal surface by marine organisms [494]. Most biocides are harmful to the submarine world [495, 496]. The usage of the organic biocide Irgarol® (cybutryn) was prohibited 2016 in the course of the biozid-directive 98/8/EC [497] and the European implementing decision 2016/107 [498]. Cybutryn was banned because its degradation products showed an increased persistence and toxicity compared to the starting compound [497, 499, 500]. The re-evaluation of biocides is based on the impact assessment of the review of priority substances under directive 2000/60/EC and will affect all biocides on the market [501]. Therefore, a growing demand challenges scientists all over the world to develop economic, ecological, and effective antifouling materials and coatings.

8.11 Biomimetic Antimicrobial and Antifouling Solutions

Following Nature's defense strategies against epibiont growth, physical and (bio)chemical antifouling strategies [486–489, 492–494, 502–507] have concentrated so far on engineered microtopographies ("shark skin-effect") [154, 155], superhydrophobic [508] or superhydrophilic [486] surfaces, electrochemical defense [509], natural product antifoulants [510, 511], and enzyme-based coatings [494]. Physical methods affect the interaction of marine fouling organisms with artificial surface topographies [155, 488]. An "unattractive" surface has only a loose connection to early-stage foulers (phase I, Fig. 8.21). The removal of bacteria is enhanced by flow mechanics and conditions [155], and a mechanical surface cleaning is facilitated [512]. Silica-based protective coating formulations (e.g., *AquaFast* formulation) reduce the strength of biofilm adhesion after ~7 weeks in a static immersion test in saltwater [512]. A grooming process yields an almost clean coating.

Substrates under static conditions (e.g., in ship piers, etc.) suffer from instant fouling. Once the biofilm has formed, other organisms follow. Fouling of static marine structures such as oil production platforms, drilling rigs or wind parks from seaweed, barnacles, mussels, etc. is prevented by securing a coated flexible sheet material (basalt or silicone rubber) to the underwater surface of the marine structure [513]. A flexible sheet material, useful as an antifouling cladding, is a fabric with an outermost coated surface that consists of a silicone rubber/silicone oil mixture. It is produced by curing a room-temperature-vulcanized silicone rubber with hydroxyl end groups in the presence of silicone oil. In an alternative approach, basalt filaments are applied as antifouling coating fabric for flow around underwater surfaces. The fabric is fixed with adhesives on the substrate or the protected marine area. The basalt fabric may also be applied on the underwater surface by covering with tissues or ligaments or by braiding, where the surface of the antifouling coating is formed predominantly by free basalt fibers [514, 515]. A related strategy is a two-part hybrid silicon-ceramic coating that infuses a "gel coat" with nanoparticular ceramic-glass. The waxes, paints, and polyurethane coatings (membranes) physically attach to the surface, while the nanoparticle "glassifier" binds covalently to the gel coat surface.

This creates a mirror finish gel coat that is self-cleaning, mold-resistant, nonfading, non-skidding, and hydrophobic.

Low-Emission Antifouling Coatings. The basic concept in antifouling with biocide-loaded coatings is a steady-state release of biocides into the surrounding water that is required for preventing the settlement of marine organisms. Here, the antifouling activity is based only on the interaction between the biocide and fouling organisms on the coating. This scenario, however, does not properly describe the state of affairs on the paint surface during barnacle colonization, because there are synergistic effects between the embedded biocide in low concentrations and the mechanical properties of the coating. This was demonstrated for preventing the deposition of barnacles. Low amounts (0.1%) of the biocide Ivermectin embedded in a soft or abrasive paint coating suppressed the attachment of adult barnacles completely, whereas no reduction in barnacle recruitment was observed when Ivermectin was embedded in a hard polystyrene-derived coating [516]. Both coatings suffered from initial fouling by settling barnacle larvae, but barnacle growth stopped after metamorphosis on the soft coating, although there was no difference in the biocide release rates (several ng/cm^2 per day). This shows that intoxication is triggered by the organism itself in contact with a coating containing an embedded biocide. Full biocide release is dispensable because the reaction is shifted from the solid/water interface to the subsurface layers of the coating.

No Emission Antifouling Coatings. Nature has solved the biofilm problem by an active, biochemical defense strategy. Communities of bacteria can coordinate their behavior with three classes of signaling molecules that are used for communication within the organism to coordinate gene expressions to (i) generate adequate response behavior, (ii) signaling between the same or related species, and (iii) signaling among different species. Bacteria generate different signal-mediated interactions for different purposes with a limited number of molecules and a limited number of combinatorial rules. Bacterial sign users share a set of syntactic rules to combine signs for a variety of interactional contexts (development, growth, mating, virulence, attack, and defense). The situational context of these interactions determines the signal semantics. The complementarity of these three levels of semiotic rules can be identified in every sign-mediated interaction within and between organisms. This leads to the generation of intra- and intercellular processes that enable bacterial communities to generate memory.

Stationary marine organisms, e.g., red and brown algae, use this strategy against epibiont growth with a group of enzymes that catalyze the synthesis of halogenated antifoulants or the oxidative halogenation of bacterial signaling compounds that prevent biofilm formation by interfering with cidal cell-to-cell communication [202]. Vanadium-dependent bromoperoxidases (VBPOs) are the best-established group of enzymes. Thus, biofouling is defended at its lowest level, the microbial colonization [159, 177, 198].

Inspired by marine red algae, HPOs [484, 494] were proposed as additives in antifouling paints more than 15 years ago [65, 196, 483]. Wever [65], ter Steeg [483], and coworkers were the first to file patents for embedding native HPOs in paint formulations to prevent biofouling [65]. Initial problems of enzyme stability and activity under non-physiological conditions could be circumvented by making specially designed mutants [138]. "Biocoats" were devised as a new active hybrid materials based on a matrix or xerogel containing living microorganisms rather than pure enzymes. The bacteria produce antifouling substrates [154], which extend the antifouling activity of the hybrid materials. However, the use of natural or recombinant enzymes or microorganisms raises the issue of production costs, long-term stability, and the reaction conditions for proper activity [253].

The use of HPO mimics in antifouling coatings is a low-cost bioinspired alternative to overcome these shortcomings. Only a few, but promising HPO mimics, were developed in the laboratory and, more significantly, tested in the field. The use of modified xerogel coatings combines the physical and chemical antifouling characteristics [154, 425] of porous silica coatings [512, 517] with the HPO-like activity of the organo-selenium compounds [425]. Leaching was encountered for phenyl chalcogenic acids, and phenyl chalcogenic chloride was formed for low-molecular-weight organo-selenium or tellurium compounds (Fig. 8.16c) [425, 426], which, however, may turn into an advantage because active components can be released during the defense process [426]. A similar xerogel containing hydrogen peroxide and artificial seawater (ASW, with halides) was successful as "hybrid" coating against the settlement of cypris larvae of the barnacle *Balanus amphitrite*, larvae of the tubeworm *Hydroides elegans* and of *Ulva zoospores* [425]. Hybrid titania/silica-derived xerogels impede the settlement of zoospores of the marine alga *Ulva linza* and sporelings (young plants) at pH 7–8 in the presence of halides (bromide, chloride) and H_2O_2 [518]. The activity of the xerogels (with Se or Ti) is connected to the HPO-like activity of the embedded compounds, i.e., the formation of hypohalous acid in combination with a submerged, durable, and smooth surface of the coating [425, 518]. Lejars [488] and Nurioglu et al. [486] compiled silica release coatings and their alternatives.

Besides HPO-like xerogels, transition metal oxides have shown oxidative bromination activity [221, 389, 390, 430, 519, 520]. Many HPOs employ vanadium(V) as cofactor, but molybdate(VI) shows higher catalytic activity toward halide oxidation in solution compared to vanadate(V). Density functional calculations revealed that the oxodiperoxo species $[MoO(O_2)_2L]$ is catalytically more active than the dioxomonoperoxo species $[MoO_2(O_2)L]$. However, the spatial environment of the VHPO active site is tailored to a monoperoxo species. Thus, the reason for the inactivity of a hypothetical molybdenum bromoperoxidase might be that no peroxo species is formed in the first place [521].

Oxides of molybdenum and tungsten exhibit good antimicrobial properties. Microbiological roll-on tests with *S. aureus*, *E. coli*, and *P. aeruginosa* showed antimicrobial activities for anhydrous samples with large specific surface area [522–524]. The proposed antimicrobial mechanism for transition metal oxides is based on local acidity, as supported by the identification of molybdenum oxide nanostructures with $Mo(=O)_2OH$ units by in situ Raman spectroscopy [525]. One might speculate that a HPO-like reactivity contributes to the observed antimicrobial properties of MoO_3 and WO_3, as indicated by antifouling steel surfaces after electrodeposition of nanoporous WO_3 films on naval construction steels and surgical instruments that showed reduced marine algal film adhesion, *E. coli* attachment, and bloodstaining [526].

The ability of V_2O_5 and CeO_{2-x} to emulate nature's chemical defense strategy against predation and epibiont growth has been demonstrated in the laboratory and in the field. V_2O_5 nanowires and CeO_{2-x} nanorods reduced bacterial adhesion and algal growth in hard (smooth) and soft (porous) paint formulations containing small amounts (2 wt%) of the metal oxide. Cultures containing the enzyme mimics, V_2O_5 or CeO_{2-x}, and both substrates, bromide and H_2O_2, showed virtually no bacterial colonization compared to negative controls. The bromination mechanism differs significantly for both oxides: Whereas the reaction for V_2O_5 involves a surface-bound η^2-peroxo group as for vanadium-dependent HPOs, the halogenation via CeO_{2-x} follows a radical mechanism as observed for iron heme HPOs. Both metal oxides showed reduced algal growth in field test. Figure 8.22 shows test plates with 2 wt% of V_2O_5 after immersing in the Atlantic ocean (stationary at a boat hull for 2 months) and with 2 wt% of CeO_{2-x} after immersing in the Maas in Belgium (stationary, at

Fig. 8.22 Representative digital images showing the antifouling activity in field test of V_2O_5 nanowires [430] and CeO_{2-x} nanorods [221]. **a** Antifouling activity in field test of V_2O_5 nanowires implemented in a commercially available paint for boat hulls [430]. Digital images of a stainless steel plates (2 × 2 cm) covered with the paint formulation without ($-V_2O_5$nw, a1, 3) and with ($+V_2O_5$nw, a2, 4) V_2O_5 nanowires: a1, 2) Immediately after fixation on a boat hull, a3, 4) after 60 days of immersion period in seawater. Severe natural biofouling was observed at the painted stainless steel plates in the absence (a3), but not in the presence of V_2O_5 nanowires (a4) (Digital images reprinted with permission from Ref. [430]), **b** Commercially available antifouling paint (b3, 5, 7 hard and b4, 6, 8 soft formulation) with (b5, 6) and without (b1–4, 7, 8) cerium dioxide nanoparticles was applied to stainless steel plates (4 × 4 cm) [221]. Digital images (b1–b8) after static field immersion (52 days). Several control plates (b1, 2 stainless steel, b3 native hard paint formulation, b4 native soft formulation, b7 hard paint formulation with copper oxide, and b8 soft paint formulations with copper oxide) are displayed as well. The plates were attached statically to a boat bridge with direct exposure to freshwater. After 52 days, the control plates without cerium dioxide nanoparticles (b1–4, 7, 8) showed heavy fouling. In contrast, the plate with the cerium dioxide coating (b5 in soft and b6 in hard paint formulation) did not and the activity appeared five times higher in comparison to copper oxide (Digital images reprinted with permission from Ref. [221])

a boat pier, 2 months). Both examples demonstrate the antifouling activity of the HPO mimics catalysts in the field and compared to cuprous oxide, the current gold standard.

The antifouling activity of CeO_{2-x} nanoparticles can be linked to native defense system of marine organisms. The use of natural intermediates such as HOBr or halogenated metabolites, which are different from conventional antifouling agents, targets specific bacterial signaling and regulatory systems like 3-oxo-HL. This represents a new strategy with the goal to emulate a natural defense system for preventing bacterial colonization or biofilm development [221]. A biomimicry approach that allows the substitution of conventional biocides or costly enzymatic preservation systems by a stable, nontoxic, and highly abundant rare earth oxide represents a sustainable antifouling solution that maximizes damage in bacterial biofilms with minimal collateral effects to the environment.

8.12 Conclusions and Outlook

Most current antifouling solutions contain copper in combination with co-biocides. Even after years of research copper-based antifouling solutions still offer the best performance. They rely on the use of seawater-insoluble cuprous oxide combined with various organic booster biocides. Many studies, however, have shown clearly that copper compounds are toxic to nontarget marine species like the embryo/larvae of the oyster, mussel, and sea urchin. Cyanobacteria are highly sensitive to copper toxicity as well. Hatchery salmons are vulnerable to the olfactory neurotoxicity caused by copper. This leads to neurobehavioral response changes which are critical for a successful migration [527]. Copper from antifouling sources has become a problem in the marine environment, especially in isolated waters like marinas and harbors that have little water exchange and high levels of boating activity. Only in San Diego's Shelter Island Yacht Basin, approx. 2.5 tons of copper leach from the hulls of two thousand boats each year [528].

The growing awareness of copper toxicity has led to changes in legislation. The Baltic Sea is an environmentally sensitive environment with a small diversity of aquatic species. Swedish authorities placed severe restrictions on the use of copper antifouling paints in the Baltic Coast of Sweden. Likewise, the copper loading in the inland aquatic environment of the Netherlands is uniquely high. The Dutch pesticide authorities decided to ban the use of copper-containing antifouling paints on all pleasure craft, especially in inland waters. In the United States, western states are leading the campaign against copper antifouling paints. In California, copper-based paints are banned for recreational boats starting in 2019.

There is a need to develop environmentally friendly antifouling solutions. The goal for biocidal antifouling coatings is to provide a surface with active biocides that are (i) stationary or (ii) deactivated shortly after leaving the surface. Thus, biocides with a short half-life, either by biodegradation or deactivation, are needed. There are several thousand fouling organisms in the waters of the world, broadly classified into hard (such as barnacles) and soft organisms (grasses, algae). An effective antifouling should have a broad activity spectrum which must be continuously available to protect against biofilm formation.

Although the interest in biofilm formation and particular in biofouling originates from its corrosive effects on manmade structures, it also occurs on the surfaces of living marine organisms and leads to problems in aquaculture of seaweeds and shellfish. Many sessile marine organisms do not possess any physical or mechanical means of defense against possible colonizers but nevertheless resist colonization and overgrowth by epibionts. Studying the natural defense strategies against fouling of organisms has led to biomimetic approaches that can be used to better understand the structure and function of biological systems as models or inspiration for the sustainable design and engineering of materials and machines. Natural antifouling strategies developed by marine organisms can be classified into three categories: (i) chemical, (ii) physical/mechanical, and (iii) behavioral. With a special focus on biofouling, two of them have been pursued for biotechnological applications, i.e., chemical defense and effects of surface topography. A successful large-scale approach to combat biofouling will have to make use of a combination of at least two of these strategies.

The ability of organisms to combat colonization of epibionts chemically is associated with the formation of secondary metabolites, commonly involved in chemical defense against predators, pathogens, and also against fouling. The discovery of potent antifouling compounds such as brominated furanones, diterpenoids [529], synoxazolidinones [530], and pulmonarins [531] demonstrates the potential of natural products as a source of new antifouling compounds. The discovery of new lead compounds with antifouling activity and their development requires large quantities of these compounds for laboratory screening, field assays, and tests of paint formulations. The challenge of natural products is the scale-up for the coating market. There are several strategies to produce larger amounts of natural products. (i) *Natural products extracted from organisms collected in the field.* Although this is the first step during development, it is prohibitive during up-scaled production due to the availability of natural resources. (ii) *Natural products extracted from cultured organisms and plants*, e.g., using aqua- or maricultures. In-sea cultures may be a cost-effective option for the supply of natural compounds. Culturing may be a successful strategy if the natural products are of microbial origin, but many organisms are very difficult to culture. (iii) *Culturing of recombinant microorganisms* that are transfected with genes involved in the synthesis of the natural product. This is only practical if the desired natural product is a direct gene product (e.g., peptides). (iv) *Chemical synthesis of natural products.* Many natural products are too complex and expensive to synthesize. The number of synthetic steps and the cost of initial reactants determine the yield of the final product and thus the success. Analogues of the natural product with similar bioactivity, but lower costs would be a viable alternative. (v) *Employing HPOs to generate the natural products* in situ. The main advantage of the biocatalyst would be that the required biocides are delivered on-demand. Many bacterial pathogens use QS to regulate the timing and extent of virulence factor production, thereby allowing them to amass until a sufficient population has been achieved to overwhelm the immune response of the hosts. As QS is dependent on small molecule signals and their relative concentration, there is substantial interest in developing chemical strategies that disable QS signaling networks or even prevent virulence. This "anti-virulence" strategy would not only be a pathway to suppress epibiont growth but also to mitigate bacterial infection in humans, animals, and plants. Several opportunistic microorganisms have become resistant to current antimicrobial and antibiotic treatment. Therefore, strategies to inhibit QS have received significant attention. Such nonbactericidal approaches could also be robust to resistance development. The (non-halogenated) starting compounds for interference with QS are supplied by the colonizers themselves. This eliminates the need to synthesize many natural products as antimicrobial agents against different colonizers on spec. Moreover, there would be a constant supply of active biocides that are stationary in low concentration. The biocides are not only complexed and used shortly after leaving the surface, but also compounds of the ecosphere, present in the biological life cycle for millions of years. (vi) *Employing nanoparticle HPO mimics to generate the natural products* in situ. This biomimicry approach to substitute conventional biocides or costly enzymatic preservation systems by cheap and very stable "heterogeneous catalysts" seems to be the most practical and sustainable solution. HPO mimics allow designing antibacterial materials with high chemical stability, low environmental toxicity, costs, constant availability of the substrates (H_2O_2 and halide) which are required for a dynamic antibacterial activity (HOX formation), and intrinsic activity at substrate concentrations comparable to those encountered in aqueous environments. This reduces time and energy and produces less polluted water.

Because of their intrinsic catalytic characteristics and environmental sustainability, oxide nanoparticles (e.g., CeO_{2-x}) may find application not only in paints, but also on bridge piers, naval vessels or in offshore windparks, in aquaculture systems [532], polymer membranes for water desalination or filters for the reclamation of wastewater [533], drinking water pipes and clarification plants, cooling systems, or on textiles [533]. Biofouling in aquacultures is a well-known problem, whose severity depends on factors like biota in the region, temperature, light, salinity, tides, and water transparency among others. Fouling on fish cages can cause a reduced water flow through the meshes, clogging of the mesh, and the increased excess weight. Water flow through nets can be reduced upon fouling, and the diameter of the net twine can increase. Consequently, oxygen levels in the cage drop, while fish wastes and ammonia levels increase. Moreover, biofouling itself can act as a pathogen vector with serious consequences to fish health. The costs associated with biofouling can be very significant.

Similarly, biofouling leads to the use of higher operating pressure, frequent chemical cleaning, and shorter lifetimes in reverse osmosis membranes. They are becoming increasingly popular for water purification applications that require high salt rejection such as brackish and seawater desalination. However, due to fouling by microorganisms, they have been unable so far to realize their full potential.

Many plants and fungi have coevolved and established symbiotic associations with bacteria. Plant-associated proteobacteria have AHL-mediated quorum-sensing systems [534]. Plants and fungi do not have active immune systems. They rely on chemical defense systems to cope with the bacteria in the environment. Since plants and fungi produce related chemical compounds to inhibit (or to stimulate) bacterial AHL-mediated communication [535], similar strategies may be successful for suppressing fungal growth.

With a movement toward environmentally friendly products, latest developments in the field of antifoulants include fouling-release coatings (silicone (PDMS), fluorsilicones), fast polishing materials, contact activity, removable foils, on-demand systems, spiky coatings, surfaces with defined microstructures, hydrogels, and metallic layers including organometallic coatings and metal claddings. Biofilms are around virtually everywhere. They are present on practically all surfaces such as food packaging, door handles, pushbuttons, keyboards, and other elements made of plastic. In medicinal applications, they develop, for example, in catheter tubes or on surgical instruments. The main problem in connection with combatting these using biocides and antibiotics is the risk of the development of resistance. The use of natural intermediates, formed in situ with nontoxic and highly abundant metal oxide enzyme mimics, which target specific bacterial signaling and regulatory systems, represents a new and efficient "green" strategy with the goal to emulate and utilize a natural defense system for preventing bacterial colonization or biofilm development.

References

1. Bruggink A, Roos EC, de Vroom E (1998) Penicillin acylase in the industrial production of β-Lactam antibiotics. Org Process Res Dev 2(2):128–133
2. Estell DA, Graycar TP, Miller JV, Powers DB, Burnier JP, Ng PG, Wells JA (1986) Probing steric and hydrophobic effects on enzyme-substrate interactions by protein engineering. Science 233(4764):659–663
3. Jensen VJ, Rugh S (1987) Industrial scale production and application of immobilised glucose isomerase. Methods Enzymol: 356–370
4. Sedlaczek K (1988) Biotransformation of steroids. Crit Rev Biotechnol 7:187–236
5. Dinçer A, Telefoncu A (2007) Improving the stability of cellulase by immobilization on modified polyvinyl alcohol coated chitosan beads. J Mol Catal B Enzym 45:10–14
6. Elleuche S, Schröder C, Sahm K, Antranikian G (2014) Extremozymes–biocatalysts with unique properties from extremophilic microorganisms. Curr Opin Biotechnol 29:116–123
7. Breslow R, Overman LE (1970) "Artificial enzyme" combining a metal catalytic group and a hydrophobic binding cavity. J Am Chem Soc 92(4):1075–1077
8. Raynal M, Ballester P, Vidal-Ferran A, van Leeuwen PWNM (2014) Supramolecular catalysis. Part 2: artificial enzyme mimics. Chem Soc Rev 43:1734–1787
9. Breslow R (2006) Artificial enzymes. Wiley-VCH, Weinheim, Germany
10. Wiester MJ, Ulmann PA, Mirkin CA (2006) Enzyme mimics based upon supramolecular coordination chemistry. Angew Chem Int Ed 50(1):114–137
11. Fan K, Cao C, Pan Y, Lu D, Yang D, Feng J, Song L, Liang M, Yan X (2012) Magnetoferritin nanoparticles for targeting and visualizing tumour tissues. Nat Nanotechnol 7(7):459–464
12. Kim CK, Kim T, Choi IY, Soh M, Kim D, Kim YJ, Jang H, Yang HS, Kim JY, Park HK, Park SP, Park S, Yu T, Yoon BW, Lee SH, Hyeon T (2012) Ceria nanoparticles that can protect against ischemic stroke. Angew Chem Int Ed 51(44):11039–11043

13. Ragg R, Natalio F, Tahir MN, Janssen H, Kashyap A, Strand D, Strand S, Tremel W (2014) Molybdenum trioxide nanoparticles with Intrinsic sulfite oxidase activity. ACS Nano 8(5):5182–5189
14. Yusop RM, Unciti-Broceta A, Johansson EM, Sanchez-Martin RM, Bradley M (2011) Palladium-mediated intracellular chemistry. Nat Chem 3(3):239–243
15. Wei H, Wang E (2013) Nanomaterials with enzyme-like characteristics (nanozymes): next-generation artificial enzymes. Chem Soc Rev 42(14):6060–6093
16. Lin Y, Ren J, Qu X (2014) Catalytically active nanomaterials: a promising candidate for artificial enzymes. Acc Chem Res 47(4):1097–1105
17. Pirmohamed T, Dowding JM, Singh S, Wasserman B, Heckert E, Karakoti AS, King JES, Seal S, Self WT (2010) Nanoceria exhibit redox state-dependent catalase mimetic activity. Chem Commun 46:2736–2738
18. Fan J, Yin JJ, Ning B, Wu X, Hu Y, Ferrari M, Anderson GJ, Wie J, Zhao Y, Nie G (2011) Direct evidence for catalase and peroxidase activities of ferritin-platinum nanoparticles. Biomaterials 32(6):1611–1618
19. Mu J, Zhang L, Zhao M, Wang Y (2014) Catalase mimic property of Co_3O_4 nanomaterials with different morphology and its application as a calcium sensor. ACS Appl Mater Interfaces 6(10):7090–7098
20. Baran T, Inanan T, Menteş A (2016) Synthesis, characterization, and catalytic activity in Suzuki coupling and catalase-like reactions of new chitosan supported Pd catalyst. Carbohydr Polym 145:20–29
21. Song L, Huang C, Zhang W, Ma M, Chen Z, Gu N, Zhang Y (2016) Graphene oxide-based Fe_2O_3 hybrid enzyme mimetic with enhanced peroxidase and catalase-like activities. Colloids Surf 506:747–755
22. Liu X, Wang Q, Zhao H, Zhang L, Su Y, Lv Y (2012) BSA-templated MnO_2 nanoparticles as both peroxidase and oxidase mimics. Analyst 137:4552–4558
23. Guo R, Wang Y, Yu S, Zhu W, Zheng F, Liu W, Zhang D, Wang J (2016) Dual role of hydrogen peroxide on the oxidase-like activity of nanoceria and its application for colorimetric hydrogen peroxide and glucose sensing. RSC Adv 6:59939–59945
24. Salvemini D, Riley DP, Cuzzocrea S (2002) SOD mimetics are coming of age. Nat Rev Drug Discov 1:367–374
25. Wang W, Jiang X, Chen K (2012) Iron phosphate microflowers as peroxidase mimic and superoxide dismutase mimic for biocatalysis and biosensing. Chem Commun 48:7289–7291
26. Schladt TD, Schneider K, Schild H, Tremel W (2011) Synthesis and bio-functionalization of magnetic nanoparticles for medical diagnosis and treatment. Dalton Trans 40:6315–6343
27. Sun T, Zhang YS, Pang B, Hyun DC, Yang M, Xia Y (2014) Engineered nanoparticles for drug delivery in cancer therapy. Angew Chem Int Ed 53:12320–12364
28. Ragg R, Tahir MN, Tremel W (2016) Solids Go Bio: inorganic nanoparticles as enzyme mimics. Eur J Inorg Chem 2016:1906–1915
29. Rosenzweig AC, Frederick CA, Lippard SJ, Nordlund P (1993) Crystal structure of a bacterial non-haem iron hydroxylase that catalyses the biological oxidation of methane. Nature 366:537–543
30. Lieberman RL, Rosenzweig AC (2005) Crystal structure of a membrane-bound metalloenzyme that catalyses the biological oxidation of methane. Nature 434:177–182
31. Hakemian AS, Rosenzweig AC (2007) The biochemistry of methane oxidation. Annu Rev Biochem 76:223–241
32. Himes RA, Karlin KD (2009) Copper-dioxygen complex mediated C–H bond oxygenation: relevance for particulate methane monooxygenase (pMMO). Curr Opin Chem Biol 13:119–131
33. Groothaert MH, Smeets PJ, Sels BF, Jacobs PA, Schoonheydt RA (2005) Selective oxidation of methane by the bis(μ-oxo)dicopper core stabilized on ZSM-5 and mordenite zeolites. J Am Chem Soc 127:1394–1395
34. Grundner S, Markovits MAC, Li G, Tromp M, Pidko EA, Hensen EJM, Jentys A, Sanchez-Sanchez M, Lercher JA (2015) Single-site trinuclear copper oxygen clusters in mordenite for selective conversion of methane to methanol. Nat Commun 6:7546

35. Wu H, Tian C, Song X, Liu C, Yang D, Jiang Z (2013) Methods for the regeneration of nicotinamidecoenzymes. Green Chem 15:1773–1789
36. Rodriguez C, Lavandera I, Gotor V (2012) Recent advances in cofactor regeneration systems applied to biocatalyzed oxidative processes. Curr Org Chem 16:2525–2541
37. Estabrook RW, Faulkner KM, Shet MS, Fisher CW (1996) Application of electrochemistry for P450-catalyzed reactions. Methods Enzymol 272:44–51
38. Reipa V, Mayhew MP, Vilker VL (1997) A direct electrode-driven P450 cycle for biocatalysis. Proc Natl Acad Sci USA 94:13554–13558
39. Kim JH, Nam DH, Park CB (2014) Nanobiocatalytic assemblies for artificial photosynthesis. Curr Opin Biotechnol 28:1–9
40. Hollmann F, Arends IWCE, Buehler K (2010) Biocatalytic redox reactions for organic synthesis: nonconventional regeneration methods. ChemCatChem 2:762–782
41. Hollmann F, Hofstetter K, Habicher T, Hauer B, Schmid B (2005) Direct electrochemical regeneration of monooxygenase subunits for biocatalytic asymmetric epoxidation. J Am Chem Soc 127:6540–6541
42. Hollmann F, Taglieber A, Schulz F, Reetz MT (2007) A light-driven stereoselective biocatalytic oxidation. Angew Chem Int Ed 46(16):2903–2906
43. Grau MM, Van Der Toorn JC, Otten LG, Macheroux P, Taglieber A, Zilly FE, Arends IWCE, Hollmann F (2009) Photoenzymatic reduction of C=C double bonds. Adv Synth Catal 351(18):3279–3286
44. Taglieber A, Schulz F, Hollman F, Rusek M, Reetz MT (2008) Light-driven biocatalytic oxidation and reduction reactions: scope and limitations. ChemBioChem 9:565–572
45. Udit AK, Hill MG, Bittner VG, Arnold FG, Gray HB (2004) Reduction of dioxygen catalyzed by pyrene-wired heme domain cytochrome P450 BM3 electrodes. J Am Chem Soc 126:10218–10219
46. Unversucht S, Hollmann F, Schmid A, van Pée KH (2005) FADH2-dependence of tryptophan 7-Halogenase. Adv Synth Catal 347:1163–1167
47. Zilly FE, Taglieber A, Schulz F, Hollmann F, Reetz MT (2009) Deazaflavins as mediators in light-driven cytochrome P450 catalyzed hydroxylations. Chem Commun 2:7152–7154
48. Peterson DH, Murray HC, Eppstein SH, Reineke LM, Weintraub A, Meister PD, Leigh HM (1952) Microbiological transformations of steroids.[1] I. Introduction of oxygen at carbon-11 of progesterone. J Am Chem Soc 74:5933–5936
49. Groves JT (2003) The bioinorganic chemistry of iron in oxygenases and supramolecular assemblies. Proc Natl Acad Sci USA 100:3569–3574
50. Liu W, Groves JT (2010) Manganese porphyrins catalyze selective C−H bond halogenations. J Am Chem Soc 132(37):12847–12849
51. Liu W, Groves JT (2015) Manganese catalyzed C–H halogenation. Acc Chem Res 48(6):1727–1735
52. Liu W, Groves JT (2013) Manganese-catalyzed oxidative benzylic C–H fluorination by fluoride ions. Angew Chem Int Ed 52(23):6024–6027
53. Liu W, Huang X, Groves JT (2013) Oxidative aliphatic C–H fluorination with manganese catalysts and fluoride ion. Nat Protoc 8(123):2348–2354
54. Zechel DL, Reid SP, Nashiru O, Mayer C, Stoll D, Jakeman DL, Warren RAJ, Wither SG (2001) Enzymatic synthesis of Carbon–Fluorine bonds. J Am Chem Soc 123(18):4350–4351
55. O'Hagan D, Deng H (2015) Enzymatic fluorination and biotechnological developments of the fluorinase. Chem Rev 115(2):634–649
56. O'Hagan D, Schaffrath C, Cobb SL, Hamilton JTG, Murphy CD (2002) Biosynthesis of an organofluorine molecule. Nature 416:279
57. Senn HM, O'Hagan D, Thiel W (2005) Insight into enzymatic C−F Bond formation from QM and QM/MM calculations. J Am Chem Soc 127(39):13643–13655
58. Neidleman SL, Geigert J (1987) Biological halogenation: roles in nature, potential in industry. Endeavour 11(1):5–15
59. Fauvarque J (1996) The chlorine industry. Pure Appl Chem 68(9):1713–1720

60. Ratnasamy P, Singh AP (2008) Organic reactions: halogenation reactions. In: Ertl G (ed) Handbook of heterogenous catalysis, vol 7, 2nd edn. Wiley-VCH, Weinheim, Germany, pp 3564–3578
61. Wittcoff HA, Reuben BG, Plotkin JS (2013) Industrial organic chemicals, 3rd edn. Wiley, Hoboken, NJ, USA
62. Purser S, Moore PR, Swallow S, Gouverneur V (2008) Fluorine in medicinal chemistry. Chem Soc Rev 37(2):320–330
63. Campbell MG, Ritter T (2014) Late-stage fluorination: from fundamentals to application. Org Process Res Dev 18(4):474–480
64. Marsden SP (1997) Organic Halides. Contemp Org Synth 4:118–135
65. Wever R, Dekker HL, Van Schijndel JWPM, Vollenbroek EGM (1995) Antifouling paint containing haloperoxidases and method to determine halide concentrations, WO/1995/027009
66. Johansen C (2004) Antimicrobial composition containing a haloperoxidase, a hydrogen peroxide source, a halide source and an ammonium source. US 6.818.212 B2
67. Christensen BE, Allesen-Holm M, Gjermansen M (2009) Methods for killing spores and disinfecting or sterilizing devices. US 2009/0104172 A1
68. André R, Natalio F, Tremel W, Wever R, Hartog A (2015) Use of vanadium pentoxide particles as a biocide. EP 2 671 449 A1
69. Fischer NF, Dobner C, Bräu M (2016) Use of vanadium-vontaining particles as biocides. US 2016/0113283 A1
70. Hofrichter M, Ullrich R (2006) Heme-thiolate haloperoxidases: versatile biocatalysts with biotechnological and environmental significance. Appl Microbiol Biotechnol 71(3):276–288
71. Littlechild J, Isupov M (2014) Haloperoxidase enzymes as redox catalysts important for industrial biocatalysis. In: Jacob C, Kirsch G, Slusarenko A, Winyard PG, Burkholz T (eds) Recent advances in redox active plant and microbial products. Springer Netherlands, Dordrecht, pp 425–446
72. Blunt JW, Copp BR, Keyzers RA, Munro MHG, Prinsep MR (2015) Marine natural products. Nat Prod Rep 32(2):116–211
73. Shannon E, Abu-Ghannam N (2016) Antibacterial derivatives of marine Algae: an overview of pharmacological mechanisms and applications. Mar Drugs 14(4):81
74. Cheung RCF, Ng TB, Wong JH, Chen Y, Chan WY (2016) Marine natural products with anti-inflammatory activity. Appl Microbiol Biotechnol 100(4):1645–1666
75. Malve H (2016) Exploring the ocean for new drug developments: marine pharmacology. J Pharm Bioallied Sci 8(2):83–91
76. Blunt JW, Copp BR, Keyzers RA, Munro MHG, Rrinsep MR (2016) Marine natural products. Nat Prod Rep 33(3):382–431
77. Mayer AMS, Rodríguez AD, Taglialatela-Scafati O, Fusetani N (2013) Marine pharmacology in 2009–2011: marine compounds with antibacterial, antidiabetic, antifungal, anti-inflammatory, antiprotozoal, antituberculosis, and antiviral activities; affecting the immune and nervous systems, and other miscellaneous mechanisms of action. Mar Drugs 11(7):2510–2573
78. Mayer AM, Nguyen M, Newman DJ, Glaser KB (2016) The marine pharmacology and pharmaceuticals pipeline in 2015. FASEB J 30(1):932–937
79. Soedjak HS, Walker JV, Butler A (1995) Inhibition and inactivation of vanadium bromoperoxidase by the substrate hydrogen peroxide and further mechanistic studies. Biochemistry 34(39):12689–12696
80. Rehder D (2010) Bioanorganische chemie des vanadiums. Leben ohne vanadium? Chem unserer Zeit 44(5):322–331
81. Leblanc C, Vilter H, Fournier JB, Delage L, Potin P, Rebuffet E, Michel G, Solari PL, Feiters MC, Czjzek M (2015) Vanadium haloperoxidases: from the discovery 30 years ago to X-ray crystallographic and V K-edge absorption spectroscopic studies. Coord Chem Rev 301–302:134–146
82. Martinez JS, Carroll GL, Tschirret-Guth RA, Altenhoff G, Little RD, Butler A (2001) On the regiospecificity of vanadium bromoperoxidase. J Am Chem Soc 123(14):3289–3294

83. Vaillancourt FH, Yeh E, Vosburg DA, Garneau-Tsodikova S, Walsh CT (2006) Nature's inventory of halogenation catalysts: oxidative strategies predominate. Chem Rev 106(8):3364–3378
84. von Gunten U (2003) Ozonation of drinking water: part I. Oxidation kinetics and product formation. Water Res 37 (7):1443–1467
85. Soedjak HS, Butler A (1990) Chlorination catalyzed by vanadium bromoperoxidase. Inorg Chem 29(25):5015–5017
86. Crans DC, Smee JJ, Gaidamauskas E, Yang L (2004) The chemistry and biochemistry of vanadium and the biological activities exerted by vanadium compounds. Chem Rev 104(2):849–902
87. Xu G, Wang BG (2016) Independent evolution of six families of halogenating enzymes. PLoS ONE 11:e0154619
88. Furtmüller PG, Zederbauer M, Jantschko W, Helm J, Bogner M, Jakopitsch C, Obinger C (2006) Active site structure and catalytic mechanisms of human peroxidases. Arch Biochem Biophys 445(2):199–213
89. Henderson JP, Heinecke JW (2003) Myeloperoxidase and eosinophil peroxidase: phagocyte enzymes for halogenation in humans. In: Hutzinger O (ed) The handbook of environmental chemistry, vol 3, pp 201–214
90. Banerjee S, Furtmüller PG, Obinger C (2011) Bovine lactoperoxidase—a versatile one- and two-electron catalyst of high structural and thermal stability. Biotechnol J 6(2):231–243
91. Porterfield SP, Hendrich CE (1993) The role of thyroid hormones in prenatal and neonatal neurological development–current perspectives. Endocr Rev 14(1):94–106
92. Arnhold J, Flemmig J (2010) Human myeloperoxidase in innate and acquired immunity. Arch Biochem Biophys 500(1):92–106
93. Winterbourn CC, Kettle AJ (2000) Biomarkers of myeloperoxidase-derived hypochlorous acid. Free Radic Biol Med 29(5):403–409
94. Winterbourn CC (2002) Biological reactivity and biomarkers of the neutrophil oxidant, hypochlorous acid. Toxicology 181–182:223–227
95. Thomas EL, Grisham MB, Jefferson MM (1986) Cytotoxicity of chloramines. Methods Enzymol 118:585–593
96. Thomas EL (1979) Myeloperoxidase-hydrogen peroxide-chloride antimicrobial system: effect of exogenous amines on antibacterial action against *Escherichia coli*. Infect Immun 25(1):110–116
97. Ligtenbarg AGJ, Hage R, Feringa BL (2003) Catalytic oxidations by vanadium complexes. Coord Chem Rev 237(1–2):89–101
98. Renirie R, Pierlot C, Aubry JM, Hartog AF, Schoemaker HE, Alsters PL, Wever R (2003) Vanadium chloroperoxidase as a catalyst for hydrogen peroxide disproportionation to singlet oxygen in mildly acidic aqueous environment. Adv Synth Catal 345(6–7):849–858
99. Khan AU (1984) Myeloperoxidase singlet molecular oxygen generation detected by direct infrared electronic emission. Biochem Biophys Res Commu. 122(2):668–675
100. Kanofsky JR, Hoogland H, Wever R, Weiss SJ (1988) Singlet oxygen production by human eosinophils. J Biol Chem 263(20):9692–9696
101. Rosen H, Klebanoff SJ (1977) Formation of singlet oxygen by the myeloperoxidase-mediated antimicrobial system. J Biol Chem 252(14):4803–4810
102. Butler A, Walker JV (1993) Marine haloperoxidases. Chem Rev 93(5):1937–1944
103. Weyand M, Hecht HJ, Kieß M, Liaud MF, Vilter H, Schomburg D (1999) X-ray structure determination of a vanadium-dependent haloperoxidase from Ascophyllum nodosum at 2.0 a resolution. J Mol Biol 293(3):595–611
104. Everett RR, Soedjak HS, Butler A (1990) Mechanism of dioxygen formation catalyzed by vanadium bromoperoxidase. J Biol Chem 265(26):15671–15679
105. Dembitsky VM (2003) Oxidation, epoxidation and sulfoxidation reactions catalysed by haloperoxidases. Tetrahedron 59:4701–4720
106. Neumann CS, Fujimori DG, Walsh CT (2008) Halogenation strategies in natural product biosynthesis. Chem Biol 15(2):99–109

107. Butler A, Carter-Franklin JN (2004) The role of vanadium bromoperoxidase in the biosynthesis of halogenated marine natural products. Nat Prod Rep 21:180–188
108. Wagner C, El Omari M, König GM (2009) Biohalogenation: nature's way to synthesize halogenated metabolites. J Nat Prod 72(3):540–553
109. Geigert J, Lee TD, Dalietos DJ, Hirano DS, Neidleman SL (1986) Epoxidation of alkenes by chloroperoxidase catalysis. Biochem Biophys Res Commun 136(2):778–782
110. Lund H, Kalum L, Hofrichter M, Sebastian P (2013) Epoxidation using peroxygenase, WO 2013/144105 A1
111. Albrich JM, McCarthy CA, Hurst JK (1981) Biological reactivity of hypochlorous acid: implications for microbicidal mechanisms of leukocyte myeloperoxidase. Proc Natl Acad Sci USA 78(1):210–214
112. Winterbourn CC (1985) Comparative reactivities of various biological compounds with myeloperoxidase-hydrogen peroxide-chloride, and similarity of the oxidant to hypochlorite. Biochim Biophys Acta 840(2):204–210
113. Lesser MP (2006) Oxidative stress in marine environments: biochemistry and physiological ecology. Annu Rev Physiol 68:253–278
114. Leblanc C, Colin C, Cosse A, Delage L, La Barre S, Morin P, Fiévet B, Voiseux C, Ambroise Y, Verhaeghe E, Amouroux D, Donard O, Tessier E, Potin P (2006) Iodine transfers in the coastal marine environment: the key role of brown algae and of their vanadium-dependent haloperoxidases. Biochimie 88(11):1773–1785
115. Wuosmaa AM, Hager LP (1990) Methyl chloride transferase: a carbocation route for biosynthesis of halometabolites. Science 249(4965):160–162
116. Wever R, van der Horst MA (2013) The role of vanadium haloperoxidases in the formation of volatile brominated compounds and their impact on the environment. Dalton Trans 42(33):11778–11786
117. Paul C, Pohnert G (2011) Production and role of volatile halogenated compounds from marine algae. Nat Prod Rep 28:186–195
118. Hager LP, Morris DR, Brown FS, Eberwein H (1966) Chloroperoxidase II. Utilization of halogen anions. J Biol Chem 241(8):1769–1777
119. van Pée KH (2001) Microbial biosynthesis of halometabolites. Arch Microbiol 175(4):250–258
120. Ohsawa N, Ogata Y, Okada N, Itoh N (2001) Physiological function of bromoperoxidase in the red marine alga, Corallina pilulifera: production of bromoform as an allelochemical and the simultaneous elimination of hydrogen peroxide. Phytochemistry 58(5):683–692
121. Thorenz UR, Kundel M, Müller L, Hoffmann T (2012) Generation of standard gas mixtures of halogenated, aliphatic, and aromatic compounds and prediction of the individual output rates based on molecular formula and boiling point. Anal Bioanal Chem 404(8):2177–2183
122. Carter-Franklin JN, Parrish JD, Tschirret-Guth RA, Little RD, Butler A (2003) Vanadium haloperoxidase-catalyzed bromination and cyclization of terpenes. J Am Chem Soc 125(13):3688–3689
123. Faulkner DJ (1998) Marine natural products. Nat Prod Rep 15(2):113–158
124. Carter-Franklin JN, Butler A (2004) Vanadium bromoperoxidase-catalyzed biosynthesis of halogenated marine natural products. J Am Chem Soc 126(46):15060–15066
125. Fujimori DG, Walsh CT (2007) What's new in enzymatic halogenations. Curr Opin Chem Biol 11:553–560
126. Gribble GW (2003) The diversity of naturally produced organohalogens. Chemosphere 52(2):289–297
127. Crockford SJ (2009) Evolutionary roots of iodine and thyroid hormones in cell-cell signaling. Integr Comp Biol 49(2):155–166
128. Wang L, Zhou X, Fredimoses M, Liao S, Liu Y (2014) Naturally occurring organoiodines. RSC Adv 4(101):57350–57376
129. Ihssen J, Schubert M, Thöny-Meyer L, Richter M (2014) Laccase catalyzed synthesis of iodinated phenolic compounds with antifungal activity. PLoS ONE 9(3):e89924

130. Reddy CM, Xu L, Drenzek NJ, Sturchio NC, Heraty LJ, Kimblin C, Butler A (2002) A chlorine isotope effect for enzyme-catalyzed chlorination. J Am Chem Soc 124(49):14526–14527
131. Sundaramoorthy M, Terner J, Poulos TL (1998) Stereochemistry of the chloroperoxidase active site: crystallographic and molecular-modeling studies. Chem Biol 5(9):461–473
132. Isupov MN, Dalby AR, Brindley AA, Izumi Y, Tanabe T, Murshudov GN, Littlechild JA (2000) Crystal structure of dodecameric vanadium-dependent bromoperoxidase from the red algae Corallina officinalis. J Mol Biol 299(4):1035–1049
133. Itoh N, Hasan AKMQ, Izumi Y, Yamada H (1988) Substrate specificity, regiospecificity and stereospecificity of halogenation reactions catalyzed by non-heme-type bromoperoxidase of Corallina pilulifera. Eur J Biochem 172(2):477–484
134. de Boer E, Wever R (1988) The reaction mechanism of the novel vanadium-bromoperoxidase. A steady-state kinetic analysis. J Biol Chem 263:12326–12332
135. Manoj KM (2006) Chlorinations catalyzed by chloroperoxidase occur via diffusible intermediate(s) and the reaction components play multiple roles in the overall process. Biochim Biophys Acta 1764(8):1325–1339
136. Tschirret-Guth RA, Butler A (1994) Evidence for organic substrate binding to vanadium bromoperoxidase. J Am Chem Soc 116(1):411–412
137. Dunford HB (2010) Peroxidases and catalases: Biochemistry, biophysics, biotechnology and physiology. Wiley-VCH, Hoboken, USA
138. Schijndel JWPM, Barnett P, Roelse J, Vollenbroek EGM, Wever R (1994) The stability and steady-state kinetics of vanadium chloroperoxidase from the fungus Curvularia inaequalis. Eur J Biochem 225(1):151–157
139. Vardhaman AK, Barman P, Kumar S, Sastri CV, Kumar D, de Visser SP (2013) Mechanistic insight into halide oxidation by non-heme iron complexes. Haloperoxidase versus halogenase activity. Chem Commun 49(93):10926–10928
140. Sandy M, Carter-Franklin JN, Martin JD, Butler A (2011) Vanadium bromoperoxidase from *Delisea pulchra*: enzyme-catalyzed formation of bromofuranone and attendant disruption of quorum sensing. Chem Commun 47(44):12086–12088
141. Winter JM, Moore BS (2009) Exploring the chemistry and biology of vanadium-dependent haloperoxidases. J Biol Chem 284(28):18577–18581
142. Hasan Z, Renirie R, Kerkman R, Ruijssenaars HJ, Hartog AF, Wever R (2006) Laboratory-evolved vanadium chloroperoxidase exhibits 100-fold higher halogenating activity at alkaline pH. J Biol Chem 281:9738–9744
143. Carter JN, Beatty KE, Simpson MT, Butler A (2002) Reactivity of recombinant and mutant vanadium bromoperoxidase from the red alga Corallina officinalis. J Inorg Biochem 91(1):59–69
144. Hemrika W, Renirie R, Macedo-Ribeiro S, Messerschmidt A, Wever R (1999) Heterologous expression of the vanadium-containing chloroperoxidase from curvularia inaequalis in saccharomyces cerevisiae and site-directed mutagenesis of the active site residues His^{496}, Lys^{353}, Arg^{360}, and Arg^{490}. J Biol Chem 274:23820–23827
145. Wever R, Renirie R (2010) Application of peroxidases, in peroxidases catalases biochemisrty, biophysics, biotechnology, physiology, 2nd ed. Wiley, Hoboken, pp 363–385
146. Gupta R, Hou G, Renirie R, Wever R, Polenova T (2015) ^{51}V NMR Crystallography of vanadium chloroperoxidase and its directed evolution P395D/L241 V/T343A Mutant: Protonation environments of the active site. J Am Chem Soc 137(16):5618–5628
147. Zampella G, Fantucci P, Pecoraro VL, De Gioia L (2006) Oxidation of organic sulfides by vanadium haloperoxidase model complexes. Inorg Chem 45(18):7133–7143
148. Zampella G, Kravitz JY, Webster CE, Fantucci P, Hall MB, Carlson HA, Pecoraro VL, De Gioia L (2004) Quantum mechanical models of the resting state of the vanadium-dependent haloperoxidase. Inorg Chem 43(14):4127–4136
149. von Gunten U, Oliveras Y (1998) Advanced oxidation of bromide-containing waters: Bromate formation mechanisms. Environ Sci Technol 32(1):63–70
150. Clague MJ, Butler A (1995) On the mechanism of cis-Dioxovanadium(V)-catalyzed oxidation of bromide by hydrogen peroxide: evidence for a reactive, binuclear vanadium(V) peroxo complex. J Am Chem Soc 117(12):3475–3484

151. BarnettP, Hondmann DH, Lambertus Henricus S, Ter Steeg PF, Wever R (1995) Enzymatic antimicrobial compositions, WO 1995/27046
152. Johansen C (1997) Antimicrobial peroxidase compositions, WO 1997/42825
153. Johansen C, Fuglsang CC (2006) Enzymatic preservation of water based paints. US 7.063.970 B1
154. Detty MR, Ciriminna R, Bright RV, Pagliaro M (2014) Environmentally benign sol-gel antifouling and foul-releasing coatings. Acc Chem Res 47(2):678–687
155. Myan FWY, Walker J, Paramor O (2013) The interaction of marine fouling organisms with topography of varied scale and geometry: a review. Biointerphases 8:30
156. Rasmussen TB, Manefield M, Andersen JB, Eberl L, Anthoni U, Christophersen C, Steinberg P, Kjelleberg S, Givskov M (2000) How Delisea pulchra furanones affect quorum sensing and swarming motility in Serratia liquefaciens MG1. Microbiology 146:3237–3244
157. Gram L, de Nys R, Maximilien R, Givskov M, Steinberg P, Kjelleberg S (1996) Inhibitory effects of secondary metabolites from the red alga delisea pulchra on swarming motility of proteus mirabilis. Appl Environ Microbiol 62(11):4284–4287
158. Annous BA, Fratamico PM, Smith JL (2009) Scientific status summary. J Food Sci 74(1):R24–37
159. Dobretsov S, Teplitski M, Paul V (2009) Mini-review: quorum sensing in the marine environment and its relationship to biofouling. Biofouling 25(5):413–427
160. Kjelleberg S, Steinberg P, Givskov M, Gram L, Manefield M, De Nys R (1997) Do marine natural products interfere with prokaryotic AHL regulatory systems? Aquat Microb Ecol 13:85–93
161. Smith D, Wang JH, Swatton JE, Davenport P, Price B, Mikkelsen H, Stickland H, Nishikawa K, Gardiol N, Spring DR, Welch M (2006) Variations on a theme: diverse *N*-acyl homoserine lactone-mediated quorum sensing mechanisms in gram-negative bacteria. Sci Prog 89:167–211
162. Decho AW, Frey RL, Ferry JL (2011) Chemical challenges to bacterial AHL signaling in the environment. Chem Rev 111(1):86–99
163. Chhabra SR, Philipp B, Eberl L, Givskov M, Williams P, Cámara M (2005) Extracellular communication in bacteria, in the chemistry of pheromones and other semiochemicals II. In: Schulz S (ed) Topics in current chemistry, vol 240, pp 279–315
164. Shirtliff ME, Mader JT, Camper AK (2002) Molecular interactions in biofilms. Chem Biol 9(8):859–871
165. Schwartz T, Walter S, Marten SM, Kirschhöfer F, Nusser M, Obst U (2007) Use of quantitative real-time RT-PCR to analyse the expression of some quorum-sensing regulated genes in Pseudomonas aeruginosa. Anal Bioanal Chem 387(2):513–521
166. Fuqua C, Winans SC, Greenberg EP (1996) Census and consensus in bacterial ecosystems: the LuxR-LuxI family of quorum-sensing transcriptional regulators. Annu Rev Microbiol 50:727–751
167. Bassler BL, Wright M, Silverman MR (1994) Sequence and function of LuxO, a negative regulator of luminescence in Vibrio harveyi. Mol Microbiol 12(3):403–412
168. Thoendel M, Kavanaugh JS, Flack CE, Horswill AR (2011) Peptide signaling in the Staphylococci. Chem Rev 111(1):117–151
169. Nakayama J, Cao Y, Horii T, Sakuda S, Akkermans ADL, De Vos WM, Nagasawa H (2001) Gelatinase biosynthesis-activating pheromone: a peptide lactone that mediates a quorum sensing in Enterococcus faecalis. Mol Microbiol 41(1):145–154
170. Håvarstein LS, Coomaraswamy G, Morrison DA (1995) An unmodified heptadecapeptide pheromone induces competence for genetic transformation in Streptococcus pneumoniae. Proc Natl Acad Sci USA 92(24):11140–11144
171. Rocha-Estrada J, Aceves-Diez AE, Guarneros G, De La Torre M (2010) The RNPP family of quorum-sensing proteins in Gram-positive bacteria. Appl Microbiol Biotechnol 87(3):913–923
172. Schauder S, Shokat K, Surette MG, Bassler BL (2001) The LuxS family of bacterial autoinducers: biosynthesis of a novel quorum-sensing signal molecule. Mol Microbiol 41(2):463–476

173. Chen X, Schauder S, Potier N, Van Dorsselaer A, Pelczer I, Bassler BL, Hughson FM (2002) Structural identification of a bacterial quorum-sensing signal containing boron. Nature 415(6871):545–549
174. Neiditch MB, Federle MJ, Pompeani AJ, Kelly RC, Swem DL, Jeffrey PD, Bassler BL, Hughson FM (2006) Ligand-induced asymmetry in histidine sensor kinase complex regulates quorum sensing. Cell 126(6):1095–1108
175. Taga ME, Miller ST, Bassler BL (2003) Lsr-mediated transport and processing of AI-2 in Salmonella typhimurium. Mol Microbiol 50(4):1411–1427
176. Dobretsov S, Abed RMM, Teplitski M (2013) Mini-review: Inhibition of biofouling by marine microorganisms. Biofouling 29(4):423–441
177. Borchardt SA, Allain EJ, Michels JJ, Stearns GW, Kelly RF, McCoy WF (2001) Reaction of acylated homoserine lactone bacterial signaling molecules with oxidized halogen antimicrobials. Appl Environ Microbiol 67(7):3174–3179
178. Manefield M, de Nys R, Kumar N, Read R, Givskov M, Steinberg P, Kjelleberg S (1999) Evidence that halogenated furanones from Delisea pulchra inhibit acylated homoserine lactone (AHL)-mediated gene expression by displacing the AHL signal from its receptor protein. Microbiology 145(2):283–291
179. Ren D, Bedzyk LA, Ye RW, Thomas SM, Wood TK (2004) Differential gene expression shows natural brominated furanones interfere with the autoinducer-2 bacterial signaling system of Escherichia coli. Biotechnol Bioeng 88(5):630–642
180. Zang T, Lee BWK, Cannon LM, Ritter KA, Dai S, Ren D, Wood TK, Zhou ZS (2009) A naturally occurring brominated furanone covalently modifies and inactivates LuxS. Bioorg Med Chem Lett 19(21):6200–6204
181. Michels JJ, Allain EJ, Borchardt SA, Hu P, McCoy WF (2000) Degradation pathway of homoserine lactone bacterial signal molecules by halogen antimicrobials identified by liquid chromatography with photodiode array and mass spectrometric detection. J Chromatogr A 898(2):153–165
182. Slaughter JC (1999) The naturally occurring furanones: formation and function from pheromone to food. Biol Rev 74(3):259–276
183. Martinelli D, Grossmann G, Séquin U, Brandl H, Bachofen R (2004) Effects of natural and chemically synthesized furanones on quorum sensing in Chromobacterium violaceum. BMC Microbiol 4:25
184. Xavier KB, Bassler BL (2003) LuxS quorum sensing: more than just a numbers game. Curr Opin Microbiol 6(2):191–197
185. Manefield M, Rasmussen TB, Henzter M, Andersen JB, Steinberg P, Kjelleberg S, Givskov M (2002) Halogenated furanones inhibit quorum sensing through accelerated LuxR turnover. Microbiology 148(4):1119–1127
186. Pantanella F, Berlutti F, Passariello C, Sarli S, Morea C, Schippa S (2007) Violacein and biofilm production in Janthinobacterium lividum. J Appl Microbiol 102(4):992–999
187. McClean KH, Winson MK, Fish L, Taylor A, Chhabra SR, Camara M, Daykin M, Lamb JH, Swift S, Bycroft BW, Stewart GSAB, Williams P (1997) Quorum sensing and Chromobacterium violaceum: exploitation of violacein production and inhibition for the detection of N-acylhomoserine lactones. Microbiology 143(12):3703–3711
188. Burt SA, Ojo-Fakunle VTA, Woertman J, Veldhuizen EJA (2014) The natural antimicrobial carvacrol inhibits quorum sensing in Chromobacterium violaceum and reduces bacterial biofilm formation at sub-lethal concentrations. PLoS ONE 9(4):e393414
189. Steenackers HP, Levin J, Janssens JC, De Weerdt A, Balzarini J, Vanderleyden J, De Vos DE, De Keersmaecker SC (2010) Structure–activity relationship of brominated 3-alkyl-5-methylene-2(5*H*)-furanones and alkylmaleic anhydrides as inhibitors of *Salmonella* biofilm formation and quorum sensing regulated bioluminescence in *Vibrio harveyi*. Bioorg Med Chem 18(14):5224–5233
190. Butler A, Sandy M (2009) Mechanistic considerations of halogenating enzymes. Nature 460(7257):848–854

191. Syrpas M, Ruysbergh E, Blommaert L, Vanelslander B, Sabbe K, Vyverman W, De Kimpe N, Mangelinckx S (2014) Haloperoxidase mediated quorum quenching by *Nitzschia cf pellucida*: study of the metabolization of *N*-Acyl homoserine lactones by a benthic diatom. Mar Drugs 12(1):352–367
192. Kalia VC (2015) Quorum sensing vs quorum quenching: a battle with no end in sight. Springer, New Delhi, India
193. Fusetani N, Clare AS (2006) Antifouling compounds. Springer, Berlin, Heidelberg
194. van Pée KH (2012) Enzymatic chlorination and bromination. Methods Enzymol 516:237–257
195. Dobretsov S, Dahms HU, Qian PY (2006) Inhibition of biofouling by marine microorganisms and their metabolites. Biofouling 22(1–2):43–54
196. Dobretsov S, Teplitski M, Bayer M, Gunasekera S, Proksch P, Paul VJ (2011) Inhibition of marine biofouling by bacterial quorum sensing inhibitors. Biofouling 27(8):893–905
197. Dobretsov S, Dahms HU, YiLi H, Wahl M, Qian PY (2007) The effect of quorum-sensing blockers on the formation of marine microbial communities and larval attachment. FEMS Microbiol Ecol 60(2):177–188
198. Camilli A, Bassler BL (2006) Bacterial small-molecule signaling pathways. Science 311(5764):1113–1116
199. Rasmussen TB, Givskov M (2006) Quorum-sensing inhibitors as anti-pathogenic drugs. Int J Med Microbiol 296(2–3):149–161
200. Worthington RJ, Richards JJ, Melander C (2012) Small molecule control of bacterial biofilms. Org Biomol Chem 10(37):7457–7474
201. Pal A, Kamthania MC, Kumar A (2014) Bioactive compounds and properties of seaweeds—a review. Open Access Libr J 1(4):1–17
202. Kalia VC (2013) Quorum sensing inhibitors: an overview. Biotechnol Adv 31(2):224–245
203. Cartagena E, Colom OA, Neske A, Valdez JC, Bardón A (2007) Effects of plant lactones on the production of biofilm of Pseudomonas aeruginosa. Chem Pharm Bull (Tokyo) 55(1):22–25
204. Palmer AG, Senechal AC, Mukherjee A, Ané JM, Blackwell HE (2014) Plant responses to bacterial N-acyl L-homoserine lactones are dependent on enzymatic degradation to L-homoserine. ACS Chem Biol 9(8):1834–1845
205. Hughes DT, Sperandio V (2008) Inter-kingdom signalling: communication between bacteria and their hosts. Nat Rev Microbiol 6(2):111–120
206. Uroz S, Heinonsalo J (2008) Degradation of *N*-acyl homoserine lactone quorum sensing signal molecules by forest root-associated fungi. FEMS Microbiol Ecol 65(2):271–278
207. Verhaeghe E, Buisson D, Zekri E, Leblanc C, Potin P, Ambroise Y (2008) A colorimetric assay for steady-state analyses of iodo- and bromoperoxidase activities. Anal Biochem 379(1):60–65
208. Jerlich A, Tschabuschnig S, Fabjan JS, Schaur RJ (2000) Kinetics of chlorination of monochlorodimedone by myeloperoxidase. Int J Clin Lab Res 30(1):33–37
209. Bakkenist ARJ, De Boer JEG, Plat H, Wever R (1980) The halide complexes of myeloperoxidase and the mechanism of the halogenation reactions. Biochim Biophys Acta 613(2):337–348
210. Wagner C, Molitor IM, König GM (2008) Critical view on the monochlorodimedone assay utilized to detect haloperoxidase activity. Phytochemistry 9(2):323–332
211. Hansen PJ, Espenson JH (1995) Oxidation of chloride ions by hydrogen peroxide, catalyzed by methylrhenium trioxide. Inorg Chem 34(23):5839–5844
212. Griffin BW, Ashley PL (1984) Evidence for a radical mechanism of halogenation of monochlorodimedone catalyzed by chloroperoxidase. Arch Biochem Biophys 233(1):188–196
213. Griffin BW (1983) Mechanism of halide-stimulated activity of chloroperoxidase evidence for enzymatic formation of free hypohalous acid. Biochem Biophys Res Commun 116(3):873–879
214. Dypbukt JM, Bishop C, Brooks WM, Thong B, Eriksson H, Kettle AJ (2005) A sensitive and selective assay for chloramine production by myeloperoxidase. Free Radic Biol Med 39(11):1468–1477

215. http://www.sigmaaldrich.com/catalog/product/fluka/32661?lang=de&region=DE, Accessed 06 Nov 2016 (2016)
216. http://www.sigmaaldrich.com/catalog/product/sial/114545?lang=de&region=DE, Accessed 06 Nov 2016 (2016)
217. "http://www.sigmaaldrich.com/catalog/product/sial/114391?lang=de&region=DE, Accessed 06 Nov 2016 (2016)
218. "http://www.sigmaaldrich.com/catalog/product/sial/114413?lang=de&region=DE, Accessed 06 Nov 2016 (2016)
219. Terrón MC, Verhagen FJM, Franssen MCR, Field JA (1998) Chemical bromination of phenol red by hydrogen peroxide is possible in the absence of haloperoxidases. Chemosphere 36(6):1445–1452
220. Rao AVS, Ravishankar HN, Ramasarma T (1996) Vanadium catalysis in bromoperoxidation reaction. Arch Biochem Biophys 334(1):121–134
221. Herget K, Hubach P, Pusch S, Deglmann P, Götz H, Gorelik TE, Gural'skiy IA, Pfitzner F, Link T, Schenk S, Panthöfer M, Ksenofontov V, Kolb U, Opatz T, André R, Tremel W (2017) Haloperoxidase mimicry by CeO_{2-x} nanorods combats biofouling. Adv Mater 29(4):1603823
222. Debowska K, Debski D, Hardy M, Jakubowska M, Kalyanaraman B, Marcinek A, Michalski R, Michalowski B, Ouari O, Sikora A, Smulik R, Zielonka J (2015) Toward selective detection of reactive oxygen and nitrogen species with the use of fluorogenic probes–limitations, progress, and perspectives. Pharmacol Rep 67(4):756–764
223. Wardman P (2007) Fluorescent and luminescent probes for measurement of oxidative and nitrosative species in cells and tissues: progress, pitfalls, and prospects. Free Radic Biol Med 43(7):995–1022
224. Winterbourn CC (2014) The challenges of using fluorescent probes to detect and quantify specific reactive oxygen species in living cells. Biochim Biophys Acta 1840(2):730–738
225. Flemmig J, Zschaler J, Remmler J, Arnhold J (2012) The fluorescein-derived dye aminophenyl fluorescein is a suitable tool to detect hypobromous acid (HOBr)-producing activity in eosinophils. J Biol Chem 287(33):27913–27923
226. Martínez VM, De Cremer G, Roeffaers MBJ, Sliwa M, Baruah M, De Vos DE, Hofkens J, Sels BF (2008) Exploration of single molecule events in a haloperoxidase and its biomimic: localization of halogenation activity. J Am Chem Soc 130(40):13192–13193
227. Jin X, Hao L, Hu Y, She M, Shi Y, Obst M, Li J, Shi Z (2013) Two novel fluorescein-based fluorescent probes for hypochlorite and its real applications in tap water and biological imaging. Sens Actuat B 186:56–60
228. Chen X, Lee KA, Ha EM, Lee KM, Seo YY, Choi HK, Kim HN, Kim MJ, Cho CS, Lee SY, Lee WJ, Yoon J (2011) A specific and sensitive method for detection of hypochlorous acidfor the imaging of microbe-induced HOCl production. Chem Commun 47(15):4373–4375
229. Lee KA, Kim SH, Kim EK, Ha EM, You H, Kim B, Kim MJ, Kwon Y, Ryu JH, Lee WJ (2013) Bacterial-derived uracil as a modulator of mucosal immunity and gut-microbe homeostasis in Drosophila. Cell 153(4):797–811
230. Lin W, Long L, Chen B, Tan W (2009) A ratiometric fluorescent probe for hypochlorite based on a deoximation reaction. Chem Eur J 15(10):2305–2309
231. Allen RC (1991) Haloperoxidase acid optimum chemiluminescence assay system, WO 91/05063 (A1)
232. Zheng MH, Hu X, Wang XW, Liu XL, Jin JY (2016) Fluorescence-enhanced sensing of hypochlorous acid based on 2-Pyridylthiazole Unit. J Fluoresc 26(2):593–598
233. Kalyanaraman B, Darley-Usmar V, Davies KJA, Dennery PA, Forman HJ, Grisham MB, Mann GE, Moore K, Roberts LJ, Ischiropoulos H (2012) Measuring reactive oxygen and nitrogen species with fluorescent probes: challenges and limitations. Free Radic Biol Med 52(1):1–6
234. Flemmig J, Arnhold J (2010) Interaction of hypochlorous acid and myeloperoxidase with phosphatidylserine in the presence of ammonium ions. J Inorg Biochem 104(7):759–764
235. Thomas EL, Bozeman PM, Jefferson MM, King CC (1995) Oxidation of bromide by the human leukocyte enzymes myeloperoxidase and eosinophil peroxidase. J Biol Chem 270(7):2906–2913

236. Koelsch M, Mallak R, Graham GG, Kajer T, Milligan MK, Nguyen LQ, Newsham DW, Keh JS, Kettle AS, Scott KF, Ziegler JB, Pattison DI, Fu S, Hawkins CL, Rees MD, Davies MJ (2010) Acetaminophen (paracetamol) inhibits myeloperoxidase-catalyzed oxidant production and biological damage at therapeutically achievable concentrations. Biochem Pharmacol 79(8):1156–1164
237. Peskin AV, Winterbourn CC (2001) Kinetics of the reactions of hypochlorous acid and amino acid chloramines with thiols, methionine, and ascorbate. Free Radic Biol Med 30(5):572–579
238. Kettle AJ (1996) Neutrophils convert tyrosyl residues in albumin to chlorotyrosine. FEBS Lett 379(1):103–106
239. Kettle AJ, Albrett AM, Chapman AL, Dickerhof N, Forbes LV, Khalilova I, Turner R (2014) Measuring chlorine bleach in biology and medicine. Biochim Biophys Acta 1840(2):781–793
240. Sokolov AV, Kostevich VA, Kozlov SO, Donskyi IS, Vlasova II, Rudenko AO, Zakharova ET, Vasilyev VB, Panasenko OM (2015) Kinetic method for assaying the halogenating activity of myeloperoxidase based on reaction of celestine blue B with taurine halogenamines. Free Radic Res 49(6):777–789
241. Colpas GJ, Hamstra BJ, Kampf JW, Pecoraro VL (1996) Functional models for vanadium haloperoxidase: reactivity and mechanism of halide oxidation. J Am Chem Soc 118(14):3469–3478
242. Wischang D, Brücher O, Hartung J (2011) Bromoperoxidases and functional enzyme mimics for oxidative bromination - a sustainable synthetic approach. Coord Chem Rev 255(19):2204–2217
243. Yonehara K, Kamata K, Yamaguchi K, Mizuno N (2011) An efficient H_2O_2-based oxidative bromination of alkenes, alkynes, and aromatics by a divanadium-substituted phosphotungstate. Chem Commun 47(6):1692–1694
244. Kato N, Hayashi Y (2013) Discrete spherical hexadecavanadates incorporating a bromide with oxidative bromination activity. Dalton Trans 42(33):11804–11811
245. Licini G, Conte V, Coletti A, Mba M, Zonta C (2011) Recent advances in vanadium catalyzed oxygen transfer reactions. Coord Chem Rev 255(19–20):2345–2357
246. da Silva JAL, da Silva JJRF, Pombeiro AJL (2011) Oxovanadium complexes in catalytic oxidations. Coord Chem Rev 255(19–20):2232–2248
247. Clague MJ, Keder NL, Butler A (1993) Biomimics of vanadium bromoperoxidase: Vanadium(V)-Schiff base catalyzed oxidation of bromide by hydrogen peroxide. Inorg Chem 32(22):4754–4761
248. Rana S, Haque R, Santosh G, Maiti D (2011) Decarbonylative halogenation by a vanadium complex. Inorg Chem 52(6):2927–2932
249. Smith TS, Pecoraro VL (2002) Oxidation of organic sulfides by vanadium haloperoxidase model complexes. Inorg Chem 41(25):6754–6760
250. Ten Brink HB, Schoemaker HE, Wever R (2001) Sulfoxidation mechanism of vanadium bromoperoxidase from Ascophyllum nodosum. Evidence for direct oxygen transfer catalysis. Eur J Biochem 268(1):131–382
251. en Brink HB, Tuynman A, Dekker HL, Hemrika W, Izumi Y, Oshiro T, Schoemaker HE, Wever R (1998) Enantioselective sulfoxidation catalyzed by vanadium haloperoxidases. Inorg Chem 37(26):6780–6784
252. Strassberger Z, Ramos-Fernandez EV, Boonstra A, Jorna R, Tanase S, Rothenberg G (2013) Synthesis, characterization and testing of a new V_2O_5/Al_2O_3–MgO catalyst for butane dehydrogenation and limonene oxidation. Dalton Trans 42(15):5546–5553
253. Mimoun H, Saussine L, Daire E, Postel M, Fischer J, Weiss R (1983) Vanadium(V) peroxy complexes. New versatile biomimetic reagents for epoxidation of olefins and hydroxylation of alkanes and aromatic hydrocarbons. J Am Chem Soc 105(10):3101–3110
254. Velusamy S, Punniyamurthy T (2004) Novel vanadium-catalyzed oxidation of alcohols to aldehydes and ketones under atmospheric oxygen. Org Lett 6(2):217–219
255. Si TK, Drew MGB, Mukherjea KK (2011) Peroxidative bromination and oxygenation of organic compounds. Polyhedron 30(13):2286–2293

256. Chen C, Sun Q, Ren DX, Zhang R, Bai FY, Xing YH, Shi SH (2013) Bromoperoxidase mimic as catalysts for oxidative bromination - synthesis, structures and properties of the diversified oxidation state of vanadium(III, IV and V) complexes with pincer *N*-heterocycle ligands. CrystEngComm 15(27):5561–5573
257. Zhang R, Liu J, Chen C, Xing YH, Guan QL, Hou YN, Wang X, Zhang XX, Bai FY (2013) Synthesis, structures and properties of the catalytic bromination reaction of a series of novel scorperate oxidovanadium complexes with the potential detection of hydrogen peroxide in water. Spectrochim. Acta Part A Mol Biomol Spectrosc 115:476–482
258. Fernández TL, Souza ET, Visentin LC, Santos JV, Mangrich AS, Faria RB, Antunes OAC, Scarpellini M (2009) A new oxo-vanadium complex employing an imidazole-rich tripodal ligand: a bioinspired bromide and hydrocarbon oxidation catalyst. J Inorg Biochem 103(4):474–479
259. Cao YZ, Zhao HY, Bai FY, Xing YH, Wei DM, Niu SY, Shi Z (2011) Amino acid-derivatized oxidovanadium complexes: Synthesis, structure and bromination reaction activity. Inorg Chim Acta 368(1):223–230
260. Tótaro RM, Williams PAM, Apella MC, Blesa MA, Baran EM (2000) Bromination of phenol red mediated by vanadium(V) peroxo complexes at pH 6.5. Dalton Trans 4403–4406
261. Khan AT, Ali S (2012) A useful and convenient synthetic protocol for iodination of organic substrates using a combination of vanadyl acetylacetonate, hydrogen peroxide, and sodium iodide. Bull Chem Soc Jpn 85(11):1239–1243
262. Saikia L, Talukdar D, Deka RC, Thakur AJ (2013) KI-VO(acac)2–H_2O_2-AcOH, A new iodinating system for selective iodination at C-5 position of activated pyrimidinediones: a combined experimental and density functional study. J Heterocycl Chem 50(5):1031–1038
263. Dinesh CU, Kumar R, Pandey B, Kumar P (1995) Catalytic halogenation of selected organic compounds mimicking vanadate-dependent marine metalloenzymes. Chem Commun 20(6):611–612
264. Kikushima K, Moriuchi T, Hirao T (2009) Vanadium-catalyzed oxidative bromination under atmospheric oxygen. Chem Asian J 4(8):1213–1216
265. Kikushima K, Moriuchi T, Hirao T (2010) Vanadium-catalyzed oxidative bromination promoted by Brønsted acid or Lewis acid. Tetrahedron 66(34):6906–6911
266. Podgoršek A, Zupan M, Iskra J (2009) Oxidative halogenation with "green" oxidants: oxygen and hydrogen peroxide. Angew Chem Int Ed 48(45):8424–8450
267. Meister GE, Butler A (1994) Molybdenum(VI)- and Tungsten(VI)-mediated biomimetic chemistry of vanadium. bromoperoxidase. Inorg Chem 33(15):3269–3275
268. Islam NS, Boruah JJ (2015) Macromolecular peroxo complexes of Vanadium(V) and Molybdenum(VI): Catalytic activities and biochemical relevance. J Chem Sci 127(5):777–795
269. Assem FL, Levy LS (2009) A review of current toxicological concerns on vanadium pentoxide and other vanadium compounds: gaps in knowledge and directions for future research. J Toxicol Environ Heal B Crit Rev 12(4):289–306
270. ECHA, https://echa.europa.eu/home. Accessed 12 Dec 2016
271. Choudary BM, Sudha Y, Reddy PN (1994) Regioselective oxybromination of activated aromatic compounds catalysed by ammonium molybdate. Synlett 1994(6):450
272. Conte V, Floris B, Galloni P, Silvagni A (2005) Oxybromination of ethynylbenzene catalysed by molybdenum complexes in organic solvent and in ionic liquids. Adv Synth Catal 347(10):1341–1344
273. Romano F, Linden A, Mba M, Zonta C, Licini G (2010) Adv Synth Catal 352 (17). 2937–2942
274. Kurapati SK, Pal S (2016) cis-Dioxomolybdenum(VI) complexes with unsymmetric linear tetradentate ligands: syntheses, structures and bromoperoxidase activities. Appl Organomet Chem 30(3):116–124
275. Pasayat S, Dash SP, Roy S, Dinda R, Dhaka S, Maurya MR, Kaminsky W, Patil YP, Nethaji M (2014) Synthesis, structural studies and catalytic activity of dioxidomolybdenum(VI) complexes with aroylhydrazones of naphthol-derivative. Polyhedron 67(1):1–10
276. Beinker P, Hanson JR, Meindl N, Medina ICR (1998) Oxidative iodination of aromatic amides using sodium perborate or hydrogen peroxide with sodium tungstate. J Chem Res 4:204–205

277. Medina ICR, Hanson JR (2003) The oxidative bromination and iodination of dimethylacetanilides. J Chem Res 2003(7):62, 428
278. Hazarika P, Kalita D, Sarmah S, Borah R, Islam NS (2006) New oxo-bridged dinuclear peroxotungsten(VI) complexes: synthesis, stability and activity in bromoperoxidation. Polyhedron 25(18):3501–3508
279. Mizuno N, Kamata K, Yamaguchi K (2012) Oxidative functional group transformations with hydrogen peroxide catalyzed by a divanadium-substituted phosphotungstate. Catal Today 185(1):157–161
280. Yamaura T, Kamata K, Yamaguchi K, Mizuno N (2013) Efficient sulfoxidation with hydrogen peroxide catalyzed by a divanadium-substituted phosphotungstate. Catal Today 203:76–80
281. Wang SS, Yang GY (2015) Recent advances in polyoxometalate-catalyzed reactions. Chem Rev 115(11):4893–4962
282. Maurya MR, Rana L, Avecilla F (2016) Oxidoperoxidotungsten(VI) and dioxidotungsten(VI) complexes catalyzed oxidative bromination of thymol in presence of H_2O_2–KBr–$HClO_4$. Inorg Chim Acta 440:172–180
283. Badetti E, Romano F, Marchiò L, Taşkesenlioğlu S, Daştan A, Zonta C, Licini G (2016) Effective bromo and chloro peroxidation catalysed by tungsten(VI) amino triphenolate complexes. Dalton Trans 45(37):14603–14608
284. Jafarzadeh M, Amani K, Nikpour F (2005) Efficient one-pot transformation of aminoarenes to haloarenes using halodimethylisulfonium halides generated in situ. Can J Chem 83(3):1808–1819
285. Espenson JH, Zhu Z, Zauche TH, Bromide ions and methyltrioxorhenium as cocatalysts for hydrogen peroxide oxidations and brominations. J Org Chem 64(4):1191–1196
286. Adam W, Mock-Knoblauch C, Saha-Möller CR, Herderich M (2000) Are MnIV species involved in Mn(Salen)-catalyzed Jacobsen−Katsuki epoxidations? A mechanistic elucidation of their formation and reaction modes by EPR spectroscopy, mass-spectral analysis, and product studies: chlorination versus oxygen transfer. J Am Chem Soc 122(40):9685–9691
287. Carney JR, Dillon BR, Thomas SP (2016) Recent advances of manganese catalysis for organic Synthesis. Eur J Org Chem 23:3912–3923
288. Feng XD, Zhang R, Wang XY, Zhang XX, Wang JX, Xing YH, Sun LX (2015) Mimicing bromoperoxidase for copper complexes: Synthesis, structures and properties of Cu(II)–triazine pyrazolyl complex. Polyhedron 90:69–76
289. Paraskevas SM, Paraskevas MS (2004) Chlorination and oxidation of some aldehydes by H_2O_2 and diphenic acid·$CuCl_2$ complex. Catal Commun 5(11):687–690
290. Menini L, Gusevskaya EV (2006) Aerobic oxychlorination of phenols catalyzed by copper(II) chloride. Appl Catal A Gen 309(1):122–128
291. Menini L, Parreira LA, Gusevskaya EV (2007) A practical highly selective oxybromination of phenols with dioxygen. Tetrahedron Lett 48(36):6401–6404
292. Suess AM, Ertem MZ, Cramer CJ, Stahl SS divergence between organometallic and single-electron-transfer mechanisms in copper(II)-mediated aerobic C–H oxidation. J Am Chem Soc 135(26):9797–9804
293. Yang L, Lu Z (2009) Stahl SS (2009) Regioselective copper-catalyzed chlorination and bromination of arenes with O_2 as the oxidant. Chem Commun 42:6460–6462
294. Hao W, Liu Y (2015) C–H bond halogenation catalyzed or mediated by copper: an overview. Beilstein J Org Chem 11:2132–2144
295. van der Werf N, Selander N (2015) Para-selective halogenation of nitrosoarenes with Copper(II) Halides. Org Lett 17(24):6210–6213
296. Zhu X, Chiba S (2016) Copper-catalyzed oxidative carbon–heteroatom bond formation: a recent update. Chem Soc Rev 45(16):4504–4523
297. Liu AH, He LN, Hua F, Yang ZZ, Huang CB, Yu B, Li B (2011) In situ acidic carbon dioxide/ethanol system for selective oxybromination of aromatic ethers catalyzed by copper chloride. Adv Synth Catal 353(17):3187–3195
298. Liu AH, Ma R, Zhang M, He LN (2012) In situ acidic carbon dioxide/water system for selective oxybromination of electron-rich aromatics catalyzed by copper bromide. Catal Today 194(1):38–43

299. Luo Y, Pan X, Wu J (2010) Silver-catalyzed decarboxylative halogenation of carboxylic acids. Tetrahedron Lett 51(50):6646–6648
300. Huang FQ, Xie J, Sun JG, Wang YW, Dong X, Qi LW, Zhang B (2016) Regioselective synthesis of carbonyl-containing alkyl chlorides via silver-catalyzed ring-opening chlorination of cycloalkanols. Org Lett 18(4):684–687
301. Limberg C, Teles JH (2001) The activation of O_2 at ruthenium complexes: catalytic chlorination of unsaturated organic substrates within the system $O_2/HCl/H_2O$. Adv Synth Catal 343(5):447–449
302. Bonaccorsi C, Althaus M, Becker C, Togni A, Mezzetti A (2006) Chiral Ru/PNNP complexes in catalytic and stoichiometric electrophilic O- and F-atom transfer to 1,3-dicarbonyl compounds. Pure Appl Chem 78(2):391–396
303. Wang Y, Li GX, Yang G, He G, Chen G (2016) A visible-light-promoted radical reaction system for azidation and halogenation of tertiary aliphatic C–H bonds. Chem Sci 7(4):2679–2683
304. Field RJ, Körös E, Noyes RM (1972) Oscillations in chemical systems. II. Thorough analysis of temporal oscillation in the bromate-cerium-malonic acid system. J Am Chem Soc 94(25):8649–8664
305. Noyes RM, Jwo JJ (1975) Oscillations in chemical systems. X.1 Implications of cerium oxidation mechanisms for the belousov-zhabotinskii reaction. J Am Chem Soc 97(19):5431–5433
306. Asakura J, Robins MJ (1990) Cerium(IV)-mediated halogenation at C-5 of uracil derivatives. J Org Chem 55(16):4928–4933
307. Nair V, Panicker SB, Augustine A, George TG, Thomas S, Vairamani M (2001) An efficient bromination of alkenes using cerium(IV) ammonium nitrate (CAN) and potassium bromide. Tetrahedron 57(34):7417–7422
308. Schweizer AE, Jones ME, Hickman DA (2004) Oxidative halogenation and optional dehydrogenation of C3 + hydrocarbons (US 2004/0158110 A1)
309. Sridharan V, Menéndez JC (2010) Cerium(IV) ammonium nitrate as a catalyst in organic synthesis. Chem Rev 110(6):3805–3849
310. Babu RS (2002) Cerium(III) chloride heptahydrate: $CeCl_3{\cdot}7H_2O$. Synlett 11:1935–1936
311. Das B, Krishnaiah M, Venkateswarlu K, Reddy VS (2007) A mild and simple regioselective iodination of activated aromatics with iodine and catalytic ceric ammonium nitrate. Tetrahedron Lett 48(1):81–83
312. Firouzabadi H, Iranpoor N, Kazemi S, Ghaderi A, Garzan A (2009) Highly efficient halogenation of organic compounds with halides catalyzed by cerium(III) chloride heptahydrate using hydrogen peroxide as the terminal oxidant in water. Adv Synth Catal 351(11–12):1925–1932
313. Vardhaman AK, Sastri CV, Kumar D, de Visser SP (2011) Nonheme ferric hydroperoxo intermediates are efficient oxidants of bromide oxidation. Chem Commun 47(39):11044–11046
314. Vione D, Maurino V, Man SC, Khanra SC, Arsene C, Olariu RI, Minero C (2008) Formation of organobrominated compounds in the presence of bromide under simulated atmospheric aerosol conditions. Chemsuschem 1(3):197–204
315. Draksharapu A, Angelone D, Quesne MG, Padamati SK, Gómez L, Hage R, Costas, Browne WR, de Visser SP (2015) Identification and spectroscopic characterization of nonheme iron(III) hypochlorite intermediates. Angew Chem Int Ed 54(14):4357–4361
316. Ma R, Huang CB, Liu AH, Li XD, He LN (2014) An *in situ* acidic carbon dioxide/glycol system for aerobic oxidative iodination of electron-rich aromatics catalyzed by $Fe(NO_3)_3{\cdot}9H_2O$. Catal Sci Technol 4(12):4308–4312
317. Miyake T, Hanaya M (1995) Screening of metal chloride catalysts for oxychlorination of propene. Appl Catal A 121(1):L13–L17
318. Kalyani D, Dick AR, Anani WR, Sanford MS (2006) Scope and selectivity in palladium-catalyzed directed C–H bond halogenation reactions. Tetrahedron 62(49):11483–11498
319. Lyons TW, Sanford MS (2010) Palladium-catalyzed ligand-directed C–H functionalization reactions. Chem Rev 110(2):1147–1169

320. Du B, Jiang X, Sun P (2013) Pd(II)-catalyzed bromo- and chlorodecarboxylation of electron-rich arenecarboxylic acids. J Org Chem 78(6):2786–2791
321. Peng X, Shao XF, Liu ZQ (2013) Palladium-catalyzed highly selective *ortho*-Halogenation (I, Br, Cl) of arylnitriles via sp^2 C–H bond activation using Cyano as directing group. Tetrahedron Lett 54(24):3079–3081
322. Liu S, Chen J, Zhang R, Zhao F, Deng GJ (2014) Palladium-catalyzed desulfinative iodination of sodium sulfinates and sulfonyl hydrazides. Asian J Org Chem 3(11):1150–1153
323. Xiong HY, Cahard D, Pannecoucke X, Besset T (2016) Pd-catalyzed directed chlorination of unactivated C(sp3)–H Bonds at room temperature. Eur J Org Chem 21:3625–3630
324. Wang L, Wang SS, VO-Thanh G, Liu Y (2013) The oxidative halogenations of arenes in water using hydrogen peroxide and halide salts over an ionic catalyst containing sulfo group and hexafluorotitanate. J Mol Catal A Chem 371:56–62
325. Das DP, Parida K (2006) Liquid phase bromination of phenol over titania pillared zirconium phosphate and titanium phosphate. Catal Commun 7(2):68–72
326. Maniatakou A, Parsons S, Karaliota A (2007) Photosensitized oxidation of bromide to bromine catalyzed by niobium pentachloride in methanol solution: Formation and crystal structure of the tetraphenylphosphonium tribromide. J Photochem Photobiol A Chem 192(1):29–35
327. Bora U, Chaudhuri MK, Dey D, Dhar SS (2001) Peroxometal-mediated environmentally favorable route to brominating agents and protocols for bromination of organics. Pure Appl Chem 73(1):93–102
328. Liu C, Zhang H, Shi W, Lei A (2011) Bond formations between two nucleophiles: Transition metal catalyzed oxidative cross-coupling reactions. Chem Rev 111(3):1780–1824
329. Beletskaya IP, Makhon'kov DI (1981) Oxidation of alkyl derivatives of aromatic hydrocarbons by transition metal salts. Russ Chem Rev 50(6):534–552
330. Petrone DA, Ye J, Lautens M (2016) Modern transition-metal-catalyzed carbon–halogen bond formation. Chem Rev 116(14):8003–8104
331. Nath J, Chaudhuri MK (2008) Boric acid catalyzed bromination of a variety of organic substrates: an eco-friendly and practical protocol. Green Chem Lett Rev 1(4):223–230
332. Lin R, Ding Y, Gong L, Li J, Chen W, Yan L, Lu Y (2009) Oxidative bromination of methane on silica-supported non-noble metal oxide catalysts. Appl Catal A Gen 353(1):87–92
333. Wan S, Wang SR, Lu W (2006) One-Pot preparation of arylalkynes by a tandem catalytic iodination of arenes and palladium-catalyzed coupling of iodoarenes with terminal alkynes. J Org Chem 71(11):4349–4352
334. Earle MJ, Katdare SB (2006) Oxidative halogenation of aromatic compound. US 7084317 B2
335. Leonard KA, Zhou F, Detty MR (1996) Chalcogen(IV)–Chalcogen(II) redox cycles. 1. Halogenation of organic substrates with Dihaloselenium(IV) and -tellurium(IV) derivatives. Dehalogenation of vicinal dibromides with diaryl tellurides. Organometallics 15(20):4285–4292
336. Drake MD, Bateman MA, Detty MR (2003) Substituent effects in arylseleninic acid-catalyzed bromination of organic substrates with sodium bromide and hydrogen peroxide. Organometallics 22(20):4158–4162
337. Mellegaard-Waetzig SR, Wang C, Tunge JA (2006) Selenium-catalyzed oxidative halogenation. Tetrahedron 62(30):7191–7198
338. Balkrishna SJ, Prasad CD, Panini P, Detty MR, Chopra D, Kumar S (2012) Isoselenazolones as catalysts for the activation of bromine: bromolactonization of alkenoic acids and oxidation of alcohols. J Org Chem 77(21):9541–9552
339. Detty MR, Zhou F, Friedman AE (1996) Positive halogens from halides and hydrogen peroxide with organotellurium catalysts. J Am Chem Soc 118(2):313–318
340. Detty MR, Higgs DE, Nelen MI (2001) Iodination of organic substrates with halide salts and h_2o_2 using an organotelluride catalyst. Org Lett 3(3):349–352
341. Abe M, You Y, Detty MR (2002) 21-Telluraporphyrins. 2. Catalysts for bromination reactions with hydrogen peroxide and sodium bromide. Organometallics 21(21):4546–4551

342. Alberto EE, Muller LM, Detty MR (2014) Synthesis and properties of tellurinic anhydride-tellurone adducts. Organometallics 33(19):5571–5578
343. Shul'pin G (2016) New trends in oxidative functionalization of carbon–hydrogen bonds: a review. Catalysts 6(4):50
344. Hartwig JF, Larsen MA (2016) Undirected, homogeneous C–H bond functionalization: challenges and opportunities. ACS Cent Sci 2(5):281
345. Smith AMR, Hii KK (2011) Transition metal catalyzed enantioselective α-heterofunctionalization of carbonyl compounds. Chem Rev 111(3):1637–1656
346. Hagen J (2015) Industrial catalysis: a practical approach, 3rd edn. Wiley-VCH, Weinheim, Germany
347. Woodley JM (2008) New opportunities for biocatalysis: making pharmaceutical processes greener. Trends Biotechnol 26(6):321–327
348. Patel RN (2008) Synthesis of chiral pharmaceutical intermediates by biocatalysis. Coord Chem Rev 252(5–7):659–701
349. Ansari SA, Husain Q (2012) Potential applications of enzymes immobilized on/in nano materials: A review. Biotechnol Adv 30(3):512–523
350. Lee CH, Lin TS, Mou CY (2009) Mesoporous materials for encapsulating enzymes. Nano Today 4(3):165–168
351. Betancor L, Luckarift HR (2008) Bioinspired enzyme encapsulation for biocatalysis. Trends Biotechnol 26(10):566–572
352. Andre R, Tahir MN, Natalio F, Tremel W (2012) Bioinspired synthesis of multifunctional inorganic and bio-organic hybrid materials. FEBS J 279(10):1737–1749
353. Drauz K, Gröger H, May O (2012) Enzyme catalysis in organic synthesis, 3rd edn. Wiley-VCH, Weinheim, Germany
354. Kirk O, Borchert TV, Fuglsang CC (2002) Industrial enzyme applications. Curr Opin Biotechnol 13(4):345–351
355. Rasor JP, Voss E (2001) Enzyme-catalyzed processes in pharmaceutical industry. Appl Catal A Gen. 221(1–2):145–158
356. Adrio JL, Demain AL (2014) Microbial enzymes: tools for biotechnological processes. Biomolecules 4(1):117–139
357. Smith DR, Grüschow S, Goss RJ (2013) Scope and potential of halogenases in biosynthetic applications. Curr Opin Chem Biol 17(2):276–283
358. Brown S, O'Connor SE (2015) Halogenase engineering for the generation of new natural product analogues. ChemBioChem 16(15):2129–2135
359. Schmid A, Dordick JS, Hauer B, Kiener A, Wubbolts M, Witholt B (2001) Industrial biocatalysis today and tomorrow. Nature 409(6817):258–268
360. Mateo C, Palomo JM, Fernandez-Lorente G, Guisan JM, Fernandez-Lafuente R (2007) Improvement of enzyme activity, stability and selectivity via immobilization techniques. Enzyme Microb Technol 40(6):1451–1644
361. Iyer PV, Ananthanarayan L (2008) Enzyme stability and stabilization—aqueous and non-aqueous environment. Process Biochem 43(10):1019–1032
362. Que L, Tolman WB (2008) Biologically inspired oxidation catalysis. Nature 455(7211):333–340
363. Lu Y, Yeung N, Sieracki N, Marshall NM (2009) Design of functional metalloproteins. Nature 460(7252):855–862
364. Kirby AJ, Hollfelder F (2009) From enzyme models to model enzymes. RSC Publishing, Cambridge UK, pp 1–273
365. Bezzu CG, Helliwell M, Warren JE, Allan DR, McKeown NB (2010) Heme-like coordination chemistry within nanoporous molecular crystals. Science 327(5973):1627–1630
366. Grzybowski BA, Huck WTS (2016) The nanotechnology of life-inspired systems. Nat Nanotechnol 11(7):585–592
367. Rothenberg G (2008) Catalysis—concepts and green applications. Wiley-VCH, Weinheim, Germany, pp 1–38

368. André R, Natálio F, Tremel W (2013) in New future developments in catalysis. In: Suib S (ed), 1st ed. Elsevier, pp 149–173
369. Deutschmann O, Knözinger H, Kochloefl K, Turek T (2009) Ullmann's encyclopedia of industrial chemistry. Wiley-VCH, Weinheim, Germany, pp 1–110
370. Wang Z, Ding K, Uozumi Y (2008) Handbook of asymmetric heterogeous catalalysis. Wiley-VCH, Weinheim, Germany, pp 1–24
371. Callender R, Dyer RB (2015) Molecular designs for controlling the local environments around metal Ions. Acc Chem Res 48(8):407–414
372. Roduner E (2014) Understanding catalysis. Chem Soc Rev 43(24):8226–8239
373. del Pino P, Pelaz B, Zhang Q, Maffre P, Nienhaus GU, Parak WJ (2014) Protein corona formation around nanoparticles—from the past to the future. Mater. Horiz. 1(3):301–313
374. André R, Natálio F, Humanes M, Leppin J, Heinze K, Wever R, Schröder HC, Müller WEG, Tremel W (2011) V_2O_5 nanowires with an intrinsic peroxidase-like activity. Adv Funct Mater 21(3):501–509
375. Tahir MN, André R, Sahoo JK, Jochum FD, Theato P, Natalio F, Berger R, Branscheid R, Kolb U, Tremel W (2011) Hydrogen peroxide sensors for cellular imaging based on horse radish peroxidase reconstituted on polymer-functionalized TiO_2 nanorods. Nanoscale 3(9):3907–3914
376. Kluenker M, Tahir MN, Ragg R, Korschelt K, Simon P, Gorelik TE, Barton B, Shylin SI, Panthöfer M, Herzberger J, Frey H, Ksenofontov V, Möller A, Kolb U, Grin J, Tremel W (2017) Pd@Fe_2O_3 superparticles with enhanced peroxidase activity by solution phase epitaxial growth. Chem Mater 29(3):1134–1146
377. Shin HY, Park TJ, Kim MI (2015) Recent research trends and future prospects in nanozymes. J Nanomater 2015:1
378. Solomon EI, Decker A, Lehnert N (2003) Non-heme iron enzymes: contrasts to heme catalysis. Proc Natl Acad Sci USA 100(7):3589–3594
379. Jittam P, Boonsiri P, Promptmas C, Sriwattanarothai N, Archavarungson N, Ruenwongsa P, Panijpan B (2009) Red seaweed enzyme-catalyzed bromination of bromophenol red: an inquiry-based kinetics laboratory experiment for undergraduates. Biochem Mol Biol Educ 37(2):99–105
380. Lee Y, von Gunten U (2010) Oxidative transformation of micropollutants during municipal wastewater treatment: comparison of kinetic aspects of selective (chlorine, chlorine dioxide, ferrate VI, and ozone) and non-selective oxidants (hydroxyl radical). Water Res 44(2):555–566
381. Gallard H, Pellizzari F, Croué JP, Legube B (2003) Rate constants of reactions of bromine with phenols in aqueous solution. Water Res 37(12):2883–2892
382. Gazda M, Margerum DW (1994) Reactions of monochloramine with bromine, tribromide, hypobromous acid and hypobromite: formation of bromochloramines. Inorg Chem 33(1):118–123
383. Acero JL, Piriou P, von Gunten U (2005) Kinetics and mechanisms of formation of bromophenols during drinking water chlorination: assessment of taste and odor development. Water Res 39(13):2979–2993
384. von Gunten U, Pinkernell U (2000) Ozonation of bromide-containing drinking waters:a delicate balance between disinfection and bromate formation. Water Sci Technol 41(7):53–59
385. von Gunten U (2003) Ozonation of drinking water: part II. Disinfection and by-product formation in presence of bromide, iodide or chlorine. Water Res 37(7):1469–1487
386. Deborde M, von Gunten U (2008) Reactions of chlorine with inorganic and organic compounds during water treatment-Kinetics and mechanisms: a critical review. Water Res 42(1–2):13–51
387. Goodwin JGJ, Kim S, Rhodes WD (2007) Turnover frequencies in metal catalysis: Meanings, functionalities and relationships. In: Spivey JJ, Roberts GW (eds) Catalysis, vol 17. RSC Publishing, pp 320–347
388. Korschelt K, Ragg R, Metzger CS, Kluenker M, Oster M, Barton B, Panthöfer M, Strand D, Kolb U, Mondeshki M, Strand S, Brieger J, Tahir MN, Tremel W (2017) Glycine-functionalized copper(II) hydroxide nanoparticles with high intrinsic superoxide dismutase activity. Nanoscale 9(11):3952–3960

389. Sels B, De Vos D, Buntinx M, Pierard F, Kirsch-De Mesmaeker A, Jacobs P (1999) Layered double hydroxides exchanged with tungstate as biomimetic catalysts for mild oxidative bromination. Nature 400(6747):855–857
390. Sels BF, De Vos DE, Buntinx M, Jacobs PA (2003) Transition metal anion exchanged layered double hydroxides as a bioinspired model of vanadium bromoperoxidase. J Catal 216(1–2):288–297
391. Shannon RD (1976) Revised effective ionic radii and systematic studies of interatomic distances in halides and chalcogenides. Acta Crystallogr A 32(5):751–767
392. Pearson RG (1997) Chemical hardness applications from molecules to solids. Wiley-VCH, Weinheim, Germany
393. Livage J (1991) Vanadium pentoxide gels. Chem Mater 3(4):578–593
394. Mallick S, Parida KM (2007) Studies on heteropoly acid supported zirconia II. Liquid phase bromination of phenol and various organic substrates Catal Commun 8(6):889–893
395. Choi H, Chang YY, Kwon YU, Han OH (2003) Incorporation of decavanadate ions into silica gels and mesostructured silica walls. Chem Mater 15(17):3261–3267
396. Narender N, Mohan K, Vinod Reddy R, Srinivasu P, Kulkarni SJ, Raghavan KV (2003) Liquid phase bromination of phenols using potassium bromide and hydrogen peroxide over zeolites. J Mol Catal A: Chem 192(1–2):73–77
397. Maurya MR, Kumar A, Manikandan P, Chand S (2004) Synthesis, characterisation and catalytic potential of oxovanadium(IV) based coordination polymers having a bridging methylene group. Appl Catal A Gen 277(1–2):45–53
398. Maurya MR, Kumar U, Manikandan P (2006) Polymer supported vanadium and molybdenum complexes as potential catalysts for the oxidation and oxidative bromination of organic substrates. Dalton Trans 29:3561–3575
399. Maurya MR, Chaudhary N, Avecilla F (2014) Polymer-grafted and neat vanadium(V) complexes as functional mimics of haloperoxidases. Polyhedron 67(1):436–448
400. Maurya MR, Kumar M, Arya A (2008) Model dioxovanadium(V) complexes through direct immobilization on polymer support, their characterization and catalytic activities. Catal Commun 10(2):187–191
401. Maurya MR, Chaudhary N, Kumar A, Avecilla F, Costa Pessoa J (2014) Polystyrene bound dioxidovanadium(V) complexes of 2-acetylpyridine derived ligands for catalytic oxidations. Inorg Chim Acta 420:24–38
402. Bhunia S, Saha D, Koner S (2011) MCM-41-supported Oxo-vanadium(IV) complex: a highly selective heterogeneous catalyst for the bromination of hydroxy aromatic compounds in water. Langmuir 27(24):15322–15329
403. Maurya MR, Saklani H, Agarwal S (2004) Oxidative bromination of salicylaldehyde by potassium bromide/H_2O_2 catalysed by dioxovanadium(V) complexes encapsulated in zeolite–Y: a functional model of haloperoxidases. Catal Commun 5(10):563–568
404. Sels BF, Levecque P, Brosius R, De Vos DE, Jacobs PA, Gammon DW, Kinfe HH (2005) A new catalytic route for the oxidative halogenation of cyclic enol ethers using tungstate exchanged on takovite. Adv Synth Catal 347(1):93–104
405. Choudary BM, Someshwar T, Kantam ML, Reddy CV (2004) Molybdate-exchanged Mg–Al–LDH catalyst: an eco-compatible route for the synthesis of β-bromostyrenes in aqueous medium. Catal Commun 5(5):215–219
406. Sels BF, De Vos DE, Jacobs PA (1997) Catalytic oxidations by in situ generated peroxotungsten complexes immobilized on layered double hydroxides (LDH): relation between catalytic properties and peroxotungstate micro-environment. In: Grasselli RK, Oyama ST, Gaffney AM, Lyons JE (eds) 3rd world congress on oxidation catalysis, vol 110, pp 1051–1059
407. Molinari A, Varani G, Polo E, Vaccari S, Maldotti A (2007) Photocatalytic and catalytic activity of heterogenized $W_{10}O_{32}{}^{4-}$ in the bromide-assisted bromination of arenes and alkenes in the presence of oxygen. J Mol Catal A 262(1–2):156–163
408. Doggali P, Waghmare S, Rayalu S, Teraoka Y, Labhsetwar N (2011) Transition metals supported on mesoporous ZrO_2 for the catalytic control of indoor CO and PM emissions. J Mol Catal A 347(1–2):52–59

409. Thomas JM, Raja R (2004) Catalytic significance of organometallic compounds immobilized on mesoporous silica: economically and environmentally important examples. J Organomet Chem 689(24):4110–4124
410. Rajagopal R, Siddiqui SA, Daniel T, Lahoti RJ, Srinivasan KV (2004) Regioselective side-chain as well as nuclear monobromination of aromatic substrates with*N*-bromosuccinimide using phosphotungstic acid supported on zirconia as a heterogeneous catalyst. J Mol Catal A 210(1–2):165–169
411. Firouzabadi H, Iranpoor N, Amani K (2003) Heteropoly acid cesium salt/cetyltrimethylammonium bromide a catalytic heterogeneous system which highly controls regioselective bromination of aromatic compounds with bromine. J Mol Catal A 195(1–2):289–294
412. Mallik S, Parida KM, Dash SS (2007) Studies on heteropoly acid supported zirconia: III: oxidative bromination of phenol using phosphotungstic acid supported on zirconia. J Mol Catal A 261(2):1721–1779
413. Wong ST, Hwang CC, Mou CY (2006) Tungstated zirconia catalyzed bromination of phenol red under nearly neutral solution. Appl Catal B 63(1–2):1–8
414. Das DP, Parida KM (2006) Liquid phase bromination of phenol: III. Over heteropoly acid (HPA)-impregnated titanium phosphate (TiP). J Mol Catal A Chem 253(1–2):70–78
415. Peng L, Liu Y, Kim N, Readman JE, Grey CP (2005) Detection of Brønsted acid sites in zeolite HY with high-field 17O-MAS-NMR techniques. Nat Mater 4(3):216–219
416. Walker JV, Morey M, Carlsson H, Davidson A, Stucky GD, Butler A (1997) Peroxidative halogenation catalyzed by transition-metal-ion-grafted mesoporous silicate materials. J Am Chem Soc 119(29):6921–6922
417. Chen AJ, Chen XR, Mou CY (2010) Highly regioselective oxybromination in an aqueous system using SBA-15 supported sulfated zirconia catalyst. J Chin Chem Soc 57(4B):820–828
418. Riley DP (1999) Functional mimics of superoxide dismutase enzymes as therapeutic agents. Chem Rev 99(9):2573–2588
419. Jameson GB, Ibers JA (1994) Bioinorganic chemistry. In: Bertini I, Gray HB, Lippard SJ, Valentine JS (eds). University Science Books, Mill Valley, CA, pp 167–252
420. Islam SM, Ghosh K, Roy AS, Salam N, Chatterjee T (2014) Oxidation and oxidative bromination reactions catalyzed by a reusable polymer-anchored iron(III) complex in water at room temperature. J Inorg Organomet Polym Mater 24(2):457–467
421. Islam SM, Roy AS, Mondal P, Salam N, Paul S (2013) Polymer-anchored Cu(II) complex as an efficient catalyst for selective and mild oxidation of sulfides and oxidative bromination reaction. Catal Lett 143(2):225–233
422. Sharma RK, Sharma C (2010) Oxidative bromination reaction using Cu^{2+}-perfluorophthalocyanine-immobilized silica gel catalyst under mild reaction conditions. Tetrahedron Lett 51(33):4415–4418
423. Raja R, Ratnasamy RP (1997) Oxyhalogenation of aromatics over copper phthalocyanines encapsulated in zeolites. J Catal 170(2):244–253
424. Chen AJ, Wong ST, Hwang CC, Mou CY (2011) Highly efficient and regioselective halogenation over well dispersed rhenium-promoted mesoporous zirconia. ACS Catal 1(7):786–793
425. McMaster DM, Bennett SM, Tang Y, Finlay JA, Kowalke GL, Nedved B, Bright FV, Callow ME, Callow JA, Wendt DE, Hadfield MG, Detty MR (2009) Antifouling character of 'active' hybrid xerogel coatings with sequestered catalysts for the activation of hydrogen peroxide. Biofouling 25(1):21–33
426. Gatley C, Muller L, Lang M, Alberto E, Detty M (2015) Xerogel-sequestered silanated organochalcogenide catalysts for bromination with hydrogen peroxide and sodium bromide. Molecules 20(6):9616–9639
427. Das B, Venkateswarlu K, Mahender G, Mahender I (2005) A simple and efficient method for α-bromination of carbonyl compounds using *N*-bromosuccinimide in the presence of silica-supported sodium hydrogen sulfate as a heterogeneous catalyst. Tetrahedron Lett 46(17):3041–3044

428. Das B, Venkateswarlu K, Krishnaiah M, Holla H (2006) An efficient, rapid and regioselective nuclear bromination of aromatics and heteroaromatics with NBS using sulfonic-acid-functionalized silica as a heterogeneous recyclable catalyst. Tetrahedron Lett 47(49):8693–8697
429. Paul V, Sudalai A, Daniel T, Srinivasan KV (1994) Regioselective bromination of activated aromatic substrates with N-bromosuccinimide over HZSM-5. Tetrahedron Lett 35(38):7055–7056
430. Natalio F, André R, Hartog AF, Stoll B, Jochum KP, Wever R, Tremel W (2012) Vanadium pentoxide nanoparticles mimic vanadium haloperoxidases and thwart biofilm formation. Nat Nanotechnol 7(7):530–535
431. Rothenberg G, Clark JH (2000) On oxyhalogenation, acids, and non-mimics of bromoperoxidase enzymes. Green Chem 2(5):248–251
432. Natalio F, André R, Pihan SA, Humanes M, Wever R, Tremel R (2011) V_2O_5 nanowires with an intrinsic iodination activity leading to the formation of self-assembled melanin-like biopolymers. J Mater Chem 21(32):11923–11929
433. Bora U, Bose G, Chaudhuri MK, Dhar SS, Gopinath R, Khan AT, Patel BK (2000) Regioselective bromination of organic substrates by tetrabutylammonium bromide promoted by $V_2O_5{-}H_2O_2$: an environmentally favorable synthetic protocol. Org Lett 2(3):247–249
434. Tilley JD (2008) Defects in solids. Wiley, Hoboken, NJ
435. Smith DM (2000) The defect chemistry of metal oxides. Oxford University Press, Oxford, UK
436. Mehonic A, Kenyon AJ (2015) Defects at oxide surfaces. Springer, Cham, Germany
437. Leyva-Pérez A, Cómbita-Merchán D, Cabrero-Antonino JR, Al-Resayes SI, Corma A (2013) Oxyhalogenation of activated arenes with nanocrystalline Ceria. ACS Catal 3(2):250–258
438. Huang W, Gao Y (2014) Morphology-dependent surface chemistry and catalysis of CeO_2 nanocrystals. Catal Sci Technol 4(11):3772–3784
439. Herget K, Tremel W (2016) Biozide zusammensetzung enthaltende ceroxide und damit hergestellte produkte. 2017-10-30PCT/EP2017/077827
440. Rodríguez-López JN, Lowe DJ, Hernández-Ruiz J, Hiner ANP, García-Cánovas F, Thorneley RN (2001) Mechanism of reaction of hydrogen peroxide with horseradish peroxidase: identification of intermediates in the catalytic cycle. J Am Chem Soc 123(48):11838–11847
441. Kahandal SS, Kale SR, Gawande MB, Zboril R, Varma RS, Jayaram RV (2014) Greener iodination of arenes using sulphated ceria–zirconia catalysts in polyethylene glycol. RSC Adv 4(12):6267–6274
442. Zhang P, Sun D, Wen M, Yang J, Zhou K, Wang Z (2012) Hydroxyl radical promotes the direct iodination of aromatic compounds with iodine in water: A combined experimental and theoretical study. Adv Synth Catal 354(4):720–729
443. Wever R (2010) Application of peroxidases. In: Dunford HB, Jones PA (eds) Peroxidases and catalases: biochemistry, biophysics, biotechnology and physiology. Wiley, pp 403–424
444. Wever R (1997) Mechanismsm of biological halogenation and dehalogenation. In: Jansen DB, Soda K, Wever R (eds) Proc Colloq Royal Netherlands Academie of Arts and Science, Amsterdam, pp 89–100
445. Chung W, Vanderwal CD (2016) Stereoselective halogenation in natural product synthesis. Angew Chem Int Ed 55(14):4396–4434
446. Li JJ, Gribble GW (2007) Palladium in heterocyclic chemistry: a guide for the synthetic chemist, 2nd edn. Elsevier, Oxford, UK
447. Kürti L, Czakó B (2005) Strategic applications of named reactions in organic synthesis, 1st edn. Elsevier, Amsterdam, Netherlands
448. Hassan J, Sévignon M, Gozzi C, Schulz E, Lemaire M (2002) Aryl−Aryl bond formation one century after the discovery of the ullmann reaction. Chem Rev 102(5):1359–1470
449. Littlechild J (1999) Haloperoxidases and their role in biotransformation reaction. Curr Opin Chem Biol 3(1):28–34
450. Hollmann F, Arends IWCE, Buehler K, Schallmey A, Bühler B (2011) Enzyme-mediated oxidations for the chemist. Green Chem 13(2):226–265

451. Franssen MCR, Steunenberg P, Scott EL, Zuilhof H, Sanders JPM (2013) Immobilised enzymes in biorenewables production. Chem Soc Rev 42(15):6491–6533
452. Ayala M, Verdin J, Vazquez-Duhalt R (2007) The prospects for peroxidase-based biorefining of petroleum fuels. Biocatal Biotrans 25(2–4):114–129
453. Sassolas A, Blum KJ, Leca-Bouvier BD (2012) Immobilization strategies to develop enzymatic biosensors. Biotechnol Adv 30(3):489–511
454. Uraisin K, Nacapricha D, Lapanantnoppakhun S, Grudpan K, Motomizu S (2005) Determination of trace amounts of bromide by flow injection/stopped-flow detection technique using kinetic-spectrophotometric method. Talanta 68(2):274–280
455. Uraisin K, Takayanagi T, Oshima M, Nacapricha D, Motomizu S (2006) Kinetic-spectrophotometric method for the determination of trace amounts of bromide in seawater. Talanta 68(3):951–956
456. Borges EP, Lavorante AF, dos Reis BF (2005) Determination of bromide ions in seawater using flow system with chemiluminescence detection. Anal Chim Acta 528(1):115–119
457. Shi L, Liu X, Niu W, Li H, Han S, Chen J, Xu G (2009) Hydrogen peroxide biosensor based on direct electrochemistry of soybean peroxidase immobilized on single-walled carbon nanohorn modified electrode. Biosens Bioelectron 24(5):1159–1163
458. Yin H, Ai S, Shi W, Zhu L (2009) A novel hydrogen peroxide biosensor based on horseradish peroxidase immobilized on gold nanoparticles–silk fibroin modified glassy carbon electrode and direct electrochemistry of horseradish peroxidase. Sensors Actuators B Chem 137(2):747–753
459. Guedes EAC, da Silva TG, Aguiar JS, de Barros LD, Pinotti LM, Sant'Ana AEG (2013) Cytotoxic activity of marine algae against cancerous cells. Rev Bras Farmacogn 23(4):668–673
460. Clark RA, Klebanoff SJ, Einstein AB, Fefer A (1975) Peroxidase-H_2O_2-halide system: cytotoxic effect on mammalian tumor cells. Blood 45(2):161–170
461. Mohammed KA, Hossain CF, Zhang L, Bruick RK, Zhou YD, Nagle DG (2004) Laurenditerpenol, a new diterpene from the tropical marine alga *Laurencia intricata* that potently inhibits HIF-1 mediated hypoxic signaling in breast tumor cells. J Nat Prod 67(12):2002–2007
462. Smit AJ (2004) Medicinal and pharmaceutical uses of seaweed natural products: a review. J Appl Phycol 16(4):245–262
463. Ilah Al-Saif SSA, Abdel-Raouf N, El-Wazanani HA, Aref IA (2014) Antibacterial substances from marine algae isolated from Jeddah coast of Red sea, Saudi Arabia. Saudi J Biol Sci 21(1):57–64
464. Oh KB, Lee JH, Chung SC, Shin J, Shin HJ, Kim HK, Lee HS (2008) Antimicrobial activities of the bromophenols from the red alga *Odonthalia corymbifera* and some synthetic derivatives. Bioorg Med Chem Lett 18(1):104–108
465. Vairappan CS (2003) Potent antibacterial activity of halogenated metabolites from Malaysian red algae, Laurencia majuscula (Rhodomelaceae, Ceramiales). Biomol Eng 20(4–6):255–259
466. Cragg GM (1830) Newman DJ (2013) Natural products: a continuing source of novel drug leads. Biochim Biophys Acta 6:3670–3695
467. Manivasagan P, Venkatesan J, Sivakumar K, Kim SK (2014) Pharmaceutically active secondary metabolites of marine actinobacteria. Microbiol Res 169(4):262–278
468. Cabrita MT, Vale C, Rauter AP (2010) Halogenated compounds from marine algae. Mar Drugs 8(8):2301–2317
469. Wanke T, Philippus AC, Zatelli GA, Vieira LFO, Lhullier C, Falkenberg M (2015) C_{15} acetogenins from the *Laurencia* complex: 50 years of research—an overview. Rev Bras Farmacogn 25(6):569–587
470. Rosenbaum RS, Kessler J (1984) Bactericidal compositions and methods. US Patent 4.473.550
471. Torres FAE, Passalacqua TG, Velásquez AMA, de Souza RA, Colepicolo P, Graminha MAS (2014) New drugs with antiprotozoal activity from marine algae: a review. Rev Bras Farmacogn 24(3):265–276
472. Wang BG, Gloer JB, Ji NY, Zhao JC (2013) Halogenated organic molecules of rhodomelaceae origin: chemistry and biology. Chem Rev 113(5):3632–3685

473. Vairappan CS, Suzuki M, Ishii T, Okino T, Abe T, Masuda M (2008) Antibacterial activity of halogenated sesquiterpenes from Malaysian Laurencia spp. Phytochemistry 69(13):2490–2494
474. Renirie R, Dewilde A, Pierlot C, Wever R, Hober D, Aubry JM (2008) Bactericidal and virucidal activity of the alkalophilic P395D/L241 V/T343A mutant of vanadium chloroperoxidase. J Appl Microbiol 105(1):264–270
475. Lane AL, Moore BS (2011) A sea of biosynthesis: marine natural products meet the molecular age. Nat Prod Rep 28(2):411–428
476. Snelgrove PVR (2016) An ocean of discovery: biodiversity beyond the census of marine life. Planta Med 82(9–10):790–799
477. Benitez FJ, Acero JL, Real FJ, Roldan G, Casas F (2011) Bromination of selected pharmaceuticals in water matrices. Chemosphere 85(9):1430–1437
478. Rubiolo J, Alonso E, Cagide E (2014) Marine compounds as a starting point to drugs. In: Seafood and freshwater toxins. CRC Press, pp 1141–1178
479. Benaglia M (2009) Recoverable and recyclable catalysts. Wiley, Chichester, UK
480. Hansen EH, Albertsen L, Schäfer T, Frisvad JC, Molin S, Gram L, Scha T, Johansen C (2003) Curvularia haloperoxidase: antimicrobial activity and potential application as a surface disinfectant. Appl Environ Microbiol 69(8):4611–4617
481. Johansson J, Christensen BE, Marie A, Danielsen S, Gjermansen M (2010) A method of obtaining high level disinfection in a washer disinfector, and a washer disinfector, WO 2010/046142 A2
482. Rothenberg G, Clark JH (2000) Vanadium-catalysed oxidative bromination using dilute mineral acids and hydrogen peroxide: an option for recycling waste acid streams. Org Process Res Dev 4(4):270–274
483. Barnett P, Hondmann DH, Henricus SL, Ter Steeg PF, Wever R (1995) Enzymatic antimicrobial compositions, WO 1995/27046
484. Talbert JN, Goddard JM (2012) Enzymes on material surfaces. Colloids Surf B 93:8–19
485. Railkin AI (2005) Marine biofouling: colonization processes and defenses. CRC Press
486. Nurioglu AG, Esteves ACC, de With G (2015) Non-toxic, non-biocide-release antifouling coatings based on molecular structure design for marine applications J Mater Chem B 3(32):6547–6570
487. Mohammed JS (2015) Micro- and nanotechnologies in plankton research. Prog Oceanogr 134(5):451–473
488. Lejars M, Margaillan A, Bressy C (2012) Fouling release coatings: a nontoxic alternative to biocidal antifouling coatings. Chem Rev 112(8):4347–4390
489. Banerjee I, Pangule RC, Kane RS (2011) Antifouling coatings: recent developments in the design of surfaces that prevent fouling by proteins, bacteria, and marine organisms. Adv Mater 23(6):690–718
490. Dürr S, Thomason JC (2009) Biofouling. Wiley-Blackwell
491. Flemming HC, Murthy PS, Venkatesan R, Cooksey KE (2009) Marine and industrial biofouling. Springer, Berlin, Heidelberg
492. Dafforn KA, Lewis JA, Johnston EL (2011) Antifouling strategies: history and regulation, ecological impacts and mitigation. Mar Pollut Bull 62(3):453–465
493. Yebra DM, Kill S, Dam-Johansen K (2004) Antifouling technology—past, present and future steps towards efficient and environmentally friendly antifouling coatings. Prog Org Coatings 50(2):75–104
494. Kristensen JB, Meyer RL, Laursen BS, Shipovskov S, Besenbacher F, Poulsen CH (2008) Antifouling enzymes and the biochemistry of marine settlement. Biotechnol Adv 26(5):471–481
495. Tornero V, Hanke G (2016) Identification of marine chemical contaminants released from sea-based sources. JRC Technical Reports. ISBN 978–92-79-60688-5
496. Konstantinou IK, Albanis TA (2004) Worldwide occurrence and effects of antifouling paint booster biocides in the aquatic environment: a review. Environ Int 30(2):235–248

497. Umwelt Bundesamt (2016) http://www.Umweltbundesamt.De/Themen/Chemikalien/Chemikalienforschung-Im-Uba/Fliess-Stillgewaesser-Simulationsanlage/Projekte/Fate-Studie-Irgarol. Accessed 14 Dec 2016
498. Europäische Kommission, Durchführungsbeschluss (EU) 2016/107 Der Kommission vom 27. Januar 2016 zur Nichtgenehmigung von Cybutryn als altem Wirkstoff zur Verwendung in Biozidprodukten der Produktart 21
499. Mohr S, Berghahn R, Mailahn W, Schmiediche R, Feibicke M, Schmidt R (2009) Toxic and accumulative potential of the antifouling biocide and TBT successor irgarol on freshwater macrophytes: a pond mesocosm study. Environ Sci Technol 43(17):6838–6843
500. Zhou JL (2008) Occurrence and persistence of antifouling biocide Irgarol 1051 and its main metabolite in the coastal waters of Southern England. Sci Total Environ 406(1–2):239–246
501. European Commission (DG, Environment) (2011) Technical support for the impact assessment of the review of priority substances under directive 2000/60/EC. Substance Assessment: Cybutryne
502. Scardino AJ, de Nys R (2011) Mini review: biomimetic models and bioinspired surfaces for fouling control. Biofouling 27(1):73–86
503. Callow JA, Callow ME (2011) Trends in the development of environmentally friendly fouling-resistant marine coatings. Nat Commun 2:244
504. Chambers LD, Stokes KR, Walsh FC, Wood RJK (2006) Modern approaches to marine antifouling coatings. Surf Coatings Technol 201(6):3642–3652
505. Almeida E, Diamantino TC, de Sousa O (2007) Marine paints: the particular case of antifouling paints. Prog Org Coat 59(1):2–20
506. Kill S, Dam-Johansen K, Weinell CE, Pedersen MS (2002) Seawater-soluble pigments and their potential use in self-polishing antifouling paints: simulation-based screening tool. Prog Org Coat 45(4):423–434
507. Grozea CM, Walker GC (2009) Approaches in designing non-toxic polymer surfaces to deter marine biofouling. Soft Matter 5(21):4088–4100
508. Genzer J, Efimenko K (2006) Recent developments in superhydrophobic surfaces and their relevance to marine fouling: a review. Biofouling 22(5–6):339–360
509. Perez-Roa RE, Anderson MA, Rittschof D, Hunt CG, Noguera DR (2009) Involvement of reactive oxygen species in the electrochemical inhibition of barnacle (Amphibalanus amphitrite) settlement. Biofouling 25(6):63–571
510. Rittschof D (2000) Natural product antifoulants: One perspective on the challenges related to coatings development. Biofouling 15(1–3):119–127
511. Raveendran TV, Mol VPL (2009) Natural product antifoulants. Curr Sci 97(4):508–520
512. Detty MR, Ciriminna R, Bright FV, Pagliaro M (2015) Xerogel coatings produced by the Sol-Gel process as anti-fouling, fouling-release surfaces: from lab bench to commercial reality. ChemNanoMat 1(1):148–154
513. Sykes B, Neild J (1980) Preventing Fouling on Marine Structures, WO 000554 A1
514. Sakrowski KD (2005) Verwendung von Basaltfasern oder Basaltfilamenten Als Textiles Flächengebilde Zur Biozidfreien Antifouling-Beschichtung Für Umströmte Unterwasserflächen, DE 10353185 B4
515. Sakrowski KD (2005) Biocide-free antifouling coating consisting of a fabric based on basalt fibres, WO2005047403 A1
516. Pinori E, Berglin ME, Brive LM, Hulander M, Dahlström M, Elwing H (2011) Multi-seasonal barnacle (Balanus improvisus) protection achieved by trace amounts of a macrocyclic lactone (ivermectin) included in rosin-based coatings. Biofouling 27(9):941–953
517. Destino JF, Gatley CM, Craft AK, Detty MR, Bright FV (2015) Probing nanoscale chemical segregation and surface properties of antifouling hybrid xerogel films. Langmuir 31(11):3510–3517
518. Damon CA, Gatley CM, Beres JJ, Finlay JA, Franco SC, Clare AS, Detty MR (2016) The performance of hybrid titania/silica-derived xerogels as active antifouling/fouling-release surfaces against the marine alga *Ulva linza: in situ* generation of hypohalous acids. Biofouling 32(8):883–896

519. Singh PP, Thatikonda T, Kumar KAA, Sawant SD, Singh B, Sharma AK, Sharma PR, Singh D, Vishwakarma RA (2012) Cu–Mn spinel oxide catalyzed regioselective halogenation of phenols and *N*-Heteroarenes. J Org Chem 77(13):5823–5828
520. Amrute AP, Mondelli C, Hevia MAG, Pérez-Ramírez J (2011) Temporal analysis of products study of HCl oxidation on Copper- and Ruthenium-based catalysts. J Phys Chem C 115(4):1056–1063
521. NIC Series Vol 39, Münster G, Wolf D, Kremer M (Eds) NIC symposium 2008—proceedings symposium, 20–21 February 2008, Jülich, Germany
522. Guggenbichler JP, Eberhardt N, Martinz HP, Wildner H (2008) Antimicrobial agent, WO 2008058707 A3
523. Zollfrank C, Gutbrod K, Wechsler P, Guggenbichler JP (2012) Antimicrobial activity of transition metal acid MoO(3) prevents microbial growth on material surfaces. Mater Sci Eng C 32(1):47–54
524. Shafaei S, Van Opdenbosch D, Fey T, Koch M, Kraus T, Guggenbichler JP, Zollfrank C (2016) Enhancement of the antimicrobial properties of orthorhombic molybdenum trioxide by thermal induced fracturing of the hydrates. Mater Sci Eng C 58(1):1064–1070
525. Gao J, Zheng Y, Jehng JM, Tang Y, Wachs IE, Podkolzin SG (2015) Identification of molybdenum oxide nanostructures on zeolites for natural gas conversion. Science 348(6235):686–690
526. Tesler AB, Kim P, Kolle S, Howell C, Ahanotu O, Aizenberg J (2015) Extremely durable biofouling-resistant metallic surfaces based on electrodeposited nanoporous tungstite films on steel. Nat Commun 6:8649
527. Baldwin DH, Tatara CP, Scholz NL (2011) Copper-induced olfactory toxicity in salmon and steelhead: extrapolation across species and rearing environments. Aquat Toxicol 101(1):295–297
528. Shelter island yacht basin dissolved copper total maximum daily load 2011 monitoring and progress final report. Weston solutions, Inc. 2433 Impala Drive Carlsbad, California 92010 March 2012
529. Lai D, Liu D, Deng Z, Van Ofwegen L, Proksch P, Lin W (2012) Antifouling eunicellin-type diterpenoids from the gorgonian *Astrogorgia* sp. J Nat Prod 75(9):1595–1602
530. Trepos R, Cervin G, Hellio C, Pavia H, Stensen W, Stensvåg K, Svendsen JS, Haug T, Svenson J (2014) Antifouling compounds from the sub-arctic ascidian *Synoicum pulmonaria*: Synoxazolidinones A and C, Pulmonarins A and B, and synthetic analogues. J Nat Prod 77(9):2105–2113
531. Qian PY, Li Z, Xu Y, Li Y, Fusetani N (2015) Mini-review: marine natural products and their synthetic analogs as antifouling compounds: 2009–2014. Biofouling 31(1):101–122
532. Fitridge I, Dempster T, Guenther J, de Nys R (2012) The impact and control of biofouling in marine aquaculture: a review. Biofouling 28(7):649–669
533. Nguyen T, Roddick FA, Fan L (2012) Biofouling of water treatment membranes: a review of the underlying causes, monitoring techniques and control measures. Membranes 2(4):804–840
534. Cha C, Gao P, Chen YC, Shaw PD, Farrand SK (1998) Production of acyl-homoserine lactone quorum-sensing signals by gram-negative plant-associated bacteria. Mol Plant Microbe Interact 11(11):1119–1129
535. Dixon EF, Hall RA (2015) Noisy neighbourhoods: quorum sensing in fungal-polymicrobial infections. Cell Microbiol 17(10):1431–1441
536. Gribble GW (2015) Biological activity of recently discovered halogenated marine natural products. Mar. Drugs 13(7):4044–4136
537. Winter JM, Moffitt MC, Zazopoulos E, McAlpine JB, Dorrestein PC, Moore BS (2007) Molecular basis for chloronium-mediated meroterpene cyclization. Cloning, sequencing, and heterologous expression of the napyradiomycin biosynthetic gene cluster. J Biol Chem 282(22):16362–16368
538. Blasiak LC, Vaillancourt FH, Walsh CT, Drennan CL (2006) Crystal structure of the non-haem iron halogenase SyrB2 in syringomycin biosynthesis. Nature 440(7082):368–371
539. Sundaramoorthy M, Terner J, Poulos TL (1995) The crystal structure of chloroperoxidase: a heme peroxidase–cytochrome P450 functional hybrid. Structure 3(12):1367–1377

540. Kühnel K, Blankenfeldt W, Terner J, Schlichting I (2006) Crystal structures of chloroperoxidase with its bound substrates and Complexed with Formate, Acetate, and Nitrate. J Biol Chem 281(33):23990–23994
541. Boruah JJ, Das SP, Borah R, Gogoi SR, Islam NS (2013) Polymer-anchored peroxo compounds of molybdenum and tungsten as efficient and versatile catalysts for mild oxidative bromination. Polyhedron 52:246–254
542. Parida KM, Parija S, Das J, Mukherjee PS (2006) Preparation, characterisation of molybdophosphoric and tungstophosphoric acid intercalated zinc aluminium hydrotalcite-like compounds and their catalytic evaluation towards the oxidative bromination of phenol. Catal Commun 7(11):913–919
543. Maurya MR, Kumar N, Avecilla F (2014) Polymer and non-polymer-grafted dioxidomolybdenum(VI) complexes having ONO donor ligand and their catalytic activities for the oxidative bromination of organic substrates. J Mol Catal A Chem 392:50–60

Chapter 9
Cerium Oxide Based Nanozymes

Ruofei Zhang, Kelong Fan and Xiyun Yan

Abbreviations

AA	Ascorbic acid
adRP	Autosomal dominant retinitis pigmentosa
AMD	Age-related macular degeneration
BBB	Blood–brain barrier
BDNF	Brain-derived neurotrophic factor
CAT	Catalase
CML	Chronic myelogenous leukemia
CMP	Chemical–mechanical planarization
CNPs	Ceria nanoparticles
CPC	Cardiac progenitor cells
DAO	Diamine oxidase
DM	Diabetes mellitus
DR	Diabetic retinopathy
EAE	Experimental autoimmune encephalomyelitis
ECM	Extracellular matrix
ERG	Electroretinogram recordings
FAM	Carboxyfluorescein
GNPs	Gold nanoparticles
GS	Graphene oxide nanosheets
GSH-Px	Glutathione peroxidase

R. Zhang · K. Fan (✉) · X. Yan (✉)
CAS Engineering Laboratory for Nanozyme, Key Laboratory of Protein and Peptide Pharmaceutical, Institute of Biophysics, Chinese Academy of Sciences, 15 Datun Road, Beijing 100101, China
e-mail: fankelong@ibp.ac.cn

X. Yan
e-mail: yanxy@ibp.ac.cn

X. Yan (ed.), *Nanozymology*, Nanostructure Science and Technology,
https://doi.org/10.1007/978-981-15-1490-6_9

H2TCPP	5,10,15,20-tetrakis 4-carboxyl phenyl-porphyrin
HMTA	Hexamethylenetetraamine
iNOS	Inducible nitric oxide synthase
LOD	Limits of detection
LOX	Lactate oxidase
LPS	Lipopolysaccharides
MPO	Myeloperoxidase
MV	Methyl violet
MWCNTs	Multi-walled carbon nanotubes
NOS	Nitric oxide synthase
PAAN	A polyacrylic acid sodium salt
PANI	Polyaniline
PEP	Case phosphoenolpyruvate carboxylase
RNS	Reactive nitrogen species
ROS	Reactive oxygen species
RT	Radiation
SIRS	System inflammatory response syndrome
SOD	Superoxide dismutase
TMB	3,3,5,5-tetramethylbiphenyl dihydrochloride
UA	Uric acid
XPS	X-ray photoelectron spectroscopy

9.1 History of Nanoceria Nanozymes Development

Ceria (CeO_2) is an oxide of the rare-earth metal cerium with fluorite structure. Cerium blends with other rare-earth elements in its natural sources bastnaesite and monazite. After cerium metals are extracted into ion state and oxidized by addition of an oxidant, ceria is separated from mixture based on its low solubility in aqueous solution [1]. Ceria has gradually become the most widely used compound of cerium after being discovered in 1803. Ceria is mainly used for polishing applications in traditional industry, especially for chemical–mechanical planarization (CMP), a process of smoothing surfaces with the combination of chemical and mechanical forces [2, 3]. Because of its excellent efficiency on producing high-quality optical surfaces, ceria has supplanted other metal oxides (such as iron oxide and zirconia) in CMP [1]. Ceria is also broadly used as glass decolorizer by converting blue-green Fe^{2+} impurities to almost colorless Fe^{3+} oxides [1].

With the development of nanoscience and nanotechnology, nanoceria with striking catalytic potential and excellent oxygen storage capacity has attracted scientists' attention. In industry, the most important application of nanoceria is as the key component in three-way catalysts for treating exhaust gas from automobiles [4, 5]. Recently, nanoceria has been found to exhibit a variety of enzyme-like activities. These findings allow the application of nanoceria to extend from industrial catalysis to biological fields, such as biocatalysts, biopharmaceutics, drug delivery, and biological scaffold [6]. This chapter mainly focuses on the enzymatic activity of nanoceria and its catalytic mechanism and recent advances in biomedical applications.

9.2 Biological Enzyme-like Activities of Nanoceria Nanozymes

During the aerobic metabolic process of cells, a series of reactive oxygen species (ROS) will be produced, such as superoxide radical anion ($O_2^{\bullet -}$), hydrogen peroxide (H_2O_2), and hydroxyl radical (•OH). High levels of ROS disrupt the dynamic balance of cells, which can cause oxidative stress [7]. Oxidative stress is a critical factor associated with many disorders and diseases which is directly caused by the imbalance between antioxidant defenses and ROS. There are two kinds of antioxidant defenses in organism, named small molecule antioxidants (e.g., vitamin C, vitamin E, glutathione) and redox enzymes [e.g., superoxide dismutase (SOD), catalase (CAT), glutathione peroxidase (GSH-Px)] [7]. Nanoceria have been revealed to reduce the ROS levels in vitro (cells) and in vivo models (animals). This section mainly focuses on the biological enzyme-like activities of nanoceria and the mechanisms of these nanoceria nanozyme activities.

9.2.1 Structural Basis of Nanoceria Enzyme-like Activities

To understand nanoceria nanozymes' catalytic mechanism in depth, a certain understanding of their specific atomic structure in advance is needed. Cerium is the second element in the lanthanide series with atomic number 58, also traditionally known as one of the rare-earth elements. The electron configuration of cerium is [Xe]4f15d16s2, of which the four external electrons are valence electrons [8]. As the remaining 4f electrons are too strongly bound to be removed, most lanthanides possess only three valence electrons except for cerium. Cerium is the exception owing to the fact that it has a smaller atomic number in the lanthanide series, whose nuclear charge is still too low to tightly hold the fourth valence electron [9]. The specific atomic structure renders cerium exhibiting dual valence states, whereas the cerium not only form CeO_2 with a stable +4 state, but also form Ce_2O_3, which is unstable and easy to be oxidized to CeO_2 [10]. In addition, stable Ce^{4+} typically forms CeO_2 crystallization with a structure of fluorite in which one cerium atom is surrounded by eight oxygen atoms in average (Fig. 9.1a) [11]. However, because the internal vacancy defects are easy to form in CeO_2 crystallization and a part of Ce^{4+} turn into Ce^{3+} in the meantime, nanoceria nanozymes usually perform as nonstoichiometric CeO_{2-x} (Fig. 9.1b) [10]. The surface oxygen vacancies are proposed to mediate catalysis on ceria nanostructures. Moreover, the higher ratio of Ce^{3+}/Ce^{4+} in nanoceria

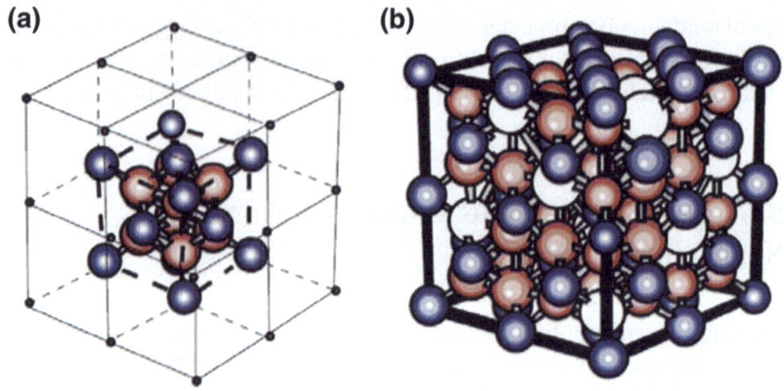

Fig. 9.1 Lattice unit cells for CeO_2 (**a**) and CeO_{2-x} (**b**) [10]. Blue, red, and white spheres indicate the cerium, oxygen atoms, and vacancies, respectively. Copyright 2002 American Physical Society

nanozymes preparation correlates with higher oxygen and electron vacancy [12, 13]. In the autocatalytic reaction of nanoceria, Ce^{4+} and low surface defect formation energy play a key role in the oxidation process, while Ce^{3+} and free mobile electrons in the oxygen vacancy lattice play a key role in the reduction process [10].

9.2.2 Superoxide Dismutase-like and Catalase-like Activities

Superoxide radicals are produced in the process of aerobic metabolism and act as signaling molecules in the signal transduction. However, superoxide radicals increase remarkably via activation of NADPH oxidases in the case of inflammatory response, resulting in the disruption of mitochondrial electron transfer chain and the interference of the production of ATP, which will damage the cell if not removed in time [14].

Superoxide dismutase is an enzyme that consecutively decompose superoxide ($O_2^{\bullet-}$) radical into either molecular oxygen (O_2) or hydrogen peroxide (H_2O_2) [15]. The catalytic dismutation reaction of copper-containing SOD is as follows [16]:

$$O_2^{\bullet-} + 2H^+ + (Cu^+) - SOD \rightarrow H_2O_2 + (Cu^{2+}) - SOD \tag{9.1}$$

$$O_2^{\bullet-} + (Cu^{2+}) - SOD \rightarrow O_2 + (Cu^+) - SOD \tag{9.2}$$

In the first step of this process, a molecule of $O_2^{\bullet-}$ is reduced to H_2O_2 by receiving an electron from (Cu^+)-SOD. While in the second step, another molecule of $O_2^{\bullet-}$ is oxidized to O_2 as the oxidized (Cu^{2+})-SOD returns to (Cu^+)-SOD. All in all, two molecules of $O_2^{\bullet-}$ are dismutated into one molecule of H_2O_2 and one molecule of O_2 during the time that SOD is circulating between oxidized state and reduced state. The overall reaction process is given by Ivanov et al. [16]:

$$O_2^{\bullet -} + O_2^{\bullet -} + 2H^+ \rightarrow H_2O_2 + O_2 \quad (9.3)$$

The SOD-like activity of nanoceria was first discovered by Self et al. in 2007 [17]. Before identification of the SOD-mimetic activity of nanoceria, several evidences have already been shown that oxygen vacancies can mediate catalysis on nanoceria surface, and high level of oxygen vacancies is correlated with high Ce^{3+}/Ce^{4+} ratio [12, 13]. Based on the previous study, Self et al. synthesized two kinds of ceria nanoparticles with mixed-valence states and found that the nanoceria with higher Ce^{3+}/Ce^{4+} ratio exhibited SOD-mimetic activity [17]. Similar as the catalytical process of SOD, the reaction mechanism of nanoceria catalyzing the dismutation of superoxide is speculated to be like follows [17]:

$$O_2^{\bullet -} + Ce^{3+} + 2H^+ \rightarrow H_2O_2 + Ce^{4+} \quad (9.4)$$

$$O_2^{\bullet -} + Ce^{4+} \rightarrow O_2 + Ce^{3+} \quad (9.5)$$

It is worth noting that when cerium oxide nanoparticles are used as SOD enzyme mimetics, H_2O_2 is one of the final products. But in the organism, excess H_2O_2 has been considered more harmful than high level of $O_2^{\bullet -}$ as it can be transferred into destructive $^{\bullet}OH$ by Fenton reaction. Catalase is one of the most important enzymes in protecting cell from oxidative damage by ROS. Through the catalysis of catalase, H_2O_2 can be decomposed into H_2O and O_2. The CAT-mimetic activity of nanoceria was also first reported by Self and his partners [18]. In this pioneering study, the Amplex Red assay was applied to detect low levels of peroxide (2 μM). Results from the Amplex Red assay showed that nanoceria with lower Ce^{3+}/Cc^{4+} ratio rather than nanoceria with higher Ce^{3+}/Ce^{4+} ratios exhibited significant CAT-mimetic activity. The hypothetical mechanism of nanoceria CAT mimetics has been proposed by investigators to be as follows [16, 19]:

$$H_2O_2 + 2Ce^{4+} + 2OH^- \rightarrow 2H_2O + O_2 + 2Ce^{3+} \quad (9.6)$$

To clarify the mechanisms of the SOD-mimetic and CAT-mimetic activities of nanoceria, Celardo et al. proposed a plausible mechanism of nanoceria redox self-regeneration [20]. When H_2O_2 approaches the surface of nanoceria, it may attach to oxygen vacancy site. Then, the Ce^{4+} around the oxygen vacancy is reduced to Ce^{3+} while H_2O_2 is decomposed into O_2 and H_2O. If there are $O_2^{\bullet -}$ in the environment, the $O_2^{\bullet -}$ can be reduced to H_2O_2 and regenerate the oxygen vacancy site in the meantime. In this progress, nanoceria acts as SOD-mimetic antioxidant (Fig. 9.2). Alternatively, another molecule of H_2O_2 can attach to the reduced vacancy site and be decomposed into H_2O, in which case nanoceria acts as CAT-mimetic catalyst (Fig. 9.3).

Nanoceria nanozymes are typically reported to exhibit SOD-mimetic or CAT-mimetic activity in different works. Systematically studies revealed the enzyme-like activities of nanoceria are mainly controlled by the ratio of Ce^{3+}/Ce^{4+} on the surface

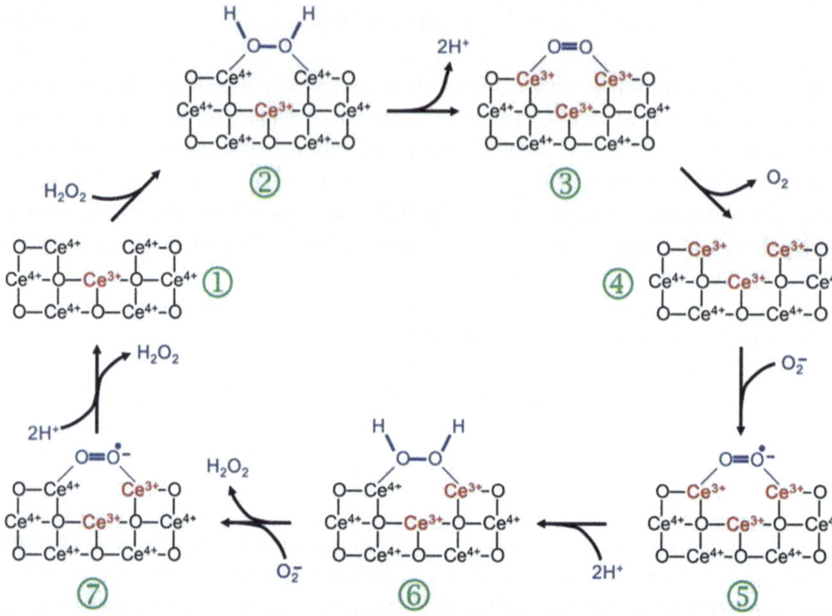

Fig. 9.2 A model of the reaction mechanism for the oxidation of hydrogen peroxide by nanoceria and the regeneration via reduction by superoxide [20]. An oxygen vacancy site on the nanoceria surface ① presents a $2Ce^{4+}$ binding site for H_2O_2 ② after the release of protons and two-electron transfer to the two cerium ions ③ oxygen is released from the now fully reduced oxygen vacancy site ④. Subsequently, superoxide can bind to this site ⑤, and after the transfer of a single electron from one Ce^{3+}, and uptake of two protons from the solution, H_2O_2 is formed ⑥ and can be released. After repeating this reaction with a second superoxide molecule ⑦ the oxygen vacancy site returns to the initial $2Ce^{4+}$ state ①. It is also possible that the third Ce^{4+} indicated, which gives rise to the oxygen vacancy, could participate directly in the reaction mechanism. The square Ce–O matrix is shown here only to illustrate the model and does not correspond to the actual spatial arrangement of the atoms in the crystal structure. Copyright 2011 RSC Pub

of nanoceria, the size and morphology of nanoceria, the components, and the pH values of the reaction solutions [21–26].

The surface oxidation state of nanoceria plays a fundamental role in the redox enzyme-mimetic activities of nanoceria. Self et al. used X-ray photoelectron spectroscopy (XPS) and UV–visible spectroscopy to analyze the nanoceria treated with H_2O_2. The results demonstrated that nanoceria was able to be oxidized from Ce^{3+} to Ce^{4+} by H_2O_2 with a loss of SOD-mimetic activity. After H_2O_2 was completely degraded, the SOD-mimetic activity of nanoceria restored slowly within 15 days [27]. Thus, SOD-mimetic activity of nanoceria depends upon a high Ce^{3+}/Ce^{4+} ratio, because only Ce^{3+} acts as electron donor to transfer electron to $O_2^{\bullet-}$ to produce peroxide [20]. In contrast, further studies have confirmed that lower level of Ce^{3+} (low Ce^{3+}/Ce^{4+} ratio) on nanoceria surface is more beneficial to CAT-mimetic activity [18, 28]. More generally, nanoceria with 40 to 60% of Ce^{3+} in their composition are more like to show SOD-mimetic behaviors, while nanoceria with 70 to 80% of Ce^{4+} in

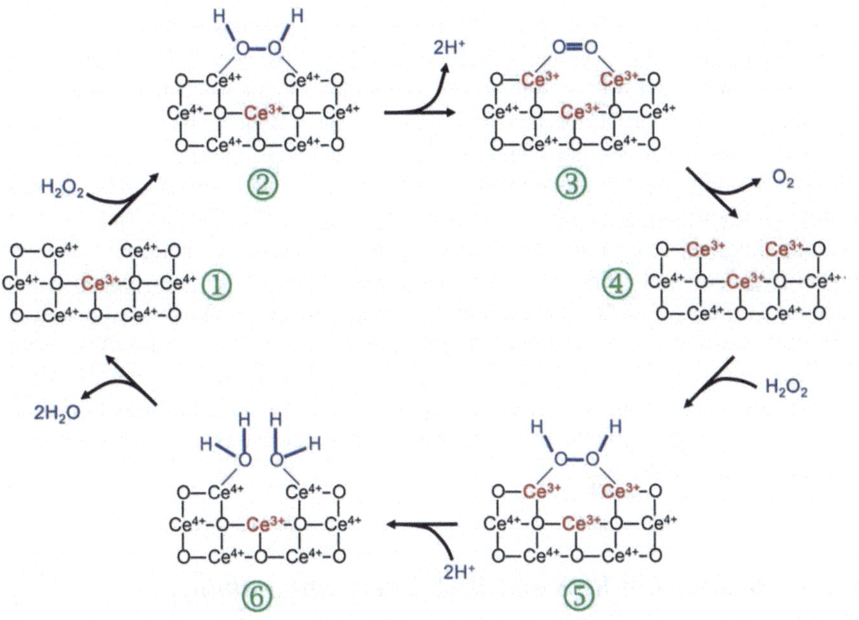

Fig. 9.3 A model of the reaction mechanism for the complete dismutation of hydrogen peroxide [20]. The oxidative half-reaction is identical to the sequence shown in Fig. 9.2 (①–④) The reductive half involves binding of H_2O_2 to the 2 Ce^{3+} site ⑤, uptake of two protons and homolysis of the O–O bond with transfer of electrons to the two Ce^{3+} ⑥, and release of the water molecules to regenerate the initial Ce^{4+} site ①. This reaction sequence would be analogous to the one found in catalases. Copyright 2011 RSC Pub

their composition prefer to be CAT-mimetic catalysts [19]. The nanosize effect is an important factor affecting the physical properties of nanoceria. Seal et al. found that with the decrease in the particle size from 30 to 3 nm, the Ce^{3+} ions concentration in nanoceria increased from 17 to 44% [12]. A similar result obtained recently found that the Ce^{3+} fractions decreased from 43 to 9.5% with the increase of nanoceria's diameter from 4.5 to 28 nm [29].

The surface Ce^{3+}/Ce^{4+} ratio is determined by the synthesis method and microenvironment which plays an important role in tuning enzyme-like activities of nanoceria. Numerous methods for the synthesis of nanoceria have been reported. These methods include solution precipitation, hydrothermal, solvothermal, ball milling, thermal decomposition, spray pyrolysis, thermal hydrolysis, sol–gel methods, and so on [30]. Different synthesis methods give nanoceria diverse physical properties (e.g., particle size, agglomeration status, surface charge, and coating or residual contamination of the surfactant on the surface), which subsequently influence the biological activity and toxicity of nanoceria [31]. Temperature, pH, and synthesis materials are some of the most important parameters in determining different synthetic products. Briefly, nanoceria synthesized by high-temperature methods usually have larger heterogeneous size and are easier to form agglomeration. By contrast, nanoceria formed under room temperature conditions generally possess homogeneous particle size. Moreover, room temperature synthesis methods produced nanoceria have higher Ce^{3+} concentration [32]. However, some room temperature methods such as microemulsion cannot remove surface surfactant as high-temperature methods, and the residual contaminations on the surface could interfere the nano–bio-interface. Also, nanoceria with a stable unagglomerated (3–5 nm) or loosely agglomerated (10–12 nm) nanoparticle suspension can be synthesized under acidic pH. The materials used for synthesizing nanoceria could also matters. Nanoceria prepared using H_2O_2 possess higher Ce^{3+}/Ce^{4+} ratio (55–65%) than that prepared using hexamethylenetetramine or base (21–30%) [31]. Besides, nanoceria doped with europium showed increased Ce^{3+}/Ce^{4+} ratio, while samarium doping decreases the Ce^{3+}/Ce^{4+} ratio [33]. Moreover, a recent study suggested that the ligand modified on the surface may also affect the oxidation state of surface cerium ions [34]. For more detailed information about nanoceria synthesis methods, there are some comprehensive reviews recommended to readers [30, 31, 35].

9.2.3 *Antioxidant ROS and RNS Eliminating Ability*

As shown in Sect. 9.2.2, nanoceria successfully scavenge $O_2^{\bullet-}$ and H_2O_2 with their SOD-mimetic and CAT-mimetic activities, respectively. However, some studies also indicated that nanoceria nanozymes directly eliminate reactive oxygen species (ROS) such as $^{\bullet}OH$ [36–39] and reactive nitrogen species (RNS) such as $^{\bullet}NO$ [40, 41] and O_2NO^- [42]. Hickman et al. found that ultra-small nanoceria (2–5 nm) protected adult rat spinal cord neurons from H_2O_2 induced damage in vitro [37]. Because H_2O_2 generates damaging $^{\bullet}OH$ via Fenton reaction, the detailed mechanism of nanoceria protecting the spinal cord cells from H_2O_2-induced damage was due to the free radical scavenging capacity of nanoceria nanozymes. In this study, the autocatalytic ability of nanoceria was tested using UV–visible spectroscopy. Treated nanoceria with 10 mM H_2O_2, a shift to the lower energy portion of the spectrum was observed, which represents a change in the redox state from Ce^{3+} to Ce^{4+}. After then, a gradual shift to the higher energy portion of the spectrum was seen over 30 days which reflects a regeneration of nanoceria (Ce^{4+} to Ce^{3+}). Based on these clues, an assumed mechanism of nanoceria's ROS-scavenging ability was proposed (Fig. 9.4). In another study, Perez et al. used water-soluble dextran-coated nanoceria (DNC) to study its autocatalytic property [39]. They suggested that the acidic condition (pH = 4) disrupted the regeneration of nanoceria by disturbing the conversion from Ce^{4+} to Ce^{3+}. Thus, they modified the previously supposed mechanism of the reaction between nanoceria and ROS (Fig. 9.5).

Although the above studies have proposed a hypothetical ROS-scavenging mechanism, it is still lacking direct evidences to deeply understand the antioxidant role of nanoceria in biological system. To solve this problem, Xue et al. established a simple photometric assay using methyl violet (MV), a chromogenic reagent, to directly show the level of hydroxyl radical [43]. Because nanoceria could competitively react

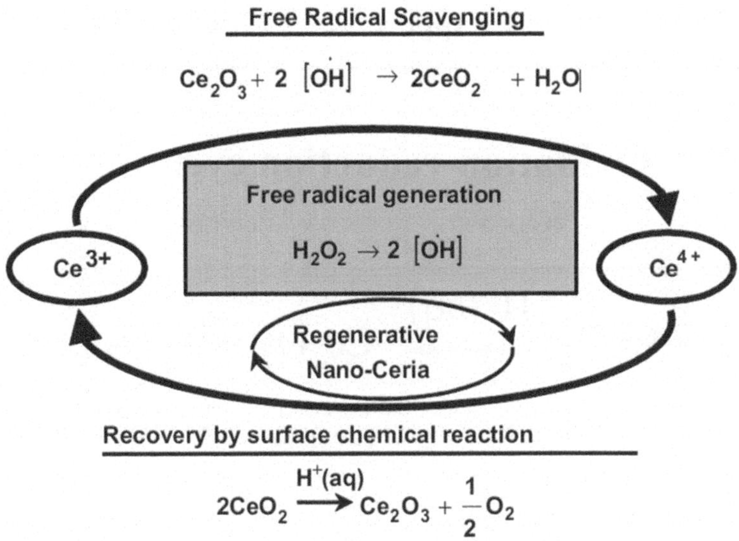

Fig. 9.4 Schematic detailing the proposed regenerative properties of nanoceria and probable mechanism of nanoceria's free radical scavenging property and autocatalytic behavior [37]. Copyright 2007 Elsevier

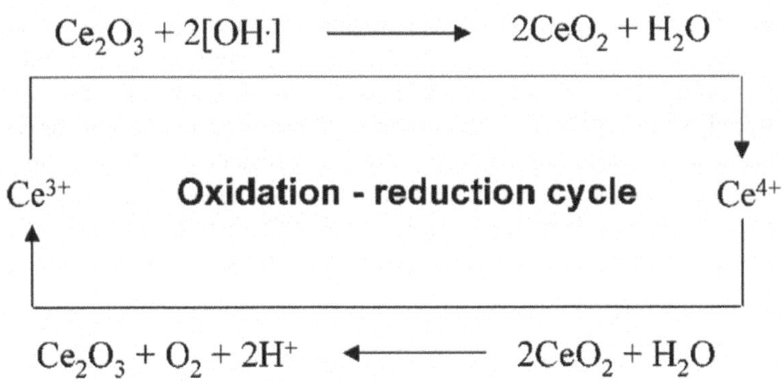

Fig. 9.5 Schematic diagram of proposed surface chemistries on nanoceria [39]. Copyright 2008 John Wiley and Sons

with •OH, the absorbance change of MV reflected the level of nanoceria's •OH eliminating ability. Interestingly, these investigators noted that nanoceria nanozymes with smaller size and higher Ce^{3+}/Ce^{4+} ratio on the surface were more efficient in scavenging •OH. Based on their intuitive observations, Xue et al. give an assumed mechanism to elucidate the redox cycle reaction (Fig. 9.6).

Nitric oxide is a broad and unique signaling molecule in organism. NO is produced from L-arginine by different nitric oxide synthase (NOS) subtypes, which owns many normal physiological purposes, such as promoting vasodilation and regulating

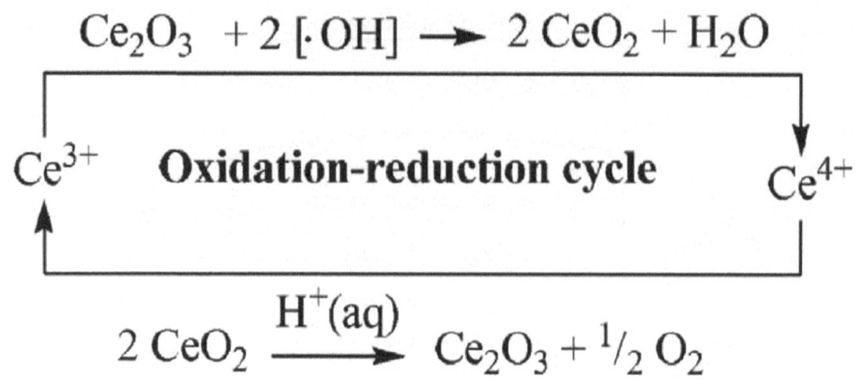

Fig. 9.6 Schematic diagram detailing the oxidation–reduction cycle as the probable mechanism of hydroxyl radical scavenging activity of CeO_2 nanoparticles [43]. Copyright 2011 American Chemical Society

communication between nervous system cells [44]. However, excessive production of reactive nitrogen species (RNS), such as $^\bullet NO$ and O_2NO^-, can cause a damaging phenomenon known as nitrosative stress [45]. Unlike extremely damaging $^\bullet OH$, $^\bullet NO$ is not a particularly toxic molecule in itself, but it reacts with O_2 or $O_2^{\bullet -}$ to produce highly reactive and destructive molecules, such as O_2NO^- [19]. Many diseases (e.g., neurodegenerative diseases) exhibit abnormal high $^\bullet NO$ contents, and thus it is necessary to remove excess $^\bullet NO$ in time [45]. Inspired by the ability that nanoceria nanozymes absorb and decompose NO in the industrial waste gas, Self et al. studied the effects of nanoceria on the $^\bullet NO$ in organism. These researchers found that nanoceria with low Ce^{3+}/Ce^{4+} ratios are efficient to scavenge nitric oxide radical. They also postulated a reaction mechanism that nanoceria eliminate $^\bullet NO$ by electropositive nitrosyl ligand between Ce^{4+} and NO (or between Ce^{3+} and NO^+) [41]:

$$Ce^{4+} + {}^\bullet NO \rightarrow [Ce^{4+}NO \longleftrightarrow Ce^{3+}NO^+] \quad (9.7)$$

Self et al. have also studied the ability of nanoceria nanozymes to directly scavenge peroxynitrite (O_2NO^-) [42]. Different from $^\bullet NO$ scavenging activity, the reaction between nanoceria and O_2NO^- is independent from Ce^{3+}/Ce^{4+} ratio on the surface. Because the reaction is too complex to be detailly described, no mechanism was proposed by these investigators.

9.2.4 Other Enzyme-Mimetic Activities

Beyond the systematically studied catalytic functions of nanoceria nanozymes mentioned in Sects. 9.2.2 and 9.2.3, nanoceria nanozymes still exhibit some other enzyme-mimetic activities. Up to now, in addition to SOD-like and CAT-like activities, nanoceria have been found to exhibit peroxidase-like [46], oxidase-like [47–50], phosphatase-like [40, 47, 51–54], ATPase-like [40], and urease-like [55] enzyme-mimetic activities.

Self et al. found that cerium metal catalyzes H_2O_2 to produce harmful oxygen radicals based on a three-step Fenton-like reaction (Eqs. 9.8–9.10) [56]. This finding is promising to the discovery of the peroxidase-like activity of nanoceria. Later on, Lv et al. synthesized a kind of well-dispersed ceria nanoparticles and observed that this nanoceria nanozyme exhibited wonderful catalytic ability toward 3,3,5,5-tetramethylbiphenyl dihydrochloride (TMB, a traditional peroxidase substrate) with H_2O_2 [46]. Based on this finding, they developed a colorimetric method for the detection of glucose which shown excellent selectivity and sensitivity.

$$Ce^{3+} + H_2O_2 + H^+ \rightarrow Ce^{4+} + {}^{\bullet}OH + H_2O \tag{9.8}$$

$$^{\bullet}OH + H_2O_2 \rightarrow HO_2^- + H_2O \tag{9.9}$$

$$Ce^{4+} + HO_2^- \rightarrow O_2 + Ce^{3+} + H^+ \tag{9.10}$$

The oxidase-like activity of nanoceria was founded by Perez et al. for the first time [49]. In this study, a series of dextran-coated or acrylic acid-coated nanoceria with different sizes were synthesized. These nanoceria could quickly oxidize organic substrates in the absence of H_2O_2 in acidic conditions, which represented the inherent oxidase-like activity of nanoceria. What is more interesting is that the oxidase-mimetic feature of these nanoceria is dependent on the pH value of the solution, the size of nanoceria and the thickness of the polymer coating. Based on these findings, an immunoassay was developed for targeted detection of folate-expressing cancer cells using folate-modified nanoceria in this study.

Nowadays, it is not surprising for investigators to find another redox enzyme-mimetic activity of nanoceria. However, it is amazing to know that nanoceria could also exhibit phosphatase-like properties. Phosphatase is a class of enzymes that can hydrolyze a phosphoric acid monoester into a phosphate ion and an alcohol [57]. Phosphatase is critical to organism, because phosphorylation and dephosphorylation of organisms play important roles in many biological processes, including signal transduction, cell differentiation, proliferation, and metabolism [58]. Qian et al. described a method for dephosphorylation of phosphor-peptides using nanoceria [53]. The results showed that the process of dephosphorylation for all the phosphor-peptides was complete within 10 min with the catalysis of nanoceria. In addition, the temperature had no influence on this reaction. These excellent properties provide the possibility for applicating nanoceria as phosphatase mimetics in biomedicine.

Unexpectedly, several studies demonstrated that nanoceria nanozymes own ATPase-like activity [40, 51]. Self et al. synthesized a series of ceria nanoparticles (CNPs) using identical precursors by a wet chemical method but with dissimilar oxidizer/reducer: H_2O_2 (CNP1), NH_4OH (CNP2), or hexamethylenetetramine

(HMT-CNP1) [40]. These different nanoceria exhibited distinct ATPase-like activities, which noted that the preparation methods of nanoceria are important to its enzyme-mimetic functions.

Recently, nanoceria has also been reported to have urease-like activity. Tremel et al. found that nanoceria with a rod-like shape efficiently catalyzed the hydrolysis of urea under ambient conditions and had catalytic activity comparable to that of natural enzymes [55]. The urease-like activity ($K_{cat} = 9.58 \times 10^1\ s^{-1}$) of nanoceria was about one order of magnitude lower than that of the native jack bean urease. At the same time, the urease-like activity of nanoceria decreased with the decrease of Ce^{4+} ratio. This indicates that the catalytic activity of urease is still related to the valence state of Cerium on its surface. In addition, heavy metal ion Cu^{2+} can inactivate natural urease but has no effect on the urease activity of cerium nanoparticles. This Cu^{2+} tolerance of nanoceria, combined with its excellent properties of stability, biocompatibility, and low cost may make it a more valuable alternative to natural enzymes.

9.3 Applications of Nanoceria Nanozyme in Disease Treatment

Nanoceria nanozymes possess diverse redox enzyme-like activities in different conditions, resulting in difficult to control their functions in biological organism. Many findings have demonstrated that nanoceria nanozymes own the ability to protect cells from reactive radicals (Table 9.1). While in other studies, these nanoparticles exhibit cell-killing effects by increasing the level of cell oxidative stress and disrupting the homeostasis of cells (Table 9.1). A recent study also reported that nanoceria at a low concentration exhibited protective effects [59]. While at a higher concentration, nanoceria exhibited oxidase-like activity which produce damaged radicals. These paradoxical phenomena make nanoceria nanozymes difficult to be applied in biomedical field. However, if we can control the enzymatic activities of nanoceria nanozymes in different scenarios, the biomedical application of nanoceria nanozymes would be extended and diversified. This section reviews the recent progress of biological applications of nanoceria nanozyme in vitro and in vivo.

9.3.1 Nanoceria as Antioxidants

In recent years, many studies have already confirmed the antioxidant function of nanoceria nanozymes in several cell and animal models including human breast cancer cell lines [60], cardiac cells [61, 62], photoreceptor cells [63–68], hepatic cells [69–71], gastrointestinal epithelium [72], endometrium [73], neuronal cells [74–76], and marrow stromal cells [77].

Table 9.1 Summary of nanoceria nanozymes used as antioxidant or prooxidant in cell culture and animal models

Nanoceria nanozymes	Type of activity	Experimental model	Dosage	Results/remarks	References
Bare nanoceria (3–5 nm)	Antioxidant activity	CRL8798 and MCF-7 cell lines	10 nM	Nanoceria confers protection to the CRL8798 cell line from radiation-induced cell death but not to the MCF-7 cell line	[60]
Bare nanoceria (5–8 nm)	Antioxidant activity	Cardiac progenitor cells (CPCs)	5–50 μg/mL	Nanoceria protects CPCs from H_2O_2-induced cytotoxicity for at least 7 days	[62]
Bare (FITC) nanoceria (7 and 94 nm)	Antioxidant activity	SKOV3 and WiDr cell lines	50 μg/mL	Larger sized nanoceria (94 nm) exhibits stronger ROS scavenging effects than that of the smaller nanoceria (7 nm) in these cell lines	[81]
Bare nanoceria (3–5 nm)	Antioxidant activity	J774A.1 murine macrophages	0.001–10 μM	Nanoceria suppresses the production of iNOS protein and mRNA as well as decreasing ROS	[87]
Spherical nanoceria (7–10 nm)	Antioxidant activity	Male Sprague Dawley rats	0.5 μg/kg body weight/week for 5 weeks	Nanoceria protects the lungs by reducing noxious free radicals during hypobaric hypoxia	[89]

(continued)

Table 9.1 (continued)

Nanoceria nanozymes	Type of activity	Experimental model	Dosage	Results/remarks	References
Rod-like nanoceria (65 ± 10 nm in length and 8 ± 1 nm in width)	Antioxidant activity	Chondrocyte-seeded hydrogel constructs	100–1000 μg/mL	Nanoceria improves the performance of engineered cartilage and protects the ECM from the IL-1α-induced inflammatory damage	[91]
Bare nanoceria (7 nm)	Antioxidant activity	MCP-1 transgenic mice	15 nmoles	Nanoceria protects against the progression of cardiac dysfunction and remodeling by attenuation of myocardial oxidative stress, ER stress, and inflammatory processes	[61]
Agglomerated spherical nanoceria (37 nm)	Antioxidant activity	Adult male Sprague Dawley rats	0.5 mg/kg body weight	Nanoceria reduces the levels of TNF-α and ROS induced by LPS	[99]
Bare nanoceria (3–5 nm)	Antioxidant activity	*Tubby* mutant mouse	20 μL of 1 mM	Nanoceria protects the retina from oxidative stress by scavenging ROS and modulating apoptosis/survival signaling pathways	[64, 66]
Water-dispersed nanoceria (3–5 nm)	Antioxidant activity	P23H-1 rats	1 mM/eye	Nanoceria relieves oxidative damage in the P23H-1 retinas by reducing ROS level	[65]

(continued)

Table 9.1 (continued)

Nanoceria nanozymes	Type of activity	Experimental model	Dosage	Results/remarks	References
Non-agglomerated nanoceria (2–5 nm)	Antioxidant activity	Adult rat spinal cord cells	10 nM	Nanoceria improves the cell survival upon H_2O_2-induced oxidative stress	[37]
Bare nanoceria (10 nm)	Antioxidant activity	Mouse hippocampal brain slice model	0.1–2 μg/mL	Nanoceria mitigates ischemic brain injury mainly by scavenging peroxynitrite	[106]
Citrate/EDTA-coated nanoceria (2.9 nm)	Antioxidant activity	EAE mouse model	10–30 mg/kg	Nanoceria penetrates the brain, reduces ROS levels, and alleviates clinical symptoms and motor deficits in EAE mice	[109]
Bare nanoceria (3–5 nm)	Antioxidant activity	Human colon cells (CRL 1541)	1–100 nM	Nanoceria protects the gastrointestinal epithelium from radiation-induced damage by reducing ROS and upregulating SOD2	[72]
Bare nanoceria (3–5 nm)	Antioxidant activity	SKOV3, HUVEC, A2780, and C200 cell lines	25–200 μM	Nanoceria inhibits the proliferation of HUVEC and the production of ROS in A2780 cells, attenuates the growth factor-mediated invasion of SKOV3 cells	[112]
Bare nanoceria (3–5 nm)	Antioxidant activity	A2780 cell bearing nude mice	0.1 mg/kg body weight	Nanoceria reduces the tumor growth in mice by attenuation of angiogenesis	[112]

(continued)

Table 9.1 (continued)

Nanoceria nanozymes	Type of activity	Experimental model	Dosage	Results/remarks	References
Bare nanoceria (5–8 nm)	Prooxidant activity	hTERT-HPNE and L3.6pl cell lines	10–200 μM	Nanoceria enhances RT-induced ROS production and cell death selectively in human pancreatic tumor cells while protecting normal tissues from the toxic side effect of RT depending upon the environmental acidity	[113]
Dextran-coated nanoceria (5 nm)	Prooxidant activity	A375 cell line and human dermal fibroblasts (HDF)	50–300 μM	Nanoceria enhances the antitumor activity of doxorubicin in human melanoma cells by causing oxidative damage while protects human dermal fibroblasts from doxorubicin-induced cytotoxicity	[116]
Dextran-coated nanoceria (5 nm)	Prooxidant activity	A375 cell bearing nude mice	50–150 μM	Nanoceria prevents tumor growth in vivo based on selective prooxidative and antioxidative activities	[117]

(continued)

Table 9.1 (continued)

Nanoceria nanozymes	Type of activity	Experimental model	Dosage	Results/remarks	References
Bare nanoceria (N/A)	Antioxidant activity	A murine model of diabetes	Nanoceria (60 mg/kg per day) combined with sodium selenite (5 μmol/kg per day)	Nanoceria combined with sodium selenite improves diabetes-induced oxidative stress	[123]
Bare nanoceria (N/A)	Antioxidant activity	Isolated rat pancreatic islets	10–1000 nM	Combination of nanoceria/sodium selenite improves transplantation outcome and graft function by control of oxidative stress damage	[124]

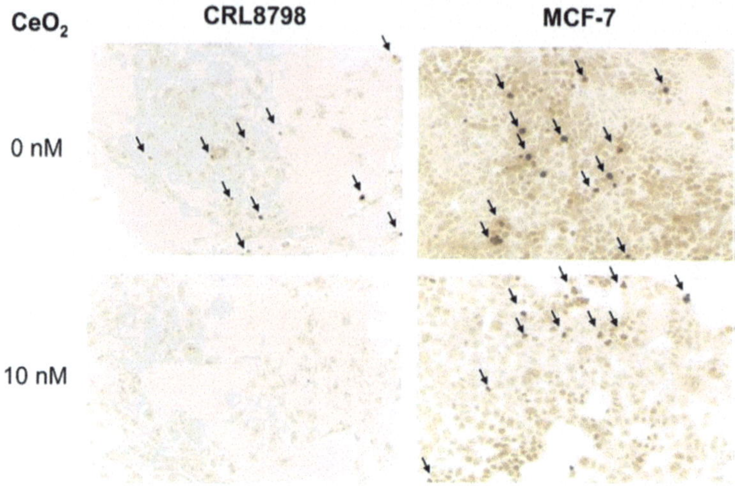

Fig. 9.7 TUNEL staining of breast cells at 48 h following 10 Gy irradiation and protection by cerium oxide nanoparticles [60]. The arrows denote TUNEL-positive apoptotic nuclei. Copyright 2005 American Chemical Society

Before the identification of enzyme-mimetic activities of nanoceria, Seal et al. have already found that nanoceria conferred protection to normal human breast cell line CRL8798 from radiation-induced cellular damage [60]. In this study, normal human breast cell line (CRL8798) and human breast tumor cell line (MCF-7) were all treated with nanoceria and radiation under the same condition. However, while almost 99% normal human breast cells were protected by nanoceria from radiation-induced apoptosis, there showed no obvious protection to human breast tumor cells (Fig. 9.7). This amazing observation makes it possible for using radiation to selectively killing tumor cells with the help of nanoceria. Based on the unique physicochemical properties of nanoceria, these researchers proposed that the antioxidant activity of nanoceria may be attributed to the valence state conversion of Ce^{3+} and Ce^{4+} on the surface of the ceria nanoparticles. In view of their hypothesis, the reactive radicals produced by radiation could be scavenged in the process of converting Ce^{3+} into Ce^{4+}. In addition; they also predicted the regeneration reaction of Ce^{4+} to Ce^{3+} on the surface of nanoceria, so that nanoceria will have extraordinary autocatalytic function [60].

Traversa et al. evaluated the capability of nanoceria nanozymes to balance the oxidative stress in isolated cardiac progenitor cells (CPCs) [62]. As a promising autologous source of cells for cardiac regenerative medicine, CPCs require to maintain ROS within physiological levels when cultured in vitro [78, 79]. Their results showed a time-dependent antioxidant activity in CPCs after a single 24 h pulse of 50 μg/mL nanoceria nanozymes against to oxidative stress induced by H_2O_2 (Fig. 9.8). In addition, these nanoceria nanozymes also exhibited dose-dependent (10, 25, and 50 μg/mL) ROS-scavenging capability. While 25 μg/mL nanoceria

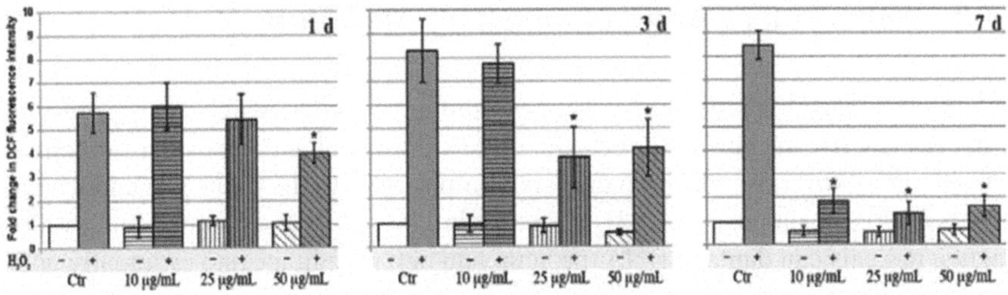

Fig. 9.8 ROS production in Sca-1pos CPCs exposed to H_2O_2 [62]. The results indicated a time-dependent decrease in ROS production when the cells were pretreated with different concentrations of CeO_2, as assessed by DCF fluorescence (*: H_2O_2-treated cells vs. CeO_2 H_2O_2-treated cells). ROS levels in the control group were arbitrarily set at fluorescence value of 1. All data are expressed as means (SD of three independent experiments and each repeated in triplicate. The relative SD of the samples has been calculated after normalization of the values compared to the mean values of the CeO_2 untreated control (negative) (*$p < 0.05$). Copyright 2012 American Chemical Society

nanozymes revealed 80% antioxidant activity at 7 days after a single pulse in the presence of H_2O_2, 10 μg/mL nanoceria nanozymes only showed 70% antioxidant capability. Besides, this study also indicated that the internalized nanoceria in the cytosol exhibited no obvious disturbance to the structure and function of CPCs after 7 days of incubation. The authors noted that these observations may refresh the understanding of the relationship between nanoparticles and organism. However, further studies are needed to understand the detailed antioxidant mechanism [62].

Lord et al. reported that nanoceria nanozymes with a large diameter (94 nm) showed better uptake and ROS-eliminating activities than that of nanoceria with a small diameter (7 nm) in activated U973 monocytes [80]. Later on, this group systematically investigated the effects of these nanoparticles on the cell proliferation, uptake mechanisms, intracellular trafficking, and ROS scavenging in SKOV3 human ovarian and WiDr human colon cancer cells [81]. Results indicated that nanoceria nanozymes with size of 7 and 94 nm reduced the levels of intracellular ROS in both cell models. Besides, as the previous study, larger sized nanoceria exhibited stronger ROS-scavenging effects than that of the smaller nanoceria in these cell lines. The uptake of nanoceria nanozymes was found to be energy-dependent and executed by nonspecific pathways and clathrin-mediated endocytosis, as well as caveolar endocytosis. These investigators concluded that the behavior of nanoceria nanozymes in ovarian and colon cancer cells was tightly associated with their size. In addition, Lord and colleagues appealed researchers to pay attention to the importance of testing new nanoparticle systems.

Taken together, considerable evidences have been reported for the antioxidant activity of nanoceria nanozymes. Since the imbalance of oxidative stress is the cause of many diseases, the antioxidant function of nanoceria nanozymes provides a promising tool for treatment of various oxidative stress-related disorders.

9.3.2 Nanoceria as Anti-inflammatory Mediators

Inflammatory reaction is a complex state of immune disorder, which can be activated by excess reactive species in tissues and organs. Frequent and persistent inflammatory responses will raise macrophages and lymphocytes to the tissue site and lead to irreversible tissue damage [82, 83]. As major inflammatory mediators, high level of reactive radicals can damage DNA, protein, and cell membrane and eventually activate apoptotic processes [84, 85]. Thus, scavenging reactive radical species may contribute to allocate the progression of inflammation. Depending on their antioxidant function in cells, nanoceria nanozymes have been found to trigger anti-inflammatory effects in a variety of chronic inflammation.

Excess radical nitric oxide (NO) produced by inducible nitric oxide synthase (iNOS) is tightly associated with the development of inflammation [82, 86]. Reilly et al. treated J774A.1 murine macrophages with nanoceria nanozymes to access the role of nanoceria in inhibiting NO-induced inflammatory [87]. The results of this study showed that the levels of iNOS protein and mRNA as well as ROS were all suppressed by the treatment of J77A.1 cells with nanoceria nanozymes (Fig. 9.9). Another important finding of this investigation is that nanoceria did not exhibit any significant toxicity in mice for up to 30 days after injection with a high concentration (0.5 mg/ml) of nanoceria nanozymes. Besides, these intravenously administered nanoceria nanozymes detained in tissues for at least 30 days post-injection, which indicated that nanoceria nanozymes play a long-term role in anti-inflammation treatment.

Hypoxia conditions typically raise the levels of reactive species and activate inflammatory effects, resulting in inducing lung pathological processes which include increase of pulmonary, epithelial malfunction, edema, and acute inflammation [88]. Bhargava et al. evaluated the anti-inflammation efficacy of nanoceria nanozymes in rat lung tissue during hypobaric hypoxia [89]. A dose of 0.5 μg/kg per week of nanoceria nanozymes was administered intraperitoneally into animals for 5 weeks. After the final injection, these animals were treated with hypobaric hypoxia condition. Then, the level of reactive species and inflammatory cytokines were estimated using the homogenate, and the plasma of lungs was isolated. Results showed that the lung-deposited nanoceria nanozymes scavenged harmful free reactive radicals and reduced hypoxia-induced inflammation without evoking any negative effect.

Inflammatory environment limited the application of engineered cartilage for the restoration of articular cartilage in injured and arthritic joints [90]. Somasundaran et al. reported the anti-inflammation effects of nanoceria nanozymes in an interleukin-1α (IL-1α) induced inflammatory context, and they accessed the function of nanoceria nanozymes in promoting the use of engineered cartilage [91]. They used bovine chondrocytes as study model and found that this model is well tolerated to a high dose of nanoceria nanozymes (60,000 cells/g). In this model, engineered cartilage was embedded in the agarose scaffold, and IL-1α was added to the culture media for inducing inflammatory action, which could cause matrix

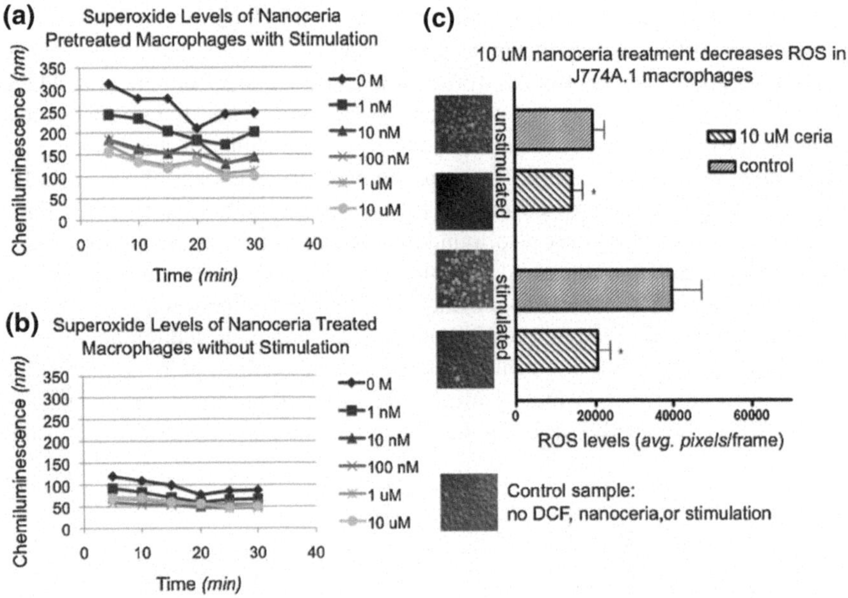

Fig. 9.9 Chemiluminescence and DCF fluorescence to determine ROS levels [87]. Cells were combined with Diogenes assay solution to measure luminescence directly proportional to superoxide production. **a** Stimulated J774A.1 macrophages without a nanoceria pretreatment exhibited about twice the levels of luminescence as compared to nanoceria-treated cells of varying concentrations. **b** Nanoceria-treated cells without stimulation also exhibited a decrease in superoxide production as the concentration of nanoceria treatment increased. Results are representative of four independent trials. ROS levels directly proportionate to DCF fluorescence levels were viewed. **c** 24 h, 10 μM nanoceria pretreated, LPS/IFN-g stimulated J774A.1 macrophages showed less DCF fluorescence than stimulated cells without nanoceria pretreatment. 10 μM pretreated, non-stimulated cells also showed less DCF fluorescence than cells with neither stimulation nor pretreatment. Control cells exhibited no DCF fluorescence. Five random images of each sample were visualized on a Zeiss LSM510 Confocal microscope (excitation 488 nm, emission 515–555 nm) under identical parameters for each sample. Copyright 2008 John Wiley and Sons

degeneration. Raman microspectroscopy revealed that internalized nanoceria evaluated production of proline, procollagen, and glycogen in chondrocytes compared with that of untreated cells. The Raman spectroscopy results indicate that nanoceria nanozymes improve synthesis of the elaborated extracellular matrix (ECM). In conclusion, these researchers demonstrated that nanoceria nanozyme-containing constructs improve mechanical and biochemical properties of engineered cartilage and showed a long-term performance against inflammatory reactions.

Recently, inflammation and oxidative stress have been closely tied with the pathogenesis of ischemic heart disease [92, 93]. Encouraged by the cell protection function of nanoceria in culture, Kolattukudy et al. evaluated the cardioprotective effects of nanoceria nanozymes in a transgenic mouse model of cardiomyopathy [61]. They constructed a transgenic murine model (MCP mice) which perform a distinct syndrome of ischemic cardiomyopathy with the features of myocardial inflammation, cardiac remodeling, and ventricular dysfunction at the age of about 6 months. Then, 15 nmol of nanoceria nanozymes were administered intravenously into MCP mice and wild-type controls twice a week for 2 weeks. Treated with nanoceria nanozymes significantly reduced the expression of proinflammatory cytokines (TNF-α, IL-1β, IL-6, and MCP1) and repressed monocyte/macrophage infiltration in MCP mice (Fig. 9.10). Besides, this study also found that nanoceria nanozymes inhibited peroxynitrite formation and repressed the initiation of ER stress-related genes in the hearts. In conclusion, nanoceria nanozymes' autocatalytic properties may contribute to the protection against the progression of cardiac dysfunction and remodeling.

Sepsis is a major life-threatening disease related with system inflammatory response syndrome (SIRS) which results in injury of multiple organisms and tissues [94, 95]. Sepsis is typically caused by the affection of lipopolysaccharides (LPS) and

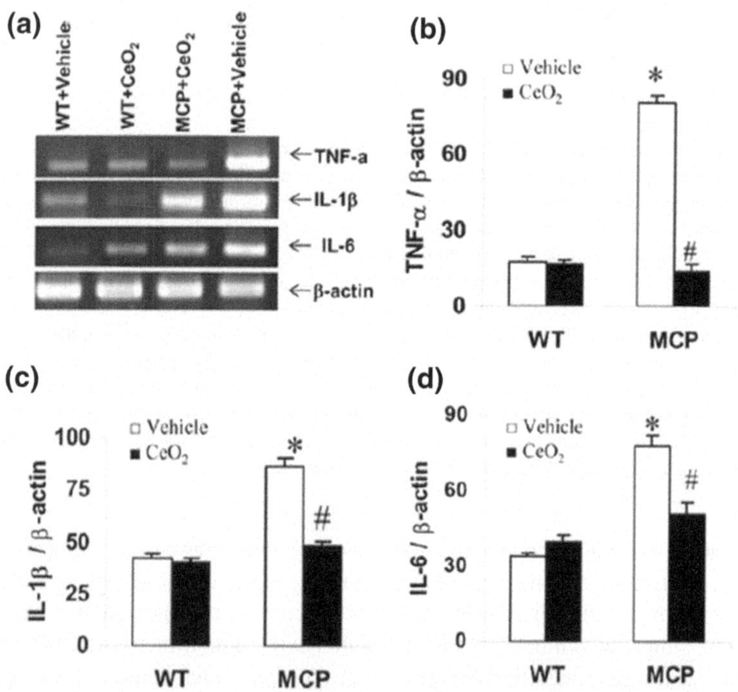

Fig. 9.10 Effects of CeO_2 nanoparticles on TNF-α, IL-1β, and IL-6 gene expression in the myocardium of MCP mice [61]. **a** Expression of TNF-α, IL-1β, and IL-6 mRNA in the myocardium of wild-type control, vehicle- and CeO_2-treated MCP mice were assayed by RT-PCR. **b–d** Bands were quantified by densitometric analysis and normalized by β-actin. *$P < 0.001$ versus wild-type controls; $^{\#}P < 0.05$ versus vehicle-treated MCP mice; n = 5 per group. Copyright 2007 Oxford University Press

gram-negative bacteria, and subsequently lead to the release of several factors including IL-6, ROS, and TNF-α [96]. Some investigators have tried to employ antioxidants to alleviate the damage of sepsis in some reported studies [97, 98]. Enlighted by the excellent antioxidant and anti-inflammatory activities, Zhu et al. synthesized a type of nanoceria nanozyme with a size of 37 nm. They analyzed the effects of nanoceria nanozymes in LPS-induced sepsis [99]. Nanoceria nanozymes were injected into sepsis rat model through tail nerves in this study. Results suggested that the administered nanoceria nanozymes blocked the release of TNF-α and ROS, which indicated the remission of sepsis.

In summary, this section summarizes the studies that explored the function of nanoceria as anti-inflammation agents. Some diseases associated with inflammation and oxidative stress have been evaluated to be alleviated under the treatment of nanoceria nanozymes. Further studies of detailed anti-inflammatory mechanisms and potential anti-inflammatory pathways of nanoceria nanozymes in biological conditions need to be conducted.

9.3.3 Nanoceria as Potential Therapies for Ocular Diseases

Oxidative stress has been considered as a major cause of many ocular diseases such as diabetic retinopathy (DR) [100], age-related macular degeneration (AMD) [101], and glaucoma [102]. McGinnis et al. have conducted a series of works in recent years to illustrate the role of nanoceria nanozymes in protecting of photoreceptor cells from oxidative damage [64–68]. In their early works, a homozygous *tubby* mutant mouse was used as a light damage model to access the ability of nanoceria to relieve the progression of retinal degeneration. After 20 μL of 1 mM nanoceria nanozymes were injected into tubby mice intracardially at postnatal day 10, day 20, and day 30, the protection effects of nanoceria to the retina were observed, including scavenged ROS, inhibited apoptosis signaling pathways, and up-regulated antioxidant-associated genes, which delayed the retinal degeneration for up to 2 weeks [64]. In a subsequent study, these *tubby* mutant mice were injected with 1 μL of 1 mM nanoceria nanozymes intravitreally and consecutively at postnatal day 28, day 49, day 80, and day 120 for analyzing the long-term effects of nanoceria nanozymes on the structure and function of retina [66]. The results showed that the structure and functions of retina are preserved for 6 weeks after a single intravitreal administration of nanoceria, while untreated tubby mutant mice typically lose about 2/3 of photoreceptor cells in the retina. Interestingly, this study also found that the function of rods (account for 95 to 97% of photoreceptor cells) is reserved longer than the function of cones (account for 3 to 5% of photoreceptor cells) [66].

While the antioxidant activity of nanoceria nanozymes has been confirmed in rodent models of retinal light damage, the potential cytotoxicity of nanoceria in retinal tissues is still obscure. Based on above reports, McGinnis and coworkers conducted a systematic study to access the toxic effects and the temporal and spatial distributions of nanoceria in retina of healthy rats [67]. After a single intravitreal

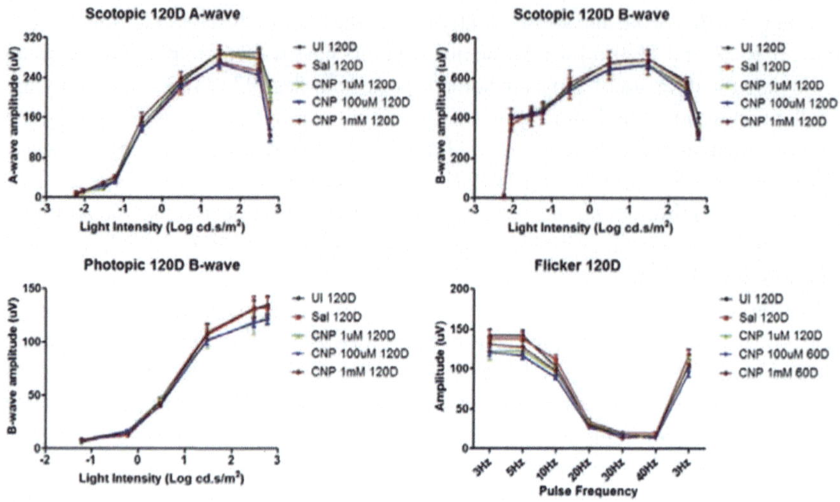

Fig. 9.11 Nanoceria (1–1000 μM) had no negative effects on retinal function, even 4 months after injection, as measured by ERG recordings [67]. Scotopic a-wave amplitude reflects primarily the function of rod cells. Scotopic b-wave amplitude reflects the function of neurons in the inner retina, post-synaptic to the rod photoreceptor cells. Photopic b-wave amplitude and flicker ERG reflect the function of cone cells. No changes were detected in amplitudes of scotopic a- and b-waves, photopic b-wave, and flicker at the frequencies indicated. Copyright 2013 Wong et al.

administration, nanoceria were rapidly absorbed by retina tissue and retained in there for up to 120 days. These nanoparticles distributed mainly in the retina, but also in the lens and eyecup at 1 h after administration. Investigators used electroretinogram recordings (ERGs) to evaluate the functions of retinal cells including rod cells, cone cells, and retinal neurons in this research. Surprisingly, almost no disfunction effects of ocular tissue were detected after such a long-term exposure (120 days), even though the used concentration (1000 μM) of nanoceria nanozymes was much higher than the effective concentration (1 μM) (Fig. 9.11) [67].

In order to illustrate the intrinsic cellular mechanism of nanoceria nanozymes' catalytic properties in protecting retinal photoreceptor cells from light damage, these researchers subsequently used an autosomal dominant retinitis pigmentosa (adRP) model, named P23H-1 rats, as their study subject [65]. A single intravitreal injection of 1 mM nanoceria (per eye) was conducted to P23H-1 rats at 2–3 weeks of age in this study. The results demonstrated that nanoceria exhibited a time-dependent regeneration activity, which was confirmed by the fact that the number of damaged cells were reduced by 46%, 56%, 21%, and 24% compared to saline-treated group at 3, 7, 14, and 21 days post-injection, respectively.

Taken together, the functions of nanoceria nanozymes in retina protection are to eliminate reactive radicals. The persistent antioxidant functions of nanoceria nanozymes have also been confirmed in several rodent models in vivo. Importantly, nanoceria nanozymes exhibited little toxic effects on ocular tissues. These systematic works demonstrated that nanoceria nanozymes are promising antioxidant agents for ophthalmic treatments.

9.3.4 Nanoceria as Potential Therapies for Neurodegenerative Diseases

Oxidative stress involves in the initiation and progression of neurodegenerative diseases [103, 104]. To date, few antioxidants have been successfully developed to treat these pathological disorders [105]. In principle, nanoceria nanozymes are hard to be used as potential therapies for brain diseases, because many studies have already demonstrated that nanoceria nanozymes could not easily across blood–brain barrier (BBB) and distribute in brain. However, attracted by the unbelievable power of nanoceria nanozymes in reducing oxidative stress, some investigators tried to explore the potential of nanoceria in treating neurogenerative diseases.

One of the pioneer studies to announce the protective function of nanoceria in neurosystem was published by Hickman et al. [37]. In this study, adult rat spinal cord cells were isolated from the whole rat cord and then cultured on neurobasal medium before treated with nanoceria (Fig. 9.12). Results showed that a single dose of 10 nM nanoceria nanozymes (2–5 nm) at the time of plate culture greatly improved the cell survival upon H_2O_2-induced oxidative stress. Besides, as we previously mentioned in Sect. 9.2.3, this work is also the pioneer studies to indirectly reveal the $^{\bullet}OH$ scavenging ability of nanoceria nanozymes.

Another in vitro study conducted by Erlichman et al. also suggested that nanoceria nanozymes possess neuroprotective function [106]. A mouse hippocampal brain slice of cerebral ischemia was used as in vitro model to evaluate the profile of nanoceria in reducing ROS and protecting nervous system in this study. Within the dose range of 0.2–1 μg/ml, nanoceria significantly reduced ischemic cell death by >50% (Fig. 9.13). In addition, these investigators demonstrated that the reduction (~15%) of ROS is the reason of the neuroprotective effects in general. Interestingly, the level of ischemia-induced 3-nitrotyrosine, a modification to proteins induced by peroxynitrite, was decreased remarkedly (~70% reduction) after treated with nanoceria nanozymes, which indicated that reduction of peroxynitrite may be a key point in nanoceria neuroprotective activity.

Employing a human AD in vitro model, Cimini et al. have confirmed that nanoceria nanozymes not only scavenge free radicals as antioxidant agents but also regulate signaling pathways associated with neuroprotection [107]. In this study, cells were treated with aggregated Aβ to mimetic an AD situation with plaque presence. The effects of nanoceria nanozymes were then evaluated by analyzing the parameters involved in signal transduction pathways of neuroprotection. Results showed that nanoceria nanozymes improved neuronal survival by modulating brain-derived neurotrophic factor (BDNF) pathway. Based on this work, Cimini et al. synthesized PEG-coated and anti-Aβ antibody-conjugated nanoceria (Aβ-CNPs-PEG) and assessed the

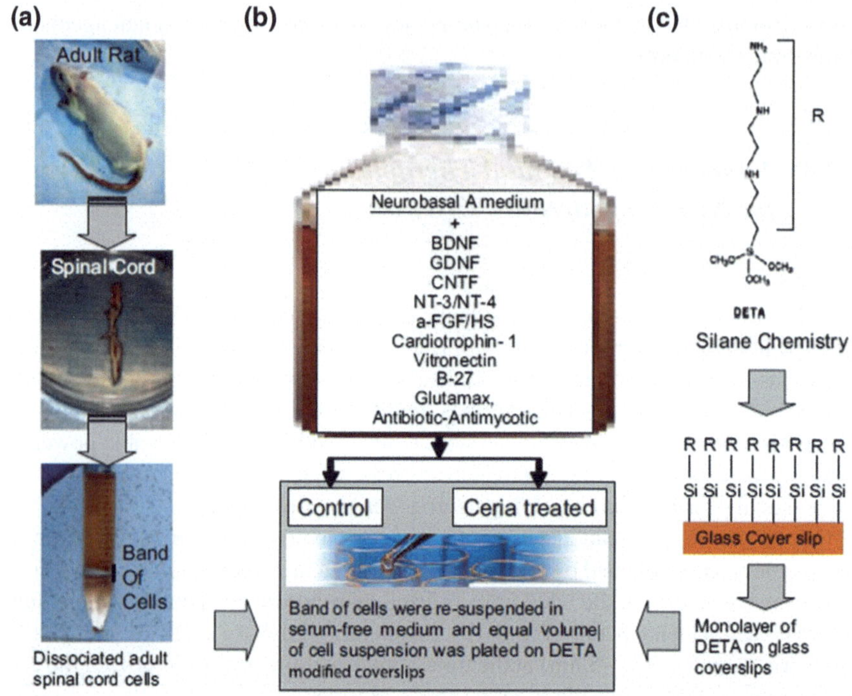

Fig. 9.12 Adult rat spinal cord culture [37]. **a** Isolation of adult rat spinal cord cells from the whole cord. **b** Development of serum-free culture medium using various growth factors. **c** Surface modification of the glass coverslips for cell culture. Copyright 2007 Elsevier

effect of this nanoceria on neuronal survival and BDNF signaling pathway [108]. Results showed that Aβ-CNPs-PEG exhibits better neuroprotection ability than control groups by selectively targeting to the Aβ plagues and regulating the BDNF signaling transduction pathway (Fig. 9.17).

Although abovementioned works have demonstrated the neuroprotective role of nanoceria nanozymes in neural cell models, further studies need to demonstrate the neuroprotective efficiency of nanoceria nanozymes on animal models. In addition, the capability of nanoceria to across the BBB in vivo should be improved. Erlichman et al. first reported the characteristics of nanoceria nanozymes in the treatment of a neurodegenerative disease animal model [109]. These researchers synthesized citrate/EDTA-coated nanoceria with an average diameter of 2.9 nm. The plasma half-time and tissue distribution of nanoceria were analyzed after administered intravenously into healthy rodent models. The half-life of these nanoparticles is about 4.0 h in plasma. Compared to previous nanoceria formulations, the nanoceria used in this study were less taken up by the liver and spleen, but successfully traversed the BBB and distributed in the brain. Nanoceria nanozymes were detected to be accumulated in the cerebellum, a typical region with significant damage in experimental autoimmune encephalomyelitis (EAE) animal models. A murine model of EAE was

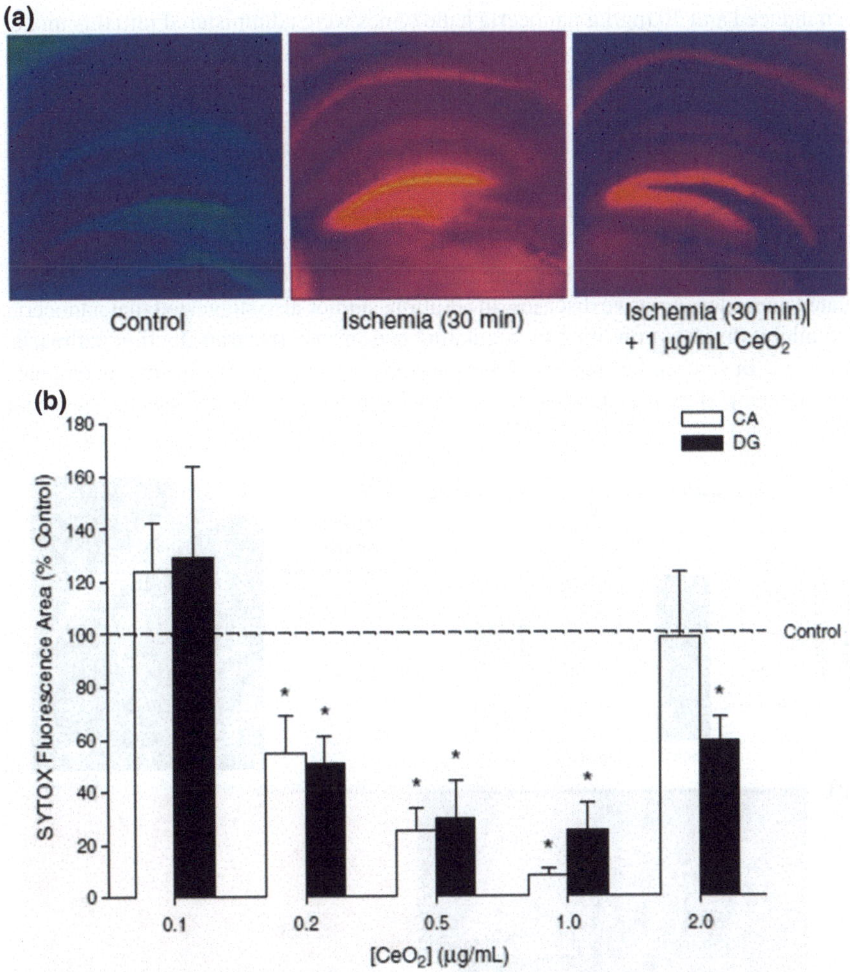

Fig. 9.13 Neuroprotective effect of nanoceria treatment after ischemia. Nanoceria significantly decreased the area of ischemia-induced cell death in hippocampal slices [106]. **a** Pseudocolor images of brain slices loaded with Sytox blue. **b** Quantification of the area of Sytox fluorescence in mouse hippocampal slices after 30 min of ischemia and treatment with varying doses of nanoceria (added to the solution at the onset of the ischemic insult). All Sytox measurements were made 1 h postischemia. Data are presented as mean percentage ± SEM of control (percentages defined by the ratio of anatomically matched test and control slices from same-sex littermates sliced and imaged on the same day). Statistical significance was determined using paired t tests and $*p < 0.01$ denotes significant differences compared to vehicle controls ($n = 3$–6 pairs). Copyright 2011 Elsevier

then induced and 30 mg/kg nanoceria nanozymes were administered into this model to evaluate the therapeutic efficiency of nanoceria. Results showed that nanoceria nanozymes were able to decrease the level of ROS and relieve clinical symptoms in EAE model. The efficacy of 30 mg/kg nanoceria nanozymes was comparable to the effectiveness of fingolimod (an immunomodulatory agent used to treat patients with relapsing MS) at the same concentration (Fig. 9.14).

Overall, the neuroprotective function of nanoceria nanozymes has been systematically studied in vitro and in vivo. As antioxidant agents, nanoceria nanozymes effectively reduce oxidative stress and therefore alleviate the progress of oxidative-related neurodegenerative diseases. In addition, studies also suggested that nanoceria also inhibit diseases condition by regulating neuroprotective transduction pathways. Moreover, in vivo studies indicated that nanoceria nanozymes are promising reagents to be developed as regenerative treatments for brain disorders. However, a recent

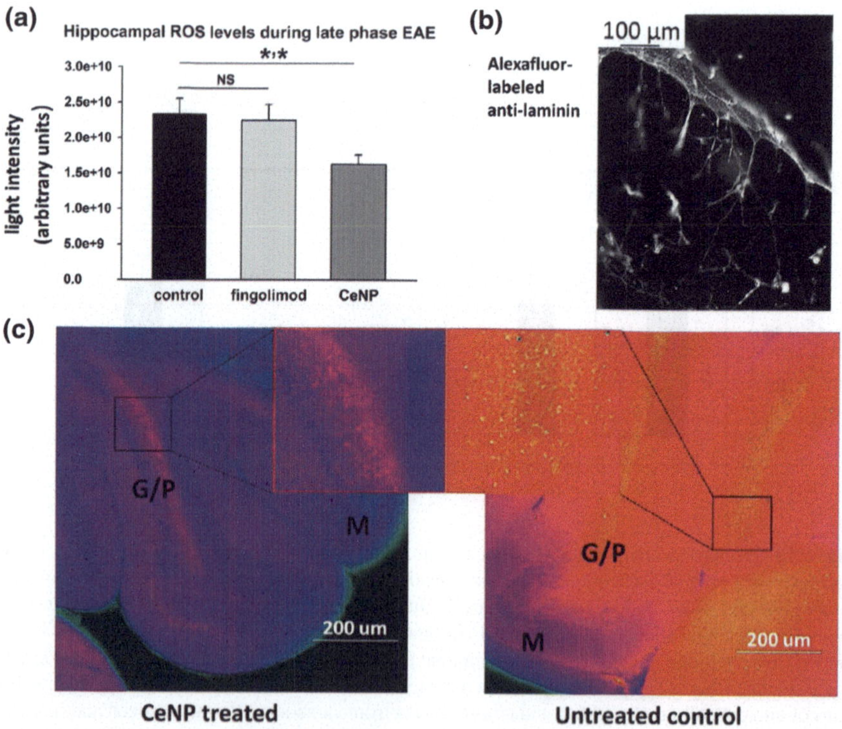

Fig. 9.14 CeNPs decrease cerebellum ROS levels in EAE mice [109]. **a** ROS levels were measured with a nonspecific free radical fluorophore (CM-H2DCFDA) in brain slices prepared from 30 mg/kg CeNP-treated (preventative regimen) or fingolimod-treated mice beginning on day 42 after induction of EAE. This time point fell at least 7 days after the last CeNP treatment. **b** The cerebellum was stained with an antilaminin AlexaFluor reagent to illustrate the location of the detected ROS staining (**c**) relative to vasculature in brain slices. **c** Representative images of ROS detected in brain slices from a CeNP-treated animal (left) and a control animal (right), which was injected with saline only. G/P: granular/Purkinje layer; M: molecular layer. Copyright 2013 American Chemical Society

study suggested that nanoceria may cause neurotoxicity by inhibiting differentiation of neural stem cells [110]. Thus, the toxicity of nanoceria nanozymes to brain tissues and the capability of nanoceria nanozymes to cross the BBB are still misty which needs investigators to make more efforts into this place.

9.3.5 Nanoceria as Potential Therapies for Cancers

From the previous sections, we have witnessed the extraordinary protective function to normal cells of ceria nanoparticles. Surprisingly, nanoceria have been demonstrated to protect normal cells but not cancer cells from radiation-induced damage in some studies [60, 72, 111, 112], which indirectly improved the efficacy of radiation cancer therapy. Besides, there are also some studies found that nanoceria directly damaged tumor cells in some conditions by improving ROS production [113].

In radiation therapy, high levels of ROS will be produced in biological environment by ionizing radiation. While these damaging ROSs enhance the efficiency of cancer cell killing, it also be harmful to normal cells. The pioneer study of nanoceria nanozymes in protecting cells from radiation-induced cellular damage was conducted by Seal et al. as previously described [60] (see Sect. 9.3.1). In this study, almost 99% of normal cells were protected under the treatment of nanoceria, while cancer cells treated with nanoceria exhibited no improvement of survival rate. Another study conducted by Baker et al. suggested that nanoceria nanozymes protected gastrointestinal epithelium against radiation-induced damage by scavenging free radicals [72]. In addition, this research also reported that nanoceria nanozymes protected cells by elevating the level of SOD2 before radiation insult.

Ovarian cancer is one of the most common and malignant diseases in women. Because of the high mortality rate and the poor prognosis, novel therapies strategies are in urgent demand [114, 115]. Inspired by the fabulous catalytical activity of nanoceria nanozymes, Shridhar et al. tested the efficiency of nanoceria to treat ovarian cancer in vitro and in vivo [112]. In vitro nanoceria nanozymes significantly reduced the cellular ROS in A2780 cells and inhibited growth factor-mediated cell migration and invasion of SKOV3 cells without affecting the cell proliferation in both cells. Besides, $VEGF_{165}$-induced proliferation, capillary tube formation, and activation of VEGFR2 in HUVEC cells were all inhibited by nanoceria nanozymes. In vivo, A2780 ovarian cancer cells treated with nanoceria nanozymes (0.1 mg/kg body weight) were intraperitoneally administered into nude mice. Results showed that nanoceria reduced the tumor size and decreased angiogenesis of ovarian cancer in mice (Fig. 9.15). This research suggested that nanoceria nanozymes are potentially to be developed as anti-angiogenic therapies in ovarian cancer.

Different from most other studies using the antioxidant function of nanoceria nanozymes, Baker et al. introduced the prooxidant property of nanoceria nanozymes to cancer treatment [113]. Employing a pancreatic cancer cell model, they found that nanoceria nanozymes preferentially improved radiation-induced ROS production in acidic condition, as compared to neutral condition. Because the pH in cancer cells is

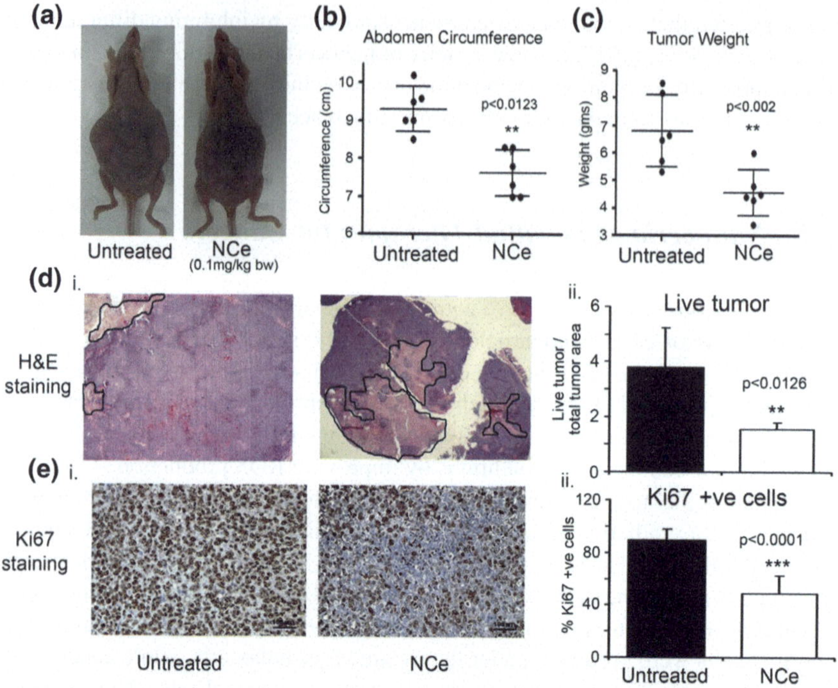

Fig. 9.15 Nanoceria (NCe) treatment inhibited ovarian tumor growth in vivo [112]. **a** Gross morphology of representative mouse with tumors at day 30 ($n = 6$). **b** Cumulative abdominal circumference at the end of the study. **c** Excised tumor weight from vehicle (PBS) treated and NCe (0.1 mg/kg bd wt; every third day). **d** (i) Representative H and E (×20) photomicrographs exhibiting live (purple) and necrotic (pink, encircled) areas in untreated and treated xenografts. (ii) Graphical representation of viable tumor size measured as described in Materials and methods. **e** (i) Representative Ki-67 staining (×200) of excised A2780 xenografts at day 30. (ii) Count of positive Ki-67 cells from five high-powered fields (×400) in three different xenografts from each group. Copyright 2013 Giri et al.

typically lower than that in normal cells, nanoceria nanozymes rendered cancer cells more sensitive to radiation (RT)-induced damage than normal cells. Taken together, it is a promising idea for using nanoceria nanozymes as a novel RT-sensitizer for improving cancer treatment.

In addition to enhancing the therapeutic effect of radiation therapy, nanoceria nanozymes also combined with chemotherapeutic drugs to kill cancer. Sack et al. demonstrated that nanoceria nanozymes improved the anticancer property of doxorubicin, a classical chemotherapeutic agent, in human melanoma cells [116]. This study found that nanoceria nanozymes enhanced the apoptosis in A375 cells, while protected normal cells from cytotoxicity of doxorubicin. Before this report, some evidences have already demonstrated that nanoceria nanozymes suppress the growth of tumor in a xenograft mouse model with A357 melanoma cells [117]. Nanoceria nanozymes exerted prooxidant activity in tumor cells which significantly increased the level of ROS and activated the apoptosis of tumor cells. However, in normal cells such as stromal fibroblasts, nanoceria exhibited antioxidant properties. One possible explanation for different enzymatic activities of nanoceria nanozymes is the pH difference between the tumor cells and normal cells. Compared with healthy cells, tumor cells typically exhibit an enhanced glycolysis rate (Warburg effect) which leading to the rapid production of lactate and eventually resulting in lowed cellular pH value [118–120]. Nanoceria nanozymes exhibit different enzyme-mimetic activities in different pH conditions. Previous studies have shown that nanoceria nanozymes are able to exhibit both as SOD-like and CAT-like enzyme mimetics (see Sect. 9.2.2). In acidic cancer cells, nanoceria nanozymes exhibit enhanced SOD-mimetic activity and suppressed CAT-mimetic activity, which subsequently elevate the production of H_2O_2 in cancer cells. Excess of H_2O_2 is transformed into highly dangerous $^{\bullet}OH$ by Fenton reaction and eventually destroy cancer cells. However, in normal cells, nanoceria usually distribute in the neutral plasma and show both SOD-like and CAT-like enzyme-mimetic properties. Under the synergistic effects of these two enzyme-mimetic activities, ROS including H_2O_2 and $O_2^{\bullet-}$ in normal cells can be eliminated simultaneously [113]. In a word, based on the above principles, nanoceria nanozymes act as both antioxidant in normal cells and prooxidant in tumor cells, which provides a broad prospect for its application in the treatment of cancer.

Recently, a study conducted by Kim et al. also suggested that nanoceria nanozymes enhanced the anticancer effect of doxorubicin in ovarian cancer cells in vitro [121]. In this study, nanoceria nanozymes served as a delivery platform to mediate the uptake of doxorubicin by ovarian cancer cells (Fig. 9.16). After conjugated onto the surface of nanoceria, doxorubicin showed superior drug-loading concentration (22.41%) and loading efficiency (99.51%). This CeO_2/Dox system improved cellular uptake and drug release performances of doxorubicin when compared to free doxorubicin. Interestingly, the concentration of the initial intracellular doxorubicin was lower in the nanoceria–doxorubicin groups than that of in the free doxorubicin groups. However, nanoceria–doxorubicin exhibited constant drug release over time and kept a high intracellular drug content for up to 72 h. The results of cytotoxicity assay showed that nanoceria-conjugated doxorubicin exhibit stronger lethality for ovarian cancer cells. This research suggested that nanoceria nanozymes is a promising drug delivery agent for cancer treatment.

9.3.6 Nanoceria as potential therapies for diabetes

Diabetes mellitus (DM) is an endocrine–metabolic disorder with high morbidity and mortality rates, in which high blood sugar level exists over a prolonged period. Studies have suggested that oxidative stress play a critical role in the development and progression of diabetes mellitus and its complications [122]. Abdollahi et al. used a combination of nanoceria and selenium selenite to treat diabetes in a murine model [123]. Nanoceria nanozymes (60 mg/kg, body weight) and sodium selenite (5 μg/kg, body weight) were together administered intraperitoneally into diabetic rats per week

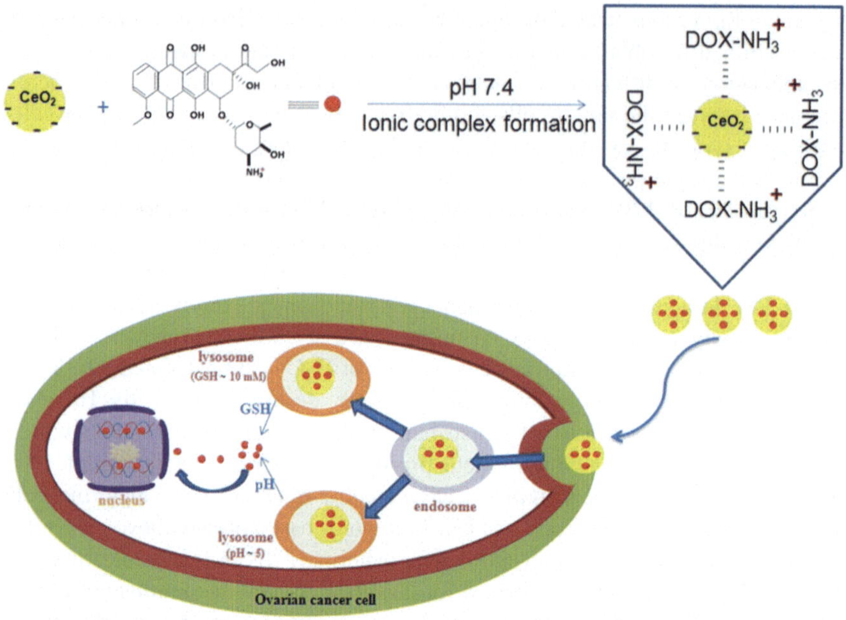

Fig. 9.16 Schematic diagram of CeO_2/DOX nanoparticle preparation, uptake in ovarian cancer cells and release of DOX [121]. Copyright 2017 Nature Publishing Group

for 2 weeks. The parameters associated with diabetes including level of oxidative stress markers, extent of energy depletion, and lipid profile were then analyzed. Results of this study showed that the combination of nanoceria and sodium selenite significantly reduced diabetes-induced oxidative stress. Patients with diabetes mellitus may suffer from destruction of pancreatic β-cells in which case pancreatic islets transplantation is the final cure. However, oxidative stress is still a serious problem during the isolation and transplantation processes of pancreatic islets. Based on their previous work, Abdollahi et al. further studied the effects of the combination of nanoceria/sodium selenite on isolated rat pancreatic islets [124]. Results suggested that nanoceria/sodium selenite combination greatly improved cell viability of isolated islets and secretion of insulin (Fig. 9.17).

9.4 Applications of Nanoceria Nanozyme in Biosensor

Biosensor is a kind of device which can recognize biological components and transform the signal of the target into measurable signal [125]. Nanoceria nanozymes have been extensively used in biosensors based on their excellent multiple enzyme-like activities. Compared with traditional biosensors, nanoceria-based diagnostic devices are fast, simple, sensitive, and low cost [126]. Up to now, nanoceria nanozymes-based

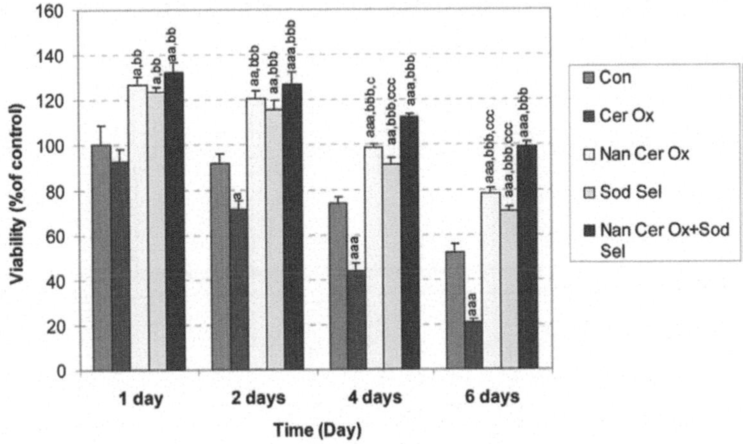

Fig. 9.17 Effects of nanoceria, nanoceria, and/or sodium selenite pretreatments on viability of isolated rat islets on days 1, 2, 4, and 6 [124]. Abbreviations: Con: Control; Cer Ox: Cerium Oxide; Nan Cer Ox: Nano Cerium Oxide; Sod Sel: Sodium Selenite; Nan Cer Ox + Sod Sel: Nano Cerium Oxide + Sodium Selenite. Copyright 2012 Taylor & Francis

biosensor have been developed to detect biochemicals (e.g., H_2O_2 and ethanol), small biomolecules (e.g., glucose and dopamine), and biomacromolecules (e.g., DNA and protein) (Table 9.2).

9.4.1 Sensing of biochemicals and small biomolecules

Reactive radical species including ROS and RNS are dangerous to organism as they could damage the components of cell, thus resulting in various diseases [84, 85]. Therefore, methods capable of detecting these highly reactive chemicals in living tissues are needed. H_2O_2 is a relative stable molecule in oxidative species. Thus, measuring the levels of H_2O_2 would be a useful strategy to identify oxidative stress-related diseases. Ujjain et al. used nanoceria-based nanozymes as electrochemical biosensors to detect the levels of H_2O_2 [127]. They synthesized two types of nanoceria, named hexamethylenetetraamine (HMTA)-coated nanoceria and fructose-coated nanoceria. These nanoceria nanozymes oxidized the H_2O_2 on their surface, which change the electric current of electrodes and hence quantify H_2O_2. The limits of detection (LOD) of HMTA–nanoceria and fructose–nanoceria nanozymes to detect H_2O_2 were assessed to be 0.6 μM and 2.0 μM, respectively. Thus, HMTA-coated nanoceria nanozyme possesses better potential in H_2O_2 detection.

To further improve the sensitivity of H_2O_2 detection, some investigations doped different nanomaterials into ceria nanoparticles. For instance, nanoceria-coated graphene oxide nanosheets (CNP-GS) were synthesized and modified with gold nanoparticles (GNPs). This nanocluster exhibited an LOD of 260 pM and a linear

Table 9.2 Summary of nanoceria-based nanozymes used as biosensors

Nanoceria-based nanozymes	Type of activity	Target molecules	Limit of detection	Linear range	References
HMTA–CeO_2	CAT- and SOD-like activities	H_2O_2	0.6 μM	1 μM–1 mM	[126]
Fructose–CeO_2	CAT- and SOD-like activities	H_2O_2	2.0 μM	1–25 mM	[126]
AuNP–GS–CNPs	N/A	H_2O_2	260 pM	3–10 mM	[128]
Pt–GS–CNPs	N/A	H_2O_2	0.43 μM	1–10 μM	[129]
CeO_2/NiO	Peroxidase-like activity	H_2O_2	0.88 μM	0.05–40 mM	[130]
PAANa–CeO_2	Oxidase-like activity	Antioxidants	N/A	N/A	[131]
Pt–CNPs–LOX	Oxidase-like activity	Lactate	100 pM	100 pM–15.5 mM	[132]
Zr–CNPs	N/A	Ammonia	3 ppm	3–40 ppm	[133]
CNPs	N/A	Ethanol	124 μM	0.17 mM–0.17 M	[134]
H2TCPP–CeO_2	Peroxidase-like activity	Glucose	33 μM	50–100 μM	[136]
CuO–CeO_2	N/A	Glucose	10 μM	N/A	[137]
PANI–CeO_2	N/A	Glucose	0.56 μM	N/A	[138]
CeO_2–HEG–nafion	N/A	Dopamine	1 μM	10 μM–780 μM	[144]
DAO–PANI–CeO_2	N/A	Histamine	48.7 μM	0.45–1.05 mM	[147]
CeO_2/TiO_2/GmOx/Chit/o–PD/Pt	N/A	L–glutamic acid	0.493 μM	5–50 μM	[148]
Ur–GLDH/CeO_2/ITO	N/A	Urea	0.1 mg/dL	10–80 mg/dL	[149]
CeO_2/GO	N/A	Influenza	0.43 pg/ml	N/A	[152]
CeO_2/Chitosan	N/A	Colorectal cancer gene	10 pM	15.9 pM–11.6 μM	[153]

(continued)

Table 9.2 (continued)

Nanoceria-based nanozymes	Type of activity	Target molecules	Limit of detection	Linear range	References
GNPs/MWCNTs/CeO_2/Chitosan	N/A	BCR/ABL fusion gene	0.5 pM	1 pM–1 nM	[154]
CeO_2/SWNTs/BMIMPF6/GCE	N/A	phosphoenolpyruvate carboxylase gene	230 fM	1 pM–0.1 μM	[155]
Nanoceria	N/A	Target ssDNA	0.12 nM	1.1–37 nM	[156]
Nanoceria	N/A	ATP	54 pM	0.1 nM–1.5 μM	[158]
MPO IgGs–nanoceria/nanogold	N/A	Myeloperoxidase	0.06 ng/mL	10–400 ng/mL	[157]

range of 3–10 mM for H_2O_2 detection [128]. Later on, a study used Pt-doped CNP-GS to detect H_2O_2 was reported, which exhibited a LOD of 0.43 μM and a linear range of 1–10 μM [129]. In a recent study, nanoceria were successfully incorporated into NiO nanostructures to form a coral-like CeO_2/NiO nanocomposites which exhibited enhanced peroxidase-mimetic activity [130]. The catalytic property was related to the percent of cerium in the CeO_2/NiO nanocomposites. When the ratio of cerium reached 2.5%, these nanocomposites performed the highest catalytic activity. CeO_2/NiO was then used to develop colorimetric H_2O_2 biosensor which showed a LOD of 0.88 μM and a wide linear range of 0.05 to 40 mM.

In addition to the detection of reactive radical species, an interesting study recently proposed a nanoceria-based assay to detect the level of antioxidants [131]. In this study, a polyacrylic acid sodium salt (PAANa)-coated nanoceria nanozyme was synthesized which was capable of oxidizing TMB in an acidic solution (pH = 4.0). When antioxidants existed in the solution, the color intensity of the nanoceria suspension would be reduced and indirectly reflected the level of antioxidants. For example, the LOD of this detection procedure for quercetin, a kind of antioxidant, was calculated as 8.25 nM.

As lactate could be decomposed into H_2O_2 and pyruvate under the catalysis of lactate oxidase (LOX), Pt-doped nanoceria nanozymes have been used to detect the levels of lactate [132]. When H_2O_2 was produced by the decomposition of lactate, H_2O_2 reacted with the oxygen vacancies in nanoceria nanozymes and indirectly showed the levels of lactate. The LOD of this biosensor for lactate was determined to be 100 pM.

A method of sensing ammonia was invented by putting zirconium (Zr) into nanoceria electrochemical detector [133]. When ammonia was approaching to this electrode, it was absorbed and catalyzed by Zr-doped nanoceria and then produced electron current (Fig. 9.18). The LOD of this biosensor for ammonia was estimated to be as low as 3 ppm. Similar with this research, nanoceria nanozymes are developed as biosensors for ethanol. Because electrical current is positively correlated with the concentration of ethanol in solution, ethanol can change the electrical response of electrode made by glassy carbon-coated nanoceria. Based on this mechanism, a

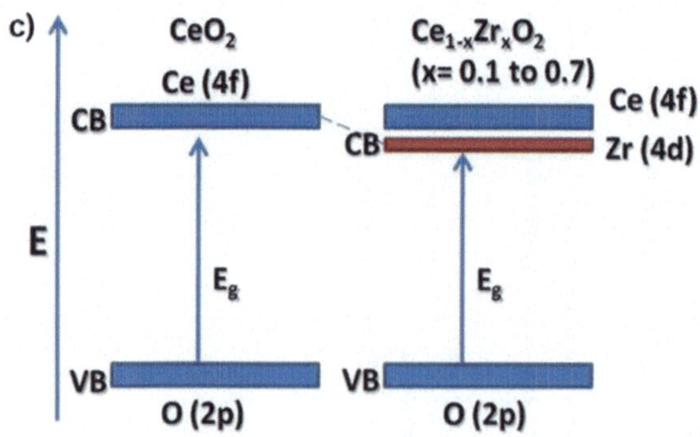

Fig. 9.18 Effect of Zr-doping on the electron energy transfer of nanoceria electrode [133]. Copyright 2017 Elsevier

biosensor of nanoceria electrode for ethanol with a linear range of 0.17 mM to 0.17 M and a LOD of 124 μM was achieved [134].

Another important application of nanoceria nanozyme-based biosensors is to detect glucose. Glucose is a small molecule but the most important component needed in a basic health [135]. Gao et al. synthesized a peroxidase-mimetic nanocomposite by modifying nanoceria with 5,10,15,20-tetrakis (4-carboxyl phenyl)-porphyrin (H2TCPP) to detect glucose [136]. This nanocomposite (H2TCPP–CeO_2) oxidizes TMB in the presence of H_2O_2 to produce a typical color reaction from colorless to blue. This method showed an LOD of 33 μM and a linear range of 50 to 100 μM for glucose detection. Guan et al. developed a glucose detect method based on a copper oxide (CuO)-modified nanoceria electrode [137]. With the levels of glucose increased, the electron current between electron increased. This biosensor exhibited an LOD of 10 μM for glucose detection. Another study used polyaniline (PANI) decorated nanoceria nanozyme as biosensor to detect glucose, and its LOD was estimated to be 0.56 μM [138]. Moreover, nanoceria combined with other peroxidase-like nanozymes (e.g., Fe_3O_4 and Co_3O_4) showed synergistic and enhanced ability to detect glucose as recently reported [139, 140].

In addition to chemical molecules, nanoceria-based biosensors are also developed to detect small biomolecules, such as dopamine and histamine. In many acute neurological diseases including Parkinson's and Schizophrenia, the concentrations of dopamine often exhibit pathological changes [141–143]. Hence, it is valuable to develop sensitive biosensors to detect dopamine. However, due to the disturbance of some small molecules, such as uric acid (UA) and ascorbic acid (AA), it is difficult to detect the low concentration of dopamine in the extracellular fluid of the central nervous system. Nayak et al. demonstrated that modified graphene nanosheets with nafion-coated nanoceria nanozymes improved LOD of dopamine detection from 2.64 to 1 μM under the interfere of disturbing species as compared with previous study in which only graphene was used [144]. Histamine is an important compound associated with immune system which acts as a mediator of itching and a biomarker of allergic responses [145, 146]. Using polyaniline as the shell of ceria nanoparticle, a histamine sensing method was developed with the presence of diamine oxidase (DAO) [147]. In the aerobic environment, histamine is catalyzed by DAO into imidazole acetaldehyde and peroxide. Peroxide is then catalyzed by nanoceria nanozyme-based electrodes and turns into oxygen in the cathodic process, and by this time the detection signal is being detected. This biosensor showed an LOD of 48.7 μM and a linearity of 0.45 to 1.05 mM for histamine detection.

In addition to the abovementioned studies, some other small biomolecules, such as L-glutamic acid [148], urea [149], thrombin [150], melamine [151], and biomarker of influenza [152], have also been detected by nanoceria nanozyme-based biosensors.

9.4.2 Sensing of Biomacromolecules

Cerium-based nanozymes have been used to develop as biosensors to recognize biomacromolecules as well, including nucleic acids and proteins. Nanoceria nanozymes immobilize DNA probe via the non-covalent interaction (π–π stacking) between DNA bases and their conjugated interface [126]. Employing this property, Feng et al. synthesized a porous CeO_2/chitosan nanocluster to immobilize single-strand DNA (ssDNA) for the recognition of colorectal cancer gene [153]. In this research, methylene blue (MB) was used as a hybridization indicator which could reduce electron current when exposed to free guanine bases. Increased current signal would be detected if MB was released by the combination of ssDNA and its target gene. Similar with this study, a research used gold nanoparticles (GNPs) modified with multi-walled carbon nanotubes (MWCNTs), nanoceria, and chitosan to detect the BCR/ABL fusion gene in chronic myelogenous leukemia (CML) [154]. MB was also used as an electroactive indicator as the previous study. When target gene was captured by the nanocomplex, the detached MB increased the electron current peak and then showed the detection signal (Fig. 9.19). The results demonstrated that the

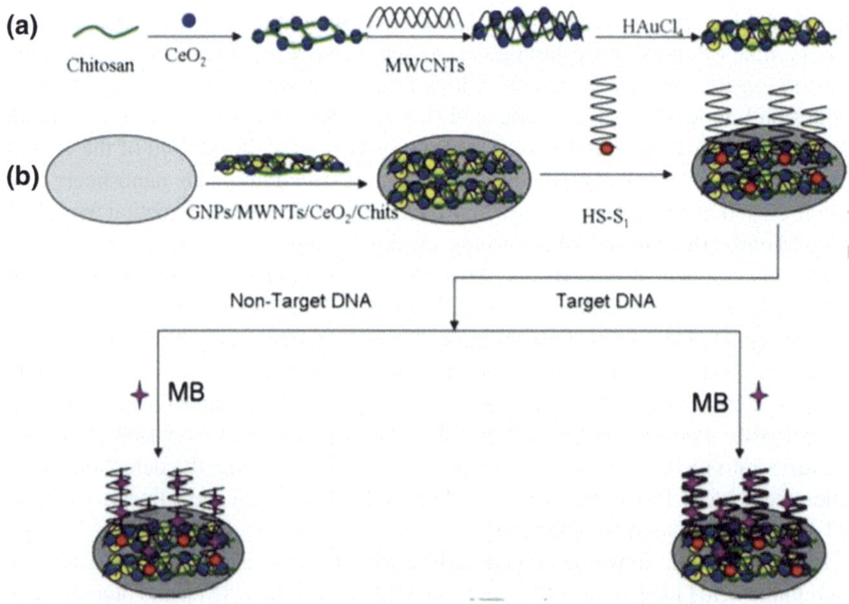

Fig. 9.19 Schematic representation of an efficient DNA electrochemical biosensor, based on the GNPs/MWCNTs/CeO_2/Chitosan nanocomposite for the detection of BCR/ABL fusion gene in chronic myelogenous leukemia (CML) [154]. In the presence of BCR/ABL fusion gene, the hybridization between target and the DNA probe occurred and the exposed free guanine bases which are the sites of MB attachment were remarkably reduced followed by reduction in the amount of MB binding in the dsDNA. Thus, detached MB increased the current peak and demonstrated the presence of BCR/ABL gene. Copyright 2013 Elsevier

presence of nanoceria nanozymes enhanced the signal up to 12 times when compared to the condition without nanoceria. In the optimized situations, this biosensor showed a linear range of 1 pM to 1 nM and an LOD of 0.5 pM for detection of the BCR/ABL fusion gene.

Zhang et al. developed a phosphoenolpyruvate carboxylase (PEPCase) gene detection method using nanoceria–carbon complex which was supported by ionic liquid (IL), a kind of nonaqueous but polar solvent [155]. It is the first study of putting ionic liquid and nanoparticles together to synthesize a DNA electrochemical biosensor. This method possesses many special advantages including advanced stabilities, negligible vapor pressure, high electron conductivity, wide electrochemical windows, reduced toxicity, and wonderful solubility. This ceria-based nanocomposite exhibited an LOD of 230 fM and a linearity of 1 pM to 0.1 μM for PEPCase gene detection.

Recently, Andreescu et al. introduced a nanoceria-based fluorescence quencher to rapidly and sensitively detect target DNA [156]. They demonstrated that nanoceria nanozymes were conjugated with oligonucleotides, which quenched their fluorescence in the absence of target DNA. However, the fluorescent tag would be released from the surface of nanoceria under the competitive binding of the target DNA. With the concentration of target DNA increased, the fluorescent signal was also enhanced (Fig. 9.20). Using a single-step method, the LOD of this nanoceria biosensor for fluorescent DNA detection was determined to be 0.12 nM.

Besides, based on the interaction of nanoceria and DNA, Xiao et al. recently developed a detection method of ATP [34]. Carboxyfluorescein (FAM)-labeled DNA was attached to the surface of nanoceria nanozymes and subsequently displaced by ATP via the stronger coordination activity of ATP with nanoceria. This method showed a wonderful liner range from 0.1 nM to 1.5 μM and an LOD of 54 pM for ATP detection.

Nanoceria nanozymes own a high isoelectric point (IEP ~ 9.2), which makes them possible to immobilize protein (such as antibody and enzyme) via electrostatic interactions. Based on this principle, the antibody of myeloperoxidase (MPO) was

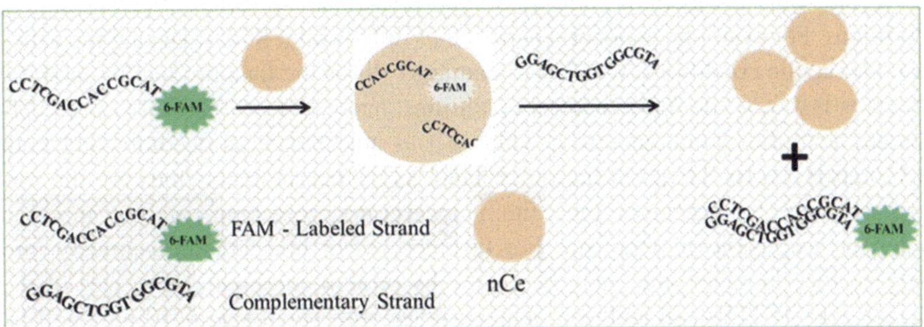

Fig. 9.20 Schematic representation of the detection mechanism showing target hybridization to nanoceria-conjugated FAM-ssDNA [156]. Upon hybridization, the fluorescent FAM probe is released from the particle surface, restoring the fluorescent signal. Copyright 2018 Nature Publishing Group

immobilized by nanoceria and nanogold-coated electrode for sensing MPO enzyme. After incubated with MPO solution for 10 min, this biosensor exhibited a reduced electron current. This method of detect MPO enzyme exhibited a linear range of 10–400 ng/mL and an LOD of 0.06 ng/mL [157].

In summary, nanoceria nanozymes possess the advantages of nano-morphological, multiple functionalization, biocompatible, and catalytic properties, which render them as promising candidates to develop ideal biosensor for detecting biochemical targets. However, the catalytic activity and stability of nanoceria nanozymes still need to be optimized when developed nanoceria nanozyme-based biosensor.

9.5 Future perspectives

This chapter reviews the studies and applications of cerium-based nanozymes. As described in Sect. 9.2, nanoceria nanozymes have been found to possess multiple enzyme-like properties including SOD, CAT, peroxidase, oxidase, phosphatase, and ATPase-mimetic activities. Based on their fabulous catalytic function, nanoceria nanozymes are already being studied as potential biosensors for detecting biochemicals and biomolecules (see Sect. 9.4), and as therapeutic compounds for a variety of ROS/NOS-related disorders and diseases including oxidative stress, chronic inflammation, ocular diseases, neurodegenerative diseases, cancers, diabetes, etc. (see Sect. 9.3).

The biomedical applications of nanoceria nanozymes are developing rapidly in the last decades. Although the catalytic mechanism of nanoceria has been demonstrated deeply in vitro, the behaviors of nanoceria in vivo are still not clear enough. On the one hand, some controversy exists when referring to the effects of physiological conditions on nanoceria enzymatic activities. Studies indicated that the enzyme-mimetic properties of nanoceria nanozymes are influenced by multiple factors such as nanoparticle diameter, chemical composition, and pH value [21–23]. Nanoceria nanozymes may act as both antioxidant and prooxidant in different physiological conditions. For example, nanoceria nanozymes only exhibit SOD-like enzyme activity under acidic conditions, which will damage cells as the final product of catalytic effects is H_2O_2, while under neutral conditions, nanoceria exhibit SOD- and CAT-mimetic activities simultaneously, which act as antioxidant as decomposing both $O_2^{\bullet -}$ and H_2O_2 [6, 113]. On the other hand, as nanoceria nanozymes are the foreign substance for organism, thus, the nanoceria may lose their catalytic function under the exclusion of organism. For instance, when injected intravenously into body, nanoceria nanozymes may be coated by protein corona and lose part of the surface activity [159]. In addition, nanoparticles with small size (<10 nm) will be eliminated by the kidney, while large nanoparticles are easier to be digest by mononuclear phagocyte system (MPS) [160, 161]. Besides, after being internalized by cell, nanoceria nanozymes may interact with cellular components including nucleic acids [162, 163], proteins [164], and ATP [165], thereby affecting their catalytic activity. At the same time, nanoceria nanozymes may also be cytotoxic by disturbing cell homeostasis. A recent study systematically evaluated the toxicity of nanoceria on multiple organs. The results demonstrated that nanoceria nanozymes may cause oxidative stress, inflammation, and DNA damage in various organs by exposing to the lung of mice [166]. Further studies should focus on the construction of novel nanoceria materials and use them to clarify the long-term effects of nanoceria in organism.

In conclusion, nanoceria nanozymes have built a solid bridge between nanotechnology and biology. It is worth expecting cerium-based nanozymes to be brilliant in biomedical applications in the future. Nanoceria nanozymes possess good biocompatibility, high regeneration ability, and excellent catalytic activity, which attract a lot of attention. Compared with other nanozymes (e.g., nano-Fe_3O_4, nanogold, nano-V_2O_5), the enzyme-like mechanism and biological applications of nanoceria nanozymes have been carried out more extensively and deeply. The research processes of nanoceria nanozymes may serve as good examples for inspiring other nanozymes research to move on. However, the catalytic activity and the long-term biological effect of nanoceria nanozymes in vivo are still unclear. From laboratory research to clinical application, there is still a long way for nanoceria nanozymes to go. Since the studies of nanoceria nanozymes are too many to be discussed here completely, we recommend some excellent reviews to help the readers get more comprehensive information [19, 20, 126, 167–175].

Acknowledgement This work was financially supported by the National Natural Science Foundation of China (No. 81930050, 31871005, 31530026, 31900981), Chinese Academy of Sciences under Grant No. YJKYYQ20180048, the Strategic Priority Research Program (No. XDB29040101), the Key Research Program of Frontier Sciences (No. QYZDY-SSW-SMC013), Chinese Academy of Sciences and National Key Research and Development Program of China (No. 2017YFA0205501), and Youth Innovation Promotion Association CAS (2019093).

References

1. Reinhardt K, Winkler H (2000) Cerium mischmetal, cerium alloys, and cerium compounds. Wiley-VCH Verlag GmbH & Co, KGaA
2. Wang L, Zhang K, Song Z, Feng S (2007) Ceria concentration effect on chemical mechanical polishing of optical glass. Appl Surf Sci 253(11):4951–4954. https://doi.org/10.1016/j.apsusc.2006.10.074
3. Zhang Z, Yu L, Liu W, Song Z (2010) Surface modification of ceria nanoparticles and their chemical mechanical polishing behavior on glass substrate. Appl Surf Sci 256(12):3856–3861. https://doi.org/10.1016/j.apsusc.2010.01.040
4. Trovarelli A, Leitenburg CD, Boaro M, Dolcetti G (1999) The utilization of ceria in industrial catalysis. Catal Today 50(2):353–367
5. Trovarelli A (1996) Catalytic properties of ceria and CeO_2-containing materials. Catal Rev Sci Eng 38(4):439–520. https://doi.org/10.1080/01614949608006464

6. Can XU, Xiaogang QU (2014) Recent progress of rare earth cerium oxide nanoparticles applied in biology. Sci Sinica 44(4):506
7. Finkel T, Holbrook NJ (2000) Oxidants, oxidative stress and the biology of ageing. Nature 408(6809):239
8. Johansson B, Luo W, Li S, Ahuja R (2014) Cerium; crystal structure and position in the periodic table. Sci Rep 4:6398
9. Holleman AF, Wiberg E, Eagleson M, Brewer W, Aylett BJ, Wiberg N (2001) Inorganic chemistry. Academic Press
10. Skorodumova NV, Simak SI, Lundqvist BI, Abrikosov IA, Johansson B (2002) Quantum origin of the oxygen storage capability of ceria. Phys Rev Lett 89(16):166601
11. Suzuki T, Kosacki I, Anderson HU, Colomban P (2001) Electrical conductivity and lattice defects in nanocrystalline cerium oxide thin films. J Am Ceram Soc 84(9):2007–2014
12. Deshpande S, Patil S, Kuchibhatla SV, Seal S (2005) Size dependency variation in lattice parameter and valency states in nanocrystalline cerium oxide. Appl Phys Lett 87(13):223–278
13. Ying JY, Tschöpe A (1996) Synthesis and characteristics of non-stoichiometric nanocrystalline cerium oxide-based catalysts. Chem Eng J Biochem Eng J 64(2):225–237
14. Hayyan M, Hashim MA, Alnashef IM (2016) Superoxide ion: generation and chemical implications. Chem Rev 116(5):3029
15. Mccord JM, Fridovich I (1969) Superoxide dismutase an enzymic function for erythrocuprein (hemocuprein). J Biol Chem 244(22):6049
16. Ivanov VK, Shcherbakov AB, Usatenko AV (2010) ChemInform abstract: structure-sensitive properties and biomedical applications of nanodispersed cerium dioxide. Cheminform 41(9):no–no
17. Korsvik C, Patil S, Seal S, Self WT (2007) Superoxide dismutase mimetic properties exhibited by vacancy engineered ceria nanoparticles. Chem Commun 10(10):1056
18. Pirmohamed T, Dowding JM, Singh S, Wasserman B, Heckert E, Karakoti AS, King JE, Seal S, Self WT (2010) Nanoceria exhibit redox state-dependent catalase mimetic activity. Chem Commun (Camb) 46(16):2736–2738. https://doi.org/10.1039/b922024k
19. Nelson BC, Johnson ME, Walker ML, Riley KR, Sims CM (2016) Antioxidant cerium oxide nanoparticles in biology and medicine. Antioxidants (Basel) 5(2). https://doi.org/10.3390/antiox5020015
20. Celardo I, Pedersen JZ, Traversa E, Ghibelli L (2011) Pharmacological potential of cerium oxide nanoparticles. Nanoscale 3(4):1411
21. Korsvik C, Patil S, Seal S, Self WT (2007) Superoxide dismutase mimetic properties exhibited by vacancy engineered ceria nanoparticles. Chem Commun (Camb) 10:1056–1058. https://doi.org/10.1039/b615134e
22. Singh S, Dosani T, Karakoti AS, Kumar A, Seal S, Self WT (2011) A phosphate-dependent shift in redox state of cerium oxide nanoparticles and its effects on catalytic properties. Biomaterials 32(28):6745–6753. https://doi.org/10.1016/j.biomaterials.2011.05.073
23. Alili L, Sack M, Karakoti AS, Teuber S, Puschmann K, Hirst SM, Reilly CM, Zanger K, Stahl W, Das S, Seal S, Brenneisen P (2011) Combined cytotoxic and anti-invasive properties of redox-active nanoparticles in tumor-stroma interactions. Biomaterials 32(11):2918–2929. https://doi.org/10.1016/j.biomaterials.2010.12.056
24. Wei X, Li X, Feng Y, Yang S (2018) Morphology- and pH-dependent peroxidase mimetic activity of nanoceria. Rsc Advances 8(21):11764–11770. https://doi.org/10.1039/c8ra00622a
25. Fisher TJ, Zhou Y, Wu T-S, Wang M, Soo Y-L, Cheung CL (2019) Structure-activity relationship of nanostructured ceria for the catalytic generation of hydroxyl radicals. Nanoscale 11(10):4552–4561. https://doi.org/10.1039/c8nr09393h

26. Zhang B, Yu H, Wang J, Wang W, Zhang Q, Zhang H (2019) Synthesis of CeO_2 nanoparticles with different morphologies and their properties as peroxidase mimic. J Am Ceram Soc 102(4):2218–2227. https://doi.org/10.1111/jace.16071
27. Heckert EG, Karakoti AS, Seal S, Self WT (2008) The role of cerium redox state in the SOD mimetic activity of nanoceria. Biomaterials 29(18):2705–2709. https://doi.org/10.1016/j.biomaterials.2008.03.014
28. Rzigalinski BA, Meehan K, Davis RM, Xu Y, Miles WC, Cohen CA (2006) Radical nanomedicine. Nanomedicine 1(4):399
29. Baldim V, Bedioui F, Mignet N, Margaill I, Berret JF (2018) The enzyme-like catalytic activity of cerium oxide nanoparticles and its dependency on Ce^{3+} surface area concentration. Nanoscale 10(15):6971–6980. https://doi.org/10.1039/c8nr00325d
30. Dhall A, Self WT (2018) Cerium oxide nanoparticles: a brief review of their synthesis methods and biomedical applications. 7(8)
31. Das S, Dowding JM, Klump KE, Mcginnis JF, Self W, Seal S (2013) Cerium oxide nanoparticles: applications and prospects in nanomedicine. Nanomedicine 8(9):1483–1508
32. Karakoti AS, Munusamy P, Hostetler K, Kodali V, Kuchibhatla S, Orr G, Pounds JG, Teeguarden JG, Thrall BD, Baer DR (2012) Preparation and characterization challenges to understanding environmental and biological impacts of nanoparticles. Surface Interface Anal Sia 44(8):882–889
33. Kumar A, Babu S, Karakoti AS, Schulte A, Seal S (2009) Luminescence properties of europium-doped cerium oxide nanoparticles: role of vacancy and oxidation states. Langmuir 25(18):10998–11007
34. Patel V, Singh M, Mayes ELH, Martinez A, Shutthanandan V, Bansal V, Singh S, Karakoti AS (2018) Ligand-mediated reversal of the oxidation state dependent ROS scavenging and enzyme mimicking activity of ceria nanoparticles. Chem Commun 54(99):13973–13976. https://doi.org/10.1039/c8cc08355j
35. Sun C, Li H, Chen L (2012) ChemInform abstract: nanostructured ceria-based materials: synthesis, properties, and applications. Cheminform 5(9):8475–8505
36. Xue Y, Zhai Y, Zhou K, Wang L, Tan H, Luan Q, Yao X (2012) The vital role of buffer anions in the antioxidant activity of CeO_2 nanoparticles. Chemistry 18(35):11115–11122
37. Das M, Patil S, Bhargava N, Kang JF, Riedel LM, Seal S, Hickman JJ (2007) Auto-catalytic ceria nanoparticles offer neuroprotection to adult rat spinal cord neurons. Biomaterials 28(10):1918–1925. https://doi.org/10.1016/j.biomaterials.2006.11.036
38. Zhang Y, Zhou K, Zhai Y, Qin F, Pan L, Yao X (2014) Crystal plane effects of nano-CeO_2 on its antioxidant activity. Rsc Advances 4(92):50325–50330
39. Perez JM, Asati A, Nath S, Kaittanis C (2008) Synthesis of biocompatible dextran-coated nanoceria with pH-dependent antioxidant properties. Small 4(5):552–556. https://doi.org/10.1002/smll.200700824
40. Dowding JM, Das S, Kumar A, Dosani T, Mccormack R, Gupta A, Sayle TXT, Sayle DC, Kalm LV, Seal S (2013) Cellular interaction and toxicity depends on physiochemical properties and surface modification of redox active nanomaterials. ACS Nano 7(6):4855–4868
41. Dowding JM, Dosani T, Kumar A, Seal S, Self WT (2012) Cerium oxide nanoparticles scavenge nitric oxide radical ($^{\cdot}NO$). Chem Commun 48(40):4896–4898
42. Dowding JM, Seal S, Self WT (2013) Cerium oxide nanoparticles accelerate the decay of peroxynitrite (ONOO(-)). Drug Deliv Transl Res 3(4):375–379
43. Xue Y, Luan Q, Yang D, Yao X, Zhou K (2011) Direct evidence for hydroxyl radical scavenging activity of cerium oxide nanoparticles. J Phys Chem C 115(11):4433–4438
44. Faassen EEV, Bahrami S, Feelisch M, Hogg N, Kelm M, Kimshapiro DB, Kozlov AV, Li H, Lundberg JO, Mason R (2009) Nitrite as regulator of hypoxic signaling in mammalian physiology. Med Res Rev 29(5):683–741

45. Yun HY, Dawson VL, Dawson TM (2009) Nitric oxide in health and disease of the nervous system. Antioxid Redox Signal 11(3):541–554
46. Jiao X, Song H, Zhao H, Bai W, Zhang L, Lv Y (2012) Well-redispersed ceria nanoparticles: promising peroxidase mimetics for H_2O_2 and glucose detection. Anal Methods 4(10):3261. https://doi.org/10.1039/c2ay25511a
47. Xu C, Liu Z, Li W, Ren J, Qu X (2014) Nucleoside triphosphates as promoters to enhance nanoceria enzyme-like activity and for single-nucleotide polymorphism typing. Adv Func Mater 24(11):1624–1630
48. Asati A, Kaittanis C, Santra S, Perez JM (2011) pH-tunable oxidase-like activity of cerium oxide nanoparticles achieving sensitive fluorigenic detection of cancer biomarkers at neutral pH. Anal Chem 83(7):2547–2553. https://doi.org/10.1021/ac102826k
49. Asati A, Santra S, Kaittanis C, Nath S, Perez JM (2009) Oxidase-like activity of polymer-coated cerium oxide nanoparticles. Angew Chem Int Ed Engl 48(13):2308–2312. https://doi.org/10.1002/anie.200805279
50. Peng Y, Chen X, Yi G, Gao Z (2011) Mechanism of the oxidation of organic dyes in the presence of nanoceria. Chem Commun 47(10):2916
51. Kuchma MH, Komanski CB, Colon J, Teblum A, Masunov AE, Alvarado B, Babu S, Seal S, Summy J, Baker CH (2010) Phosphate ester hydrolysis of biologically relevant molecules by cerium oxide nanoparticles. Nanomedicine 6(6):738–744. https://doi.org/10.1016/j.nano.2010.05.004
52. Patil AJ, Krishna KR, Barron NJ, Mann S (2012) Cerium oxide nanoparticle-mediated self-assembly of hybrid supramolecular hydrogels. Chem Commun 48(64):7934–7936
53. Feng T, Zhang Y, Wang J, Wei J, Cai Y, Qian X (2008) An efficient method for dephosphorylation of phosphopeptides by cerium oxide. J Mass Spectrom JMS 43(5):628–632
54. Yao T, Tian Z, Zhang Y, Qu Y (2019) Phosphatase-like activity of porous nanorods of CeO_2 for the highly stabilized dephosphorylation under interferences. ACS Appl Mater Interfaces 11(1):195–201. https://doi.org/10.1021/acsami.8b17086
55. Korschelt K, Schwidetzky R, Pfitzner F, Strugatchi J, Schilling C, von der Au M, Kirchhoff K, Panthoefer M, Lieberwirth I, Tahir MN, Hess C, Meermann B, Tremel W (2018) CeO_{2-x} nanorods with intrinsic urease-like activity. Nanoscale 10(27):13074–13082. https://doi.org/10.1039/c8nr03556c
56. Heckert EG, Seal S, Self WT (2008) Fenton-Like reaction catalyzed by the rare earth inner transition metal cerium. Environ Sci Technol 42(13):5014–5019
57. Li X, Wilmanns M, Thornton J, Köhn M (2013) Elucidating human phosphatase-substrate networks. Sci Signal 6(275):rs10
58. Cohen P (2002) The origins of protein phosphorylation. Nat Cell Biol 4(5):127–130
59. Vinothkumar G, Arunkumar P, Mahesh A, Dhayalan A, Babu KS (2018) Size- and defect-controlled anti-oxidant enzyme mimetic and radical scavenging properties of cerium oxide nanoparticles. New J Chem 42(23):18810–18823. https://doi.org/10.1039/c8nj04435j
60. Tarnuzzer RW, Colon J, Patil S, Seal S (2005) Vacancy engineered ceria nanostructures for protection from radiation-induced cellular damage. Nano Lett 5(12):2573
61. Niu J, Azfer A, Rogers LM, Wang X, Kolattukudy PE (2007) Cardioprotective effects of cerium oxide nanoparticles in a transgenic murine model of cardiomyopathy. Cardiovasc Res 73(3):549–559
62. Pagliari F, Mandoli C, Forte G, Magnani E, Pagliari S, Nardone G, Licoccia S, Minieri M, Nardo PD, Traversa E (2012) Cerium oxide nanoparticles protect cardiac progenitor cells from oxidative stress. ACS Nano 6(5):3767
63. Bhargava N, Shanmugaiah V, Saxena M, Sharma M, Sethy NK, Singh SK, Balakrishnan K, Bhargava K, Das M (2016) Nanocerium oxide increases the survival of adult rod and cone photoreceptor in culture by abrogating hydrogen peroxide-induced oxidative stress. Biointerphases 11(3):031016. https://doi.org/10.1116/1.4962263

64. Kong L, Cai X, Zhou X, Wong LL, Karakoti AS, Seal S, McGinnis JF (2011) Nanoceria extend photoreceptor cell lifespan in tubby mice by modulation of apoptosis/survival signaling pathways. Neurobiol Dis 42(3):514–523. https://doi.org/10.1016/j.nbd.2011.03.004
65. Wong LL, Pye QN, Chen L, Seal S, McGinnis JF (2015) Defining the catalytic activity of nanoceria in the P23H-1 rat, a photoreceptor degeneration model. PLoS ONE 10(3):e0121977. https://doi.org/10.1371/journal.pone.0121977
66. Cai X, Sezate SA, Seal S, Mcginnis JF (2012) Sustained protection against photoreceptor degeneration in tubby mice by intravitreal injection of nanoceria. Biomaterials 33(34):8771–8781
67. Wong LL, Hirst SM, Pye QN, Reilly CM, Seal S, McGinnis JF (2013) Catalytic nanoceria are preferentially retained in the rat retina and are not cytotoxic after intravitreal injection. PLoS ONE 8(3):e58431. https://doi.org/10.1371/journal.pone.0058431
68. Chen J, Patil S, Seal S, Mcginnis JF (2006) Rare earth nanoparticles prevent retinal degeneration induced by intracellular peroxides. Nat Nanotechnol 1(2):142–150
69. Singh S (2018) Investigating the role of catalase mimetic cerium oxide-based nanozyme to impart protection to hepatic cells from acatalasia. Free Radic Biol Med 128:S57–S58. https://doi.org/10.1016/j.freeradbiomed.2018.10.109
70. Singh R, Singh S (2019) Redox-dependent catalase mimetic cerium oxide-based nanozyme protect human hepatic cells from 3-AT induced acatalasemia. Colloids Surf B-Biointerfaces 175:625–635. https://doi.org/10.1016/j.colsurfb.2018.12.042
71. Singh R, Singh S (2019) Catalytically active cerium oxide nanoparticles protect mammalian cells from endogenous reactive oxygen species. Mater Today Proc 10:25–31. https://doi.org/10.1016/j.matpr.2019.02.184
72. Colon J, Hsieh N, Ferguson A, Kupelian P, Seal S, Jenkins DW, Baker CH (2010) Cerium oxide nanoparticles protect gastrointestinal epithelium from radiation-induced damage by reduction of reactive oxygen species and upregulation of superoxide dismutase 2. Nanomedicine 6(5):698–705. https://doi.org/10.1016/j.nano.2010.01.010
73. Chaudhury K, Babu KN, Singh AK, Das S, Kumar A, Seal S (2013) Mitigation of endometriosis using regenerative cerium oxide nanoparticles. Nanomedicine 9(3):439–448. https://doi.org/10.1016/j.nano.2012.08.001
74. Dowding JM, Song W, Bossy K, Karakoti A, Kumar A, Kim A, Bossy B, Seal S, Ellisman MH, Perkins G, Self WT, Bossy-Wetzel E (2014) Cerium oxide nanoparticles protect against Abeta-induced mitochondrial fragmentation and neuronal cell death. Cell Death Differ 21(10):1622–1632. https://doi.org/10.1038/cdd.2014.72
75. Schubert D, Dargusch R, Raitano J, Chan SW (2006) Cerium and yttrium oxide nanoparticles are neuroprotective. Biochem Biophys Res Commun 342(1):86–91
76. Estevez AY, Boateng Y, Lipps J, Stadler B, Erlichman JS (2018) Analysis of the antioxidant enzyme-mimetic activity and neuroprotective effects of cerium oxide nanoparticles stabilized with varying ratios of citric acid and EDTA. Faseb J 32(1)
77. Zhang Q, Ge K, Duan J, Chen S, Zhang R, Zhang C, Wang S, Zhang J (2014) Cerium oxide nanoparticles protect primary mouse bone marrow stromal cells from apoptosis induced by oxidative stress. J Nanopart Res 16(11). https://doi.org/10.1007/s11051-014-2697-3
78. Forte G, Carotenuto F, Pagliari F, Pagliari S, Cossa P, Fiaccavento R, Ahluwalia A, Vozzi G, Vinci B, Serafino A (2008) Criticality of the biological and physical stimuli array inducing resident cardiac stem cell determination. Stem Cells 26(8):2093–2103
79. Pagliari S, Vilela-Silva AC, Forte G, Pagliari F, Mandoli C, Vozzi G, Pietronave S, Prat M, Licoccia S, Ahluwalia A (2011) Cooperation of biological and mechanical signals in cardiac progenitor cell differentiation. Adv Mater 23(4):514–518

80. Lord MS, Jung M, Teoh WY, Gunawan C, Vassie JA, Amal R, Whitelock JM (2012) Cellular uptake and reactive oxygen species modulation of cerium oxide nanoparticles in human monocyte cell line U937. Biomaterials 33(31):7915–7924
81. Vassie JA, Whitelock JM, Lord MS (2017) Endocytosis of cerium oxide nanoparticles and modulation of reactive oxygen species in human ovarian and colon cancer cells. Acta Biomater 50:127–141. https://doi.org/10.1016/j.actbio.2016.12.010
82. Singh VK, Mehrotra S, Narayan P, Pandey CM, Agarwal SS (2000) Modulation of autoimmune diseases by nitric oxide. Immunol Res 22(1):1–19
83. Sabroe I, Parker LC, Calverley PM, Dower SK, Whyte MK (2007) Pathological networking: a new approach to understanding COPD. Thorax 62(8):733–738
84. Balboa MA, Balsinde J (2006) Oxidative stress and arachidonic acid mobilization. Biochem Biophys Acta 1761(4):385
85. Olmedo DG, Tasat DR, Evelson P, Guglielmotti MB, Cabrini RL (2008) Biological response of tissues with macrophagic activity to titanium dioxide. J Biomed Mater Res Part A 84a(4):1087–1093
86. Ahsan H, Ali A, Ali R (2003) Oxygen free radicals and systemic autoimmunity. Clin Exp Immunol 131(3):398–404
87. Hirst SM, Karakoti AS, Tyler RD, Sriranganathan N, Seal S, Reilly CM (2009) Anti-inflammatory properties of cerium oxide nanoparticles. Small 5(24):2848–2856. https://doi.org/10.1002/smll.200901048
88. Bärtsch P, Straub PW, Haeberli A (2001) Hypobaric hypoxia. Lancet 357(9260):955–955
89. Arya A, Sethy NK, Singh SK, Das M, Bhargava K (2013) Cerium oxide nanoparticles protect rodent lungs from hypobaric hypoxia-induced oxidative stress and inflammation. Int J Nanomed 8:4507–4520. https://doi.org/10.2147/IJN.S53032
90. Rotter N, Ung F, Roy AK, Vacanti M, Eavey RD, Vacanti CA, Bonassar LJ (2005) Role for interleukin 1alpha in the inhibition of chondrogenesis in autologous implants using polyglycolic acid-polylactic acid scaffolds. Tissue Eng Part A 11(1–2):192–200
91. Ponnurangam S, O'Connell GD, Chernyshova IV, Wood K, Hung CT, Somasundaran P (2014) Beneficial effects of cerium oxide nanoparticles in development of chondrocyte-seeded hydrogel constructs and cellular response to interleukin insults. Tissue Eng Part A 20(21–22):2908–2919. https://doi.org/10.1089/ten.TEA.2013.0592
92. Aukrust P, Gullestad L, Ueland T, Damås JK, Yndestad A (2005) Inflammatory and anti-inflammatory cytokines in chronic heart failure: potential therapeutic implications. Ann Med 37(2):74
93. Grieve DJ, Byrne JA, Cave AC, Shah AM (2004) Role of oxidative stress in cardiac remodelling after myocardial infarction. Heart Lung Circ 13(2):132–138
94. Singer M, Deutschman CS, Seymour CW, Shankar-Hari M, Annane D, Bauer M, Bellomo R, Bernard GR, Chiche JD, Coopersmith CM, Hotchkiss RS, Levy MM, Marshall JC, Martin GS, Opal SM, Rubenfeld GD, van der Poll T, Vincent JL, Angus DC (2016) The third international consensus definitions for sepsis and septic shock (Sepsis-3). JAMA 315(8):801–810. https://doi.org/10.1001/jama.2016.0287
95. Annane D, Bellissant E, Cavaillon JM (2015) Septic shock. Lancet 365(9453):63–78
96. Larrosa M, Azorínortuño M, Yañezgascón MJ, Garcíaconesa MT, Tomásbarberán F, Espín JC (2011) Lack of effect of oral administration of resveratrol in LPS-induced systemic inflammation. Eur J Nutr 50(8):673–680
97. Madhumitha G, Saral AM (2011) Preliminary phytochemical analysis, antibacterial, antifungal and anticandidal activities of successive extracts of Crossandra infundibuliformis. Asian Pac J Trop Med 4(3):192–195

98. Roopan SM, Rohit Madhumitha G, Rahuman AA, Kamaraj C, Bharathi A, Surendra TV (2013) Low-cost and eco-friendly phyto-synthesis of silver nanoparticles using Cocos nucifera coir extract and its larvicidal activity. Ind Crops Prod 43(5):631–635
99. Ge L, Hu Q, Chen J, Shi M, Yang H, Zhu G (2017) Inhibition of TNF-alpha sepsis of lipopolysaccharide induction using nano cerium oxide system. Mater Sci Eng C, Mater Biol Appl 77:405–410. https://doi.org/10.1016/j.msec.2017.03.207
100. Liao Y (2007) Oxidative stress and diabetic retinopathy, vol 25
101. Hollyfield J, Bonilha V, Me Yang X, Shadrach K, Lu L, Ufret R, Salomon R, Perez V (2008) Oxidative damage-induced inflammation initiates age-related macular degeneration. Nat Med 14(2):194–198
102. Aslan M, Cort A, Yucel I (2008) Oxidative and nitrative stress markers in glaucoma. Free Radic Biol Med 45(4):367–376
103. Halliwell B (2006) Oxidative stress and neurodegeneration: where are we now? J Neurochem 97(6):1634–1658
104. Harman D (2003) The free radical theory of aging. Antioxid Redox Signal 5(5):557
105. Knott AB, Perkins G, Schwarzenbacher R, Bossy-Wetzel E (2008) Mitochondrial fragmentation in neurodegeneration. Nat Rev Neurosci 9(7):505–518
106. Estevez AY, Pritchard S, Harper K, Aston JW, Lynch A, Lucky JJ, Ludington JS, Chatani P, Mosenthal WP, Leiter JC (2011) Neuroprotective mechanisms of cerium oxide nanoparticles in a mouse hippocampal brain slice model of ischemia. Free Radic Biol Med 51(6):1155–1163
107. D'Angelo B, Santucci S, Benedetti E, Loreto SD, Phani R, Falone S, Amicarelli F, Ceru M, Cimini A (2009) Cerium oxide nanoparticles trigger neuronal survival in a human Alzheimer disease model by modulating BDNF pathway. Curr Nanosci 5(2)
108. Cimini A, D'Angelo B, Das S, Gentile R, Benedetti E, Singh V, Monaco AM, Santucci S, Seal S (2012) Antibody-conjugated PEGylated cerium oxide nanoparticles for specific targeting of Aβ aggregates modulate neuronal survival pathways. Acta Biomater 8(6):2056–2067
109. Heckman KL, DeCoteau W, Estevez A, Reed KJ, Costanzo W, Sanford D, Leiter JC, Clauss J, Knapp K, Gomez C, Mullen P, Rathbun E, Prime K, Marini J, Patchefsky J, Patchefsky AS, Hailstone RK, Erlichman JS (2013) Custom cerium oxide nanoparticles protect against a free radical mediated autoimmune degenerative disease in the brain. ACS Nano 7(12):10582–10596. https://doi.org/10.1021/nn403743b
110. Gliga AR, Edoff K, Caputo F, Kallman T, Blom H, Karlsson HL, Ghibelli L, Traversa E, Ceccatelli S, Fadeel B (2017) Cerium oxide nanoparticles inhibit differentiation of neural stem cells. Sci Rep 7(1):9284. https://doi.org/10.1038/s41598-017-09430-8
111. Colon J, Herrera L, Smith J, Patil S, Komanski C, Kupelian P, Seal S, Jenkins DW, Baker CH (2009) Protection from radiation-induced pneumonitis using cerium oxide nanoparticles. Nanomed Nanotechnol Biol Med 5(2):225–231
112. Giri S, Karakoti A, Graham RP, Maguire JL, Reilly CM, Seal S, Rattan R, Shridhar V (2013) Nanoceria: a rare-earth nanoparticle as a novel anti-angiogenic therapeutic agent in ovarian cancer. PLoS ONE 8(1):e54578. https://doi.org/10.1371/journal.pone.0054578
113. Wason MS, Colon J, Das S, Seal S, Turkson J, Zhao J, Baker CH (2013) Sensitization of pancreatic cancer cells to radiation by cerium oxide nanoparticle-induced ROS production. Nanomed Nanotechnol Biol Med 9(4):558–569
114. Bast RC, Urban N, Shridhar V, Smith D, Zhen Z, Skates S, Lu K, Liu J, Fishman D, Mills G (2002) Early detection of ovarian cancer: promise and reality. Cancer Treat Res 107(2):61
115. Friedlander ML (1998) Prognostic factors in ovarian cancer. Semin Oncol 25(3):305–314

116. Sack M, Alili L, Karaman E, Das S, Gupta A, Seal S, Brenneisen P (2014) Combination of conventional chemotherapeutics with redox-active cerium oxide nanoparticles–a novel aspect in cancer therapy. Mol Cancer Ther 13(7):1740–1749
117. Alili L, Sack M, Montfort CV, Carroll K, Giri S, Das S, Zanger K, Seal S, Brenneisen P (2013) Downregulation of tumor growth and invasion by redox-active nanoparticles. Antioxid Redox Signal 19(8):765–778
118. Denko NC (2008) Hypoxia, HIF1 and glucose metabolism in the solid tumour. Nat Rev Cancer 8(9):705
119. Gatenby RA, Gillies RJ (2004) Why do cancers have high aerobic glycolysis? Cancer 4(11):891–899
120. Kuphal S, Winklmeier A, Warnecke C, Bosserhoff AK (2010) Constitutive HIF-1 activity in malignant melanoma. Eur J Cancer 46(6):1159
121. Das J, Choi YJ, Han JW, Reza A, Kim JH (2017) Nanoceria-mediated delivery of doxorubicin enhances the anti-tumour efficiency in ovarian cancer cells via apoptosis. Sci Rep 7(1):9513. https://doi.org/10.1038/s41598-017-09876-w
122. Rahimi R, Nikfar S, Larijani B, Abdollahi M (2005) A review on the role of antioxidants in the management of diabetes and its complications. Biomed Pharmacother 59(7):365–373
123. Pourkhalili N, Hosseini A, Nili-Ahmadabadi A, Hassani S, Pakzad M, Baeeri M, Mohammadi-rad A, Abdollahi M (2011) Biochemical and cellular evidence of the benefit of a combination of cerium oxide nanoparticles and selenium to diabetic rats. World J Diabetes 2(11):204–210. https://doi.org/10.4239/wjd.v2.i11.204
124. Pourkhalili N, Hosseini A, Nili-Ahmadabadi A, Rahimifard M, Navaei-Nigjeh M, Hassani S, Baeeri M, Abdollahi M (2012) Improvement of isolated rat pancreatic islets function by combination of cerium oxide nanoparticles/sodium selenite through reduction of oxidative stress. Toxicol Methods 22(6):476
125. Charbgoo F, Soltani F, Taghdisi SM, Abnous K, Ramezani M (2016) Nanoparticles application in high sensitive aptasensor design. TrAC Trends Anal Chem 85:85–97
126. Charbgoo F, Ramezani M, Darroudi M (2017) Bio-sensing applications of cerium oxide nanoparticles: advantages and disadvantages. Biosens Bioelectron 96:33–43. https://doi.org/10.1016/j.bios.2017.04.037
127. Ujjain SK, Das A, Srivastava G, Ahuja P, Roy M, Arya A, Bhargava K, Sethy N, Singh SK, Sharma RK (2014) Nanoceria based electrochemical sensor for hydrogen peroxide detection. Biointerphases 9(3):031011
128. Yang X, Ouyang Y, Wu F, Hu Y, Ji Y, Wu Z (2017) Size controllable preparation of gold nanoparticles loading on graphene sheets@cerium oxide nanocomposites modified gold electrode for nonenzymatic hydrogen peroxide detection. Sens Actuat B Chem 238:40–47
129. Yang X, Ouyang Y, Wu F, Hu Y, Zhang H, Wu Z (2016) In situ & controlled preparation of platinum nanoparticles dopping into graphene sheets@cerium oxide nanocomposites sensitized screen printed electrode for nonenzymatic electrochemical sensing of hydrogen peroxide. J Electroanal Chem 777:85–91
130. Mu J, Zhao X, Li J, Yang EC, Zhao XJ (2017) Coral-like CeO_2/NiO nanocomposites with efficient enzyme-mimetic activity for biosensing application. Mater Sci Eng C, Mater Biol Appl Sens 74:434–442. https://doi.org/10.1016/j.msec.2016.12.037
131. Ozdemir Olgun FA, Uzer A, Ozturk BD, Apak R (2018) A novel cerium oxide nanoparticles-based colorimetric sensor using tetramethyl benzidine reagent for antioxidant activity assay. Talanta 182:55–61. https://doi.org/10.1016/j.talanta.2018.01.047
132. Sardesai NP, Ganesana M, Karimi A, Leiter JC, Andreescu S, Chem A (2015) Platinum-doped ceria based biosensor for in vitro and in vivo monitoring of lactate during hypoxia. Anal Chem 87(5):2996–3003
133. Sharan R, Dutta A (2017) Structural analysis of Zr^{4+} doped ceria, a possible material for ammonia detection in ppm level. J Alloy Compd 693:936–944

134. Khan SB, Faisal M, Rahman MM, Jamal A (2011) Exploration of CeO_2 nanoparticles as a chemi-sensor and photo-catalyst for environmental applications. Sci Total Environ 409(15):2987–2992
135. Library WP (2011) WHO model list of essential medicines
136. Liu Q, Ding Y, Yang Y, Zhang L, Sun L, Chen P, Gao C (2016) Enhanced peroxidase-like activity of porphyrin functionalized ceria nanorods for sensitive and selective colorimetric detection of glucose. Mater Sci Eng C, Mater Biol Appl Sens 59:445–453. https://doi.org/10.1016/j.msec.2015.10.046
137. Guan P, Li Y, Zhang J, Li W (2016) Non-enzymatic glucose biosensor based on CuO-decorated CeO_2 nanoparticles. Nanomaterials 6(9):159
138. Zhang J, Guan P, Li Y, Li W, Guo Q (2016) Polyaniline/cerium oxide hybrid modified carbon paste electrode for non-enzymatic glucose detection. Bull Korean Chem Soc 37(7):985–986
139. Huang F, Wang J, Chen W, Wan Y, Wang X, Cai N, Liu J, Yu F (2018) Synergistic peroxidase-like activity of CeO_2-coated hollow Fe_3O_4 nanocomposites as an enzymatic mimic for low detection limit of glucose. J Taiwan Inst Chem Eng 83:40–49. https://doi.org/10.1016/j.jtice.2017.12.011
140. Alizadeh N, Salimi A, Hallaj R (2019) Mimicking peroxidase-like activity of Co_3O_4–CeO_2 nanosheets integrated paper-based analytical devices for detection of glucose with smartphone. Sens Actuat B-Chem 288:44–52. https://doi.org/10.1016/j.snb.2019.01.068
141. Weinberger DR (1987) Implications of normal brain-development for the pathogenesis of schizophrenia. Arch Gen Psychiatry 44(7):660–669
142. Cummings JL (1992) Depression and Parkinsons-disease—a review. Am J Psychiatry 149(4):443–454
143. Spanagel R, Weiss F (1999) The dopamine hypothesis of reward: past and current status. Trends Neurosci 22(11):521–527. https://doi.org/10.1016/s0166-2236(99)01447-2
144. Nayak P, Santhosh PN, Ramaprabhu S (2015) Cerium oxide nanoparticles decorated graphene nanosheets for selective detection of dopamine. J Nanosci Nanotechnol 15(7):4855
145. Crespo LM, Oliveira NDD, Damatta RA, Nascimento VVD, Soares TP, Machado OLT (2016) Identification of IgE-binding peptide and critical amino acids ofJatropha curcasallergen involved in allergenic response. Springerplus 5(1):454
146. Andersen HH, Elberling J, Arendt-Nielsen L (2015) Human surrogate models of histaminergic and non-histaminergic itch. Acta dermato-venereologica 95(7):771
147. Gumpu MB, Nesakumar N, Sethuraman S, Krishnan UM, Rayappan JBB (2014) Development of electrochemical biosensor with ceria–PANI core–shell nano-interface for the detection of histamine. Sens Actuat B Chem 199(6):330–338
148. Re Ö, Ispas C, Ganesana M, Leiter JC, Andreescu S (2014) Glutamate oxidase biosensor based on mixed ceria and titania nanoparticles for the detection of glutamate in hypoxic environments. Biosens Bioelectron 52(4):397–402
149. Ansari AA, Azahar M, Malhotra BD (2012) Electrochemical urea biosensor based on sol-gel derived nanostructured cerium oxide, pp 5490–5498
150. Song HP, Jang JY, Bae SH, Choi SB, Yu BJ, Kim MI (2018) Convenient colorimetric detection of thrombin via aptamer-mediated inhibition and restoration of the oxidase activity of nanoceria. J Nanosci Nanotechnol 18(9):6570–6574. https://doi.org/10.1166/jnn.2018.15696
151. Jin X, Yin W, Ni G, Peng J (2018) Hydrogen-bonding-induced colorimetric detection of melamine based on the peroxidase activity of gelatin-coated cerium oxide nanospheres. Anal Methods 10(8):841–847. https://doi.org/10.1039/c7ay02296d
152. Nair H, Brooks WA, Katz M, Roca A, Berkley JA, Madhi SA, Simmerman JM, Gordon A, Sato M, Howie S (2011) Global burden of respiratory infections due to seasonal influenza in young children: a systematic review and meta-analysis. Lancet 378(9807):1917
153. Feng KJ, Yang YH, Wang ZJ, Jiang JH, Shen GL, Yu RQ (2006) A nano-porous CeO_2/Chitosan composite film as the immobilization matrix for colorectal cancer DNA sequence-selective electrochemical biosensor. Talanta 70(3):561–565

154. Li S, Wang L, Li Y, Zhu X, Liang Z, Lu L, Zhang W, Liu B, Xie G, Feng W (2013) Electrochemical determination of BCR/ABL fusion gene based on in situ synthesized gold nanoparticles and cerium dioxide nanoparticles. Colloids Surf, B 112(12):344–349
155. Zhang W, Yang T, Zhuang X, Guo Z, Jiao K (2009) An ionic liquid supported CeO_2 nanoshuttles-carbon nanotubes composite as a platform for impedance DNA hybridization sensing. Biosens Bioelectron 24(8):2417
156. Bulbul G, Hayat A, Mustafa F, Andreescu S (2018) DNA assay based on nanoceria as fluorescence quenchers (NanoCeracQ DNA assay). Sci Rep 8(1):2426. https://doi.org/10.1038/s41598-018-20659-9
157. Lu L, Liu B, Li S, Zhang W, Xie G (2011) Improved electrochemical immunosensor for myeloperoxidase in human serum based on nanogold/cerium dioxide-BMIMPF6/L-Cysteine composite film. Colloids Surf B Biointerfaces 86(2):339–344
158. Li F, Hu X, Wang F, Zheng B, Du J, Xiao D (2018) A fluorescent "on-off-on" probe for sensitive detection of ATP based on ATP displacing DNA from nanoceria. Talanta
159. Bargheer D, Nielsen J, Gã©Bel G, Heine M, Salmen SC, Stauber R, Weller H, Heeren J, Nielsen P (2015) The fate of a designed protein corona on nanoparticles in vitro and in vivo. Beilstein J Nanotechnol 6(1):36–46
160. Feliu N, Docter D, Heine M, Del PP, Ashraf S, Kolosnjaj-Tabi J, Macchiarini P, Nielsen P, Alloyeau D, Gazeau F (2016) In vivo degeneration and the fate of inorganic nanoparticles. Chem Soc Rev 45(9):2440
161. Arami H, Khandhar A, Liggitt D, Krishnan KM (2015) In vivo delivery, pharmacokinetics, biodistribution and toxicity of iron oxide nanoparticles. Chem Soc Rev 44(23):8576–8607
162. Lu C, Huang Z, Liu B, Liu Y, Ying Y, Liu J (2017) Poly-cytosine DNA as a high-affinity ligand for inorganic nanomaterials. Ange Chemie 56(22)
163. Yang D, Fa M, Gao L, Zhao R, Luo Y, Yao X (2018) The effect of DNA on the oxidase activity of nanoceria with different morphologies. Nanotechnology 29(38). https://doi.org/10.1088/1361-6528/aacf86
164. Zhao F (2011) Cellular uptake, intracellular trafficking, and cytotoxicity of nanomaterials. Small 7(10):1322–1337
165. Shah J, Purohit R, Singh R, Karakoti AS, Singh S (2015) ATP-enhanced peroxidase-like activity of gold nanoparticles. J Colloid Interface Sci 456:100–107
166. Nemmar A, Yuvaraju P, Beegam S, Fahim MA, Ali BH (2017) Cerium oxide nanoparticles in lung acutely induce oxidative stress, inflammation, and DNA damage in various organs of mice. Oxid Med Cell Longev 2017:9639035. https://doi.org/10.1155/2017/9639035
167. Karakoti A, Singh S, Dowding JM, Seal S, Self WT (2010) Redox-active radical scavenging nanomaterials. Chem Soc Rev 39(11):4422–4432. https://doi.org/10.1039/b919677n
168. Celardo I, Traversa E, Ghibelli L (2011) Cerium oxide nanoparticles: a promise for applications in therapy. J Exp Therap Oncol 9(1):47
169. Estevez AY, Erlichman JS (2014) The potential of cerium oxide nanoparticles (nanoceria) for neurodegenerative disease therapy. Nanomedicine 9(10):1437–1440
170. Wong LL, McGinnis JF (2014) Nanoceria as bona fide catalytic antioxidants in medicine: what we know and what we want to know. Adv Exp Med Biol 801:821–828. https://doi.org/10.1007/978-1-4614-3209-8_103
171. Yokel RA, Hussain S, Garantziotis S, Demokritou P, Castranova V, Cassee FR (2014) The Yin: an adverse health perspective of nanoceria: uptake, distribution, accumulation, and mechanisms of its toxicity. Environ Sci Nano 1(5):406–428. https://doi.org/10.1039/C4EN00039K
172. Walkey C, Das S, Seal S, Erlichman J, Heckman K, Ghibelli L, Traversa E, McGinnis JF, Self WT (2015) Catalytic properties and biomedical applications of cerium oxide nanoparticles. Environ Sci Nano 2(1):33–53. https://doi.org/10.1039/C4EN00138A
173. Cai X, McGinnis JF (2016) Nanoceria: a potential therapeutic for dry AMD. Adv Exp Med Biol 854:111–118. https://doi.org/10.1007/978-3-319-17121-0_16

174. Singh S (2016) Cerium oxide based nanozymes: Redox phenomenon at biointerfaces. Biointerphases 11(4):04B202. https://doi.org/10.1116/1.4966535
175. Naz S, Beach J, Heckert B, Tummala T, Pashchenko O, Banerjee T, Santra S (2017) Cerium oxide nanoparticles: a 'radical' approach to neurodegenerative disease treatment. Nanomedicine (London, England) 12(5):545–553. https://doi.org/10.2217/nnm-2016-0399

Chapter 10
Noble Metal-Based Nanozymes

Shuangfei Cai and Rong Yang

Abbreviations

AA	Ascorbic acid
ABTS	2,2′-Azino-bis(3-ethylbenzo-thiazoline-6-sulfonic acid) diammonium salt
ACP	Acid phosphatase
ADP	Adenosine diphosphate disodium salt
apoFT	Apoferritin
ATP	Adenosine triphosphate disodium salt
BMPO	5-tert-bytoxycarbonyl-5-methyl-1-pyrroline-N-oxide
BNMs	Bimetallic nanomaterials
BNPs	Bimetallic nanoparticles
BSA	Bovine serum albumin
CEA	Carcinoembryonic antigen
CTAB	Cetyltrimethylammonium bromide
CTPO	3-Carbamoyl-2,2,5,5-tetramethyl-3-pyrroline-1-yloxyl
D	Dimensional
DMPO	Dimethyl pyridine N-oxide
DNA	Deoxyribonucleic acid
ELISA	Enzyme linked immunosorbent assay
ESR	Electron spin resonance
FDA	Food and Drug Administration

S. Cai · R. Yang (✉)
CAS Key Lab for Biomedical Effects of Nanomaterials and Nanosafety, Center of Materials Science and Optoelectronics Engineering, CAS Center for Excellence in Nanoscience, National Center for Nanoscience and Technology, Beijing, China
e-mail: yangr@nanoctr.cn

R. Yang
Sino-Danish College, Sino-Danish Center for Education and Research, University of Chinese Academy of Sciences, Beijing, China

X. Yan (ed.), *Nanozymology*, Nanostructure Science and Technology,
https://doi.org/10.1007/978-981-15-1490-6_10

GO	Graphene oxide
GOx	Glucose oxidase
GSH	Glutathione
H_2O_2	Hydrogen peroxide
HRP	Horseradish peroxidase
ICA	Immunochromatographic assay
LDR	Linear detection range
LOD	Limit of detection
MOF	Metal–organic framework
NDs	Nanodiamonds
NMs	Nanomaterials
NRs	Nanorods
NSs	Nanosheets
$O_2^{\bullet -}$	Superoxide radicals
•OH	Hydroxyl radicals
OPD	O-Phenylenediamine
ORR	Oxygen reduction reaction
Ox-TMB	Oxidized TMB
PAH	Polyallylamine hydrochloride
PSA	Prostate-specific antigen
RGO	Reduced graphene oxide
ROS	Reactive oxygen species
SDM	Sulfadimethoxine
SOD	Superoxide dismutase
SWCNT	Single-walled carbon nanotube
TA	Terephthalic acid
TMB	3,3′,5,5′-Tetramethylbenzidine
TMDs	Transition metal dichalcogenides
XPS	X-ray photoelectron spectroscopy

10.1 Introduction of Noble-Metal Nanozymes

In general, noble metals are transition metals with high resistance to corrosion and oxidation in moist air. Noble-metal nanomaterials (NMs), which contain one/a few noble metal elements like Pt, Au, Pd, Ag, Rh, Ru, and Ir (Fig. 10.1), can be classified as unsupported NMs (monometallic, bimetallic, and multimetallic NMs) and immobilized NMs on various substrates. While other transition metals (e.g., Ti, V, Cr, Mn, Fe, Co, Ni, Zn, Al, Mo, and W) are generally stabilized in the form of their complex ions (e.g., oxides, halides, sulfides, and hydroxides), which are beyond the scope of this chapter and thus will not be covered. The first NM-based enzyme mimic was metal oxide nanoparticles (Fe_3O_4 NPs) reported in 2007 [1]. Since then, an increasing number of NM-based nanozymes have been reported within the last decade. We

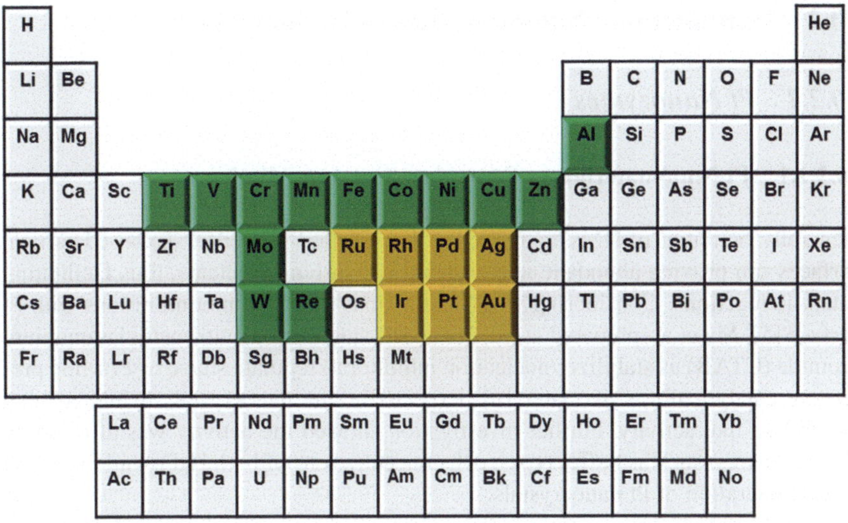

Fig. 10.1 Highlighted elements in the periodic table show noble metals in the yellow boxes, and other transition metals in the green boxes are often present as their complex ions

herein discuss noble-metal NMs showing enzymatic properties, with special focus on Pt- and Au-based ones.

In the past few decades, noble-metal NMs have been extensively studied in catalysis due to many unique attributes, as well-reviewed by several representative literatures [2–4]. To allow for the development of an in-depth insight of origins of their catalytic properties, numerous mechanistic studies have been directed toward surface catalysis, thus paving the way for their applications. As green catalysts, the emerging field of noble metal-based nanozymes represents a burgeoning area with increasing use in biosensing, environmental monitoring, clinical diagnosis, and drug discovery. Undoubtedly, this rapidly expanding research subject was built upon fruitful multidisciplinary collaborations between material, chemistry, and biology. As promising candidates for natural enzymes, noble-metal nanozymes demonstrated good tunability in activities and high stability. They were found to display multiple enzyme-like properties including oxidase-, peroxidase-, catalase-, and/or superoxide dismutase-(SOD) like activities, depending on the intrinsic nature of catalytically active metals (composition, structure, size, and morphology), capping molecules, surface charge, support materials, and specific experimental conditions (e.g., pH, temperature, and substrate). To develop noble metal-based nanozymes with facile preparation, excellent activity, high stability, good biocompatibility as well as low cost still remains an active topic in this important yet exciting field.

10.2 Monometallic Nanomaterials as Nanozymes

10.2.1 Pt Nanozymes

10.2.1.1 Pt Nanomaterials as Peroxidase/Catalase Mimics

Generally, activities of NMs are related to their surface properties. Exposed particle surfaces can provide abundant active sites for absorbing reactants, thus facilitating catalytic reactions. Considering weak interactions of alkylammonium ions with Pt surface [5], Ma et al. prepared cubic Pt nanocrystals with cetyltrimethylammonium bromide (CTAB) as stabilizer/nucleation inhibitor to regulate shape of Pt while preserving catalytically active sites [6]. Expectedly, the nanocrystals exhibited high peroxidase-like activity. Further investigation showed the activity was affected by several factors, such as buffer types, pH, concentrations of both H_2O_2 and Pt as well as agglomeration of Pt nanocrystals.

In another study, Fu et al. revealed peroxidase-like activities of the DNA-stabilized Pt NPs were greatly affected by surface charge state and particle size [7], which could be conveniently modulated by programming DNA sequences, precursor ions, and molar ratios of [precursor]/[DNA]. Compared to small clusters, Pt NPs containing more metallic Pt^0 species had higher activity. Specifically, the i-motif RET2-stabilized Pt with about 66% Pt^0 and an average diameter of 2.9 nm was most efficient.

The activity of Pt nanozymes is also affected by external metal ions. A typical example is that Li et al. found Hg^{2+} ions could downregulate peroxidase-like activity of bovine serum albumin (BSA)-stabilized 2 nm-Pt NPs (BSA-Pt), mainly through interactions between Hg^{2+} and Pt^0 [8]. On a basis of this finding, a colorimetric Hg^{2+} ions sensing system based on BSA-Pt was established, with a limit of detection (LOD) of 7.2 nM of Hg^{2+} and a linear detection range (LDR) of 0–120 nM. The sensing system also displayed excellent selectivity toward Hg^{2+} over a variety of competing ions like Na^+, Mg^{2+}, Ca^{2+}, Mn^{2+}, Ni^{2+}, Zn^{2+}, Co^{2+}, Cu^{2+}, Pd^{2+}, Cd^{2+}, Fe^{2+}, and Au^{3+}. It was further applied for quantitative determination of Hg^{2+} in drinking water samples.

Interestingly, some studies showed Pt nanozymes possessed dual enzymatic properties (peroxidase- and catalase-like activities), since both have H_2O_2 as substrate. Fan et al. fabricated 1–2 nm-Pt NPs using apoferritin (apoFt) protein shell as a nanoreactor (Fig. 10.2) [9]. The as-synthesized Pt-Ft NPs were highly stable and had both catalase- and peroxidase-like activities, showing differential response to pH and temperature.

Similar observation was reported by Liu et al. [10], in which Pt NPs showed catalase-like activity under neutral/alkaline conditions and peroxidase-like activity under acidic conditions. Besides, Li et al. investigated origins of both activities and pH-switchability for several noble metals involving Pt by means of calculations and experiments [11]. The peroxidase-like activity of Pt in low-pH conditions could be

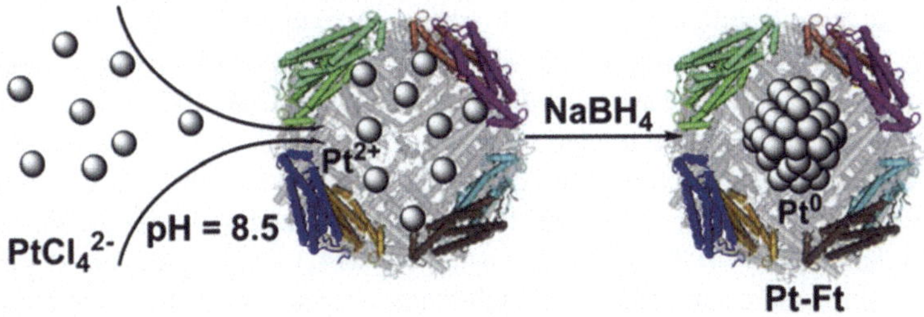

Fig. 10.2 Schematic illustration of synthesis of Pt-Ft [9]. Copyright (2011) Elsevier

explained from the base-like decomposition of hydrogen peroxide (H_2O_2) on Pt surfaces, whereas catalase-like activity of Pt in high-pH conditions could be attributed to the acid-like decomposition of H_2O_2 on Pt surfaces. Either experimentally or theoretically, these studies not only provided useful guidance to design Pt NPs with desired catalytic properties, but also showed Pt NPs were potentially versatile biocatalysts for multiple biomedical applications, such as for various human diseases involved in oxidative stress damage.

Currently, utilization of peroxidase- and catalase-like activities of monometallic Pt nanozymes for practical biological applications is still at the beginning, mainly due to challenges in controlled size and structures of Pt NMs to achieve superior activities. Ultrasmall Pt nanoclusters have evoked considerable interest in bioanalysis, owing to many favorable attributes including small size, excellent activity, and good biocompatibility. The recent report [12] of the use of yeast extract as a reductant/stabilizer to synthesize ultrasmall Pt clusters is interesting in this context. Since yeast extract contains abundant active sites from biomolecules (amino acids and proteins), they can efficiently bind and reduce metal ions. The yeast extract-stabilized Pt clusters possessed both high water solubility and peroxidase-like activity. With these observations, a colorimetric method for glucose detection was developed with LOD of 0.28 μM.

10.2.1.2 Pt Nanomaterials as Oxidase Mimics

Peroxidase-mimic-catalyzed reactions require involvement of H_2O_2, which potentially cause some problems: (1) some analytes might be damaged by H_2O_2 due to its strong oxidative capacity and (2) H_2O_2 can readily decompose and lose its oxidative capacity, limiting its practical applications. Contrarily, H_2O_2 is unnecessary for oxidase-mimic-catalyzed reactions, which can be completed with oxygen. Thus, the distinct advantage makes oxidase mimics attractive alternatives for biomedical applications with operational simplicity, good compatibility, and high reliability.

As is known, surface ligands not only enhance the dispersion and stability of NMs, but also facilitate further surface functionalization and biological applications. Interestingly, Deng et al. demonstrated the superiority of chitosan as a capping molecule for simultaneous stabilizing effect and enhancing oxidase-like activity of Pt NPs [13]. On a basis of inhibitory effect of ascorbic acid (AA) on activity of chitosan-stabilized Pt, a colorimetric approach for acid phosphatase (ACP) determination was developed with good sensitivity and selectivity.

Sometimes the capping molecules can block active sites of the catalytic core and then decrease their activity. Yu et al. found a way to deal with this problem [14]. They synthesized Lys-stabilized Pt_4 clusters (Lys-Pt_4) using lysozyme (Lys) as a template. Since the use of additional stabilizers is unnecessary for template-based direct synthesis, the Pt clusters located in templates can provide non-block and active surfaces for catalysis. Several typical chromogenic substrates, such as 2,2′-azino-bis(3-ethylbenzo-thiazoline-6-sulfonic acid) diammonium salt (ABTS), TMB, and dopamine, could be catalytically oxidized by the clusters with dissolved oxygen, following a Michaelis–Menten mechanism. Compared to horseradish peroxidase (HRP) and Fe_3O_4 NP-catalyzed oxidation of TMB with H_2O_2 as well as other NM-based oxidase mimics like CeO_2 NPs, the Lys-Pt_4 clusters gave a lower K_m value and a larger V_{max} value, indicating high affinity for the substrate and high catalytic efficiency. The authors also found the oxidase-like activities of these nanozymes were highly dependent on their particle size by comparison with Pt NPs of 5 and 30 nm, of which Lys-Pt_4 displayed the best catalytic performance due to a high surface area-to-volume ratio. With above findings, the Lys-Pt_4 clusters were used as oxidase mimics to oxidize methylene blue, demonstrating their practical applicability in pollutant treatment.

10.2.1.3 Pt Nanomaterials as SOD Mimics

As revealed, environmental pH plays an important role in both catalase- and peroxidase-like activities of Pt nanozymes. Liu et al. found Pt NPs displayed SOD-like property of scavenging superoxide radicals ($O_2^{\bullet-}$) under neutral conditions, i.e., catalytic dismutation of $O_2^{\bullet-}$ radicals into H_2O_2 and O_2 [10]. Similar observations for Pt NPs were reported by Clark et al. [15] and Zhang et al. [16]. $O_2^{\bullet-}$ radicals are physiologically relevant reactive oxygen species (ROS), and overproduction of $O_2^{\bullet-}$ can denature enzymes and oxidize lipids and fragment DNA. Thus, the studies of Pt-based SOD mimics to scavenge $O_2^{\bullet-}$ are of great significance.

10.2.2 Au Nanozymes

10.2.2.1 Au Nanomaterials as Peroxidase Mimics

Au NMs stabilized by various chemical or biological molecules displayed peroxidase-like activities. Wang et al. found BSA-stabilized Au clusters (BSA-Au) as peroxidase mimics could be used in colorimetric detection of H_2O_2 [17]. In their study, as low as 0.02 μM H_2O_2 could be detected with LDR of 0.5–20 μM. Then, a colorimetric assay for xanthine detection using xanthine oxidase (XOD) and BSA-Au was developed with LOD of 0.5 μM. The assay could be further applied to determine xanthine in urine and human serum sample.

Similar to Pt nanozymes, surface charge state of Au NMs is also crucial to their peroxidase-like activities. For example, Ahmed et al. found positively charged Au NPs stabilized by cysteamine displayed approximately sevenfold higher than that of positively charged ones capped by citrate [18]. As peroxidase mimics, Au NPs displayed ultra-sensitivity for the colorimetric detection of influenza viruses. A level concentration of 10.79 pg mL^{-1} and 11.62 plaque-forming units (PFU) mL^{-1} could be detected for influenza virus A (H_1N_1) and clinically isolated influenza virus A (H_3N_2), respectively. The sensing system was found to be 100- and 500-fold more sensitive than the conventional enzyme-linked immunosorbent assay (ELISA) and commercial virus kit, respectively.

Particle size also plays an important role in the activity of Au nanozymes. Unlike small molecule stabilizers, apoFT can be used not only as a stabilizer but also as a size-constrained reaction vessel to direct nanostructure synthesis. This is demonstrated in the study of apoFT-paired Au clusters (Au-Ft) with a diameter of less than 2 nm [19]. Au-Ft displayed dramatically higher activity than larger colloidal Au NPs with 6 or 20 nm. On the basis of these observations, a simple and selective assay for glucose detection was developed.

Besides, the surface properties of Au atoms are essential to the performance of Au-based nanozymes, yet little is known about the effect of biologically relevant molecules and ions on activity. In this context, Shah et al. investigated the effect of adenosine triphosphate disodium salt (ATP), adenosine diphosphate disodium salt (ADP) and ions on peroxidase-like activity of citrate-capped Au NPs [20]. The authors found ATP and ADP addition enhanced peroxidase-like activity of Au NPs due to a synergistic effect of ATP and Au NPs, in which ATP as an H_2O_2 activator facilitated electron transfer (between H_2O_2 and TMB) during the catalytic reaction. On the contrary, ions like free phosphate, sulfate, and carbonate anions did not change the peroxidase-like activity of Au NPs. Wang et al. also investigated the impact of surface modification on peroxidase-like activity of Au NPs [21]. Compared with those amino-modified or citrate-capped Au NPs, the unmodified Au ones exhibited significantly higher activity, which could be ascribed to the superficial Au atoms. Additionally, charge characteristics of NPs and substrates were also essential to the activity of Au NPs.

In another study, Ni et al. found the addition of melamine can improve peroxidase-like activity of bare Au NPs (curve 2, Fig. 10.3a), since Au NPs could conjugate with melamine to form Au-melamine aggregates [22]. Benefiting from the melamine concentration-dependent enhancement of Au, the authors further developed a sensitive, accurate, and reliable colorimetric method for melamine determination (Fig. 10.3b). LOD of the assay was as low as 0.2 nM by UV–Vis spectroscopy and 0.5 μM by naked-eye observation, both of which are far below the US Food and Drug Administration (FDA) estimated melamine safety limit of 20 μM. The proposed method was successfully applied to detect melamine in raw milk and milk powder.

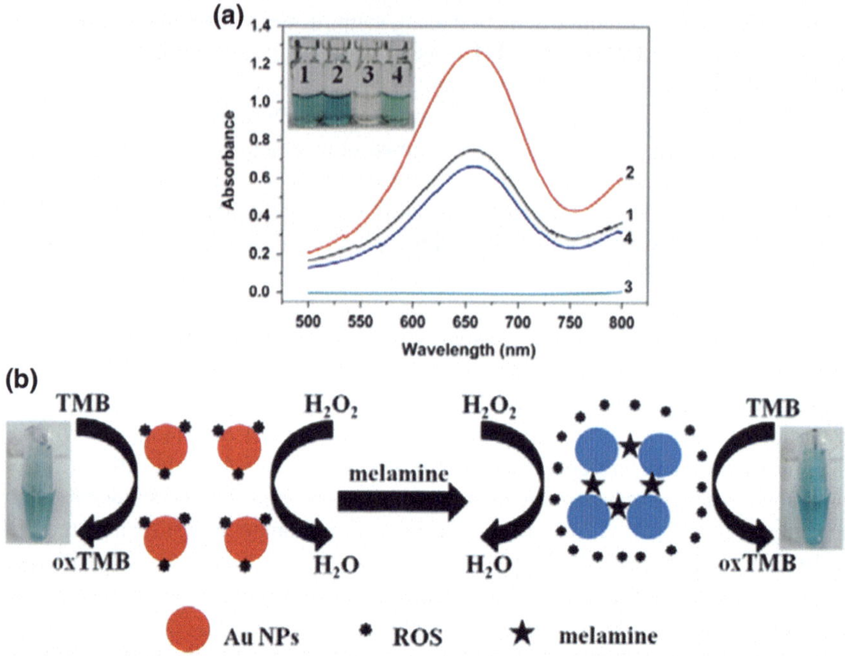

Fig. 10.3 Enhanced peroxidase-like activity of Au NPs conjugated with melamine [22]. **a** Absorption spectra of (1) Au NPs (160 μL)/TMB/H_2O_2; (2) Au NPs (160 μL)/melamine (0.5 μM)/TMB/H_2O_2; (3) melamine (0.5 μM)/TMB/H_2O_2; (4) 170 μL of supernatant of the mixture of Au NPs and melamine (800 μL of Au NPs and 50 μL of melamine (0.5 mM))/TMB/H_2O_2. The concentration of TMB and H_2O_2 was 2.5 mM and 0.28 mM, respectively. Inset shows corresponding digital images. **b** Schematic illustration of the proposed detection system. Copyright (2014) Elsevier

Other strategies to enhance activity of Au nanozymes have been reported. For example, Wang et al. demonstrated peroxidase-like activity of Au NPs could be significantly enhanced by visible light stimulation [23]. Hizir et al. proposed another approach to tune peroxidase-like activity of citrate-capped Au NPs by integrating oligonucleotides on Au NP surface [24].

However, several biologically relevant molecules could inhibit catalytic properties of Au nanozymes' activity. Feng et al. found glutathione (GSH) can suppress peroxidase-like activity of the GSH-stabilized Au nanoclusters (GSH-Au) [25]. The inhibition of GSH can be explained from two factors: (1) GSH can strongly bind with Au atom by Au–S bond, thus decreasing the possibility of collision between H_2O_2 and Au to generate hydroxyl radicals (•OH) and (2) GSH as a strong reducing agent can scavenge •OH arising from the decomposition of H_2O_2. On the basis of these findings, the authors established a colorimetric protocol based on GSH-Au for facile sensing of GSH in cells (Fig. 10.4). Similar observation was reported for the malathion-inhibited peroxidase-like activity of CTAB-capped Au nanorods (NRs) [26]. The as-prepared NRs were found to possess higher activity than HRP and other

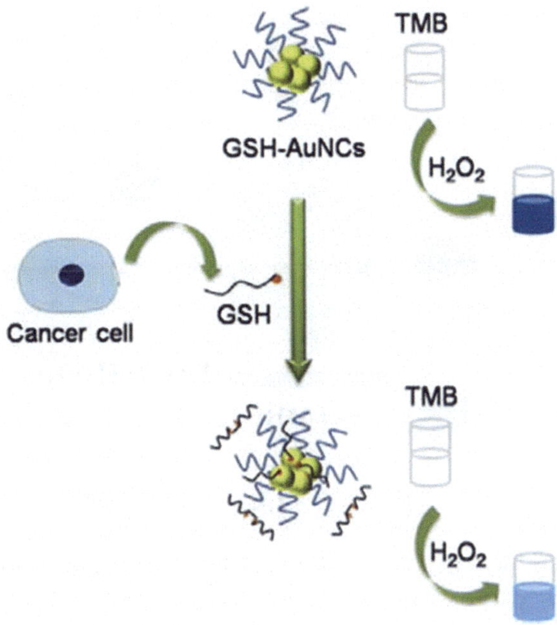

Fig. 10.4 Schematic diagram of colorimetric detection of GSH in cancer cells based on the inhibition effect of GSH on the peroxidase-like activity of GSH-Au [25]. Copyright (2017) Elsevier

Au nanostructures, as a result of charged surface of Au NRs which might contribute to the generation and stabilization of peroxide free radical, and partial electron-transfer process. In the following, the authors established a simple, colorimetric inhibition assay for malathion based on Au NR peroxidase mimics.

Like Pt nanozymes, Au nanozymes' activity may also be affected by the introduction of metal ions. For example, on the basis of inhibition effect of Ag^+ ions on peroxidase-like activity of BSA-Au, Chang et al. developed a simple and feasible method for the detection of Ag^+ ions [27]. Once Ag^+ was introduced, Ag^+ ions selectively reacted with Au^0 through redox reaction, evidenced by the X-ray photoelectron spectroscopy (XPS) analysis. The high specificity of Ag–Au interaction gave excellent selectivity over the interfering metal ions. The LOD of Ag^+ sensing was 0.204 μM with LDR of 0.5–10 μM. The proposed method has been further applied for rapid analysis of Ag^+ in lake water with satisfactory results. For another example, Zhu et al. developed a method for Hg^{2+} detection based on BSA-Au clusters (Fig. 10.5) [28]. Owing to specific interaction between Hg^{2+} and Au^+, Hg^{2+} induced an inhibition effect on peroxidase-like activity of BSA-Au. In their study, as low as 3 nM Hg^{2+} could be detected with LDR of 10 nM to 10 μM. The method has been applied for colorimetric determination of total mercury content in skin-lightening products.

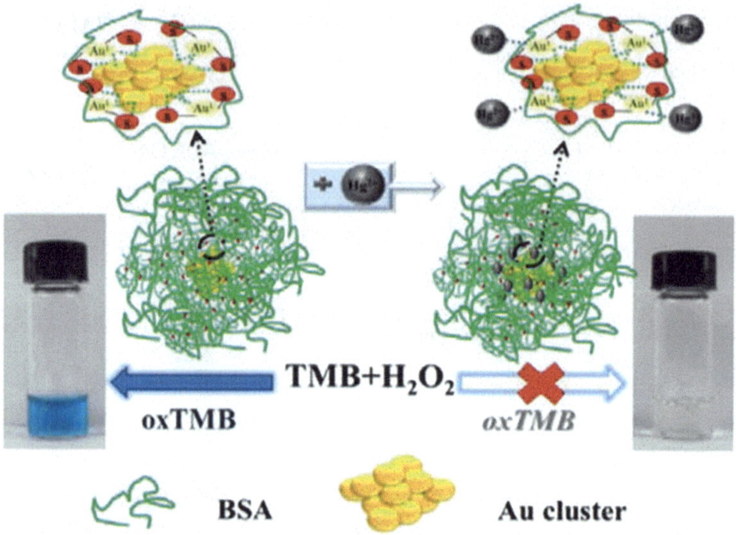

Fig. 10.5 Schematic diagram of Hg^{2+} detection based on the inhibition effect of Hg^{2+} on the peroxidase-like activity of BSA-Au [28]. Copyright (2013) Elsevier

10.2.2.2 Au Nanomaterials as Oxidase Mimics

Despite numerous examples of monometallic Au peroxidase mimics, few studies of Au oxidase mimics are available. Therefore, to explore Au NM-based oxidase mimics is of special interest. As active catalysts, colloid Au NPs without any solid supports displayed intrinsic catalytic activity for many organic reactions (e.g., CO oxidation, borohydride reductions, aerobic oxidation of alcohols and diols as well as C-C coupling reactions, etc.) [29]. The naked Au NPs were also efficient for the aerobic oxidation of glucose with dissolved oxygen [30]. The reaction was similar to that catalyzed by glucose oxidase (GOx), indicating the unsupported Au NPs could be a promising candidate for GOx. By contrast, no significant oxidase-like activity for other metal-based NPs (e.g., Ag, Cu, Pd, and Pt NPs) was exhibited. An Eley–Rideal mechanism was proposed based on the experiment results, i.e., glucose may be first adsorbed onto Au NPs; then oxygen reacted with glucose, yielding gluconic acid and H_2O_2 as products. Although native enzyme was 55-fold more active than Au nanozymes, Au colloids demonstrated potential for practical application in glucose oxidation, considering their easy preparation, nontoxicity, and possibility of recyclability. By carrying out a series of kinetic measurements, the same group further made a comparison between enzymatic and Au-based catalysis in a later study [31].

Inspired by above findings, Luo et al. designed an interesting self-limiting system for controllable growth of Au NPs by taking advantage of the GOx-like activity of Au NPs [32]. The size, shape, and catalytic activity of Au NPs could be simultaneously controlled in this system, thus providing a new strategy for their controllable synthesis. With GOx-like property of Au NPs, the same group further developed a sensing strategy for the detections of DNA and microRNA [33]. In principle, with different affinities of ssDNA and dsDNA toward Au seeds, the growth of Au NPs could be finely regulated and thus the oxidase-like activity of Au NPs was adjusted. Consequently, facilitated by DNA hybridization, enzymatic activity of Au NPs could be exploited to detect target DNA or microRNA.

10.2.3 Pd Nanozymes

Like Pt and Au NMs, Pd NMs with lower cost also have emerged as promising peroxidase mimics in visible detection of metal ions and small biomolecules [34–36]. A representative example is the GSH-capped Pd NPs with 2.6 nm, which were used as a peroxidase mimic for the colorimetric detection of Ag^+ ions, on a basis of the inhibition effect of Ag^+ on the activity [36].

DNA oligonucleotides can be used as templates to synthesize and stabilize metal NMs with tunable physiochemical properties. Recently, Li et al. fabricated the oligonucleotide-stabilized Pd NPs with an average diameter of 4.8 nm [37]. On the basis of thiol-induced shielding of peroxidase-like activity of Pd NPs, they were employed for biothiol detection with LOD of 3.7 nM for cysteine and 4.3 nM for homocysteine, respectively.

More recently, Fang et al. reported that different crystal planes of Pd NMs had an influence on their peroxidase-like activity and antibacterial effect [38]. Higher efficiency of the {100}-faceted Pd cubes against Gram-positive bacteria was observed, corresponding to the extent of their enzymatic activity, while the {111}-faceted Pd octahedra against Gram-negative bacteria was efficient.

10.2.4 Other Monometallic Nanozymes

Despite a variety of Pt-, Au-, and Pd-containing monometallic nanozymes, examples of other noble metal-containing ones are extremely scarce. Ru, compared with other noble metals, is rather unique, in that it is not only ultrastable but also the only one that normally takes the hexagonal close-packed (hcp) crystal structure. Thus, to explore enzymatic properties of the Ru-based NMs is appealing.

This was first demonstrated in the Ru nanoframes synthesized via seeded growth and chemical etching [39]. The Ru nanoframes were chemically and thermally stable, and displayed approximate threefold more active peroxidase-like activity than HRP. Ru nanoframes were also found to possess a stronger binding affinity to TMB yet a weaker binding affinity to H_2O_2. The study provided a type of novel noble-metal nanozymes with excellent activity and high stability. Besides, Cui et al. found the tannin acid-stabilized Ir NPs also had superior peroxidase-like activity [40]. Rh, another member of the platinum group with corrosion resistance and chemical inertness, has been explored for nanozyme catalysis until recently [41, 42].

10.3 Bimetallic Nanozymes

Bimetallic nanomaterials (BNMs), which are composed of two metals, often display enhanced catalytic performance than their monometallic counterparts because of a synergistic effect [43–45]. Among them, Pt-based BNMs have been extensively studied in catalysis field for many years. Recent studies showed they had multiple enzyme-mimic properties (e.g., oxidase-, peroxidase-, catalase-, and SOD-like activities), demonstrating great potential in biomedical applications such as biosensing, biomedicine, and immunohistochemistry.

10.3.1 Au_{core}- and Pt_{core}-Based Bimetallic Nanostructures

Currently, for bimetallic nanozymes, Au_{core}- and Pt_{core}-based ones are most studied. One of the typical examples is $Au_{core}Pt_{shell}$ nanozymes. By epitaxial growth of Pt nanodots on the premade Au NRs, He et al. synthesized $Au_{core}Pt_{shell}$ nanostructures, which exhibited oxidase-, peroxidase-, and catalase-like activities [46]. An ELISA based for detection of mouse interleukin 2 (IL-2) was further established.

Using a similar strategy, Liu et al. prepared $Au_{core}Pt_{shell}$ nanostructures [47]. The materials were found to exhibit dual enzyme-like (peroxidase- and oxidase-like) activities. As peroxidase mimics, $Au_{core}Pt_{shell}$ NPs were further used to detect H_2O_2, giving LOD of 45 μM and LDR of 45–1000 μM. Moreover, in combination with natural enzymes (GOx and cholesterol oxidase), similar LODs and LDRs were demonstrated in selective detection of hydrophilic glucose and lipophilic cholesterol. Notably, some endogenous reducing species like AA could be eliminated in advance by oxidase-like activity of $Au_{core}Pt_{shell}$, thus improving the sensitivity for detection.

$Au_{core}Pd_{shell}$ NPs fabricated by growing Pd NPs on Au NR surface also had peroxidase-like activity [48]. $Au_{core}Pd_{shell}$ was found to be stable over a broad temperature range (4–70 °C). Apparent kinetic study showed $Au_{core}Pd_{shell}$ gave a low K_m value and a high V_{max} value, better than HRP. On a basis of selective quenching of malathion on the peroxidase-like activity of $Au_{core}Pd_{shell}$ NPs, the authors developed a sensitive assay for malathion detection with LOD of 60 ng mL^{-1}, lower than Au NR-based assay previously reported by same group (1.78 μg mL^{-1}).

Different from Au NR core in above studies, Gao et al. prepared urchin-like $Au_{core}Pt_{shell}$ NPs with Au NPs as core [49]. The as-prepared BNPs could be used for in situ amplified colorimetric immunoassay of prostate-specific antigen (PSA) based on their peroxidase-like activity. The developed immunoassay displayed a good response toward PSA in the LDR of 5–500 pg mL^{-1} with LOD of 2.9 pg mL^{-1}.

The urchin-like $Au_{core}Pt_{shell}$ NPs also could be utilized for the colorimetric detection of Cu^{2+} by taking advantage of their peroxidase-like activity [50]. The developed Cu^{2+} sensor was based on following principle: catalytic ability of $Au_{core}Pt_{shell}$ could be inhibited by L-cysteine in the absence of Cu^{2+}; however, the case in the presence of Cu^{2+} was just the opposite, since L-cysteine could be oxidized by O_2 with Cu^{2+}. By measuring colorimetric signal variation of the H_2O_2-TMB reaction, a colorimetric detection method for Cu^{2+} was developed with high sensitivity and selectivity.

Besides, Au NPs coated with Pd dots ($Au_{core}Pd_{shell}$) were synthesized by Nangia et al, which were further modified via electrostatic interaction between the surface-terminated anionic groups on $Au_{core}Pd_{shell}$ NPs and the positively charged lysine residues of the antibody [51]. Benefiting from the peroxidase-like activity of $Au_{core}Pd_{shell}$, the authors developed a sensitive fluoroimmunoassay for the herbicide bensulfuron-methyl with LDR of 0.001–100 ng mL^{-1} and LOD of 0.01 ng mL^{-1}.

Studies on the Pt_{core}-based bimetallic nanozymes are scarce. Meanwhile, few studies of catalase mimics on the bioanalysis are available. Wu et al. reported a cost-effective, portable capillary-based biosensor based on the catalase-like activity of $Pt_{core}Au_{shell}$ NPs for quantitative detection of biomarkers by the naked eyes [52]. The developed biosensor was tested by measuring the distance of blue ink movement, which was directly correlated with O_2 produced by $Pt_{core}Au_{shell}$ NP-catalyzed H_2O_2 decomposition. By linking $Pt_{core}Au_{shell}$ NPs with antibodies, capillary-based biosensor sandwich immunoassays were further constructed. The biosensor showed high sensitivity and selectivity for the cancer biomarker prostate-specific antigen (PSA) and carcinoembryonic antigen (CEA), well correlated with those by chemiluminescence immunoassays. In another study, Lee et al. demonstrated the uncoated $Pt_{core}Au_{shell}$ NPs had catalase- and SOD-like activities, and in vitro experiments revealed cellular protection against oxidative stress induced by H_2O_2 [53].

Additionally, Wu et al. synthesize BSA-capped $Pt_{core}Bi_{shell}$ NPs (BSA-$Pt_{core}Bi_{shell}$), which exhibited high stability and enhanced peroxidase-like activity than BSA-capped Pt NPs, resulting from formation of monolayer or submonolayer of Bi on Pt surface [54]. Several bioassays (serum glucose detection, extracellular H_2O_2 monitor, and cancer cells labeling) based on peroxidase-like activity of BSA-$Pt_{core}Bi_{shell}$ NPs had been further realized.

10.3.2 Pt-Based Alloy Nanostructures

Different to bimetallic core-shell NPs typically prepared by a two-step procedure, some alloy-based nanozymes can be simply fabricated via a one-pot strategy. This is demonstrated in the preparation of AuPt bimetallic NPs (BNPs) by co-reduction of Au and Pt precursors. He et al. found the peroxidase- and oxidase-like activities of AuPt NPs prepared by this facile strategy were highly composition-dependent [55]. By studying the inhibiting effects of HS on their activities, the authors established a method for colorimetric detection of HS and evaluation of the inhibitory effects of inhibitors on AuPt nanozymes. The responsive ability of this method was found to be largely influenced by the composition, i.e., AuPt NPs showed much lower LOD for HS than Pt NPs, whereas Pt NPs showed wider LDR than AuPt NPs. The study paved the way for tuning catalytic activities of alloy nanostructure for biomedical applications.

Sun et al. demonstrated the peroxidase-like activity of C-rich oligonucleotide-stabilized AuPt BNPs also showed composition dependence [56]. A colorimetric assay for biothiols with high sensitivity and selectivity was further established based on thiol-induced inhibition of activities of AuPt nanozymes. Besides, benefiting from peroxidase-like activity of AuPt NPs, Jiang et al. developed an immunochromatographic assay (ICA) with enhanced sensitivity for the visual and quantitative detection of *E. coli* O157:H7 with LOD of 10^2 cell mL^{-1}, about 1/1000 of the recent conventional colloidal-based ICA [57].

PtAg BNMs could be utilized as another type of enzyme mimics. By co-reduction of two metal ions in aqueous solutions, He et al. developed a general and facile way to prepare a series of Ag-based BNPs AgM (M = Pt, Au, and Pd) with hollow/porous structures [58]. By controlling Ag/M ratio and/or surfactant concentration, the morphology of AgM NPs could be tailored. Like PtAu nanoalloys, PtAg NPs also exhibited the composition-dependent peroxidase-like activity.

Wu et al. found the integration of both Hg^0 and Hg^{2+} to deoxyribonucleic acid (DNA)-stabilized AgPt (DNA-AgPt) nanoclusters resulted in DNA-AgPt aggregation and the decrease of Pt^{2+} content on DNA-AgPt nanocluster surface, which finally impaired their catalytic activity. Based on this observation, colorimetric detection of Hg^{2+} was developed [59]. In another study, the same group developed a highly sensitive and selective colorimetric detection for L-cysteine, on the basis of its inhibiting effect on the peroxidase-like activity of DNA-AgPt nanoclusters [60].

Nanoporous materials have found broad catalytic applications because of many attributes, such as large surface area, low density, and high catalytic efficiency [61]. Through a facile and economical dealloying process of regular Pt_xAg_{100-x} ($x = 25$, 50, 75) octahedra, our group fabricated a series of porous PtAg BNPs [62]. Owing to the resulted Pt-rich surface structure, increased surface area, and a synergistic effect of Pt and Ag atoms in the dealloyed $Pt_{50}Ag_{50}$ NPs, a remarkable enhancement in multifunctional enzyme-like activities (oxidase-, peroxidase-, and catalase-like activities) was observed. Besides, the peroxidase-like activities of the dealloyed NPs were composition-dependent, in which the highest content of Ag in the porous NPs catalyst coincided well with its highest activity. The dealloyed $Pt_{50}Ag_{50}$ NPs also exhibited excellent antibacterial effects for Gram-negative *E. coli* and Gram-positive *S. aureus*. The optimized H_2O_2 concentrations for them were 10 and 200 μM in the presence of 20 $\mu g\ mL^{-1}$ of NPs, respectively, much lower than those antibacterial systems containing graphene quantum dots [63] and mesoporous silica-supported Au NPs [64]. The work represented a significant step in the exploration between controlled synthesis of high-quality nanoalloys and their novel catalytic properties for potential biomedical applications.

In addition to PtAu and PtAg, PtPd bimetallic nanozymes with different morphologies have been studied. Through a bromide-induced galvanic replacement reaction between Pd nanowires and K_2PtCl_6, Ge et al., prepared one-dimensional (1D) PtPd porous NRs [65]. The as-prepared PtPd NRs could be used as peroxidase mimics for the colorimetric detection of the flux of H_2O_2 released from living cells. Similarly, Zhu et al. illustrated 3D PtPd nanosponges had better oxidase- and peroxidase-like activities than 1D PtPd nanowires [66], benefiting from their unique structures (high porosity, good flexibility, large specific surface area, and interconnected branches).

Contrary to above studies, introduction of non-precious metals into Pt-based alloy nanozymes is of special interest. In this regard, Lu et al. prepared 3D hierarchical porous PtCu dendrites via galvanic replacement, followed by chemical dealloying to selectively dissolve Cu atoms from PtCu alloys [67]. The resulted porous PtCu dendrites as peroxidase mimics have been applied to the determination of H_2O_2 in daily milk products with satisfactory results. Besides, incorporation of magnetic metals (e.g., Fe, Co, and Ni) into alloys not only enhances catalytic activity, but also endows materials with magnetic property. However, there are very limited reports concerning Pt-based magnetic nanozymes. In 2016, our group prepared PtCo alloys by a co-reduction method. With several advantages including high oxidase-like activity, good water solubility, low cell toxicity, and superparamagnetic properties, a magnetic-enhanced colorimetric assay based on aptamer-conjugated $Pt_{72}Co_{28}$ NPs was further constructed (Fig. 10.6) [68]. Notably, incorporation of Co into alloys not only resulted in enhanced oxidase-like activity of NPs for cancer-cell detection without destructive H_2O_2, but facilitated magnetic separation for improving detection sensitivity.

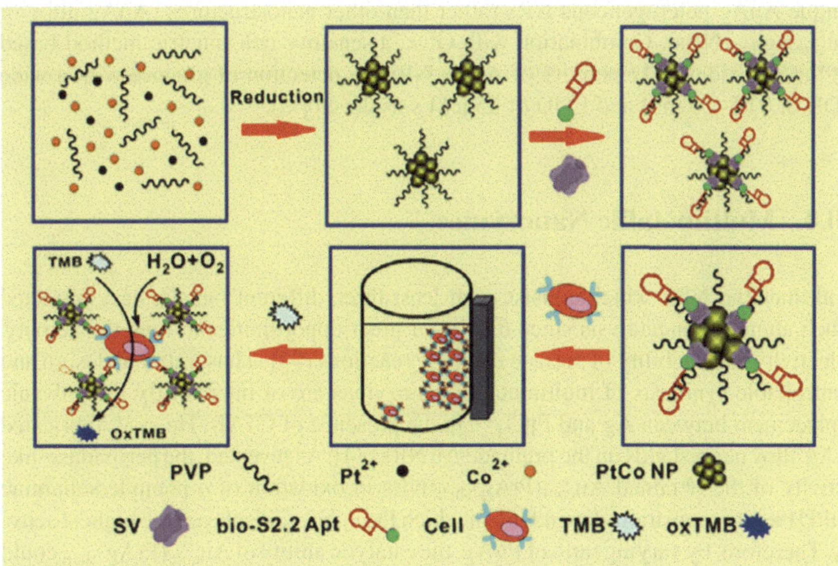

Fig. 10.6 Schematic illustration of synthesis of PtCo-aptamer nanoconjugates for magnetic separation and detection of target cancer cells [68]. Copyright (2016) Royal Society of Chemistry

10.3.3 Other Bimetallic Nanostructures

In addition to Pt-based bimetallic nanozymes, a few Pd-based ones are reported. One typical example is $Pd_{core}Ir_{shell}$ nanocubes prepared by depositing Ir atoms as ultrathin skins on Pd nanocubes [69]. The authors found the $Pd_{core}Ir_{shell}$ displayed significantly enhanced efficiency, with catalytic constants more than 20- and 400-fold higher than those of Pd nanocubes and HRP, respectively. $Pd_{core}Ir_{shell}$ nanocubes were further applied to the colorimetric ELISA of human PSA with LOD of 0.67 pg mL^{-1}, 100-fold lower than that of the conventional HRP-based ELISA. Another example is $Pd_{core}Pt_{shell}$ mesoporous NPs synthesized by galvanic replacement reaction [70], which could be used as peroxidase mimics for colorimetric and sensitive immunoassay of p53 with LDR of 0.1–10 ng mL^{-1} and LOD of 0.05 ng mL^{-1}.

Like PtCo alloy nanozymes discussed previously, Pd-based ones also provide an attractive platform for biosensing. Wang et al. prepared porous PdNi nanoalloys by a galvanic replacement method [71]. The as-prepared PdNi NPs were found to exhibit triple enzyme-like activities: oxidase-, peroxidase-, and catalase-like activities. Taking advantage of peroxidase-like activity of PdNi, the authors developed a simple glucose biosensor with a wide LDR (0.005–0.5 mM) and low LOD (4.2 μM).

A more interesting example is AuAg heterogeneous NRs fabricated by co-reduction of Au and Ag sources [72], different from multistep seed-mediated methods for synthesis of Au_{core}-based BNMs in above studies. Notably, Au was totally buried in Ag to emerge with a clear boundary between two metals, forming the unique AuAg heterogeneous NRs rather than other nanostructures (AuAg alloy or $Au_{core}Ag_{shell}$ NRs). Combination with GOx, a sensitive colorimetric method based on the peroxidase-like activity of AuAg NRs for detection of glucose with a wide LDR of 0.05–20 mM and LOD of 25 μM was developed.

10.4 Multimetallic Nanozymes

Multimetallic NPs, which consist of at least three different metals, have attracted much attention in catalysis since they offer great opportunities to tune the activity, selectivity, and stability of surface catalytic reactions [73]. Thus, rational design and controllable synthesis of multimetallic nanozymes are of interest. By the galvanic replacement between Ag and $PtCl_4^{2-}$ in the presence of CTAB, He et al. fabricated PtAg alloy nanoislands on the premade Au NRs [74]. As revealed, the peroxidase-like activity of the obtained $Au_{core}(PtAg)_{shell}$ NRs in oxidation of *o*-phenylenediamine (OPD) was composition-dependent, of which $Pt_{0.62}Ag_{0.38}$ displayed the highest activity. Therefore, by varying ratio of Pt/Ag, the catalytic ability of $Au_{core}(PtAg)_{shell}$ could be finely adjusted.

Despite peroxidase-like activities of AgM (M = Pt and Pd) nanoalloys, a high Ag percentage in alloys is unstable in the presence of H_2O_2, thus restricting the tunability of catalytic activity. To explore stable substitutes of Ag and design porous nanostructure with large surface area facilitated for catalysis, Zhang et al. studied the combination of Pd and Pt to construct PdPt alloy nanodots as a shell with nanoscale pores on Au NRs [75]. The successful formation of $Au_{core}(PdPt)_{shell}$ nanostructure was related to the unique Pt-guided island growth mode. Interestingly, different from Ag, alloying with Pd enhanced oxidase-mimic activities. On a basis of competitive oxidation of TMB and Fe^{2+} with dissolved oxygen, selective and sensitive detection of Fe^{2+} ions in air was realized with LDR of 1.5–21 μM and LOD of 1.5 μM. For another kinetic study on the reduction of *p*-nitrophenol, the apparent rate constant (k_{app}) showed a nice linear relationship with Pd/Pt ratio.

Considering high price and scarcity of Pt, reducing Pt usage in the Pt-based nanozymes is of importance. In this context, through a chitosan-mediated epitaxial growth of Pt multi-branches occurs on cresyl violet-labeled $Au_{core}Ag_{shell}$ NPs, Li et al. prepared Fructus *Broussonetia*-like $(AuAg)_{core}Pt_{shell}$ nanostructures [76]. The as-prepared nanostructures as peroxidase mimics together with GOx were used for the colorimetric detection of glucose, giving LDR of 1 nM to 200 μM and LOD of 1 nM. In a recent study, Mu et al. synthesized the PtPdMo nanocubes by doping Pt-based nanocrystals with Mo atoms, aiming to boost nanozyme selectivity in the physiological environment and exploring potential applications of multimetallic nanostructures in brain science [77]. The as-obtained nanocubes exhibited improved activity for H_2O_2 and O_2 reductions due to lattice distortion and exposure of more active sites. Meanwhile, the nanocubes as nanozymes possessed multi-antioxidative properties, with special neutral environment preference for removing the reactive oxygen and nitrogen species radicals like •OH, $^1O_2$, and •NO. The in vitro experiments further showed that PtPdMo could protect the H_2O_2- and lipopolysaccharide-treated neutral cells from oxidative damage of these radicals. Moreover, the PtPdMo nanozyme treatment could remarkably improve the survival rate, neuroinflammation, and reference memory of injured mice by scavenging the overproduced free radicals.

10.5 Nanocomposite Enzyme Mimics

10.5.1 Noble-Metal Nanozymes Supported on Graphene and Its Derivatives

Since the discovery of 2D graphene in 2004 [78], graphene and its derivatives such as graphene oxide (GO) and reduced graphene oxide (rGO) have been widely used as promising carbon matrices to immobilize metal NPs for nanozyme catalysis. For example, Ahmed et al. prepared the graphene/Au nanocomposites [79] with HCOONa as a reductant and graphene sheets as support materials. After modification of graphene/Au with antibody, a colorimetric immunoassay for the detection of norovirus-like NP was established, on a basis of peroxidase-like activity of the as-prepared nanohybrids. For another example, using a two-step procedure involving a rapid hydrothermal method and a replacement method, Wang et al. synthesized graphene/($Ni_{core}Pd_{shell}$) nanocomposites [80]. In the following, the authors developed a colorimetric aptasensor for sulfadimethoxine (SDM) detection based on the peroxidase-like activity of graphene/($Ni_{core}Pd_{shell}$). The addition of SDM aptamer could inhibit their peroxidase-like activity, whereas the target SDM and aptamer could be triggered tightly and recover the activity of the nanocomposites. Besides, Chen et al. synthesized PtPd nanodendrites on graphene sheets with ethanol as a reductant [81]. Compared with monometallic analogues, graphene/PtPd showed a stronger affinity for TMB. Taking advantage of high peroxidase-like activity, a colorimetric sensor for H_2O_2 based on graphene/PtPd was developed with fast response, a wide LDR (0.5–150 μM), and low LOD (0.1 μM).

Unlike graphene, GO has abundant surface oxygen-functional groups, which facilitates the in situ reduction of metal precursors on GO sheets. For example, with tannic acid as a stabilizer, Chen et al. prepared graphene/Au nanocomposites by the in situ reduction of Au^{3+} on GO sheets [82]. By taking advantage of the peroxidase-like activity of graphene/Au flexibly controlled by hairpin assembly, the authors developed a sensitive and specific platform for DNA detection. As illustrated in Fig. 10.7, without target DNA, the assembly of H1 and H2 could not be triggered, and the activity of graphene/Au was inhibited due to the surface adsorption of H1 and H2. However, the case in the presence of target was just the opposite. The target-catalyzed hairpin assembly could produce plenty of H1–H2 duplex, inducing a weak binding affinity with graphene/Au. Therefore, the protected interface of graphene/Au became catalytically active for reaction.

By similar in situ growth of metal NPs on GO sheets, Tao et al. [83] and Zhang et al. [84] prepared GO/Au and GO/Pt nanocomposites, respectively. After conjugation with folic acid, the authors proposed two simple, cheap, and highly selective

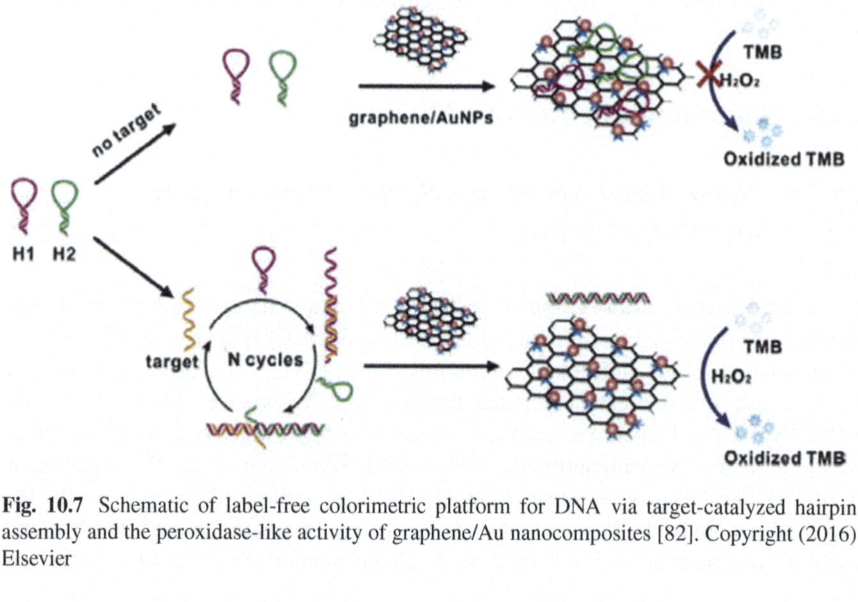

Fig. 10.7 Schematic of label-free colorimetric platform for DNA via target-catalyzed hairpin assembly and the peroxidase-like activity of graphene/Au nanocomposites [82]. Copyright (2016) Elsevier

and sensitive colorimetric assays to detect cancer cells based on the peroxidase-like activity of the as-synthesized nanocomposites. As low as 125 MCF-7 cells could be detected based on GO/Pt, lower than that of GO/Au (1000 MCF-7 cells). More recently, Zhang et al. fabricated GO/Au via the covalent modification of the pre-made Au@BSA NPs with GO by the amide bond [85]. Interestingly, the as-obtained nanocomposites showed both GOx- and peroxidase-like activities at same pH, which acted as a "tandem nanozyme" for the one-pot glucose detection, with LOD of 0.6 μM lower than that of the conventional two-step assay. Besides, Darabdhara et al. synthesized CuAg NPs supported on rGO sheets (rGO/CuAg) via a solvothermal route, in which GO was reduced to rGO [86]. A combination of rGO/CuAg as peroxidase mimics with GOx allowed the construction of a sensitive and a selective colorimetric assay for glucose detection in blood serum with a low LOD of 3.8 μM. The as-prepared rGO/CuAg nanocomposites were also applied for sensing AA with low LOD of 3.6 μM.

10.5.2 Transition Metal Dichalcogenides Supported Noble-Metal NPs

Transition metal dichalcogenides (TMDs), especially molybdenum disulfide (MoS_2) nanosheets (NSs) as typical 2D graphene-like layered compounds, have received considerable attention because of their tunable energy bandgap and natural richness [87–89]. With abundant active edges and large surface area, MoS_2 provides a promising support material for potential catalytic performance. To extend the functionalities of MoS_2 NSs as versatile materials, the modification of MoS_2 with metal NPs is of great and widespread interest [90–94].

The MoS_2/BNP nanocomposites are emerging as immobilized nanozymes. Sun et al. synthesized $Au_{core}Pd_{shell}$ NPs anchored on chemical exfoliated MoS_2 NSs through a facile co-reduction method [95]. The resulted nanocomposites with a Au/Pd mass ratio of 1:2 (MoS_2/Au_1Pd_2) exhibited synergistically enhanced peroxidase-like activity, owing to not only intrinsic catalytic activity of MoS_2 but also strong interactions of metal–metal and metal–support.

Notably, despite numerous examples of MoS_2/BNP nanocomposites in literatures, the synthesis of MoS_2/BNP with high quality and different heterostructures still remains challenging. By decorating ultrathin MoS_2 NSs with octahedral $Pt_{74}Ag_{26}$ nanoalloys, our group synthesized the MoS_2/PtAg nanocomposites [96]. As illustrated in Fig. 10.8, the few-layer MoS_2 NSs were obtained with a H_2SO_4-intercalation/exfoliation method, followed by modification of polyallylamine hydrochloride (PAH). Then PtAg octahedra were hydrothermally fabricated on the as-obtained MoS_2 NSs. The resulted $MoS_2/Pt_{74}Ag_{26}$ nanocomposites were investigated as peroxidase mimics to catalyze the TMB oxidation, producing a blue-colored reaction and exhibiting typical Michaelis–Menten kinetics. The nanocomposites had

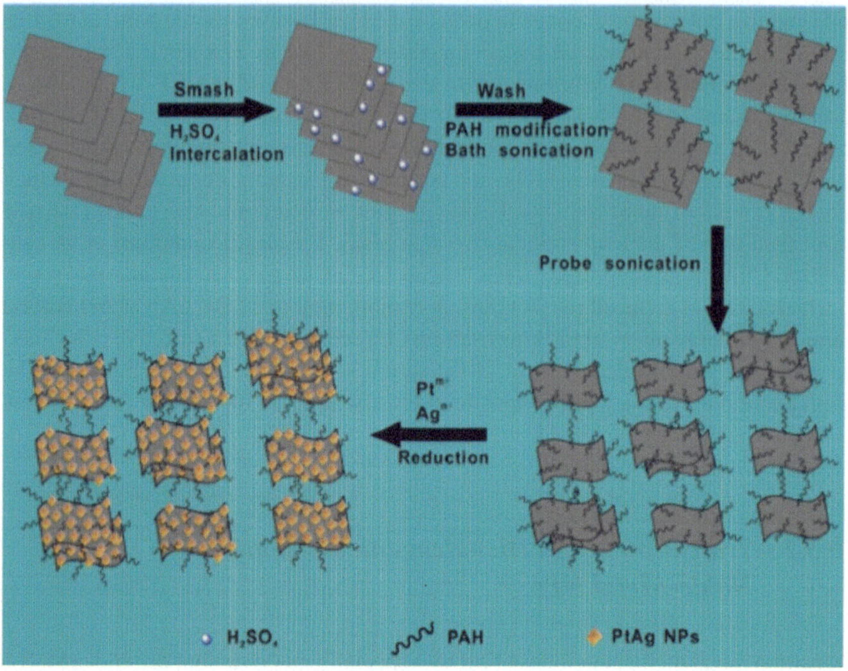

Fig. 10.8 Schematic illustration of synthesis of MoS_2-PtAg nanocomposites [96]. Copyright (2016) Royal Society of Chemistry

higher affinity for H_2O_2 than HRP and higher V_{max} value with TMB as the substrate than MoS_2. The improved catalytic activity of the nanocomposites for TMB oxidation could be attributed to synergistic effects of octahedral $Pt_{74}Ag_{26}$ NPs and ultrathin MoS_2 NSs as supports.

Similarly, uniform Pt_3Au_1 NPs decorated few-layer MoS_2 NSs as novel peroxidase mimics were studied by our group [97]. Compared to pure MoS_2 NSs, the as-prepared nanocomposites displayed enhanced peroxidase-like activity. The nanocomposites also possessed well dispersibility and high stability in water. With these findings, a simple, fast, and low-cost colorimetric detection of phenol with high sensitivity and selectivity was developed, on the basis of oxidative coupling reaction of phenol and 4-aminoantipyine with H_2O_2 as an oxidant to form pink color products.

Additionally, our group prepared MoS_2/PtCu nanocomposites with good oxidase-like activity [98]. Facilitated with their oxidase-like activity, a colorimetric aptasensor (apt-MoS_2/PtCu) was constructed and its application in the colorimetric detection of cancer cells was demonstrated, with good sensitivity and selectivity to target cells.

10.5.3 Metal Oxide-Supported Noble-Metal NPs

Some studies revealed noble-metal NPs immobilized on the reducible metal oxides could exhibit enhanced activity owing to the strong metal–support interaction. For example, Ma et al. prepared Pt-modified Fe_3O_4 NPs by a two-step procedure [99], which exhibited stronger affinity toward substrates (TMB and H_2O_2) and higher peroxidase-like activity than Fe_3O_4 NPs. Our group synthesized Pt NPs supported on 2D CuO NSs [100] and CeO_2 nanorods [101] by the liquid-phase reduction method. Both immobilized nanozymes displayed synergistically enhanced peroxidase-like activity. Benefiting from the inhibition of AA on the nanozyme activity, the colorimetric detection of AA was developed with high selectivity against amino acids, carbohydrates, and normal ions. In a similar study, Li et al. synthesized Pt NPs supported on the CeO_2 cubes, which acted as a peroxidase mimic for colorimetric detection of H_2O_2 and glucose [102].

Unlike metal NPs grown on support materials described above, Kluenker et al. demonstrated 2D Pd nanotetrahedra and nanoplates could be used to direct epitaxial growth of γ-Fe_2O_3/Pd NRs [103], providing a facile metal-templated epitaxial growth and high-yield synthesis of functional materials. With formation of a Fe_xPd "buffer phase", the interfacial strain at γ-Fe_2O_3 interface was minimized, facilitating growth of γ-Fe_2O_3 NRs. The nanocomposites exhibited 10-fold higher peroxidase-like activity compared with that of Fe_2O_3 NRs, resulting from a synergistic effect for the charge separation and electron transport. Similarly, by growing MnO_2 NSs from $KMnO_4$ and 2-(n-morpholino) ethanesulfonic acid on the surface of preformed Pt NPs as a template under sonication, Liu et al. prepared the $Pt_{core}MnO_{2shell}$ nanostructures, in which the MnO_2 NSs formed a 3D scaffold adhered to the Pt NPs [104]. The as-obtained 3D Wonton-like nanostructures displayed synergistically enhanced oxidase-like activity, owing to Pt NPs and MnO_2 NSs as well as the changes to their electronic properties in the nanocomposites. In the following studies, the $Pt_{core}MnO_{2shell}$ as an oxidase mimic was successfully employed for the colorimetric detection of GSH and dopamine.

10.5.4 Other Nanocomposites

In this section, nanozymes immobilized on other support materials are discussed. As a new star in 2D materials, metallic NSs have attracted significantly increasing attention in catalysis in recent years [105–107]. Owing to their unique properties, such as large specific surface area, high-density unsaturated atoms, and anisotropic structure, metallic NSs as promising support materials provide enormous opportunities to develop active nanozymes. To explore the enzyme-like properties of metallic NS-based nanocomposites, Wei et al. synthesized Pd NSs supported metallic nanostructures, including Au NPs and Pt nanodots [108]. The as-prepared bimetallic nanoplates possessed intrinsic peroxidase-, oxidase-, and catalase-like activities.

Compared with Pd NSs and Pd/Au, the resulted Pd/Pt nanoplates displayed the highest peroxidase-like activity because of a synergistic effect between Pt nanodots and Pd NSs. Most recently, different from the traditional two-step procedure to access the supported metallic nanostructure, our group reported the one-pot synthesis of PtRh NPs supported on ultrathin Rh NSs by a modified seed-mediated growth method [109]. The successive processes including generation of PtRh NPs, formation of Rh NSs, and immobilization of PtRh NPs on Rh NSs were involved in the synthesis. Owing to both structural and electronic effects, the as-obtained nanocomposites as an oxidase mimic exhibited synergistically enhanced activity.

Similarly, our group reported a facile protocol for one-pot preparation of Pd NPs supported on ultrathin $NiCl_2$ NSs [110]. The effective strategy can be described as a three-step process of "in situ reduction–oxidation assembly," constructing $NiCl_2$/Pd nanocomposites, which displayed synergistically enhanced peroxidase-like activity for TMB oxidation. The in situ nucleation and growth of NS around Pd NPs guaranteed cleanness of the metal–support interface, and greatly facilitated the catalysis.

The 2D metal–organic framework (MOF) NSs are another interesting material to support metallic NPs. By structurally rational design, Huang et al. synthesized ultrasmall Au NP-decorated metalloporphyrinic MOF NSs, which were used as dual nanozymes (peroxidase and GOx) for cascade reactions in glucose detection [111]. A good linear relationship between the absorbance of oxidized TMB at 652 nm and glucose concentration in the range of 10–300 μM with LOD of 8.5 μM was observed. The work provides an avenue to design nanozymes with multiple complex functions.

Other types of support materials were also studied. For example, Haider et al. found Au NPs decorated single-walled carbon nanotube (SWCNT) exhibited synergistically enhanced peroxidase-like activity and good affinity toward H_2O_2 oxidation [112]. The obtained SWCNT/Au was used as a robust nanoprobe for D-alanine detection with LDR of 0.1–25 μM and LOD of 0.05 μM.

Kim et al. prepared Au NPs deposited on carboxylated nanodiamonds (NDs) by in situ chemical reduction of Au precursor ions, which displayed peroxidase-like activity in OPD oxidation [113]. Interestingly, the remarkable catalytic activity was exhibited only by ND/Au and was not observed for either Au NPs or NDs separately. Although the sp^2 graphene-like layer was likely to exist on the surface, ND itself did not show intrinsic catalytic properties, while graphene was intrinsically active as a peroxidase. The enhanced activity of ND/Au could be attributed to the synergistic effect between Au NPs and the sp^2 carbon layer. The ND/Au could be utilized to detect xanthine using cascade catalysis.

Besides, Tao et al. found Au NPs supported on the bifunctionalized mesoporous silica displayed enhanced peroxidase-like activities and demonstrated potential in the antibacterial application [64]. Most recently, Xu et al. presented a more interesting example for the bioenergy supply system, by taking advantage of the GOx- and peroxidase-like activities of Au NPs immobilized on the hollow silica microspheres (SiO_2/Au) [114]. The nanocomposites exhibited superior cascade catalysis for mitochondria-mimicking oxidative phosphorylation, with a high efficiency of 4920 nmol mg^{-1} h^{-1} comparable to that in natural systems. As illustrated in Fig. 10.9,

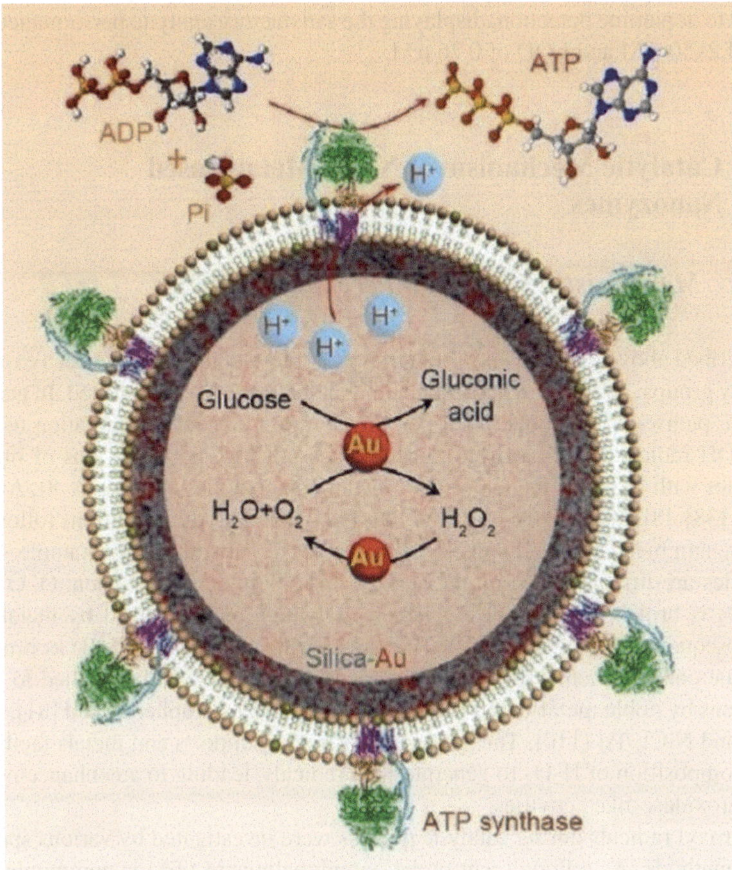

Fig. 10.9 Illustration of SiO_2/Au for cascade reactions in the mitochondria-mimicking oxidative phosphorylation process [114]. Copyright (2019) Wiley-VCH Verlag GmbH & Co. KGaA, Weinheim

the SiO_2/Au catalyzes the aerobic oxidation of glucose to gluconic acid, accompanied by the decrease of surrounding pH value and generation of a proton gradient. Subsequently, the protons diffuse across the proteoliposome membrane coating on the silica surface, facilitating the ultimate synthesis of ATP by ATP synthase. At the same time, the produced H_2O_2 is captured by the reformation of Au NPs, which protects ATP synthase from possible oxidation. The study provides a prototype of natural–artificial hybrids for in vitro applications.

Additionally, Ivanova et al. synthesized Pt NPs anchored on the chemically modified hydrophilic hexagonal boron nitride NSs [115], a new type of support that has only recently been recognized to have potential in nanozyme catalysis. The as-prepared nanocomposites possessed good peroxidase-like activity and were further applied to dopamine detection, displaying the satisfactory analytic performance with LDR of 2–50 μM and LOD of 0.76 μM.

10.6 Catalytic Mechanism of Noble Metal-Based Nanozymes

10.6.1 Mechanistic Studies on Peroxidase Mimics

As described above, enzyme-like properties of noble-metal NMs have been explored by many groups, and several key influencing factors have been identified. In general, catalytic pathways for noble-metal peroxidase mimics involve generation of ROS (e.g., •OH radicals) and electron-transfer process. Catalytic oxidations of organic substrates with H_2O_2 by the majority of metallic nanozymes (e.g., Pt [6–8], Au [19, 20], Pd [38], PtPd [65], PtCu [67], and PdIr [69]) under acidic conditions followed a radical chain mechanism. Taking TMB molecules as substrates, for example, H_2O_2 molecules are first absorbed on the surface of noble-metal NPs. Then, O–O bond of H_2O_2 is broken to generate •OH radicals, which are stabilized by metal NPs and subsequently oxidize substrates, yielding *ox*idized TMB (*ox*TMB) as products with blue color (Scheme 10.1). The similar mechanism could be applied to TMB oxidations by noble metal-based nanocomposites such as graphene/PtPd [81], Pd/Pt [108], and $NiCl_2$/Pd [110]. The synergistic effect of supports and metals facilitated the decomposition of H_2O_2 to generate •OH radicals, leading to an enhancement in their peroxidase-like activities.

Hydroxyl radicals during catalytic process were investigated by various spectroscopic methods. As a fluorescent probe, terephthalic acid (TA) is commonly used to investigate •OH, since TA easily reacts with •OH to form highly fluorescent 2-hydroxy terephthalic acid with an emission peak at about 430 nm [23, 38, 63–65]. Isopropanol and *t*-butyl alcohol [13] also acted as •OH quenchers by inhibiting peroxidase-like activity. Notably, ESR spectroscopy provides a direct and reliable method to identify and quantify •OH radicals. Dimethyl pyridine *N*-oxide (DMPO) is a commonly used trap probe [59, 110]. A four-line signal with a 1:2:2:1 peak-to-peak intensity pattern and the isotropic hyperfine coupling constants ($A^N = A^H \approx$

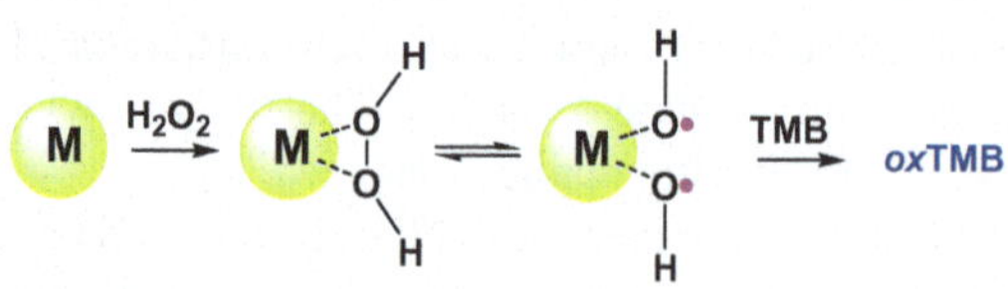

Scheme 10.1 Noble metal-based peroxidase-mimic-catalyzed TMB oxidation proceeds via a free radical chain mechanism

14.9 G) is characteristic of DMPO-HO• adduct. Besides •OH radicals, $O_2^{\bullet -}$/HOO• radicals were also found in Au NPs-melamine-TMB-H_2O_2 system [24].

Meanwhile, some studies revealed the nature of catalytic behavior of noble-metal peroxidase mimics was attributed to electron transfer between substrate and oxidant, instead of •OH radical generation. For example, Jin et al. found the yeast extract-stabilized Pt clusters consumed •OH radicals in TA oxidation experiment [12]. The electron-transfer ability of Pt clusters was further verified by electrocatalytic reduction of H_2O_2, and the reaction kinetics followed typical Michaelis–Menten theory. Cui et al. [40] also showed that no •OH formed during the TMB oxidation with tannin acid-stabilized Ir NPs, evidenced by the ESR experiment. The authors speculated Ir NPs could facilitate electron transfer between TMB and H_2O_2, i.e., TMB was absorbed on Ir NP surface and then donated lone-pair electrons from amino groups to Ir NPs, resulting in the increase of electron density of Ir NPs. The electrons subsequently transferred from Ir NPs to H_2O_2 and facilitated oxidation. Similar observation was reported for hollow PdNi BNPs by Wang et al. [71].

However, not all peroxidase-mimic-catalyzed oxidations occur with a single pathway. Biswas et al. [26] revealed the peroxidase-like activity of CTAB-stabilized Au NRs was attributed to multiple processes, including generation of •OH and partial electron transfer. Regardless of the mechanism involved, one of the important characteristics of peroxidase mimics is the accessibility of reactants.

10.6.2 Mechanistic Studies on Oxidase Mimics

As revealed, Pt-based oxidase mimics [13, 68] have good capacity of activating oxygen for the aerobic oxidation of organic substrates at low pH. Similar to the cases of noble-metal peroxidase mimics, catalytic pathways for noble-metal oxidase mimics can also be conveniently divided into two categories: generation of ROS (e.g., $O_2^{\bullet -}$ radicals) and electron-transfer process. For the former, it is believed that the first step is adsorption of dissolved O_2 on nanozyme surface (Scheme 10.2). The successive one-electron transfer results in reduction of O_2 molecules, releasing $O_2^{\bullet -}$ radicals to oxidize substrates eventually. A typical example is the chitosan-stabilized Pt NPs for oxidase-mimic catalysis [13]. $O_2^{\bullet -}$ can be detected by several spectroscopic methods. A common $O_2^{\bullet -}$ dye is nitro blue tetrazolium (NBT), which reacts with $O_2^{\bullet -}$ to form a blue compound with a maximum absorbance at 680 nm [68]. Besides, as a highly selective fluorescent probe for $O_2^{\bullet -}$ radicals, hydroethidine is often used by the increase in the intensity of its fluorescence [64]. AA can also be employed to capture $O_2^{\bullet -}$ [13].

Scheme 10.2 Noble metal-based oxidase-mimic-catalyzed TMB oxidation by $O_2^{\bullet -}$

On the other hand, it is known Pt is the highly efficient electrocatalyst for oxygen reduction reaction (ORR). The TMB oxidation with $Au_{core}Pt_{shell}$ nanostructures [46] was also believed to go through a similar pathway to the electrocatalytic reduction of oxygen, where electrons were provided by TMB molecules. Yu et al. [14] revealed that TMB oxidation with lysozyme-stabilized Pt as an oxidase mimic involved a four-electron reduction process [$O_2(g) + 4H^+(aq) + 4e^- \rightarrow 2H_2O(l)$], and TMB acted as an electron donor, rather than generation of $O_2^{\bullet -}$ radicals. Similar experimental observation was obtained for the $Pt_{core}MnO_{2shell}$ nanostructures as an oxidase mimic, reported by Liu et al. [104]. Despite above findings, it is noteworthy that the actual active sites during the catalysis are still unclear. For a specific oxidase mimic, it is also hard to accurately predict the mechanism, considering different natures of catalytically active metals (composition, structure, size, and morphology), coating agents, surface charge, support materials, and reaction conditions. Thus, a deep understanding of catalytic mechanism remains challenging yet highly desirable.

10.6.3 Mechanistic Studies on Catalase Mimics

As is known, the decomposition of H_2O_2 has received extensive studies in water treatment. It has been also used as a probe reaction for the redox catalytic activity of noble-metal nanozymes. To study mechanism of the dealloyed $Pt_{50}Ag_{50}$ (D-$Pt_{50}Ag_{50}$) NPs for catalase-mimic catalysis under alkaline conditions [62], the ESR experiments using 3-carbamoyl-2,2,5,5-tetramethyl-3-pyrroline-1-yloxyl (CTPO) as a trap in a closed chamber with samples were carried out by our group. To quantify the evolution of oxygen concentration, a parameter K that reflects the change in super hyperfine structure in ESR spectra [9] was calculated by the equation $K = (h_1 + h_2)/2h$, where h_1, h_2, and h are determined as shown in Fig. 10.10. The comparison of activities of native SOD, $Pt_{50}Ag_{50}$, and D-$Pt_{50}Ag_{50}$ NPs using K values showed D-$Pt_{50}Ag_{50}$ NPs had the highest activity.

10.6.4 Mechanistic Studies on SOD Mimics

The SOD-like activity of Pt NPs under neutral conditions was studied by ESR spectroscopy [10]. Xanthine/XOD was first chosen as the generator of $O_2^{\bullet -}$ radicals, which were then captured by 5-tert-bytoxycarbonyl-5-methyl-1-pyrroline-*N*-oxide (BMPO) to form spin adducts (BMPO/HOO•). A typical ESR spectrum of BMPO/HOO• has four lines with relative intensities of 1:1:1:1 (Fig. 10.11a) and the hyperfine splitting parameters of the lines are $A^H = 13.4$, $A_H^\beta = 12.1$. Upon the addition of Pt NPs, ESR signal intensity of adducts gradually decreased (Fig. 10.11b), suggesting their ability to scavenge $O_2^{\bullet -}$ radicals similar to SOD.

Although noble-metal nanozymes provide a valuable platform to probe fundamental processes at nanoscale surface, it should be noted that active intermediates during

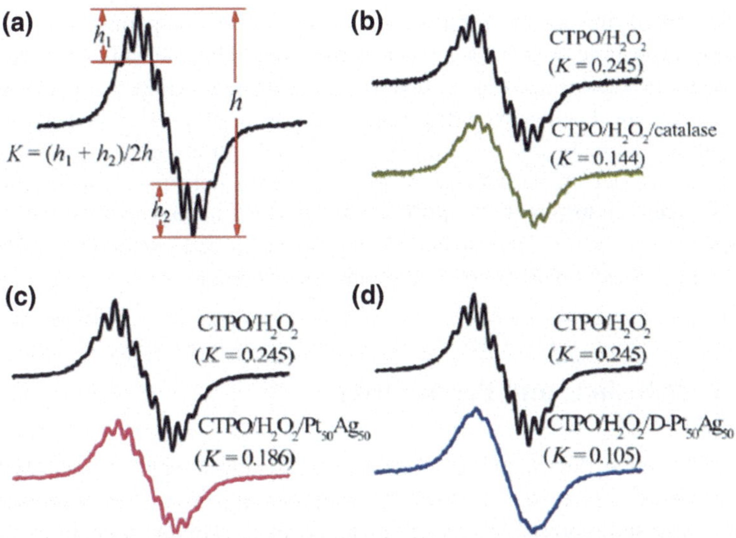

Fig. 10.10 Enhanced catalase-like activity of the dealloyed $Pt_{50}Ag_{50}$ NPs [62]. Copyright (2017) Tsinghua University Press and Springer-Verlag GmbH Germany, Part of Springer Nature

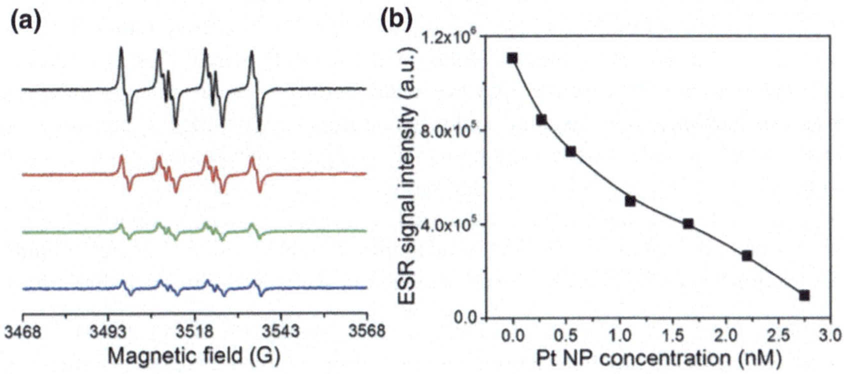

Fig. 10.11 Evidence for SOD-like activity of Pt NPs [10]. **a** ESR spectra of BMPO/HOO• adducts obtained from samples containing 1 mM xanthine, 0.05 unit per mL XOD, 0.05 mM DTPA, 25 mM BMP, 10 mM PBS buffer (pH 7.27), and different concentrations of Pt NPs: 0 mM (black); 1.1 mM (red); 2.75 mM (green); 0.5 unit per mL SOD (blue). **b** Effect of Pt NP concentration on the activity. Copyright (2014) Royal Society of Chemistry

actual reactions are complex and competition catalysis might exist. Some studies showed catalase-like activity might suppress peroxidase-like activity of nanozymes, since both have H_2O_2 as a substrate. The type of mechanism is closely associated with H_2O_2 concentration [46]. Similarly, owing to presence of oxygen, the oxidase-like activity of nanozymes might be inhibited by their SOD-like activity. Thus, to attain high nanozyme activity for a specific catalysis, reaction parameters (e.g., pH, temperature, substrate, and H_2O_2 concentration) should be precisely adjusted.

Although there have been many investigations on nanozymes, some unresolved issues still remain. For instance, for a bimetallic or multimetallic nanozyme, it is challenging to elucidate how and why introducing other metal components has such a significant impact on its activity. Moreover, we do not know which structural feature of nanocomposites is the most beneficial to the nanozymes' activity or the identities of true reactive intermediate during the catalytic reaction cycle. Thus, to design highly efficient nanozymes still remains a challenge.

10.7 Conclusions and Perspectives

As discussed above, this chapter gives a summary of current noble-metal NM-based enzyme mimics for bioanalysis. Associated with their intrinsic nature (composition, structure, size, and morphology), surface charge state, capping molecules, support materials, and specific reaction conditions (e.g., pH, temperature, and substrate), several noble-metal NMs displayed different enzymatic properties. By varying these key influencing factors, as well as introducing appropriate metal ions or biological relevant molecules, their catalytic activities could be adjusted. From these observations, a number of biological applications were also demonstrated, mostly in the analytical chemistry and biomedical fields. With tunability in activities and high stability, noble metal NM-based nanozymes could be utilized as promising candidates for natural enzymes. Besides, they could also be attractive for studying fundamental processes at nanoscale surface, facilitating the in-depth understanding of mechanism and providing valuable guidance for the rational design of nanozymes.

To date, although many noble metal-based nanozymes have been identified, some basic issues still need to be fully addressed in future studies, which ultimately require interdisciplinary collaborations. Briefly, we herein speculate a few such directions.

(1) *Reducing nanozyme cost.* The literature is currently full of Pt- and Au-based nanozymes. For practical applications, nanozymes with high catalytic efficiency yet low cost would be more desirable and appealing. In this goal, a solution may be offered by constructing bimetallic nanozymes. Besides, the single-atom nanozymes, a very new concept with little reported works [116–118], could provide great opportunities to develop advanced enzyme mimics at the atomic scale.

(2) *Enhancing catalytic activity and stability.* Although nanozymes' activities could be tuned by several strategies, some of them are still low compared with those of natural enzymes. Thus, the utilization of enzymatic properties of noble-metal NMs for practical biological applications still remains largely unexplored. We believe the controllable synthesis of noble-metal NMs may be benefited from the deep understanding of their nucleation and growth mechanism. On the other hand, although surface modification or enzyme immobilization on appropriate

matrixes could improve their stability, the catalytically active surface would be blocked simultaneously. Thus, it is critical to find a balance between colloidal stability and catalytic activity of nanozymes.

(3) *Exploring novel reactions*. Although numerous noble-metal NMs were found to possess enzymatic properties, current types of enzyme-like activities studied are mostly redox reactions and focus on peroxidase-like activity. While, studies on oxidase-mimic-catalyzed reactions, which do not require the use of constructive and unstable H_2O_2, are few. In a broader sense, to expand activity of noble-metal NMs as multifunctional nanozymes to other reactions is of great significance. This new front, however, has not been explored to a large degree.

(4) *Improving substrate specificity*. Although most studies focused on catalytic activity of nanozymes, they lack substrate specificity compared with most natural enzymes. Some attractive strategies such as by introducing specific ligands are now available [119, 120], which could be a potential direction for future research.

Acknowledgements This work was supported by the National Key Research and Development Program from the Ministry of Science and Technology of China (2016YFC0207102), National Natural Science Foundation of China (21501034, 21573050), and Strategic Priority Research Program of the Chinese Academy of Sciences (Grant No. XDA09030303).

References

1. Gao LZ, Zhuang J, Nie L, Zhang JB, Zhang Y, Gu N, Wang TH, Feng J, Yang DL, Perrett S, Yan XY (2007) Intrinsic peroxidase-like activity of ferromagnetic nanoparticles. Nat Nanotechnol 2(9):577–583
2. Hashmi ASK, Hutchings GJ (2006) Gold catalysis. Angew Chem Int Ed 45(47):7896–7936
3. Johansson Seechurn CC, Kitching MO, Colacot TJ, Snieckus V (2012) Palladium-catalyzed cross-coupling: a historical contextual perspective to the 2010 Nobel prize. Angew Chem Int Ed 51(21):5062–5085
4. Pushkarev VV, Zhu ZW, An K, Hervier A, Somorjai GA (2012) Monodisperse metal nanoparticle catalysts: synthesis, characterizations, and molecular studies under reaction conditions. Top Catal 55(19–20):1257–1275
5. Borodko Y, Jones L, Lee H, Frei H, Somorjai GA (2009) Spectroscopic study of tetradecyltrimethylammonium bromide Pt-C_{14}TAB nanoparticles: structure and stability. Langmuir 25(12):6665–6671
6. Ma M, Zhang Y, Gu N (2011) Peroxidase-like catalytic activity of cubic Pt nanocrystals. Colloid Surf A 373(1):6–10
7. Fu Y, Zhao XY, Zhang JL, Li W (2014) DNA-based platinum nanozymes for peroxidase mimetics. J Phys Chem C 118(31):18116–18125
8. Li W, Chen B, Zhang HX, Sun YH, Wang J, Zhang JL, Fu Y (2015) BSA-stabilized Pt nanozyme for peroxidase mimetics and its application on colorimetric detection of mercury(II) ions. Biosens Bioelectron 66(66c):251–258
9. Fan J, Yin JJ, Ning B, Wu XC, Hu Y, Ferrari M, Anderson GJ, Wei JY, Zhao YL, Nie GJ (2011) Direct evidence for catalase and peroxidase activities of ferritin-platinum nanoparticles. Biomaterials 32(6):1611–1618

10. Liu Y, Wu HH, Li M, Yin JJ, Nie ZH (2014) pH dependent catalytic activities of platinum nanoparticles with respect to the decomposition of hydrogen peroxidase and scavenging of superoxide and singlet oxygen. Nanoscale 6(20):11904–11910
11. Li JN, Liu WQ, Wu XC, Gao XF (2015) Mechanism of pH-switchable peroxidase and catalase-like activities of gold, silver, platinum and palladium. Biomaterials 48:37–44
12. Jin LH, Meng Z, Zhang YQ, Cai SJ, Zhang ZH, Li C, Shang L, Shen YH (2017) Ultrasmall Pt nanoclusters as robust peroxidase mimics for colorimetric detection of glucose in human serum. ACS Appl Mater Inter 9(11):10027–10033
13. Deng HH, Lin XL, Liu YH, Li KL, Zhang QQ, Peng HP, Liu AL, Xia XH, Chen W (2017) Chitosan-stabilized platinum nanoparticles as effective oxidase mimics for colorimetric detection of acid phosphate. Nanoscale 9(29):10292–10300
14. Yu CJ, Chen TH, Jiang JY, Tseng WL (2014) Lysozyme-directed synthesis of platinum nanoclusters as a mimic oxidase. Nanoscale 6(16):9618–9624
15. Clark A, Zhu AP, Sun K, Petty HR (2011) Cerium oxide and platinum nanoparticles protect cells from oxidant-mediated apoptosis. J Nanopart Res 13(10):5547–5555
16. Zhang LB, Laug L, Münchgesang W, Pippel E, Gösele U, Brandsch M, Knez M (2010) Reducing stress on cells with apoferritin-encapsulated platinum nanoparticles. Nano Lett 10(1):219–223
17. Wang XX, Wu Q, Shan Z, Huang QM (2011) BSA-stabilized Au clusters as peroxidase mimetics for use in xanthine detection. Biosens Bioelectron 26(8):3614–3619
18. Ahmed SR, Kim J, Suzuki T, Lee J, Park EY (2016) Detection of influenza virus using peroxidase-mimic of gold nanoparticles. Biotech Bioeng 113(10):2298–2303
19. Jiang X, Sun CJ, Guo Y, Nie GJ, Xu L (2015) Peroxidase-like activity of apoferritin paired gold clusters for glucose detection. Biosens Bioelectron 64:165–170
20. Shah J, Purohit R, Singh R, Karakoti AS, Singh S (2015) ATP-enhanced peroxidase-like activity of gold nanoparticles. J Colloid Interface Sci 456:100–107
21. Wang S, Chen W, Liu AL, Hong L, Deng HH, Lin XH (2012) Comparison of the peroxidase-like activity of unmodified, amino-modified, and citrate-capped gold nanoparticles. ChemPhysChem 13:1199–1204
22. Ni PJ, Dai HC, Wang YL, Sun YJ, Shi Y, Hu JT, Li Z (2014) Visual detection of melamine based on the peroxidase-like activity enhancement of bare gold nanoparticles. Biosens Bioelectron 60:286–291
23. Wang C, Shi Y, Dan YY, Nie XG, Li J, Xia XH (2017) Enhanced peroxidase-like performance of gold nanoparticles by hot electrons. Chem Eur J 23(28):6717–6723
24. Hizir MS, Top M, Balcioglu M, Rana M, Robertson NM, Shen FS, Sheng J, Yigit MV (2016) Multiplexed activity of perAuxidase: DNA-capped AuNPs act as adjustable peroxidase. Anal Chem 88:600–605
25. Feng JY, Huang PC, Shi SZ, Deng KY, Wu FY (2017) Colorimetric detection of glutathione in cells based on peroxidase-like activity of gold nanoclusters: a promising powerful tool for identifying cancer cells. Anal Chim Acta 967:64–69
26. Biswas S, Tripathi P, Kumar N, Nara S (2016) Gold nanorods as peroxidase mimetics and its application for colorimetric biosensing of malathion. Sensor Actuat B-Chem 231:584–592
27. Chang YQ, Zhang Z, Hao JH, Yang WS, Tang JL (2016) BSA-stabilized Au clusters as peroxidase mimetic for colorimetric detection of Ag^{+}. Sensor Actuat B-Chem 232:692–697
28. Zhu R, Zhou Y, Wang XL, Liang LP, Long YJ, Wang QL, Zhang HJ, Huang XX, Zheng HZ (2013) Detection of Hg^{2+} based on the selective inhibition of peroxidase mimetic activity of BSA-Au clusters. Talanta 117(22):127–132
29. Mikami Y, Dhakshinamoorthy A, Alvaro M, García H (2013) Catalytic activity of unsupported gold nanoparticles. Catal Sci Technol 3:58–69
30. Comotti M, Pina CD, Matarrese R, Rossi M (2004) The catalytic activity of "naked" gold particles. Angew Chem Int Ed 43(43):5812–5815
31. Beltrame P, Comotti M, Pina CD, Rossi M (2006) Aerobic oxidation of glucose II. Catalysis by colloidal gold. Appl Catal A Gen 297 (1):1–7

32. Luo WJ, Zhu CF, Su S, Li D, He Y, Huang Q, Fan CH (2010) Self-catalyzed, self-limiting growth of glucose oxidase-mimicking gold nanoparticles. ACS Nano 4(12):7451–7458
33. Zheng XX, Liu Q, Jing C, Li Y, Li D, Luo WJ, Wen YQ, He Y, Huang Q, Long YT, Fan CH (2011) Catalytic gold nanoparticles for nanoplasmonic detection of DNA hybridization. Angew Chem Int Ed 50(50):11994–11998
34. Lan J, Xu W, Wan Q, Zhang X, Lin J, Chen J, Chen J (2014) Colorimetric determination of sarcosine in urine samples of prostatic carcinoma by mimic enzyme palladium nanoparticles. Anal Chim Acta 825(2):63–68
35. Shi W, Fan H, Ai S, Zhu L (2015) Pd nanoparticles supported on nitrogen, sulfur-doped three-dimensional hierarchical nanostructures as peroxidase-like catalysts for colorimetric detection of xanthine. RSC Adv 5(41):32183–32190
36. Fu Y, Zhang H, Dai S, Zhi X, Zhang J, Li W (2015) Glutathione-stabilized palladium nanozyme for colorimetric assay of silver(I) ions. Analyst 140(19):6676–6683
37. Li W, Zhi X, Yang JJ, Zhang JL, Fu Y (2016) Colorimetric detection of cysteine and homocysteine based on an oligonucleotide-stabilized Pd nanzyme. Anal Methods 8(25):5111–5116
38. Fang G, Li WF, Shen XM, Perez-Aguilar JM, Chong Y, Gao XF, Chai ZF, ChenCY Ge CC, Zhou RH (2018) Differential Pd-nanocrystal facets demonstrate distinct antibacterial activity against Gram-positive and Gram-negative bacteria. Nat Commun 9:129
39. Ye HH, Mohar J, Wang QX, Catalano M, Kim MJ, Xia XH (2016) Peroxidase-like properties of ruthenium nanoframes. Sci Bull 61(22):1739–1745
40. Cui ML, Zhou JD, Zhao Y, Song QJ (2017) Facile synthesis of iridium nanoparticles with superior peroxidase-like activity for colorimetric detection of H_2O_2 and xanthine. Sens Actuat B 243:203–210
41. Choleva TG, Gatselou VA, Tsogas GZ, Giokas DL (2018) Intrinsic peroxidase-like activity of rhodium nanoparticles, and their application to the colorimetric determination of hydrogen peroxide and glucose. Microchim Acta 185(1):22–30
42. Cai SF, Xiao W, Duan HH, Liang XX, Wang C, Yang R (2018) Single-layer Rh nanosheets with ultrahigh peroxidase-like activity for colorimetric biosensing. Nano Res 11(12):6304–6315
43. Wang DS, Li YD (2011) Bimetallic nanocrystals: liquid-phase synthesis and catalytic applications. Adv Mater 23(9):1044–1060
44. Gu J, Zhang YW, Tao F (2012) Shape control of bimetallic nanocatalysts through well-designed colloidal chemistry approaches. Chem Soc Rev 41(24):8050–8065
45. Tao F (2012) Synthesis, catalysis, surface chemistry and structure of bimetallic nanocatalysts. Chem Soc Rev 41(24):7977–7979
46. He WW, Liu Y, Yuan JS, Yin JJ, Wu XC, Hu XN, Zhang K, Liu JB, Chen CY, J YL, Guo YT (2011) Au@Pt nanostructures as oxidase and peroxidase mimetics for use in immunoassays. Biomaterials 32:1139–1147
47. Liu JB, Hu XN, Hou S, Wen T, Lu WQ, Zhu X, Yin JJ, Wu XC (2012) Au@Pt core/shell nanorods with peroxidase- and ascorbate oxidase-like activities for improved detection of glucose. Sensor Actuat B-Chem 166–167(6):708–714
48. Singh S, Tripathi P, Kumar N, Nara S (2017) Colorimetric sensing of malathion using palladium-gold bimetallic nanozyme. Biosens Bioelectron 92:280–286
49. Gao ZQ, Xu MD, Lu MH, Chen GN, Tang DP (2015) Urchin-like (gold core)@(platinum shell) nanohybrids: a highly efficient peroxidase-mimetic system for in situ amplified colorimetric immunoassay. Biosens Bioelectron 70:194–201
50. Pan N, Zhu Y, Wu LL, Xie ZJ, Xue F, Peng CF (2016) Highly sensitive colorimetric detection of copper ions based on regulating the peroxidase-like activity of Au@Pt nanohybrids. Anal Methods 8:7531–7536
51. Nangia Y, Kumar B, Kaushal J, Suri CR (2012) Palladium@gold bimetallic nanostructures as peroxidase mimic for development of sensitive fluoroimmunoassay. Anal Chim Acta 751:140–145
52. Wu Z, Fu QQ, Yu ST, Sheng LR, Xu M, Yao CZ, Xiao W, Li XQ, Tang Y (2016) Pt@AuNPs integrated quantitative capillary-based biosensors for point-of-care testing application. Biosens Bioelectron 85:657–663

53. Lee JW, Son J, Yoo KM, Lo M, Moon B (2014) Characterization of the antioxidant activity of gold@platinum nanoparticles. RSC Adv 4:19824–19830
54. Wu GW, Shen YM, Shi XQ, Deng HH, Zheng XQ, Peng HP, Liu AL, Xia XH, Chen W (2017) Bimetallic Bi/Pt peroxidase mimic and its bioanalytical applications. Anal Chim Acta 971(8):88–96
55. He WW, Han XN, Jia HM, Cai JH, Zhou YL, Zheng Z (2016) AuPt alloy nanostructures with tunable composition and enzyme-like activities for colorimetric detection of bisulfide. Sci Rep 7:40103
56. Sun YH, Wang J, Li W, Zhang JL, Zhang YD, Fu Y (2015) DNA-stabilized bimetallic nanozyme and its application on colorimetric assay of biothiols. Biosens Bioelectron 74:1038–1046
57. Jiang T, Song Y, Wei TX, Li H, Du D, Zhu MJ, Lin YH (2016) Sensitive detection of Escherichia coli O157:H7 using Pt-Au bimetal nanoparticles with peroxidase-like amplification. Biosens Bioelectron 77:687–694
58. He WW, Wu XC, Liu JB, Hu XN, Zhang K, Hou S, Zhou WY, Xie SS (2010) Design of AgM bimetallic alloy nanostructures (M = Au, Pd, Pt) with tunable morphology and peroxidase-like activity. Chem Mater 22(9):2988–2994
59. Wu LL, Wang LY, Xie ZJ, Xue F, Peng CF (2016) Colorimetric detection of Hg^{2+} based on inhibiting the peroxidase-like activity of DNA-Ag/Pt nanoclusters. RSC Adv 6:75384–75389
60. Wu LL, Wang LY, Xie ZJ, Pan N, Peng CF (2016) Colorimetric assay of L-cysteine based on peroxidase-mimicking DNA-Ag/Pt nanoclusters. Sensor Actuat B-Chem 235:110–116
61. Polarz S, Smarsly B (2002) Nanoporous materials. J Nanosci Nanotechno 2(6):581–612
62. Cai SF, Jia XH, Han QS, Yan XY, Yang R, Wang C (2017) Porous Pt/Ag nanoparticles with excellent multifunctional enzyme mimic activities and antibacterial effects. Nano Res 10(6):2056–2069
63. Sun HJ, Gao N, Dong K, Ren JS, Qu XG (2014) Graphene quantum dots-band-acids used for wound disinfection. ACS Nano 8(6):6202–6210
64. Tao Y, Ju EG, Ren JS, Qu XG (2015) Bifunctionalized mesoporous silica-supported gold nanoparticles: intrinsic oxidase and peroxidase catalytic activities for antibacterial applications. Adv Mater 27(6):1097–1104
65. Ge SG, Liu WY, Liu HY, Liu F, Yu JH, Yan M, Huang JD (2015) Colorimetric detection of the flux of hydrogen peroxide released from living cells based on the high peroxidase-like catalytic performance of porous PtPd nanorods. Biosens Bioelectron 71:456–462
66. Zhu ZJ, Zhai YL, Zhu CZ, Wang ZX, Dong SJ (2013) Bimetallic alloy nanowires and nanosponges: a comparative study of peroxidase mimetics and as enhanced catalysts for oxygen reduction reaction. Electrochem Commun 36(6):22–25
67. Lu Y, Ye WC, Yang Q, Yu J, Wang Q, Zhou PP, Wang CM, Xue DS, Zhao SQ (2016) Three-dimensional hierarchical porous PtCu dendrites: a highly efficient peroxidase nanozyme for colorimetric detection of H_2O_2. Sensor Actuat B-Chem 230:721–730
68. Cai SF, Qi C, Li YD, Han QS, Yang R, Wang C (2016) PtCo bimetallic nanoparticles with high oxidase-like catalytic activity and their applications for magnetic-enhanced colorimetric biosensing. J Mater Chem B 4(10):1869–1877
69. Xia XH, Zhang JT, Lu N, Kim MJ, Ghale K, Xu Y, McKenzie E, Liu JB, Ye HH (2015) Pd-Ir core-shell nanocubes: a type of highly efficient and versatile peroxidase mimic. ACS Nano 9(10):9994–10004
70. Jiang T, Song Y, Du D, Liu XT, Lin YH (2016) Detection of p53 protein based on mesoporous Pt-Pd nanoparticles with enhanced peroxidase-like catalysis. ACS Sensors 1(6):717–724
71. Wang QQ, Zhang LL, Shang CS, Zhang ZQ, Dong SJ (2016) Triple-enzyme mimetic activity of nickel-palladium hollow nanoparticles and their application in colorimetric biosensing of glucose. Chem Commun 52(31):5410–5413
72. Han L, Li CC, Zhang T, Lang QL, Liu AH (2015) Au@Ag heterogeneous nanorods as nanozyme interfaces with peroxidase-like activity and their application for one-pot analysis of glucose at nearly neutral pH. ACS Appl Mater Inter 7(26):14463–14470

73. Cai Z, Kuang Y, Qi XH, Wang P, Zhang Y, Zhang ZC, Sun XM (2015) Ultrathin branched PtFe and PtRuFe nanodendrites with enhanced electrocatalytic activity. J Mater Chem A 3(3):1182–1187
74. He WW, Wu XC, Liu JB, Zhang K, Chu WG, Feng LL, Hu XN, Zhou WY, Xie SS (2010) Formation of AgPt alloy nanoislands *via* chemical etching with tunable optical and catalytic properties. Langmuir 26(6):4443–4448
75. Zhang K, Hu XN, Liu JB, Yin JJ, Hou S, Wen T, He WW, Ji YL, Guo YT, Wang Q, Wu XC (2011) Formation of PdPt alloy nanodots on gold nanorods: tuning oxidase-like activities *via* composition. Langmuir 27(6):2796–2803
76. Li JR, Lv L, Zhang GN, Zhou XD, Shen AG, Hu JM (2016) Core-shell Fructus Broussonetia-like Au@Ag@Pt nanoparticles as highly efficient peroxidase mimetics for supersensitive resonance-enhanced Raman sensing. Anal Methods 8(9):2097–2105
77. Mu XY, Wang JY, Li YH, Xu FJ, Long W, Ouyang LF, Liu HL, Jing YQ, Wang JY, Dai HT, Liu Q, Sun YM, Liu CL (2019) Redox trimetallic nanozyme with neutral environment preference for brain injury. ACS Nano 13(2):1870–1884
78. Novoselov KS, Geim AK, Morozov SV, Jiang D, Zhang Y, Dubonos SV, Grigorieva IV, Firsov AA (2004) Electric field in atomically thin carbon films. Science 306:666–669
79. Ahmed SR, Takemeura K, Li TC, Kitamoto N, Tanaka T, Suzuki T, Park EY (2017) Size-controlled preparation of peroxidase-like graphene-gold nanoparticle hybrids for the visible detection of norovirus-like particles. Biosens Bioelectron 87:558–565
80. Wang AC, Zhao HM, Chen XC, Tan B, Zhang YB, Quan X (2017) A colorimetric aptasensor for sulfadimethoxine detection based on peroxidase-like activity of graphenene/nickel@palladium hybrids. Anal Biochem 525(15):92–99
81. Chen XM, Su BY, Cai ZX, Chen X, Oyama M (2014) PtPd nanodendrites supported on graphene nanosheets: a peroxidase-like catalyst for colorimetric detection of H_2O_2. Sensor Actuat B-Chem 201(4):286–292
82. Chen CH, Li NX, Lan JW, Ji XH, He ZK (2016) A label-free colorimetric platform for DNA *via* target-catalyzed hairpin assembly and the peroxidase-like catalytic of graphene/Au-NPs hybrids. Anal Chim Acta 902:154–159
83. Tao Y, Lin YH, Huang ZZ, Ren JS, Qu XG (2013) Incorporating graphene oxide and gold nanoclusters: a synergistic catalyst with surprisingly high peroxidase-like activity over a broad pH range and its application for cancer cell detection. Adv Mater 25(18):2594–2599
84. Zhang LN, Deng HH, Lin FL, Xu XW, Weng SH, Liu AL, Lin XH, Xia XH, Chen W (2014) In situ growth of porous platinum nanoparticles on graphene oxide for colorimetric detection of cancer cells. Anal Chem 86(5):2711–2718
85. Zhang HJ, Liang X, Han L, Li F (2018) "Non-naked" gold with glucose oxidase-like activity: a nanozyme for tandem catalysis. Small 14(44):1803256
86. Darabdhara G, Sharma B, Das MR, Boukherroub R, Szunerits S (2017) Cu-Ag bimetallic nanoparticles on reduced graphene oxide nanosheets as peroxidase mimic for glucose and ascorbic acid detection. Sensor Actuat B-Chem 238:842–851
87. Ganatra R, Zhang Q (2014) Few-layer MoS_2: a promising layered semiconductor. ACS Nano 8(5):4074–4099
88. Huang YX, Guo JH, Kang YJ, Ai Y, Li MC (2015) Two dimensional atomically thin MoS_2 nanosheets and their sensing applications. Nanoscale 7(46):19358–19376
89. Zhang X, Lai ZC, Tan CL, Zhang H (2016) Solution-processed two-dimensional MoS_2 nanosheets: preparation, hybridization, and applications. Angew Chem Int Ed 55(31):8816–8838
90. Singha SS, Nandi D, Singha A (2015) Tuning the photoluminescence and ultrasensitive trace detection properties of few-layer MoS_2 by decoration with gold nanoparticles. RSC Adv 5(31):24188–24193
91. Yuwen L, Xu F, Xue B, Luo Z, Zhang Q, Bao B, Su S, Weng L, Huang W, Wang L (2014) General synthesis of noble metal (Au, Ag, Pd, Pt) nanocrystals modified MoS_2 nanosheets and the enhanced catalytic activity of Pd-MoS_2 for methanol oxidation. Nanoscale 6(11):5762–5769

92. Deng J, Li H, Xiao J, Tu Y, Deng D, Yang H, Tian H, Li J, Ren P, Bao X (2015) Arousing the electrocatalytic hydrogen evolution activity of inert wo-dimensional MoS_2 surface *via* single-atom metal doping. Energy Environ Sci 8(5):1594–1601
93. Zhong X, Yang H, Guo S, Li S, Gou G, Niu Z, Dong Z, Lei Y, Jin J, Li R, Ma J (2012) In situ growth of Ni-Fe alloy on graphene-like MoS_2 for catalysis of hydrazine oxidation. J Mater Chem 22(28):13925–13927
94. Su S, Zhang C, Yuwen LH, Liu XF, Wang LH, Fan CH, Wang LH (2016) Uniform Au@Pt core-shell nanodendrites supported on molybdenum disulfide nanosheets for the methanol oxidation reaction. Nanoscale 8(1):602–608
95. Sun Z, Zhao QS, Zhang GH, Li Y, Zhang GL, Zhang FB, Fan XB (2015) Exfoliated MoS_2 supported Au-Pd bimetallic nanoparticles with core-shell structures and superior peroxidase-like activities. RSC Adv 5(14):10352–10357
96. Cai SF, Han QS, Qi C, Lian Z, Jia XH, Yang R, Wang C (2016) $Pt_{74}Ag_{26}$ nanoparticle-decorated ultrathin MoS_2 nanosheets as novel peroxidase mimics for highly selective colorimetric detection of H_2O_2 and glucose. Nanoscale 8(6):3685–3693
97. Cai SF, Han QS, Qi C, Wang XH, Wang T, Jia XH, Yang R, Wang C (2017) MoS_2-Pt_3Au_1 nanocomposites with enhanced peroxidase-like activities for selective colorimetric detection of phenol. Chin J Chem 35(5):605–612
98. Qi C, Cai SF, Wang XH, Li JY, Lian Z, Sun SS, Yang R, Wang C (2016) Enhanced oxidase/peroxidase-like activities of aptamer conjuagated MoS2/PtCu nanocomposites and their biosensing application. RSC Adv 6(60):54949–54955
99. Ma M, Xie J, Zhang Y, Chen ZP, Gu N (2013) Fe_3O_4@Pt nanoparticles with enhanced peroxidase-like catalytic activity. Mater Lett 105(16):36–39
100. Wang XH, Han QS, Cai SF, Wang T, Qi C, Yang R, Wang C (2017) Analyst 142(13):2500–2506
101. Liu XL, Wang XH, Qi C, Han QS, Xiao W, Cai SF, Wang C, Yang R (2019) Sensitive colorimetric detection of ascorbic acid using Pt/CeO_2 nanocomposites as peroxidase mimics. Appl Surf Sci 479:532–539
102. Li ZH, Yang XD, Yang YB, Tan YN, He Y, Liu M, Liu XW, Yuan Q (2018) Peroxidase-mimicking nanozyme with enhanced activity and high stability based on metal-support interactions. Chem Eur J 24(2):409–415
103. Kluenker M, Tahir MN, Ragg R, Korschelt K, Simon P, Gorelik TE, Barton B, Shylin SI, Panthöfer M, Herzberger J, Frey H, Ksenofontov V, Möller A, Kolb U, Grin J, Tremel W (2017) Pd@Fe_2O_3 superparticles with enhanced peroxidase activity by solution phase epitaxial growth. Chem Mater 29(3):1134–1146
104. Liu J, Meng LJ, Fei ZF, Dyson PJ, Zhang L (2018) On the origin of the synergy between the Pt nanoparticles and MnO_2 nanosheets in Wonton-like 3D nanozyme oxidase mimics. Biosens Bioelectron 121:159–165
105. Lin T, Wang JJ, Zhang H, Song ST, Zhou YZ, Zhao J, Du XW (2015) Freestanding ultrathin metallic nanosheets: materials, synthesis, and applications. Adv Mater 27(36):5396–5402
106. Chen Y, Fan ZX, Zhang ZC, Niu WX, Li CL, Yang NL, Chen B, Zhang H (2018) Two-dimensional metal nanomaterials: synthesis, properties, and applications. Chem Rev 118(13):6409–6455
107. Sial MAZG, Din MAU, Wang X (2018) Multimetalllic nanosheets: synthesis and applications in fuel cells. Chem Soc Rev 47(16):6175–6200
108. Wei JP, Chen XL, Shi SG, Mo SG, Zheng NF (2015) An investigation of the mimetic enzyme activity of two-dimensional Pd-based nanostructures. Nanoscale 7(45):19018–19026
109. Cai SF, Lian C, Duan HH, Xiao W, Han QS, Qi C, Wang C, Yang R (2019) Facile strategy to prepare Rh nanosheet-supported PtRh nanoparticles with synergistically enhanced catalysis in oxidation. Chem Mater 31(3):808–818
110. Cai SF, Liu XL, Han QS, Qi C, Yang R, Wang C (2018) A novel strategy to construct supported Pd nanocomposites with synergistically enhanced catalytic performances. Nano Res 11(6):3272–3281
111. Huang Y, Zhao MT, Han SK, Lai ZC, Yang J, Tan CL, Ma QL, Lu QP, Chen JZ, Zhang X, Zhang ZC, Li B, Chen B, Zong Y, Zhang H (2017) Growth of Au nanoparticles on 2D

metalloporphyrinic metal-organic framework nanosheets used as biomimetic catalysts for cascade reactions. Adv Mater 29(32):1700102
112. Haider W, Hayat A, Raza Y, Chaudhry AA, Rehman IU, Marty JL (2015) Gold nanopariticle decorated single walled carbon nanotube nanocomposite with synergistic peroxidase like activity for D-alanine detection. RSC Adv 5(32):24853–24858
113. Kim MC, Lee D, Jeong SH, Lee SY, Kang E (2016) Nanodiamond-gold nanocomposites with the peroxidase-like oxidative catalytic activity. ACS Appl Mater Inter 8(50):34317–34326
114. Xu YQ, Fei JB, Li GL, Yuan TT, Xu X, Li JB (2019) Nanozyme-catalyzed cascade reactions for mitochondria-mimicking oxidative phosphorylation. Angew Chem Int Ed 58(17):5572–5576
115. Ivanova MN, Grayfer ED, Plotnikova EE, Kibis LS, Darabdhara G, Boruah PK, Das MR, Fedorov VE (2019) Pt-decorated boron nitride nanosheets as artificial nanozyme for detection of dopamine. ACS Appl Mater Interfaces 11(25):22102–22112
116. Zhao C, Xiong C, Liu XK, Qiao M, Li ZJ, Yuan TW, Wang J, Qu YT, Wang XQ, Zhou FY, Xu Q, Wang SQ, Chen M, Wang WY, Li YF, Yao T, Wu YE, Li YD (2019) Unraveling the enzyme-like activity of heterogeneous single atom catalyst. Chem Commun 55(16):2285–2288
117. Huang L, Chen JX, Gan LF, Wang J, Dong SJ (2019) Single-atom nanozymes. Sci Adv 5(5):eaav5490
118. Xu BL, Wang H, Wang WW, Gao LZ, Li SS, Pan XT, Wang HY, Yang HL, Meng XQ, Wu QW, Zheng LR, Chen SM, Shen XH, Fan KL, Yan XY, Liu HY (2019) A single-atom nanozyme for wound disinfection applications. Angew Chem Int Ed 58(15):4911–4916
119. Zhang Z, Liu B, Liu J (2016) Molecular imprinting for substrate selectivity and enhanced activity of enzyme mimics. Small 13(7):1602730
120. Golub E, Albada HB, Liao WC, Biniuri Y, Willner I (2016) Nucleoapzymes: Hemin/G-quadruplex DNAzyme-aptamer binding site conjugates with superior enzyme-like catalytic functions. J Am Chem Soc 138(1):164–172

Chapter 11
Hybrid Nanozyme: More Than One Plus One

Aipeng Li, Yao Chen and Lianbing Zhang

Abbreviations

ABTS	2,2′-azino-bis(3-ethylbenzothiazoline-6-sulfonic acid) diammonium salt
Aβ	Amyloid-β peptides
AD	Alzheimer's disease
ALP	Alkaline phosphatase
ATP	Adenosine triphosphate
bR	Bacteriorhodopsin
CTP	Cytidine triphosphate
DNA	Deoxyribonucleic acid
ELISA	Enzyme-linked immuno-sorbent assay
GO	Graphene oxide
GOx	Glucose oxidase
GTP	Guanosine triphosphate
HRP	Horseradish peroxidase
NaAsc	Sodium ascorbate
NMNPs	Noble metal nanoparticles
NPs	Nanoparticles
NTP	Nucleoside triphosphates
PC	Plastocyanin (PC)
PEG	Polyethylene glycol
POMs	Polyoxometalates
PSI	Photosystem I
rGO	Reduced graphene oxide
RNA	Ribonucleic acid

A. Li · Y. Chen · L. Zhang (✉)
School of Life Sciences, Northwestern Polytechnical University,
127 West Youyi Road, Xi'an 710072, China
e-mail: lbzhang@nwpu.edu.cn

X. Yan (ed.), *Nanozymology*, Nanostructure Science and Technology,
https://doi.org/10.1007/978-981-15-1490-6_11

ROS	Reactive oxygen species
ssDNA	Single-strand DNA
TMB	3,3′,5,5′-tetramethylbenzidine
UTP	Uridine triphosphate

11.1 Introduction

If you read through the previous chapters, you must be surprised that significant progress has been made in the research field of "nanozyme", although the concept was defined only a few years ago. You may also have noticed that the type of reactions catalyzed by nanozymes, comparing with that of natural enzymes, however, is still limited. Current nanozymes are, if we try to fit them into the shoes of natural enzymes, mostly oxidoreductases with a few exceptions of hydrolases [1]. This is still a bit far from the six major groups of its biological counterpart. As artificial enzymes consisting mostly of inorganic materials, nanozymes have several unique advantages over biological macromolecules, such as stability in both ambient and harsh conditions, more convenient mass production and so on. Applications of nanozymes in fabrication, in health care have been already in development focus and the front of basic research, in order to remove several hampers. For example, the activity of most nanozymes is relatively low compared to that of their protein counterparts. More perishing is that most nanozymes have poor substrate selectivity.

Since current nanozymes are inorganic nanomaterials consisting of metals, metal oxides, metal sulfides, or carbon, the catalytic activity depends mainly on their surface morphology and chemical states, which play a critical role in free radical generation, electron transfer, as well as substrates or products adsorption/desorption. Substrates recognition and their binding on the nanomaterials surfaces are unlike that in the binding pocket and active site of an enzyme. However, facing these challenges, little progress has been made in activity and selectivity enhancement by mimicking natural enzymes. One approach is through specific surface modifications to increase the activity or selectivity toward desired substrates. This is similar or derived from the commonly used surface modification methods by colloidal synthesis, which is aimed to enhance the solubility, modulate size, morphology, and crystal structure of nanomaterials. However, surface modifications handled in this chapter are not for nanozymes synthesis, but their catalytic property oriented. Since organic molecules or polymers are used for those modifications, the final nanozymes are hybrids of organic and inorganic materials, which fit to the classic definition of hybrid materials. Due to this reason, we include those nanozymes in hybrid nanozymes. In addition, we would like to cover another type of "hybrid" nanozymes in this chapter, which use nanomaterials to modify biological enzymes or substitute part of enzymes, and the final combinations show, in many cases, novel catalytic properties which cannot be performed by either part alone.

Some of the research works presented in this chapter may have been already mentioned in other chapters of this book. However, we try to put them under a different scope in this chapter and to analyze them from a different point of view. We think this approach of combining inorganic nanozymes with organic enzymes may be interesting to open a new venue for nanozyme design, fabrication, and application.

11.2 Hybrid Nanozymes: Surface Modifications for Performance Enhancement

Derived from commonly used surface modification methods for nanomaterial synthesis, hybridization with organic molecules and polymers are used to enhance the performances of nanozymes, either activities or substrate selectivity. Those hybridizations lead to formation of complex materials either showing enhanced characteristics in between the two original phases or even new properties. Taking advantage of this approach, copious efforts have been devoted. In some cases, researches even find inspirations from natural enzymes.

11.2.1 Hybridization for Selectivity Enhancement

One simple mechanism to bind specific substrate by natural enzymes is the "lock and key" principle, by which the binding pocket is the "lock" and the substrate is the "key". Only the right key which fits exactly into the lock can unlock. Compared to natural protein enzymes, inorganic nanozymes are relatively poor in substrate selectivity due to lacking of those specific bonding pockets for substrates. Can we borrow this idea for nanozyme design by engineering a binding pocket on the surface of nanozymes, which only a defined substrate can fit into? If it works, a better substrate selectivity might be achieved! How can we make such a pocket for substrate on nanozymes?

Indeed, binding pockets can be accomplished by molecular imprinting or capping with specific molecules on surfaces of nanozymes. Molecular imprinting is created by polymerizing functional monomers with template molecules (here could be the desired substrate of nanozymes) imbedded in the polymer matrix. And a cavity complementary to the shape and size of template, eventually for rebinding the template, will be left behind after removing the template. With this technique, Liu and his coworkers demonstrated that moderate specificity for oxidization of 3,3′,5,5′-tetramethylbenzidine (TMB) or 2,2′-azino-bis(3-ethylbenzothiazoline-6-sulfonic acid) diammonium salt (ABTS) can be achieved by imprinting with an adsorbed TMB or ABTS on the surfaces of Fe_3O_4 nanoparticles [2]. Moreover, further improvement was obtained by imprinting with monomers having opposite charge to the substrate, which makes binding pockets not only complementary in shape

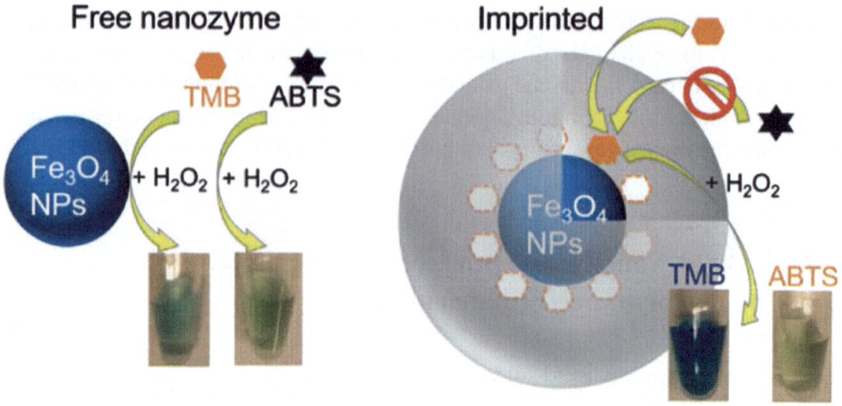

Fig. 11.1 Hundred-fold nanozyme specificity improvement by molecular imprinting on Fe_3O_4 nanoparticles. Reprinted with permission from Ref. [2]. Copyright (2017) American Chemical Society

to substrate molecules, but also electrostatically attractive. These double improvements increased almost 100-fold binding specificity compared to that of bare Fe_3O_4 nanoparticles (Fig. 11.1). This genius method inspired by the well-known "lock and key" model is also feasible for gold nanoparticle and nanoceria [2], paving the way for application of nanozymes in biosensing, drug delivering, environmental remediation, and so on.

Substrate selectivity between TMB and ABTS is based on the fact that these two molecules are different significantly in size and shape. However, biological enzymes can distinguish and recognize substrate molecules with similar structures, such as isomers or even enantiomers which are mirror images of each other. Can we go so far with nanozymes? Endowing nanozymes with a chiral selectivity is beyond the limit of current molecular imprinting technique. However, capping nanozymes with specific ligands is another way to achieve desired selectivity, especially for substrate with special structures. For example, DNA can be stimulated externally to switch between random-coiled and multi-stranded structures, and hence it can serve as a chiral layer adsorbed on surfaces of gold nanoparticles (AuNPs) to engineer and adjust the catalytic selectivity for chiral ligands like glucose enantiomers. Single-stranded DNA (ssDNA) has a special stereo orientation that arises from chiral stacking of nucleosides. The surface-adsorbed ssDNA has much stronger interaction with L-glucose. These effects endow ssDNA-capped AuNPs surface with a stronger affinity to L-glucose than D-glucose, which resulted in a higher oxidation rate of L-glucose than D-glucose with AuNPs as glucose oxidases [3]. Interestingly, the chiral selectivity of ssDNA-capped AuNPs can be reversed by hybridization or folding of the capping DNA, which is not an easy task by natural enzymes. It has never been reported to change or shut down an enzyme's selectivity without tedious engineering. Here you see not only a clever design of nanozymes, but also an amazing technological potential of nanozymes as artificial enzymes. Furthermore, this simple method is not restricted to AuNPs and can also be applied for other nanozymes.

11.2.2 Hybridization for Activity Enhancement

Despite the great progress that has been made in improving the catalytic activities of nanozymes, it is still low compared to that of natural enzymes. In catalytic reaction, the binding affinity of enzymes to substrates is one of the key factors that determines the catalytic efficiency. For natural enzymes, the excellent activity originates from not only the core of active sites, but also the protein corona around the core, which helps binding and positioning substrate molecules. Obviously, the inorganic nanoparticles are poor in substrate binding affinity, due to the lacking of a coordinated binding site to bind substrate molecules. Mimicking the binding site in natural enzymes, functional groups, or protein has been introduced to modify the surfaces of nanozymes, which has been demonstrated to be effective to improve the affinity for substrate binding and hence the reaction activity.

It is reported that small molecules like nucleoside triphosphates (NTP), including adenosine triphosphate (ATP), guanosine triphosphate (GTP), cytidine triphosphate (CTP), and uridine triphosphate (UTP) can be incorporated with nanoceria or gold nanoparticle as an effective modulator to improve their biocatalyst-mimic activities [4, 5]. ATP, present in all known forms of life, can form charge-transfer complexes with aromatic molecules and participate in free radical redox chemistry. Therefore, an incorporation of ATP with Fe_3O_4 nanoparticle is beneficial for promoting free radical generation and stabilizing the intermediate products. Thus, peroxidase-like activity, as well as cycling stability of Fe_3O_4 nanoparticles can be greatly improved (Fig. 11.2a) [6]. Such activity enhancement can also be observed in polymer-hybridized nanozymes. For example, some small catechol organic ligands, such as dopamine and catecholamine, can bind to nanoparticles and modulate their catalytic performance, finding wide applications in protein sensing, H_2O_2 assay, and so on [7]. Moreover, it was found that ferric ions can accelerate the mimic enzymatic ability of 2, 6-diaminopurine (DAP)-modified AuNPs, which could be ascribed to the synergistic reaction of AuNPs and ferric ions [8].

The large ratio of surface area to volume and the strong dipole–dipole attraction between particles make the Fe_3O_4 nanozymes easy to aggregate, which seriously hamper their activity in some certain conditions [9]. Promoting the catalytic activity and the stability of Fe_3O_4 nanozymes is an urgent challenge. Besides the outstanding inherent performances of two-dimensional graphene nanosheets, the chemical vapor deposition-grown three-dimensional graphene foams (3D GF) exhibit fascinating properties such as strong mechanical strengths, large surface areas, rapid charge transfer, and mass transport kinetics [9]. Therefore, 3D GF/metal oxide nanohybrids attracted great attention for the development of advanced composite materials for multiple applications [10, 11]. A three-dimensional graphene/mesoporous Fe_3O_4

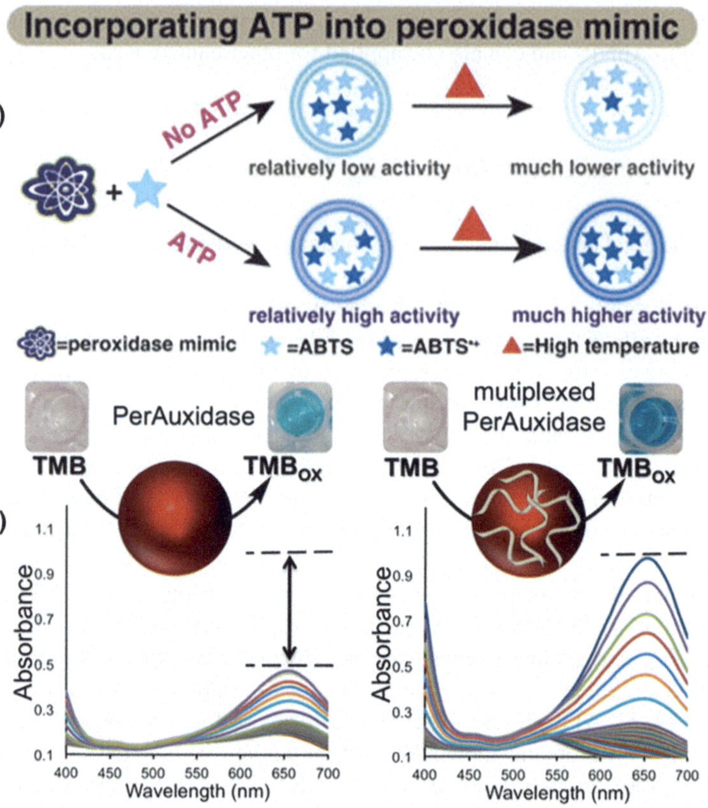

Fig. 11.2 **a** The catalytic activity within a fairly broad range of operating temperatures by incorporating adenosine triphosphate (ATP) with the peroxidase mimic. Reprinted with permission from Ref. [6]. Copyright (2014) Springer Nature. **b** The intrinsic peroxidase-like activity of citrate-capped AuNPs (perAuxidase) fabricated by integrating oligonucleotides on a nanoparticle surface. Reprinted with permission from Ref. [17]. Copyright (2016) American Chemical Society

(3D GF/m-Fe_3O_4) heterogeneous nanozyme was prepared using a simple hydrothermal procedure, in which mesoporous Fe_3O_4 nanoparticles were in situ synthesized on 3D GF. Compared with individual m-Fe_3O_4 nanoparticles, the 3D GF/m-Fe_3O_4 nanohybrid possesses high peroxidase mimicking activity due to the synergistic effect of 3D GF and m-Fe_3O_4. Moreover, the 3D GF/m-Fe_3O_4 nanohybrid shows to be extremely robust, as it still retains its original response to PNP after being reused for 10 times [9]. It was also reported that a nanozyme assembled by polyethylenimine (PEI)-protected reduced graphene oxide anchoring iron oxide (PRGI) and Pt nanoparticle (PRGI/Pt nanohybrid) showed higher catalytic activity than the sum of PRGI nanocomposite and Pt nanoparticle [12].

Compared to small molecules, polymers are macromolecules that can adsorb tightly onto nanoparticle surface via various binding interactions. The adsorbed molecules can, besides afford better biocompatibility and colloidal stability, facilitate the substrate binding via charge attraction or organic bind interaction leading to an activity enhancement [13, 14]. For example, the negatively charged DNA, especially longer DNA attached on surface of nanoceria, prevents the nanoparticle agglomeration and facilitates binding with TMB. The reason lies in the phosphate groups of DNA backbone which convert the surface charge of nanoceria from positive to negative, and hence enhance the binding interaction with positively charged TMB and accelerate its oxidation by nanoceria as peroxidase. Interestingly, the longer the DNA molecule is, the higher the rate enhancement becomes [6]. Similar improvement in activity has been also achieved for other DNA hybridized nanozymes [4, 13, 15, 16]. Moreover, by altering the oligonucleotide sequences or changing the reaction parameters, the activity of ssDNA-integrated gold nanocatalysts can be precisely tuned (Fig. 11.2b) [17]. The oxidase-like activity of spinel manganese cobalt oxide ($MnCo_2O_4$) submicrospheres exert also can be reversibly regulated by ssDNA strands [18].

To clear how DNA molecules with different structures interact with nanozymes and influence catalytic activities of nanozymes, a systematic and comprehensive investigation on the interactions between typical nanozymes and different structured DNAs was performed [19]. Various structured DNA ligands, for example, ssDNA, double-stranded DNA, hairpin DNA (H-DNA), and long dsDNA of hybridization chain reaction (HCR), were utilized to regulate the catalytic activity of metal nanoparticles (Fe_3O_4 NPs and AuNPs) toward substrate oxidation (Fig. 11.3). HCR product-treated nanoparticles showed the highest peroxidase-like activities among four DNA structures. The treatment of stiff double-stranded DNA only induced a negligible enhancement of the catalytic activity, while both coil-like single-stranded DNA and hairpin DNA-capped nanozymes produced a medium signal increase. These

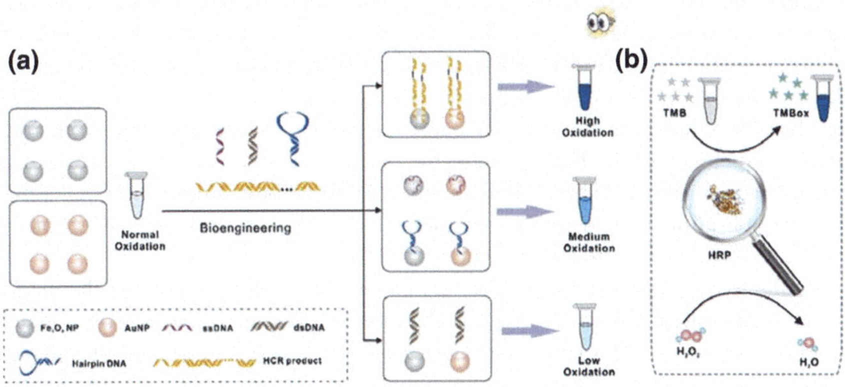

Fig. 11.3 Schematic illustration of high-precision regulation of catalytic nanozymes; **a** Construction of versatile bio-nanointerfaces on two nanozymes, including Fe_3O_4 NPs and AuNPs, by using ssDNA, dsDNA, H-DNA, and HCR products, respectively; **b** Oxidation of TMB using peroxidase mimics in the presence of H_2O_2. Reprinted with permission from Ref. [19]. Copyright (2018) American Chemical Society

results demonstrated that the DNA-engineered catalytic regulation of nanozymes was strongly dependent on their structural and characteristic differences.

Other organic polymers like dextran, hyaluronan, folic acid, polylysine, and polyethyleneimine can also be coated on surfaces of nanozymes via encapsulation/adsorption or covalent binding to promote functional enhancement [20–23]. In general, the redox chemistry of nanoceria as enzyme mimics depends on its oxidative state, especially the Ce^{3+} oxidization state. Hence, the controlling on generation of Ce^{3+} plays a key role in the application of nanoceria. Since PEG has been reported to increase the residence time of nanoparticles, the PEGylated nanoceria created by complexation cerium ion with PEG exhibited the ability to adjust the redox chemistry as function of PEG concentrations, owing to the reason that the regeneration of Ce^{3+} oxidation state of nanoceria can be tuned by PEG [24]. Polymer hybridization can also lead to a root and branch change of nanozyme-catalyzed reaction. For example, one study even found that poly(acrylic acid)-coated nanoceria conjugated to folic by click chemistry can quickly oxidize organic dyes and small molecules without the assistance of any oxidizing agent, which makes the hybrid nanozyme a more robust nanoprobe for immunoassay [25].

Hybridization with polymer macromolecules presents an effective and applicable approach to enhance nanozyme activities. Investigations demonstrated that the thickness, the packing density, and the molecular weight of polymer coatings can have effects on the catalytic activity. In general, a relatively thinner layer and an increasing molecular weight with decreasing surface coverage are correlated directly with increasing rate constant and better performance [22, 26]. In addition, Lin and his coworkers reported that charge characteristics of the polymer-hybridized nanozymes and the substrates also play an important role in the catalytic performances [27].

As for selectivity enhancement, the idea of mimicking the binding environment of natural enzymes is also applied to improve the activity of nanozymes. For example, Fe_3O_4 nanoparticle was modified with histidine to mimic the architecture of natural horseradish peroxidase (HRP) which contains a heme cofactor in the active site with two distal histidine residues involving in the substrate positioning. The introduction of histidine significantly improves the peroxidase-like activity and the catalytic efficiency of the Fe_3O_4 nanozyme by enhancing the affinity for H_2O_2 via hydrogen bond formation between the imidazole group of histidine and H_2O_2 [28]. This enzyme-mimetic approach paves a way for the rational design/construction of artificial enzymes with improved functions [29].

Similar hybrid nanozymes can also be fabricated by mixing inorganic chemicals directly with protein molecules [27]. For example, the mixture of copper (II) ions and proteins leads to growth of micrometer-size particles with flower-petal-like shape. When using an enzyme as the protein component, a nanoflower confined hybrid nanozyme was fabricated and exhibited enhanced activity and stability due to the high surface area of nanomaterial and confinement effect of the flower-like structure [30]. Some similar work was carried out by encapsulation of nanoceria into a protein cage of apoferritin via a dissociation/reconstruction route, which provides the nanocomplex (apoferritin-CeO_2) with greatly enhanced mimetic SOD activity and changes the cellular uptake route of nanoceria at the same time. By encapsulation in apoferritin, the electron localization at the surface of nanoparticle was manipulated; therefore, the reactive oxygen species (ROS)-scavenging ability of the hybrid nanozyme was improved. In this case, the hybridization with a function protein or enzyme can be viewed as an extremity, since each component itself has its own enzymatic functionality. Therefore, we would like to analyze similar cases in the following sub-chapter for a better overview.

11.3 Hybrid Nanozymes: Coupling Functional Building Blocks

As described above, nanozymes involved hybrid systems represent new complex materials revealing improved or even new catalytic characteristics. However, in addition to simply improving their catalytic performance, a particular challenge is to combine the intrinsic functionality of nanozymes for example, unique electronic, optical, or catalytic properties, and other functional entities into a functional cluster with combined and synergistic, or totally novel properties originating from the two original building blocks. To this end, several impressive developments have been made over past years on various fields.

11.3.1 Hybridization with Inorganic Functional Building Blocks

(1) **Coupling of Cerium Oxide/Gold Nanoparticles with POMs**

The key pathological markers in Alzheimer's disease (AD) are cerebral extracellular amyloid plaques which are composed mainly of 39–43 amino acid amyloid-β peptides (Aβ), accompanied by local inflammation and the production of reactive oxygen species (ROS) [31, 32]. Recent advances have demonstrated that the onset of neurotoxicity is strongly associated with the aggregation process of monomer Aβ to oligomers and successively to fibrils in the pathogenesis. Owing to the critical role of Aβ in AD pathogenesis, proteolytic degradation of Aβ aggregates and clearance of Aβ-induced ROS have been regarded as promising therapeutic approach to treat AD [33].

Due to the synergistic activity between both Ce^{3+} and Ce^{4+} ions, cerium oxide nanoparticles (CeONP) are in possession of antioxidant activity so that it has the potential for protection of the brain against oxidative stress in AD [34]. Moreover, rare earth metal-substituted polyoxometalates (POMs) possessed ability to hydrolyze peptides [35, 36]. For above reasons, CeONP and POMs were chosen as scaffolds to design and construct an artificial Ceria/Polyoxometalates hybrid (CeONP@POMs) with both proteolytic and superoxide dismutase activities (Fig. 11.4a). The results indicated that CeONP@POMs could efficiently

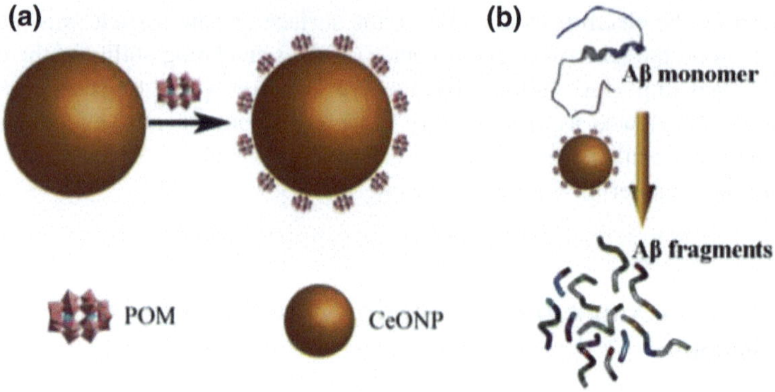

Fig. 11.4 **a** Schematic drawing of synthesis of CeONP@POMs; **b** scheme of CeONP@POMD as hydrolytic enzyme to degrade Aβ40 monomers. Reprinted with permission from Ref. [37]. Copyright (2016) Elsevier Ltd.

degrade Aβ aggregates and reduce intracellular reactive oxygen species (Fig. 11.4b). More importantly, CeONP@POMD (POMD, POMs with Wells–Dawson structure ($K_6[P_2W_{18}O_{62}]$) could not only promote PC12 cell proliferation and can cross blood–brain barrier, but also inhibit Aβ-induced BV2 microglial cell activation. In vivo studies further indicated that CeONP@POMD as nanozyme hybrid possessed good biocompatibility [37]. Similar nanozyme hybrid, which composed of the POMD, gold nanoparticles (AuNP), and N-Ac-Cys-heptapeptide complex (AuNPs@POMD-8pep), also exhibits protease activities, SOD-like functionality, and metal-ion chelation capabilities. By virtue of these properties, the nanozyme may act as a multifunctional anti-Aβ agent for the treatment of AD [38].

(2) **Coupling of Palladium Nanoparticles with Iron Oxide/Graphene Nanosheets**

Noble-metal nanoparticles (NMNPs) with small particle size in the reaction conditions tend to aggregate and lose their catalytic activities [39]. Moreover, the high cost and limited resources also hinder their practical application. To solve these problems, the nanoparticles are often immobilized on various solid supports. Among all solid supports studied so far, magnetic nanoparticles (such as Fe_3O_4 nanoparticles) and graphene have attracted particular interest [40–43]. Magnetic NPs have efficient magnetic separation capacity for easy separation of the catalyst from the reaction solution and graphene has a high surface area, which offers an effective protocol to minimize aggregation and maximize the chemically accessible area. Liu and coworkers present a general strategy for the preparation of strongly coupled Fe_3O_4 and palladium nanoparticles (PdNPs) to graphene sheets by employing polyethyleneimine as the coupling linker. These ternary nanocomposites have a well-defined 2D morphology, controllable size, high dispersion of PdNPs and Fe_3O_4 NPs, and high specific surface area. The novel nanocatalyst exhibits synergistic catalysis by Pd nanoparticles supported on reduced graphene oxide (rGO) and a tertiary amine of polyethyleneimine

($Pd/Fe_3O_4/PEI/rGO$) for the Tsuji–Trost reaction in water and air. For example, the reaction of ethyl acetoacetate with allyl ethyl carbonate afforded the allylated product in more than 99% isolated yield, and the turnover frequency reached 2200 h^{-1}. The yield of allylated products was 66% for Pd/rGO without polyethyleneimine. And the results indicate that the tertiary amine on $Pd/Fe_3O_4/PEI/rGO$ accelerates the Pd-catalyzed Tsuji–Trost reaction. More importantly, the noble-metal nanocatalyst on the magnetic support has a very low leaching loss and allows one to reach at least 92% yield in Tsuji–Trost catalysis over 30 cycling tests [44].

(3) PEG Derivatives Coupled $Au/Fe_3O_4/GO$ Hybrid Material

Mercury has been widely recognized as one of the most hazardous pollutants and highly dangerous elements for both the environment and human health [45–47]. Typically, Hg^{2+} is one of the most common and stable forms of mercury pollution. Therefore, accurately and selectively recognizing and rapidly removing Hg^{2+} is of great significance in the aqueous environment [48–50].

In this regard, Zhang and coworkers developed an efficient strategy for the synthesis of a strongly coupled $Au/Fe_3O_4/GO$ hybrid material by using polyethylene glycol (PEG) derivatives as the coupling linkers between the three components (Fig. 11.5). The unique thin nanosheet structure and the abundance of functional groups in GO offer a large surface area and anchoring sites for the assembly of the as-synthesized Au (nanozyme with mercury-stimulated peroxidase activity) and Fe_3O_4 NPs (magnetic nanoparticles). This enables the full utilization of the catalyst surface by minimizing the agglomeration of Au NPs and facilitates the magnetic separation of $Hg^0@Au/Fe_3O_4/GO$ from biological solutions, leading to the selective colorimetric detection of Hg^{2+} at ultralow concentrations. The mercury-stimulated peroxidase activity studies show that the $Au/Fe_3O_4/GO$ hybrid material exhibits not

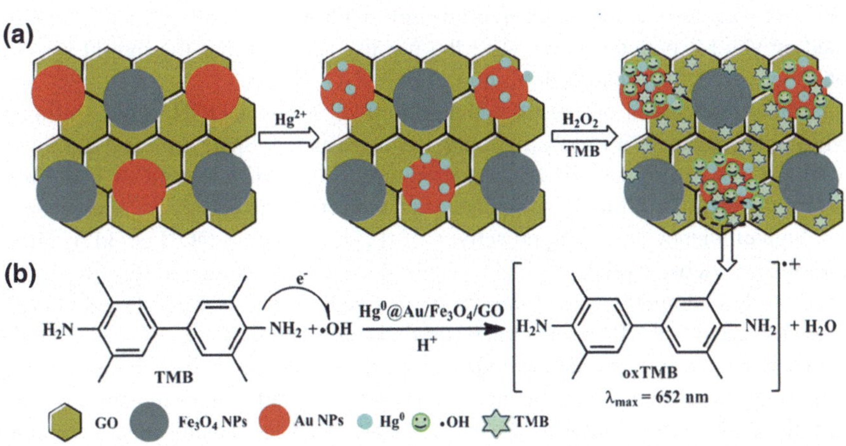

Fig. 11.5 **a** The mercury-enhanced peroxidase-like activity of the $Au/Fe_3O_4/GO$ hybrid material; **b** The reaction principles in this system. Reprinted with permission from Ref. [51]. Copyright (2015) Royal Society of Chemistry

only an enhanced mercury-stimulated peroxidase activity but also superior stability compared to Au in catalytic reaction media, which renders it a high-performance catalyst for mercury-stimulated peroxidase reactions. The Au/Fe_3O_4/GO hybrid material could be applied successfully for the rapid (within 30 min) and efficient (>99% elimination efficiency) removal of Hg^{2+} from water with a high reusability (>15 cycles) [51].

(4) **Coupling of $[Fe_2L_3]^{4+}$ with GO–COOH**

$[Fe_2L_3]^{4+}$ is a chiral metallo-supramolecular cylinder with similar size and shape like zinc finger protein [52, 53]. And it was found that $[Fe_2L_3]^{4+}$ possesses peroxidase-like activity and has enantioselectivity for catalysis reaction. On the other hand, it is known that graphene-family materials represent an interesting geometrical support for improving molecular catalysts [54, 55]. GO–COOH was chosen because it had better water solubility than graphene nanosheet or reduced graphene oxide, and a higher electron transport capacity than graphite oxide [56]. In addition, the negatively charged carboxyl on the surface of GO–COOH could enhance the adsorption of positively charged $[Fe_2L_3]^{4+}$ through electrostatic interactions. Furthermore, it is known that the GO–COOH also possesses intrinsic peroxidase-like activity. The combination of $[Fe_2L_3]^{4+}$ and GO–COOH was expected to have mutual-promoted effect. Therefore, Xu and coworkers functionalized GO with the chiral supramolecular complex $[Fe_2L_3]^{4+}$ via electrostatic interaction to create novel nanocatalysts ($[Fe_2L_3]^{4+}$–GO–COOH) [57].

Surprisingly, different from other enzyme mimics, the novel peroxidase mimic designed could be not only simply duplicating and imitating the properties of natural peroxidase, but also introducing additionally new features such as enantioselectivity and near-infrared photothermic effect. Based on the excellent peroxidase-like property of $[Fe_2L_3]^{4+}$–GO–COOH, a colorimetric method for intracellular H_2O_2 detection was designed through hybridization with metal ions (e.g., Cu^{2+}) conjugated amyloid-β peptides (Aβ). And the results indicated that the hybrid system could be successfully detection of intracellular H_2O_2 [57].

Furthermore, based on the photothermic effect of GO–COOH, a near-infrared (NIR) light-activated biocatalyst system was constructed. Since the catalytic activity of $[Fe_2L_3]^{4+}$–GO–COOH was temperature-dependent, while GO–COOH has absorption in near-infrared region and can convert light energy to heat energy with high efficiency, the catalytic activity of $[Fe_2L_3]^{4+}$–GO–COOH could be NIR-controlled. Glucose oxidase (GOx) was introduced for reducing the reaction background at low temperature and the enzyme cascade reactions were designed. The results indicated that NIR laser could be used to control GOx-$[Fe_2L_3]^{4+}$–GO–COOH cascade reactions and served as a model system for remote control of biological reactions using the designed novel nanocatalyst and biofriendly, high transparent, and adaptable NIR laser. Moreover, $[Fe_2L_3]^{4+}$–GO–COOH had enantio-discrimination ability for the chiral drug dopa. Although the discrimination mechanism needs further studies to clear up, this is important because L-dopa is the most effective drug at present used to combat Parkinson's disease while D-dopa is inactive and can even cause side effects [57].

(5) **Hybrid of Different Types of Nanozyme**

The use of pure nanomaterials to construct a system capable of mimicking an enzyme cascade reaction also attracted the attention of researchers. Qu and coworkers constructed a novel nanocomposite consisting of V_2O_5 nanowires, polydopamine, and gold nanoparticles (AuNPs), in which V_2O_5 nanowires and AuNPs possess intrinsic peroxidase and glucose oxidase (GOx)-like activity, respectively [58]. The V_2O_5-PDA-AuNP nanocomposite could be applied as a robust nanoreactor for mimicking an enzyme cascade reaction without the need for the natural enzymes (horseradish peroxidase (HRP) and GOx). Based on this nanozyme cascade reaction system, a direct and selective colorimetric method for the detection of glucose was successfully developed.

Furthermore, by taking advantage of the different deactivating effects of single-stranded (ssDNA) and double-stranded DNA (dsDNA) on the GOx-like activity of AuNPs, by which ssDNA strongly binds to AuNPs, leading to effective suppression of their catalytic activity, whereas dsDNA binds only weakly to AuNPs and thus only slightly perturbs their catalytic activity, the sensing of target complementary DNA can be realized and disease-associated single-nucleotide polymorphism of DNA can be easily distinguished by the V_2O_5-PDA-AuNP nanocomposite [58].

11.3.2 Hybridization with Functional Biomacromolecules

If aforementioned hybridization is inorganic nanozymes with other inorganic components, it is reasonable that there are also trials of hybridizations of nanozymes with enzymes. Interestingly, several experiments have shown those hybrid nanozymes are not simple superposition of activities of the natural and the artificial enzymes. Novel catalytic active hybrids were created in this way.

(1) **Pt Nanozymes Hybridize with Ferritin Light Chain as Artificial Ferroxidases**

Ferritin is a protein highly conserved in all living forms [59]. In mammals, two distinct subunits of ferritin have emerged as a result of evolution. The heavy-chain (H-chain) subunit conserves the common di-iron binding site and catalyzes the oxidation of Fe^{2+} to Fe^{3+}, being a ferroxidase. Although similar in sequence and size, the light-chain (L-chain) subunit lacks ferroxidase activity as well as further enzymatic activities [60]. The detailed mechanisms of involvement of the L-chains in the mineralization process remain elusive. Therefore, to confirm that the L-chains are the only entities responsible for the electron transfer, a hybrid system, which is composed of L-chain apoferritins and Pt nanoparticles, was developed (Fig. 11.6). Pt nanoparticles act as artificial ferroxidases of H-chain for Fe^{2+} oxidation in this hybrid. The result demonstrated that the H-chain proteins are not required for the electron transfer and the electron transfer is a specific property of the L-chains [60].

Unlike the apoferritin-encapsulated CeO_2 (apoferritin-CeO_2) which enhances the activity of CeO_2, Pt/L-chain must be together to function as an active ferritin with

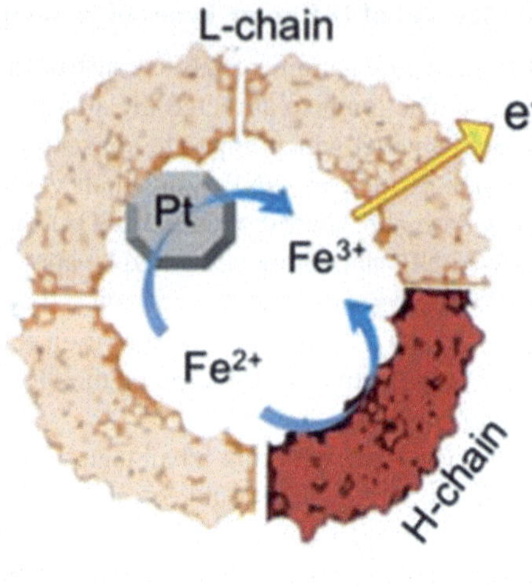

Fig. 11.6 Hybrid system which composed of L-chain apoferritins and Pt nanoparticles. Reprinted with permission from Ref. [60]. Copyright (2014) Royal Society of Chemistry

ferroxidase activity. No enzymatic activity will exist without Pt. Equally, Pt alone cannot function as a ferritin in a cellular environment to effectively involve in the iron homeostasis [61].

(2) **Nanoceria Nanozymes Hybridize with Apoferritin or CuZn-SOD Enzyme**

The construction of hybrid system in which nanoparticles are encapsulated by functional proteins has been proven to be an efficient strategy to alter the electronic structure and antioxidant activity of nanoceria. Liu et al. prepared apoferritin-encapsulated nanoceria (AFt-CeO_2) (Fig. 11.7). The SOD activity of this conjugate is higher than that of free apoferritin, free CeO_2, and even their simple mixture, suggesting a synergic effect. In addition, the cage proteins, apoferritin, carry their biological identities to the AFt-CeO_2 to initiate an endocytosis process and improve its biocompatibility. Further analysis showed that the surface Ce(III) content of CeO_2 increased from 0 to 70% after hybridization with apoferritin. Ce(III) is known to be important for the

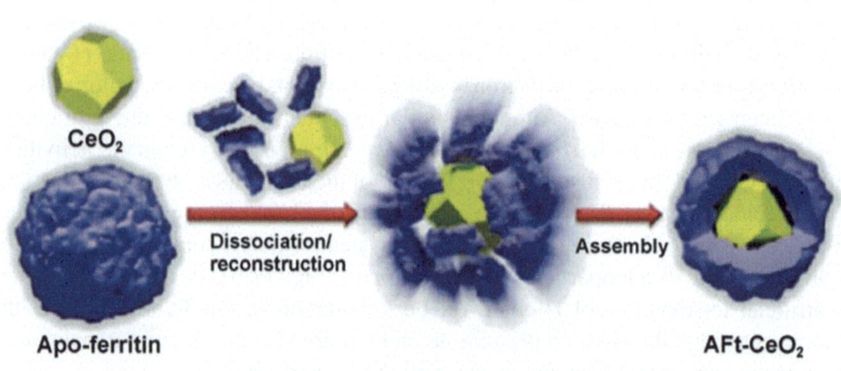

Fig. 11.7 A schematic showing the preparation of apoferritin-encapsulated nanoceria. Reprinted with permission from Ref. [62]. Copyright (2011) Royal Society of Chemistry

activity of the nanoceria. In this system, apoferritin not only improved the biocompatibility of the nano-CeO_2, but also manipulated the electron localization at the surface of the nanoparticle (apoferritin reduced Ce(IV) to Ce(III) on the nanoceria surface), thereby ameliorating the ROS-scavenging activity of the apoferritin–CeO_2 hybrid nanocomposite [62].

Zhang et al. found that interfacing nanoceria with a redox-active CuZn-SOD enzyme or a sensitizing dye, $[Ru(dcbpy)_2(NCS)_2](RuN_3)$, also can significantly enhance the SOD activity. And it was confirmed that nanoceria acquire their superoxide-scavenging ability after interfacial electron transfer [63].

(3) **Protein–Pt Nanoparticles@Mesoporous Iron Oxide Hybrid**

Inspired by the delicate structure and prominent efficiency of natural multiple-enzyme systems, integrating nanozymes and natural enzymes has been seen as an effective approach to developing enzyme cascade platforms [64–66]. Liu and coworkers adopted a facile and convenient structural-design approach to create an organized cascade platform based on a dual-functionalized β-casein-Pt nanoparticles@mesoporous-Fe_3O_4 (CM-PtNP@m-Fe_3O_4) hybrid nanozyme [67]. In their design, the CM-PtNP@m-Fe_3O_4 hybrid acts as not only a nanozyme with outstanding peroxidase-like activity but also a scaffold to immobilize and stabilize the natural oxidase. The hybrid provides a favored microenvironment that can immobilize oxidase and preserve the oxidase activity, which resulted in significantly enhanced peroxidase-like activity. The nanohybrid-triggered cascade platform exhibited excellent sensitivity with a wide linear range of 0.1–400 mM and a detection limit of 0.05 mM for cholesterol [67]. The highly rationally designed protein/inorganic hybrid and dual-functional strategy provide a facile one-pot and effective high-performance organized enzyme cascade bioplatform.

(4) **Co_3O_4 and CeO_2 Nanoparticles Hybridize with Natural Enzymes**

Zhang et al. synthesized Co_3O_4 magnetic nanoparticles (MNPs) with intrinsic peroxidase-like activity by using bull serum albumin (BSA) as templates at mild condition (Fig. 11.8) [68]. It was found that the immobilization of glucose oxidase (GOx) and glucoamylase (GA) on Co_3O_4 MNPs could regulate pH-dependent activity of enzymes. Further, based on the overlapped pH range between peroxidase-like activity of Co_3O_4 MNPs and catalytic activity of GOx, GOx–Co_3O_4 bioinorganic hybrid catalysts were constructed and applied on the one-pot colorimetric glucose sensing. The bioinorganic cascade catalysts exhibited advantages of good pH stability, high catalytic activity, strong magnetism, wide detection range, good selectivity, and excellent recyclability. Furthermore, the coimmobilization of GA and GOx on Co_3O_4 MNPs was successfully performed and applied on the one-pot colorimetric starch detection, which further confirms the universality of the proposed strategy. An additional bonus of the hybrid cascade catalysts is that it could be expediently separated from the reaction solution under the magnetic field in order to rapidly terminate the reaction and recycle enzymes [68]. This work opens an avenue for the various one-pot enzyme–nanozyme coupled assay.

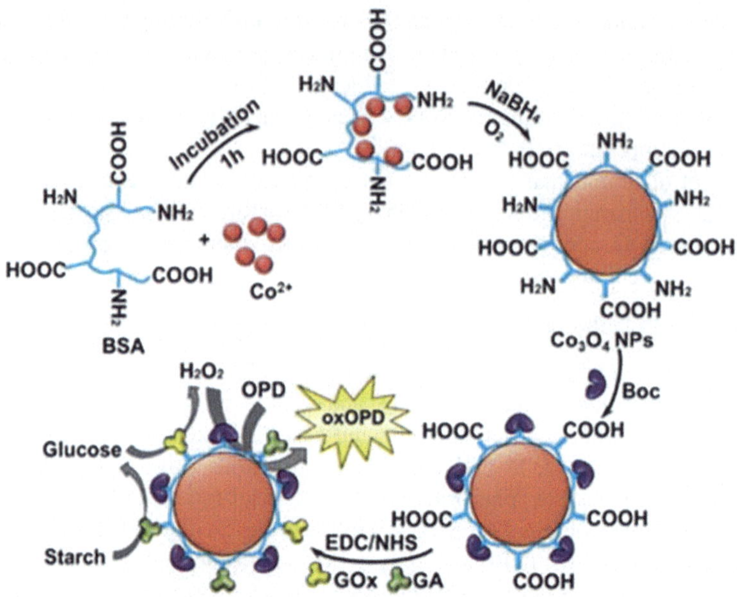

Fig. 11.8 Schematic of BSA-directed synthesis of Co_3O_4 MNPs, immobilization of GA/GOx on Co_3O_4 MNPs and one-pot colorimetric starch detection. Reprinted with permission from Ref. [68]. Copyright (2019) Elsevier B.V.

Similarly, a CeO_2/glucose oxidase (GOx) hybrid nanozyme, which exhibited excellent catalytic activity toward cascade reactions, was developed through self-assembly [69]. Due to the minimal diffusion of intermediate in the nanocomplexes system, nanocomplexes displayed enhanced efficiency compared to mixed GOx and CeO_2. Moreover, the CeO_2/GOx nanocomplexes exhibited outstanding long-term stability and excellent recyclability.

(5) **Integrating Gold Nanoparticles and Natural ATP Synthase**

Artificial photophosphorylation systems have attached much attention in recent years [70]. Multiple enzymes involved cascade reactions are necessary in natural oxidative phosphorylation. However, the natural enzymes usually suffered from environmental sensitivity, high cost of preparation and purification, difficulties in recovery and recycling, and low stability. Nanozymes possess enzyme-like activities are a good compensatory option. Xu and coworkers developed a mitochondria-mimicking oxidative phosphorylation scenario by integrating gold nanoparticles (AuNPs) and natural ATP synthase as a compartmentalized architecture [70]. In this hybrid system, AuNPs immobilized on hollow silica microspheres (HSMs) possess two enzyme-like activities (glucose oxidase and peroxidase). In the presence of oxygen, glucose was converted into gluconic acid by AuNPs catalyzed cascade reactions, which resulted in a decrease of the surrounding pH value and a generation of a proton gradient. Afterward, ATP was synthesized through rotational catalysis of ATP synthase, which was driven by a gluconic acid-mediated vectoral proton gradient.

(6) **MOF-Based Nanozymes Hybridize with Glucose Oxidase**

Metal–organic frameworks (MOFs) are a class of porous crystalline materials constructed by connecting metal clusters via organic linkers [71]. Owing to their finely tunable chemical composition, morphology, and robust structure, MOFs have generally been explored to use as protective networks for immobilizing enzymes [72]. However, the regular arrangement of metal nodes and organic ligands in MOFs can produce high-density biomimetic active material centers with significant enzyme-like catalytic activity [73]. Thus, a very recent advanced application in MOF chemistry was the use of MOFs as nanozymes. Therefore, the MOF could perform a dual function in acting as an enzyme mimic, and as a solid support for immobilization of enzymes.

Zhao and coworkers immobilized glucose oxidase (GOx) into a MOF of type Fe(III)-BTC with peroxidase-like activity (Pox) by a one-pot method (Fig. 11.9) [74]. In this work, Fe-BTC is used not only as a support for immobilizing GOx, but also as a nanozyme. The resulted GOx@Fe-BTC could produce H_2O_2 by GOx catalyzed oxidation of glucose, while the POx mimic catalytically oxidizes 3,3′,5,5′-tetramethylbenzidine (TMB) to form a blue-green product in the presence of H_2O_2. Therefore, the GOx@Fe-BTC MOF was successfully applied to the one-step colorimetric determination of glucose in serum with the detection limit 2.4 μM [74]. The

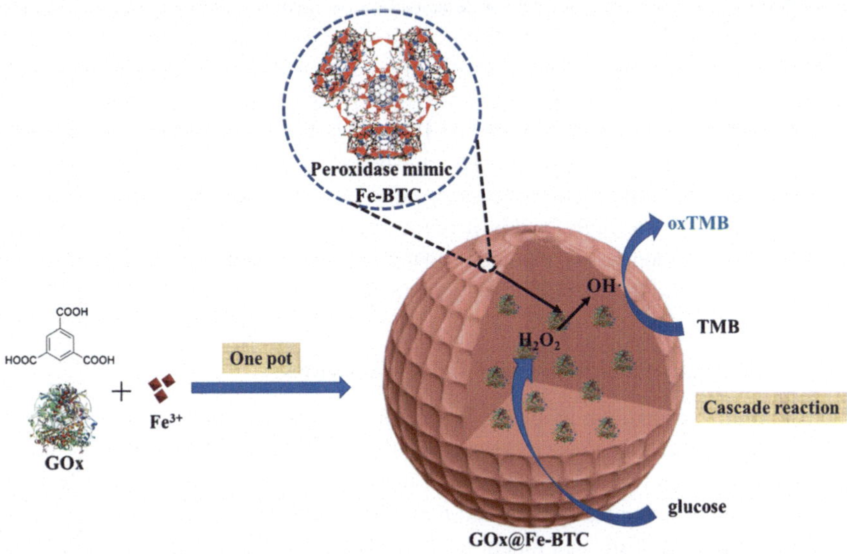

Fig. 11.9 Schematic illustration of the synthesis of GOx@Fe-BTC and the catalytic oxidation of glucose. Reprinted with permission from Ref. [74]. Copyright (2019) Springer-Verlag GmbH

enzymatic and enzyme-mimicking system exhibits advantages, such as low cost, easy preparation, and good reusability. This strategy provides a cost-effective integrated enzyme cascade system for colorimetric assay.

(7) **Alkaline Phosphatase Coupled Generation of Photoresponsive Nanozyme**

The alkaline phosphatase (ALP) is one of the most widely used labeling tracer in enzyme-linked immuno-sorbent assay (ELISA) [75–77]. Therefore, highly sensitive detection of ALP activity is needed in various diagnostic applications. The majorities of the currently developed methods for ALP activity detection [78–80] or ALP-based immunoassays [75, 81] merely rely on the catalytic reaction of ALP for signal readout, which are simple but suffer from poor sensitivity due to limited signal amplification strategy. Thus, ALP biocatalysis followed by the in situ enzymatic generation of a visible light-responsive nanozyme is coupled to elucidate a novel amplification strategy by enzymatic cascade reaction for versatile biosensing. Based on this, a new concept to produce a photo-triggered nanozyme through the ALP-based in situ biocatalytic process was introduced. The dephosphorylation process of o-phosphon-oxyphenol (OPP) by ALP generated the product of catechol (CA) containing enediol-ligand, which had specificity and high affinity for the uncoordinated Ti atoms on the surface of TiO_2 NPs and formed the CA–Ti (IV) charge-transfer (CT) complex. Upon the stimuli by CA generated from ALP, the inert TiO_2 NPs are activated, which demonstrates highly efficient oxidase-mimicking activity for catalyzing the oxidation of the typical substrate of 3,3′,5,5′-tetramethylbenzidine (TMB) under visible light ($\lambda \geq 400$ nm) irradiation utilizing dissolved oxygen as an electron acceptor. Also, this photoactive nanozyme possessed advantages of high catalytic activity, controllable and tailored activities by light irradiation, and mild reaction conditions without destructive H_2O_2. Therefore, a cascade reaction was coupled for probing ALP activity with high sensitivity: ALP catalyzed the generation of CA and, in turn, activated a second nanozyme (TiO_2–CA), resulting in efficient signal amplification. As a proof of concept, ALP-based colorimetric immunoassay using mouse IgG as a model analyte was constructed based on the signal-transduction tag by coupling ALP and in situ generated nanozyme of TiO_2–CA, which was also proved to be highly sensitive. The high sensitivity of the protocols for probing ALP activity and the ALP-based immunoassay resulted from the exquisite consolidation of the cascade reaction of ALP and the enzymatically generated nanozyme (TiO_2–CA)(Fig. 11.10) [82].

(8) **Hybrid Nanozymes with Light-Driven Activities**

Photocatalytic production of clean hydrogen fuels using sunlight has attracted remarkable attention due to the increasing global energy demand. One of the promising means to this end is the construction of biomimetic photosynthetic hybrids. Due to their unique electronic, optical, and catalytic properties, nanozymes seemed as important components of such functional hybrid clusters, for example, platinum (Pt) nanoparticles. Metallic platinum can be precipitated at the reducing end of Photosystems I (PSI) in photosynthetic membranes and isolated free PSI so that emergent electrons drive the photocatalytic evolution of molecular hydrogen [83]. Therefore,

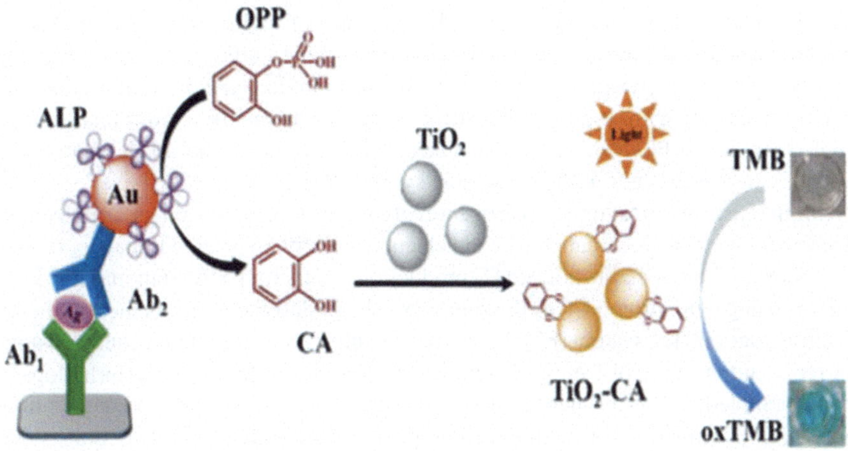

Fig. 11.10 Proposed mechanism of catalytic oxidation of TMB by photo-activated TiO_2–CA and proposed immunodetection process for mouse IgG by coupling the cascade reaction of ALP and the enzymatically in situ generated photoresponsive nanozyme of TiO_2–CA. Reprinted with permission from Ref. [82]. Copyright (2015) American Chemical Society

a simple photocatalytic hydrogen-evolving system, which is composed of sodium ascorbate (NaAsc), plastocyanin (PC), and platinized Photosystem I, was developed. The hybrid systems are illustrated schematically in Fig. 11.11a. The upper panel of Fig. 11.11a illustrates free NaAsc + PC + platinized PSI, whereas the lower panel illustrates the system with PC cross-linked to PSI. In all cases, the platinization of PSI was achieved by photochemical deposition of metallic platinum according to the reaction $PtCl_6{}^{2-} + 4e^- \rightarrow Pt\downarrow + 6Cl^-$ where the electrons were photogenerated by PSI with NaAsc as the reductant. In this system, the photochemistry of platinum

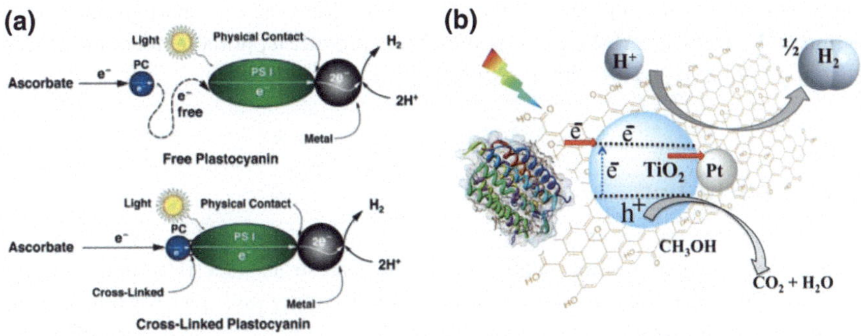

Fig. 11.11 **a** Schematic illustration of photocatalytic hydrogen-evolving system composed of NaAsc + PC + platinized PSI. Upper panel: free PC; lower panel: PC cross-linked to PSI. Reprinted with permission from Ref. [84]. Copyright (2004) American Chemical Society; **b** Depiction of photocatalytic hydrogen evolution using Pt/TiO_2 interfaced with rGO and bR. Reprinted with permission from Ref. [87]. Copyright (2014) American Chemical Society

deposition and catalyst formation precedes the onset of hydrogen evolution. And it was confirmed that photoprecipitated platinum and the natural electron-accepting enzyme system composed of ferredoxin plus ferredoxin: $NADP^+$ oxidoreductase behave differently with respect to the fate of the photogenerated electrons that emerge from cross-linked PC-PSI. The cross-linked PC-PSI is functional in the platinized system, whereas it is not in the enzymatic system [84].

Similarly, a photocatalytic hydrogen-evolving system based on intermolecular electron transfer between native PSI and electrostatically associated Pt nanoparticles was reported. Visible light-induced H_2 production occurs for PSI/Pt nanoparticle biohybrids using cytochrome C6 as the soluble mediator and ascorbate as the sacrificial electron donor in this system [85]. Moreover, Shankar Balasubramanian and coworkers report upon the application of light harvesting proton pump bacteriorhodopsin (bR) assembled on Pt/TiO_2 nanocatalyst for visible light-driven hydrogen generation [86]. And hybridization of the reduced graphene oxide (rGO) obviously boosts the above nanobiocatalyst performance (Fig. 11.11b) [87].

The splitting of dinitrogen (N_2) and reduction to ammonia (NH_3) are a kinetically complex and energetically challenging multistep reaction. In the Haber–Bosch process, N_2 reduction is accomplished at high temperature and pressure [88]. In nitrogen-fixing bacteria, the enzymatic reduction of N_2 to NH_3 is catalyzed by nitrogenase enzymes and proceeds via the hydrogenation of N_2 through metal-hydride intermediates. The Mo-dependent nitrogenase is a multiprotein complex composed of MoFe and Fe proteins. Although nitrogenase functions under ambient conditions, it requires a large input of chemical energy provided by the hydrolysis of 5′-triphosphate (ATP) [89].

Different from the role of nanoparticles described in above photocatalytic hybrid systems, cadmium sulfide (CdS) nanocrystals act as semiconductor nanocrystals and hybridized with MoFe to construct a light-driven dinitrogen reduction system. In this system, CdS nanocrystals which deliver photogenerated electrons to the MoFe protein were used to photosensitize the nitrogenase molybdenum–iron (MoFe) protein, where light harvesting replaces ATP hydrolysis to drive the enzymatic reduction of N_2 into NH_3. The CdS:MoFe protein biohybrids provide a photochemical model for achieving light-driven N_2 reduction to NH_3 [90].

11.4 Conclusion

Facing the challenges of limit reaction types, poor substrate selectivity, and catalytic activities, nanozymology as a research field is still young and, just for this reason, very active. Hybridization with organic materials has been proved as an effective way to enhance the performances of nanozymes. It is not uncommon that we can be inspired by natural enzymes to get ideas for improvements. Just as the "lock and key" model can be mimicked with molecular imprinting to design nanozymes with better substrate selectivity [2]. With chiral molecules as hybrid partners, even enantiomer selectivity is possible for nanozymes. Could the next trial and success be the "induced fit" model?

Coupling different nanozymes can be a way to fabricate multifunctional nanozymes, as in the case of CeONP@POMD system which combined SOD activity with proteolytic activity [37, 38]. Sometimes, those couplings reveal totally novel properties beyond simple synergetic combination as the enantiomer selectivity emerged from the $[Fe_2L_3]^{4+}$–GO–COOH system [57]. With the continuing developments of nanozyme research, high-efficiency catalysis by proximity with multi-nanozyme systems may become true.

It may take some time to have a nanozyme family which covers all types of reaction catalyzed by natural enzymes. But we can combine natural enzymes with current nanozymes. Just like to substitute the ferroxidase active sites with PtNPs and to turn an ATP-driven nitrogenase into a light-driven enzyme with CdS [60, 90]. The hybrid of an inorganic nanozyme with an organic enzyme may not be a "pure" nanozyme. However, it can integrate the advantages from both sides, which may unfold the promising potentials of a hybrid system.

Acknowledgements We appreciate the financial support of the National Natural Science Foundation of China (31771577 and 31700700), the Natural Science Basic Research Plan in Shaanxi Province of China (2018JM3027 and 2018JQ2038), and the Fundamental Research Funds for the Central Universities (3102017OQD047, 3102017OQD048, and 3102017OQD049).

References

1. Wei H, Wang E (2013) Nanomaterials with enzyme-like characteristics (nanozymes): next-generation artificial enzymes. Chem Soc Rev 42(14):6060–6093
2. Zhang Z, Zhang X, Liu B, Liu J (2017) Molecular imprinting on inorganic nanozymes for hundred-fold enzyme specificity. J Am Chem Soc 139(15):5412–5419
3. Zhan P, Wang ZG, Li N, Ding B (2015) Engineering gold nanoparticles with DNA ligands for selective catalytic oxidation of chiral substrates. ACS Catalysis 5(3):1489–1498
4. Shah J, Purohit R, Singh R, Karakoti AS, Singh S (2015) ATP-enhanced peroxidase-like activity of gold nanoparticles. J Colloid Interface Sci 456:100–107
5. Xu C, Liu Z, Wu L, Ren J, Qu X (2014) Nucleoside triphosphates as promoters to enhance nanoceria enzyme-like activity and for single-nucleotide polymorphism typing. Adv Funct Mater 24(11):1624–1630
6. Lin Y, Huang Y, Ren J, Qu X (2014) Incorporating ATP into biomimetic catalysts for realizing exceptional enzymatic performance over a broad temperature range. NPG Asia Mater 6(7):e114
7. Li X, Wen F, Creran B, Jeong Y, Zhang X, Rotello VM (2012) Colorimetric protein sensing using catalytically amplified sensor arrays. Small 8(23):3589–3592
8. Wu Y, Chen Y, Li Y, Huang J, Yu H, Wang Z (2018) Accelerating peroxidase-like activity of gold nanozymes using purine derivatives and its application for monitoring of occult blood in urine. Sensor Actuat B-Chem 270:443–451
9. Qiu N, Liu Y, Xiang M, Lu X, Yang Q, Guo R (2018) A facile and stable colorimetric sensor based on three-dimensional graphene/mesoporous Fe_3O_4 nanohybrid for highly sensitive and selective detection of p-nitrophenol. Sensor Actuat B-Chem 266:86–94
10. Yue HY, Zhang H, Huang S, Lin XY, Gao X, Chang J, Yao LH, Guo EJ (2017) Synthesis of ZnO nanowire arrays/3D graphene foam and application for determination of levodopa in the presence of uric acid. Biosens Bioelectron 89(Pt 1):592–597

11. Ma Y, Zhao M, Cai B, Wang W, Ye Z, Huang J (2014) 3D graphene network@WO_3 nanowire composites: a multifunctional colorimetric and electrochemical biosensing platform. Chem Commun 50:11135–11138
12. Liu L, Du B, Shang C, Wang J, Wang E (2018) Construction of surface charge-controlled reduced graphene oxide-loaded Fe_3O_4 and Pt nanohybrid for peroxidase mimic with enhanced catalytic activity. Anal Chim Acta 1014:77–84
13. Pautler R, Kelly EY, Huang PJ, Cao J, Liu B, Liu J (2013) Attaching DNA to nanoceria: regulating oxidase activity and fluorescence quenching. ACS Appl Mater Inter 5(15):6820–6825
14. Gao W, Wei X, Wang X, Cui G, Liu Z, Tang B (2016) A competitive coordination-based CeO_2 nanowire-DNA nanosensor: fast and selective detection of hydrogen peroxide in living cells and in vivo. Chem Commun (Camb) 52(18):3643–3646
15. Liu B, Sun Z, Huang PJ, Liu J (2015) Hydrogen peroxide displacing DNA from nanoceria: mechanism and detection of glucose in serum. J Am Chem Soc 137(3):1290–1295
16. Bulbul G, Hayat A, Andreescu S (2016) ssDNA-functionalized nanoceria: a redox-active aptaswitch for biomolecular recognition. Adv Healthc Mater 5(7):822–828
17. Hizir MS, Top M, Balcioglu M, Rana M, Robertson NM, Shen F, Sheng J, Yigit MV (2016) Multiplexed activity of perAuxidase: DNA-capped AuNPs act as adjustable peroxidase. Anal Chem 88(1):600–605
18. Huang L, Chen K, Zhang W, Zhu W, Liu X, Wang J, Wang R, Hu N, Suo Y, Wang J (2018) ssDNA-tailorable oxidase-mimicking activity of spinel $MnCo_2O_4$ for sensitive biomolecular detection in food sample. Sensor Actuat B-Chem 269:79–87
19. Zeng C, Lu N, Wen Y, Liu G, Zhang R, Zhang J, Wang F, Liu X, Li Q, Tang Z, Zhang M (2019) Engineering nanozymes using DNA for catalytic regulation. ACS Appl Mater Inter 11(2):1790–1799
20. Perez JM, Asati A, Nath S, Kaittanis C (2008) Synthesis of biocompatible dextran-coated nanoceria with pH-dependent antioxidant properties. Small 4(5):552–556
21. Nath S, Kaittanis C, Ramachandran V, Dalal N, Perez JM (2009) Synthesis, magnetic characterization and sensing applications of novel dextran-coated iron oxide nanorods. Chem Mater 21(8):1761–1767
22. Yu F, Huang Y, Cole AJ, Yang VC (2009) The artificial peroxidase activity of magnetic iron oxide nanoparticles and its application to glucose detection. Biomaterials 30(27):4716–4722
23. Lord MS, Farrugia BL, Yan CM, Vassie JA, Whitelock JM (2016) Hyaluronan coated cerium oxide nanoparticles modulate CD44 and reactive oxygen species expression in human fibroblasts. J Biomed Mater Res A 104(7):1736–1746
24. Karakoti AS, Singh S, Kumar A, Malinska M, Kuchibhatla SVNT, Wozniak K, Self WT, Seal S (2009) PEGylated nanoceria as radical scavenger with tunable redox chemistry. J Am Chem Soc 131(40):14144–14145
25. Asati A, Santra S, Kaittanis C, Nath S, Perez JM (2009) Oxidase-like activity of polymer-coated cerium oxide nanoparticles. Angew Chem Int Ed Engl 48(13):2308–2312
26. Ansar SM, Kitchens CL (2016) Impact of gold nanoparticle Stabilizing ligands on the colloidal catalytic reduction of 4-nitrophenol. ACS Catalysis 6(8):5553–5560
27. Wang XX, Wu Q, Shan Z, Huang QM (2011) BSA-stabilized Au clusters as peroxidase mimetics for use in xanthine detection. Biosens Bioelectron 26(8):3614–3619
28. Fan K, Wang H, Xi J, Liu Q, Meng X, Duan D, Gao L, Yan X (2016) Optimization of Fe_3O_4 nanozyme activity via single amino acid modification mimicking an enzyme active site. Chem Commun (Camb) 53(2):424–427
29. Liu X, Wei W, Yuan Q, Zhang X, Li N, Du Y, Ma G, Yan C, Ma D (2012) Apoferritin-CeO_2 nano-truffle that has excellent artificial redox enzyme activity. Chem Commun (Camb) 48(26):3155–3157
30. Ge J, Lei J, Zare RN (2012) Protein-inorganic hybrid nanoflowers. Nat Nanotechnol 7(7):428–432
31. Cimini A, D'Angelo B, Das S, Gentile R, Benedetti E, Singh V, Monaco AM, Santucci S, Seal S (2012) Antibody-conjugated PEGylated cerium oxide nanoparticles for specific targeting of Abeta aggregates modulate neuronal survival pathways. Acta Biomater 8(6):2056–2067

32. Mw R (1997) Oxidative stress hypothesis in Alzheimer's disease. Free Radical Bio Med 23(1):134–147
33. Hamley IW (2012) The amyloid beta peptide: a chemist's perspective. Role in Alzheimer's and fibrillization. Chem Rev 112(10):5147–5192
34. Heckert EG, Karakoti AS, Seal S, Self WT (2008) The role of cerium redox state in the SOD mimetic activity of nanoceria. Biomaterials 29(18):2705–2709
35. Ly HG, Absillis G, Janssens R, Proost P, Parac-Vogt TN (2015) Highly amino acid selective hydrolysis of myoglobin at aspartate residues as promoted by Zirconium(IV)-substituted polyoxometalates. Angew Chem Int Ed Engl 54(25):7391–7394
36. Stroobants K, Moelants E, Ly HG, Proost P, Bartik K, Parac-Vogt TN (2013) Polyoxometalates as a novel class of artificial proteases: selective hydrolysis of lysozyme under physiological pH and temperature promoted by a cerium(IV) Keggin-type polyoxometalate. Chem-Eur J 19(8):2848–2858
37. Guan Y, Li M, Dong K, Gao N, Ren J, Zheng Y, Qu X (2016) Ceria/POMs hybrid nanoparticles as a mimicking metallopeptidase for treatment of neurotoxicity of amyloid-β peptide. Biomaterials 98:92–102
38. Gao N, Dong K, Zhao A, Sun H, Wang Y, Ren J, Qu X (2016) Polyoxometalate-based nanozyme: design of a multifunctional enzyme for multi-faceted treatment of Alzheimer's disease. Nano Res 9(4):1079–1090
39. Jin Z, Nackashi D, Lu W, Kittrell C, Tour JM (2010) Decoration, migration, and aggregation of palladium nanoparticles on graphene sheets. Chem Mater 22(20):5695–5699
40. Shylesh S, Schünemann V, Thiel WR (2010) Magnetisch abtrennbare Nanokatalysatoren: Brücken zwischen homogener und heterogener Katalyse. Angew Chem 122(20):3504–3537
41. Shylesh S, Schünemann V, Thiel WR (2010) Magnetically separable nanocatalysts: bridges between homogeneous and heterogeneous catalysis. Angew Chem Int Edit 49(20):3428–3459
42. Huo X, Liu J, Wang B, Zhang H, Yang Z, She X, Xi P (2013) A one-step method to produce graphene-Fe_3O_4 composites and their excellent catalytic activities for three-component coupling of aldehyde, alkyne and amine. J Mater Chem A 1(3):651–656
43. Venkateswara Rao C, Cabrera CR, Ishikawa Y (2011) Graphene-supported Pt–Au alloy nanoparticles: a highly efficient anode for direct formic acid fuel cells. J Phys Chem C 115(44):21963–21970
44. Liu J, Huo X, Li T, Yang Z, Xi P, Wang Z, Wang B (2014) Palladium nanoparticles bonded to two-dimensional iron oxide graphene nanosheets: a synergistic and highly reusable catalyst for the Tsuji-Trost reaction in water and air. Chem-Eur J 20(36):11549–11555
45. Reichhardt T (2001) Hunt for Earth-like planets edges towards launch pad. Nature 409(6817):124–125
46. Nolan EM, Lippard SJ (2008) Tools and tactics for the optical detection of mercuric ion. Chem Rev 108(9):3443–3480
47. Cai S, Lao K, Lau C, Lu J (2011) "Turn-On" chemiluminescence sensor for the highly selective and ultrasensitive detection of Hg^{2+} ions based on interstrand cooperative coordination and catalytic formation of gold nanoparticles. Anal Chem 83(24):9702–9708
48. Dressler VL, Antes FG, Moreira CM, Pozebon D, Duarte FA (2011) As, Hg, I, Sb, Se and Sn speciation in body fluids and biological tissues using hyphenated-ICP-MS techniques: a review. Int J Mass Spectrom 307(1):149–162
49. Xiang Y, Wang Z, Xing H, Wong NY, Lu Y (2010) Label-free fluorescent functional DNA sensors using unmodified DNA: a vacant site approach. Anal Chem 82(10):4122–4129
50. Mor-Piperberg G, Tel-Vered R, Elbaz J, Willner I (2010) Nanoengineered electrically contacted enzymes on DNA scaffolds: functional assemblies for the selective analysis of Hg^{2+} ions. J Am Chem Soc 132(20):6878–6879
51. Zhang S, Li H, Wang Z, Liu J, Zhang H, Wang B, Yang Z (2015) A strongly coupled Au/Fe_3O_4/GO hybrid material with enhanced nanozyme activity for highly sensitive colorimetric detection, and rapid and efficient removal of Hg(2+) in aqueous solutions. Nanoscale 7(18):8495–8502

52. Zhao C, Geng J, Feng L, Ren J, Qu X (2011) Chiral Metallo–supramolecular complexes selectively induce human telomeric G-quadruplex formation under salt-deficient conditions. Chem-Eur J 17(29):8209–8215
53. Yu H, Wang X, Fu M, Ren J, Qu X (2008) Chiral metallo-supramolecular complexes selectively recognize human telomeric G-quadruplex DNA. Nucleic Acids Res 36(17):5695–5703
54. Geim AK, Novoselov KS (2007) The rise of graphene. Nat Mater 6(3):183–191
55. Geim AK (2009) Graphene: status and prospects. Science 324(5934):1530–1534
56. Song Y, Qu K, Zhao C, Ren J, Qu X (2010) Graphene oxide: intrinsic peroxidase catalytic activity and its application to glucose detection. Adv Mater 22(19):2206–2210
57. Xu C, Zhao C, Li M, Wu L, Ren J, Qu X (2014) Artificial evolution of graphene oxide chemzyme with enantioselectivity and near-infrared photothermal effect for cascade biocatalysis reactions. Small 10(9):1841–1847
58. Qu K, Shi P, Ren J, Qu X (2014) Nanocomposite incorporating V_2O_5 nanowires and gold nanoparticles for mimicking an enzyme cascade reaction and its application in the detection of biomolecules. Chem-Eur J 20(24):7501–7506
59. Theil EC, Turano P (2013) Metalloenzymes: cage redesign explains assembly. Nat Chem Biol 9(3):143–144
60. Carmona U, Li L, Zhang L, Knez M (2014) Ferritin light-chain subunits: key elements for the electron transfer across the protein cage. Chem Commun 50(97):15358–15361
61. Li L, Zhang L, Carmona U, Knez M (2014) Semi-artificial and bioactive ferroxidase with nanoparticles as the active sites. Chem Commun 50(59):8021–8023
62. Liu X, Wei W, Yuan Q, Zhang X, Li N, Du Y, Ma G, Yan C, Ma D (2012) Apoferritin-CeO_2 nano-truffle that has excellent artificial redox enzyme activity. Chem Commun 48(26):3155–3157
63. Li Y, He X, Yin JJ, Ma Y, Zhang P, Li J, Ding Y, Zhang J, Zhao Y, Chai Z, Zhang Z (2015) Acquired superoxide-scavenging ability of ceria nanoparticles. Angew Chem Int Ed Engl 54(6):1832–1835
64. Huo M, Wang L, Chen Y, Shi J (2017) Tumor-selective catalytic nanomedicine by nanocatalyst delivery. Nat Commun 8(1):357
65. Liang H, Liu B, Yuan Q, Liu J (2016) Magnetic iron oxide nanoparticle seeded growth of nucleotide coordinated polymers. ACS Appl Mater Inter 8(24):15615–15622
66. Cheng H, Zhang L, He J, Guo W, Zhou Z, Zhang X, Nie S, Wei H (2016) Integrated nanozymes with nanoscale proximity for in vivo neurochemical monitoring in living brains. Anal Chem 88(10):5489–5497
67. Liu Y, Qin Y, Zheng Y, Qin Y, Cheng M, Guo R (2019) A one-pot and modular self-assembly strategy for high-performance organized enzyme cascade bioplatforms based on dual-functionalized protein–PtNP@mesoporous iron oxide hybrid. J Mater Chem B 7:43–52
68. Zhang H, Han L, Li F (2019) A universal one-pot assay strategy based on bio-inorganic cascade catalysts for different analytes by changing pH-dependent activity of enzymes on enzyme mimics. Sensor Actuat B-Chem 286:460–467
69. Liu M, Li Z, Li Y, Chen J, Yuan Q (2019) Self-assembled nanozyme complexes with enhanced cascade activity and high stability for colorimetric detection of glucose. Chin Chem Lett 30(5):1009–1012
70. Xu Y, Fei J, Li G, Yuan T, Xu X, Li J (2019) Nanozyme-catalyzed cascade reactions for mitochondria-mimicking oxidative phosphorylation. Angew Chem Int Ed Engl 58(17):5572–5576
71. Furukawa H, Cordova KE, O'Keeffe M, Yaghi OM (2013) The chemistry and applications of metal-organic frameworks. Science 341(6149):1230444
72. Chen G, Huang S, Kou X, Wei S, Jiang S, Shen J, Zhu F, Ouyang G (2019) A convenient and versatile amino-acid-boosted biomimetic strategy for the nondestructive encapsulation of biomacromolecules within metal-organic frameworks. Angew Chem Int Ed Engl 58(5):1463–1467
73. Li S, Liu X, Chai H, Huang Y (2018) Recent advances in the construction and analytical applications of metal-organic frameworks-based nanozymes. TRAC-Trend Anal Chem 105:391–403

74. Zhao Z, Pang J, Liu W, Lin T, Ye F, Zhao S (2019) A bifunctional metal organic framework of type Fe(III)-BTC for cascade (enzymatic and enzyme-mimicking) colorimetric determination of glucose. Microchim Acta 186(5):295
75. Akanda MR, Tamilavan V, Park S, Jo K, Hyun MH, Yang H (2013) Hydroquinone diphosphate as a phosphatase substrate in enzymatic amplification combined with electrochemical–chemical–chemical redox cycling for the detection of *E. coli* O157: H7. Anal Chem 85(3):1631–1636
76. Xianyu Y, Wang Z, Jiang X (2014) A plasmonic nanosensor for immunoassay via enzyme-triggered click chemistry. ACS Nano 8(12):12741–12747
77. Geraud E, Prevot V, Forano C, Mousty C (2008) Spongy gel-like layered double hydroxide–alkaline phosphatase nanohybrid as a biosensing material. Chem Commun 13:1554–1556
78. Qian Z, Chai L, Tang C, Huang Y, Chen J, Feng H (2015) Carbon quantum dots-based recyclable real-time fluorescence assay for alkaline phosphatase with adenosine triphosphate as substrate. Anal Chem 87(5):2966–2973
79. Zhang L, Zhao J, Duan M, Zhang H, Jiang J, Yu R (2013) Inhibition of dsDNA-templated copper nanoparticles by pyrophosphate as a label-free fluorescent strategy for alkaline phosphatase assay. Anal Chem 85(8):3797–3801
80. Wei H, Chen C, Han B, Wang E (2008) Enzyme colorimetric assay using unmodified silver nanoparticles. Anal Chem 80(18):7051–7055
81. Gao Z, Xu M, Hou L, Chen G, Tang D (2013) Magnetic bead-based reverse colorimetric immunoassay strategy for sensing biomolecules. Anal Chem 85(14):6945–6952
82. Jin LY, Dong YM, Wu XM, Cao GX, Wang GL (2015) Versatile and amplified biosensing through enzymatic cascade reaction by coupling alkaline phosphatase in situ generation of photoresponsive nanozyme. Anal Chem 87(20):10429–10436
83. Greenbaum E (1985) Platinized chloroplasts: a novel photocatalytic material. Science 230:1373–1376
84. Evans BR, O'Neill HM, Hutchens SA, Bruce BD, Greenbaum E (2004) Enhanced photocatalytic hydrogen evolution by covalent attachment of plastocyanin to photosystem I. Nano Lett 4(10):1815–1819
85. Utschig LM, Dimitrijevic NM, Poluektov OG, Chemerisov SD, Mulfort KL, Tiede DM (2011) Photocatalytic hydrogen production from noncovalent biohybrid photosystem I/Pt nanoparticle complexes. J Phys Chem Lett 2(3):236–241
86. Balasubramanian S, Wang P, Schaller RD, Rajh T, Rozhkova EA (2013) High-performance bioassisted nanophotocatalyst for hydrogen production. Nano Lett 13(7):3365–3371
87. Wang P, Dimitrijevic NM, Chang AY, Schaller RD, Liu Y, Rajh T, Rozhkova EA (2014) Photoinduced electron transfer pathways in hydrogen-evolving reduced graphene oxide-boosted hybrid nano-bio catalyst. ACS Nano 8(8):7995–8002
88. Kim D, Sakimoto KK, Hong D, Yang P (2015) Artificial photosynthesis for sustainable fuel and chemical production. Angew Chem Int Edit 54(11):3259–3266
89. Hoffman BM, Lukoyanov D, Yang ZY, Dean DR, Seefeldt LC (2014) Mechanism of nitrogen fixation by nitrogenase: the next stage. Chem Rev 114(8):4041–4062
90. Brown KA, Harris DF, Wilker MB, Rasmussen A, Khadka N, Hamby H, King PW (2016) Light-driven dinitrogen reduction catalyzed by a CdS: nitrogenase MoFe protein biohybrid. Science 352(6284):448–450

Part III
Promising Applications of Nanozymes

Chapter 12
Molecular Detection Using Nanozymes

Biwu Liu and Juewen Liu

Abbreviations

β-gal	β-galactosidase
ABTS	2,2′-azino-bis(3-ethylbenzothiazoline-6-sulfonic acid)
AP	Alkaline phosphatase
AR	Amplex red
ATP	Adenosine triphosphate
AuNPs	Gold nanoparticles
AUR	Amplex ultrared
BSA	Bovine serum albumin
CTP	Cytidine triphosphate
ELISA	Enzyme-linked immunosorbent assay
GOx	Glucose oxidase
GTP	Guanosine triphosphate
HRP	Horseradish peroxidase
LOD	Limit of detection
MOF	Metal–organic frameworks
NP(s)	Nanoparticle(s)
NTPs	Nucleoside triphosphates
rGO	Reduced graphene oxide
SOD	Superoxide dismutase
SWNTs	Single-walled carbon nanotubes
TMB	3,3′,5,5′-Tetramethylbenzidine

B. Liu · J. Liu (✉)
Department of Chemistry, Waterloo Institute for Nanotechnology, Waterloo, ON N2L 3G1, Canada
e-mail: liujw@uwaterloo.ca

X. Yan (ed.), *Nanozymology*, Nanostructure Science and Technology,
https://doi.org/10.1007/978-981-15-1490-6_12

UTP	Uridine triphosphate
WHO	World Health Organization
XPS	X-ray photoelectron spectroscopy

Nanozymes, a new type of enzyme mimics, are nanomaterials with intrinsic enzyme-like activities. Nanozymes can efficiently catalyze the conversion of substrates and follow similar kinetics and mechanisms of natural enzymes under near-physiological conditions. Many applications of nanozymes have already been identified as discussed in other chapters of this book. This chapter is focused on its analytical application for molecular detection. The scope of this chapter is to focus on the intrinsic enzyme-like activities of nanomaterials, rather than conjugating protein enzymes onto nanomaterials [1–7].

12.1 Biosensors

Biosensors use a biomolecule for target recognition. Since the inception of this concept, biosensors have played an increasingly important role complementary to traditional analytical instruments for the measurement of chemicals and providing on-site and real-time information. Due to its cost-effectiveness and simplicity, biosensors are highly attractive for biomedical diagnosis, environmental monitoring, food quality testing, and the pharmaceutical industry. The most successful biosensors are probably the glucose meters, which have occupied more than half of the total biosensor market. The pregnancy test strips are another success story. Given this impressive progression, there is still a lot of room for further improvement. For example, the biomolecule components in sensors are mainly proteins that are prone to irreversible denaturation, and it is critically important to increase their stability for a longer shelf life and to ensure more robust sensors. In addition, most biosensors only work under ambient or physiological conditions, while it is nearly impossible to detect intended targets under harsh conditions such as extreme pH, high temperature, or high ionic strength. These nonphysiological conditions may quickly denature proteins. Finally, the targets detectable by biosensors are mostly limited to metabolites and those that can elicit an immune response. For many biosensors, enzyme is a critical component, either for target recognition or for producing signals for detection.

Some of these problems might be solved by replacing enzymes with more robust and artificially developed enzyme mimicking molecules or materials. In this chapter, we focus our discussion on nanozymes: enzyme mimicking inorganic nanomaterials. This strategy, if successful, can further drop the cost and increase the stability of biosensors. Nanozymes can be prepared in bulk quantities at a very low cost, while enzymes have to be extracted from cells. At the simplest level, nanozymes might replace protein enzymes if they perform exactly the same function. In addition, nanozymes might be used at nonphysiological conditions, thus broadening the scope of sensing applications. Finally, nanozymes might be able to detect targets that are difficult for typical biosensors. Therefore, nanozymes can also go much beyond simply replacing protein enzymes.

The other chapters in this book have already covered the synthesis and property of various nanozymes, and this chapter only focuses on its analytical applications. We first give a brief summary of the roles of enzymes in biosensors and then review representative examples of nanozymes.

12.2 Enzymes in Biosensors

Enzymes have a number of basic properties including excellent substrate specificity and fast turnover rates. These properties make enzymes ideal as biosensor components. A biosensor contains a biomolecule for target recognition and also a signaling moiety (Fig. 12.1); enzymes can serve in both components. For example, the excellent substrate specificity allows enzymes to be used for target recognition (e.g., in glucose sensors). Its fast turnover rate is ideal for signal production and amplification. As such, enzymes have played a central role in biosensor development. To better illustrate the role of nanozymes in biosensors, we first introduce a few representative enzyme-based biosensors.

12.2.1 Enzyme as a Label for Signaling

Most enzymes have extremely high catalytic rates. Since each enzyme molecule can turn over thousands to millions of substrates, enzymes are highly sensitive and popular for signal generation and amplification. The most popular format of using enzyme labels is probably the enzyme-linked immunosorbent assay (ELISA). Three types of ELISAs are illustrated in Fig. 12.2a. Typically, the first step of an ELSIA is immobilization of the target (antigen), which can be adsorbed directly onto a

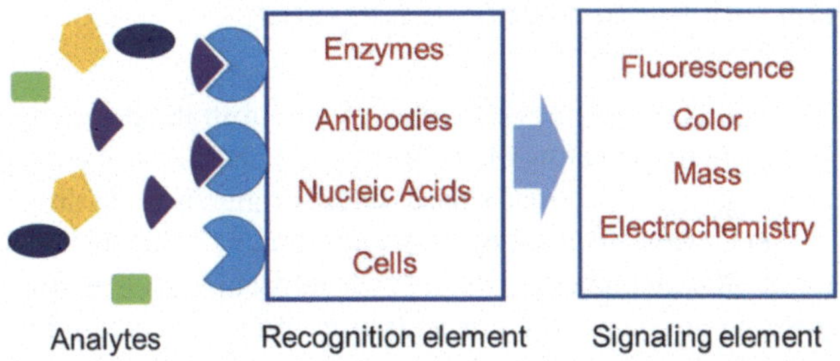

Fig. 12.1 A scheme showing the typical components of a biosensor: a target recognition element and a signal transduction element. Enzymes can be used in both components

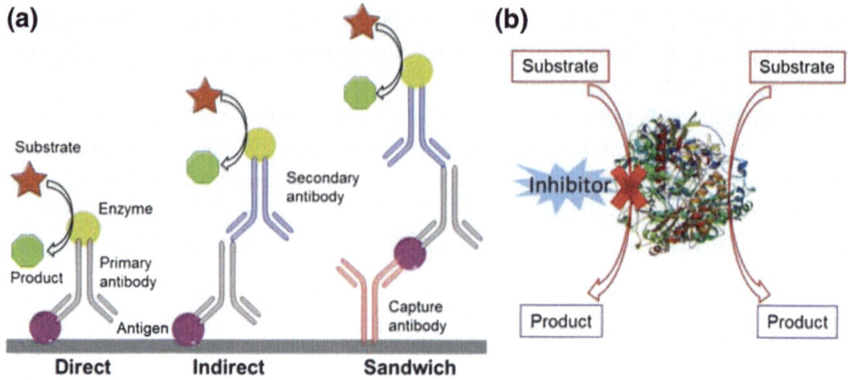

Fig. 12.2 Design of biosensors based on enzymes. **a** Three types of ELISA strategies. The role of the enzyme is to convert antibody-induced recognition events to observable signals by catalyzing the conversion of a substrate. **b** Biosensors can be designed by measuring the effect of an inhibitor. Figure b was adapted from Ref. [9]. Copyright (2016) Nature Publishing Group

plate (e.g., direct and indirect format) or captured by an antibody (sandwich format). Then, detection of the antigen is achieved by the attachment of its antibody conjugated with an enzyme. In a direct assay, the enzyme is directly conjugated to the primary antibody, while in an indirect or sandwich assay, the enzyme is bound to a secondary antibody. Besides these three formats, other ELISA assays also exist, such as competitive ELISA and in-cell ELISA. All of these assays are now commercially available.

While different ELISA formats exist, the final signaling step relies on the catalytic conversion of a substrate using a linked enzyme. A few enzymes with high catalytic activities have been used in ELISA. The most popular enzyme is probably horseradish peroxidase (HRP), followed by alkaline phosphatase (AP) and β-galactosidase (β-gal) [8]. Some other enzymes with a high catalytic rate (e.g., urease, carbonic anhydrase) have also been used.

12.2.2 Detecting the Substrate of Enzymes

A fundamental property of enzymes is their excellent substrate specificity. Therefore, enzymes are excellent for detecting their substrates, and a classic example is sensors for glucose detection using glucose oxidase (GOx) or glucose dehydrogenases [10]. Glucose can be selectively oxidized by oxygen to produce gluconic acid and H_2O_2 in the presence of GOx, while other small sugar molecules such as fructose, galactose, and sucrose do not react. The concentration of O_2 is decreased while that of H_2O_2 is increased; both molecules can be monitored using electrochemical or optical methods. Alternatively, oxidation of glucose is accompanied by the reduction of an electron mediator (e.g., a redox dye) rather than O_2. The reduced mediator can also be monitored by colorimetric or electrochemical methods. Lastly, the glucose oxidation can be detected by directly monitoring electron transfer to an electrode. Using this direct strategy, possible variations of oxygen concentration in blood samples can be avoided.

12.2.3 Detecting Enzyme Inhibitors

Many biosensors are designed based on the detection of enzyme inhibitors (Fig. 12.2b) [11, 12]. In 1962, Guilbault et al. reported that organophosphorous compounds inhibited cholinesterase activity, allowing for electrochemical detection with a high sensitivity [13]. Since then, enzyme inhibition-based sensors have been developed for the detection of drugs, nerve agents, and pesticides. Depending on the interaction between the inhibitor and the enzyme, sensors can work by irreversible or reversible inhibition. Irreversible inhibition typically requires the formation of a covalent bond between the inhibitor and the active site of an enzyme. This type of interaction results in a permanent deactivation of enzyme activity. In the case of reversible inhibition, the activities of enzymes can be recovered by washing away inhibitors. To measure the inhibition effect quantitatively, the degree of inhibition is defined as

$$I\% = 100 \times \frac{A_0 - A_i}{A_0} \quad (12.1)$$

where A_0 is the activity of enzyme without inhibitors and A_i is the activity in the presence of the inhibitor.

Using the enzyme cholinesterase as an example, Moscone and co-workers introduced three types of protocols to measure the inhibitors [14]. In the first protocol, enzymatic reactions were initialized by mixing the substrate and enzyme. An inhibitor was then added after a certain period of time. The activities before and after adding the inhibitor were then compared. The authors pointed out that this type of sensor was not very popular since the inhibitor did not fully react with the enzyme. In the second protocol, the enzyme is incubated with the inhibitor for a certain period before adding the substrate. In a control group, the substrate is mixed with the enzyme in the absence of any inhibitor. The difference between the two groups is then compared to quantify the inhibitor. In the third method, after incubating an inhibitor with the enzyme, the biosensor is rinsed with water and immersed in a new buffer solution with its substrate. This so-called "medium-exchange method" is intended to avoid interferences from residual components.

Using the methods described above, detection of nerve agents was reported. Lee et al. reported the detection of organophosphorous compounds based on the inhibition of the enzyme acetylcholinesterase using planar waveguide absorbance spectroscopy [15]. The limits of detection for soman and sarin were reported to be ca. 100 pM and

600 pM, respectively. Using the same enzyme, sensitive detection of many other enzyme inhibitors was also demonstrated, which was reviewed by other groups [12, 16], and thus is not repeated here in detail.

12.3 Nanozymes in Biosensors

The above section briefly summarized a few typical ways of using protein-based enzymes for biosensor development. Since the first reports in the 1990s, nanozymes have been widely employed as functional materials in analytical chemistry, environmental monitoring, colloidal chemistry, and biomedicine. In this section, we describe efforts using nanozymes for biosensor design in parallel to the strategies described above. Various sensing methods with representative examples in the literatures are discussed.

12.3.1 Typical Nanozymes Reactions Used in Biosensors

Before discussing specific examples, we first introduce some typical nanozyme reactions reported so far that may be useful for producing detectable signals (Fig. 12.3a). The most common reaction is probably the activation of H_2O_2 to oxidize a diverse range of chromogenic or fluorescence substrates (e.g., peroxidase-mimicking nanozymes). Many types of nanomaterials have been found to possess such activities. Commonly used substrates include the 3,3′,5,5′-tetramethylbenzidine

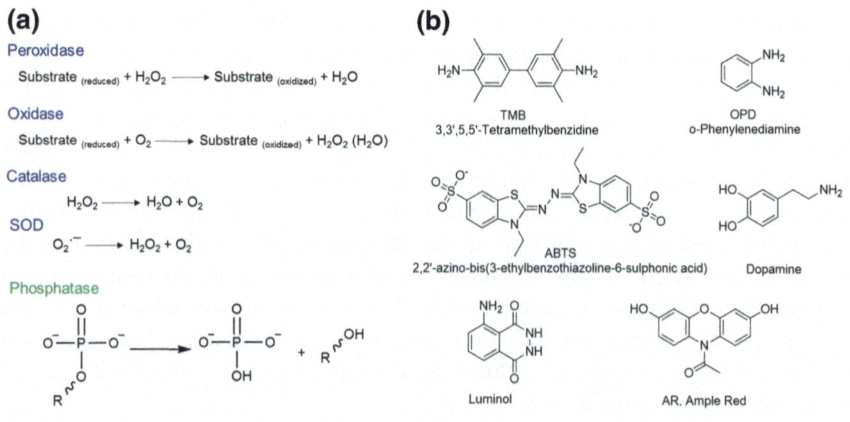

Fig. 12.3 **a** Typical reactions catalyzed by nanozymes. **b** Some commonly used chromogenic and fluorescence substrates for oxidase/peroxidase reactions

(TMB), 2,2′-azino-bis(3-ethylbenzothiazoline-6-sulfonic acid) (ABTS) for colorimetric assays, and Amplex red (AR), Amplex ultrared (AUR) for fluorescent assays (Fig. 12.3b). On one hand, H_2O_2 is an important small molecule involved in many biological reactions. Thus, detection of H_2O_2 becomes a routine for any newly identified peroxidase mimics. On the other hand, many natural enzymes can generate H_2O_2 as a by-product. In this way, nanozymes can be combined with protein enzymes to expand the range of detectable analytes. Besides the peroxidase activity, nanozymes are also able to catalyze several other redox reactions that are outlined in Fig. 12.3a. For example, oxidase can directly use O_2 as the oxidizing agent to oxidize the substrates mentioned above. Catalase mimics catalyze the decomposition of H_2O_2 into H_2O and O_2. Superoxide dismutase (SOD) mimics are able to accelerate the disproportion of $O_2^{\bullet-}$ into H_2O_2 and O_2. Hydrolysis of phosphate ester bonds catalyzed by nanozymes has also been reported.

12.3.2 Nanozymes Replacing Protein Enzymes in Signal Amplification

A straightforward application of nanozymes is to replace protein enzymes for ELISA. Compared to their protein counterparts, nanozymes have higher stability, lower cost, and are easier to prepare. A representative example was demonstrated by Yan and co-workers in their seminal work [9]. The intrinsic-peroxidase activity of Fe_3O_4 nanoparticles (NPs) allowed the design of an ELISA in the absence of any protein enzymes. In their initial design, Yan et al. immobilized an antigen (hepatitis B virus surface antigen, preS1) on a plate, and attached an antibody (anti-HBV preS1 antibody) with an incubation step (Fig. 12.4a). Afterward, the Fe_3O_4 NPs modified with protein A and a substrate were added. Basically, the role of protein A-modified Fe_3O_4 NPs was to replace an enzyme-conjugated secondary antibody in the indirect ELISA (Fig. 12.2a). The bound Fe_3O_4 NPs were able to catalyze the oxidation of TMB in the presence of H_2O_2. In their second immunoassay, antibody-modified Fe_3O_4 NPs were used to capture biomarkers in serum. The intrinsic magnetic property of Fe_3O_4 NPs made the separation step easy to operate. Then, these antigen–antibody Fe_3O_4 conjugates were attached to another antibody on the plate. Again, oxidation of TMB by H_2O_2 was then catalyzed by bound Fe_3O_4 NPs as a peroxidase mimic (Fig. 12.4b). With the developed assays, the authors successfully detected preS1 (Fig. 12.4c) and cardiac troponin I (TnI) (Fig. 12.4d), and selectivity against BSA was demonstrated.

Nanoceria, as an oxidase mimic, is able to catalyze the oxidation of various substrates using dissolved O_2 as the oxidizing agent (no H_2O_2 needed). Asati et al. [17] proposed to use the oxidase-mimicking nanoceria to replace signaling enzymes (Fig. 12.4e). Poly(acrylic acid)-modified nanoceria was functionalized with folic acid, which is the ligand for the overexpressed folate receptor in cancer cell lines. After incubating the ligand-modified nanoceria with target cells, the authors introduced a chromogenic substrate (i.e., TMB). Since nanoceria is an oxidase mimic, the

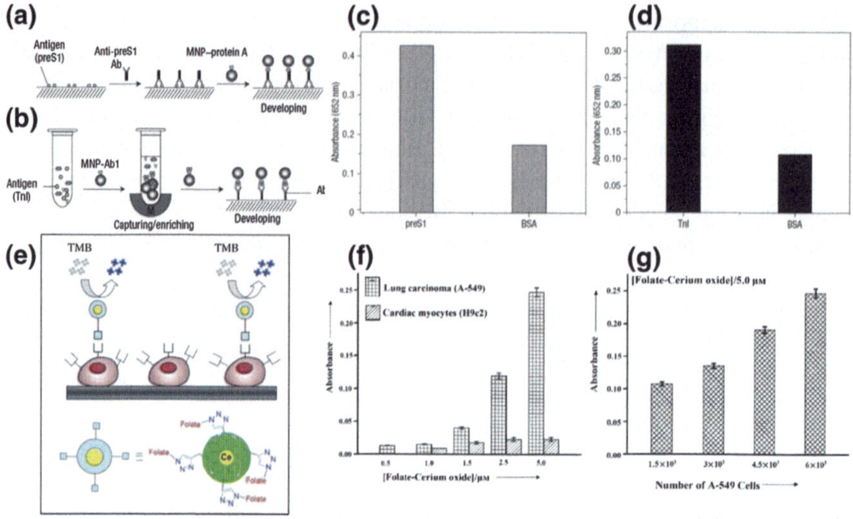

Fig. 12.4 Design of assays using nanozymes for signal amplification. **a** An indirect ELISA method by immobilizing the antigen on a plate, followed by attaching an antibody. Fe_3O_4 NPs modified with proteins were further linked to the antibody for signal production. **b** Design of a "capture-and-detection" strategy by combining the magnetic property and nanozyme activity of Fe_3O_4 NPs. **c** and **d** are the analytical performance of sensors shown in (**a**) and (**b**), respectively. **e** Using nanoceria in replacing enzymes in signaling. Oxidation of TMB by cell bound nanoceria as a function of **f** nanoceria concentration and **g** target cell concentration. Figure a–d was adapted from Ref. [9]. Copyright (2007) Nature Publishing Group. Figure e–g was adapted from Ref. [17]. Copyright (2009) Wiley–VCH Verlag GmbH & Co. KGaA, Weinheim

unstable H_2O_2 was not required to initiate the oxidation. To examine the feasibility of the assay, a lung cancer line (A-549) was tested, and cardiac myocytes (H9c2) were used as a negative control. Results showed that folic acid-modified nanoceria prefers to bind to the lung cancer cell line (A-549) over H9c2, as indicated by the faster oxidation of TMB (Fig. 12.4f). Furthermore, the TMB oxidation was correlated to the number of cells (A549). With more A549 cells, more nanoceria particles were attached on the cell, resulting in faster TMB oxidation (Fig. 12.4g).

Instead of using organic dyes, Zhao et al. demonstrated that the catalytic growth of gold nanoclusters (AuNCs, diameter < 2 nm) induced by H_2O_2 for signaling [18]. A sandwich-type assay was used by sequentially adding a biotin-modified goat IgG, avidin, and biotin-modified AuNCs to the plate. Then, $HAuCl_4$ and H_2O_2 were added as precursors to grow AuNCs for the color change. An obvious color change was observed in the presence of 1.0×10^{-20} M avidin. This strategy was further demonstrated to detect several other analytes by using different probes. For example, CA15-3, a breast cancer biomarker, was successfully detected with a limit of detection of 7.52×10^{-14} U/mL.

12.3.3 Sensors Based on Regulating the Activity of Nanozymes

Protein enzymes usually have excellent catalytic activities, leaving a large room for measuring their inhibitor concentrations. The activity of nanozymes varies quite a lot depending on the type of material, surface chemistry, and the type of catalytic reaction. As a result, both inhibitors and promoters of nanozymes can be identified and detected (Fig. 12.5). Some analytes may interact with the surface-active species of nanozymes to accelerate the reaction. Alternatively, some surface capping ligands may block the accessibility of a substrate, which slows down the reaction. In addition, analytes may affect the activity by interacting with the substrate and/or the product (e.g., by removing the product from the nanozyme surface). In any case, if the degree of inhibition or promotion can be correlated with the concentration of an analyte sensitively and selectively, a sensor can be designed. In this section, some typical strategies are introduced along with representative examples reported in literatures. Three types of analytes are summarized, including inorganic ions (cations and anions), small molecules, and biomacromolecules. In each part, examples of the activity promoting effects are introduced first followed by those acting as inhibitors.

Detecting Inorganic Anions by Regulating Nanozymes Activity

The oxidase-like activity of nanoceria has been extensively used for biosensor development since this reaction does not require H_2O_2 [17, 19]. In general, smaller nanoceria particles have higher activity [17], and most nanozyme-related works used nanoceria are below 5 nm. The oxidase activity of nanoceria is affected by various inorganic anions that might be adsorbed on the particle surface [20]. Our lab found that fluoride ions can bind to nanoceria, resulting in an increase of its oxidase activity up to 100-fold (Fig. 12.6a). The oxidation of both TMB and ABTS was enhanced in the presence of fluoride. The catalytic efficiency (k_{cat}/K_m) was increased by ca. 15-fold for ABTS and ca. 100-fold for TMB. The surface charge of nanoceria inverted from positive to negative after fluoride adsorption, while the same concentration of chloride had little effect (Fig. 12.6b). In addition, fluoride has a strong electron-withdrawing property. These factors may contribute to the enhanced activity. The

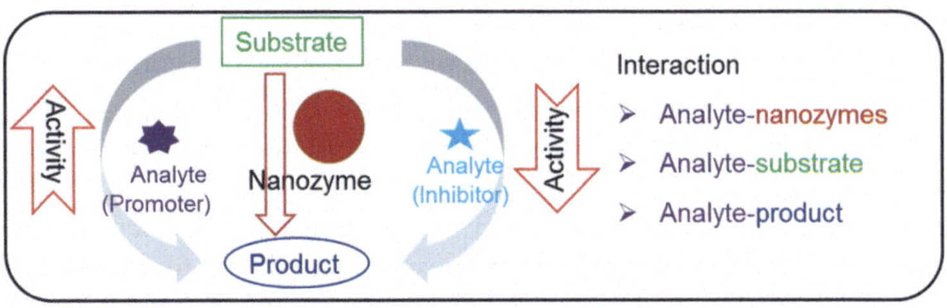

Fig. 12.5 A scheme showing the detection of analytes by modulating the activity nanozymes. The modulation effect can be from the interaction of analytes with the nanozyme, the substrate used, or the product

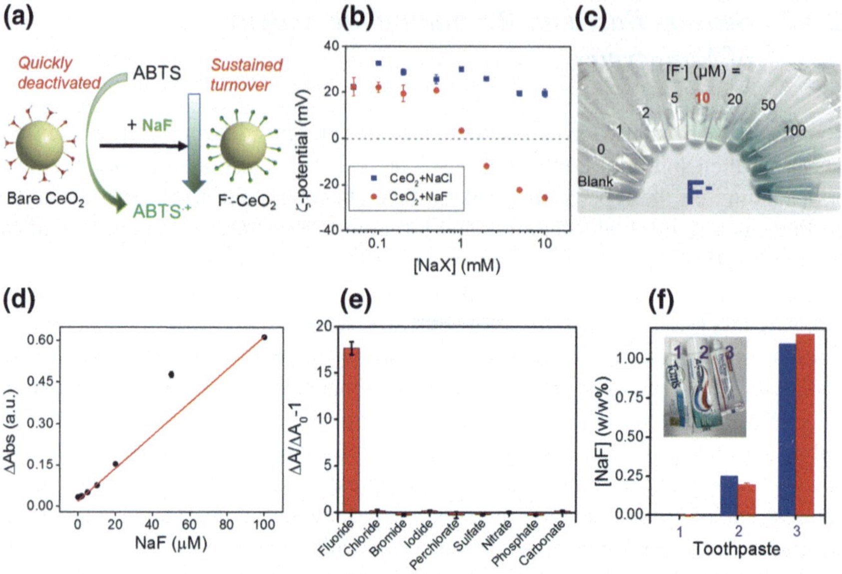

Fig. 12.6 Detection of fluoride ions based on its promoter effect on the oxidase-like activity of nanoceria. **a** A scheme showing that fluoride capping allows sustained turnover of ABTS by nanoceria. **b** ζ-potential of nanoceria in acetate buffer (pH 4) as a function of F^- and Cl^- concentration. **c** Photographs showing the F^- enhanced oxidase-like activity using ABTS as a substrate. **d** Calibration curve and **e** selectivity test for F^- detection. **f** Detection of fluoride in toothpaste samples using our sensor (red bars) and the results are comparable with those on the label (blue bars). Inset: a photograph of the three toothpastes used. This figure is adapted from Ref. [21] with permission. Copyright (2016) Royal Society of Chemistry

unique promoting effect of fluoride allows a colorimetric assay for its detection. As low as 10 μM of F^- can be detected by the naked eye (Fig. 12.6c), and the limit of detection was calculated to be 640 nM F^- based on the calibration curve using a spectrophotometer (Fig. 12.6d). Selectivity test showed that compared to fluoride (0.1 mM), other anions (1 mM), including chloride, bromide, iodide, perchlorate, sulfate, nitrate, phosphate, and carbonate, did not induce a significant enhancing effect (Fig. 12.6e). In addition, this sensor was robust enough to accurately detect fluoride in toothpaste samples (Fig. 12.6f).

Inhibition effects of anions on certain nanozymes were also reported [22, 23]. For example, phosphate was detected based on its inhibition effect on the peroxidase-like activity of Fe_3O_4 NPs. Chen and co-workers [22] examined the TMB oxidation by H_2O_2 and Fe_3O_4 NPs after adsorbing different anions. They found that only phosphate can significantly decrease the TMB oxidation, leading to a sensitive method with a limit of detection (LOD) of 110 nM phosphate. Furthermore, phosphate in real drinking water samples was detected and the recovery efficiency was 97.8–113.6% for two levels of phosphate spiking. The results were comparable to the standard molybdenum-blue method for phosphate detection. In another work, Qin et al. [24] reported that sulfite was able to decrease the peroxidase-like activity of the Co_3O_4

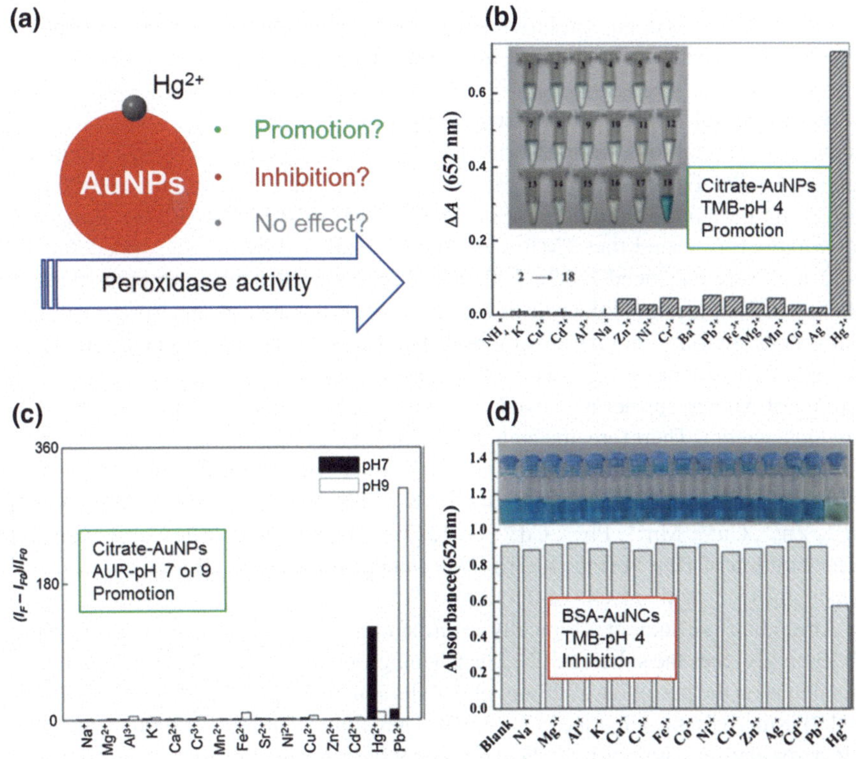

Fig. 12.7 Detection of Hg^{2+} based on its modulation of the peroxidase-like activity of AuNPs. **a** Hg^{2+} may play different roles in regulating the activity of AuNPs. Selectivity test of metal ions on **b** TMB oxidation at pH 4 with citrate-capped AuNPs, **c** AUR oxidation at pH 7 and pH 9 with citrate-capped AuNPs, and **d** TMB oxidation at pH 4 with BSA-AuNCs. Figure b from Ref. [26]. Copyright (2011) Royal Society of Chemistry. Figure c from Ref. [27]. Copyright (2012) Elsevier Ltd. Figure D from Ref. [28]. Copyright (2013) Elsevier Ltd.

nanozyme, leading to a sensitive sensor. As low as 200 nM sulfite could be detected with this method, although the mechanism of inhibition was not discussed.

Anions may also bind to gold nanoparticles (AuNPs) and affect their enzyme-like performance. For example, sulfide (S^{2-}) was found to form Au_2S species on the surface of bare AuNPs [25]. The authors reasoned that such coordination shielded the active sites of bare AuNPs. The effect of S^{2-} was then quantitatively examined. With the naked eye, as low as 2 μM sulfide could be distinguished. The LOD for sulfide was 80 nM with a linear range from 0.5 to 10 μM. This LOD was much lower than the World Health Organization (WHO) recommended maximum level in drinking water (15 μM). The selectivity was further evaluated using a diverse range of anions (100 μM, Cl^-, NO_3^-, NO_2^{2-}, SO_4^{2-}, SO_3^{2-}, Ac^-, PO_4^{3-}, CO_3^{2-}, and $EDTA^{2-}$) and metal ions (10 μM, Na^+, K^+, Mg^{2+}, Ca^{2+}, Al^{3+}, Fe^{2+}, Fe^{3+}, Co^{2+}, Ni^{2+}, Cu^{2+}, Zn^{2+}), and no interference was observed. It needs to be noted that detecting promoters are analytically preferred since detecting inhibitors are more susceptible to interference with a limited room for signal change.

Detecting Metal Ions by Regulating Nanozymes Activity

Some heavy metal ions were reported to promote enzyme-like activities of citrate-capped AuNPs. For example, several groups reported that mercury ions (Hg^{2+}) promoted the peroxidase-like activity of AuNPs and AuNCs (Fig. 12.7a). Long and co-workers [26] found that TMB peroxidation catalyzed by citrate-capped AuNPs (~30 nm) was enhanced by 25-fold in the presence of a trace amount of Hg^{2+} (400 nM). To understand the role of Hg^{2+}, X-ray photoelectron spectroscopy (XPS) was used to analyze the surface species of Hg. It was found that both Hg^0 and Hg^{2+} existed after incubating Hg^{2+} with citrate-capped AuNPs. Also, the peroxidase-like activity of Au was further improved by mixing the Hg-AuNPs sample with $NaBH_4$, a reducing agent. Therefore, the authors concluded that Hg^0 on the surface of AuNPs might be responsible for promoting H_2O_2 activation. The enhancement was selective for Hg^{2+}, while other metal ions, including K^+, Na^+, NH_4^+, Ag^+, Ba^{2+}, Mg^{2+}, Co^{2+}, Ni^{2+}, Zn^{2+}, Cu^{2+}, Mn^{2+}, Pb^{2+}, Cd^{2+}, Al^{3+}, Cr^{3+}, and Fe^{3+} had little effect even at a concentration of 10 μM (Fig. 12.7b). This enhancement effect allowed colorimetric sensing of Hg^{2+} with a LOD of 0.3 nM.

An acidic environment is typically required for the peroxidase activity assay using TMB or ABTS as the substrate [2]. The Chang group investigated the effect of heavy metal ions on the peroxidase activity of AuNPs at neutral (pH 7) and basic conditions (pH 9) using AR as a substrate [27, 29, 30]. They found that Hg^{2+} and Pb^{2+} enhanced AR peroxidation with optimal pH at 7.0 and 9.0, respectively (Fig. 12.7c) [27].

Besides the peroxidase activity, the oxidase- and catalase-like activity of AuNPs is also affected by metal ions [30]. In the absence of H_2O_2, oxidation of AR was enhanced by 117-fold by Ag^+ and 31-fold by Hg^{2+}. The decomposition of H_2O_2 (e.g., catalase activity) catalyzed by AuNPs was also enhanced by Hg^{2+} and Bi^{3+}. Spectroscopic investigations showed that these metal ions can bind to the AuNP surface and form metal–Au alloys, which are responsible for the altered catalytic performance. These modulation effects of metal ions can be used for sensor designs.

The effect of metal ions is quite sensitive to the surface chemistry of gold. For example, bovine serum albumin (BSA)-encapsulated AuNCs exhibited a decreased activity toward TMB oxidation at pH 4 in the presence of Hg^{2+} [28]. Note that without the high-affinity ligand BSA, the activity of citrate-capped AuNPs was increased as mentioned above. However, as low as 1 μM of Hg^{2+} inhibited the activity of BSA-AuNCs by 30%, while other metal ions, including 50 μM of Na^+, Fe^{3+}, Co^{2+}, Ag^+, 100 μM of Al^{3+}, K^+, Ca^{2+}, Cr^{3+}, Ni^{2+}, Cu^{2+}, Zn^{2+}, Cd^{2+}, Pb^{2+}, and Mg^{2+} did not show any effect (Fig. 12.7d). Aggregation of AuNCs by Hg^{2+}-induced fusion was proposed to account for the decreased activity. With this inhibition effect, detection of Hg^{2+} down to 3 nM was achieved. Another metal ion, uranyl (UO_2^{2+}), also inhibited the peroxidase activity of BSA-AuNCs [31]. It was suggested that binding of uranyl to the BSA ligand resulted in the fusion of AuNCs and formation of larger NPs. Based on such an inhibition effect, a UO_2^{2+} sensor was then designed, and the LOD was 1.86 μM. The selectivity test showed that other metal ions (Ag^+ and Hg^{2+}) also inhibited the reaction, while their interference was masked by adding iodide.

Similar inhibition effects were also observed by Fu and co-workers [32]. Among different metal ions, Hg^{2+} selectively inhibited the peroxidase activity of BSA-capped Pt NPs, leading to a sensitive sensor with a LOD of 7.2 nM Hg^{2+}. In an earlier study, Liu et al. [33] examined the role of several metal ions on the oxidase activity of Au@Pt NPs in order to obtain inhibitors of the nanozyme. Three metal ions (Fe^{2+}, Cu^{2+}, and Hg^{2+}) among others (Ba^{2+}, Co^{2+}, Cr^{3+}, Ca^{2+}, Ni^{2+}, Zn^{2+}, Pb^{2+}, Mn^{2+}, and Fe^{3+}) were found to inhibit the oxidation of OPD. A Hg^{2+} sensor was then constructed with a LOD of 550 nM and the linear range was from 1 to 10 μM Hg^{2+}.

Detecting Small Molecules by Regulating Nanozymes Activity

In addition to cations and anions, the surface chemistry and activity of nanozymes can also be modulated by adsorbing small molecules, many of which are metabolites playing important roles in biological reactions. Small molecules with functional groups may cap the surface of nanozymes, which may in turn affect their catalytic activity. For example, compared to bare AuNPs, melamine-capped AuNPs exhibited enhanced peroxidase-like activity (~two fold) [34]. Based on this observation, the authors successfully detected melamine down to 0.5 μM with the naked eyes. Using a spectrometer, the authors obtained a LOD of 0.2 nM with a linear range of 1–800 nM.

Nucleoside triphosphates (NTPs), including adenosine triphosphate (ATP), guanosine triphosphate (GTP), cytidine triphosphate (CTP), and uridine triphosphate (UTP), are important molecules in biology. With both phosphate groups and the nitrogen-containing nucleobases, NTPs are very versatile in interacting with various nanomaterials. For example, Xu et al. [20] reported that these NTPs, serving as a coenzyme, enhanced the oxidase-like activity of nanoceria. Among the four NTPs, GTP was the most efficient. The enhanced oxidase-like activity was believed to be related to the hydrolysis of NTP, although the actual activator might still be the phosphate anion part. While direct sensing of NTPs has not been studied, the promotion effect of NTPs on nanoceria was used to construct colorimetric assays to analyze single-nucleotide polymorphism in DNA.

In a recent work, Yan and co-workers reported that histidine was able to modulate the peroxidase-like activity of Fe_3O_4 NPs [35]. They intended to mimic the microenvironment of the protein counterpart (HRP), which has a histidine near the active site. Such a simple modification improved the apparent affinity of the substrate, H_2O_2, by 10-fold and the catalytic efficiency by 20-fold. The catalase-like activity of Fe_3O_4 NPs was also enhanced with such a simple modification. While analytical detection was not demonstrated in this initial communication, the significantly improved peroxidase activity should facilitate the more sensitive detection of H_2O_2 and provide a new strategy in rational design of nanozyme-based sensors.

Catecholamine participates in many brain functions and its abnormal concentrations are believed to correlate with Parkinson's disease. Liu et al. studied the peroxidase-like activity of Fe_3O_4 NPs using AUR as a fluorescent substrate at pH 7.0 [36]. Dopamine and other catecholamines (epinephrine, norepinephrine) were able to bind to Fe^{3+} on the surface Fe_3O_4 NPs via their catechol moiety and such binding inhibited the oxidation of AUR. Therefore, the concentration of catecholamine was correlated to the inhibited enzyme activity. The LOD for dopamine and other catecholamines was ca. 3 nM. Other metabolites, such as ascorbic acid, amino acids, carbohydrates, uric acid (1 mM for each), and phosphate (100 mM), did not cause a significant inhibition of AUR oxidation. Redox-active small molecules may also react with nanozymes, sometimes resulting in a permanent loss of its structure and activity [37].

Small molecules may not only affect catalytic activity, but also the stereoselectivity of nanozymes [38]. The Qu group reported a chiral nanozyme by modifying nanoceria with D- or L-amino acids. After screening a series of amino acids, they found that phenylalanine-nanoceria exhibited excellent stereoselectivity toward 3,4-dihydroxyphenylalanine enantiomers.

Detecting Biomacromolecules by Regulating Nanozymes Activity
Compared to simple ions, biomacromolecules or biopolymers such as antibodies and aptamers have more sophisticated structures and functions such as molecular recognition. To design nanozyme-based biosensors, conjugating nanomaterials with biopolymer probes (e.g., antibody, DNA) is often performed. Such surface modifications may also modulate the activity of nanozymes. On one hand, surface-bound biopolymers may impede the access of substrate due to steric effects or by occupying surface-active sites. On the other hand, surface modification may improve the colloidal stability of nanozymes, which in turn may increase the number of active sites for catalysis. Other surface forces may also come into play by surface modification, for example, electrostatic interactions, hydrophobic interactions, π-π stacking, and hydrogen bonding. By introducing such interactions, the affinity between substrates and nanozymes could be enhanced.

Citrate-capped AuNPs prefer to adsorb single-stranded DNA (ssDNA) over double-stranded DNA (dsDNA), since DNA uses its bases for interacting with AuNPs, and only ssDNA has exposed bases [39, 40]. After DNA adsorption, the colloidal stability of AuNPs against salt is increased. This affinity difference between ssDNA and dsDNA could be reflected on the stability and thus color of the system, allowing the design of colorimetric sensors for DNA detection [41].

Using the intrinsic glucose oxidase-like activity of AuNPs, Fan and co-workers demonstrated multiple biosensors for DNA detection [42]. As shown in Fig. 12.8a, ssDNA rather than dsDNA binds to AuNPs, inhibiting their activity for catalyzing oxidation of glucose. Compared to the as-prepared bare AuNPs, both ssDNA-AuNPs and dsDNA-AuNPs showed decreased activity, which was attributed to the reduced affinity of glucose to AuNPs due to the adsorbed DNA. The drop in pH due to the production of gluconic acid might also affect the kinetics of the glucose oxidation reaction [43].

Furthermore, the other product, H_2O_2, can be detected by multiple methods: (1) HRP was used to catalyze the oxidation of a chromogenic substrate (ABTS) (Fig. 12.8b) or a chemiluminescence probe (luminol) (Fig. 12.8c). Using this method, sensitive DNA detection was achieved, with a LOD of 14 nM for the target DNA and 0.75 nM for the chromogenic substrates, respectively. Detection of miRNA and K^+

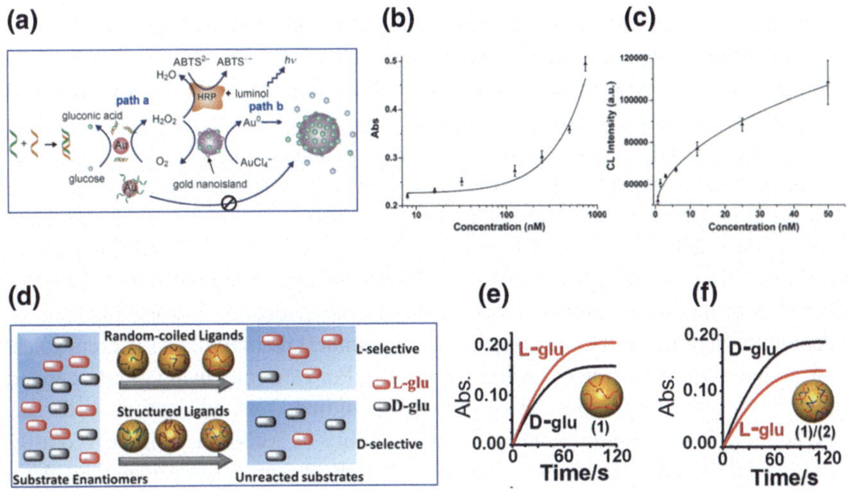

Fig. 12.8 DNA detection by modulating the GOx-like activity of AuNPs. **a** Different affinities of binding of ssDNA and dsDNA to AuNPs resulting in different concentrations of H_2O_2, which can be detected either by the enzyme (HRP)-catalyzed method or self-growth of AuNPs. Calibration curves of DNA detection using **b** ABTS and **c** luminol for signal production. **d** A scheme showing that DNA with different structures can alter the selectivity of AuNP-based GOx activity. Kinetics showing, **e** L-glucose oxidation is preferred by ssDNA-capped AuNPs, while **f** D-glucose is preferred by dsDNA-capped AuNPs. Figure a–c adapted from Ref. [42]. Copyright (2011) Wiley-VCH Verlag GmbH & Co. KGaA, Weinheim. Figure d–f from Ref. [45]. Copyright (2015) American Chemical Society

was also demonstrated using the HPR-assisted method. (2) The self-catalyzed and self-limited growth of AuNPs was reported [44]. In this method, the H_2O_2 produced during glucose oxidation reacted with $HAuCl_4$ to form gold nanoislands on the surface of AuNPs. This growth process was limited by two factors: the decreased catalytic activity of larger AuNP and gluconic acid-induced surface passivation. In the presence of ssDNA, this growth process was inhibited due to the inhibited enzyme-like activity. The size and shape variation of AuNPs was directly observed using a dark-field microscope [42].

DNA adsorbed on AuNPs not only affected the GOx-like activity, but also the selectivity to different substrates. Zhan and co-workers adsorbed various DNAs (ssDNA, dsDNA, and ssDNA with secondary structures) onto AuNPs and each DNA had a different affinity [45]. Such a difference was due to different binding affinities of glucose with different chiralities. They claimed that AuNPs modified with a random ssDNA prefer to catalyze the oxidation of L-glucose (Fig. 12.8d–f). In contrast, AuNPs modified with dsDNA, i-motif, or G-quadruplex DNA exhibited a higher catalytic activity toward the oxidation of D-glucose. The observed selectivity was attributed to the fact that DNA with different conformations bind to the two chiral substrates differently. However, the selectivity induced by the DNA on AuNPs was quite moderate. Also, the underlying chemical basis was not very clear, since glucose lacks functional groups for strong binding with DNA [46].

In addition to relying on the ssDNA-to-dsDNA transition, the conformation of DNA can also be controlled by small molecules if the DNA is an aptamer. For example, Bansal and co-workers [47] found that the peroxidase activity of tyrosine-reduced AuNPs dropped by ~90% after adsorbing a 21-mer ssDNA (the kanamycin aptamer). When the target kanamycin was added to form a complex with the aptamer, the peroxidase activity of AuNPs was recovered back to ~75%. With this activity modulation effect, a selective kanamycin biosensor was designed with a LOD of 1.49 nM. Similar strategies were used to detect a pesticide, acetamiprid [48].

As mentioned above, DNA adsorbed on nanomaterials may have different roles, resulting in either inhibition or promotion effects. For example, using citrate-capped AuNPs as a peroxidase mimic, Hizir et al. [49] studied DNA adsorption on its catalytic activity. They found that the colloidal stability of AuNPs was improved after adsorbing DNA, and the surface charge of AuNPs became more negative. As a result, binding of the positively charged substrate TMB was enhanced, and the oxidation reaction was accelerated. Biosensors based on this enhancement effect can be designed when DNA aptamers are used (Fig. 12.9a).

Abrin, a protein, is a natural toxin found in the seeds of rosary peas. It is recognized as a biological threat due to the high toxicity, and fast detection methods are needed. Based on the enhancement effect of peroxidase activity of AuNPs, Hu et al. designed a highly selective and sensitive biosensor to detect abrin [50]. In the absence of abrin, the free aptamer effectively adsorbed to AuNPs, increasing the peroxidase-like activity of the AuNPs (Fig. 12.9b). In the presence of abrin, an aptamer binding complex was formed, which had a lower affinity to AuNPs. As a result, the activity of AuNPs in catalyzing the TMB oxidation decreased. Figure 12.9c shows a calibration curve for the abrin detection, and the LOD was calculated to be 0.05 nM. Selectivity tests showed that the designed sensor was very specific to abrin (Fig. 12.9d). It is interesting to note that DNA adsorption promoted TMB peroxidation here, while it inhibited the same reaction for ABTS in Fig. 12.8a. This could be attributed to the different charges of the substrates (e.g., cationic TMB was attracted by DNA while anionic ABTS was repelled), and a similar effect on these substrates was also observed with iron oxide nanoparticles [51].

Carbon-based nanomaterials (e.g., carbon nanotube, graphene oxide) are also able to discriminate ssDNA and dsDNA [52]. While most previous work used fluorescence to detect this affinity difference, biosensors have also been developed based on the intrinsic-peroxidase activity of such nanocarbons. For example, Qu and co-workers designed a label-free colorimetric DNA biosensor as depicted in Fig. 12.10a [53]. Single-walled carbon nanotubes (SWNTs) bind to ssDNA more strongly than dsDNA. Without the ssDNA, the bare SWNTs or dsDNA-SWNTs easily aggregated and precipitated upon adding NaCl (0.3 M). After centrifugation and collecting the precipitate, the TMB assay was performed. The reaction kinetics as a function of added target DNA in Fig. 12.10b showed that with more target DNA, more SWNTs were precipitated, and the TMB oxidation was then faster. A disease-related DNA was detected down to 1 nM.

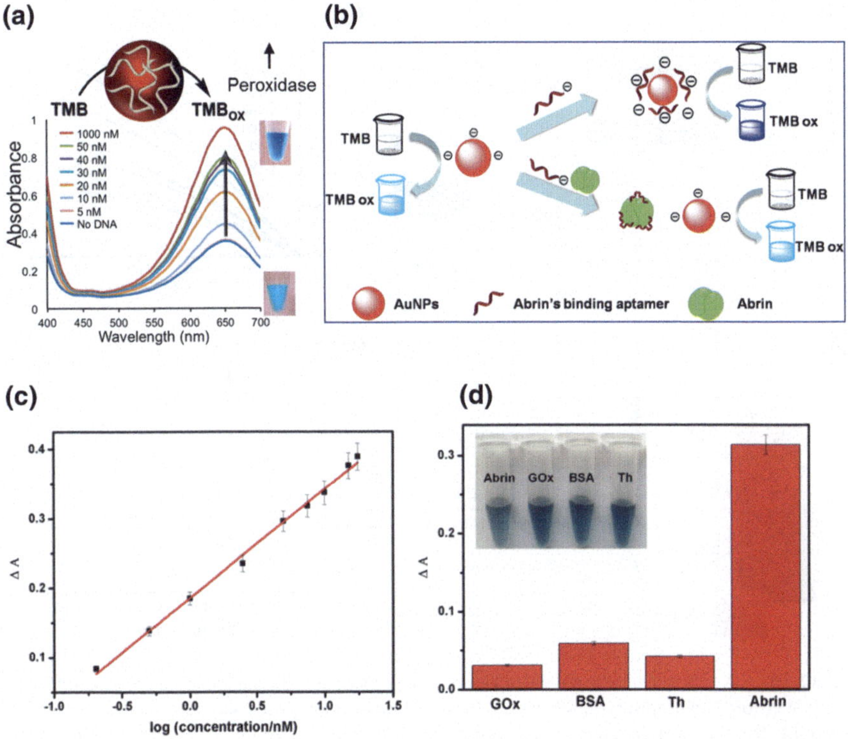

Fig. 12.9 **a** A scheme and UV–vis absorbance spectra showing the enhanced oxidation of TMB catalyzed by citrate-AuNPs in the presence of different concentrations of DNA. **b** A scheme showing the design of a protein biosensor based on the modulation of peroxidase-like activity of AuNPs. The free aptamer has higher affinity to AuNPs compared to the aptamer–protein complex. **c** A calibration curve for the detection of protein arbrin. **d** Selectivity test of abrin against other proteins (GOx, BSA, and thrombin, 5 nM each). Figure a from Ref. [49]. Copyright (2016) American Chemical Society. Figure b–d from Ref. [50]. Copyright (2015) Royal Society of Chemistry

In another work, a slightly different strategy was used for the sensor design [54]. A conjugate composed of hemin and reduced graphene oxide (hemin/rGO) was able to function as a peroxidase mimic (Fig. 12.10c). Similarly, ssDNA adsorbed more strongly to the conjugate, and protected it from salt-induced aggregation. However, instead of collecting the precipitate, the authors used the supernatant to catalyze TMB oxidation. A target concentration-dependent reaction kinetics was also observed (Fig. 12.10d). Using this sensor, a LOD of 2 nM DNA was achieved with a linear range from 5 to100 nM.

Besides DNA and aptamers, peptide affinity ligands have also been coupled with nanozyme for this purpose. Wei and co-workers synthesized a series of MOF nanosheets serving as peroxidase mimics [55]. The adsorption of heparin-specific AG73 peptides inhibited the enzymatic properties of MOF nanozymes, while desorption of the peptide by heparin binding recovered the MOF activity. With this

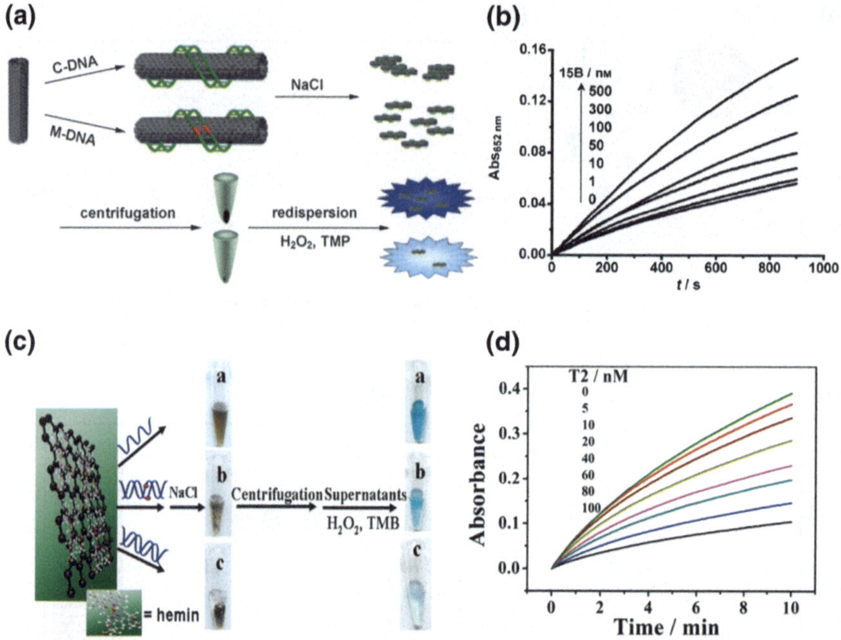

Fig. 12.10 **a** A scheme showing the discrimination of matched and mismatch dsDNA using their different adsorption affinities on SWNTs. After DNA adsorption, NaCl was added to precipitate the less stable SWNTs with perfectly matched dsDNA. **b** The activity of the pellet after centrifugation and re-dispersion. **c** Adsorption of ssDNA, matched dsDNA, and mismatch dsDNA onto hemin/rGO. The supernatant after centrifugation was tested. **d** The activity of the supernatant as a function of target DNA concentration. Figure a, b from Ref. [53]. Copyright (2010) Wiley-VCH Verlag GmbH & Co. KGaA, Weinheim. Figure c, d from Ref. [58]. Copyright (2011) American Chemical Society

strategy, the elimination of heparin in live rats was successfully detected. Interestingly, in another report [56], Hu et al. found that heparin adsorption on AuNCs enhanced the peroxidase-like activity at neutral pH, which was used to monitor the heparin and heparinase activity.

The different affinities for biopolymers on nanozymes were also used to prepare array-based sensors [57]. For example, Fe_3O_4 NPs were modified with two types of small molecules, dopamine and trimethylammonium, respectively. Proteins of varying sizes, charges, and surface hydrophobicity interacted with these two modified Fe_3O_4 NPs differentially. As a result, the oxidation of ABTS by H_2O_2 was different in each case, and a pattern of response was formed. Ten proteins (50 nM each) were differentiated using the linear discrimination analysis method.

12.3.4 Detecting the Substrate of Nanozymes

Many nanozymes identified so far mimic the activity of peroxidases, i.e., activating the co-substrate H_2O_2 to oxidize various substrate molecules. As a result, H_2O_2 has been a very popular analyte, which can be quantified by various colorimetric, fluorescent, and electrochemical methods. In 2008, Wei and Wang used ABTS as a substrate, and Fe_3O_4 as a peroxidase mimic to detect H_2O_2 down to 3 μM with a linear range from 5 to 100 μM [59]. In the same year, the Yan group also reported the detection of H_2O_2 using TMB as the substrate, and the limit of detection was 175 nM with a linear range from 0 to 70 μM [60]. After this early work, many other peroxidase-mimicking nanozymes have been demonstrated to detect H_2O_2 based on this strategy. The role of nanozymes here is simply to catalyze the redox reaction, which is similar to natural protein enzymes (e.g., HRP). Compared to nature enzymes, nanozymes are more controllable by tuning the composition, structure, and surface chemistry of nanomaterials.

Aside from enzyme-like activities, many nanomaterials possess unique optical, electrochemical, and magnetic properties, which may facilitate the signaling of H_2O_2 consumption. For example, nanoceria has been recognized as a catalase mimic, which catalyzes the decomposition of H_2O_2 into H_2O and O_2. Typical assays detect the H_2O_2 or O_2 concentration by reacting with other substrates as we discussed above. Interestingly, substrate-free methods have also been reported. Ornatska et al. developed a colorimetric assay using the intrinsic color change of nanoceria upon reacting with H_2O_2 [61]. In the absence of H_2O_2, the nanoceria dispersion (20 nm, 0.2% w/v) was almost colorless (Fig. 12.11a, first spot). Upon adding H_2O_2, its color immediately changed to reddish orange. The intensity of the color was positively correlated to the concentration of H_2O_2 added with a linear range from 10 to 150 μM. Furthermore, the authors developed a paper-based sensor (Fig. 12.11b). Nanoceria was trapped within the cellulose paper with the help of a silica layer. The prepared paper sensor can still respond to H_2O_2 (Fig. 12.11c). The calibration curve enabled the quantitative detection of H_2O_2 in the range of 2.5–100 mM (Fig. 12.11d). By doping nanoceria with Eu^{3+}, Pratsinis et al. obtained luminescent nanozymes [62]. Here, the interaction of H_2O_2 with nanoceria affected its luminescence, allowing sensitive H_2O_2 detection down to 150 nM.

Our lab developed a fluorescent assay using the DNA adsorption and fluorescence quenching ability of nanoceria [63, 64]. As shown in Fig. 12.11e, nanoceria of ~5 nm in size could adsorb a fluorophore-labeled ssDNA and fully quench the fluorescence. Upon introducing H_2O_2, the surface-bound DNA was displaced and the fluorescence was fully recovered (Fig. 12.11f). With this “turn-on” sensor, the reaction kinetics were examined (Fig. 12.11g). The results showed that the signal recovery was fast, approaching a plateau within 5 min. The H_2O_2-concentration-dependent fluorescence recovery was studied to obtain a calibration curve (Fig. 12.11h). The LOD for H_2O_2 was calculated to be 130 nM (S/N = 3). The developed sensor was also very selective, since most tested small biological molecules failed to produce a fluorescent signal, except for a high concentration of ascorbate (Fig. 12.11i).

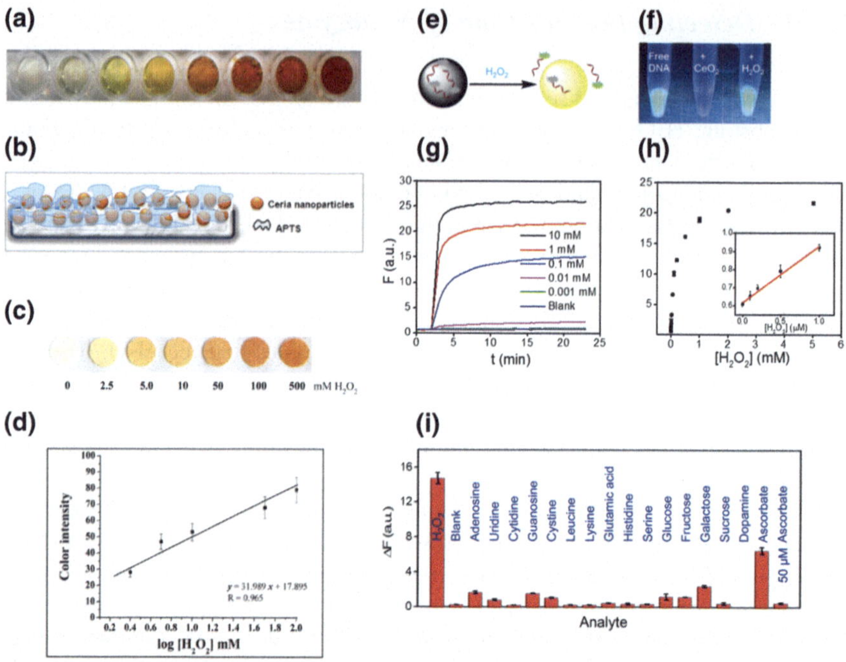

Fig. 12.11 **a** Color change of nanoceria dispersion (20 nm) upon adding different concentrations of H_2O_2. **b** Preparation of a paper-based sensor by immobilizing nanoceria with a silica layer. **c** The color change of a paper-based sensor as a function of H_2O_2 concentration. **d** A calibration curve of the paper sensor. **e** The scheme and **f** photos showing DNA desorption and fluorescence recovery after adding H_2O_2. The performance of the DNA-nanoceria sensor in detecting H_2O_2: **g** reaction kinetics, **h** calibration curve, and **i** selectivity test. Figure a–d from Ref. [61]. Copyright (2011) American Chemical Society. Figure e–i from Ref. [64]. Copyright (2015) American Chemical Society

Aside from H_2O_2, many other substrates of nanozymes have also been detected. Epinephrine, the main hormone in the adrenal medulla, is a substrate of laccase. Cu-coordinated GMP nanoparticles were able to catalyze epinephrine oxidation, mimicking the laccase activity [65]. The initial reaction rate of Cu/GMP was 11-fold faster than that of laccase. With such a high activity, highly sensitive detection of epinephrine was achieved with a limit of detection of 0.41 μg/mL. In another study, Tremel and co-workers reported that molybdenum trioxide (MoO_3) was able to catalyze the oxidation of sulfite by O_2 [66], and they also demonstrated the potential of MoO_3 as therapeutic agents for sulfite oxidase deficiency.

Hydrolysis reactions can also be accelerated by nanozymes. Several groups reported that nanoceria can catalyze the hydrolysis of phosphate esters [67–69]. Although such reactions have not yet been used for analytical applications, sensors based on them may be possible in the future.

Overall, since most nanozymes do not have a specific substrate-binding site, currently demonstrated examples are special cases than general observations. For H_2O_2, it can help produce colored products, while the product of epinephrine is colored. Achieving the same sensing mechanism like protein-based enzymes would require the introduction of specific substrate-binding pockets. We have explored some initial work in this direction by using molecularly imprinted polymers [70, 71], and aptamers may also be used for selective oxidation [72].

12.3.5 Hybrid Sensing Systems

As discussed above, chemical reactions catalyzed by nanozymes are mainly redox related, along with a few hydrolysis reactions. Therefore, the range of analytes that can be detected using nanozymes is quite limited. To expand the range of analytes, a popular strategy is to combine protein enzymes with nanozymes. For example, GOx has been combined with many nanozymes (Fig. 12.12a). GOx has excellent activity and selectivity toward glucose, and it generates H_2O_2 as a by-product. H_2O_2 is then detected using the nanozyme strategies discussed above. Many such papers were published using different nanozymes, and a typical example was reported by Wei and Wang [59]. GOx was combined with Fe_3O_4 NPs to detect glucose with a

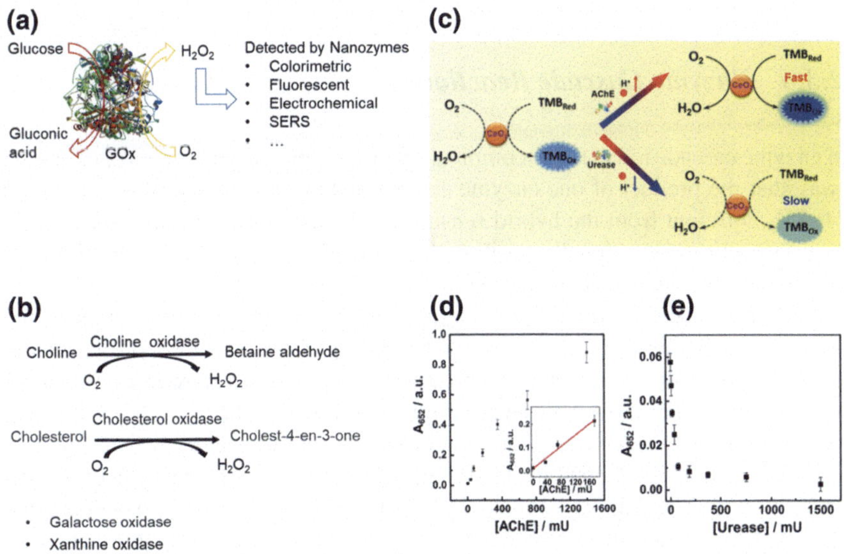

Fig. 12.12 **a** GOx-catalyzed glucose oxidation producing H_2O_2, which can be detected using nanozyme-based assays. **b** Examples of other enzymes also generating H_2O_2 as a by-product. **c** A scheme illustrating the regulation of nanoceria activity by enzymes producing/consuming protons. The activity of nanoceria toward TMB oxidation as a function of the concentration of **d** proton-producing enzyme AChE, and **e** proton-consuming enzyme urease. Figure c–e from Ref. [73]. Copyright (2016) American Chemical Society

LOD of 30 μM and a linear range of 50 μM–1 mM. Aside from GOx, some other enzymes also produce H_2O_2, such as choline oxidase, cholesterol oxidase, galactose oxidase, and xanthine oxidase (Fig. 12.12b). Detection of enzyme activity as well as the substrate can be realized by monitoring the H_2O_2 level.

Another strategy is to regulate the activity of nanozymes by enzyme-catalyzed reactions. For example, many nanozymes are sensitive to pH. At the same time, several natural enzymes are able to produce/consume protons. Therefore, it is possible to design self-regulated bioassays (Fig. 12.12c) [73]. To prove this concept, Cheng et al. designed an assay to detect the enzyme acetylcholinesterase (AChE), which catalyzed the hydrolysis of acetylcholine to acetate and choline. The initial pH of nanoceria/TMB assay was 7.0 (buffered by 5 mM PBS). In the absence of the target enzyme AChE, the oxidase activity of nanoceria was very low. Upon adding the enzyme, the generated acetic acid lowered the pH, enhancing the oxidase activity of nanoceria. By monitoring the oxidation of TMB, the concentration of AChE was quantified (Fig. 12.12d). The LOD was 25 mU with a linear range of 35–175 mU. This assay was successfully extended to other proton generation enzymes (e.g., esterase). Similarly, a proton-consuming enzyme, such as urease, was also used. Urease can increase the pH and thus deactivate nanoceria (Fig. 12.12e). The authors also proved that the developed assays provided a convenient and efficient strategy to screen enzyme inhibitors.

12.3.6 Enzyme Cascade Reactions

An enzyme cascade refers to the combination of multiple enzymes in one pot in such a way that the product of one enzyme can act as the substrate of other enzyme(s) [74, 75]. Different from the hybrid sensing mechanism mentioned above, enzyme cascade reactions intentionally position the coupled enzymes in close proximity to achieve the substrate channeling effect.

While most work used only protein enzymes for this purpose, nanozyme-based cascade systems have also been reported [76–79]. Three examples are discussed here to illustrate the design strategy. In the first example, a peroxidase-mimicking nanozyme was combined with GOx as illustrated in Fig. 12.13a. The Wei group developed an integrated nanozyme system by assembling nanozymes and protein enzymes in a MOF. With combined enzymatic property and surface-enhanced Raman scattering property, AuNPs, coupled with GOx, were assembled in metal–organic framework to form an "integrative nanozyme" [78]. This platform was able to monitor the variation of glucose and lactate in living brains, showing a great potential in biomedical and catalytic applications. Soft material networks such as hydrogel can also be used to anchor nanozymes and protein enzymes [79].

In a second example, two nanozymes were combined in the absence of natural enzymes. Qu and co-workers constructed a hybrid material composed of V_2O_5, polydopamine (PDA), and AuNPs [77]. As depicted in Fig. 12.13b, after modifying V_2O_5 nanowires with a PDA layer, AuNPs were formed in situ by reducing $HAuCl_4$

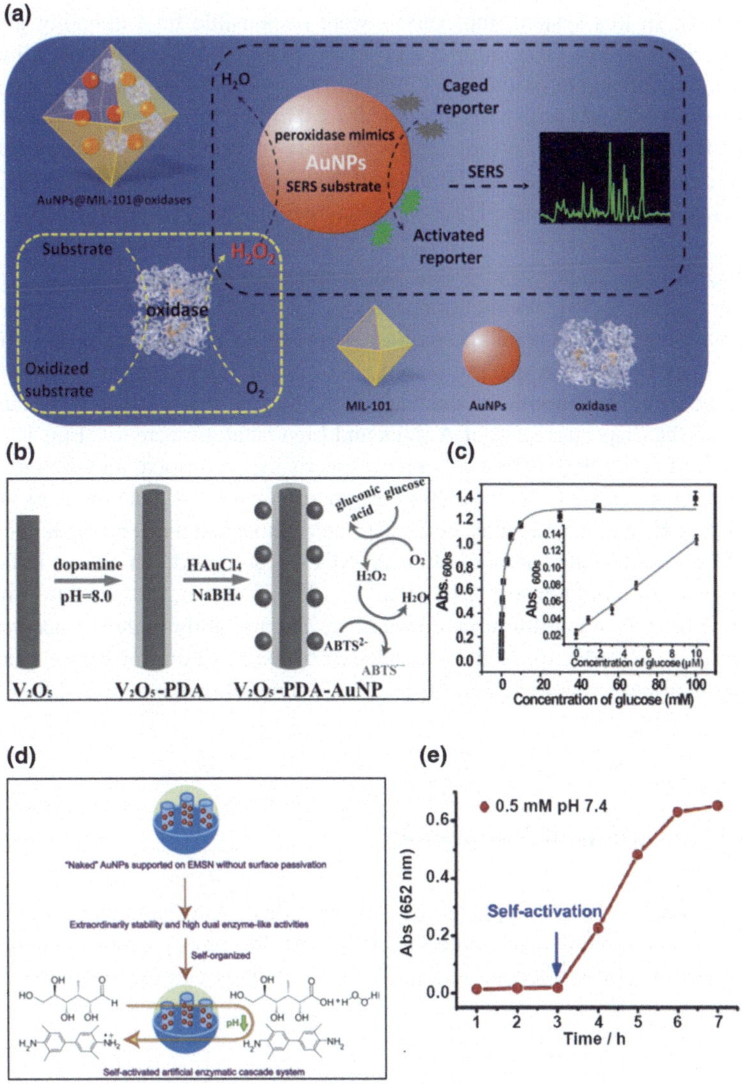

Fig. 12.13 Design of cascade enzyme systems containing nanozymes. **a** Construction of an integrated nanozyme assembling protein enzyme and nanozyme with MOF. **b** A scheme illustrating the preparation of V_2O_5@PDA@AuNP cascade system, and **c** the calibration curve for glucose detection. **d** A scheme showing the design of mesoporous silica supported AuNPs as both GOx and peroxidase. **e** Kinetics of TMB oxidation in pH 7.4 buffer (only 0.5 mM buffer concentration). Figure a from Ref. [78]. Copyright (2015) American Chemical Society. Figure b, c from Ref. [77]. Copyright (2014) Wiley-VCH Verlag GmbH & Co. KGaA, Weinheim. Figure d, e from Ref. [76]. Copyright (2013) Elsevier Ltd.

with $NaBH_4$. In this system, the AuNPs were responsible for catalyzing glucose oxidation, serving as a GOx mimic, and V_2O_5 was a peroxidase mimic. Glucose-concentration-dependent oxidation of ABTS was obtained (Fig. 12.13c). The LOD for glucose was 0.5 μM with a linear range from 0 to 10 μM. Based on the inhibition effect of DNA adsorption on the catalytic activity of AuNPs, detection of ssDNA was also demonstrated. It should be noted that this system required two steps, each at a different pH. The optimal pH for V_2O_5 nanowires was pH 4 and for AuNPs was pH 7.

The Qu group demonstrated that one nanozyme was sufficient to perform multistep reactions [76]. As mentioned above, AuNPs alone could function as a multi-enzyme system: GOx and peroxidase. However, in the solution phase, AuNPs typically exhibit only one specific activity. To maximize the enzymatic activity of AuNPs, the authors immobilized the "naked" AuNPs on mesoporous silica support to maintain the dispersed state of AuNPs and high catalytic activity (Fig. 12.13d). The GOx and peroxidase activities were, respectively, confirmed, and then the cascade reaction was tested. The oxidation of glucose generated gluconic acid, which lowered the pH in a low capacity buffer (0.5 mM phosphate buffer). With decrease of pH, the peroxidase-like activity of the AuNPs was activated, which was indicated by the enhanced oxidation of TMB (Fig. 12.13e). However, the activation of the peroxidase activity was quite time-consuming, and a slightly higher concentration of buffer (e.g., 20 mM) made the second step of the reaction much less efficient. A similar dual-enzyme mimic activity of MnO_2 nanoflakes was reported by Li and co-workers [80].

12.4 Limitations of Nanozymes

While quite impressive developments have been made in using nanozymes as biosensor components, a number of limitations still exist. We briefly summarize them in the space below. Only after addressing these limitations can the full potential of nanozymes be reached.

12.4.1 Surface Fouling

Most of the current analytical works using nanozymes activities were demonstrated in clean buffers or relatively simple environmental water samples. However, biological samples may contain concentrated proteins, small molecules, and ions. Inorganic nanoparticles are known to adsorb many types of small and large molecules. Such reactions can sometimes promote the catalytic activity such as in the case of fluoride adsorption on nanoceria. In many other cases, however, they may inhibit the reaction. The inhibition of glucose oxidase activity of gold nanoparticles by adsorption of biopolymers is a good example [43]. In either case (promoting or inhibiting), the activity of nanozymes becomes unpredictable and thus this uncertainty will affect the analytical performance of the sensors. How to produce consistent, reliable, and robust nanozyme surfaces is a critical problem to be solved.

12.4.2 Lack of Substrate Specificity

Substrate specificity is a fundamental property of natural enzymes. Although some enzymes like peroxidases accept many substrates, most enzymes like glucose oxidase are highly specific, allowing us to detect its substrate. This substrate specificity originated from complex structures of enzymes often containing a substrate-binding pocket. However, nanozymes lack such intrinsic structural elements to specifically bind substrates. Most reported nanozymes are naked inorganic nanomaterials with catalytic properties. This lack of specificity has raised concerns on the term nanozyme, since these nanomaterials function more like a conventional catalyst in some cases. Engineering the surface with polymeric networks may improve substrate specificity. Some recent developments in this direction have been demonstrated by using molecularly imprinted polymers [70, 71].

12.4.3 Sensor Immobilization

Immobilization is a key aspect of biosensors. The chemistry of nanozyme immobilization without compromising their activity has not been well developed. This effect is different on different nanozymes. For the glucose oxidase activity of gold nanoparticles, it is very sensitive to surface modifications. A balance between surface passivation and colloidal stability needs to be considered.

12.4.4 Slow Reaction Rates

The activity of most nanozymes is sensitive to buffer conditions, especially pH. Many common nanozymes such as the peroxidase-mimicking Fe_3O_4 and the oxidase-mimicking CeO_2 work the best at an acidic pH of 4. While this is not a problem for using them for signal production, it is inconvenient to directly use them for detection in biological samples, which are often around neutral pH.

12.4.5 Limited Reaction Types

Among the six types of protein enzymes (i.e., oxidoreductase, transferases, hydrolases, lyases, isomerases, and ligases), current nanozymes are mainly capable of mimicking the oxidoreductases, for example, oxidase, peroxidase, catalase, and SOD. A few of them also possess the hydrolase activity [81]. Therefore, nanoscale materials can only mimic a small fraction of enzymes at this moment. It is still not clear if commonly used nanomaterials (e.g., AuNPs, Fe_3O_4 NPs) have other enzymatic activities or not. On one hand, it is important to fully explore the potential of current materials in combination with surface engineering. On the other hand, pursuing new materials with new activities is required.

12.5 Summary and Future Perspectives

Protein-based enzymes have been quite successful in developing biosensors with many commercial products already in a rapidly growing market. To further improve such sensors by reducing the cost and increasing stability, a recent direction is to involve inorganic nanomaterials with enzyme-like activities. In this chapter, parallel examples of using nanozymes and protein enzymes for developing biosensors have been reviewed.

The study of nanozymes is a relatively young field. As detailed here, extensive applications in molecular detection have already been found. Essentially the ideas used in developing enzyme-based sensors can also be realized by using nanozymes. However, as outlined above, nanozymes also have a number of limitations that need to be addressed.

To push this field forward, one or a few high impact applications need to be demonstrated with nanozymes. In this regard, nanozymes may need to compete with protein enzymes based on existing products. To do this, the nanozyme products need to have at least a similar performance, yet a lower cost. Another possibility is to develop new applications where protein enzymes have not been very successful. This may require the researchers in the nanozyme field to communicate with clinicians to know the important problems that are possible to be measured by nanozymes.

Acknowledgements The work from the Liu lab described in this chapter was supported by The Natural Sciences and Engineering Research Council of Canada (NSERC).

References

1. Gao L, Yan X (2016) Nanozymes: an emerging field bridging nanotechnology and biology. Sci China Life Sci 59(4):400–402
2. Wei H, Wang E (2013) Nanomaterials with enzyme-like characteristics (nanozymes): next-generation artificial enzymes. Chem Soc Rev 42(14):6060–6093
3. Wang X, Hu Y, Wei H (2016) Nanozymes in bionanotechnology: from sensing to therapeutics and beyond. Inorg Chem Front 3(1):41–60
4. Ragg R, Tahir MN (2016) Tremel W (2016) Solids go bio: inorganic nanoparticles as enzyme mimics. Eur J Inorg Chem 13–14:1906–1915
5. Lin Y, Ren J, Qu X (2014) Catalytically active nanomaterials: a promising candidate for artificial enzymes. Acc Chem Res 47(4):1097–1105
6. Wu J, Wang X, Wang Q, Lou Z, Li S, Zhu Y, Qin L, Wei H (2019) Nanomaterials with enzyme-like characteristics (nanozymes): next-generation artificial enzymes (II). Chem Soc Rev 48(4):1004–1076
7. Huang Y, Ren J, Qu X (2019) Nanozymes: classification, catalytic mechanisms, activity regulation, and applications. Chem Rev 119(6):4357–4412
8. Khatkhatay MI, Desai M (1999) A comparison of performances of four enzymes used in ELISA with special reference to β-lactamase. J Immunoassay 20(3):151–183
9. Gao L, Zhuang J, Nie L, Zhang J, Zhang Y, Gu N, Wang T, Feng J, Yang D, Perrett S, Yan X (2007) Intrinsic peroxidase-like activity of ferromagnetic nanoparticles. Nat Nanotechnol 2(9):577–583
10. Ferri S, Kojima K, Sode K (2011) Review of glucose oxidases and glucose dehydrogenases: a bird's eye view of glucose sensing enzymes. J Diabetes Sci Technol 5(5):1068–1076
11. Arduini F, Amine A (2014) Biosensors based on enzyme inhibition. In: Gu MB, Kim H-S (eds) Biosensors based on aptamers and enzymes. Springer, Berlin, pp 299–326. https://doi.org/10.1007/10_2013_224
12. Amine A, Arduini F, Moscone D, Palleschi G (2016) Recent advances in biosensors based on enzyme inhibition. Biosens Bioelectron 76:180–194
13. Guilbault GG, Kramer DN, Cannon PL (1962) Electrical determination of organophosphorous compounds. Anal Chem 34(11):1437–1439
14. Arduini F, Amine A, Moscone D, Palleschi G (2010) Biosensors based on cholinesterase inhibition for insecticides, nerve agents and aflatoxin b1 detection (review). Microchim Acta 170(3):193–214
15. Lee WE, Thompson HG, Hall JG, Bader DE (2000) Rapid detection and identification of biological and chemical agents by immunoassay, gene probe assay and enzyme inhibition using a silicon-based biosensor. Biosens Bioelectron 14(10):795–804
16. Van Dyk JS, Pletschke B (2011) Review on the use of enzymes for the detection of organochlorine, organophosphate and carbamate pesticides in the environment. Chemosphere 82(3):291–307
17. Asati A, Santra S, Kaittanis C, Nath S, Perez JM (2009) Oxidase-like activity of polymer-coated cerium oxide nanoparticles. Angew Chem Int Ed 121(13):2344–2348
18. Zhao Q, Huang H, Zhang L, Wang L, Zeng Y, Xia X, Liu F, Chen Y (2016) Strategy to fabricate naked-eye readout ultrasensitive plasmonic nanosensor based on enzyme mimetic gold nanoclusters. Anal Chem 88(2):1412–1418
19. Xu C, Qu X (2014) Cerium oxide nanoparticle: a remarkably versatile rare earth nanomaterial for biological applications. NPG Asia Mater 6:e90
20. Xu C, Liu Z, Wu L, Ren J, Qu X (2014) Nucleoside triphosphates as promoters to enhance nanoceria enzyme-like activity and for single-nucleotide polymorphism typing. Adv Funct Mater 24(11):1624–1630
21. Liu B, Huang Z, Liu J (2016) Boosting the oxidase mimicking activity of nanoceria by fluoride capping: rivaling protein enzymes and ultrasensitive f-detection. Nanoscale 8(28):13562–13567

22. Chen C, Lu L, Zheng Y, Zhao D, Yang F, Yang X (2015) A new colorimetric protocol for selective detection of phosphate based on the inhibition of peroxidase-like activity of magnetite nanoparticles. Anal Methods 7(1):161–167
23. Wen S-H, Zhong X-L, Wu Y-D, Liang R-P, Zhang L, Qiu J-D (2019) Colorimetric assay conversion to highly sensitive electrochemical assay for bimodal detection of arsenate based on cobalt oxyhydroxide nanozyme via arsenate absorption. Anal Chem 91(10):6487–6497
24. Qin W, Su L, Yang C, Ma Y, Zhang H, Chen X (2014) Colorimetric detection of sulfite in foods by a TMB–O_2–Co_3O_4 nanoparticles detection system. J Agric Food Chem 62(25):5827–5834
25. Deng H-H, Weng S-H, Huang S-L, Zhang L-N, Liu A-L, Lin X-H, Chen W (2014) Colorimetric detection of sulfide based on target-induced shielding against the peroxidase-like activity of gold nanoparticles. Anal Chim Acta 852:218–222
26. Long YJ, Li YF, Liu Y, Zheng JJ, Tang J, Huang CZ (2011) Visual observation of the mercury-stimulated peroxidase mimetic activity of gold nanoparticles. Chem Commun 47(43):11939–11941
27. Wang C-I, Huang C-C, Lin Y-W, Chen W-T, Chang H-T (2012) Catalytic gold nanoparticles for fluorescent detection of mercury(II) and lead(II) ions. Anal Chim Acta 745:124–130
28. Zhu R, Zhou Y, Wang X-L, Liang L-P, Long Y-J, Wang Q-L, Zhang H-J, Huang X-X, Zheng H-Z (2013) Detection of Hg^{2+} based on the selective inhibition of peroxidase mimetic activity of BSA-Au clusters. Talanta 117:127–132
29. Lien C-W, Huang C-C, Chang H-T (2012) Peroxidase-mimic bismuth-gold nanoparticles for determining the activity of thrombin and drug screening. Chem Commun 48(64):7952–7954
30. Lien C-W, Chen Y-C, Chang H-T, Huang C-C (2013) Logical regulation of the enzyme-like activity of gold nanoparticles by using heavy metal ions. Nanoscale 5(17):8227–8234
31. Zhang D, Chen Z, Omar H, Deng L, Khashab NM (2015) Colorimetric peroxidase mimetic assay for uranyl detection in sea water. ACS Appl Mater Interfaces 7(8):4589–4594
32. Li W, Chen B, Zhang H, Sun Y, Wang J, Zhang J, Fu Y (2015) BSA-stabilized Pt nanozyme for peroxidase mimetics and its application on colorimetric detection of mercury(II) ions. Biosens Bioelectron 66:251–258
33. Liu J, Hu X, Hou S, Wen T, Liu W, Zhu X, Wu X (2011) Screening of inhibitors for oxidase mimics of Au@Pt nanorods by catalytic oxidation of OPD. Chem Commun 47(39):10981–10983
34. Ni P, Dai H, Wang Y, Sun Y, Shi Y, Hu J, Li Z (2014) Visual detection of melamine based on the peroxidase-like activity enhancement of bare gold nanoparticles. Biosens Bioelectron 60:286–291
35. Fan K, Wang H, Xi J, Liu Q, Meng X, Duan D, Gao L, Yan X (2017) Optimization of Fe_3O_4 nanozyme activity via single amino acid modification mimicking an enzyme active site. Chem Commun 53(2):424–427
36. Liu C-H, Yu C-J, Tseng W-L (2012) Fluorescence assay of catecholamines based on the inhibition of peroxidase-like activity of magnetite nanoparticles. Anal Chim Acta 745:143–148
37. Liu SG, Han L, Li N, Xiao N, Ju YJ, Li NB, Luo HQ (2018) A fluorescence and colorimetric dual-mode assay of alkaline phosphatase activity via destroying oxidase-like CoOOH nanoflakes. J Mater Chem B 6(18):2843–2850
38. Sun Y, Zhao C, Gao N, Ren J, Qu X (2017) Stereoselective nanozyme based on ceria nanoparticles engineered with amino acids. Chem Eur J 23(72):18146–18150
39. Herne TM, Tarlov MJ (1997) Characterization of DNA probes immobilized on gold surfaces. J Am Chem Soc 119(38):8916–8920
40. Liu J (2012) Adsorption of DNA onto gold nanoparticles and graphene oxide: Surface science and applications. Phys Chem Chem Phys 14(30):10485–10496
41. Li H, Rothberg L (2004) Colorimetric detection of DNA sequences based on electrostatic interactions with unmodified gold nanoparticles. Proc Natl Acad Sci USA 101(39):14036–14039
42. Zheng X, Liu Q, Jing C, Li Y, Li D, Luo W, Wen Y, He Y, Huang Q, Long Y-T, Fan C (2011) Catalytic gold nanoparticles for nanoplasmonic detection of DNA hybridization. Angew Chem Int Ed 50(50):11994–11998

43. Lang NJ, Liu B, Liu J (2014) Characterization of glucose oxidation by gold nanoparticles using nanoceria. J Colloid Interface Sci 428:78–83
44. Luo W, Zhu C, Su S, Li D, He Y, Huang Q, Fan C (2010) Self-catalyzed, self-limiting growth of glucose oxidase-mimicking gold nanoparticles. ACS Nano 4(12):7451–7458
45. Zhan P, Wang Z-G, Li N, Ding B (2015) Engineering gold nanoparticles with DNA ligands for selective catalytic oxidation of chiral substrates. ACS Catal 5(3):1489–1498
46. Yang K-A, Barbu M, Halim M, Pallavi P, Kim B, Kolpashchikov DM, Pecic S, Taylor S, Worgall TS, Stojanovic MN (2014) Recognition and sensing of low-epitope targets via ternary complexes with oligonucleotides and synthetic receptors. Nat Chem 6(11):1003–1008
47. Sharma TK, Ramanathan R, Weerathunge P, Mohammadtaheri M, Daima HK, Shukla R, Bansal V (2014) Aptamer-mediated 'turn-off/turn-on' nanozyme activity of gold nanoparticles for kanamycin detection. Chem Commun 50(100):15856–15859
48. Weerathunge P, Ramanathan R, Shukla R, Sharma TK, Bansal V (2014) Aptamer-controlled reversible inhibition of gold nanozyme activity for pesticide sensing. Anal Chem 86(24):11937–11941
49. Hizir MS, Top M, Balcioglu M, Rana M, Robertson NM, Shen F, Sheng J, Yigit MV (2016) Multiplexed activity of perAuxidase: DNA-capped AuNPs act as adjustable peroxidase. Anal Chem 88(1):600–605
50. Hu J, Ni P, Dai H, Sun Y, Wang Y, Jiang S, Li Z (2015) Aptamer-based colorimetric biosensing of abrin using catalytic gold nanoparticles. Analyst 140(10):3581–3586
51. Liu B, Liu J (2015) Accelerating peroxidase mimicking nanozymes using DNA. Nanoscale 7(33):13831–13835
52. Lu C-H, Yang H-H, Zhu C-L, Chen X, Chen G-N (2009) A graphene platform for sensing biomolecules. Angew Chem Int Ed 48(26):4785–4787
53. Song Y, Wang X, Zhao C, Qu K, Ren J, Qu X (2010) Label-free colorimetric detection of single nucleotide polymorphism by using single-walled carbon nanotube intrinsic peroxidase-like activity. Chem Eur J 16(12):3617–3621
54. Guo Y, Deng L, Li J, Guo S, Wang E, Dong S (2011) Hemin-graphene hybrid nanosheets with intrinsic peroxidase-like activity for label-free colorimetric detection of single-nucleotide polymorphism. ACS Nano 5(2):1282–1290
55. Cheng H, Liu Y, Hu Y, Ding Y, Lin S, Cao W, Wang Q, Wu J, Muhammad F, Zhao X, Zhao D, Li Z, Xing H, Wei H (2017) Monitoring of heparin activity in live rats using metal–organic framework nanosheets as peroxidase mimics. Anal Chem 89(21):11552–11559
56. Hu L, Liao H, Feng L, Wang M, Fu W (2018) Accelerating the peroxidase-like activity of gold nanoclusters at neutral pH for colorimetric detection of heparin and heparinase activity. Anal Chem 90(10):6247–6252
57. Li X, Wen F, Creran B, Jeong Y, Zhang X, Rotello VM (2012) Colorimetric protein sensing using catalytically amplified sensor arrays. Small 8(23):3589–3592
58. Guo Y, Li J, Dong S (2011) Hemin functionalized graphene nanosheets-based dual biosensor platforms for hydrogen peroxide and glucose. Sens Actuat B 160(1):295–300
59. Wei H, Wang E (2008) Fe_3O_4 magnetic nanoparticles as peroxidase mimetics and their applications in H_2O_2 and glucose detection. Anal Chem 80(6):2250–2254
60. Zhuang J, Zhang J, Gao L, Zhang Y, Gu N, Feng J, Yang D, Yan X (2008) A novel application of iron oxide nanoparticles for detection of hydrogen peroxide in acid rain. Mater Lett 62(24):3972–3974
61. Ornatska M, Sharpe E, Andreescu D, Andreescu S (2011) Paper bioassay based on ceria nanoparticles as colorimetric probes. Anal Chem 83(11):4273–4280
62. Pratsinis A, Kelesidis GA, Zuercher S, Krumeich F, Bolisetty S, Mezzenga R, Leroux J-C, Sotiriou GA (2017) Enzyme-mimetic antioxidant luminescent nanoparticles for highly sensitive hydrogen peroxide biosensing. ACS Nano 11(12):12210–12218
63. Pautler R, Kelly EY, Huang P-JJ, Cao J, Liu B, Liu J (2013) Attaching DNA to nanoceria: regulating oxidase activity and fluorescence quenching. ACS Appl Mater Interfaces 5(15):6820–6825

64. Liu B, Sun Z, Huang P-JJ, Liu J (2015) Hydrogen peroxide displacing DNA from nanoceria: mechanism and detection of glucose in serum. J Am Chem Soc 137(3):1290–1295
65. Liang H, Lin F, Zhang Z, Liu B, Jiang S, Yuan Q, Liu J (2017) Multicopper laccase mimicking nanozymes with nucleotides as ligands. ACS Appl Mater Interfaces 9(2):1352–1360
66. Ragg R, Natalio F, Tahir MN, Janssen H, Kashyap A, Strand D, Strand S, Tremel W (2014) Molybdenum trioxide nanoparticles with intrinsic sulfite oxidase activity. ACS Nano 8(5):5182–5189
67. Manto MJ, Xie P, Wang C (2017) Catalytic dephosphorylation using ceria nanocrystals. ACS Catal 7(3):1931–1938
68. Kuchma MH, Komanski CB, Colon J, Teblum A, Masunov AE, Alvarado B, Babu S, Seal S, Summy J, Baker CH (2010) Phosphate ester hydrolysis of biologically relevant molecules by cerium oxide nanoparticles. Nanomedicine: NBM 6(6):738–744
69. Vernekar AA, Das T, Mugesh G (2016) Vacancy-engineered nanoceria: enzyme mimetic hotspots for the degradation of nerve agents. Angew Chem Int Ed 55(4):1412–1416
70. Zhang Z, Liu B, Liu J (2017) Molecular imprinting for substrate selectivity and enhanced activity of enzyme mimics. Small 13(7):1602730
71. Zhang Z, Zhang X, Liu B, Liu J (2017) Molecular imprinting on inorganic nanozymes for hundred-fold enzyme specificity. J Am Chem Soc 139(15):5412–5419
72. Golub E, Albada HB, Liao W-C, Biniuri Y, Willner I (2016) Nucleoapzymes: Hemin/G-quadruplex DNAzyme–aptamer binding site conjugates with superior enzyme-like catalytic functions. J Am Chem Soc 138(1):164–172
73. Cheng H, Lin S, Muhammad F, Lin Y-W, Wei H (2016) Rationally modulate the oxidase-like activity of nanoceria for self regulated bioassays. ACS Sens 1(11):1336–1343
74. Kuchler A, Yoshimoto M, Luginbuhl S, Mavelli F, Walde P (2016) Enzymatic reactions in confined environments. Nat Nanotechnol 11(5):409–420
75. Ricca E, Brucher B, Schrittwieser JH (2011) Multi-enzymatic cascade reactions: overview and perspectives. Adv Synth Catal 353(13):2239–2262
76. Lin Y, Li Z, Chen Z, Ren J, Qu X (2013) Mesoporous silica-encapsulated gold nanoparticles as artificial enzymes for self-activated cascade catalysis. Biomaterials 34(11):2600–2610
77. Qu K, Shi P, Ren J, Qu X (2014) Nanocomposite incorporating V_2O_5 nanowires and gold nanoparticles for mimicking an enzyme cascade reaction and its application in the detection of biomolecules. Chem Eur J 20(24):7501–7506
78. Hu Y, Cheng H, Zhao X, Wu J, Muhammad F, Lin S, He J, Zhou L, Zhang C, Deng Y, Wang P, Zhou Z, Nie S, Wei H (2017) Surface-enhanced Raman scattering active gold nanoparticles with enzyme-mimicking activities for measuring glucose and lactate in living tissues. ACS Nano 11(6):5558–5566
79. Qu R, Shen L, Qu A, Wang R, An Y, Shi L (2015) Artificial peroxidase/oxidase multiple enzyme system based on supramolecular hydrogel and its application as a biocatalyst for cascade reactions. ACS Appl Mater Interfaces 7(30):16694–16705
80. Han L, Zhang H, Chen D, Li F (2018) Protein-directed metal oxide nanoflakes with tandem enzyme-like characteristics: colorimetric glucose sensing based on one-pot enzyme-free cascade catalysis. Adv Funct Mater 28(17):1800018
81. Yao T, Tian Z, Zhang Y, Qu Y (2019) Phosphatase-like activity of porous nanorods of CeO_2 for the highly stabilized dephosphorylation under interferences. ACS Appl Mater Interfaces 11(1):195–201

Chapter 13
Nanozyme-Based Tumor Theranostics

Xiangqin Meng, Lizeng Gao, Kelong Fan and Xiyun Yan

Abbreviations

AuNPs	Gold nanoparticles
Ce6	Chlorine e6
CEA	Carcinoembryonic antigen
CL	Chemiluminescence
Co_3O_4	Cobalt trioxide
CuONRs	Cupric oxide nanorods
CYFRA 21-1	Cytokeratin-19-fragment
DMSNs	Dendritic mesoporous silica nanoparticles
DNA	Deoxyribonucleic acid
ELISA	Enzyme-linked immunosorbent assay
GDs	Graphene quantum dots
GO	Graphene oxide
GOD	Glucose oxidase
GOx	Glucose oxidase
GSH	Reduced glutathione
GSSG	Glutathione disulfide
HEL	Human erythroleukemia

X. Meng · L. Gao · K. Fan (✉) · X. Yan (✉)
CAS Engineering Laboratory for Nanozyme, Key Laboratory of Protein and Peptide Pharmaceutical, Institute of Biophysics, Chinese Academy of Sciences, 15 Datun Road, Beijing 100101, China
e-mail: fankelong@ibp.ac.cn

X. Yan
e-mail: yanxy@ibp.ac.cn

X. Meng
School of Future Technology, University of Chinese Academy of Sciences, No. 380, Huaibei Village, Huaibei Town, Huairou District, Beijing 101408, China

X. Yan (ed.), *Nanozymology*, Nanostructure Science and Technology,
https://doi.org/10.1007/978-981-15-1490-6_13

HER2	Human epidermal growth factor receptor 2
HFn	Human H-ferritin
HIF	Hypoxia-inducible factor
HRP	Horseradish peroxidase
HSA	Human serum albumin
IHC	Immunohistochemistry
LOD	Limit of detection
MHFn	Magnetoferritin
MNPs	Magnetic nanoparticles
MOFs	Metal–organic frameworks
MRI	Magnetic resonance imaging
$mSiO_2$	Mesoporous silica
NPs	Nanoparticles
OMC	Ordered mesoporous carbon
OxTMB	Oxidized TMB
PBNPs	Prussian Blue nanozymes
PCR	Polymerase chain reaction
PDT	Photodynamic therapy
PHPBNs	Porous hollow Prussian blue nanoparticles
PSA	Prostate-specific antigen
PS	Photosensitizers
PtNPs	Platinum nanoparticles
RGO	Reduced graphene oxide
RNA	Ribonucleic acid
ROS	Reactive oxygen species
RT	Radiation therapy
SDT	Sonodynamic therapy
SO	Singlet oxygen
SOD	Superoxide dismutase
TMB	3,3′,5,5′-Tetramethylbenzidine
TME	Tumor microenvironment
TfR1	Transferrin receptor 1
UCL	Upconversion luminescent
UCNPs	Upconversion nanoprobes
US	Ultrasound
VEGF	Vascular endothelial growth factor
VEGFR	Vascular endothelial growth factor receptor

13.1 The Enzymatic Activity of Nanozymes Potentially Used in Medicine Practice

Owing to the remarkable catalytic efficiency and substrates specificity under mild conditions, natural enzymes have been extensively used in various biomedical applications. However, limited by the intrinsic drawbacks of natural enzymes, such as poor stability, high cost, and unrecyclable, practical biomedical applications of natural enzymes still encounter difficulties. Thus, novel artificial enzymes with high stability and low cost are regarded as the alternatives to natural enzymes to overcome those limitations.

Since the first evidence of ferromagnetic nanoparticles (magnetic nanoparticles, MNPs) as peroxidase mimetics was reported in 2007 [1], various nanomaterials have been identified that possess intrinsic enzyme-like activities [2, 3]. These materials catalyze enzymatic reactions and follow similar enzymatic kinetics and mechanisms of natural enzymes under physiological conditions. Thus, the term "Nanozyme" was coined to describe the properties of this new generation of enzyme mimetics or artificial enzymes [4]. Nanozymes exhibit a high enzymatic activity which is tunable via size control, doping and surface modification. In addition, they exhibit multiple functions, high stability, and are easy to be scaled up with low cost. These advantages make them superior to natural enzymes or traditional enzyme mimetics, presenting a new generation of artificial enzymes.

Nowadays, nanozymes have been demonstrated to exhibit multiple enzymatic activities, such as peroxidase [1], oxidase [5], catalase [6], and superoxide dismutase (SOD) [7]. Among these enzymes, the peroxidase, such as horseradish peroxidase (HRP), is extensively used in biomedical diagnosis. HRP-based bioanalytical methods, such as enzyme-linked immunosorbent assay (ELISA), immunohistochemistry (IHC), are routine techniques used in tumor diagnosis. Thus, the nanozymes with peroxidase activity can be used to develop novel tumor diagnosis methods [8]. Until now, nanozymes have been widely used in biomedical applications including immunoassays and biosensors [9].

Currently, considerable efforts have been made to explore the feasibility of applying nanozymes to in vivo clinical diagnosis and therapy [3, 10, 11]. The unique physicochemical properties of nanozymes at the nanoscale and excellent enzymatic activity have rendered them with excellent catalytic performances in vivo for biomedical applications. Nanozymes for in vivo tumor theranostics have attracted a lot of attention. However, a more systematical investigation of the relationship between biological systems and nanozyme activities, which guides the rational design of new nanozymes, is necessary for the nanozyme-based tumor theranostics.

In this chapter, we systematically summarize the recent progress on contemporarily reported nanozymes for multiple biomedical applications of cancer diagnosis and therapy. We also discuss the potential clinical practices based on nanozyme in the near future.

13.2 Nanozyme-Based In Vitro Tumor Diagnosis

While cancer seriously threatens human health, early diagnosis is a key to achieve successful treatment. Studies have found that the 5-year survival rate for renal cancer in early stage was 99%, while for patients in the second phase or higher was only 16% [12]. Therefore, the development of highly sensitive and specific early diagnostic methods is of great significance for the treatment of cancer [13, 14].

With the advantage of high catalytic activity, low cost, high stability, and multifunctionality, nanozymes can be used in the detection of cancer-related genes, molecules, cells, and tissues by combining with the early diagnosis technology, which provides a new way for rapid and accurate diagnosis of cancer [8].

13.2.1 Nanozyme for Cancer-Related Genes Detection

The occurrence of cancer is often related to changes of the genes. Proto-oncogenes and tumor suppressor genes play important roles in cell growth, proliferation, and regulation. However, under certain conditions, such as viral infection, chemical carcinogens, or radiation, these two types of genes are susceptible to mutation, loss, or inactivation, resulting in malignant cell transformation and the occurrence of tumors. Therefore, detection of cancer-related genes is one of the earliest ways for cancer diagnosis [15].

Traditionally, this detection has relied almost exclusively on polymerase chain reaction (PCR) and labeled DNA probe to convert nucleic acid recognition events into detectable signal changes such as a fluorescent, colorimetric, or electrochemical signal. However, most labeling processes are time-consuming and complex [16]. Wang et al. [16] designed a simple, label-free DNA assay for the detection of breast cancer gene BRCA1 [17] (Fig. 13.1). The nanomaterial was fabricated by embedding Pt nanoparticles (NPs) in the mesoporous silica ($mSiO_2$) matrix which denoted

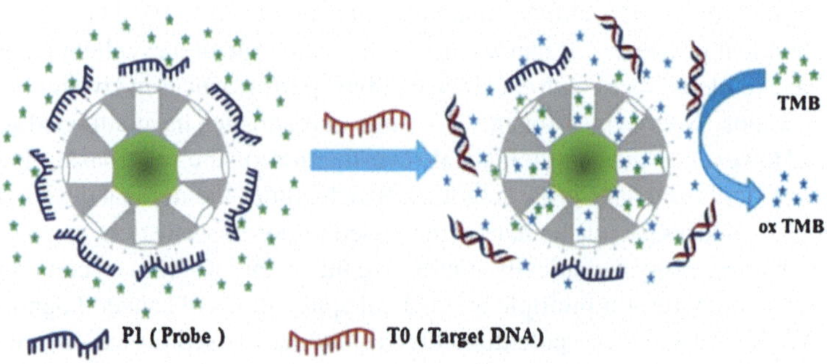

Fig. 13.1 Label-free detection of DNA by Pt@$mSiO_2$ nanozyme [16]. Copyright 2014 Royal Society of Chemistry

as Pt@mSiO_2, single strain DNA probes (P1) were then adsorbed to the surface of Pt@mSiO_2 through the electrostatic interaction. In this manner, due to the blocking by the adsorbed P1, Pt NPs encapsulated inside mSiO_2, which possess an intrinsic peroxidase-like activity, cannot catalyze the oxidation of substrate TMB. When the complementary target DNA (T0) is present in the solution, upon hybridization with T0, P1 will detach from the surface of Pt@mSiO_2. Under this condition, the oxidization reaction of TMB will be accelerated under the catalysis of exposed Pt NPs, producing the blue oxidized TMB (oxTMB). Based on the correlations of specific absorption peaks of oxTMB at OD652nm and concentrations of target DNA, the linear relationship between them was established. The results indicated that the detection sensitivity of BRCA1 can be as low as 3 nmol/L, and this method had a good single base pair mismatch-discrimination capability. Compared with other DNA detection procedures, this nanozyme-based DNA detection method not only exhibits an excellent sensitivity, but also leaves out the labeling of the probes.

13.2.2 *Nanozyme for Tumor Marker Molecules Detection*

The detection of tumor marker molecules is crucial for the diagnosis, classification, staging, and prognosis of cancers [18]. Enzyme-linked immunosorbent assay (ELISA) is a common method used in clinical practice for tumor biomarker detection, which typically employs HRP as the enzyme to catalyze the color reaction. However, several disadvantages are suffered in using natural enzymes, such as high production cost and low stability [19]. Nanozymes as an alternative to HRP have been used to develop detection methods for different tumor markers, such as human epidermal growth factor receptor 2 (HER2), prostate-specific antigen (PSA), vascular endothelial growth factor (VEGF), reduced glutathione (GSH), cytokeratin-19-fragment (CYFRA 21-1), carcinoembryonic antigen (CEA), integrin GPIIb/IIIa, etc. (Table 13.1) [19–27].

Park's group developed a novel strategy that MNPs were incorporated together with Pt NPs in ordered mesoporous carbon (OMC). The fabricated nanocomposites exhibited up to 50-fold higher catalytic efficiency than that of the MNPs. After conjugating antibodies on the surface, the nanocomposites were utilized in ultrafast colorimetric immunoassay systems for the detection of clinical biomarkers, such as HER2 (Fig. 13.2a) [19]. Results showed that the detection limit of HER2 in cell lysates of breast cancer cells achieved 1.5 ng/ml by this immunoassay system, which was sufficient to identify this biomarker at its clinical cutoff value of 15 ng/mL [28].

In addition to replacing natural enzymes in ELISA for tumor marker molecules detection, nanozymes can also be utilized to detect tumor biomarkers which can produce or consume the substrates of enzymes. GSH is the most abundant antioxidant in mammalian cells [23], which plays an important role in maintaining intracellular redox balance and protecting cells from oxidative damage. The imbalance of GSH level in cells is closely related to the occurrence of cancer [29]. In the presence of H_2O_2, GSH can be converted into glutathione disulfide (GSSG) [30]. Yang's group

Table 13.1 Detection of tumor marker molecules by nanozymes

Nanozyme	Tumor marker molecules and samples	Principle	Linear range and detection limit	Reliability in clinical tests	Reference
A nanocomposite entrapping both MNPs and Pt NPs in mesoporous carbon	HER2; Breast cancer cells lysate	Peroxidase activity; conjugating antibody, ultrafast colorimetric immunoassay	2.5–100 ng/mL; 1.5 ng/mL	The detection limit was sufficient to enable determination of this antigen at its clinical cutoff value of 15 ng/mL [28]	[19]
Urchin-like (gold core)@(platinum shell) nanohybrids	PSA; Serum	Horseradish peroxidase activity; conjugating antibody, colorimetric immunoassay	5–500 pg/mL; 2.9 pg/mL	12 human PSA serum specimens were monitored by this method and a referenced electrochemiluminescence immunoassay (ECLIA) method, the regression equation between two methods displayed a well positive correlation	[20]
Fe_3O_4 magnetic bead	PSA; Serum	Coupling highly catalytic efficient catalase with magnetic bead-based peroxidase mimics; sandwich-type reverse colorimetric immunoassay	0.05–20 ng/mL; 0.03 ng/mL	12 PSA clinical serum specimens were validated, giving results in good accordance with those obtained by the commercially available ELISA method	[21]

(continued)

Table 13.1 (continued)

Nanozyme	Tumor marker molecules and samples	Principle	Linear range and detection limit	Reliability in clinical tests	Reference
DNA-templated Ag/Pt bimetallic nanoclusters	VEGF; Human serum samples of breast cancer patients	Peroxidase activity; employing the aptamer of VEGF to design a label-free electrochemical biosensor	6.0–20 pM; 4.6 pM	Six human serum samples of breast cancer patients were tested using the sensor and ELISA Kit as a reference	[22]
Graphene dots	GSH; Hela cell lysate	Peroxidase activity; GSH consumed H_2O_2 and weakened the activity of the GDs; colorimetric assay	0.5–100 μM; 0.5 μM	A series of concentrations of GSH were spiked into cell lysate for detection, the regression equation with a correlation coefficient (R^2) of 0.9946. The concentration of GSH was estimated by this method	[23]
Carbon nanodots	GSH; The whole human blood	Peroxidase activity; the blue-colored free radical cation concentration was inversely related to the GSH concentration; indirect colorimetric detection	0–7 μM; 0.3 μM	This chemosensor was used to measure the GSH content of the whole human blood with the results consistent with that of a classical Ellman method. A key advantage is elimination of hemoglobin interference	[24]

(continued)

Table 13.1 (continued)

Nanozyme	Tumor marker molecules and samples	Principle	Linear range and detection limit	Reliability in clinical tests	Reference
Platinum nanoparticles	CYFRA 21-1; Serum	The volumetric bar-chart chip volumetrically measured the production of oxygen gas by PtNPs and can be integrated with ELISA technology to provide visible and quantitative readouts	0.5–50 ng/mL; 0.5 ng/mL;	When CYFRA 21-1 was spiked in serum at the same concentration, the bar-chart results exhibited a similar detection limit to that obtained in the buffer, thus indicating that serum matrices do not interfere significantly with PtV-Chip sensitivity	[25]
Cupric oxide nanorods	CEA; Serum	Chemiluminescent immunosensor based on cupric oxide nanorods as peroxidase mimics; when antigens were present, a large immunocomplexes were formed that prevented CL substrate access to the surface, thereby reducing the CL signal	0.1–60 ng/mL; 0.05 ng/mL	CEA levels in human serum sample were tested, the results were in good agreement with those obtained using an ECL immunoassay, the relative errors were less than 7.5%	[26]
Gold nanoparticles	GPIIb/IIIa; HELcell	Combined the peroxidase activity and two-photon luminescence of gold (Au) nanoparticles to construct a nanoprobe, and achieved accurate quantification and localization of integrin GPIIb/IIIa	31.25–375 ng/mL; 6.4×10^6/cell	Inductively coupled plasma mass spectrometry measurement verified the reliability of the new analytical method	[27]

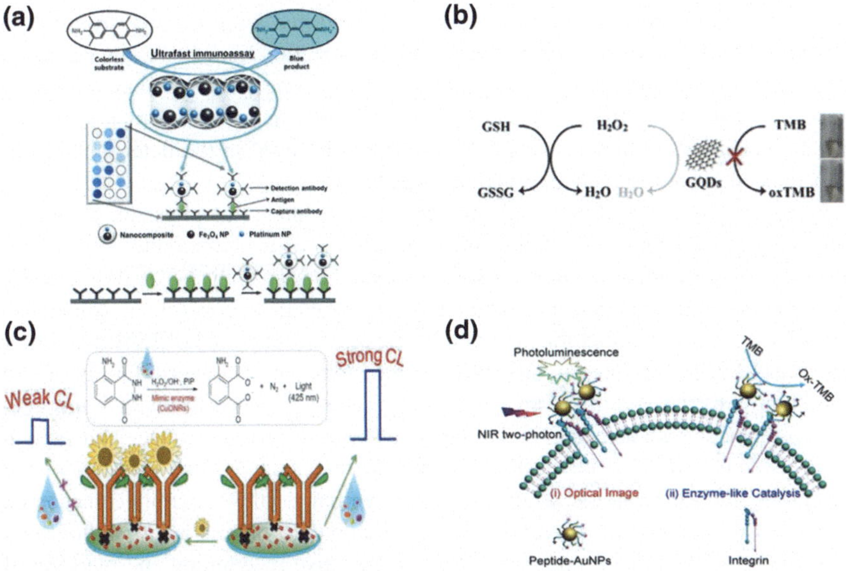

Fig. 13.2 Detection of tumor marker molecules by nanozymes. **a** Immunoassay based on a nanozyme composite in which both MNPs and Pt NPs are entrapped in mesoporous carbon [19]. Copyright 2009 John Wiley and Sons. **b** Schematic illustration of highly efficient peroxidase-like activity of graphene dots (GDs) for the detection of GSH [23]. Copyright 2013 Elsevier B.V. All rights reserved. **c** CuONR nanozyme platform for the nanoenzymatic-based label-free detection of CEA [26]. Copyright 2017 Elsevier B.V. All rights reserved. **d** Au nanozyme integrates the signal generation and amplification for display and quantify of integrin GPIIb/IIIa [27]. Copyright 2015 American Chemical Society

combined this reaction with the peroxidase activity of graphene quantum dots (GDs) to achieve highly sensitive detection of GSH (Fig. 13.2b) [23]. GSH consumed H_2O_2 in the system and thus weakened the catalytic activity of GDs nanozyme. The results indicated that the reduced absorbance value of oxTMB was linear to the concentration of GSH. Thus, the concentration of GSH in the prepared cell lysate can be estimated by this nanozyme-based method.

The catalytic activity of nanozymes can be regulated by hindering the binding of nanozyme and substrate. On the basis of this principle, Yang's team developed an efficient label-free chemiluminescence (CL) immunoassay based on the peroxidase activity of copper oxide nanorods [26], which was used to detect the broad-spectrum cancer marker—CEA [31, 32] in serum (Fig. 13.2c). Cupric oxide nanorods (CuONRs) were synthesized by a simple hydrothermal method. A CuONR-chitosan solution was first coated onto an epoxy-modified glass slide to form a solid CuONRs-chitosan support. Streptavidin was then employed to functionalize the composites for highly selective capture of biotinylated antibodies. The immunocomplexes formed on the sensing interface were shown to hinder the diffusion of the CL substrate molecules to the CuONR surface. These restrictions effectively inhibited the nanozyme-catalyzed CL reaction, thereby leading to a decrease in CL signal with increasing analyte concentration. The result showed that a linear range from 0.1 to 60 ng/ml for CEA antigen was obtained, and the detection limit was evaluated to be 0.05 ng/mL. Furthermore, the relative errors of CEA concentrations obtained by the proposed method and the reference method were less than 7.5%, which indicated that the proposed immunosensing platform could be used in the determination of CEA concentrations in clinical practical samples.

Nanozyme is a multifunctional molecule which integrates physical, chemical, and catalytic functions. Combining these properties of nanozyme effectively produces novel multifunctional materials. Gao's team combined the peroxidase activity and two-photon luminescence of gold (Au) nanoparticles to construct a nanoprobe, and achieved accurate quantification and localization of integrin GPIIb/IIIa in tumor cell (Fig. 13.2d) [27]. Integrin GPIIb/IIIa is a member of the integrin family of cell membrane receptors whose expression level is closely related to platelet aggregation and cancer pathogenesis, including metastasis in prostate cancer and human erythroleukemia (HEL) cells [33]. Utilizing intrinsic enzyme-like property of the nanoprobe, the expression level of integrin on human erythroleukemia cells can be quantitatively counted in an amplified and reliable colorimetric assay without cell lysis and protein extraction process. Through selective recognition and marking of integrin, two-photon photoluminescence of the nanoprobe was exploited for direct observation of protein spatial distribution on cell membrane. This strategy provides strong evidence for the diagnosis and early treatment of related diseases.

13.2.3 Nanozyme for Tumor Cells Detection

Rapid, sensitive, and accurate detection of tumor cells is critical in clinical practice, as it not only monitors the progression of tumors, but also helps to select effective treatment strategies and improve the clinical outcomes. At present, the methods for tumor cells detection are typically divided into three types: (a) the first type of methods is using biomarker-specific antibodies combined with flow cytometry to isolate and detect tumor biomarkers positive cells; (b) the second one is employing centrifugation and microfiltration methods to enrich and detect tumor cells based on the biophysical properties of tumor cells including size, shape, and density; and (c) the third type of methods is PCR-based methods that can detect specific genes in tumor cells. However, these methods have various disadvantages, such as complicated preprocessing steps and operations [34]. A variety of rapid and sensitive tumor cell detection methods have been developed by combining the high catalytic activity and multifunctionality of nanozymes with enzyme-linked immunosorbent assay (ELISA) [13, 14, 25, 35–40].

Park's group successfully established a novel colorimetric immunoassay for tumor cell detection by utilizing the peroxidase mimicking activity of MNPs (Fig. 13.3a). Conjugation of the MNPs with antibodies enabled MNP nanozymes specific recognition of the tumor cells with biomarkers. Results showed that the limit of detection for HER2-positive tumor cells was determined to be 341 cells [14].

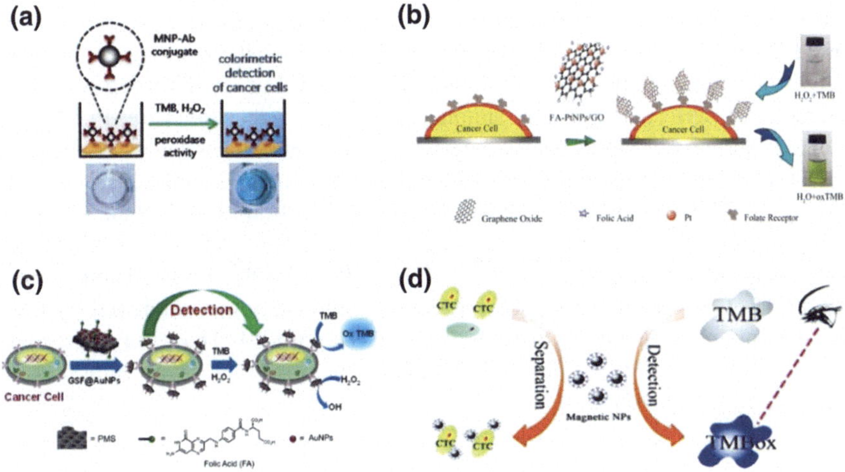

Fig. 13.3 Detection of tumor cells by nanozymes. **a** Colorimetric detection of breast cancer cells in a direct immunoassay system based on MNPs nanozyme [14]. Copyright 2013 MDPI (Basel, Switzerland). **b** Schematic representation of colorimetric direction of cancer cells by using folic acid functionalized PtNPs/GO nanozyme [40]. Copyright 2014 American Chemical Society. **c** Peroxidase activity of GSF@AuNPs for cancer cell detection [13]. Copyright 2015 American Chemical Society. **d** Simple and rapid colorimetric detection of melanoma circulating tumor cells using bifunctional magnetic nanozyme [38]. Copyright 2017 Royal Society of Chemistry

With the development of nanozyme field, scientists found that the hybrids of different types of nanozymes exhibit unique properties, such as enhanced enzyme-like activities, or increased dispersibility and stability. For example, a green approach was proposed by Chen's group for in situ growth of porous platinum nanoparticles on graphene oxide (PtNPs/GO) to synthesize a novel nanozyme. The fabricated PtNPs/GO nanozyme exhibited high peroxidase catalytic activity. By using folic acid as a recognition moiety of PtNPs/GO nanozyme, as low as 125 MCF-7 cancer cells can be detected by naked-eye observation. Using a microplate reader, the limit of detection (LOD) of this PtNPs/GO nanozyme-based method for MCF-7 cells detection was determined to be 30 cells (Fig. 13.3b) [40].

Zhao's group developed a hybrid nanozyme (GSF@AuNPs) which was prepared by the immobilization of gold nanoparticles (AuNPs) on periodic mesoporous silica (PMS)-coated nanosized reduced graphene oxide (RGO) conjugated with folic acid, a cancer cell-targeting ligand. The GSF@AuNPs hybrid nanozyme showed unprecedented peroxidase-like activity and was utilized as a selective, quantitative, and fast colorimetric detection probe for cancer cells detection (Fig. 13.3c). Results demonstrated that as low as 50 tumor cells could easily be detected by naked-eye observation [13].

Circulating tumor cells (CTCs) are cells that enter peripheral blood from primary or metastatic tumors. Recent studies have revealed that CTCs can act as a biomarker to monitor the process of tumor metastasis, cancer recurrence, and tumor therapeutic efficacy [41]. However, the rarity and inherent fragility of CTCs pose great challenges for CTC detection. An integrated platform combining magnetic property and catalytic activity of nanozymes facilitates the rapid detection of CTCs. Trau's group developed bifunctional MNPs nanozyme for simple and rapid CTC detection (Fig. 13.3d). The magnetic property of MNPs nanozyme with targeting antibodies enables the isolation and enrichment of melanoma CTCs within 5 min from blood samples. The catalytic activity of MNPs nanozyme achieves naked-eye detection of melanoma CTCs by catalyzing the oxidation of colorless TMB into blue color products. The blue color is still detectable by naked eye even the CTC cells at levels of 50 cells/mL serum. The LOD of this method for CTCs detection could be further improved by UV–Vis absorption spectroscopy to be 13 cells/mL serum [38]. Thus, multifunctional nanozyme enables the CTCs detection to be convenient and time-saving, thus making clinical CTCs detection feasible.

13.2.4 Nanozyme for Tumor Tissues Detection

Immunohistochemistry (IHC) assay is the "gold standard" of tumor pathological diagnosis [42–44]. The conventional IHC methods use primary antibodies to specific recognize tumor-associated antigens and biotin-labeled secondary antibody further to identify primary antibodies. Streptavidin is used to conjugate the HRP enzyme for the highly selective capture of biotinylated antibodies. Finally, HRP catalyzes substrate to produce color reactions so as to achieve the visualization of tumor tissue (Fig. 13.4c). However, the IHC methods involve multiple manipulation steps and long operation time. In addition, the IHC results are easily affected by the proficiency and subjectivity of manipulators. Thus, improvements of IHC methods are needed [45, 46].

Recently, Yan's group developed a new technique for tumor tissues detection by using magnetoferritin (M-HFn) nanozyme as a dual-functional reagent to simultaneous target and visualize tumor tissues (Fig. 13.4d) [46]. They found that human H-ferritin (HFn), like anti-transferrin receptor 1 (TfR1) antibody and transferrin, can be utilized to directly target tumor cells via overexpressed TfR1 [47]. In addition, M-HFn nanozymes own intrinsic peroxidase-like activity which can catalyze substrates to produce the color reaction to visualize tumor tissues (Fig. 13.4b). Therefore, magnetoferritin (M-HFn) nanozyme can be used to realize the rapid and sensitive tumor detection (Fig. 13.4b).

Compared with conventional antibody-based IHC methods for cancer detection in clinics, this M-HFn nanozyme-based method possesses the following advantages [46]: (1) high efficiency, with sensitivity of 98% and specificity of 95%, both of which are higher than conventional IHC method and (2) ease of use. As the new technology uses one reagent and one step (Fig. 13.4c), instead of traditional IHC using primary antibody, secondary antibody, or enzyme-labeled third antibody with multiple steps; (3) high speed, it only takes less than 1 h, rather than 3–4 h required for IHC method; and (4) low cost and ease for large-scale preparation. HFn can be

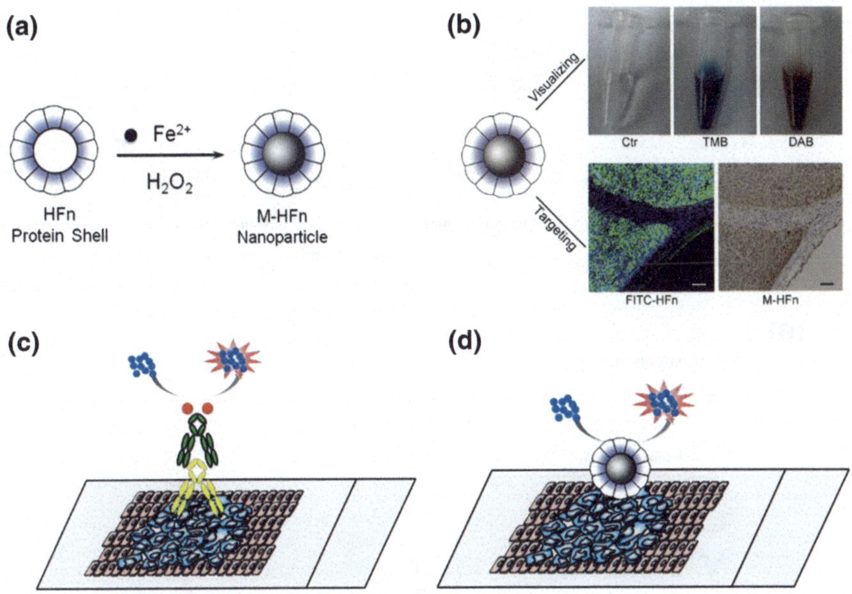

Fig. 13.4 Ferritin nanozyme-based immunohistochemical assay for clinical tumor tissue diagnosis [49]. **a** The preparation process of M-HFn nanozyme. **b** M-HFn nanozyme with intrinsic dual functions: targeting tumor tissues without any modification and giving a color signaling by its peroxidase-like activity. **c** and **d** Antibody-based immunohistochemistry and M-HFn nanozyme-based immunostaining. Copyright 2013 John Wiley and Sons

obtained by convenient production in *Escherichia coli* at high yield, avoiding the use of expensive antibodies. All these properties indicated that M-HFn nanozyme, as a new reagent, is expected to be a rapid, economical, and broad-spectrum new tool for tumor diagnosis [48].

Subsequently, Pan's group demonstrated that M-HFn nanozymes with increasing core sizes enhance the visualization of cancer tissues [50]. In addition, they found that cobalt-doped M-HFn nanozymes exhibit enhanced peroxidase activity and improve tumor tissue visualization [51]. These reports further confirm the feasibility of M-HFn nanozyme-based method for pathological diagnosis of tumor tissues.

After the report of Fe_3O_4 nanozyme exhibiting peroxidase-like activity, many scientists focused on the cobalt element which is in the same group of Fe [7, 45, 52]. Zhang's team found that, like Fe_3O_4, the cobalt trioxide (Co_3O_4) nanoparticles also exhibited strong peroxidase activity. They modified antibodies onto the surface of Co_3O_4 nanozyme by chemical conjugation to render Co_3O_4 nanozyme specifically targeting the tumor biomarker vascular endothelial growth factor receptor (VEGFR). The immunoassay based on Co_3O_4 nanozyme for clinical diagnosis of tumor tissue was successfully established (Fig. 13.5A). The results showed that this Co_3O_4 nanozyme-based method successfully differentiates VEGFR-positive tumor tissue from normal tissue [7].

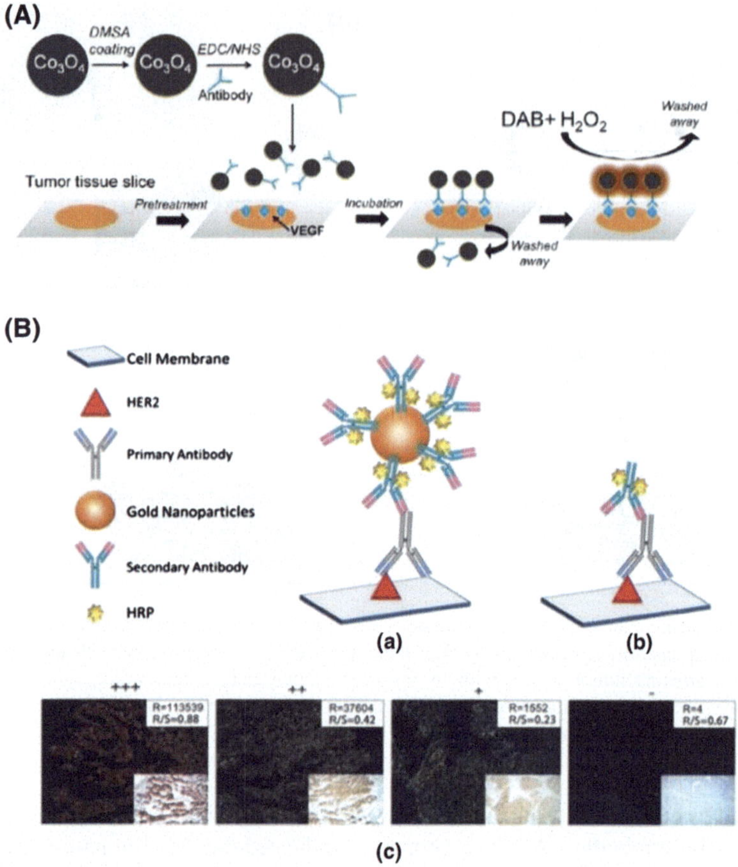

Fig. 13.5 Co_3O_4 and Au nanozyme-based immunohistochemical assay for clinical tumor tissue diagnosis. **A** Immunohistochemistry detection process based on the peroxidase-like activity of Co_3O_4 nanozyme [7]. Copyright 2014 American Chemical Society. **B** Principle of the synergistically enhanced dark-field method based on HRP-AbII-Au nanoprobes [45]. Copyright 2015 Royal Society of Chemistry

Conventional IHC method is limited to subjective judgment based on human experience, thus a quantitative IHC detection method is required [45]. DAB aggregates, a type of staining product produced by conventional IHC, were found to possess a special optical property of dark-field imaging by Zhang's team. On the basis of this finding, a novel IHC method for HER2 detection was established (Fig. 13.5B) [45]. Gold nanoparticles were used as a synergistically enhanced agent in this method due to their nanozyme activity and dark-field scattering properties. According to the scattering characteristics of DAB aggregates and gold nanoparticles under dark-field microscope, the tumor tissues stained by this method can be divided into four levels of "−, +, ++, +++". 114 cases of breast tumor tissues with different expression levels of HER2 were successfully detected with a 96.70% sensitivity and 95.65%

specificity. Compared with the traditional IHC method, enhanced dark-field imaging as a novel IHC method can be analyzed by software, thus avoiding the influence of subjective judgment.

13.3 Nanozyme for In Vivo Tumor Imaging

As we mentioned in Sect. 13.2, the development of highly sensitive and specific early diagnostic methods is of great significance for the treatment of cancer [13, 14]. Tumor imaging is an important technique to achieve in vivo tumor diagnosis in clinic. For example, magnetic resonance imaging (MRI), as a noninvasive tumor imaging technique with high spatial resolution, has been extensively used in modern clinical tumor diagnosis. Reduced superoxide dismutase (SOD) level was shown in all tumor tissues, and this makes superoxide radicals, the substrate of SOD, be associated with the development of cancer [53]. However, MR contrast agents demonstrating a superoxide radical-triggered contrast change are rare.

Tremel's group first demonstrated the intrinsic superoxide dismutase (SOD)-like activity of MnO nanozymes is capable of enhancing the MR imaging contrast (Fig. 13.6a) [54]. The T_1- and T_2-weighted MR contrast of the MnO nanozymes was investigated by evaluating the specific relaxivities r_1 and r_2 without/with the addition of superoxide generated by xanthine/xanthine oxidase. When MnO nanozymes were exposed to superoxide radicals, the T_1 and T_2 relaxation times increased significantly with r_1 and r_2 values of 0.06 ± 0.01 mM^{-1} s^{-1} and 1.90 ± 0.14 mM^{-1} s^{-1}, respectively. Increase of the relaxation times can possibly be explained by a temporary change of the oxidation states of manganese ions during the SOD reaction, leading to alterations in the paramagnetic field. In addition, MnO nanozymes are able to decompose H_2O_2, which is one of the SOD reaction products, in a catalase-like reaction subsequently. Therefore, the excellent dual enzyme-like activities of MnO nanozymes can eliminate superoxide radical via a sequential catalytic reaction. Due

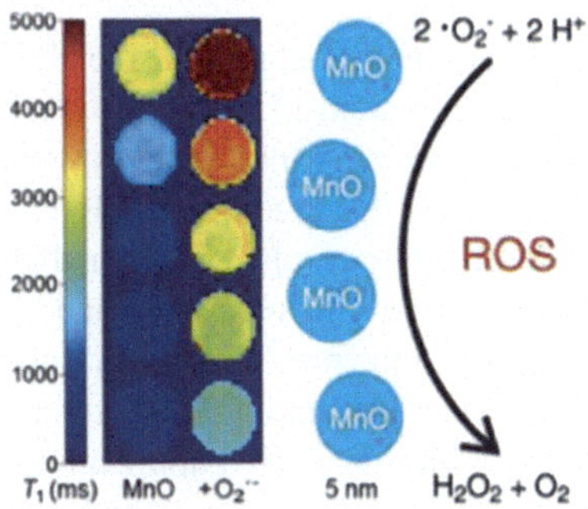

Fig. 13.6 Intrinsic superoxide dismutase activity of MnO nanozymes enhances the MR imaging contrast [54]. Copyright 2016 Royal Society of Chemistry

to the unique combination of sequential catalytic reaction with enhanced MRI contrast, MnO nanozymes could be used simultaneously for treating and imaging of tumor sites with high level of superoxide radicals.

ROS are high-energy oxygen-bearing molecules encompassing a variety of chemical species including superoxide anion, hydroxyl radical, hydrogen peroxide, and so on. As a representative ROS, H_2O_2 is diffusible and relatively long-lived. The overproduction of H_2O_2 can act as a potential diagnostic marker for the presence and progression of cancer states. However, some limitations of conventional H_2O_2 probes have hindered their further clinical applications, such as auto-oxidation, poor specificity, less sensitivity, and high cost for in vivo analysis. Therefore, it is necessary to develop novel probes for the detection of H_2O_2 with a high spatiotemporal resolution [55].

Prussian blue nanozymes ($KFe^{3+}[Fe^{2+}(CN)_6]$) were reported to be capable of catalyzing the breakdown of H_2O_2 into oxygen (O_2) molecules under the neutral pH condition. The O_2 molecules generated in this reaction can be used as an ultrasound contrast agent to enhance ultrasound (US) imaging. Moreover, an unusual structural feature that renders some Fe^{3+} centers accessible to water coordination and the paramagnetic oxygen bubbles can shorten the T_1 relaxation time, which is beneficial for developing nanoparticle-based T_1 MR imaging contrast agents. Both US and MR imaging showed enhanced contrast after PBNPs injection in the sites with overproduction of H_2O_2, evidencing the enormous potential of PBNPs for diagnostic imaging such as tumor in vivo imaging. This dual-mode diagnostic approach not only accomplishes the combination of US and MR imaging to offer complementary medical information for the diagnosis of cancer, but also provides us with much inspiration for rational design of more theranostic nanozymes [56].

13.4 Nanozyme for Tumor Therapy

13.4.1 Nanozyme Directly Used in Tumor Catalytic Therapy

High amounts of abnormal metabolic products are produced during tumor growth and development [57]. For instance, an elevated level of H_2O_2 is often exhibited in solid tumors, which renders tumor cells more resistant to therapeutic treatment. However, if the accumulated H_2O_2 can be converted instead for damaging tumor cells, effective results of tumor therapy may be obtained, with a mindset of "Give tumor a taste of its own medicine."

Natural enzymes with peroxidase activity can be used to convert H_2O_2 into toxic ROS which effectively kills tumor cells [58]. However, most natural enzymes are not suitable for in vivo applications due to their sensitivity and low stability in an unfavorable environment, which limits their practical applications.

Recently, Yan and Gao's groups reported a novel tumor catalytic therapy using nanozymes which convert O_2 and H_2O_2 to toxic ROS for tumor destruction. In this work, they developed a novel nanozyme using nitrogen-doped porous carbon nanospheres (N-PCNSs) to mimic four enzyme-like activities (oxidase, peroxidase, catalase, superoxide dismutase). They found that these nanozymes were able to regulate intracellular ROS and boost ROS generation by oxidase and peroxidase activities under acidic microenvironment. To utilize the enzymatic performance for tumor therapy, ferritin was introduced to target tumors and deliver N-PCNSs to lysosome for controlling ROS generation [60]. In vivo tests demonstrated that ferritin-N-PCNSs specifically suppress tumors in an animal model, indicating that the nanozyme activities are controllable to perform the desired purpose. This study provides evidence that N-PCNSs are powerful nanozymes capable of regulating intracellular reactive oxygen species, and ferritinylation is a promising strategy to render nanozymes to target tumor cells for in vivo tumor catalytic therapy. Importantly, as a nanomaterial, N-PCNSs own excellent biocompatibility and biodegradability under physiological conditions. In addition, N-PCNSs can be made at a large scale with low cost. These properties make them superior to natural enzymes in biomedical applications (Fig. 13.7) [59].

ROS-induced apoptosis is a promising treatment strategy for malignant neoplasms. However, current systems are highly dependent on oxygen status and/or external stimuli to generate ROS, which greatly limit their therapeutic efficacy particularly in hypoxic tumors. Qu and coworkers developed a biomimetic nanoflower based on self-assembly of nanozymes that can catalyze a cascade of intracellular biochemical reactions to produce ROS in both normoxic and hypoxic conditions without any external stimuli [61]. PtCo nanoparticles were first synthesized and used to direct the growth of MnO_2. By adjusting the ratio of reactants, highly ordered MnO_2@PtCo nanoflowers with excellent catalytic efficiency were obtained, where

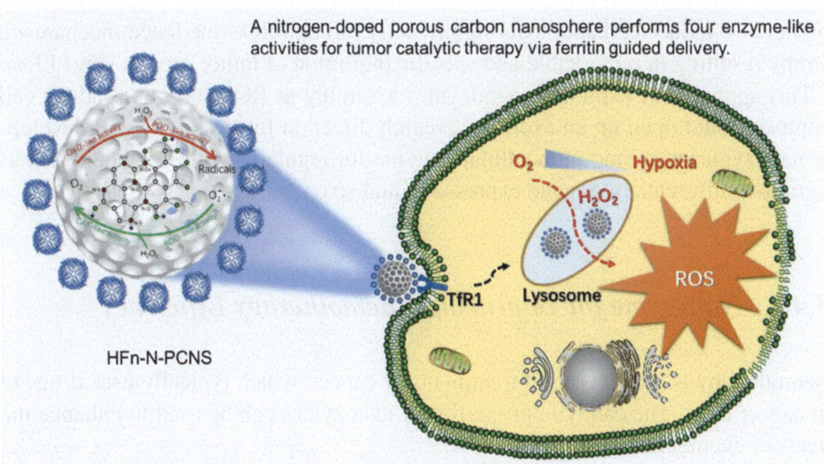

Fig. 13.7 In vivo guiding nitrogen-doped carbon nanozyme for tumor catalytic therapy. Adapted with permission from ref. [59]. Copyright 2018 Nature Publishing Group

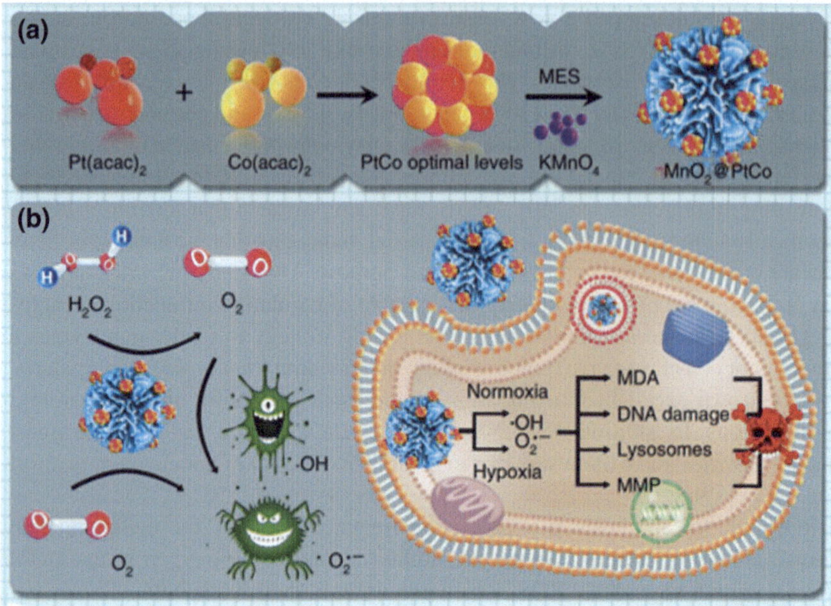

Fig. 13.8 Biomimetic nanoflowers by self-assembly of nanozymes to induce intracellular oxidative damage against normoxic and hypoxic tumors [61]. **a** Schematic illustration showing the self-assembly of nanozymes into well-defined nanoflowers. **b** MnO_2@PtCo nanoflowers generate ROS under different oxygen tensions for tumor therapy. Copyright 2018 Nature Publishing Group

PtCo behaved as oxidase mimic and MnO_2 functioned as catalase mimic. In this way, the well-defined MnO_2@PtCo nanoflowers not only can relieve hypoxic condition but also induce cell apoptosis significantly through ROS-mediated mechanism, thereby resulting in remarkable and specific inhibition of tumor growth (Fig. 13.8).

This approach of exploiting nanozymes assembly as ROS source to induce cell apoptosis could open up an exciting research direction for designing and developing nanozymes to mimic intracellular enzyme for regulating cell functions such as migration, differentiation, gene expression, and so on.

13.4.2 Nanozyme for Improving Chemotherapy Efficiency

Chemotherapy is a widely used treatment for cancer, which typically uses drugs to kill cancer cells. The catalytic properties of nanozymes can be used to enhance the effects of chemotherapy [6, 62].

In 2012, Gu and coworkers identified that Fe_3O_4 nanozymes exhibit the dual enzyme-like activity in a pH-dependent manner [6]. They systematically investigated the enzymatic activities of Fe_3O_4 nanozyme in neutral or acidic condition. Fe_3O_4

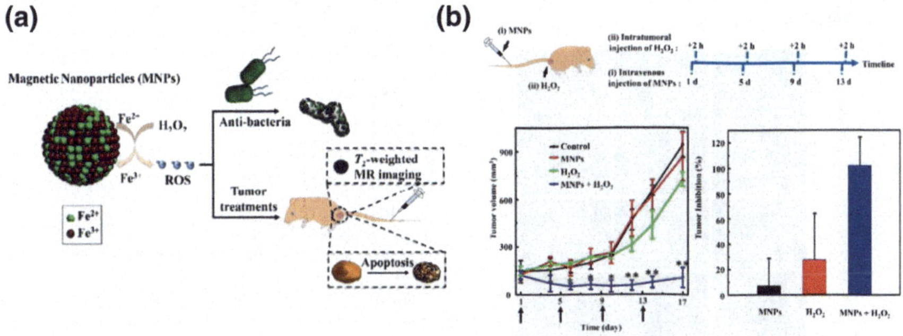

Fig. 13.9 Fe_3O_4 nanozymes for tumor therapy. **a** Schematic illustration of enzyme-mimic Fe_3O_4 nanozymes for anti-bacteria and in vivo tumor treatments. **b** Fe_3O_4 nanozymes catalyzed the intratumorally injected H_2O_2 to produce abundant ROS for tumor therapy [62]. Copyright 2013 Royal Society of Chemistry

nanozymes exhibit a catalase-like activity which is capable of catalytically decomposing H_2O_2 into nontoxic H_2O and O_2 under neutral pH condition. In contrast, Fe_3O_4 nanozymes catalyze H_2O_2 into highly toxic •OH, displaying peroxidase-like activity under acidic condition. This unique pH-dependent dual enzyme-like activity of Fe_3O_4 nanozymes makes them ideal agents to enhance H_2O_2-induced cancer cell damage dramatically because nanozymes were mostly located in the acidic environment of lysosomes.

Inspired by the peroxidase-like activity of Fe_3O_4 nanozyme, Wang's group directly used Fe_3O_4 nanozyme to catalyze the intratumorally injected H_2O_2 to produce abundant highly toxic •OH, thus resulting in significant tumor inhibition effects (Fig. 13.9) [62]. Moreover, Wang and coworkers combined the magnetic property and catalytic activity of Fe_3O_4 nanozyme to develop a nanozyme-based tumor theranostics strategy. The merits of this strategy are (i) the stability of Fe_3O_4 nanozyme is higher than natural enzymes; (ii) PEGylated Fe_3O_4 nanozyme is preferential accumulation in tumor tissue via EPR effect, and (iii) biocompatible Fe_3O_4 nanozyme can be used as a highly sensitive T_2-weighted MR imaging contrast agents. This work demonstrated that Fe_3O_4 nanozyme may open up a new avenue for cancer theranostic (Fig. 13.11).

13.4.3 Nanozyme for Improving Radiotherapy Efficiency

Radiation therapy (RT) is a conventional clinically applied treatment strategy which employs ionizing radiation to kill malignant cells. However, side effect of RT remains the most challenging issue for cancer treatment. Research to reduce the unwanted side effects of RT has yielded two categories of compounds: radiation protectants and radiation sensitizers [63]. Yet, as evidenced by the disparity between the number

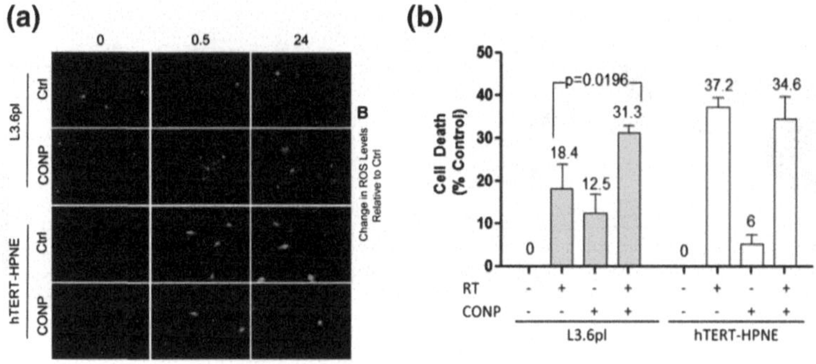

Fig. 13.10 Cerium oxide nanozyme used for improving radiotherapy efficiency [63]. **a** Cerium oxide nanozyme treatment prior to RT increases RT-induced ROS levels in acidic pancreatic cancer cells and decreases RT-induced ROS levels in neutral pancreatic normal cells. **b** The pretreatment with cerium oxide nanozyme has sensitized the cancer cells selectively to the subsequent RT without increasing the RT toxicity to the normal cells. Copyright 2013 Elsevier Inc. All rights reserved

of patients receiving RT and how few radiation protection/sensitization compounds exist, identifying viable adjuvants has proved elusive.

Cerium oxide nanoparticles (CONPs) have been shown to act as either a superoxide dismutase (SOD) mimetic, catalase mimetic, oxidase mimetic. Due to the array of radical interactions (both pro- and antioxidant) now established, CONPs can be viewed as free radical modulators. Based on these studies, Zhao and coworkers determined whether and how cerium oxide nanoparticles (CONPs) sensitize pancreatic cancer cells to RT. Notably, they found that in acidic environments of cancer cells, CONPs favor the scavenging of superoxide radical which is the product of RT over the hydroxyl peroxide resulting in accumulation of the latter and increase of ROS level, whereas in neutral pH environments of normal cells CONPs scavenge both. Therefore, CONP treatment prior to RT markedly potentiated RT-induced ROS production and cell death selectively in human pancreatic tumor cells while protecting normal tissues from the toxic side effect of RT depending upon the environmental acidity (Fig. 13.10). Taken together, these results identify cerium oxide nanozyme as a potentially novel RT-sensitizer as well as protectant for improving cancer treatment.

13.4.4 Nanozyme for Improving Photodynamic Therapy Efficiency

Photodynamic therapy (PDT) is an emerging clinically approved noninvasive therapeutic modality which exhibits great spatiotemporal selectivity for oncology [64, 65]. PDT typically composes three key components: photosensitizers (PS), light, and oxygen in tissues. The PS can be activated by light to generate singlet oxygen ($^1O_2$) in the presence of oxygen to kill tumors. Various nanomaterials have been fabricated to

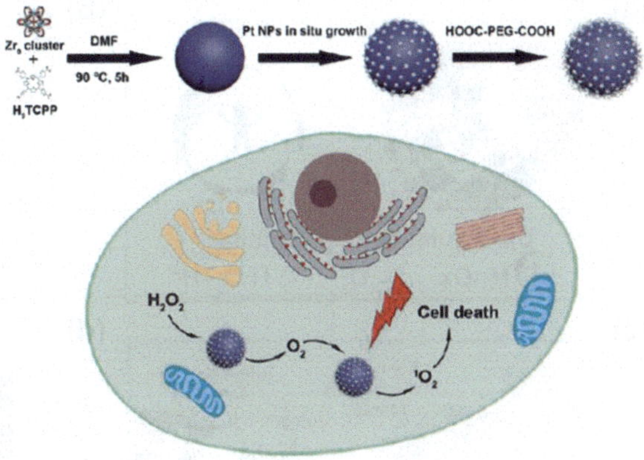

Fig. 13.11 Nanozymes with catalase activity modulate hypoxia in solid tumor microenvironment for improving photodynamic therapy [68]. Copyright 2018 American Chemical Society

investigate their photodynamic properties [66, 67]. However, PDT typically involves significant O_2 consumption, leading to drastically decrease the level of tissue oxygen which results in affecting the final antitumor efficacy for continuing treatment.

Many different nanomaterials are reported to exhibit catalase-like activity [2, 69], thus, in principle, these nanozymes can be used to modulate the hypoxia in solid tumor TME for improving PDT efficiency. Qu and coworkers fabricated a Pt nanozyme-based nanosystem by decorating Pt nanozymes on PS integrated metal–organic frameworks (MOFs). The Pt nanozymes homogeneously immobilized on the MOFs exhibit high stability and catalase-like activity. This nanosystem facilitates the formation of $^1O_2$ in hypoxic tumor site via H_2O_2-activated generation of O_2, which further causes more serious damage to cancer cells (Fig. 13.11) [68]. This study demonstrates that the O_2 generated by the catalytic activity of nanozymes for improving the antitumor efficiency of PDT is a promising strategy to overcome the current deficiency of PDT.

13.4.5 Nanozyme for Improving Sonodynamic Therapy Efficiency

Similar to light-triggered the production of ROS by photosensitizers for PDT, ultrasound (US) is also capable of triggering sonosensitizers for the generation of ROS for cancer therapy, which is thereby termed sonodynamic therapy (SDT) [70]. However, hypoxia induced the low ROS-generation efficacy remains critical issue in achieving a desirable tumor-suppression effect and further high therapeutic outcome.

The catalase activity of nanozymes was also used to augment the cancer SDT. Shi and coworkers fabricated a multifunctional nanosonosensitizer by integrating MnOx nanozymes with biodegradable hollow mesoporous organosilica nanoparticles [71]. The MnOx nanozyme in the nanosonosensitizer system acts as a catalase for converting the tumor-produced H_2O_2 molecules into oxygen, thus enhancing the

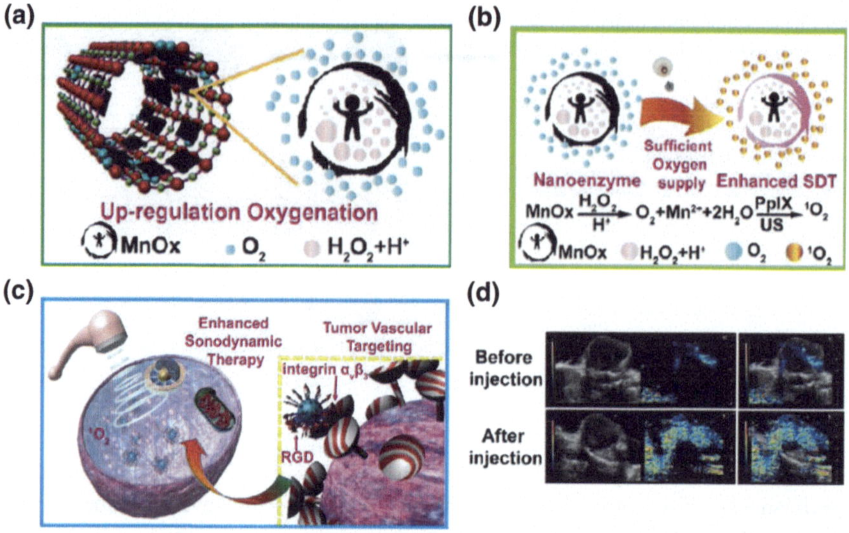

Fig. 13.12 Nanozyme-augmented cancer sonodynamic therapy by catalytic tumor oxygenation [71]. **a** Schematic illustration of alleviating tumor hypoxia and enhancing oxygenation by MnOx nanozyme catalyzed the decomposition of H_2O_2 to produce O_2 in tumor tissues. **b** Scheme of MnOx as the catalase-like nanozyme for the O_2 production and further $^1O_2$ generation upon US irradiation. **c** Schematic illustration of the augmented cancer sonodynamic therapy process. **d** Tumor oxygenation status as determined by in vivo photoacoustic imaging. Copyright 2018 American Chemical Society

tumor oxygen level subsequently (Fig. 13.12). It has been demonstrated that the involvement of MnOx nanozyme in this nanosystem facilitates SDT-induced ROS production and enhances SDT efficacy subsequently.

Taking together Sects. 13.4.4 and 13.4.5, these studies demonstrated that catalase activity of nanozymes can alleviate the tumor hypoxia via catalyzing the decomposition of tumor-produced hydrogen peroxide to generate oxygen, which provides abundant oxygen sources for enhancing the antitumor efficacy of PDT and SDT. These works highlight the application prospect of modulating unfavorable TME by taking advantage of the enzymatic activity of nanozymes to overcome current limitations of cancer therapies.

13.4.6 *Nanozyme for Improving Combination Therapy Efficiency*

Recently, combination therapy for tumor gets more attention because of its combination of the advantages of different methods and enhanced therapeutic effect. Strategies combining nanozyme-based catalytic therapy with other therapy such as photothermal therapy and starvation therapy have been reported with excellent tumor inhibition efficiency [72–75].

Starvation therapy is an emerging treatment paradigm for the clinical management of tumors and has drawn great interest [76]. It is widely accepted that the growth of tumors is highly dependent on glucose supply [77]. In most of the previous reports, tumor starvation was usually induced by cutting off the blood supply to the tumor tissues, and moderate success has been achieved in slowing down the tumor growth rate. More recently, scientists have discovered that tumor starvation may be used conjunctionally with other anticancer modalities such as chemotherapy and photothermal to improve their therapeutic efficacy [78].

Shi's group recently developed a novel Fe_3O_4 nanozyme-based tumor catalytic strategy (Fig. 13.13) [72]. The biocompatible and multifunctional GOD-Fe_3O_4@DMSNs nanosystems (GFD NCs) are fabricated by integrating glucose

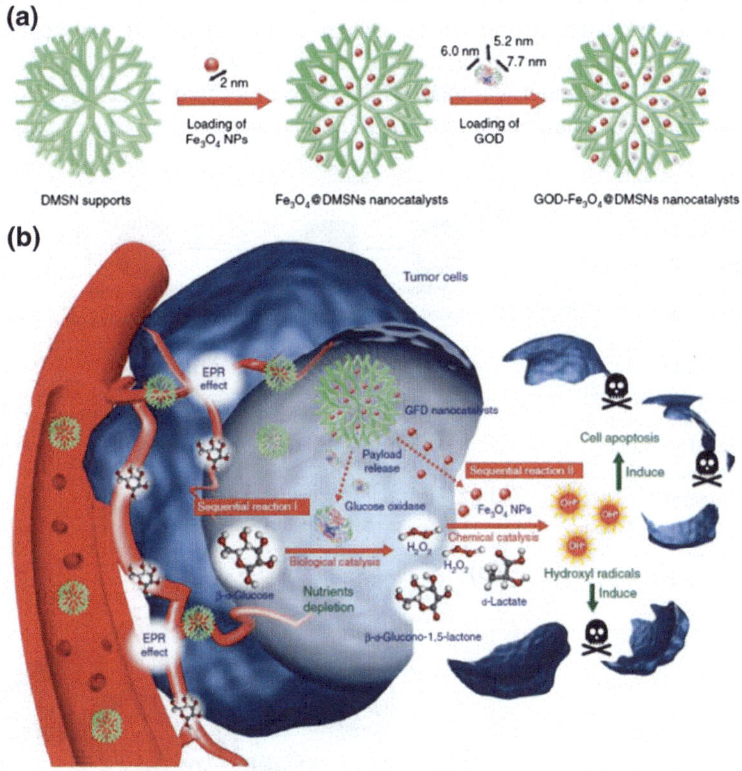

Fig. 13.13 Schematic illustrations of the fabrication process and catalytic-therapeutic methodologies of sequential GFD NCs [72]. **a** Synthetic procedure for Fe_3O_4@DMSNs and GOD-Fe_3O_4@DMSNs nanocatalysts. **b** Sequential catalytic-therapeutic mechanism of GFD NCs to generate hydroxyl radicals for cancer therapy. Copyright 2018 Nature Publishing Group

oxidase (GOD) and synthetic ultrasmall Fe_3O_4 nanozymes into the large mesopores of dendritic mesoporous silica nanoparticles (DMSNs). This Fe_3O_4 nanozyme-based system can in situ catalyze the generation of toxic ROS by a cascade enzyme reaction in response to the specific TME [56]. First, GOD catalyzes the glucose into abundant H_2O_2 in tumor region. The elevated H_2O_2 is then catalyzed by the Fe_3O_4 nanozymes via peroxidase activity to release highly toxic hydroxyl radicals, which further induce tumor cell apoptosis. Moreover, this nanosystem exhibits high biodegradability and biocompatibility. In antitumor studies, employing 4T1 breast tumor xenografts as model, this Fe_3O_4 nanozyme-based nanosystem exhibits highly desired tumor-suppression effects, concurrently with high therapeutic biosafety. This interesting study further demonstrates the feasibility of nanozymes serving as highly efficient enzyme mimics to construct an ingenious sequential catalytic reaction for desirable cancer therapy by combining with nature enzyme [56].

Employing the catalase activity and excellent photothermal effects of porous hollow Prussian blue nanoparticles (PHPBNs) nanozymes, Cai and coworkers developed a tumor-targeted redox-responsive nanosystem for enhanced tumor starvation therapy and low-temperature photothermal therapy (Fig. 13.14) [73]. In this study, GOx as a glucose consumer was loaded into the central cavity of PHPBNs to avoid biodegradation. The improved tumor starvation therapy is achieved via employing

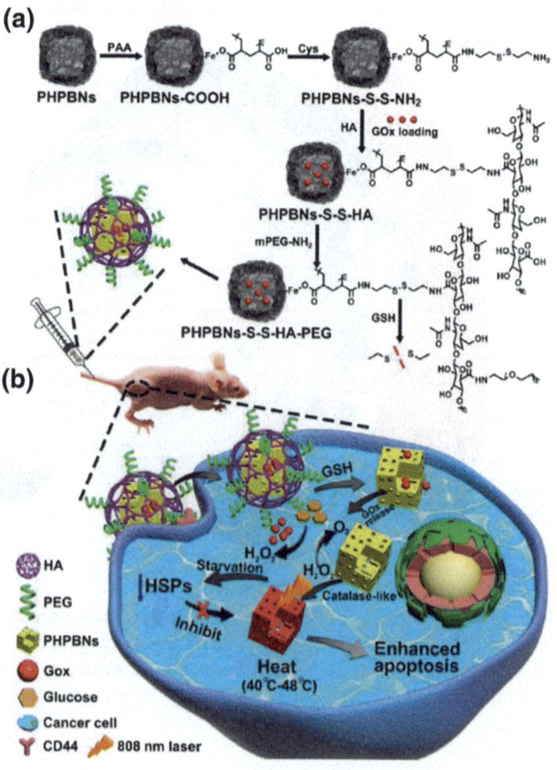

Fig. 13.14 Engineering of a nanozyme-based nanosystem for combination of tumor starvation and low-temperature photothermal therapy [73]. **a** Synthesis scheme of PHPBNs nanozyme-based nanosystem. **b** Illustration of GOx-induced starvation therapy and enhanced low-temperature photothermal therapy in a hypoxic tumor microenvironment. The hypoxia-induced photothermal resistance was circumvented via the PHPBN-mediated tumor reoxygenation. Copyright 2018 American Chemical Society

PHPBN nanozymes to generate oxygen to replenish the oxygen consumption of GOx catalyzing glucose reactions. Interestingly, Cai and coworkers also demonstrated that this nanozyme-enhanced GOx-based tumor starvation not only directly suppresses the growth of tumors, but also enhances their susceptibility to the low-temperature photothermal treatment by blocking the hyperthermia-induced expression of proteins. This work suggests that integrating the catalase activity of nanozyme into a multifunctional nanosystem is a promising strategy to improve the antitumor efficacy of GOx-mediated tumor starvation and low-hyperthermia photothermal therapy.

The improved therapeutic efficacy from the combination of tumor starvation therapy and other therapies in these studies may be indicative for the creation of nanozyme-based combinational therapeutic systems with clinical significance.

13.5 Future Perspectives

In this chapter, we discuss the applications of nanozymes in cancer diagnosis and therapy in recent years. Scientists have developed a variety of tumor diagnostic methods based on nanozymes that can detect target substances from genes, molecules to cells and tissues. These diagnostic methods are applied to the clinical sample detection, which shows high sensitivity and accuracy, and reflect the practical application value of nanozyme in the diagnosis of cancer. Moreover, employing the catalase activity of nanozymes to modulate the tumor hypoxia condition, the antitumor efficacy of various tumor therapy modalities is significantly improved. Importantly, the latest work reported by Yan and Gao's lab demonstrated that nanozymes are capable of regulating intracellular reactive oxygen species, and ferritinylation is a promising strategy to render nanozymes to target tumor cells for in vivo tumor catalytic therapy. We will discuss our future perspectives on nanozymes in tumor diagnosis and therapy in this section.

13.5.1 Future Perspectives on Nanozymes in Tumor Diagnosis

With the development of the research of nanozymes in cancer diagnosis, although much progress has been made, some problems are also emerging. Here, we summarize the problems and possible solutions as follows.

First, most of the nanozyme-based tumor diagnosis methods utilize the catalytic activity of nanozyme to amplify signals and improve the sensitivity. Therefore, the improvements of the catalytic activity and stability of nanozymes will be the key problems in nanozyme-based cancer diagnosis. According to the characteristics of nanozymes, improvements of nanozymes catalytic activities can be achieved as follows: changing the size, shape, and structure of nanozymes, making surface coating and modification, forming hybrids or developing activators and inhibitors of nanozymes [79–84].

Second, tumor-targeting ability of nanozyme probes is the precondition of accurate tumor detection. Target molecules such as antibody, polypeptide, small molecule, or aptamer are typically modified on the surface of nanozymes via chemical conjugation. However, these chemical modifications may raise nonspecific adsorption issues [85–88], which affect the specificity of the nanozyme probes. Moreover, the complex conjugation procedures increase the difficulty of clinical translation. Many effective ways have already been reported to solve these issues, such as coating nanozymes with octadecyl or carboxylate groups that could decrease nonspecific binding [85–88]; employing the intrinsic tumor-targeting molecules (e.g., ferritin) as a template to synthesize nanozymes will reduce the nonspecific adsorption and improve the accuracy of tumor diagnosis and possibility of clinical translation [46, 47, 60].

Third, taking advantage of the multifunctional properties of nanozymes, including their enzymatic activities and nanoscale physicochemical properties such as magnetics, optics, and photothermy, scientists can design dual-functional or multifunctional tumor diagnosis nanoprobes, which will expand the use of nanozymes in cancer diagnosis and other applications.

Finally, given the excellent performance of nanozymes in the cancer diagnosis research, more efforts should be devoted to the development of nanozyme products. Preclinical study of the stability, specificity, and sensitivity of nanozyme-based tumor diagnosis methods in clinical application should be determined. Importantly, a complete production process, manufacturing procedures, and quality standards should be developed to achieve the product translation of nanozymes. We are optimistic about that nanozymes will be translated from research in laboratory to the market and be utilized for the improvement of human health and quality of life in the next few years.

13.5.2 Future Perspectives on Nanozymes in Tumor Therapy

In the last decade, nanozymes have been widely used in biomedical applications including immunoassays, biosensors, antibacterial, and antibiofilm agents [3, 89]. Currently, considerable efforts are being made to explore the feasibility of applying nanozymes to in vivo clinical diagnosis and therapy [56, 59]. Great interests have been focused on the field of nanozymes in tumor theranostics owing to their excellent enzymatic activities and unique physicochemical characteristics. Taking advantage of enzymatic chemistry, nanomedicine, bionanotechnology, and oncology, nanozymes have shown great potential in combating tumors [56].

Despite the remarkable progress that has been made in recent years, the nanozyme research for tumor therapy is still in its infancy. Therefore, to increase the possibility of clinical translation of nanozymes, several unsolved critical issues need to be addressed.

First, nanozymes typically exhibit multi-enzymatic activities [59]. A critical question for in vivo applications of nanozymes in tumor therapy is that how to control a nanozyme to selectively perform the desired activity, as the off-target activity leads to counter-productive to the main desired activity. Recently, Yan and coworkers developed an in vivo ferritin-guided nanozyme for tumor catalytic therapy strategy [59], which provides a new idea for selectively activating the enzymatic activity of nanozymes. More strategies to effectively regulate the enzymatic activity of nanozymes in vivo are in urgent demand in this field.

Second, there are six types of natural enzymes in organism that are capable of catalyzing almost all the important reactions. However, the currently employed nanozymes for tumor therapy are mainly redox nanozymes (i.e., catalase, peroxidase, oxidase, SOD mimics). Thus, more efforts should be devoted to design novel types of nanozymes that mimic other diversified enzyme catalysis for effective antitumor in the near future.

Third, the activities of natural enzymes are rationally regulated by biological systems, which inspire us to develop various strategies to tune the activities of nanozymes for optimized tumor therapeutic outcomes. The enzymatic activities of nanozymes can be regulated by controlling its surrounding environment [58] and its interaction with specific ligands [3, 90, 91]. Many biological components, including ATP [90], DNA [92], and amino acids [58, 93, 94], have been reported to affect the enzymatic activity of nanozymes. Triggered by specific biological components via rational design, either regulating the enzymatic activity of nanozymes or making nanozyme involving in specific biological process will render nanozymes potentially to be powerful tools for cancer theranostics.

Fourth, to rationally design nanozymes with high enzymatic activity and tumor-targeting performances for tumor therapy, the influences on catalytical activities of nanozymes should be more comprehensively investigated. The tumor-targeting property and enzymatic activity of nanozymes should be coordinating with each other when developing nanozyme-based tumor therapy strategy. The catalytic reactions typically occur on the surface of nanozymes, while the tumor-targeting modification would change the property of nanozyme surface. Thus, how to balance the enzymatic activities and tumor-targeting property is also a challenge for nanozyme-based tumor therapy.

Finally, also the most important point, the biosafety of nanozymes used for tumor therapy should be comprehensively and systematically investigated. The nanozymes developed for in vivo tumor therapy must meet the strict safety and efficacy requirements of the regulatory agencies. However, most of the reported nanozymes used in the tumor therapy lack their acute and long-term biosafety evaluation. Thus, for the design of nanozyme-based tumor therapy, the biodegradable and biocompatible nanomaterials are the first choice. Then, more systematical investigations should be taken into consideration while evaluating a nanozyme-based tumor therapy strategy.

In conclusion, with the development of nanozyme field, great progress has been achieved in nanozyme-based tumor theranostics. However, in order to make nanozymes translate from research in laboratory to the market, more efforts are needed from the interdisciplinary fields of biology, nanotechnology, medicine, and material science. We are positive about that nanozymes will be utilized to improve human health and quality of life in the near future.

Acknowledgements This work was supported in part by the Key Research Program of Frontier Sciences, CAS (Grant No. QYZDB-SSW-SMC013), National Key R&D Program of China (2017YFA0205501), the National Natural Science Foundation of China (Grant No. 31530026), the Strategic Priority Research Program, CAS (Grant Nos. XDB29040101, XDA09030306), Young Elite Scientist Sponsorship Program by CAST (2015QNRC001), China Postdoctoral Science Foundation (Grant No. 2015M570158), and the China Postdoctoral Science Special Foundation (Grant No. 2016T90143).

References

1. Gao L, Zhuang J, Nie L, Zhang J, Zhang Y, Gu N, Wang T, Feng J, Yang D, Perrett S, Yan X (2007) Intrinsic peroxidase-like activity of ferromagnetic nanoparticles. Nat Nanotechnol 2(9):577–583. https://doi.org/10.1038/nnano.2007.260
2. Wei H, Wang E (2013) Nanomaterials with enzyme-like characteristics (nanozymes): next-generation artificial enzymes. Chem Soc Rev 42(14):6060–6093. https://doi.org/10.1039/C3CS35486E
3. Gao L, Fan K, Yan X (2017) Iron oxide nanozyme: a multifunctional enzyme mimetic for biomedical applications. Theranostics 7(13):3207–3227. https://doi.org/10.7150/thno.19738
4. Gao L, Yan X (2016) Nanozymes: an emerging field bridging nanotechnology and biology. Sci China Life Sci 59(4):400–402. https://doi.org/10.1007/s11427-016-5044-3
5. Vernekar AA, Das T, Ghosh S, Mugesh G (2016) A remarkably efficient $MnFe_2O_4$-based oxidase nanozyme. Chemistry 11(1):72–76. https://doi.org/10.1002/asia.201500942
6. Chen Z, Yin J-J, Zhou Y-T, Zhang Y, Song L, Song M, Hu S, Gu N (2012) Dual enzyme-like activities of iron oxide nanoparticles and their implication for diminishing cytotoxicity. ACS Nano 6(5):4001–4012. https://doi.org/10.1021/nn300291r
7. Dong J, Song L, Yin JJ, He W, Wu Y, Gu N, Zhang Y (2014) Co(3)O(4) nanoparticles with multi-enzyme activities and their application in immunohistochemical assay. ACS Appl Mater Interfaces 6(3):1959–1970. https://doi.org/10.1021/am405009f
8. Meng XQ, Fan KL (2018) Application of nanozymes in disease diagnosis. Prog Biochem Biophys 45(2):218–236. https://doi.org/10.16476/j.pibb.2018.0039
9. He WW, Wamer W, Xia QS, Yin JJ, Fu PP (2014) Enzyme-like activity of nanomaterials. J Environ Sci Heal C 32(2):186–211. https://doi.org/10.1080/10590501.2014.907462
10. Behrens MM, Ali SS, Dao DN, Lucero J, Shekhtman G, Quick KL, Dugan LL (2007) Ketamine-induced loss of phenotype of fast-spiking interneurons is mediated by NADPH-oxidase. Science 318(5856):1645–1647. https://doi.org/10.1126/science.1148045
11. Dugan LL, Tian LL, Quick KL, Hardt JI, Karimi M, Brown C, Loftin S, Flores H, Moerlein SM, Polich J, Tabbal SD, Mink JW, Perlmutter JS (2014) Carboxyfullerene neuroprotection postinjury in parkinsonian nonhuman primates. Ann Neurol 76(3):393–402. https://doi.org/10.1002/ana.24220
12. Linehan WM, Bratslavsky G, Pinto PA, Schmidt LS, Neckers L, Bottaro DP, Srinivasan R (2010) Molecular diagnosis and therapy of kidney cancer. Annu Rev Med 61:329–343

13. Maji SK, Mandal AK, Nguyen KT, Borah P, Zhao Y (2015) Cancer cell detection and therapeutics using peroxidase-active nanohybrid of gold nanoparticle-loaded mesoporous silica-coated graphene. ACS Appl Mater Interfaces 7(18):9807–9816. https://doi.org/10.1021/acsami.5b01758
14. Woo M-A, Kim MI, Jung JH, Park KS, Seo TS, Park HG (2013) A novel colorimetric immunoassay utilizing the peroxidase mimicking activity of magnetic nanoparticles. Int J Mol Sci 14(5):9999–10014. https://doi.org/10.3390/ijms14059999
15. He XW, Ma N (2014) A general strategy for label-free sensitive DNA detection based on quantum dot doping. Anal Chem 86(7):3676–3681
16. Wang Z, Yang X, Feng J, Tang Y, Jiang Y, He N (2014) Label-free detection of DNA by combining gated mesoporous silica and catalytic signal amplification of platinum nanoparticles. Analyst 139(23):6088–6091. https://doi.org/10.1039/c4an01539h
17. Paul A, Paul S (2014) The breast cancer susceptibility genes (BRCA) in breast and ovarian cancers. Front Biosci-Landmrk 19(4):605–618
18. Kulasingam V, Diamandis EP (2008) Strategies for discovering novel cancer biomarkers through utilization of emerging technologies. Nat Clin Pract Oncol 5(10):588–599
19. Kim MI, Ye Y, Woo MA, Lee J, Park HG (2014) A highly efficient colorimetric immunoassay using a nanocomposite entrapping magnetic and platinum nanoparticles in ordered mesoporous carbon. Adv Healthc Mater 3(1):36–41. https://doi.org/10.1002/adhm.201300100
20. Gao Z, Xu M, Lu M, Chen G, Tang D (2015) Urchin-like (gold core)@(platinum shell) nanohybrids: a highly efficient peroxidase-mimetic system for in situ amplified colorimetric immunoassay. Biosens Bioelectron 70:194–201. https://doi.org/10.1016/j.bios.2015.03.039
21. Gao Z, Xu M, Hou L, Chen G, Tang D (2013) Magnetic bead-based reverse colorimetric immunoassay strategy for sensing biomolecules. Anal Chem 85(14):6945–6952. https://doi.org/10.1021/ac401433p
22. Fu X-M, Liu Z-J, Cai S-X, Zhao Y-P, Wu D-Z, Li C-Y, Chen J-H (2016) Electrochemical aptasensor for the detection of vascular endothelial growth factor (VEGF) based on DNA-templated Ag/Pt bimetallic nanoclusters. Chin Chem Lett 27(6):920–926. https://doi.org/10.1016/j.cclet.2016.04.014
23. Zheng AX, Cong ZX, Wang JR, Li J, Yang HH, Chen GN (2013) Highly-efficient peroxidase-like catalytic activity of graphene dots for biosensing. Biosens Bioelectron 49(11):519–524. https://doi.org/10.1016/j.bios.2013.05.038
24. Shamsipur M, Safavi A, Mohammadpour Z (2014) Indirect colorimetric detection of glutathione based on its radical restoration ability using carbon nanodots as nanozymes. Sens Actuat B Chem 199(4):463–469. https://doi.org/10.1016/j.snb.2014.04.006
25. Song Y, Xia X, Wu X, Wang P, Qin L (2014) Integration of platinum nanoparticles with a volumetric bar-chart chip for biomarker assays. Angew Chem Int Ed Engl 53(46):12451–12455. https://doi.org/10.1002/anie.201404349
26. Li J, Cao Y, Hinman SS, McKeating KS, Guan Y, Hu X, Cheng Q, Yang Z (2018) Efficient label-free chemiluminescent immunosensor based on dual functional cupric oxide nanorods as peroxidase mimics. Biosens Bioelectron 100:304–311. https://doi.org/10.1016/j.bios.2017.09.011
27. Gao L, Liu M, Ma G, Wang Y, Zhao L, Yuan Q, Gao F, Liu R, Zhai J, Chai Z, Zhao Y, Gao X (2015) Peptide-conjugated gold nanoprobe: intrinsic nanozyme-linked immunsorbant assay of integrin expression level on cell membrane. ACS Nano 9(11):10979–10990
28. Bramwell VHC, Doig GS, Tuck AB, Wilson SM, Tonkin KS, Tomiak A, Perera F, Vandenberg TA, Chambers AF (2009) Changes over time of extracellular domain of HER2 (ECD/HER2) serum levels have prognostic value in metastatic breast cancer. Breast Cancer Res Tr 114(3):503–511
29. Estrela JM, Ortega A, Obrador E (2006) Glutathione in cancer biology and therapy. Crit Rev Cl Lab Sci 43(2):143–181
30. Ma YH, Zhang ZY, Ren CL, Liu GY, Chen XG (2012) A novel colorimetric determination of reduced glutathione in A549 cells based on Fe3O4 magnetic nanoparticles as peroxidase mimetics. Analyst 137(2):485–489

31. Gao J, Guo ZK, Su FJ, Gao L, Pang XH, Cao W, Du B, Wei Q (2015) Ultrasensitive electrochemical immunoassay for CEA through host-guest interaction of beta-cyclodextrin functionalized graphene and Cu@Ag core-shell nanoparticles with adamantine-modified antibody. Biosens Bioelectron 63:465–471
32. Xu YS, Zhang XP, Luan CX, Wang H, Chen BA, Zhao YJ (2017) Hybrid hydrogel photonic barcodes for multiplex detection of tumor markers. Biosens Bioelectron 87:264–270
33. Ylanne J, Cheresh DA, Virtanen I (1990) Localization of Beta-1-Subunit, Beta-3-Subunit, Alpha-5-Subunit, Alpha-V-Subunit and Alpha-Iib-Subunit of the integrin family in spreading human erythroleukemia-cells. Blood 76(3):570–577
34. Politaki E, Agelaki S, Apostolaki S, Hatzidaki D, Strati A, Koinis F, Perraki M, Saloustrou G, Stoupis G, Kallergi G, Spiliotaki M, Skaltsi T, Lianidou E, Georgoulias V, Mavroudis D (2017) A comparison of three methods for the detection of circulating tumor cells in patients with early and metastatic breast cancer. Cell Physiol Biochem 44(2):594–606
35. Sun C, Chen X, Xu J, Wei M, Wang J, Mi X, Wang X, Wu Y, Liu Y (2013) Fabrication of an inorganic–organic hybrid based on an iron-substituted polyoxotungstate as a peroxidase for colorimetric immunoassays of H_2O_2 and cancer cells. J Mater Chem A 1(15):4699–4705. https://doi.org/10.1039/C3TA01255G
36. Asati A, Kaittanis C, Santra S, Perez JM (2011) pH-tunable oxidase-like activity of cerium oxide nanoparticles achieving sensitive fluorigenic detection of cancer biomarkers at neutral pH. Anal Chem 83(7):2547–2553. https://doi.org/10.1021/ac102826k
37. Lee YC, Kim MI, Woo MA, Park HG, Han JI (2013) Effective peroxidase-like activity of a water-solubilized Fe-aminoclay for use in immunoassay. Biosens Bioelectron 42:373–378. https://doi.org/10.1016/j.bios.2012.10.092
38. Li J, Wang J, Wang Y, Trau M (2017) Simple and rapid colorimetric detection of melanoma circulating tumor cells using bifunctional magnetic nanoparticles. Analyst 142(24):4788–4793. https://doi.org/10.1039/c7an01102d
39. Wang GL, Xu XF, Qiu L, Dong YM, Li ZJ, Zhang C (2014) Dual responsive enzyme mimicking activity of AgX (X = Cl, Br, I) nanoparticles and its application for cancer cell detection. ACS Appl Mater Interfaces 6(9):6434–6442. https://doi.org/10.1021/am501830v
40. Zhang LN, Deng HH, Lin FL, Xu XW, Weng SH, Liu AL, Lin XH, Xia XH, Chen W (2014) In situ growth of porous platinum nanoparticles on graphene oxide for colorimetric detection of cancer cells. Anal Chem 86(5):2711–2718. https://doi.org/10.1021/ac404104j
41. Nagrath S, Sequist LV, Maheswaran S, Bell DW, Irimia D, Ulkus L, Smith MR, Kwak EL, Digumarthy S, Muzikansky A, Ryan P, Balis UJ, Tompkins RG, Haber DA, Toner M (2007) Isolation of rare circulating tumour cells in cancer patients by microchip technology. Nature 450(7173):1235–1239
42. Shi SR, Liu C, Pootrakul L, Tang L, Young A, Chen R, Cote RJ, Taylor CR (2008) Evaluation of the value of frozen tissue section used as "gold standard" for immunohistochemistry. Am J Clin Pathol 129(3):358–366
43. Taylor CR (2006) Standardization in immunohistochemistry: the role of antigen retrieval in molecular morphology. Biotech Histochem 81(1):3–12
44. O'Leary TJ (2001) Standardization in immunohistochemistry. Appl Immunohisto M M 9(1):3–8
45. Fan L, Tian Y, Yin R, Lou D, Zhang X, Wang M, Ma M, Luo S, Li S, Gu N, Zhang Y (2016) Enzyme catalysis enhanced dark-field imaging as a novel immunohistochemical method. Nanoscale 8(16):8553–8558. https://doi.org/10.1039/c5nr08232c
46. Fan K, Cao C, Pan Y, Lu D, Yang D, Feng J, Song L, Liang M, Yan X (2012) Magnetoferritin nanoparticles for targeting and visualizing tumour tissues. Nat Nanotechnol 7(7):459–464. https://doi.org/10.1038/nnano.2012.90
47. Fan K, Jia X, Zhou M, Wang K, Conde J, He J, Tian J, Yan X (2018) Ferritin nanocarrier traverses the blood brain barrier and kills Glioma. ACS Nano. https://doi.org/10.1021/acsnano.7b06969
48. Fan K, Zhou M, Yan X (2014) Novel properties and applications of human ferritin expressed by Escherichia coli. Microbiol China 41(3):520–538

49. Fan KL, Gao LZ, Yan XY (2013) Human ferritin for tumor detection and therapy. Wires Nanomed Nanobi 5(4):287–298
50. Cai Y, Cao CQ, He XQ, Yang CY, Tian LX, Zhu RX, Pan YX (2015) Enhanced magnetic resonance imaging and staining of cancer cells using ferrimagnetic H-ferritin nanoparticles with increasing core size. Int J Nanomed 10:2619–2634. https://doi.org/10.2147/ijn.s80025
51. Zhang T, Cao C, Tang X, Cai Y, Yang C, Pan Y (2017) Enhanced peroxidase activity and tumour tissue visualization by cobalt-doped magnetoferritin nanoparticles. Nanotechnology 28(4):045704. https://doi.org/10.1088/1361-6528/28/4/045704
52. Zhang W, Dong J, Wu Y, Cao P, Song L, Ma M, Gu N, Zhang Y (2017) Shape-dependent enzyme-like activity of Co_3O_4 nanoparticles and their conjugation with his-tagged EGFR single-domain antibody. Colloids Surf B Biointerfaces 154:55–62. https://doi.org/10.1016/j.colsurfb.2017.02.034
53. Dhar SK, St Clair DK (2012) Manganese superoxide dismutase regulation and cancer. Free Radic Biol Med 52(11–12):2209–2222. https://doi.org/10.1016/j.freeradbiomed.2012.03.009
54. Ragg R, Schilmann AM, Korschelt K, Wieseotte C, Kluenker M, Viel M, Volker L, Preiss S, Herzberger J, Frey H, Heinze K, Blumler P, Tahir MN, Natalio F, Tremel W (2016) Intrinsic superoxide dismutase activity of MnO nanoparticles enhances the magnetic resonance imaging contrast. J Mater Chem B 4(46):7423–7428. https://doi.org/10.1039/c6tb02078j
55. Yang F, Hu S, Zhang Y, Cai X, Huang Y, Wang F, Wen S, Teng G, Gu N (2012) A hydrogen peroxide-responsive O(2) nanogenerator for ultrasound and magnetic-resonance dual modality imaging. Adv Mater 24(38):5205–5211. https://doi.org/10.1002/adma.201202367
56. Yang BW, Chen Y, Shi JL (2018) Nanozymes in catalytic cancer theranostics. Prog Biochem Biophys 45(2):237–255. https://doi.org/10.16476/j.pibb.2017.0466
57. Kalyanaraman B (2017) Teaching the basics of cancer metabolism: Developing antitumor strategies by exploiting the differences between normal and cancer cell metabolism. Redox Biology 12:833–842. https://doi.org/10.1016/j.redox.2017.04.018
58. Fan K, Wang H, Xi J, Liu Q, Meng X, Duan D, Gao L, Yan X (2016) Optimization of Fe_3O_4 nanozyme activity via single amino acid modification mimicking an enzyme active site. Chem Commun (Camb) 53(2):424–427. https://doi.org/10.1039/c6cc08542c
59. Fan K, Xi J, Fan L, Wang P, Zhu C, Tang Y, Xu X, Liang M, Jiang B, Yan X, Gao L (2018) In vivo guiding nitrogen-doped carbon nanozyme for tumor catalytic therapy. Nat Commun 9(1):1440. https://doi.org/10.1038/s41467-018-03903-8
60. Liang M, Fan K, Zhou M, Duan D, Zheng J, Yang D, Feng J, Yan X (2014) H-ferritin–nanocaged doxorubicin nanoparticles specifically target and kill tumors with a single-dose injection. Proc Natl Acad Sci 111(41):14900–14905. https://doi.org/10.1073/pnas.1407808111
61. Wang ZZ, Zhang Y, Ju EG, Liu Z, Cao FF, Chen ZW, Ren JS, Qu XG (2018) Biomimetic nanoflowers by self-assembly of nanozymes to induce intracellular oxidative damage against hypoxic tumors. Nat Commun 9
62. Zhang D, Zhao Y-X, Gao Y-J, Gao F-P, Fan Y-S, Li X-J, Duan Z-Y, Wang H (2013) Anti-bacterial and in vivo tumor treatment by reactive oxygen species generated by magnetic nanoparticles. J Mater Chem B 1(38). https://doi.org/10.1039/c3tb20907e
63. Wason MS, Colon J, Das S, Seal S, Turkson J, Zhao JH, Baker CH (2013) Sensitization of pancreatic cancer cells to radiation by cerium oxide nanoparticle-induced ROS production. Nanomed-Nanotechnol 9(4):558–569
64. Agostinis P, Berg K, Cengel KA, Foster TH, Girotti AW, Gollnick SO, Hahn SM, Hamblin MR, Juzeniene A, Kessel D, Korbelik M, Moan J, Mroz P, Nowis D, Piette J, Wilson BC, Golab J (2011) Photodynamic therapy of cancer: an update. CA Cancer J Clin 61(4):250–281. https://doi.org/10.3322/caac.20114
65. Wen X, Li Y, Hamblin MR (2017) Photodynamic therapy in dermatology beyond non-melanoma cancer: an update. Photodiagnosis Photodyn Ther 19:140–152. https://doi.org/10.1016/j.pdpdt.2017.06.010
66. Kamkaew A, Chen F, Zhan Y, Majewski RL, Cai W (2016) Scintillating nanoparticles as energy mediators for enhanced photodynamic therapy. ACS Nano 10(4):3918–3935. https://doi.org/10.1021/acsnano.6b01401

67. Ethirajan M, Chen Y, Joshi P, Pandey RK (2011) The role of porphyrin chemistry in tumor imaging and photodynamic therapy. Chem Soc Rev 40(1):340–362. https://doi.org/10.1039/B915149B
68. Zhang Y, Wang F, Liu C, Wang Z, Kang L, Huang Y, Dong K, Ren J, Qu X (2018) Nanozyme decorated metal-organic frameworks for enhanced photodynamic therapy. ACS Nano 12(1):651–661. https://doi.org/10.1021/acsnano.7b07746
69. Ragg R, Tahir MN, Tremel W (2016) Solids Go Bio: inorganic nanoparticles as enzyme mimics. Eur J Inorg Chem 2016 (13–14):1906–1915. https://doi.org/10.1002/ejic.201501237
70. Chen HJ, Zhou XB, Gao Y, Zheng BY, Tang FX, Huang JD (2014) Recent progress in development of new sonosensitizers for sonodynamic cancer therapy. Drug Discov Today 19(4):502–509
71. Zhu P, Chen Y, Shi J (2018) Nanoenzyme-augmented cancer sonodynamic therapy by catalytic tumor oxygenation. ACS Nano. https://doi.org/10.1021/acsnano.8b00999
72. Huo M, Wang L, Chen Y, Shi J (2017) Tumor-selective catalytic nanomedicine by nanocatalyst delivery. Nat Commun 8(1):357. https://doi.org/10.1038/s41467-017-00424-8
73. Zhou J, Li M, Hou Y, Luo Z, Chen Q, Cao H, Huo R, Xue C, Sutrisno L, Hao L, Cao Y, Ran H, Lu L, Li K, Cai K (2018) Engineering of a nanosized biocatalyst for combined tumor starvation and low-temperature photothermal therapy. ACS Nano 12(3):2858–2872. https://doi.org/10.1021/acsnano.8b00309
74. Liu F, Lin L, Zhang Y, Wang Y, Sheng S, Xu C, Tian H, Chen X (2019) A tumor-microenvironment-activated nanozyme-mediated theranostic nanoreactor for imaging-guided combined tumor therapy. Adv Mater e1902885. https://doi.org/10.1002/adma.201902885
75. Li S, Shang L, Xu B, Wang S, Gu K, Wu Q, Sun Y, Zhang Q, Yang H, Zhang F, Gu L, Zhang T, Liu H (2019) Nanozyme with photo-enhanced dual enzyme-like activities for deep pancreatic cancer therapy. Angew Chem Int Ed Engl. https://doi.org/10.1002/anie.201904751
76. Zhang C, Ni DL, Liu YY, Yao HL, Bu WB, Shi JL (2017) Magnesium silicide nanoparticles as a deoxygenation agent for cancer starvation therapy. Nat Nanotechnol 12(4):378–386
77. Marx J (2003) A boost for tumor starvation (vol 301, pg 452, 2003). Science 302(5649):1333–1333
78. Galluzzi L, Bravo-San Pedro JM, Demaria S, Formenti SC, Kroemer G (2017) Activating autophagy to potentiate immunogenic chemotherapy and radiation therapy. Nat Rev Clin Oncol 14(4):247–258
79. Asati A, Santra S, Kaittanis C, Nath S, Perez JM (2009) Oxidase-like activity of polymer-coated cerium oxide nanoparticles. Angew Chem Int Edit 48(13):2308–2312
80. Wan Y, Qi P, Zhang D, Wu JJ, Wang Y (2012) Manganese oxide nanowire-mediated enzyme-linked immunosorbent assay. Biosens Bioelectron 33(1):69–74
81. Liu JB, Hu XN, Hou S, Wen T, Liu WQ, Zhu X, Yin JJ, Wu XC (2012) Au@Pt core/shell nanorods with peroxidase- and ascorbate oxidase-like activities for improved detection of glucose. Sens Actuat B Chem 166:708–714
82. Kong L, Cai X, Zhou XH, Wong LL, Karakoti AS, Seal S, McGinnis JF (2011) Nanoceria extend photoreceptor cell lifespan in tubby mice by modulation of apoptosis/survival signaling pathways. Neurobiol Dis 42(3):514–523
83. Singh S, Dosani T, Karakoti AS, Kumar A, Seal S, Self WT (2011) A phosphate-dependent shift in redox state of cerium oxide nanoparticles and its effects on catalytic properties. Biomaterials 32(28):6745–6753
84. Lee YM, Garcia MA, Huls NAF, Sun SH (2010) Synthetic tuning of the catalytic properties of Au-Fe_3O_4 Nanoparticles. Angew Chem Int Edit 49(7):1271–1274
85. Chen HW, Wang LY, Yeh J, Wu XY, Cao ZH, Wang YA, Zhang MM, Yang L, Mao H (2010) Reducing non-specific binding and uptake of nanoparticles and improving cell targeting with an antifouling PEO-b-P gamma MPS copolymer coating. Biomaterials 31(20):5397–5407
86. Bagwe RP, Hilliard LR, Tan W (2006) Surface modification of silica nanoparticles to reduce aggregation and nonspecific binding. Langmuir 22(9):4357–4362. https://doi.org/10.1021/la052797j

87. Grubbs RB (2007) Roles of polymer ligands in nanoparticle stabilization. Polym Rev 47(2):197–215
88. Doshi N, Mitragotri S (2009) Designer biomaterials for nanomedicine. Adv Funct Mater 19(24):3843–3854
89. Wang X, Hu Y, Wei H (2016) Nanozymes in bionanotechnology: from sensing to therapeutics and beyond. Inorganic Chemistry Frontiers 3(1):41–60. https://doi.org/10.1039/C5QI00240K
90. Vallabani NVS, Karakoti AS, Singh S (2017) ATP-mediated intrinsic peroxidase-like activity of Fe_3O_4-based nanozyme: one step detection of blood glucose at physiological pH. Colloids Surf B Biointerfaces 153:52–60. https://doi.org/10.1016/j.colsurfb.2017.02.004
91. Zhang Z, Zhang X, Liu B, Liu J (2017) Molecular imprinting on inorganic nanozymes for hundred-fold enzyme specificity. J Am Chem Soc 139(15):5412–5419. https://doi.org/10.1021/jacs.7b00601
92. Sun Y, Wang J, Li W, Zhang J, Zhang Y, Fu Y (2015) DNA-stabilized bimetallic nanozyme and its application on colorimetric assay of biothiols. Biosens Bioelectron 74:1038–1046. https://doi.org/10.1016/j.bios.2015.08.001
93. Liu Y, Ding D, Zhen Y, Guo R (2017) Amino acid-mediated 'turn-off/turn-on' nanozyme activity of gold nanoclusters for sensitive and selective detection of copper ions and histidine. Biosens Bioelectron 92:140–146. https://doi.org/10.1016/j.bios.2017.01.036
94. Sun Y, Zhao C, Gao N, Ren J, Qu X (2017) Stereoselective nanozyme based on ceria nanoparticles engineered with amino acids. Chem Eur J 23(72):18146–18150. https://doi.org/10.1002/chem.201704579

Chapter 14
Nanozymes for Therapeutics

Wen Cao, Zhangping Lou, Wenjing Guo and Hui Wei

Abbreviations

AD	Alzheimer's disease
CAL	Coronary artery ligature
CeNPs	Cerium oxide nanoparticles
CPC	Cardiac progenitor cells
DSMA	2, 3-dimercaptosuccinic acid
EPCs	Endothelial progenitor cells
GSH	Glutathione
HMSCs	Human mesenchymal stem cells
HSCs	Hepatic stellate cells
HUVECs	Human umbilical vein endothelial cells
MCP	Monocyte chemoattractant protein
MCT	Monocrotaline
MFCs	Myofibroblast-like cells
NOS	Nitric oxides synthase
PBNPs	Prussian blue nanoparticles
PD	Parkinson's disease
PLGA	Poly(lactic-co-glycolic acid)
RNS	Reactive nitrogen species
ROIs	Reactive oxygen intermediates
ROS	Reactive oxygen species
SOD	Superoxide dismutase
SuOx	Sulfite oxidase

W. Cao · Z. Lou · W. Guo · H. Wei (✉)
Nanjing National Laboratory of Microstructures, Jiangsu Key Laboratory of Artificial Functional Materials, Department of Biomedical Engineering, College of Engineering and Applied Sciences, Nanjing University, Nanjing 210093, Jiangsu, China
e-mail: weihui@nju.edu.cn
URL: http://weilab.nju.edu.cn

X. Yan (ed.), *Nanozymology*, Nanostructure Science and Technology,
https://doi.org/10.1007/978-981-15-1490-6_14

TPP	Triphenylphosphonium
VEGF	Vascular endothelial growth factor
Vn	V_2O_5 nanowires

Since the discovery of peroxidase-like iron oxide nanozymes [1], a large number of nanozymes have been developed for mimicking oxidase, peroxidase, superoxide dismutase (SOD), catalase, hydrolase, etc. [2, 3]. Among them, free radical scavenging nanozymes (mainly SOD-like nanozymes) have received particular attention due to their therapeutic applications. Free radicals play important roles in the physiological and pathological processes. A moderate amount of free radicals is useful whereas too much of them can induce oxidative stress and cause damage to DNA, protein, lipid, and then the cell. Therefore, many free radical scavenging enzymes and reagents have been developed to modulate the level of free radicals. Radical scavenging nanozymes are advantages over natural counterparts because of their high stability, ease of production, multi-functionality, and tunable activity. Herein, we will discuss the applications of nanozymes in varieties therapeutics associated with free radicals, including neuroprotection, cardioprotection, hepatoprotection, cytoprotection, cancer therapy, tissue engineering, anti-inflammation, and anti-aging.

14.1 Neuroprotection

Most neurodegenerative diseases, including retinal degeneration, brain ischemia, neurodegeneration, and autoimmune degenerative disease, are associated with excessive reactive oxygen species (ROS) production [4–7]. These ROS induce oxidative stresses to cell and trigger cell death. Based on their SOD- or catalase-like enzyme activities, many nanozymes have neuroprotection functions and are used to scavenge ROS for relieving oxidative stress.

Constant light exposure can cause harmful oxidative stress to photoreceptor cells [4]. To mitigate this harmful effect, in an early study, McGinnis et al. used nanoceria particles to protect rat retina photoreceptor cells from light-induced degeneration (Fig. 14.1). Their results demonstrated that nanoceria particles inhibited the increase of intracellular reactive oxygen intermediates (ROIs) concentration after adding H_2O_2 to rat retina cell, thus preventing the ROIs-induced cell death. After injection of nanoceria particles, they found that nanoceria effectively protected rat retina photoreceptor cells from light-induced degeneration [4].

After that, they also monitored the long-term impact of nanoceria by intravitreal injection with little nanoceria into tubby mice [8]. The untreated tubby mouse showed rapid retinal degeneration. A single dose nanoceria injection prolonged photoreceptor cell life-span and kept both retinal structure and function for more than one month. Furthermore, they showed that nanoceria, acting as an antioxidant, upregulated the antioxidant associated genes as well as decreasing the mislocalization of photoreceptor-specific proteins [8]. In another study, the authors chose a very

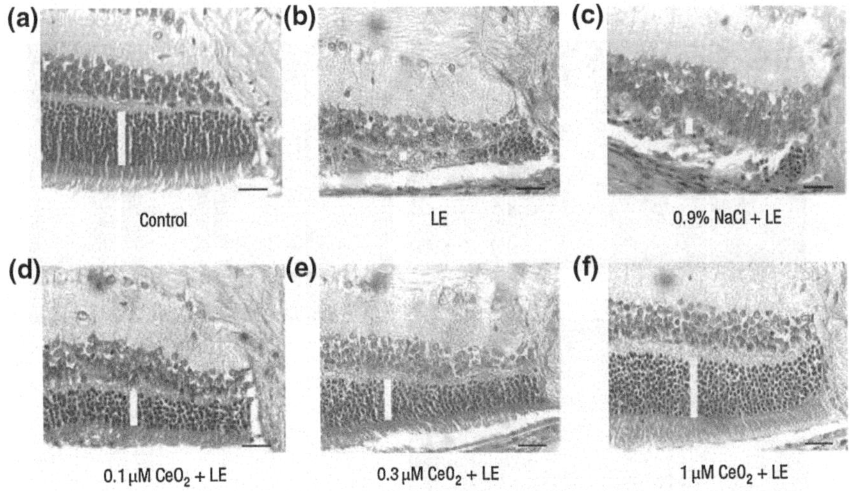

Fig. 14.1 Intravitreal injection of nanoceria particles protects rat retina photoreceptor cells from light-induced degeneration. Representative images of photomicrographs of H&E stained sections adjacent to the optic nerve are shown. The white bars indicate the thickness of the layer of nuclei of rods and cones (the outer nuclear layer, ONL). Injections were given on day 0, rats were exposed to light on day 3, and the experiment ended on day 10. **a** No light exposure (LE), no injection. **b–f** Rats were exposed to 6 h of 2,700 lux white light. No injection (**b**), injections of saline (**c**), 0.1 M CeO_2 (**d**), 0.3 M CeO_2 (**e**), and 1.0 M CeO_2 (**f**) 3 days before LE. Reprinted with permission from [4]. Copyright (2006) Nature Publishing Group

low-density lipoprotein receptor knockout (*vldlr*$^{-/-}$) mouse as a model to study the age-related macular degeneration. By intravitreal injection, nanoceria exhibited sustained inhibition of neovascularization (Fig. 14.2) [9]. Nanoceria alleviated oxidative stress, resulting in suppressing the expression of vascular endothelial growth factor (VEGF). It also demonstrated that nanoceria-inhibited vessel leakage and suppressed cell apoptosis by a way of down-regulating the ASK1-P38/JNK-NF-κB signaling pathway [9]. In their later study, McGinnis et al. used P23H-1 rat, a photoreceptor degeneration model, to research the cellular mechanism and duration of catalytic activity of nanoceria. At least within 7 days, nanoceria kept its high catalytic activity after injecting into rats. The catalytic activity of nanoceria started to decrease until the next 14 days (Fig. 14.3). They validated that nanoceria slowed down the cell death and delayed the disease progression [10].

Besides the application in preventing retinal degeneration, numerous studies demonstrated that nanozymes had great effects on brain ischemia. Brain ischemia is a common disease with a high fatality rate. During the ischemia process, ROS is rapidly produced and accumulated, which induces oxidative damage [5, 11–13]. Hickman et al. first isolated spinal cord cells from the whole cord of adult rats. Then they added hydrogen peroxide to the cell cultures to induce oxidative injury.

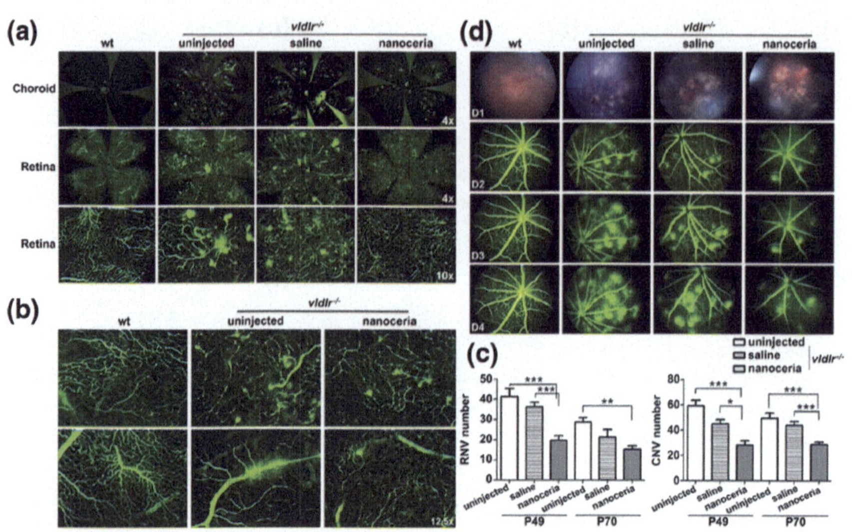

Fig. 14.2 Nanoceria regress existing neovascularization and prevent vascular leakage in $vldlr^{-/-}$ mice. **a** Uninjected $vldlr^{-/-}$ eyes at P49 exhibit numerous choroidal neovascular "tufts" (top) and intraretinal neovascular "blebs" (bottom two panels) compared to wt which have no fluorescein labeling in the choroid and have uniform organized retinal blood vessels. Nanoceria treatment markedly reduced the number of "tufts" and "blebs". **b** High magnification of the retina reveals the leakage of both the neovasculature (top) and the original larger vessels (bottom) in untreated $vldlr^{-/-}$ retina compared to wt. However, nanoceria injection not only reduced the "blebs" number but also prevented the larger vessel leakage. Representative images from 10 to 22 eyes per group are shown. **c** The quantitation of the RNV and CNV number ($N = 10$–22 eyes per group) is presented and shown as mean ± SD. $*P < 0.05$, $**P < 0.001$, $***P < 0.0001$. **d** Fundoscopy at P49 revealed that untreated $vldlr^{-/-}$ mice exhibit numerous fundus patches of neovascularization (D1) which were highlighted by fluorescent angiography (D2, D3, and D4). The blood vessel leakage in the uninjected group was progressive and severe over extended times. Saline treatment did not affect the leakage. However, nanoceria injection remarkably reduced neovascularization and neovascular leakage. Reprinted with permission from [9]. Copyright (2014) Elsevier

Obviously, a single dose of nanoceria at nanomolar concentration (10 nM) improved the cell viability and possessed great neuroprotective effects on adult rat spinal cord neurons (Fig. 14.4) [11]. Hyeon et al. validated the repairing function of nanoceria for ischemia injury in living animals by carefully studying the brain slices treated with different amounts of nanoceria. Consistent with previous results, nanoceria was effective in inhibiting cell death by decreasing ROS, resulting in the reduction of infarct volume (Fig. 14.5) [5]. In another study, using a mouse hippocampal brain slice model of ischemia, Estevez et al. showed that nanoceria reduced almost 50% ischemia cell death. The concentration of superoxide and nitric oxide was decreased by 15% and ischemia-induced 3-nitrotyrosine, which played an important role in the dissemination of oxidative injury, was significantly decreased by 70%. Thus, nanoceria was a promising therapeutic drug for stroke by scavenging both ROS and

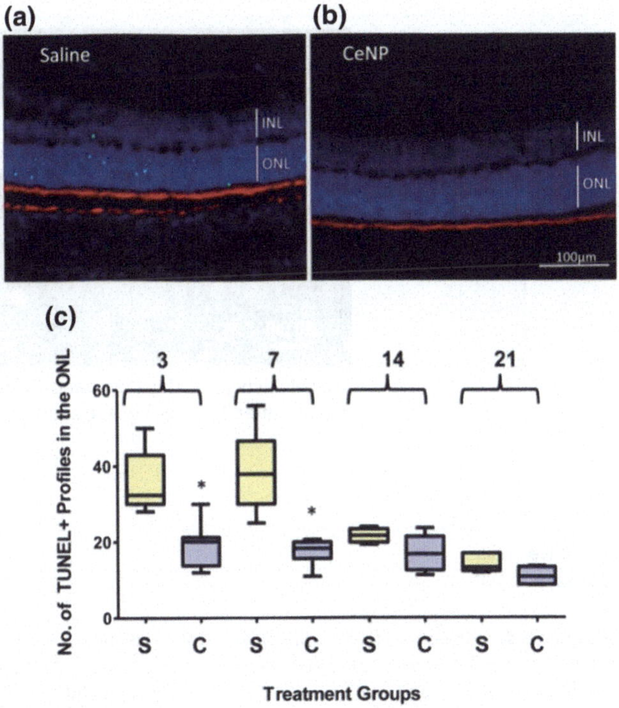

Fig. 14.3 A single application of CeNPs at P15 reduced the number of apoptotic death of photoreceptor cells for at least 21 days. **a–b** Representative photomicrographs of retinal sections from 3 dpi group with examples of TUNEL+ profiles (green). Nuclei in the ONL and INL are labeled blue. Rod outer segments are labeled red with Rhodopsin, 1D4, antibody. CeNPs-treated animals have significantly fewer TUNEL+ profiles in the ONL compared to saline-treated ones. **c** Quantification of TUNEL+ profiles in the ONL from 3, 7, 14, and 21 dpi groups. The reductions from 3 and 7 dpi groups of CeNPs injected animals were highly significant. Reprinted with permission from [10]. Copyright (2015) PLOS

nitric oxides synthase (NOS) [12]. Yang et al. utilized silica-coated superparamagnetic iron oxide nanoparticles (SiO_4@SPIONs) to label endothelial progenitor cells (EPCs). The labeled EPCs could be guided to the ischemic hemisphere of the brain with the help of external magnetic field, which significantly enhanced the therapeutic outcomes (Fig. 14.6) [13].

In addition to nanoceria, Fe_3O_4 was studied for its ROS scavenging capacity on *Drosophila* based on its catalase-like catalytic activity. These nanoparticles could alleviate oxidative stress induced by H_2O_2. Thus, they were used on two different neurodegenerative disease models, Parkinson's disease (PD) and Alzheimer's disease (AD). The climbing ability and longevity studies of *Drosophila* indicated that daily ingestion of Fe_3O_4 nanoparticles can lighten neurodegeneration both in PD and AD models (Fig. 14.7) [14].

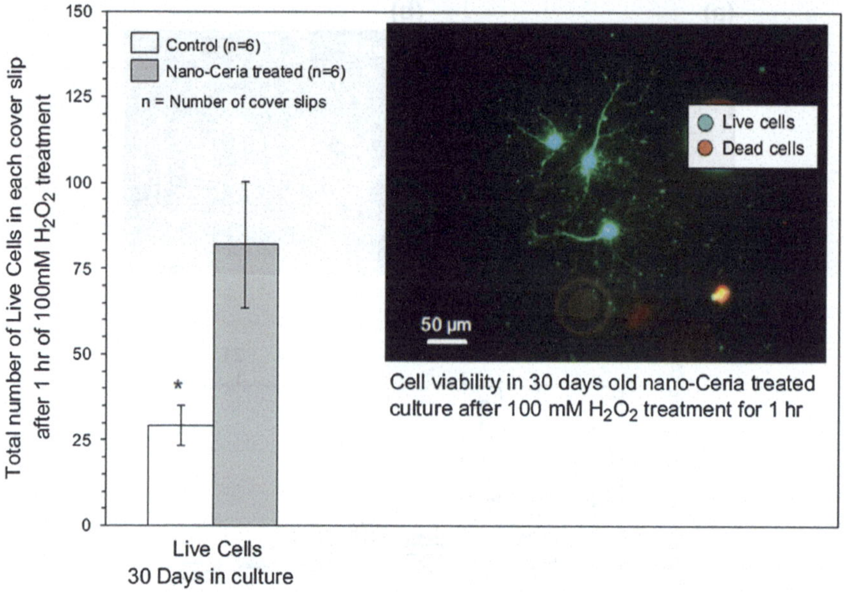

Fig. 14.4 Results after hydrogen peroxide-induced oxidative injury in control and treated cultures of adult rat spinal cord at day 30. Neuron–glial cell assays indicated a significantly high neuronal survival in treated cultures at day 15 and day 30 as compared to the control cultures. Reprinted with permission from [11]. Copyright (2007) Elsevier

Thanks to the intrinsic post-modification feature of nanozymes, nanoceria was functionalized with triphenylphosphonium (TPP) to enhance specific targeting to mitochondria. Such TPP-conjugated nanoceria effectively targeted mitochondria and thus inhibited neuronal death in an AD disease mouse model. Reactive gliosis and morphological mitochondria damage in lesion location can be seen greatly alleviated (Fig. 14.8) [6].

14.2 Cardioprotection

Similar to neuroprotection, nanozymes with ROS scavenging ability have been proved to be effective on cardioprotection [15–18]. Cardiac progenitor cells (CPCs) therapy is an emerging strategy for cardiac regenerative medicine. Traversa et al. showed that nanoceria could be internalized by CPCs and stayed inside the cytoplasm as the aggregated particles. Meanwhile, nanoceria did not affect the morphology, growth, and differentiation of CPCs. After adding H_2O_2 to mimic oxidative insults, nanoceria protected CPCs from ROS-induced cytotoxicity for at least 7 days (Fig. 14.9) [18]. Kolattukudy and coworkers treated monocyte chemoattractant protein (MCP)-1 transgenic murine model with nanoceria. Cardiac-specific expression

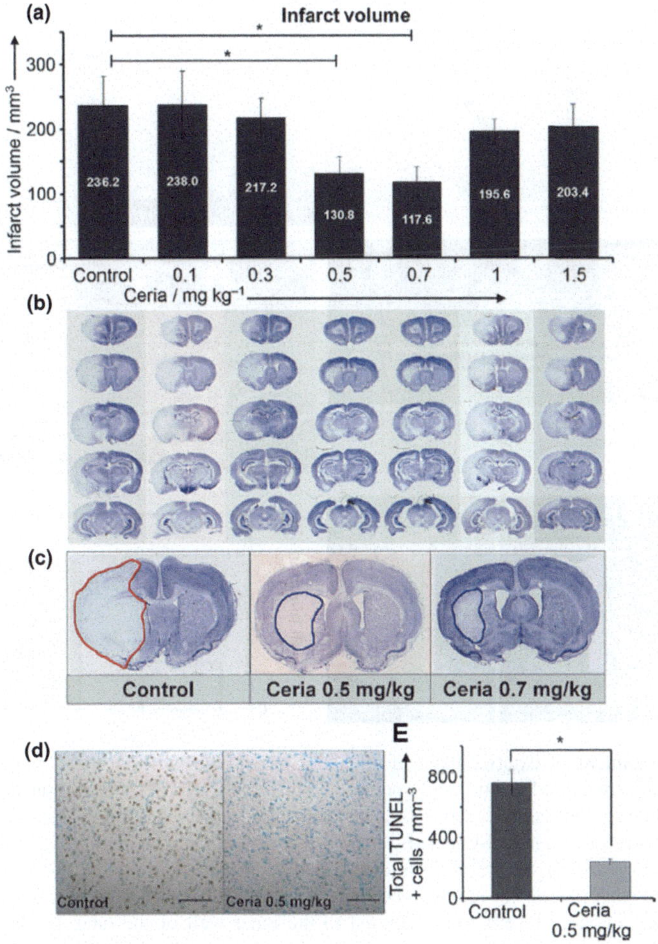

Fig. 14.5 Infarct volume and ischemic cell death in vivo. **a** Low-dose ceria nanoparticles (0.1 and 0.3 mg kg^{-1}) do not decrease infarct volumes, whereas 0.5 and 0.7 mg kg^{-1} ceria nanoparticles considerably reduce infarct volumes, to as little as 50 % of those of the control group (*, $p < 0.05$). Higher doses of ceria nanoparticles (1.0 and 1.5 mg kg^{-1}) do not exhibit protective effects against stroke ($n = 12$ for each group, except 0.1 and 1.5 mg kg^{-1}, where $n = 6$). **b** Brain slices from anterior (top) to posterior (bottom), with intervals of 2 mm. On Nissl-stained brains, infarcts are shown as pale blue-colored lesions while undamaged region are stained as deep blue. Infarct areas were maximally decreased at 0.5 and 0.7 mg kg^{-1} ceria nanoparticles. **c** Representative slices, clearly showing that 0.5 and 0.7 mg kg^{-1} ceria nanoparticles can significantly reduce infarct volumes. **d** Microscopic analysis of cell death in brain slices using TUNEL. TUNEL-positive cells are shown in brown, and TUNEL-negative cells, which were counter-stained with methyl green, are shown in blue. The number of TUNEL-positive cells was reduced in the ceria-injected group (0.5 mg kg^{-1}). Scale bar = 100 μm. **e** In our quantitative analysis, the number of TUNEL-positive cells decrease markedly in the ceria-injected group (*, $P < 0.05$; $n = 4$ each). Reprinted with permission from [5]. Copyright (2012) John Wiley & Sons

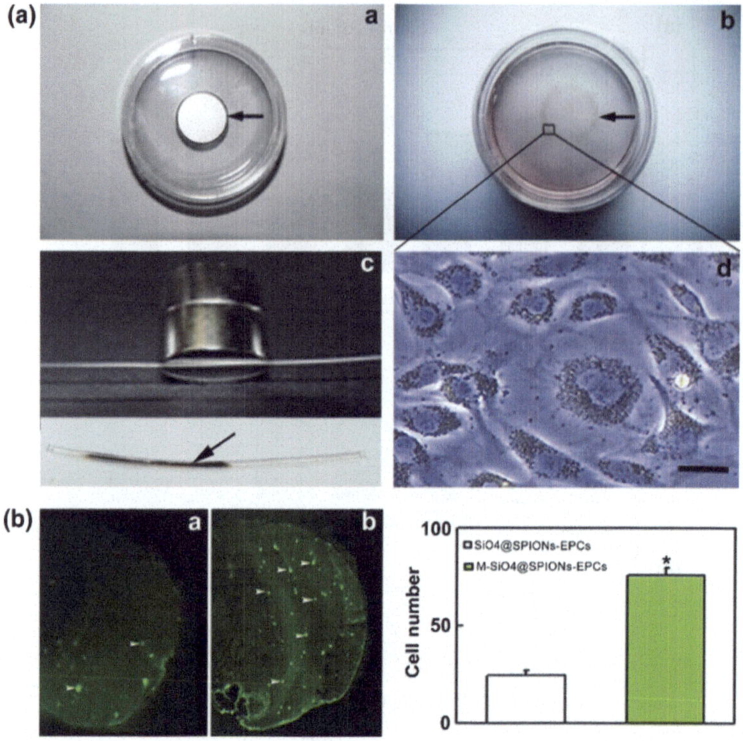

Fig. 14.6 Treatment of exterior magnetic field increased SIO_4@SPIONs-EPCs accumulation in vitro and in vivo. **A** Magnetic field applied in SIO_4@SPIONs-EPCs attraction in vitro. The picture exhibited a cell culture dish with a magnet placed under the bottom center of the dish for 24 h (**a**, arrow). A circle formed (**b**, arrow), and the rectangular region was magnified in (**d**), which showed clear SIO_4@SPIONs-EPCs. Bar = 20 μm. The picture exhibited a PE-50 tube with a magnet placed outside the tube for 1 min (**c**). SIO_4@SPIONs-EPCs flowed through tube and large amount of cells were captured to the inner wall of the tube. **B** Magnetic field applied in SIO_4@SPIONs-EPCs attraction in vivo. Images showed that GFP-SIO_4@SPIONs-EPCs (green, arrowheads) in ischemic hemisphere with magnetic field treatment (**b**) and without (**a**). Bar graph exhibited amount of cells homing in ischemic hemisphere. SIO_4@SPIONs-EPCs = MCAO mice with SIO_4@SPIONs-EPCs transplantation; M-SIO_4@SPIONs-EPCs = MCAO mice with both treatment of magnetic field and SIO_4@SPIONs-EPCs transplantation. *, $P < 0.05$, M-SIO_4@SPIONs-EPCs versus SIO_4@SPIONs-EPCs. Data is mean ± SD, $n = 3$ in each group. Reprinted with permission from [13]. Copyright (2013) Elsevier

of MCP-1 caused ischemic cardiomyopathy, coming with endoplasmic reticulum (ER) stress. The treatment of MCP-1 mice with nanoceria induced the decrease of MCP-1 and suppressed oxidative stress, ER stress, and inflammation. Nanoceria successfully protected MCP-1 mice against progressive left ventricular dysfunction [15].

Cardiac hypertrophy is associated with oxidative stress and apoptosis. Blough and coworkers used monocrotaline (MCT) to induce right ventricular hypertrophy

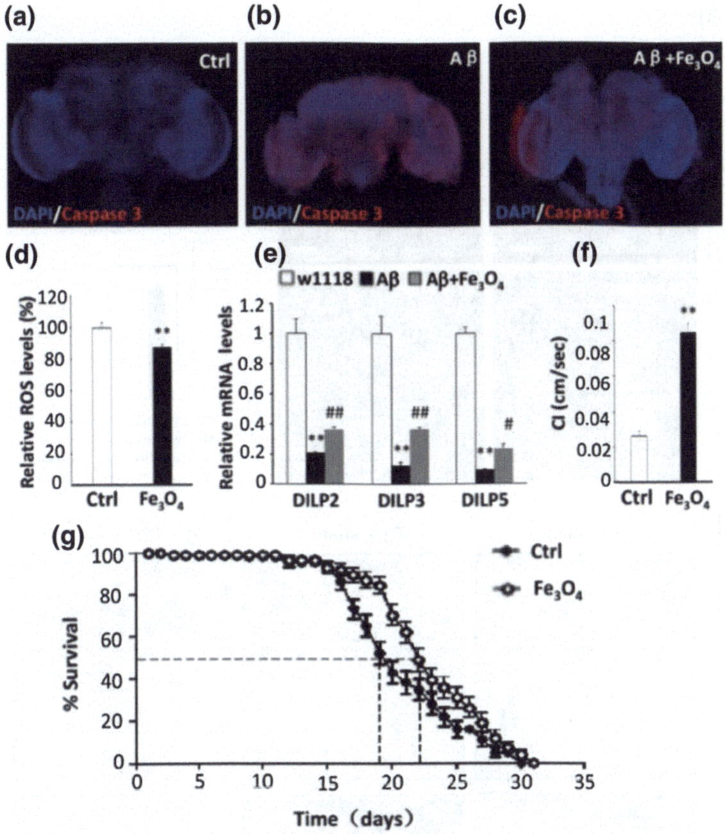

Fig. 14.7 Effects of dietary Fe_3O_4 NPs in a Drosophila AD model. **a–c** Detection of apoptosis by immunostaining in the brain of AD flies fed with control food, or food containing Fe_3O_4 NPs. *w1118* flies were used as control. **d** ROS levels of 16-day-old AD flies fed with control food, or food containing Fe_3O_4 NPs. Data represented as mean ± S.D. ($n = 3$). Student's *t*-test, $**P < 0.01$. **e** Relative DILPs expression levels in *w1118* flies, AD flies or AD flies fed with Fe_3O_4 NPs. Data represented as mean ± S.D. ($n = 3$). Student's *t*-test, AD flies compared to *w1118*: $**P < 0.01$; AD flies fed with Fe_3O_4 NPs compared to AD flies: $\#\#P < 0.01$, $\#P < 0.05$. **f** Climbing abilities of 16-day-old AD flies fed with control food, or food containing Fe_3O_4 NPs. Data represented as mean ± S.D. ($n = 3$). Student's *t*-test, $**P < 0.01$. **g** Life span of AD flies fed with control food, or food containing Fe_3O_4 NPs. Log rank test, 200 μg mL^{-1} Fe_3O_4 NPs ($n = 84$) versus control ($n = 84$): $P < 0.01$. Reprinted with permission from [14]. Copyright (2016) John Wiley & Sons

following pulmonary arterial hypertension in rat models. They observed that the administration of nanoceria increased pulmonary flow and decreased pulmonary artery remodeling, as well as reduced right ventricle wall thickness and cardiac remodeling (Fig. 14.10) [17]. Besides nanoceria, iron oxide also exhibited great cardioprotection functions both in vitro and in vivo. Gu et al. reported that the protection activity of 2, 3-dimercaptosuccinic acid (DMSA) modified Fe_2O_3 nanoparticles could only affected by their size, not by the surface molecules or charge. Compared with

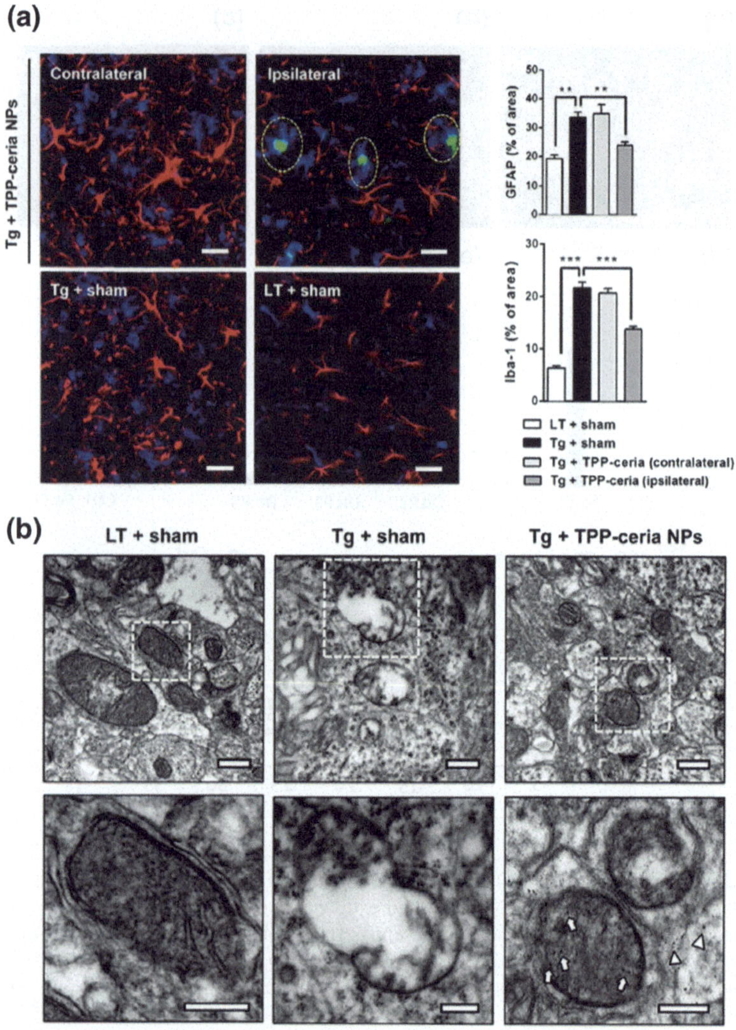

Fig. 14.8 TPP-ceria NPs reduce reactive glial activation and restore mitochondrial morphology. **a** Confocal fluorescence images (left) of gliosis in tissue sections co-labeled with GFAP (red) and Iba-1 (blue). Quantified levels of GFAP and Iba-1 in the images ($n = 4$ per group). Statistical analysis was performed using an ANOVA test. Error bars represent 95% CIs. $^{**}P < 0.01$; $^{***}P < 0.001$; LT + sham: littermate mice; Tg + sham: 5XFAD mice. Scale bar = 30 μm. **b** TEM images showing representative mitochondrial morphologies of LT + sham, Tg + sham, and Tg + TPP-ceria NPs. Scale bar = 500 nm ($n = 4$ per group). Magnified images of the boxed areas are shown below. Arrows and arrowheads indicate TPP-ceria NPs in mitochondrial matrix and cytosol, respectively. Scale bar = 250 nm. Reprinted with permission from [6]. Copyright (2016) American Chemical Society

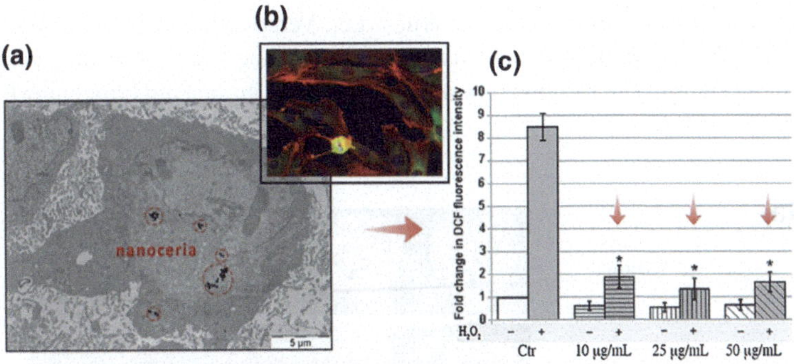

Fig. 14.9 **a** TEM image of nanoceria-treated CPCs. **b** Effect of nanoceria on CPCs morphology. **c** ROS production in CPCs exposed to H_2O_2. Reprinted with permission from [18]. Copyright (2012) American Chemical Society

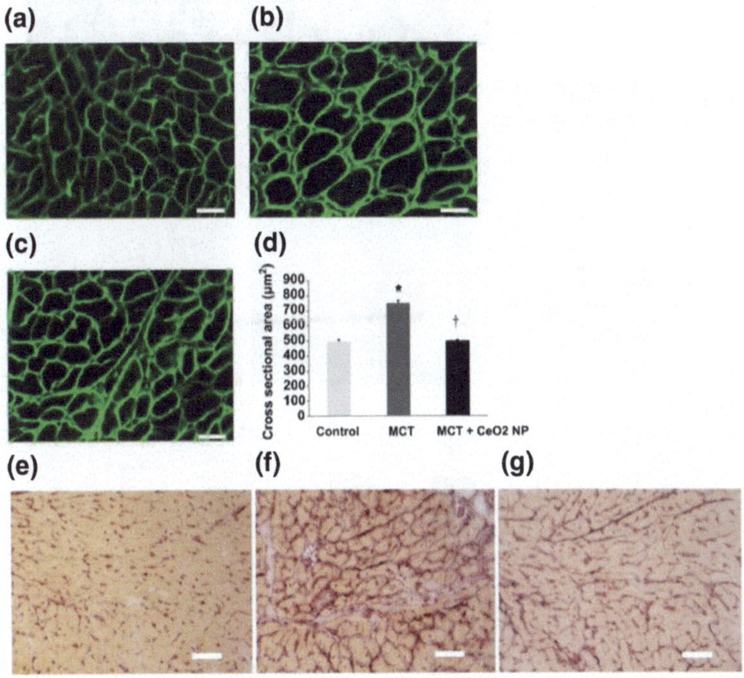

Fig. 14.10 CeO_2 nanoparticle administration attenuates monocrotaline-induced increases in cardiomyocyte cross-sectional area and cardiac fibrosis. Dystrophin-stained right ventricular sections from control (**a**), MCT only (**b**), and MCT + CeO_2 nanoparticle treatment group (**c**). Quantification of cardiomyocyte cross-sectional area (**d**). Picrosirius red staining was used to evaluate cardiac fibrosis in the right ventricles of control (**e**), MCT only (**f**), and MCT + CeO_2 nanoparticle treatment groups (**g**). Scale bar = 50 µm. Data are mean ± SEM ($n = 3$ rats/group). * Significantly different from control. † Significantly different from the MCT only group ($P < 0.05$). Reprinted with permission from [17]. Copyright (2014) Elsevier

two extensively used cardioprotective drugs (i.e., Verapamil and *Salvia miltiorrhiza* extract), Fe_2O_3 nanoparticles showed better performance in coronary artery ligature (CAL) induced injury in rats and reperfusion process in guinea pig Langendorff heart (Fig. 14.11) [15].

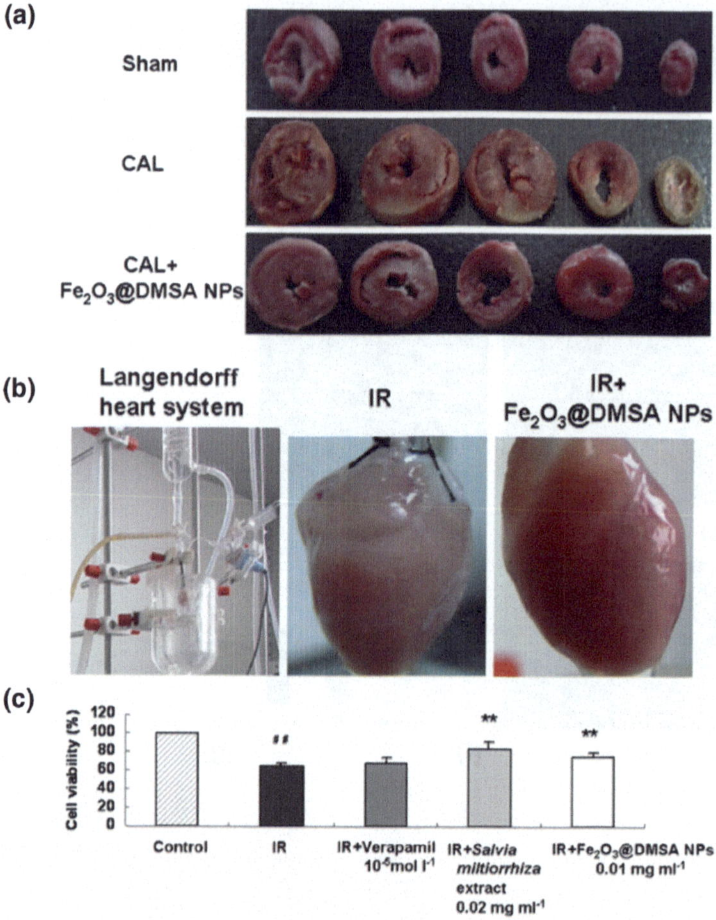

Fig. 14.11 **a** Fe_2O_3@DMSA NPs protected coronary artery ligature (CAL) induced injury in rats. **b** Fe_2O_3@DMSANPs have a caridoprotective activity in guinea pig Langendorff heart following 30 min of ischemia and 30 min of reperfusion (IR). **c** Comparison of cardioprotective effects of Fe_2O_3@DMSA NPs with Verapamil and *Salvia miltiorrhiza* extract. Reprinted with permission from [15]. Copyright (2015) Nature Publishing Group

14.3 Hepatoprotection

Livers play multiple roles in removing toxins, storing glycogen (glycogen), synthesizing secretory protein, etc. Free radical induced liver damages can lead to hepatitis and other liver diseases. Numerous studies demonstrated that excess alcohol can cause liver disease [19–21]. One possible reason is that oxidative stress appears when a liver is scavenging excess alcohol, and the liver itself may be impaired in this process. Herein, nanozymes with SOD mimicking activity have been applied to eliminate excess free radicals and thus to protect the liver.

Carbon tetrachloride (CCl_4) poisoning in mice is related to free radicals and usually regarded as a typical model of free radical induced liver injury. Moussa and coworkers applied this model to prove that fullerene could be an efficient free radical scavenger in reducing oxidative stress in liver. Hepatic stellate cells (HSCs) would transform into myofibroblast-like cells (MFCs) when they were under oxidative stress after the rats were poisoned with CCl_4. However, if the rats were pretreated with fullerene, no MFCs were observed and the number of HSCs also decreased [22]. The results indicated that fullerene was a good scavenger of free radicals. In order to ensure fullerene was nontoxic to liver, they investigated the influence of different doses and pretreatment time on the mice liver. The results were shown as macroscopic and pathological section photos (Fig. 14.12). All the mice liver pretreated with fullerene recovered to normal condition, which indicated that fullerene could be degraded by mice.

14.4 Cytoprotection

There are many factors that can cause damage or even apoptosis to cells. Among them, ROS accounts for a large part of the reasons. Previous studies showed that nanozyme could play a role in eliminating excess cellular ROS and sometimes they can also promote the production of antioxidant enzymes related to the cell itself. Therefore, nanozymes could be used to counteract the disadvantages of excessive ROS for potential anti-oxidative therapy [23–29].

To elucidate their ability to protect cells, cells are usually cultured with nanozymes first, followed by adding a source of oxidative stress like hydrogen peroxide or by adding the inhibitors of cellular antioxidant enzymes (such as Cu^{2+} ions and 3-amino-1,2,4-triazole (3-AT)). After that, the ability of the cells to defend against oxidation is studied. Compared with the control group, if the nanozyme-added cell group exhibits better resistance to ROS, it indicates that the nanozymes have the ability to protect cells from oxidative stress. In Mugesh and coworkers' study, they synthesized glutathione peroxidase-like V_2O_5 nanowires (Vn). Vn could be successfully internalized into cells and fully restored the redox balance when maintaining the cellular antioxidant defense meanwhile. Using H_2O_2-specific probe HyPer and dye Amplex Red, obvious low fluorescent intensity was observed comparing to the untreated cells

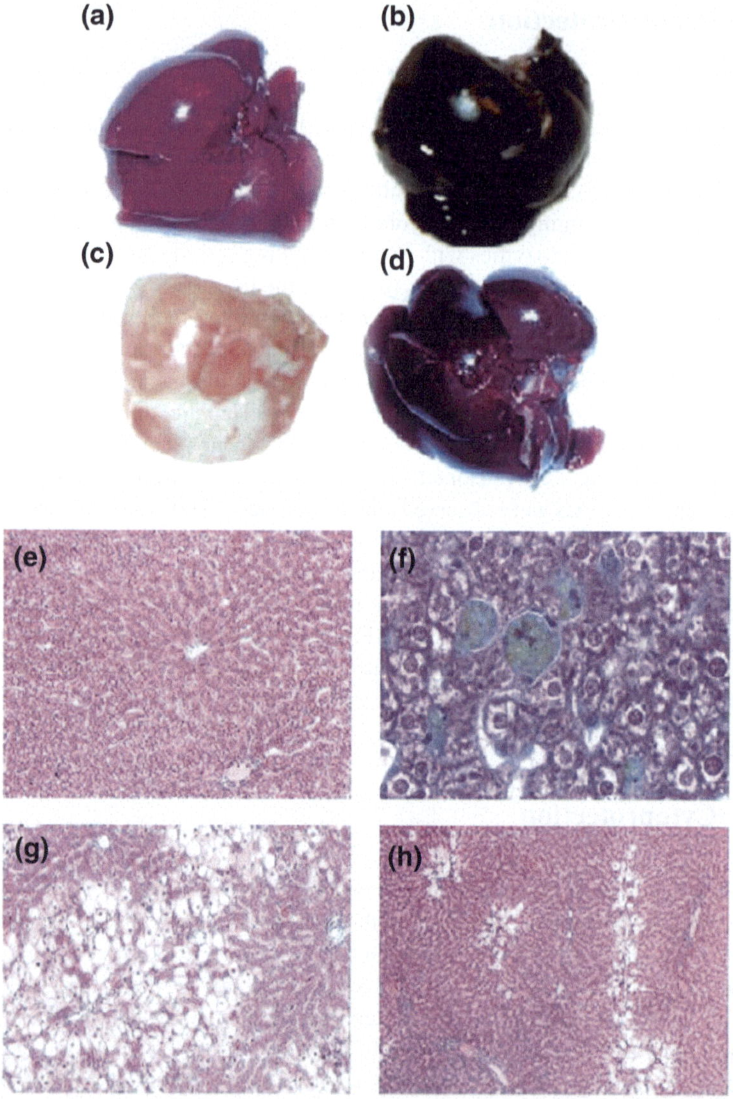

Fig. 14.12 Macroscopic and microscopic effects of C_{60} on rat livers. **a** Control liver; **b** liver after 7 days of pretreatment with C_{60} (2.0 g kg^{-1} of body weight); **c** liver of a rat intoxicated with CCl_4 (1 mL kg^{-1} of body weight); **d** liver after 14 days of pretreatment with C_{60} before CCl_4 treatment. Trichrome staining of liver sections (magnification(100x)) from: **e** C_{60}-treated rat; **f** magnification of E; **g** CCl_4-treated rat; **h** an example of C_{60}-pretreated rat before CCl_4 treatment showing a few necrotic areas limited to some cords of hepatocytes. Reprinted with permission from [22]. Copyright (2005) American Chemical Society

through microscopic analysis, confirming the cellular H_2O_2 scavenging ability of Vn (Fig. 14.13) [24].

In the case of environmental pollutants like heavy metals, acid, base, halogen, organic poisons, and so on, cells will also suffer from severe oxidative stress, resulting in a series of biological changes including viability loss, apoptotic death, the depletion of glutathione (GSH), the peroxidation of membrane lipid, and DNA damage. Tian and coworkers found that human SH-SY5Y neuroblastoma cells pretreated with C_{60}-methionine derivate (FMD) before lead exposure could decrease apoptotic death and oxidative stress without obvious toxicity. They also compared other C_{60}-amino acid derivatives, β-alanine C_{60} derivate, and cystine C_{60} derivate; the results suggested that C_{60}-methionine derivate possessed the best anti-oxidative ability among the three, which may serve as a potential anti-oxidative agent in the prevention of lead intoxication (Fig. 14.14) [26].

The defects of sulfite oxidase (SuOx), which is a disease caused by gene deficiencies, can lead to early death with no efficient or cost-effective therapy in sight. This molybdenum-containing enzyme is located at mitochondria and can functionally convert sulfite to sulfate in the amino acid and lipid metabolism, thus protecting cells from the toxicity of sulfites. In the research of Tremel et al., liver cells' SuOx activity, which was chemically knocked down, was recovered after treatment with MoO_3 nanoparticles. As shown in Fig. 14.15, the MoO_3 nanoparticles were functionalized with a TPP ligand via dopamine for targeting the mitochondria of HepG2 cells. The experiment results showed that the MoO_3 nanoparticle-treated HepG2 cells regained the sulfite oxidase activity, which may help remedy the defects induced by gene deficiencies in the future [25].

Radiation therapy is increasingly prominent in cancer therapy, which is one of the main methods to treat malignant tumor. Protecting normal cells from radiation while targeting cancer cells is an everlasting destination in the treatment process. Normal cells in close proximity to the treatment site are threatened by free radicals formed through ionizing reactions, when the protective molecules of cells themselves could not effectively eliminate these free radicals. So additional materials are needed to get rid of this dilemma. By pretreating CRL8798 cells and MCF-7 cells (as normal and breast carcinoma cells, respectively) with nanoceria before radiation, Seal et al. found contrasting results. Almost no radiation-induced apoptosis was observed in the CRL8798 cells while MCF-7 cells showed no statistically significant protection by nanoceria (Fig. 14.16) [28]. These results demonstrated the selective protection of normal cells against radiation-induced damages.

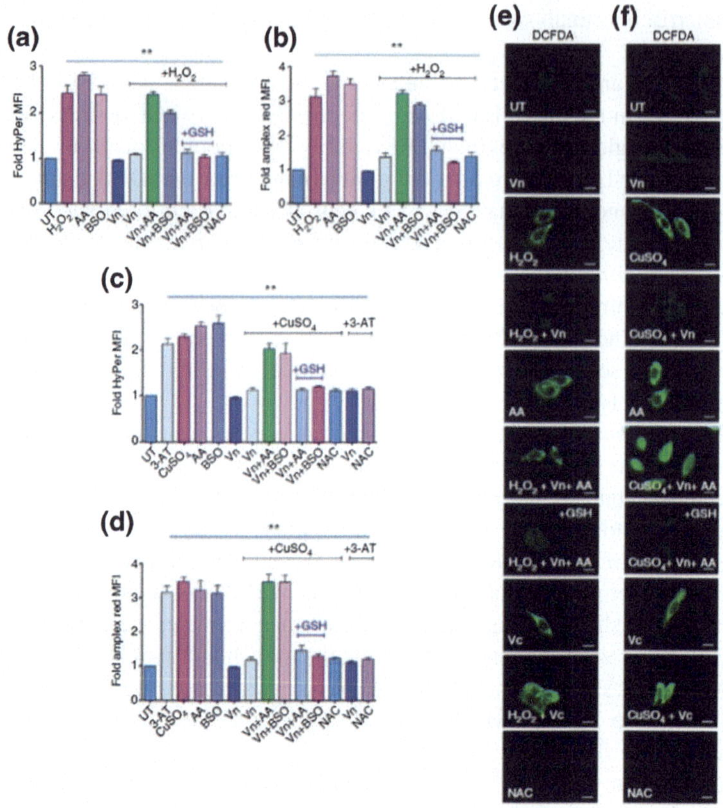

Fig. 14.13 Ability of Vn to scavenge intracellular ROS. **a–b** H_2O_2 scavenging activity of Vn was measured using genetically encoded H_2O_2-specific probe HyPer in HEK293T cells. Cells were either left untreated or pretreated with 50 ng μl^{-1} of nano-V_2O_5 (Vn) prior exposure to 200 mM H_2O_2 or 500 mM $CuSO_4$, or 20 mM 3-amino-1,2,4-triazole (3-AT). AA (0.1 mM) or 50 mM BSO was used to deplete cellular GSH level. GSH (100 mM) was used to replenish the GSH level to show the requirement of GSH as a cofactor for Vn. To measure the ROS scavenging ability of vanadium complex (Vc), 50 ng μl^{-1} Vc was used to treat the cells in the presence or absence of oxidative stress. N-acetyl cysteine-treated cells (NAC) (100 mM) were used as positive control. Data represented as mean ± s.e.m. $n = 3$, $^{**}P$ (t-test) < 0.001. **c–d** To further confirm the H_2O_2 scavenging ability of Vn, HEK293T cells were subjected to various treatments as mentioned earlier and stained with H_2O_2-specific dye Amplex Red (50 mM). The fluorescence was measured at 590 nm using spectrofluorometer and data were represented as fold change in mean fluorescent intensity (MFI) of Amplex Red over untreated cells. Data represented as mean ± s.e.m. n = 3, $^{**}P$ (t-test) < 0.001. **e–f** HeLa cells were treated with Vn before the treatment with H_2O_2 (E) or $CuSO_4$ **f** (as mentioned earlier) and then stained with 15 mM DCFDA-H2 dye. The change in fluorescent intensity was observed through microscopic analysis (scale bar, 10 mm). Reprinted with permission from [24]. Copyright (2014) Nature Publishing Group

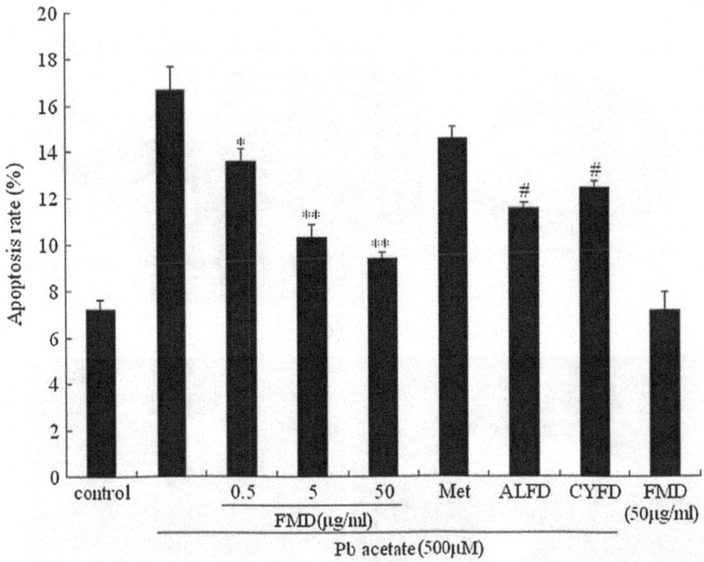

Fig. 14.14 Effect of FMD on Pb-induced apoptosis in SH-SY5Y cells. After pretreatment with either FMD or the other C_{60}-amino acid derivatives (ALFD and CYFD) for 1 h, and incubation with 500 mM Pb acetate for another 72 h, cells in culture plates were harvested and then stained with PI. Apoptosis of cells was detected by flow cytometry. Cells in the sub-G1 fraction were regarded as apoptotic. The results are representative of three independent experiments. *$P < 0.05$ and **$P < 0.01$ compared with corresponding values of the Pb^{2+}-only group. #$P < 0.05$ compared with the corresponding values of the FMD (50 mg mL^{-1}) group ($n = 3$). Reprinted with permission from [26]. Copyright (2011) John Wiley & Sons

14.5 Cancer Therapy

ROS and oxidation are also involved in cancer progression. Therefore, ROS scavenging nanozymes have been studied in the area of cancer therapy [30–33].

Tumor–stroma interaction is one of the important factors during tumor growth process. Brenneisen et al. studied the effect of nanoceria nanoparticles on myofibroblast formation, cell toxicity, and tumor invasion. They showed that nanoceria nanoparticles not only lowered the invasive capacity of tumor cells but also significantly inhibited tumor invasion compared to mock-treated SCL-1 cells (Fig. 14.17) [30]. Furthermore, their group also found that ceria nanoparticles could inhibite tumor growth and invasion (Fig. 14.18) [31].

Sudipta et al. found that nanoceria could be used as an antiangiogenic therapeutic agent. Compared with untreated mouse ovarian tumors, nanoceria-treated group showed obvious tumor inhibition. The tumor inhibition could be attributed to the fact that nanoceria effectively decreased the proliferation of tumor cells (Fig. 14.19) [32]. In addition, some researches have been focused on the modulation of the special tumor microenvironment with nanozymes. Webster et al. reported that dextran-coated

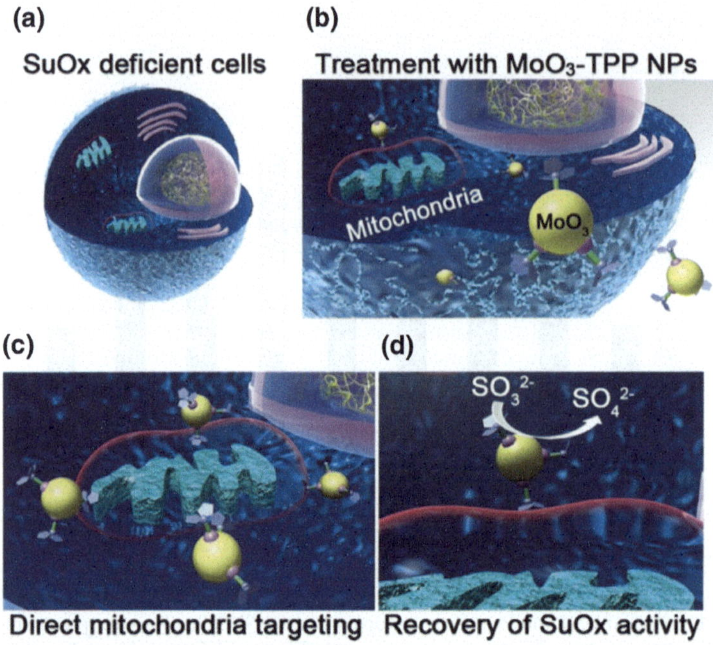

Fig. 14.15 Mode of action of MoO_3 nanoparticles. **a** Hepatocytes are treated with sodium tungstate to induce SuOx deficiency in vitro. **b** SuOx deficient cells are treated with 2 nm TPP-surface functionalized MoO_3 nanoparticles. **c** Mitochondria are directly targeted, and nanoparticles accumulate in close proximity to the membrane. **d** Sulfite is oxidized to cellular innocuous sulfate by MoO_3 nanoparticles, and therefore SuOx activity is reconstituted and cells regain their detoxifying capacity. Reprinted with permission from [25]. Copyright (2014) American Chemical Society

nanoceria showed pH-dependent activity in prohibiting osteosarcoma cell proliferation [33]. Dextran-coated nanoceria was more effective at killing bone tumor cells at acidic (pH 6) conditions compared with normal microenvironment (pH 7).

14.6 Tissue Engineering

Tissue engineering is an important discipline that combines cell biology and material science to construct tissues or organs in vitro or in vivo. Functional nanomaterials have been investigated as scaffolds for tissue engineering. Compared with the developed nanomaterial scaffolds, nanozymes could be more advantageous due to the enzyme-mimicking activities.

In tissue engineering, CeO_2 nanozymes have been demonstrated as promising auxiliary materials [34–40]. Traversa and coworkers found that CeO_2 nanoparticles in PLGA scaffolds could induce the aligned growth of stem cells. The growth of stem cells on different composite supports was compared [34]. In the presence of CeO_2

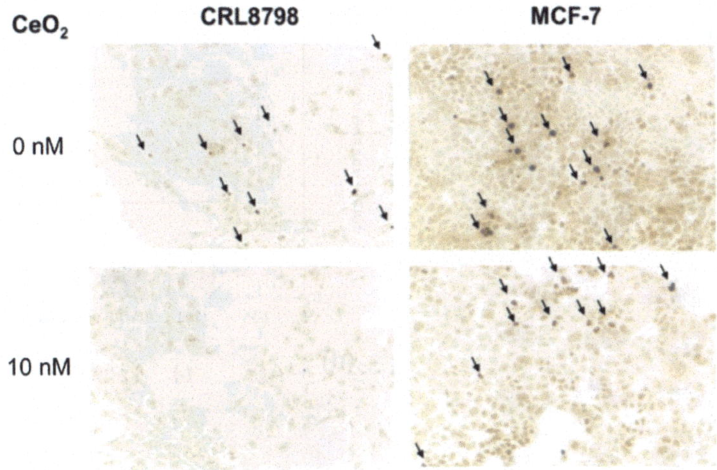

Fig. 14.16 TUNEL staining of breast cells at 48 h following 10 Gy irradiation and protection by cerium oxide nanoparticles. The arrows denote TUNEL-positive apoptotic nuclei. Reprinted with permission from [28]. Copyright (2005) American Chemical Society

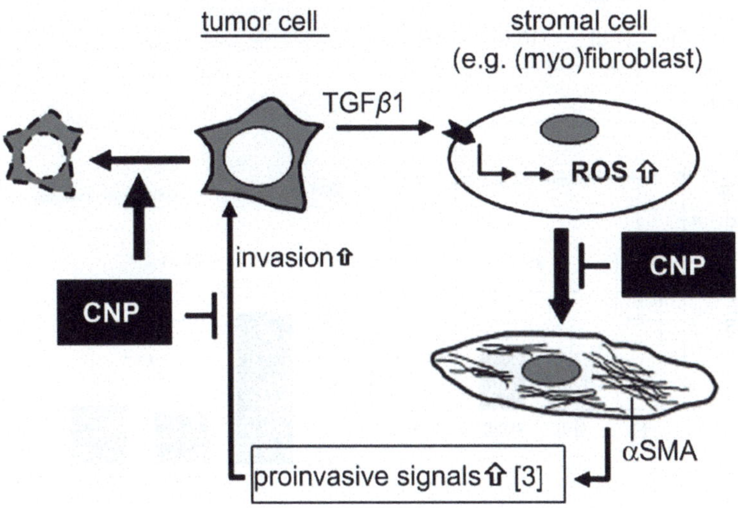

Fig. 14.17 Tumor cells release growth factors and cytokines, e.g., TGFβ1, which initiate ROS-dependent generation of αSMA-positive (stromal) myofibroblasts. Myofibroblast-derived proinvasive signals stimulate the invasive capacity of cancer cells. These signals are lowered or prevented by treatment of stromal cells with CNP. These CNP being nontoxic for stromal cells show a cytotoxic and anti-invasive effect on tumor cells, indicating a bifunctional role of CNP in tumor–stroma interactions. Reprinted with permission from [30]. Copyright (2011) Elsevier

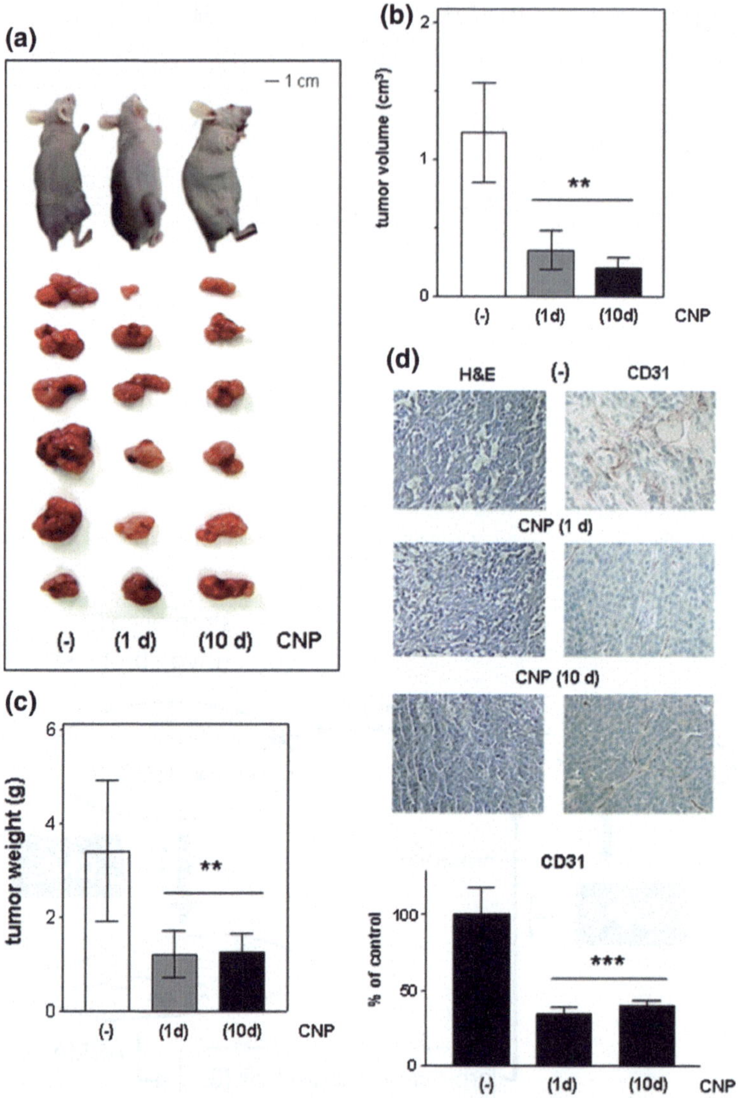

Fig. 14.18 Inhibitory effect of nanoceria nanoparticles (CNP) on the growth of A375 xenografted mice. **a** Images of dissected tumors. Three groups of mice were used. Group 1 is vehicle-treated (mock-treated, (−)); group 2 is CNP-treated starting with day 1 after tumor injection; and group 3 is CNP-treated starting with day 10 after tumor injection. After 30 days, the tumors were dissected. **b** Tumor volume. The tumor volume of six mice per group was measured. $^{**}P < 0.01$ versus ct (ANOVA and Dunnett's test). Data are presented as means ± SD. **c** Tumor weight. The tumor weight of six mice per group was measured. $^{**}P < 0.01$ versus ct (ANOVA and Dunnett's test). Data are presented as means ± SD. **d** H&E and CD31 staining. Formalin-fixed tissues were embedded in paraffin and sectioned for immunohistological analysis of angiogenesis via CD31 staining and H&E staining for morphology. The percentage of CD31 staining is presented. $^{***}P < 0.001$ versus ct (ANOVA and Dunnett's test). Data are presented as means ± SEM. Reprinted with permission from [31]. Copyright (2013) Mary Ann Liebertpub

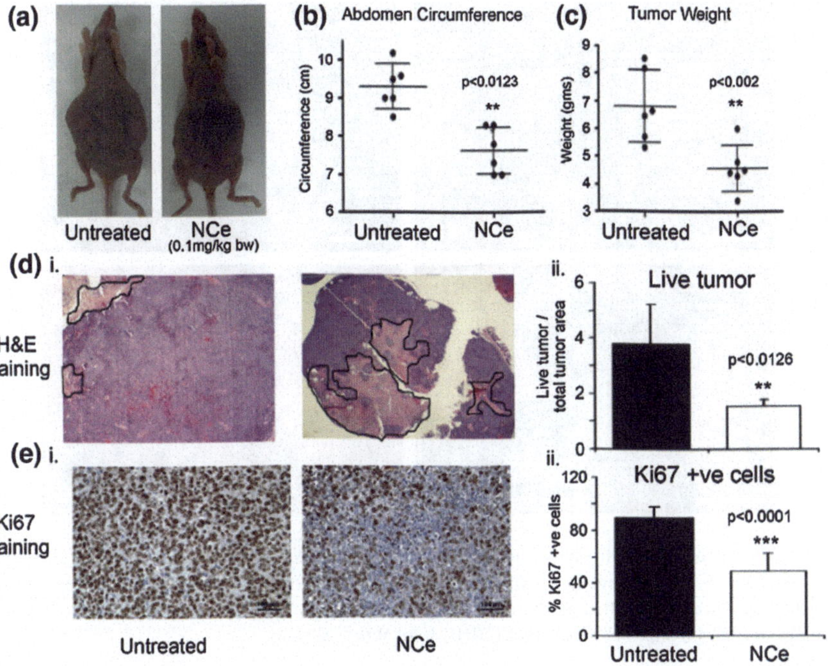

Fig. 14.19 NCe treatment inhibited ovarian tumor growth in vivo. **a** Gross morphology of representative mouse with tumors at day 30 ($n = 6$). **b** Cumulative abdominal circumference at the end of the study. **c** Excised tumor weight from vehicle (PBS) treated and NCe (0.1 mg kg^{-1} bd wt; every third day). Results are shown as mean ± S.D. of six individual animals. $^{**}P < 0.01$ NCe treated compared to untreated group using two-tailed Student's t-test (Prism). **d** (i) Representative H&E (x20) photomicrographs exhibiting live (purple) and necrotic (pink, encircled) areas in untreated and treated xenografts. (ii) Graphical representation of viable tumor size measured as described in Materials and methods. **e** (i) Representative Ki-67 staining (x200) of excised A2780 xenografts at day 30. (ii) Count of positive Ki-67 cells from five high powered fields (x400) in three different xenografts from each group. Counts are expressed as percentage of control. $^{***}P < 0.001$ and $^{**}P < 0.01$ NCe treated compared to untreated using two-tailed Student's t-test (Prism). Reprinted with permission from [32]. Copyright (2013) PLOS

nanoparticles, the growth of cells exhibited a certain orientation, which was quite different from the disorganized growth of cells on pure PLGA support (Fig. 14.20). The promoted growth of stem cells was attributed to the anti-oxidation property of CeO_2 nanoparticles.

Furthermore, CeO_2 nanoparticles have been used for the healing of wound. Mattson, Seal et al. found that CeO_2 nanoparticles could promote the proliferation and migration of fibroblasts, keratinocytes, and vascular endothelial cells, which in turn accelerated the healing of full-thickness dermal wounds of mice [35]. CeO_2 nanoparticles were found to accumulate in the wound site (Fig. 14.21C). The two main oxidative stress markers (i.e., 4-hydroxynonenal (HNE) and nitrotryosine) in the wound site from mice treated with or without CeO_2 nanoparticles were measured. As shown

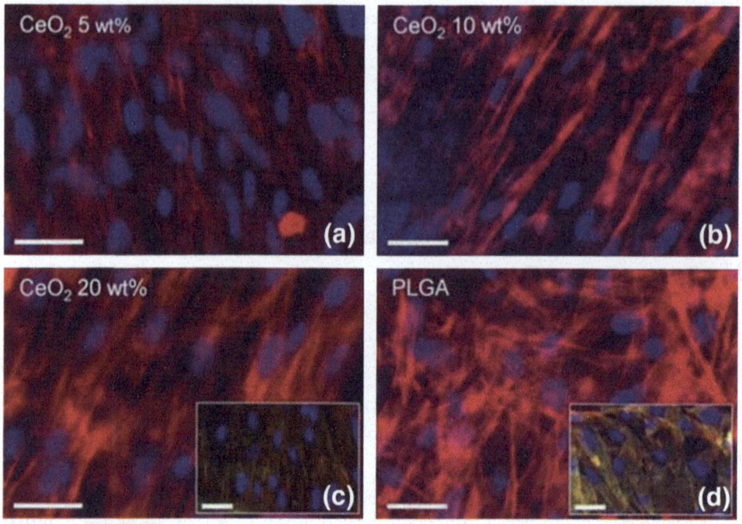

Fig. 14.20 If culture monitoring of the CSC seeded composite supports after six DIV: **a** 5, **b** 10, and **c** 20 wt% CeO_2/PLGA, together with **d** unfilled PLGA as a control. Cell nuclei, cytoskeleton, and vinculin can be distinguished in blue, red, and green, respectively (all scale bars 20 μm). Reprinted with permission from [34]. Copyright (2010) John Wiley & Sons

in Fig. 14.21A–B, the levels of the two markers were obviously lower with the CeO_2 nanoparticle-treated groups. These results suggested that nanoceria could relieve excessive oxidative stress in the wound site, therefore accelerating the healing of the wound.

More, nanoceria is considered to be helpful for bone regeneration. Seal and coworkers incorporate nanoceria in 3-D bioactive glass foam scaffolds. Compared with the bioactive glass without nanoceria, the nanoceria incorporated glass promoted the production of collagen and differentiation of osteoblast by human mesenchymal stem cells (HMSCs) [36]. HMSCs cultured onto the nanoceria contained scaffolds with osteogenic supplements exhibited distinct smooth morphology and better growth results (Fig. 14.22b, d).

14.7 Anti-inflammation

The occurrence of inflammation usually associates with the overproduction of ROS. So some researchers have focused on the anti-inflammation capacity of nanoceria due to its ROS scavenging activity. Reilly et al. reported that nanoceria nanoparticles held excellent biocompatibility [41]. They further showed that the nanoceria nanoparticles not only scavenged ROS but also inhibited the production of nitric oxide. Then, they used nanoceria as a novel therapy for inflammation. Puntes, Jimenez et al. used

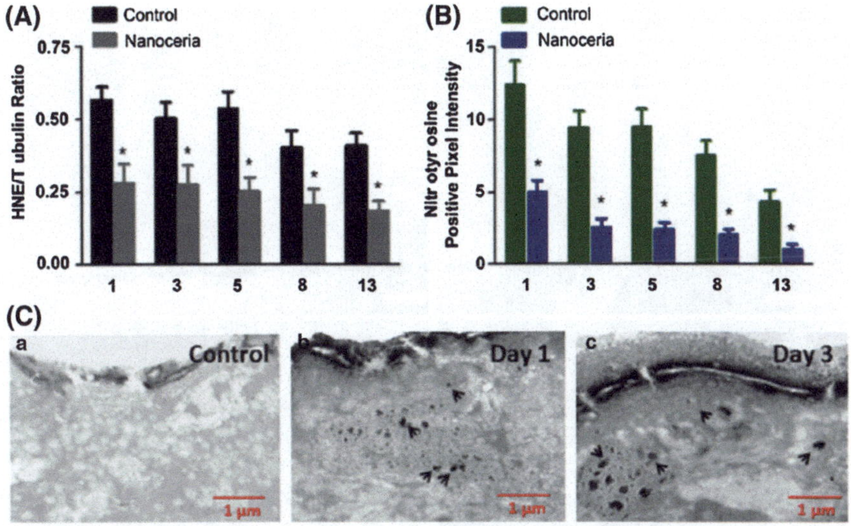

Fig. 14.21 Levels of proteins oxidatively modified by the lipid peroxidation product HNE and peroxynitrite in skin wounds are reduced in nanoceria-treated mice compared to control mice. **A** Levels of HNE-modified proteins in skin tissue samples from control and nanoceria-treated mice at 1, 3, 5, 8, 13 day after injury. *$P < 0.01$ (n ¼ 4 mice per group). **B** Levels of nitrated proteins (measured using an anti-nitrotyrosine antibody) in skin tissue samples from control and nanoceria-treated mice at 1, 3, 5, 8, 13 days after injury. *$P < 0.001$ (n ¼ 4 mice per group). **C** Nanoceria accumulate outside of and within skin cells in dermal wounds. Transmission electron micrographs of skin biopsy samples from an unwounded control mouse (**a**), and mice that had been wounded either 1 or 3 days previously (**b** and **c**, respectively). Skin samples were taken 24 h after application of nanoceria. Reprinted with permission from [35]. Copyright (2013) Elsevier

nanoceria in rats with liver disease. They found that nanoceria could help downregulate the expression of the mRNA related with inflammatory. Thereby they suggested that nanoceria significantly attenuated the intensity of inflammatory response, and it might provide therapeutic value in chronic diseases (Fig. 14.23) [42].

Gu, Zhang et al. recently studied the multiple enzyme-like activity of Prussian blue nanoparticles (PBNPs). They further evaluated their anti-inflammatory effects in vitro and in vivo [43]. The results indicated that PBNPs showed great potentials of scavenging superoxide radical and hydrogen peroxide in vitro and inhibiting inflammation in liver tissues (Fig. 14.24).

14.8 Anti-aging

Anti-aging is the eternal pursuit of human beings. Numerous hypotheses have been proposed for aging. Among them, free radical theory has gained broad support due to the high degree of fit between theory and practical aging phenomenon [44]. This

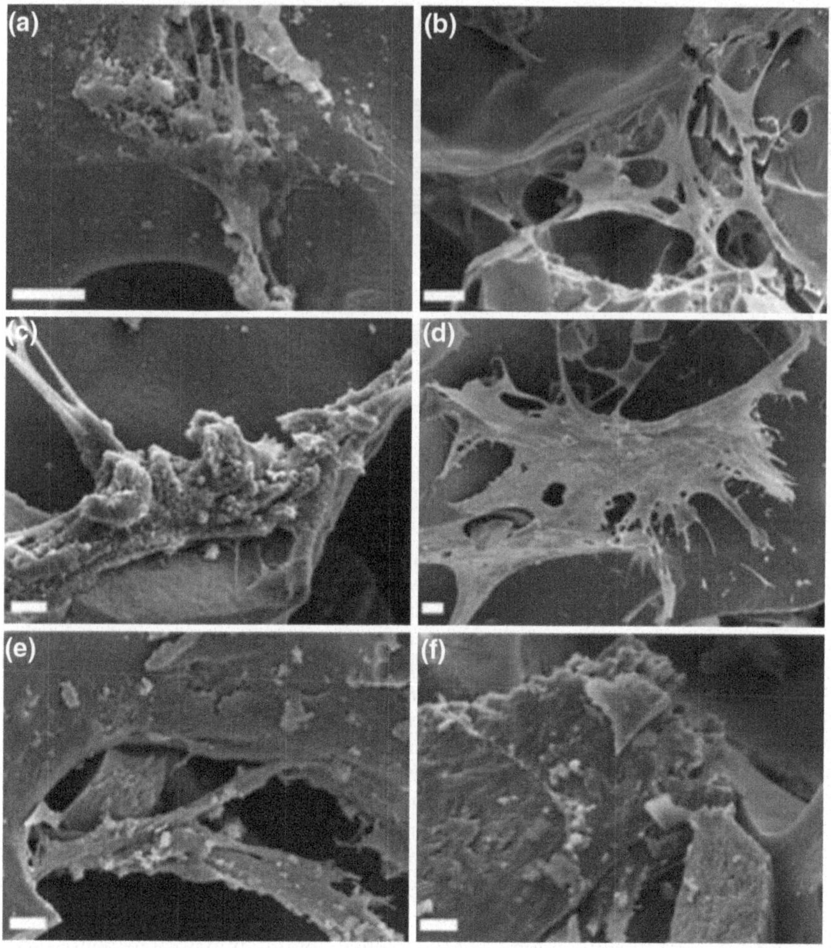

Fig. 14.22 SEM of HMSCs cultured for 10 days on the bioactive glass scaffolds doped with nanoceria synthesized in water (**a**–**b**), in dextran (**c**–**d**), and no nanoceria (**e**–**f**). Cell culture was performed in the absence (**a**, **c**, **e**) and in the presence (**b**, **d**, **f**) of osteogenic supplements. Cells attached and spread on the scaffolds' surface. Note the osteoblast-like products or minerals on the cells cultured in the absence (**a**, **c**, **e**) of osteogenic supplements (scale bar = 10 μm). Reprinted with permission from [36]. Copyright (2010) Royal Society of Chemistry

theory indicates that degenerative changes in the process of aging are due to the deleterious effects of free radicals produced and accumulated during normal cellular metabolism, which was suggested by Denham Harman in 1956 [44–46]. Free radicals may cause various pathologies (as summarized in Fig. 14.25), which can be attenuated by antioxidant nanoparticles [47]. The living systems themselves (including humans) can produce antioxidants to eliminate free radicals, but the amount of antioxidants decreases with age. Therefore, the intervention of antioxidant nanoparticles is needed

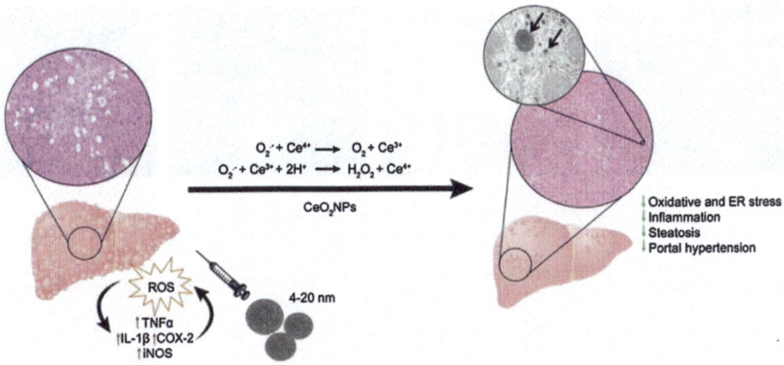

Fig. 14.23 The concept of reducing steatosis, portal hypertension, and displaying anti-inflammatory properties in rats with liver fibrosis for cerium oxide nanoparticles. Reprinted with permission from [42]. Copyright (2016) Elsevier

to compensate for this deficiency, thus to remove excess free radicals and prevent disease (Fig. 14.25).

Mitochondria, known as "biological clock of aging", are closely associated with the development of normal human aging and aging-related diseases [48–51]. Therefore, great research interests have been devoted to developing mitochondria-targeted nanozymes to mitigate aging. Halliwell and coworkers proposed a model of mitochondria-targeted nanomaterials [51], with functional groups (such as mitochondriotropics) at the surface and antioxidants in the internal space (Fig. 14.26). Due to the presence of functional groups like mitochondriotropics, these hybrid nanomaterials could specifically target mitochondria after cellular uptake. Finally, the antioxidants would be released from the internal space and achieve the purpose of reducing free radicals.

In this chapter, nanozymes with free radical scavenging properties have been applied to abundant therapeutic fields, including neuroprotection, cardioprotection, hepatoprotection, cytoprotection, cancer therapy, tissue engineering, anti-inflammation, and anti-aging. For neuroprotection and cardioprotection, nanoceria-based and Fe_3O_4-based nanozymes were widely reported for their excellent cell protection ability and have showed great effects on brain ischemia and cardiac hypertrophy. These two nanozymes have been successfully applied to two different neurodegenerative disease models, Parkinson's disease (PD) and Alzheimer's disease (AD). In the hepatoprotection section, fullerene as a representative material has shown great potential in reducing oxidative stress in liver and no toxic was found after the treatment. As for cytoprotection, numerous cell damages induced by excess ROS, pollution, intrinsic gene defects, and radiation could be relieved after the treatment of nanozymes. Cancer progression is also related to ROS and oxidation, therefore ROS scavenging nanozymes can be used in the therapy of cancer and they did inhibit tumor growth and invasion. In tissue engineering, CeO_2 nanozymes have been demonstrated as promising auxiliary, wound healing, and bone regeneration materials. Nanozymes could be more advantageous due to the enzyme-mimicking activities compared with

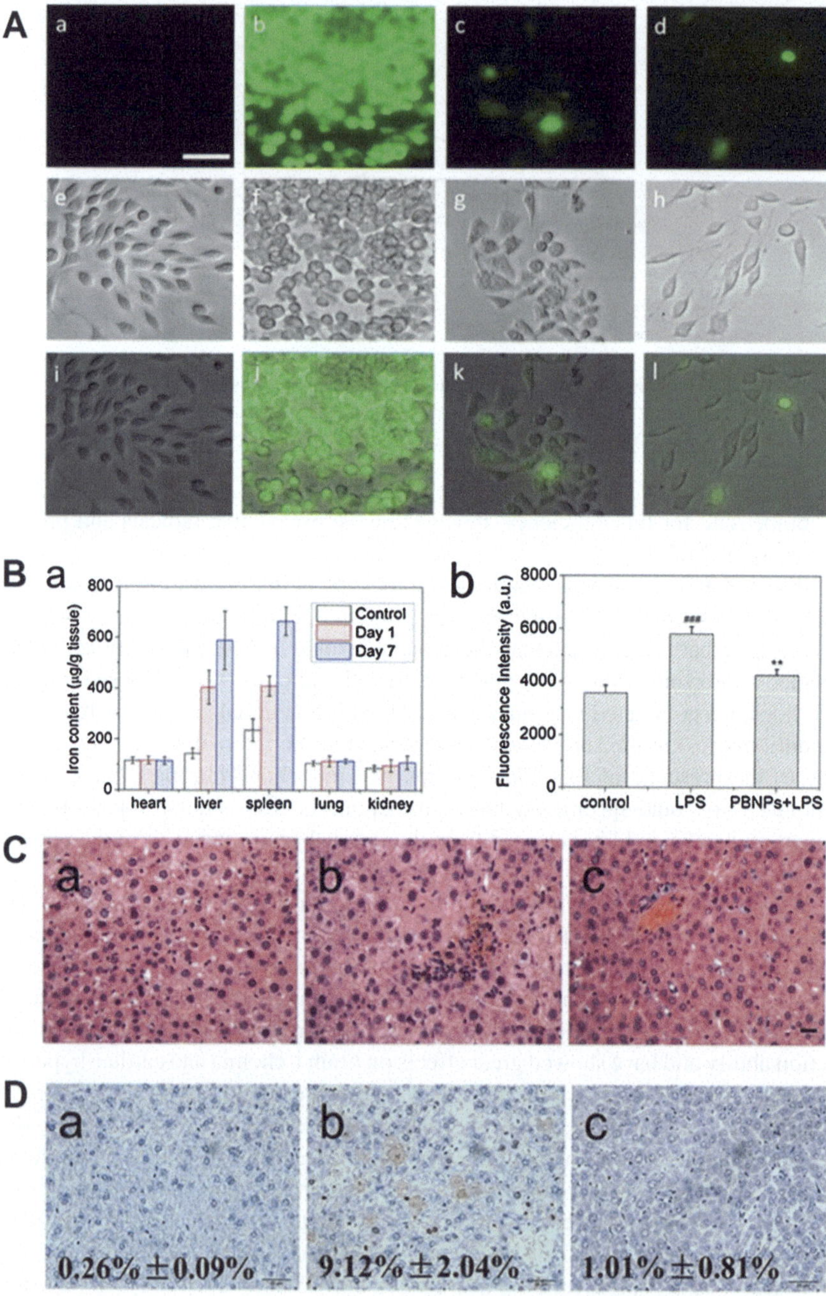
A
a
b
c
d
e
f
g
h
i
j
k
l
B a
Iron content (μg/g tissue)
0
200
400
600
800
heart
liver
spleen
lung
kidney
Control
Day 1
Day 7
b
Fluorescence Intensity (a.u.)
0
2000
4000
6000
8000
###
**
control
LPS
PBNPs+LPS
C
a
b
c
D
a
b
c
0.26%±0.09%
9.12%±2.04%
1.01%±0.81%

◂**Fig. 14.24** PBNPs can inhibit ROS generation induced by LPS. **A** Fluorescence microscopic images (**a–d**), bright-field microscopic images (**e–h**), and merged images (**i–l**) of RAW264.7 cells. (**a**) Normal RAW264.7 cells; (**b**) RAW264.7 cells were stimulated by LPS; (**c**) RAW264.7 cells were preincubated with PBNPs then stimulated by LPS; (**d**) RAW264.7 cells were preincubated with GSH and then stimulated LPS. Scale bar, 200 μm. **B** In vivo accumulation of PBNPs in mice (**a**) and ROS intensities in liver tissue (**b**). Time-dependent biodistribution measurement of iron levels in various organs including heart, liver, spleen, lung, and kidneys of mice after double injection of 20 mg kg^{-1} of PBNPs, using 0.9% saline injected ICR mice as control. The second injection was carried out on the fourth day. ROS intensities were measured in liver tissues of the same mass excised from control mice, LPS-treated mice, and LPS-treated mice with 1 week chronic injection of PBNPs. ROS intensity was expressed by fluorescence intensity; bars represent the means ± s.e.m. ($n = 5$). ### indicates $P < 0.001$ compared with the control group; ∗∗ indicates $P < 0.01$ compared with the LPS group. **C** Hematoxylin and Eosin (H&E) staining images of liver histology. Liver tissues were excised from (**a**) control, (**b**) LPS-treated mice, and (**c**) LPS-treated mice with 1 week preinjection of PBNPs. Scale bar, 20 μm. **D** TUNEL staining images of mice liver tissue sections from (**a**) control, (**b**) LPS-treated mice, and (**c**) LPS-treated mice with 1 week preinjection of PBNPs. Reprinted with permission from [43]. Copyright (2016) American Chemical Society

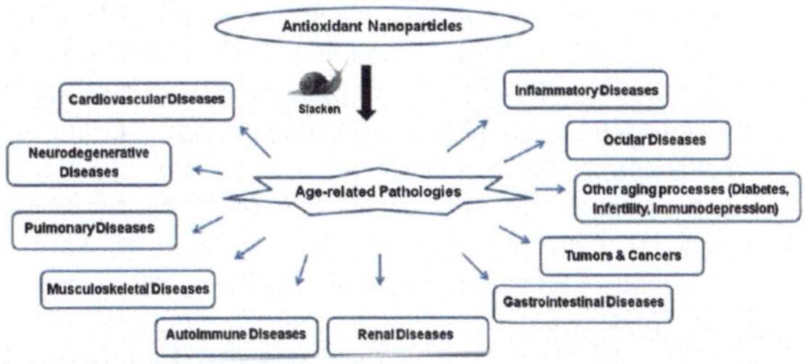

Fig. 14.25 List of age-related pathologies that can be attenuated by antioxidant nanoparticles. Reprinted with permission from [47]. Copyright (2013) Elsevier

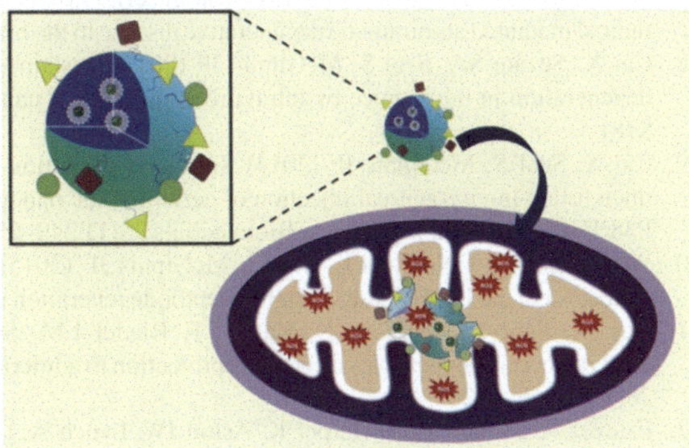

Fig. 14.26 Design of a possible mitochondria-targeted nanomaterial, bearing several functional groups (e.g., mitochondriotropics and fluorescent probes), and encapsulating antioxidants in their inner core. Reprinted with permission from [51]. Copyright (2013) Elsevier

the developed nanomaterial scaffolds. Inflammation and aging are also free radicals related processes, which provide plenty of application space for nanozymes. Though diverse successful therapeutics of nanozymes have been demonstrated, more kinds of nanozymes with novel properties for other therapies are expected, some of which will be discussed in Chap. 17.

Acknowledgements We thank National Natural Science Foundation of China (21722503 and 21874067), 973 Program (2015CB659400), PAPD program, Shuangchuang Program of Jiangsu Province, Open Funds of the State Key Laboratory of Analytical Chemistry for Life Science (SKLACLS1704), Open Funds of the State Key Laboratory of Coordination Chemistry (SKLCC1819), Fundamental Research Funds for the Central Universities (021314380103), and Thousand Talents Program for Young Researchers for financial support. We thank Xiwen Chen, Lin Lin, and Yuting Wang for proof-reading.

References

1. Gao L, Zhuang J, Nie L, Zhang J, Zhang Y, Gu N, Wang T, Feng J, Yang D, Perrett S, Yan X (2007) Intrinsic peroxidase-like activity of ferromagnetic nanoparticles. Nat Nanotechnol 2(9):577–583
2. Wang X, Guo W, Hu Y, Wu J, Wei H (2016) Nanozymes: next wave of artificial enzymes. Springer
3. Wei H, Wang E (2013) Nanomaterials with enzyme-like characteristics (nanozymes): next-generation artificial enzymes. Chem Soc Rev 42(14):6060–6093
4. Chen J, Patil S, Seal S, McGinnis JF (2006) Rare earth nanoparticles prevent retinal degeneration induced by intracellular peroxides. Nat Nanotechnol 1(2):142–150
5. Kim CK, Kim T, Choi I-Y, Soh M, Kim D, Kim Y-J, Jang H, Yang H-S, Kim JY, Park H-K, Park SP, Park S, Yu T, Yoon B-W, Lee S-H, Hyeon T (2012) Ceria nanoparticles that can protect against ischemic stroke. Angew Chem Int Ed 51(44):11039–11043
6. Kwon HJ, Cha M-Y, Kim D, Kim DK, Soh M, Shin K, Hyeon T, Mook-Jung I (2016) Mitochondria-targeting ceria nanoparticles as antioxidants for alzheimer's disease. ACS Nano 10(2):2860–2870
7. Heckman KL, DeCoteau W, Estevez A, Reed KJ, Costanzo W, Sanford D, Leiter JC, Clauss J, Knapp K, Gomez C, Mullen P, Rathbun E, Prime K, Marini J, Patchefsky J, Patchefsky AS, Hailstone RK, Erlichman JS (2013) Custom cerium oxide nanoparticles protect against a free radical mediated autoimmune degenerative disease in the brain. ACS Nano 7(12):10582–10596
8. Cai X, Sezate SA, Seal S, McGinnis JF (2012) Sustained protection against photoreceptor degeneration in tubby mice by intravitreal injection of nanoceria. Biomaterials 33(34):8771–8781
9. Cai X, Seal S, McGinnis JF (2014) Sustained inhibition of neovascularization in vldlr(-/-) mice following intravitreal injection of cerium oxide nanoparticles and the role of the ASK1-P38/JNK-NF-kappa B pathway. Biomaterials 35(1):249–258
10. Wong LL, Pye QN, Chen L, Seal S, McGinnis JF (2015) Defining the catalytic activity of nanoceria in the P23H-1 Rat, a photoreceptor degeneration model. PLoS ONE 10(3):e0121977
11. Das M, Patil S, Bhargava N, Kang J-F, Riedel LM, Seal S, Hickman JJ (2007) Auto-catalytic ceria nanoparticles offer neuroprotection to adult rat spinal cord neurons. Biomaterials 28(10):1918–1925
12. Estevez AY, Pritchard S, Harper K, Aston JW, Lynch A, Lucky JJ, Ludington JS, Chatani P, Mosenthal WP, Leiter JC, Andreescu S, Erlichman JS (2011) Neuroprotective mechanisms of cerium oxide nanoparticles in a mouse hippocampal brain slice model of ischemia. Free Radic Biol Med 51(6):1155–1163
13. Li Q, Tang G, Xue S, He X, Miao P, Li Y, Wang J, Xiong L, Wang Y, Zhang C, Yang G-Y (2013) Silica-coated superparamagnetic iron oxide nanoparticles targeting of EPCs in ischemic brain injury. Biomaterials 34(21):4982–4992
14. Zhang Y, Wang Z, Li X, Wang L, Yin M, Wang L, Chen N, Fan C, Song H (2016) Dietary iron oxide nanoparticles delay aging and ameliorate neurodegeneration in drosophila. Adv Mater 28(7):1387–1393

15. Xiong F, Wang H, Feng Y, Li Y, Hua X, Pang X, Zhang S, Song L, Zhang Y, Gu N (2015) Cardioprotective activity of iron oxide nanoparticles. Sci Rep 5:8579
16. Niu J, Azfer A, Rogers LM, Wang X, Kolattukudy PE (2007) Cardioprotective effects of cerium oxide nanoparticles in a transgenic murine model of cardiomyopathy. Cardiovasc Res 73(3):549–559
17. Kolli MB, Manne NDPK, Para R, Nalabotu SK, Nandyala G, Shokuhfar T, He K, Hamlekhan A, Ma JY, Wehner PS, Dornon L, Arvapalli R, Rice KM, Blough ER (2014) Cerium oxide nanoparticles attenuate monocrotaline induced right ventricular hypertrophy following pulmonary arterial hypertension. Biomaterials 35(37):9951–9962
18. Pagliari F, Mandoli C, Forte G, Magnani E, Pagliari S, Nardone G, Licoccia S, Minieri M, Di Nardo P, Traversa E (2012) Cerium oxide nanoparticles protect cardiac progenitor cells from oxidative stress. ACS Nano 6(5):3767–3775
19. Faremi TY, Suru SM, Fafunso MA, Obioha UE (2008) Hepatoprotective potentials of Phyllanthusamarus against ethanol-induced oxidative stress in rats. Food Chem Toxicol 46(8):2658–2664
20. Noh J-R, Kim Y-H, Gang G-T, Hwang JH, Lee H-S, Ly S-Y, Oh W-K, Song K-S, Lee C-H (2011) Hepatoprotective effects of chestnut (Castanea crenata) inner shell extract against chronic ethanol-induced oxidative stress in C57BL/6 mice. Food Chem Toxicol 49(7):1537–1543
21. Liu Y, Du J, Yan M, Lau MY, Hu J, Han H, Yang OO, Liang S, Wei W, Wang H (2013) Biomimetic enzyme nanocomplexes and their use as antidotes and preventive measures for alcohol intoxication. Nat Nanotechnol 8(3):187–192
22. Gharbi N, Pressac M, Hadchouel M, Szwarc H, Wilson SR, Moussa F (2005) [60] fullerene is a powerful antioxidant in vivo with no acute or subacute toxicity. Nano Lett 5(12):2578–2585
23. Xu ZQ, Lan JY, Jin JC, Dong P, Jiang FL, Liu Y (2015) Highly photoluminescent nitrogen-doped carbon nanodots and their protective effects against oxidative stress on cells. ACS Appl Mater Interfaces 7(51):28346–28352
24. Vernekar AA, Sinha D, Srivastava S, Paramasivam PU, D'Silva P, Mugesh G (2014) An antioxidant nanozyme that uncovers the cytoprotective potential of vanadia nanowires. Nat Commun 5:5301
25. Ragg R, Natalio F, Tahir MN, Janssen H, Kashyap A, Strand D, Strand S, Tremel W (2014) Molybdenum trioxide nanoparticles with intrinsic sulfite oxidase activity. ACS Nano 8(5):5182–5189
26. Chen T, Li YY, Zhang JL, Xu B, Lin Y, Wang CX, Guan WC, Wang YJ, Xu SQ (2011) Protective effect of C(60)-methionine derivate on lead-exposed human SH-SY5Y neuroblastoma cells. J Appl Toxicol 31(3):255–261
27. Huang Y, Liu Z, Liu C, Ju E, Zhang Y, Ren J, Qu X (2016) Self-assembly of multi-nanozymes to mimic an intracellular antioxidant defense system. Angew Chem Int Ed 55(23):6646–6650
28. Tarnuzzer RW, Colon J, Patil S, Seal S (2005) Vacancy engineered ceria nanostructures for protection from radiation-induced cellular damage. Nano Lett 5(12):2573–2577
29. Colon J, Herrera L, Smith J, Patil S, Komanski C, Kupelian P, Seal S, Jenkins DW, Baker CH (2009) Protection from radiation-induced pneumonitis using cerium oxide nanoparticles. Nanomed 5(2):225–231
30. Alili L, Sack M, Karakoti AS, Teuber S, Puschmann K, Hirst SM, Reilly CM, Zanger K, Stahl W, Das S, Seal S, Brenneisen P (2011) Combined cytotoxic and anti-invasive properties of redox-active nanoparticles in tumor-stroma interactions. Biomaterials 32(11):2918–2929
31. Alili L, Sack M, von Montfort C, Giri S, Das S, Carroll KS, Zanger K, Seal S, Brenneisen P (2013) Downregulation of tumor growth and invasion by redox-active nanoparticles. Antioxid Redox Signal 19(8):765–778
32. Giri S, Karakoti A, Graham RP, Maguire JL, Reilly CM, Seal S, Rattan R, Shridhar V (2013) Nanoceria: a rare-earth nanoparticle as a novel anti-angiogenic therapeutic agent in ovarian cancer. PLoS ONE 8(1):e54578
33. Alpaslan E, Yazici H, Golshan NH, Ziemer KS, Webster TJ (2015) pH-dependent activity of dextran-coated cerium oxide nanoparticles on prohibiting osteosarcoma cell proliferation. ACS Biomater Sci Eng 1(11):1096–1103

34. Mandoli C, Pagliari F, Pagliari S, Forte G, Di Nardo P, Licoccia S, Traversa E (2010) Stem cell aligned growth induced by CeO_2 nanoparticles in PLGA scaffolds with improved bioactivity for regenerative medicine. Adv Funct Mater 20(10):1617–1624
35. Chigurupati S, Mughal MR, Okun E, Das S, Kumar A, McCaffery M, Seal S, Mattson MP (2013) Effects of cerium oxide nanoparticles on the growth of keratinocytes, fibroblasts and vascular endothelial cells in cutaneous wound healing. Biomaterials 34(9):2194–2201
36. Karakoti AS, Tsigkou O, Yue S, Lee PD, Stevens MM, Jones JR, Seal S (2010) Rare earth oxides as nanoadditives in 3-D nanocomposite scaffolds for bone regeneration. J Mater Chem 20(40):8912
37. Kong L, Cai X, Zhou X, Wong LL, Karakoti AS, Seal S, McGinnis JF (2011) Nanoceria extend photoreceptor cell lifespan in tubby mice by modulation of apoptosis/survival signaling pathways. Neurobiol Dis 42(3):514–523
38. Martin RA, Yue S, Hanna JV, Lee PD, Newport RJ, Smith ME, Jones JR (2012) Characterizing the hierarchical structures of bioactive sol-gel silicate glass and hybrid scaffolds for bone regeneration. Philos Transact A Math Phys Eng Sci 370(1963):1422–1443
39. Naganuma T, Traversa E (2014) The effect of cerium valence states at cerium oxide nanoparticle surfaces on cell proliferation. Biomaterials 35(15):4441–4453
40. Ponnurangam S, O'Connell GD, Chernyshova IV, Wood K, Hung CT, Somasundaran P (2014) Beneficial effects of cerium oxide nanoparticles in development of chondrocyte-seeded hydrogel constructs and cellular response to interleukin insults. Tissue Eng Part A 20(21–22):2908–2919
41. Hirst SM, Karakoti AS, Tyler RD, Sriranganathan N, Seal S, Reilly CM (2009) Anti-inflammatory properties of cerium oxide nanoparticles. Small 5(24):2848–2856
42. Oro D, Yudina T, Fernandez-Varo G, Casals E, Reichenbach V, Casals G, Gonzalez de la Presa B, Sandalinas S, Carvajal S, Puntes V, Jimenez W (2016) Cerium oxide nanoparticles reduce steatosis, portal hypertension and display anti-inflammatory properties in rats with liver fibrosis. J Hepatol 64(3):691–698
43. Zhang W, Hu SL, Yin JJ, He WW, Lu W, Ma M, Gu N, Zhang Y (2016) Prussian Blue nanoparticles as multienzyme mimetics and reactive oxygen species scavengers. J Am Chem Soc 138(18):5860–5865
44. Harman D (2003) The free radical theory of aging. Antioxid Redox Signal 5(5):557–561
45. Harman D (1956) Aging: a theory based on free radical and radiation chemistry. J Gerontol 11(3):298–300
46. Harman D (1972) The biologic clock: the mitochondria? J Am Geriatr Soc 20(4):145–147
47. Narayanan KB, Park HH (2013) Pleiotropic functions of antioxidant nanoparticles for longevity and medicine. Adv Colloid Interface Sci 201–202:30–42
48. Antonenko YN, Roginsky VA, Pashkovskaya AA, Rokitskaya TI, Kotova EA, Zaspa AA, Chernyak BV, Skulachev VP (2008) Protective effects of mitochondria-targeted antioxidant SkQ in aqueous and lipid membrane environments. J Membr Biol 222(3):141–149
49. Asami DK, McDonald RB, Hagopian K, Horwitz BA, Warman D, Hsiao A, Warden C, Ramsey JJ (2008) Effect of aging, caloric restriction, and uncoupling protein 3 (UCP3) on mitochondrial proton leak in mice. Exp Gerontol 43(12):1069–1076
50. Asin-Cayuela J, Manas AR, James AM, Smith RA, Murphy MP (2004) Fine-tuning the hydrophobicity of a mitochondria-targeted antioxidant. FEBS Lett 571(1–3):9–16
51. Gruber J, Fong S, Chen CB, Yoong S, Pastorin G, Schaffer S, Cheah I, Halliwell B (2013) Mitochondria-targeted antioxidants and metabolic modulators as pharmacological interventions to slow ageing. Biotechnol Adv 31(5):563–592

Chapter 15
Nanozymes for Antimicrobes: Precision Biocide

Zhuobin Xu, Dandan Li, Zhiyue Qiu and Lizeng Gao

Abbreviations

OH	Hydroxyl radicals
ABTS	2,2′-azino-bis(3-ethylbenzothiazoline-6-sulfonic acid) diammonium salt
AgNP	Ag nanoparticles
A. niger	*Aspergillus niger*
AuNP	Au nanoparticles
B. subtilis	*Bacillus subtilis*
CAT–NP	Catalytic iron oxide nanoparticles
CNA	g-C_3N_4@AuNPs
CLSM	Confocal laser scanning microscopy
COMSTAT	Confocal microscopy and computational analysis
DMAE	DNase-mimetic artificial enzyme
DR	Drug resistance
E. coli	*Escherichia coli*
EVD	Ebola virus disease
EBOV	Ebola virus
ELISA	Enzyme-linked immunosorbent assay
EVD	Ebola virus disease
EPS	Extracellular polymeric substances
ESR	Electron spin resonance

Z. Xu · D. Li · Z. Qiu
School of Medicine, Institute of Translational Medicine, Yangzhou University, Yangzhou 225001, Jiangsu, China

L. Gao (✉)
CAS Engineering Laboratory for Nanozyme, Key Laboratory of Protein and Peptide Pharmaceutical, Institute of Biophysics, Chinese Academy of Sciences, 15 Datun Road, Beijing 100101, China
e-mail: gaolizeng@ibp.ac.cn

X. Yan (ed.), *Nanozymology*, Nanostructure Science and Technology,
https://doi.org/10.1007/978-981-15-1490-6_15

Fe–S	Iron-sulfur
GO	Graphene oxide
GQD	Graphene quantum dot
Gt	Graphite
GtfB	Glucosyltransferase B
GtO	Graphite oxide reduced
HBrO	Hypobromous acid
HRP	Horseradish peroxidase
Listeria monocytogenes	*L. monocytogenes*
MIC	Minimum inhibitory concentration
MNP	Magnetic nanoparticles
MoS_2	Molybdenum disulfide
MPNP	Magnetic polymeric nanoparticle
MRSA	Methicillin-resistant *Staphylococcus aureus*
MSN–AuNPs	AuNPs supported on biofunctionalized mesoporous silica
NIR	Near infrared
NPC	Nanoparticle cluster
O_2^-	Superoxide anion
o-CNT	Oxygenated-group-enriched carbon nanotubes
P. aeruginosa	*Pseudomonas aeruginosa*
PEG	Polyethylene glycol
PEG-MoS_2 NFs	Polyethylene glycol functionalized MoS_2 nanoflowers
PCR	Polymerase chain reaction
PDT	Photodynamic therapy
PTT	Photothermal therapy
rGO	Graphene oxide
ROS	Reactive oxide species
S. aureus	*Staphylococcus aureus*
S. epidermidis	*Staphylococcus epidermidis*
sHA	Saliva-coated hydroxyapatite
SnS_2 NPs	Tin disulfide nanoparticles
S. typhimurium	*Salmonella typhimurium*
SPION	Superparamagnetic iron oxide nanoparticles
TMB	3,3′,5,5′-tetramethylbenzidine
V_2O_5	Vanadium pentoxide
V-HPOs	Vanadium haloperoxidases
V. cholerae	*Vibrio cholerae*

15.1 Introduction

In recent years, microbial infections caused by pathogenic microorganisms are widespread. The emergence and developments of antibiotics greatly improved the treatment of microbial infection, followed by the resistance of pathogenic microorganisms become a difficult problem in clinical practice. With the increase in bacterial resistance, the dose of antibiotics used to overcome the clinical infection is also increasing, not only makes the antibiotic efficacy gradually weakened, but also leads to the emergence of uncommon clinical side effects [1–5].

With the deep understanding of the pathogenesis of bacteria, more and more researchers have found that some chronic and refractory infectious diseases, associated with the planktonic bacteria adhere to the surface of the receptor. Bacteria, once adhered on the surface of the object, began to secrete extracellular polymeric substances to the surface, wrapped and living in mucosal layer, which was called biofilm [2, 6–10].

Bacterial biofilm research has been increasingly concerned by the international medical community. The presence of bacteria biofilm, with its local barriers and altered microenvironment reducing drug access, triggers bacterial tolerance to antimicrobials while enhancing the mechanical stability of the biofilms, making them difficult to treat or remove [2, 6, 9]. A majority of human bacterial infections and bacterial biofilm, especially osteomyelitis, endocarditis, dental caries, periodontitis and many other chronic and refractory infections are associated with biofilm-related. There are studies have shown that biofilm bacteria prone to resist to antibiotics, and biofilm bacteria easily form a new species, causing greater harm to the human body [7, 8, 11].

With the study of nanotechnology, it has been found that when the size of the material reaches nanoscale, dramatic changes can be observed in their optical, biological and physicochemical properties. Nanoscale materials are unique because of their small particle size and large specific surface area, especially in the antibacterial effects. Nano antibacterial materials can overcome some shortcomings of the traditional antibacterial materials, such as continuous effectiveness, safety, high temperature resistance, and other excellent performance with super antibacterial efficiency, low cost, and easy preparations [2, 11–13]. At present, nano antibacterial materials have been widely used in textile, medical, food, sewage control and other fields.

Both metal-based nanoparticles and organic nanoparticles have been verified to possess fine antibacterial activities. The bactericidal effects depend on the nanoparticle composition, shape, size, and surface properties. Due to the diversity of antimicrobial mechanism of nanoparticles, bacterial resistance to antimicrobial nanoparticles is not easy to occur [2, 14, 15]. However, the complexity and exclusion of the interaction between nanoparticles and biological cells and tissues also arouse deep and exact researches on nano-antibacterial materials before widely used in vivo and in vitro [2, 5, 16]. Since the researches of nanozymes have been reported, scientists have explored the antimicrobial properties of different nanomaterials with catalytic functions. The antimicrobial properties of the nanomaterials that are different in composition are constantly being found [5, 13]. Currently, several types of inorganic nanozymes with antibacterial ability have arose great attention, including metal oxide [5, 17], metal nanomaterials [18–20], and non-metallic nanomaterials, such as iron oxide [21, 22], cerium oxide [23, 24], vanadium oxide [25–27], copper oxide [28, 29], titanium oxide [19, 30, 31], gold nanoparticles [1, 11, 12, 32], sliver nanoparticles [33, 34], metal sulfide [29, 35–37], fullerene [38, 39], graphene [40–42], and carbon nanotubes [38, 39]. Inhibiting several microbial specials or preventing microbial adherence and infections were studied in vitro and applied to therapeutic use.

15.2 Nanozymes for Microorganism Detection

Detection of microorganisms is important for health and safety. Due to the unique physical and chemical properties of nanozymes, they have been extensively used to develop biosensors for rapid detection of microorganisms with microbial cells. In this section, the design principles of nanozymes-based biosensors for several selected analyte categories (bacteria cells, fungus and virus), closely associated with the target analytes' properties is reviewed. Regarding peroxidase-like activity, nanozyme can be used to replace horseradish peroxidase (HRP) in HRP-related molecular detection methods such as enzyme-linked immunosorbent assay (ELISA). Nanozymes can be directly used as an HRP alternative by conjugating with antibodies to amplify the signal via a colorimetric reaction.

15.2.1 Bacteria Detection

Human *Salmonella* enteritidis infection have always been a common epidemic disease revealing fever, diarrhea symptoms within 12–72 h, which arouse persistent attention on eggs and poultry which may harbor the organism and lead to infections. There is an urgent demand for a rapid and efficient detection of food-poisoning bacteria due to the quick onset of bacterial infection diseases as well as the long periodicity of conventional approaches like polymerase chain reaction and culture methods. Contrapose to *Salmonella typhimurium*, Ji's group developed a scheme which reacts simply and rapidly with naked-eye color change [43]. Firstly, magnetic nanozymes interacted with aptamers which were able to specifically binding to the *Salmonella* species with its peroxidase activity decreasing. After adding bacterial solution, the magnetic nanozymes combined with the *S. typhimurium* with its catalytic activity increased (Fig. 15.1). According to the colorimetric reactions involving 3,3′,5,5′-tetramethylbenzidine (TMB) and H_2O_2, *S. typhimurium* could be easily detected.

Vibrio cholera, which is responsible for cholera, the fatal disease that leads to deaths annually, shift the attention to contaminated water as the transmission of this pathogen. Polymerase Chain Reaction (PCR) technique using the ctxA gene specific

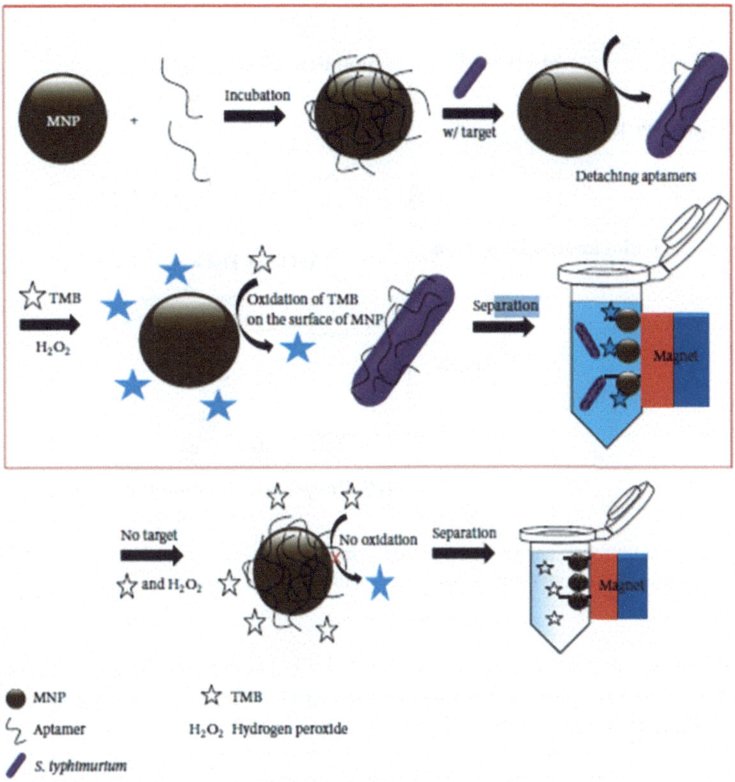

Fig. 15.1 Schematic illustration of the MNP-based colorimetric detection using label-free DNA aptamers and TMB [43]. Copyright © 2015 Ji Young Park et al.

primers performed on wastewater samples appears some limits such as the accuracy and high requirements on experimental apparatus. Raweewan's group assumed a magnetic polymeric nanoparticle (MPNP)-polymerase chain reaction-colorimetry technique to detect the target gene of *Vibrio cholerae* (*V. cholera*) [44]. MPNP was connected by an amino-modified primer specifically linked with the *ctx*AB gene. When PCR performing of the primer was finished, exert a magnet to concentrate the MPNP. Together with 2,2′-azino-bis (3-ethylbenzothiazoline-6-sulfonic acid) diammonium salt (ABTS) and H_2O_2, the amount of products could be confirmed by the colorimetric signal resulting from the peroxidase activity (Fig. 15.2) [45]. The assumption suggested a sensitive technique to identify the bacterial in daily use of water. The sensitivity of the detection was 10^3 CFU/ml.

The bacterial disease listeriosis was caused by *Listeria monocytogenes*, which were mostly resulted from contaminated food. To improve the conventional detection method of food-borne bacterial such as culture and colony counting, Xing's group developed a nanoparticle cluster (NPC) catalyzed signal amplification biosensor which took vancomycin and aptamer to recognize the cell wall of *L. monocytogenes*

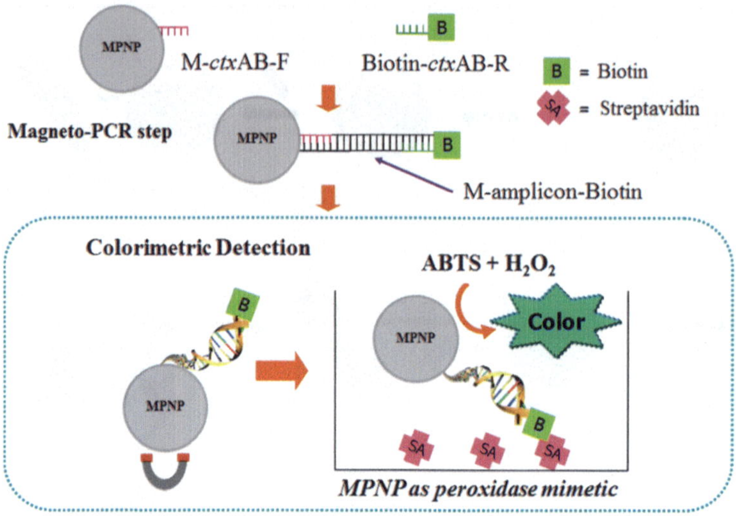

Fig. 15.2 Schematic representation of M–amplicon–biotin detection by the magneto–PCR–colorimetry technique [45]. Copyright © 2013 American Chemical Society

at distinguishing sites (Fig. 15.3) [46]. Fe_3O_4 NPC probes which cross-linked poly-L-lysine with Fe_3O_4 nanoparticles used a three-step method containing recognition, visualization and signal amplification.

15.2.2 Virus Detection

Rotavirus infection poses a great threat to infants and young children among developing and developed countries, leading to severe diarrhea diseases which badly impact on childhood morbidity and mortality. Vaccination is the only valid coping strategy to combat severe dehydrating rotavirus disease. These viruses are transmitted by the fecal-oral pathway or polluted food or water [47]. Therefore, it is necessary for joint efforts to develop a timely sensitive detection of rotaviruses, contributing to disease control and prevention. As Park's group did, when conjugated to antibodies against rotavirus, Magnetic Nanoparticles (MNP) appeared color signals through the mixture of the selected peroxidase substrate, 3,3′,5,5′-tetramethylbenzidine (TMB) and H_2O_2. The target molecules are easily quantified by measuring the absorbance of the blue color which induced the reaction between color developing agent and H_2O_2, proposing a feasible method to replace the immunoassay relied on the instabilities of enzymes (Fig. 15.4). Rotavirus antivirus were bound to the immobilized antibodies first and the sandwich-like MNP-antibody conjugates were added to the former solution later to capture the rotaviruses. In company with H_2O_2, the production of the blue colored signal demonstrated the target rotavirus exist or not [48].

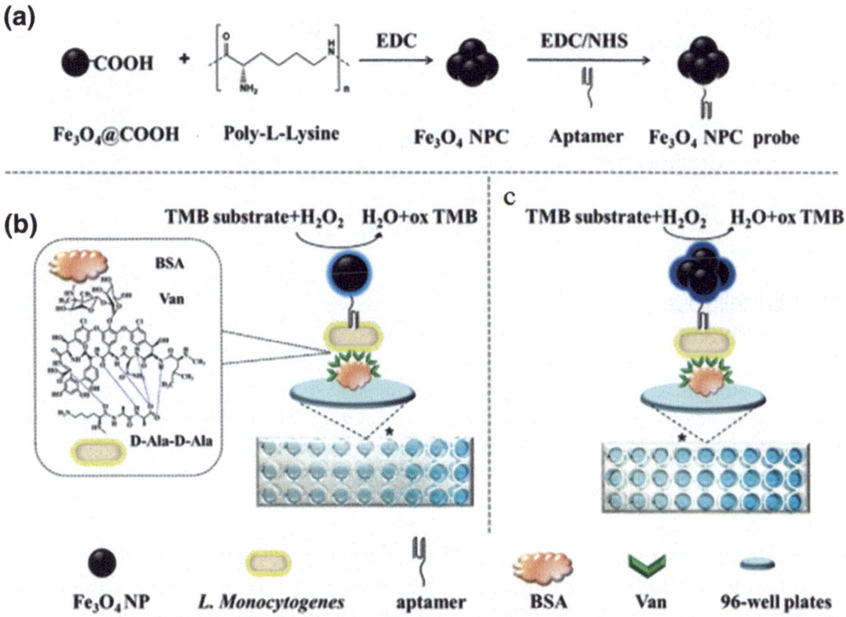

Fig. 15.3 **a** Schematic representation for the preparation of Fe_3O_4 NPC [46]. **b** The principle of the Fe_3O_4 NP-based biosensor. **c** The Fe_3O_4 NPC catalyzed signal amplification biosensor. Copyright © 2018 Elsevier B.V.

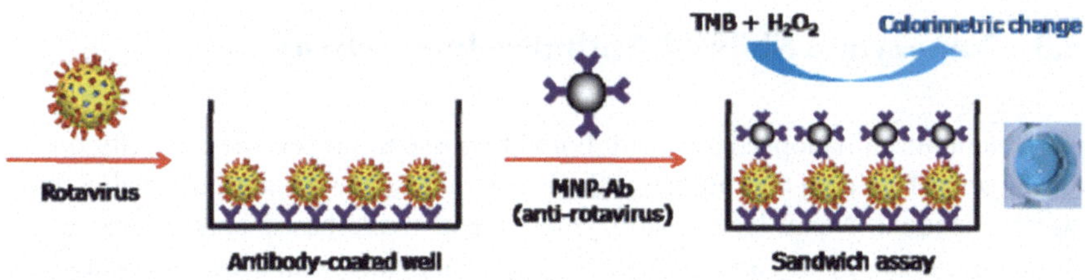

Fig. 15.4 Scheme for colorimetric detection of rotavirus in a sandwich immunoassay system. Rotaviruses are captured by antibodies immobilized on the surface of the well. MNP-Abs are then applied to the well and bind to the captured rotaviruses [48]. Copyright © 2013 by the authors; licensee MDPI, Basel, Switzerland

The outbreak of Ebola virus disease (EVD) in 2014 caused a panic among the whole society. The high fatal and rapid epidemicity of EVD led to high attention paid on the timely approach to identify and quarantine patients from health individuals [49]. Yan and her co-workers transferred Fe_3O_4 nanozymes labeling anti-Ebola virus (EBOV) antibody into a immunochromato-graphic strip based on its intrinsic peroxidase-like activity so that there were obvious color signals when it recognized EBOV [50]. The capability of detecting 1 ng/ml glycoprotein of EBOV is more than

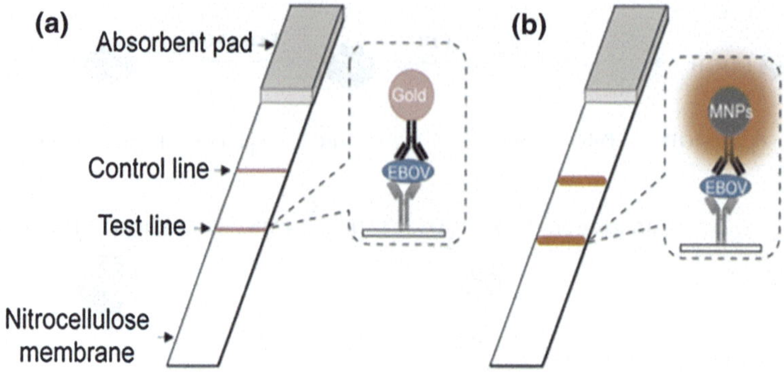

Fig. 15.5 Nanozyme-strip design [50]. **a** Standard colloidal gold strip. **b** Nanozyme-strip employing MNPs in place of colloidal gold to form a novel nanozyme probe. The probe with nanozyme activity generates a color reaction with substrates, which significantly enhances the signal so that it can be visualized by the naked-eye. Copyright © 2015 Elsevier B.V.

100-fold more sensitive than the standard colloidal gold strip. This work demonstrated that the 240 pfu/ml sensitivity of Nanozyme-strip is adequate for the detection. What's more, it also suggested a possibility that the high sensitive strip could be a platform with changeable antibodies to diagnosis particular viruses or pathogens (Fig. 15.5).

15.3 Nanozymes Showed Antimicrobial Activity

A large amount of nanoparticles were found to posse mimic enzyme activity due to their large surface area and unique crystal structure, while the antibacterial activity was also evident. Metal-based nanozymes, metal oxide-based nanozymes, metal sulfide-based nanozymes and graphene-based nanozymes with diverse biochemical and physiological properties provide an alternative route for bacteria inhibition [51–53]. It is worth applying these metal nanozymes and metal oxide nanozymes to eliminate bacterial infections.

15.3.1 Gold Nanozymes for Antibacterial Activity

Gold nanoparticles have been reported for applications in biomedicine due to their biocompatibility and photothermal or chemical activity after engineered. They are widely used for antimicrobial chemotherapy via photothermal heating under near infrared (NIR) radiation or generating reactive oxygen species (ROS) to induce damage to cells [1, 12, 32, 54–56]. AuNPs can be conjugated to several of antimicrobial drugs, including ampicillin, vancomycin, Ciprofloxacin, streptomycin and kanamycin [13, 40, 55, 57]. In addition to combined with antimicrobial drugs, AuNPs were also used for photodynamic therapy (PDT) and photothermal therapy (PTT), which were effective in cancer therapy, as well as killing pathogenic organisms. Besides, AuNPs can be decorated with bioactive substances for improvement of existing therapeutics.

Yu Tao et al. described a method for preparation of AuNPs supported on bio-functionalized mesoporous silica (MSN–AuNPs) [58]. The authors reported that this MSN–AuNPs possess intrinsic oxidase and peroxidase catalytic activities. MSN–AuNPs could catalyze the oxidation of TMB with H_2O_2 into oxidized products with maximum absorbance at 652 nm (Fig. 15.6a). Figure 15.6c showed the absorbance changes in the presence of MSN–AuNPs in different concentrations.

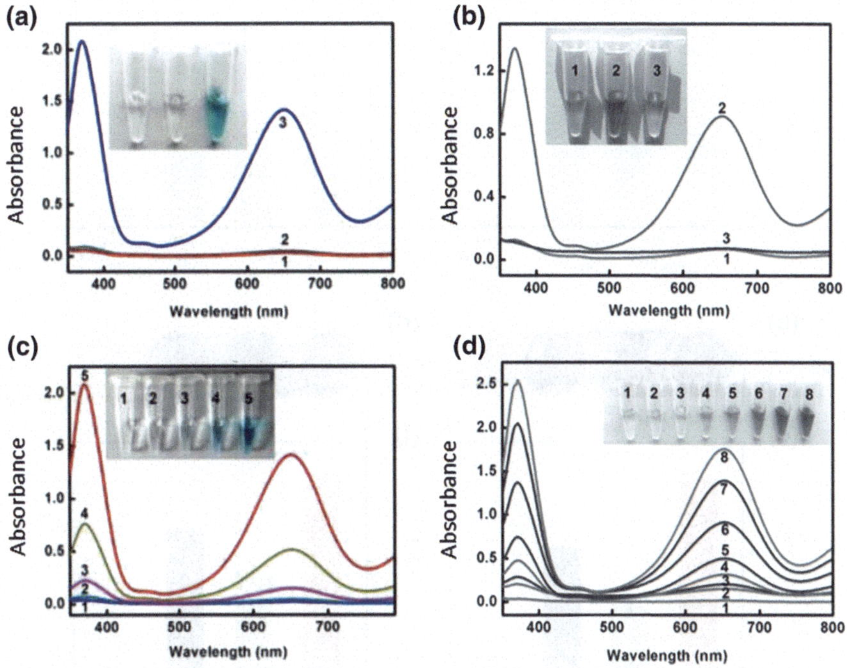

Fig. 15.6 **a** The absorbance spectra and visual color changes of TMB in different reaction systems [58]: (1) TMB + H_2O_2, (2) TMB + H_2O_2 + MSN and (3) TMB + H_2O_2 + MSN–AuNPs in a pH 4 PB buffer (25 mM) at 35 °C after 10 min incubation. **b** The absorbance spectra and visual color changes of TMB in different reaction systems: (1) TMB, (2) TMB + MSN–AuNPs, and (3) TMB + MSN in a pH 4 PB buffer (25 mM) at 35 °C after 30 min incubation. **c** The absorbance spectra and visual color changes of TMB in presence of different concentrations of MSN–AuNPs after 10 min incubation. (1) 0 μg/mL, (2) 10 μg/mL (10 μg/mL), (3) 50 μg/mL; (4) 100 μg/mL; (5) 250 μg/mL. **d** The absorbance spectra and visual color changes of TMB in presence of different concentrations of MSN–AuNPs after 30 min incubation. (1) 0 μg/mL, (2) 25 μg/mL, (3) 50 μg/mL; (4) 100 μg/mL; (5) 150 μg/mL; (6) 250 μg/mL; (7) 400 μg/mL; (8) 500 μg/mL. Copyright © 2014 Wiley-VCH Verlag GmbH & Co. KGaA, Weinheim

Furthermore, MSN–AuNPs could act as an oxidase mimic to facilitate the oxidation of TMB, apart from the intrinsic peroxidase catalytic activity (Fig. 15.6b). Meanwhile, the absorbance changes against the increasing concentrations of MSN–AuNPs (Fig. 15.6d). The authors pointed out that MSN–AuNPs showed dual enzyme activities similar to those of peroxidase and oxidase.

The MSN–AuNPs showed antibacterial activities against both Gram-negative and Gram-positive bacteria (Fig. 15.7a, b). Furthermore, the effect of the MSN–AuNPs-based antibacterial properties on preventing biofilm formation and destruction of the existing biofilm was also explored, which exhibited notable performance in biofilm elimination (Fig. 15.7c–f). These results revealed that MSN–AuNPs could generate

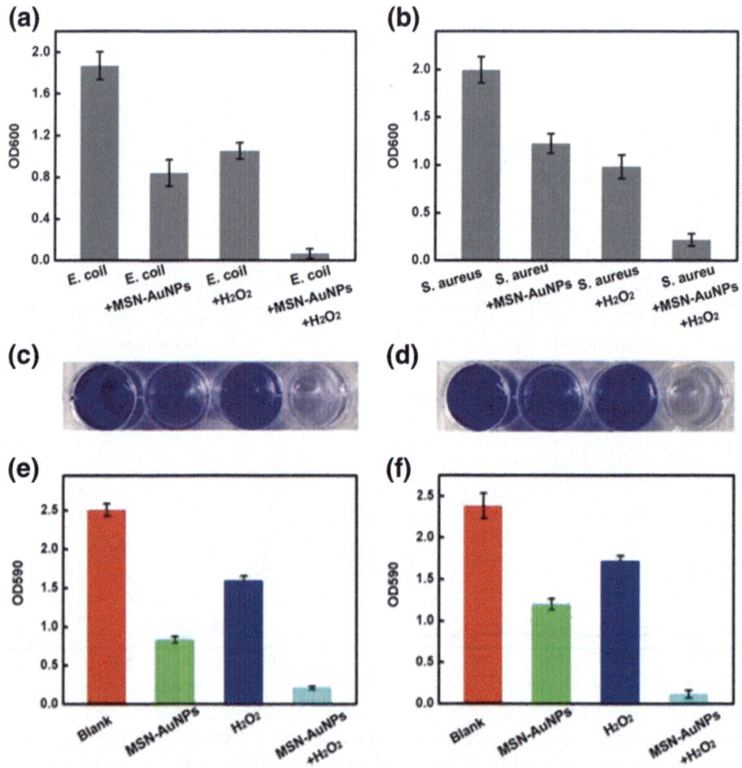

Fig. 15.7 **a** Optical density at 600 nm of bacterial suspension of *Escherichia coli* (*E. coli*) treated with MSN–AuNPs (250 μg/mL) and/or H_2O_2 (1 mM) [58]. **b** Optical density at 600 nm of bacterial suspension of Staphylococcus aureus (treated with MSN–AuNPs (250 μg/mL) and/or H_2O_2 (10 mM). **c, e** The effect of the MSN–AuNPs based antibacterial system on the biofilm destruction of *Bacillus subtilis* (*B. subtilis*). **c** Pictures of crystal-violet-stained the remaining biofilms. **e** The remaining biofilms were quantified by crystal-violet staining. **d, f** The effect of the MSN–AuNPs based antibacterial system on the biofilm formation of *B. subtilis*. **d** Pictures of crystal-violet-stained the generated biofilms. **f** The generated biofilms were quantified by crystal-violet staining. Copyright © 2014 Wiley-VCH Verlag GmbH & Co. KGaA, Weinheim

ROS and promote the antibacterial activity of H_2O_2 through converting H_2O_2 into •OH, resulting in the inhibition and destruction of biofilm.

Wang's group explored the highly efficient nanozyme hybrids of AuNPs with ultrathin graphitic carbon nitride (g-C_3N_4), which provides excellent peroxidase-activity, and catalyze the decomposition of H_2O_2 to OH radicals much more efficiently (Fig. 15.8) [51].

Furthermore, the hybrid nanoparticles not only exhibited striking bactericidal performance against both drug resistance (DR) Gram-negative and DR Gram-positive bacteria, but also showed high efficiency in breaking down the existing DR-biofilms and prevented the formation of new biofilms in vitro (Fig. 15.9). The bacterial surface

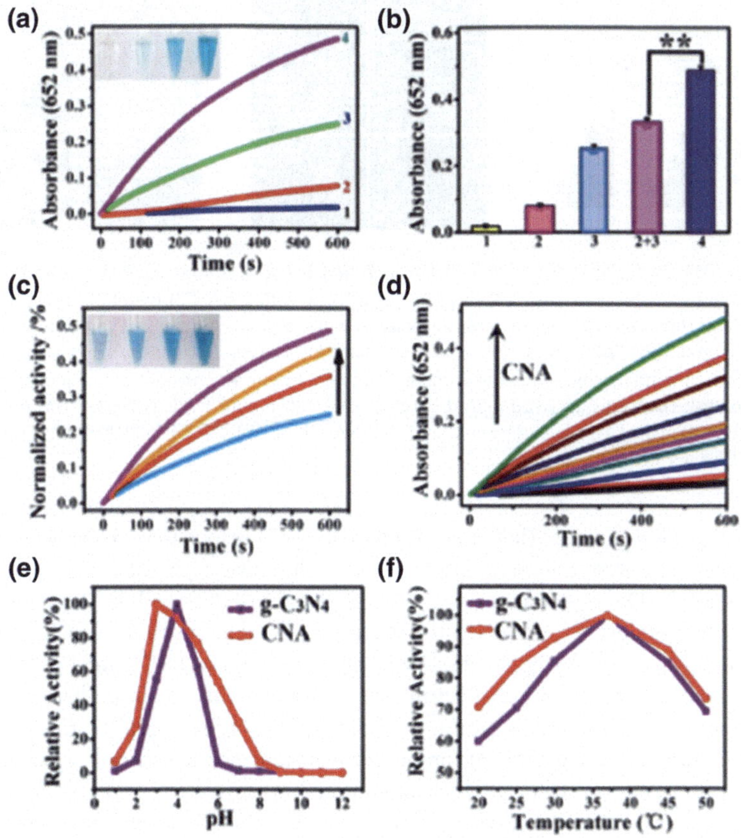

Fig. 15.8 **a** Time-dependent absorbance changes at 652 nm [51]. (1) control, (2) AuNPs, (3) g-C_3N_4 nanosheets, (4) CNA nanohybrids. **b** The absorption values at 652 nm after 600 s: 1, control; 2, AuNPs; 3, g-C_3N_4 nanosheets; 2 + 3, the sum of the absorption values of AuNPs and g-C_3N_4; 4, CNA nanohybrids. **c** Relative activities of CNA nanohybrids with various Au content (0.1, 0.2, 0.3, 0.4 mM). **d** Time-dependent absorbance changes at 652 nm of TMB reaction solutions catalyzed by the different concentrations of CNA nanohybrids. **e**, **f** The peroxidase-like activity of the CNA nanohybrids were dependent on pH and temperature. Copyright © 2016 Elsevier Ltd.

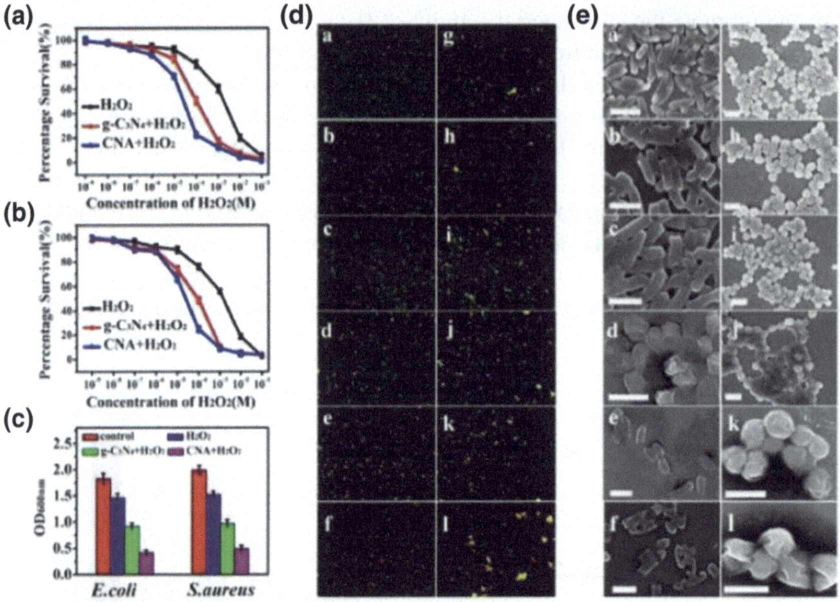

Fig. 15.9 The bacterial viability of **a** DR-*E. coli* and **b** DR-*S. aureus* incubated with different concentrations of H_2O_2 or with g-C_3N_4 (20 mg/mL) or with CNA (20 mg/mL) [51]. **c** Optical density at 600 nm ($OD_{600\ nm}$) of bacterial suspension with different treatments. Typical fluorescence images (**d**) and SEM images (**e**) of DR-*E. coli* (left) and DR-*S. aureus* (right) by various treatments. All fluorescence images were obtained under magnification of 40. All the scale bars are 2 mm. (**a**, **g**) control; (**b**, **h**) g-C_3N_4 (20 mg/mL); (**c**, **i**) CNA (20 mg/mL); (**d**, **j**) H_2O_2 (100 mM); (**e**, **k**) g-C_3N_4 (20 mg/mL) + H_2O_2 (100 mM); (**f**, **l**) CNA (20 mg/mL) + H_2O_2 (100 mM). Copyright © 2016 Elsevier Ltd

became distorted and wrinkled after exposure to H_2O_2 dispersions (100 mM), While DR-*E. coli* cells were collapsed after treated with both g-C_3N_4 and H_2O_2. In stark contrast, all the DR-*E. coli* cells were collapsed and lost their cellular integrity after treated with both g-C_3N_4@AuNPs (CNA) and H_2O_2, demonstrating much stronger antibacterial activity. A similar trend was also observed to DR-*S. aureus* cells. In order to demonstrate the broad-spectrum antibacterial ability of the antibacterial system, Other three DR-bacterial strains including multidrug-resistant *Bacillus subtilis* (MDR-*B. subtilis* ATCC 27853), ampicillin-resistant *Staphylococcus epidermidis* (DR-*S. epiddermidis* ATCC 700567) and ampicillin-resistant *E. coli* (ATCC BAA-2193) were tested. Similar results to DR-*E. coli* and DR-*S. aureus* were obtained, indicating that the CNA-based system possessed broad-spectrum antibacterial ability.

The authors also investigated the potential of their system for eradicating DR-biofilm. As shown in Fig. 15.10, when antibiotics or g-C_3N_4 or CNA was used, the DR-biofilms still integrally fixed on the surface of the wells due to the high drug resistance. Compared with control groups, H_2O_2 (1 mM) alone only induced a moderate effect for biofilm disruption. The treatment with both g-C_3N_4 (20 mg/mL) and H_2O_2 (100 mM) displayed certain effect for DR-biofilm destruction, While the

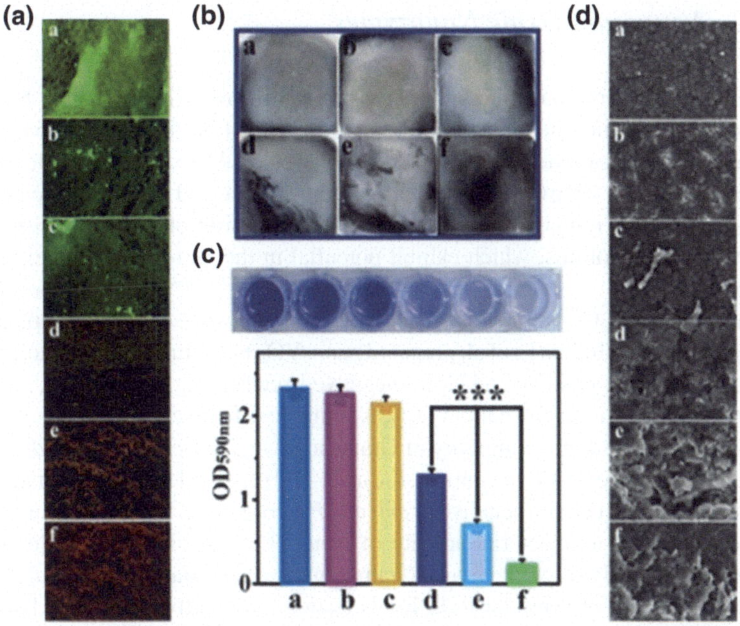

Fig. 15.10 The effect of the CNA-based antibacterial system on disrupting antibiotic-resistant biofilms of *S. aureus* [51]. Fluorescence images (**a**), representative photograph (**b**) of antibiotic-resistant biofilms with different treatments. All fluorescence images were obtained under magnification of 20. **c** The remaining biofilms were quantified by CV staining. **d** SEM images of antibiotic-resistant biofilms with treatments. All SEM images were obtained under magnification of 5 K. (**a**) control; (**b**) g-C_3N_4 (20 mg/mL); (**c**) CNA (20 mg/mL); (**d**) H_2O_2 (100 mM); (**e**) g-C_3N_4 (20 mg/mL) + H_2O_2 (100 mM); (**f**) CNA (20 mg/mL) + H_2O_2 (100 mM). Copyright © 2016 Elsevier Ltd

remaining clear biofilm bands were also observed. However, DR-bacterial biofilm was almost completely eradicated in the presence of CNA (20 mg/mL) combined H_2O_2 (100 mM).

Mohamed' group evaluated the efficiency of gold nanoparticles synthesised by co-precipitation method as an antibacterial approach against *C. pseudotuberculosis* bacteria in vitro [59]. Results revealed that minimum inhibitory concentration (MIC) of AuNPs and AuNPs–laser combined therapy were 200 μg/mL and 100 μg/mL respectively. The authors suggested that the antibacterial activity of the gold nanoparticles might be attributed to generation of ROS that increases the oxidative stress of microbial cells in form of vacuole formation as an indication of potent activity. A more attractive work was done by Jiang et al. that they creatively combined two metallic element into one which displayed significant catalytic activities, while neither single one can do [60]. Besides, the inhibition to the growth of bacteria without producing ROS also reveals another direction to prevent the bacterial infection. The existence of AuPt not only resulted in damage to the membrane potential but also elevated ATP levels inside cells.

15.3.2 Vanadium Oxide Nanozymes

Wolfgang et al. found that vanadium pentoxide nanowires with vanadium haloperoxidase activity could effectively inhibit the formation of biofilms. In company with hydrogen peroxide, vanadium pentoxide nanowires could oxidize bromide ions to hypobromous acid (HBrO) and singlet oxygen, which had a strong bactericidal activity. There is a remarkable prevention of the adhesion of microorganisms and the formation of biofilms, which exhibit potential in the anti-fouling of ship shells [25, 26].

In addition to V_2O_5 nanowires, vanadium dioxide possesses great antimicrobial activity selectively. Jinhua Li et al. prepared nano-VO_2 to evaluate their antimicrobial effect against Gram-positive and Gram-negative bacteria (Fig. 15.11) [26]. According to the investigations, the nano-VO_2 was found to effectively disrupt the bacteria morphology and membrane integrity of Gram-positive *S. aureus* and *S. epidermidis*, and eventually cause death. In contrast, the nano-VO_2 showed no significant antibacterial activity against Gram-negative *E. coli* and *P. aeruginosa*. The authors suggested that the selectively antibacterial activity of nano-VO_2 was attributed to the different sensitivity of Gram-positive and Gram-negative bacteria to intracellular ROS level, and elevated intracellular ROS levels leaded to oxidative stress and bacterial inactivation subsequently (Fig. 15.12).

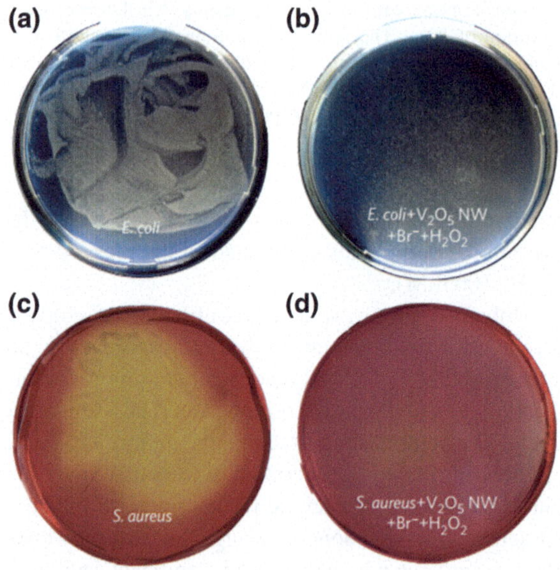

Fig. 15.11 Representative digital images showing the influence of the catalytic activity of V_2O_5 nanowires on the growth of gram-positive (*S. aureus*) and gram-negative (*E. coli*) bacteria [25]. **a** Gram-negative: *E. coli* alone. **b** *E. coli* co-incubated with V_2O_5 nanowires (0.075 mg/ml), Br_2 (1 mM) and H_2O_2 (10 mM). **C** Gram-positive: *S. aureus* only. **d** *S. aureus* co-incubated with V_2O_5 nanowires (0.075 mg/ml), Br_2 (1 mM) and H_2O_2 (10 mM). Copyright © Royal Society of Chemistry 2018

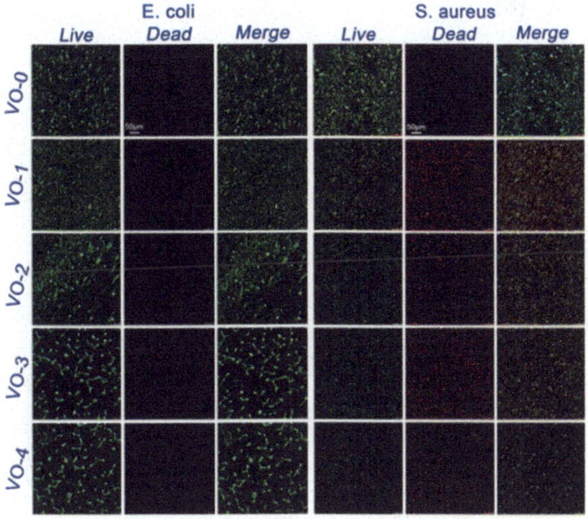

Fig. 15.12 Live/dead staining results of Gram-negative *E. coli* (left panel) and Gram-positive *S. aureus* (right panel), visualizing the bacterial viability and membrane integrity after 24 h of culture on different samples of VO_2 [26]. Copyright © Royal Society of Chemistry 2018

15.3.3 Fe_3O_4 Nanozymes

Since the great discovery of Gao and Yan's research [50, 52, 61, 62], ferromagnetic nanoparticles has been widely used to mimic natural enzymes in biomedical field. As •HO was generated by iron oxide nanoparticles in the presence of H_2O_2, peroxidase-mimicking activity helped Fe_3O_4 nanozymes to degrade nucleic acids, proteins, and polysaccharides effectively, eradicating the bacteria biofilm formation even in low level of H_2O_2 [22].

Gao investigated the ferromagnetic nanoparticles with peroxidase-like activity in biofilm degradation and prevention [22]. According to their results, the ferromagnetic nanoparticles enhanced oxidative cleavage of biofilm components, prevented new biofilms developing and efficiently destroyed the existing biofilm, with killing both planktonic bacteria and those within the biofilm (Fig. 15.13).

As bacterial die, they release a complex mixture of cellular components including nucleic acids and proteins. These released biomass might accumulate around resistant cells and contribute to biofilm formation and protection from disinfectants or antimicrobial treatments. To test whether the MNP–H_2O_2 system would degrade these complex mixtures, the authors collected the released nucleic acids and proteins and incubated them with MNP–H_2O_2 or controls, including either MNPs or H_2O_2. The results showed that treatment with MNP–H_2O_2 was successful at degrading these released products (Fig. 15.14), thus not only efficiently killing bacteria, but also degrading the biomass released from the dead cells.

Gao et al. further discussed the application of catalytic iron oxide nanoparticles (CAT–NP) used in dental biofilm-associated oral disease [21]. CAT–NP could trigger extracellular matrix degradation and cause bacterial death within acidic niches of caries-causing biofilm. It has been proved that, CAT–NP in combination with H_2O_2 can be a good idea to overcome the onset and severity of dental caries (Fig. 15.15).

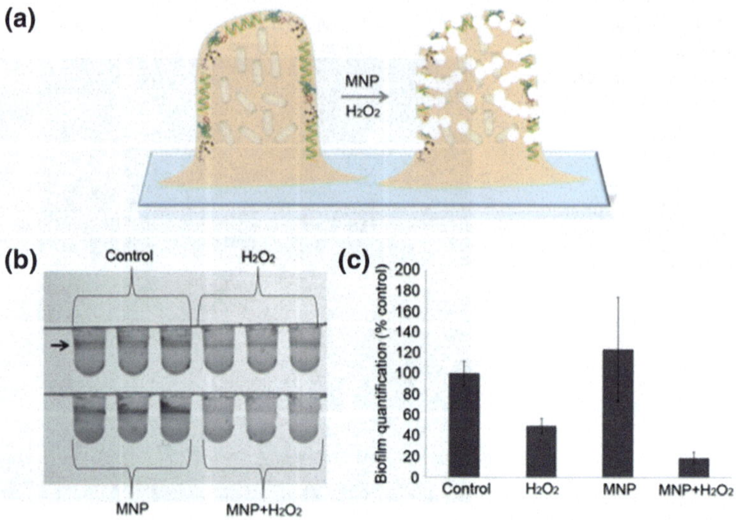

Fig. 15.13 Biofilm elimination by MNP-enhanced oxidative cleavage [22]. **a** Schematic of MNP–H_2O_2 eliminating the biofilm by cleaving nucleic acids, proteins and polysaccharides in the biofilm. **b** *P. aeruginosa* biofilm elimination with MNP–H_2O_2. **c** Quantification of the biofilm remaining after MNP–H_2O_2 treatment. Copyright © Royal Society of Chemistry 2018

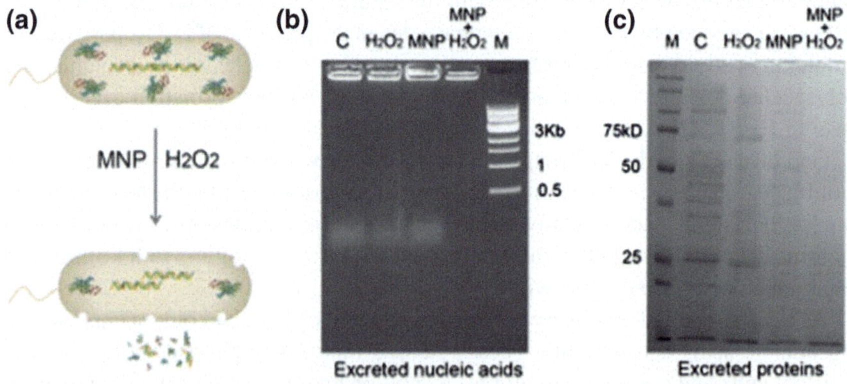

Fig. 15.14 Killing of *E. coli* and cleavage of released nucleic acids and proteins by the MNP–H_2O_2 system [22]. **a** Schematic of MNP–H_2O_2 killing of *E. coli* and cleavage of released nucleic acids and proteins. **b** MNP–H_2O_2 cleaved nucleic acids released from killed *E. coli* cells. "C" denotes control with nucleic acids only. "M" denotes the DNA marker. **c** MNP–H_2O_2 cleaved proteins released from killed *E. coli* cells. "M" denotes the protein marker. Copyright © Royal Society of Chemistry 2018

(a)

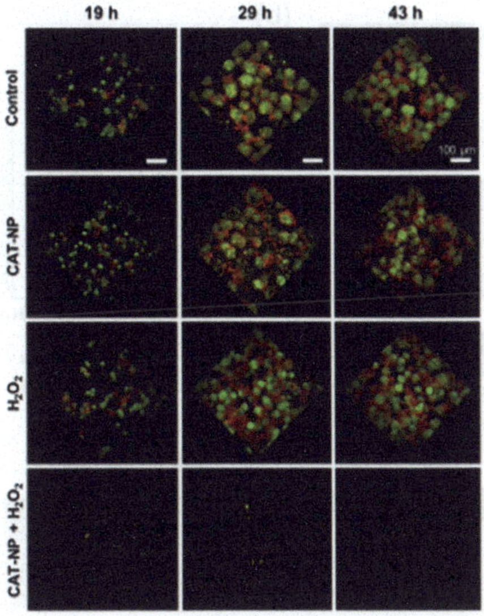

(b)

COMSTAT Analysis			
Biofilm 43h	Total Biovolume (μm³/μm²)	Cell Biovolume (μm³/μm²)	EPS Biovolume (μm³/μm²)
Control	204.2 ± 17.4	98.9 ± 6.8	105.3 ± 11.1
CAT-NP	229.5 ± 11.3	113.1 ± 12.3	116.5 ± 6.3
H_2O_2	194.8 ± 7.8	89.8 ± 6.4	105.0 ± 2.3
CAT-NP+H_2O_2	2.6 ± 0.2*	1.7 ± 0.1*	0.9 ± 0.3*

Fig. 15.15 Dynamics of biofilm disruption after topical treatments with CAT–NP + H_2O_2 [21]. **a** Confocal microscopy images at different time points. Biofilms received topical treatment by CAT–NP followed immediately by H_2O_2 exposure (CAT–NP + H_2O_2) or sodium acetate buffer (CAT–NP alone) twice daily. For H_2O_2, biofilms were treated with sodium acetate buffer followed immediately by H_2O_2 exposure. The control group consisted of biofilms treated with buffer only. Bacterial cells were stained with SYTO 9 (in green) and EPS were labeled with Alexa Fluor 647 (in red). **b** Confocal microscopy and computational analysis (COMSTAT) of total, cell and EPS biovolume for biofilm at 43 h. Copyright © 2016 Elsevier Ltd.

In other words, the CAT–NP with peroxidase-like activity can catalyze H_2O_2 to generate free-radicals that effectively breakdown extracellular polymeric substances (EPS) matrix of dental biofilm and rapidly kill bacteria for dental biofilm control and caries prevention. CAT–NP/H_2O_2 treatments were highly effective in disrupting caries development (Fig. 15.16).

Conjugation of superparamagnetic iron oxide nanoparticles (SPION) with antibacterial metal ions of $FeCl_3$, $ZnCl_2$, and $AgNO_3$ is another approach for the better eradication of persistent biofilms and antibiotic-resistant biofilms [63, 64]. In particular, this study reveals that the use of SPION could outperform antibiotics against both gram-positive and gram-negative bacteria, and even antibiotic-resistant *S. aureus* biofilms. The authors suggested that the antimicrobial activity might be

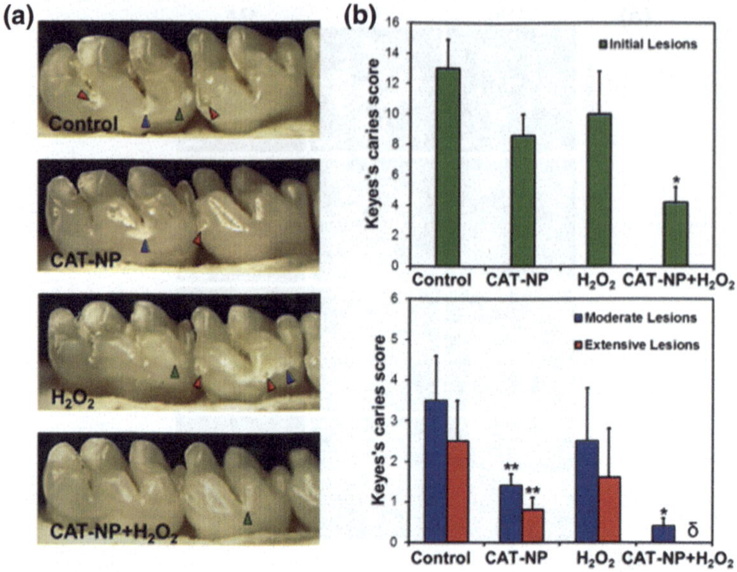

Fig. 15.16 Protection against development of carious lesions by CAT–NP/H_2O_2 treatment [21]. **a** Images of teeth from rats treated as noted. Green arrows indicate initial lesion formation where areas of the enamel is demineralized and become white; blue arrows show moderate carious lesions where areas of enamel are white-opaque or damaged. Caries scores are recorded as stages and extent of carious lesion severity according to Larson's modification of Keyes' scoring system: **b** Initial lesion (surface enamel white); **c** Moderate lesion (enamel white-opaque) and extensive (cavitation with enamel eroded and underlying dentin exposed). Copyright © 2016 Elsevier Ltd.

due to the oxidative stress since ROS formation is increased. The electron transport chain encourages the formation of Superoxide (O_2^-) and then attack iron-sulfur (Fe–S) clusters, making massive iron available for oxidation involved in the Fenton reaction. Then DNA damage, lipids and proteins destruction occur due to the formation of hydroxyl radicals (•OH) (Fig. 15.17) [65].

15.3.4 TiO_2 Nanozymes

Antimicrobial property of TiO_2 was researched for antibacterial property related to its catalytic activity and optical activity [19, 66]. TiO_2 has been widely investigated in bactericidal application for bacteria, fungi, and viruses. Jirapat A studied the antibacterial activity of undoped, Ni-doped, and N-doped TiO_2 (Fig. 15.18) [67]. The results showed that the N-doped TiO_2 nanoparticles exhibited better antibacterial activity than the Ni-doped TiO_2. K. Vijayalakshmi compared the antibacterial activity of TiO_2 with Ba doped TiO_2 nanoparticles by a microwave assisted method [68]. The antibacterial activity of Ba doped TiO_2 nanoparticles was found to be higher than

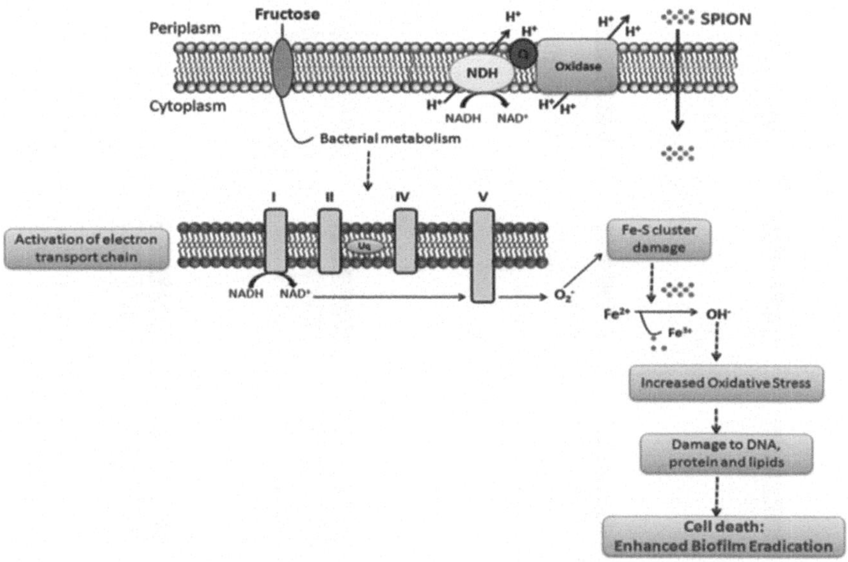

Fig. 15.17 Possible mechanism for the enhanced anti-biofilm activity when fructose metabolites were added to SPION treatment [65]. Enhanced activity is due to the increased penetration of nanoparticles within the biofilm which leads to increases in ROS formation. The uptake of fructose metabolites might induce the catabolism of metabolites and the generation of NADH. The bacterial electron transport chain oxidizes NADH and contributes to PMF. Then, bacterial cell death might occur due to the oxidative stress from ROS formation. First, superoxide (O_2^-) formation could be stimulated by the electron transport chain and O_2^- might damage Fe–S clusters, making more ferrous iron available for oxidation by the Fenton reaction. The Fenton reaction might lead to the formation of hydroxyl radicals (•OH), which can damage DNA, lipids and proteins. All of these events eventually cause bacterial cell death and the better eradication of biofilms. Copyright © 1999–2018 John Wiley & Sons, Inc.

pure TiO_2 nanoparticles. In the above two literatures, it seems that higher amount of hydroxyl radicals and ROS and more visible light absorption were generated within the N-doped TiO_2 and Ba-doped TiO_2 nanoparticles compared to other groups.

15.3.5 CeO_2 Nanozymes

As a notable oxide metal nanomaterial, CeO_2 could catalyze ROS formation which provides great portion in killing bacterial. CeO_2 nanomaterials were also known for its storage ability and high oxygen transport [23, 24]. Besides, there were also evidences indicating that CeO_2 nanoparticles could adhere on the *E. coli* cell membrane. According to Wang's work, Ag/CeO_2 were synthesized with various shape and size such as nanocubes, nanorods, and nanoparticles [69]. The electron spin resonance (ESR) result of •OH and •O_2 formation on the surface of CeO_2 and Ag/CeO_2

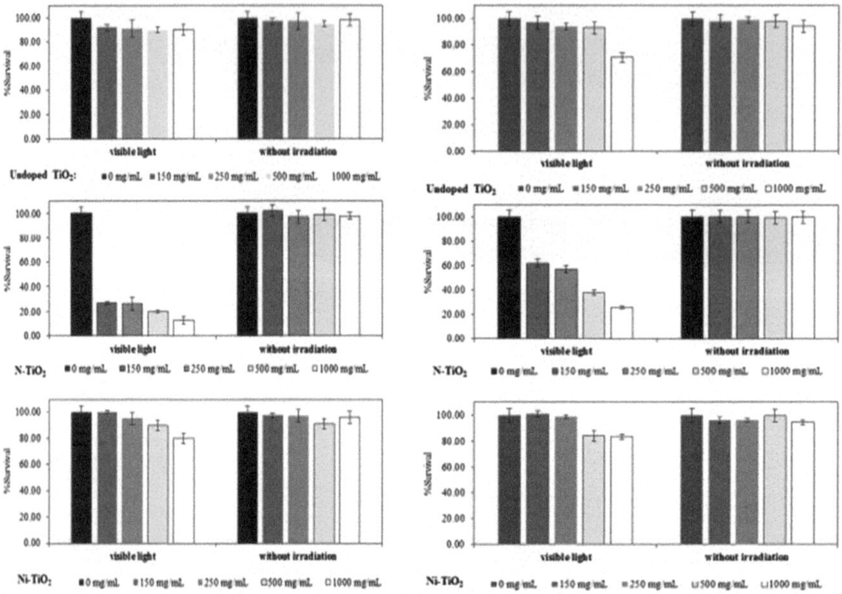

Fig. 15.18 Photocatalytic inactivation of *S. aureus* and *E. coli* with different concentrations of TiO_2 under visible light [67]. Copyright © 2017 Springer International Publishing AG. Part of Springer Nature

nanomaterials provided enough evidence that the generated various of active oxygen species result in the direct and adequate contact between *E. coli* and ROS, leading to destruction of the cell wall and membrane. As a catalyst, the inactivation of *E. coli* cells mostly attribute to the strong oxidative capabilities of extracellular ROS produced by CeO_2 while the toxicity of Ag^+ play a minor role of antibacterial activity.

Chen and co-workers constructed a DNase-mimetic artificial enzyme (DMAE) by confining passivated gold nanoparticles with multiple cerium (IV) complexes on the surface of colloidal magnetic Fe_3O_4/SiO_2 core/shell particles, a robust and recoverable artificial enzyme with DNase-like activity was obtained, which exhibited high cleavage ability towards both model substrates and eDNA [53].

Significantly reduced adhesion (>90%) and much thinner biofilms (<5 mm) were observed for the DMAE-coated surfaces even after 120 h (Fig. 15.19). However, for DNase I, the amount as well as the thickness of the adherent biofilm were low only up to 24 h of cultivation and increased gradually afterwards. Overall, by taking advantage of the eDNA degradation ability and inherent stability of DMAE, prolonged inhibition of bacterial adhesion was achieved, which avoided the continuously adding fresh natural DNase to new generations of biofilms. The authors demonstrated that DMAE could inhibit biofilm formation for prolonged periods of time and eliminate biofilms in the presence of antibiotics. By degrading the eDNA present in the EPS, the DMAE not only strongly reduced bacterial adhesion and prevented biofilm formation over 120 h, but also effectively dispersed preformed biofilms of varying ages

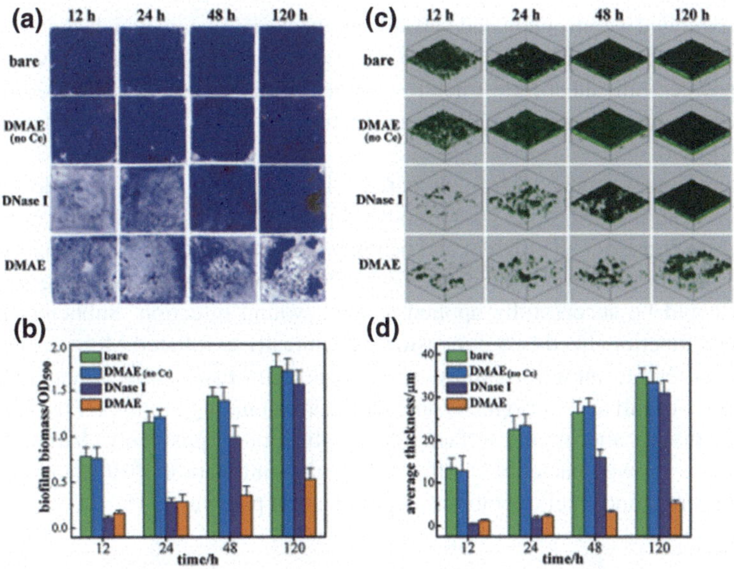

Fig. 15.19 Inhibited biofilm formation on DMAE-modified surfaces [53]. **a** Biofilms formed on bare, DMAE(no Ce)-, DNase I-, and DMAE-modified surfaces that were left to grow for different periods of time were stained with crystal violet. **b** Quantitative analysis of the crystal-violet-stained biofilms in (**a**) by measuring the absorbance at 590 nm; each bar represents the average of three experiments. **c** 3D confocal laser scanning microscopy (CLSM) of biofilms formed on bare, DMAE(no Ce)-, DNase I-, and DMAE-modified surfaces after different incubation times. The biofilms were stained with Calcein-AM. **d** Average thickness of the biofilms shown in (**c**) as calculated by the COMASTAT software. Copyright © 1999-2018 John Wiley & Sons, Inc.

and could be magnetically recovered and recycled for five runs. Moreover, disrupting the integrity of EPS by DMAE greatly increased the activity of antibiotics towards killing the enclosed bacteria, finally eliminating the biofilms [53].

15.3.6 Molybdenum Disulfide (MoS_2) Nanozymes

Since the peroxidase-mimic ability has been found on molybdenum disulfide (MoS_2) nanomaterials, numerous tests and papers proved its application for not only colorimetric detection of H_2O_2 and glucose, but also bactericidal assay and wound healing. MoS_2 nanomaterials and its functionalized materials were capable of generating hydroxyl radicals (•OH) to prevent bacterial infection [35, 70, 71]. Yin et al. demonstrated that polyethylene glycol functionalized molybdenum disulfide (PEG-MoS_2) nanoflowers, which possess peroxidase-like activity and PTT efficacy, could damage the bacterial cell wall in company with low concentration of H_2O_2 and result in wrinkles and other physical damage of cell membrane [35]. Although the antibacterial ability was only evidenced on *E. coli* and *B. Subtilis*, both the in vitro and in vivo results exhibited a convenient and versatile antibacterial system which assembles the PEG-MoS_2 nanoflowers, H_2O_2 and 808 nm laser exposure. According to Xu's work, MoS_2 powder with sheet-like structure enlarged the surface of MoS_2 materials, and provided adequate contact for ROS and the bacteria through high-specific surface area [72].

In some other studies, MoS_2 prepared with metal ion attracted widely attention. Cao et al. infused the cysteine-modified molybdenum disulfide with minimum Ag^+ ions to construct an efficient and benign antimicrobial depot [37]. Their results showed that this system exhibited much enhanced broad-spectrum antibacterial activity and could be successfully applied to treat wound infection. Subhendu Pandit reported a functionalized two-dimensional chemically exfoliated MoS_2 (ce-MoS_2), which was able to inhibit a gram-positive methicillin-resistant *S. aureus* (MRSA) and a Gram-negative *P. aeruginosa*, and their corresponding biofilms (Fig. 15.20). It is proved that by altering the surface functionalization of positively charged MoS_2, the products showed bactericidal effects through combination of ROS-independent oxidative stress and a cell membrane depolarization pathway [71].

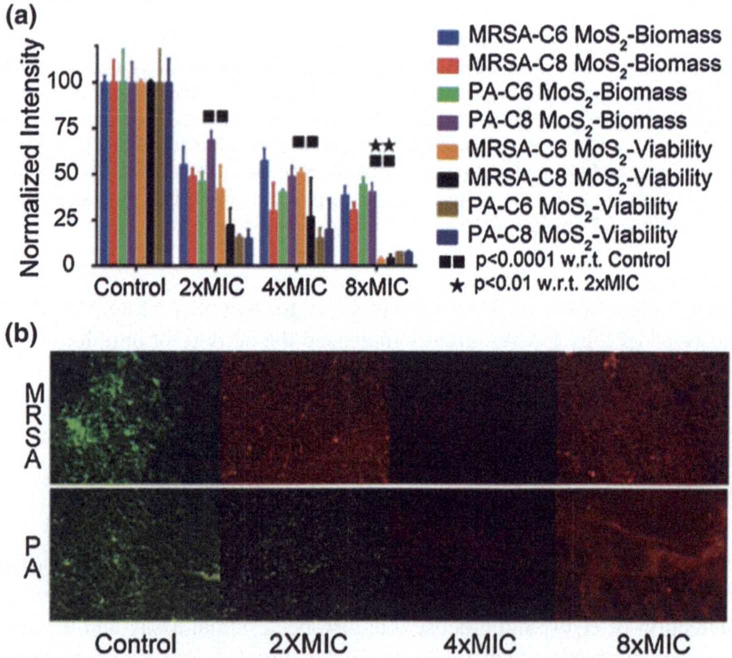

Fig. 15.20 Antibiofilm properties of functionalized ce-MoS_2 against 48 h grown mature biofilm of MRSA and *P. aeruginosa* [71]. **a** Biomass and viability of the functionalized ce-MoS2-treated biofilm. **b** Live-dead stained fluorescence microscope image of the treated biofilms. Green indicates live bacteria stained by Syto9. Copyright © 2016 American Chemical Society

15.3.7 Iron Sulfide Nanomaterials

It is a long history since people used plants in *Allium genus* to treat microbial infection. Based on the traditional health care methods, the idea of transforming organosulfur compounds into nano-sized iron sulfide was translated into reality by Gao et al. Except for the release of polysulfanes, the enzyme mimic activity possessed by nFeS itself which accelerated the hydrogen polysulfanes dispersion attribute to the destruction to several kinds of microbe including gram-negative bacteria such as *E. coli*, *P. aeruginosa*, *S. enteritidis* and gram-positive bacteria such as *S. mutans*, *S. aureus* as well as drug-resistant strains.

Apart from the marvelous bacteria killing effect, these series of nFeS presented exact cariogenic biofilm elimination on both saliva-coated hydroxyapatite (sHA) surface models and human tooth surface models as well (Fig. 15.21). The decrease of biofilm dry weight together with the cell viability demonstrated the feasibility of nFeS as an agent for dental biofilm therapy. The confocal 3D images of *S. mutans* biofilm demonstrated that nFeS restrained the development of extracellular matrix as well as the growth of the bacterial flora. The assessment of glucosyltransferase B (GtfB) exoenzyme activity showed more evidence of the destruction [73].

15.3.8 Graphene-Based Nanozymes for Antimicrobial Therapy

As a kind of non-metallic nanomaterials, graphene and its derivatives have been found multifunctional biological applications in many fields due to their superior features inherit from the nanosheet structure, unique thermal quality and lower toxicity.

Graphene related nanomaterials were reported that its capable of produce superoxide anion (O_2^-) could induce ROS production, which leads to cytotoxicity to bacterial cells [16, 74, 75]. The results from Chen's work obviously revealed that graphite (Gt), graphite oxide (GtO), graphene oxide (GO) and reduced graphene oxide (rGO) all exhibit distinct inhibition of *E. coli* viability [76]. Despite of some differences in these materials, there still exist time-dependent and concentration-dependent bacteria inactivation activities between GO and rGO, two materials with better antibacterial behaviors. Except the antibacterial ability, the functionalized graphene quantum dots also showed obvious antimycoplasma properties, together with advantages of environmental friendly and low cost.

Recently, there were researches developed by Qu's group proved that oxygenated-group-enriched carbon nanotubes (o-CNTs) performed high peroxidase activity for biocatalytic reaction. With the confirmation of carbonyl groups as the enzymatic activity center, this kind of nanozymes deactive carboxyl groups to make the antibacterial function more prominent. Generation of reactive oxygen species with high efficiency, the platform based on o-CNTs grew into a practical against both gram-negative and gram-positive bacteria. The successful animal model of curing the

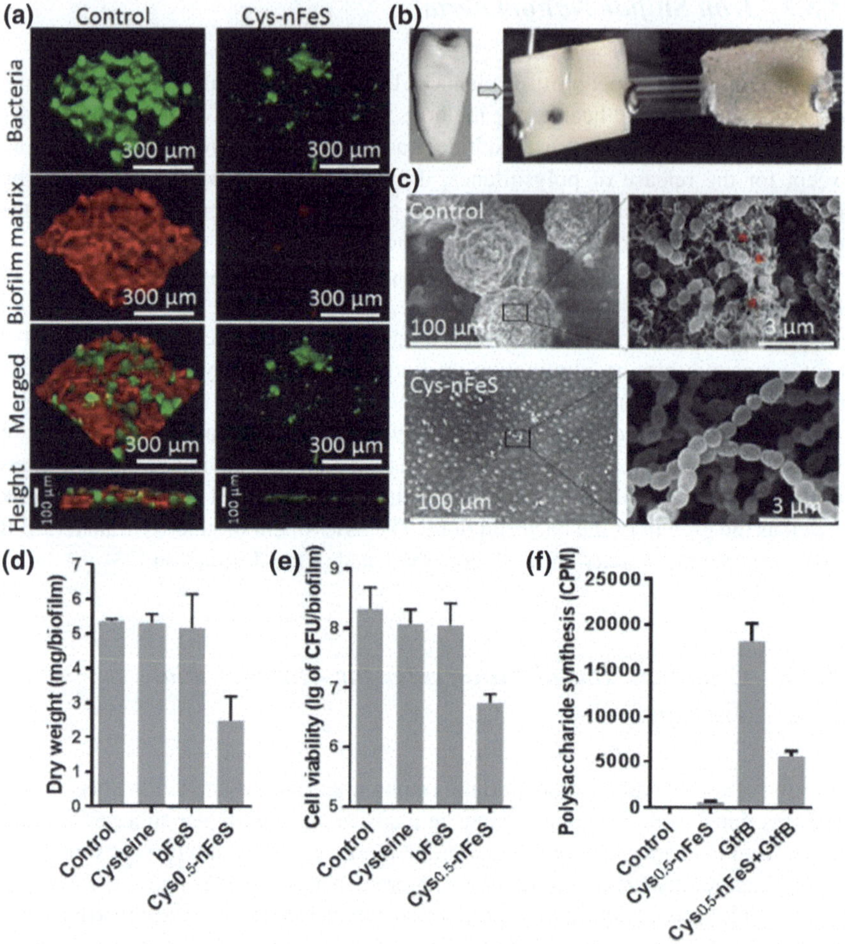

Fig. 15.21 Biofilm elimination on tooth surface using nFeS regimen. **a** Confocal 3D image of a *S. mutans* UA159 biofilm treated by Cys-nFeS. Scale bars: 100 μm. **b** *S. mutans* biofilm formed on a dentin surface sectioned from human tooth. **c** SEM image of a S. mutans biofilm treated by Cys-nFeS. The red arrows indicate EPS. Left scale bars: 100 μm. Right scale bars: 3 μm. **d**, **e** Dry weight and cell viability of a *S. mutans* biofilm treated with Cys-nFeS. **f** Inhibition on GtfB activity for polysaccharide synthesis by Cys-nFeS. CPM counts per minute. Data are shown as the mean ± S.D. Statistical significance was assessed using an unpaired Student's two-sided t-test compared to the control group

infected wound with the therapy further proves that o-CNTs-based nanomaterials would be a promising antibacterial material [77].

Qu's group reported that graphene quantum dots (GQDs) possess outstanding performance in inhibiting wound infection, assembling both peroxidase-like activity and biocompatibility [78]. GODs could catalyze H_2O_2 into •OH which greatly enhance the effect of killing both gram-negative and gram-positive bacteria while enormously

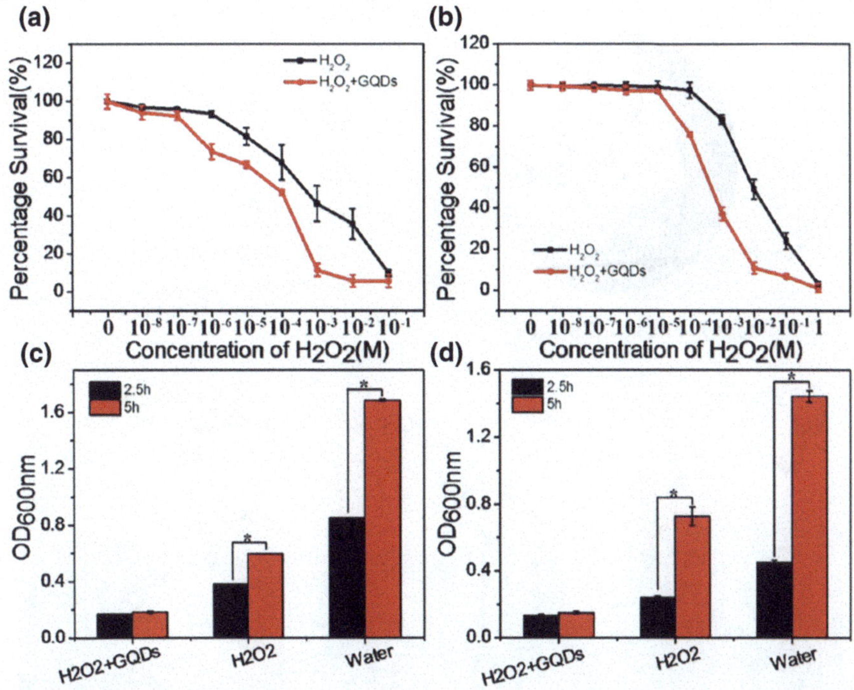

Fig. 15.22 Survival rates of *E. coli* (**a**) and *S. aureus* (**b**) treated with H_2O_2 at different concentration coincubation with or without GQDs (100 μg/mL). Optical density at 600 nm ($OD_{600\ nm}$) of bacterial suspension treated with 100 μg/mL GQDs and 1 mM H_2O_2 (*E. coli*) (**c**) and 10 mM H_2O_2 (*S. aureus*) (**d**) at 2.5 and 5 h. The original bacterial concentration is 10^6 cfu/mL [78]. Copyright © 2014 American Chemical Society

decrease the toxicity of high concentration H_2O_2 (Fig. 15.22). Furthermore, due to the treatment with GQDs and H_2O_2, the smooth and intact cell walls were impaired and became rough and wrinkled. More experimental results demonstrated that the GQD-Band-Aids could sterilize significantly during the wound infection. According to the evidence above, it is strongly believed that graphene and its derivatives may be a promising agent for antibiotic clinical application.

Through further transformation, a H_2O_2-free depots were designed by gathering the catalyze capability of graphene and hemin and the ability of producing H_2O_2 while reacting with water (Fig. 15.23). The formed ROS damaged the biofilm by destroying the components such as bacteria, proteins, nucleic acids and polysaccharides [79].

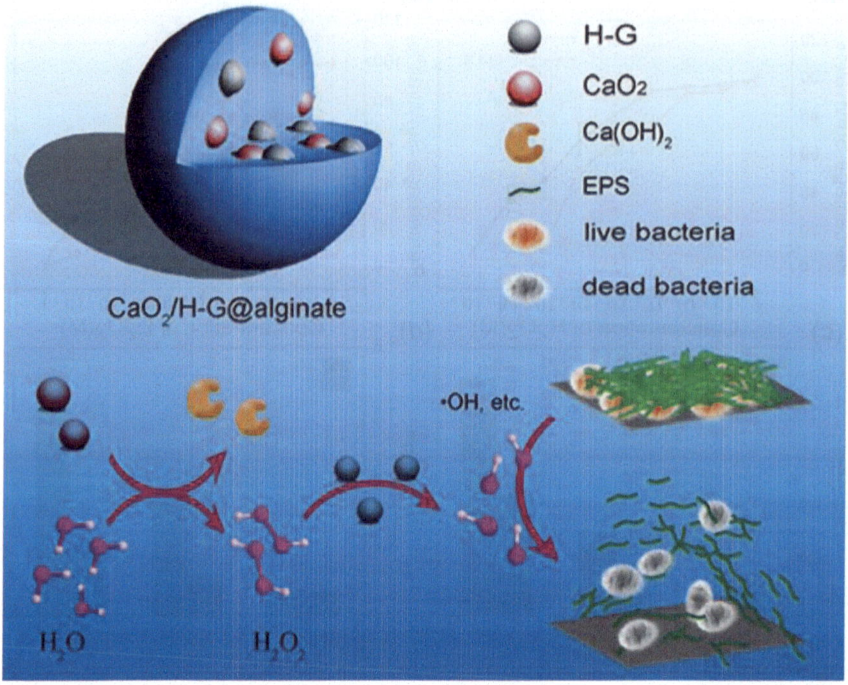

Fig. 15.23 Schematic illustration of CaO_2/H-G@alginate depots to treat bacterial infection [79]. Copyright © Royal Society of Chemistry 2018

15.4 Nanozyme for Wound Disinfection and Healing

Skin is a necessary natural barrier for protecting biological internal organs, regulating body temperature and maintaining metabolism [51, 80, 81]. Once skin was hurt, microorganisms would break the skin barrier easily, and subsequently form colonies thereby leading to different levels of wound infection and even causing life-threatening infections. With the prominent antibacterial ability and outstanding biocompatibility in vitro, several highlighted antimicrobial nanozymes were evaluated its ability in healing of skin wounds of rats suffered from the infection of pathogenic microorganism [37, 51, 54, 82].

Graphene quantum dots (GQDs) used to enhance the antibacterial efficiency of H_2O_2 for bacterial infection prevention in wound disinfection due to their high peroxidase-like activity and low toxicity in vivo were reported by Sun H et al. [78]. The presence of GQDs could convert H_2O_2 (low antibacterial efficiency) into OH radicals (high antibacterial activity), thus improving the antibacterial activity and avoiding the high concentration of H_2O_2 used in wound healing (Fig. 15.24).

Researchers utilized Kunming mice with wound on their back as a model to assess the antibacterial and wound healing efficacy of GQD-Band-Aid. The wounds of mice

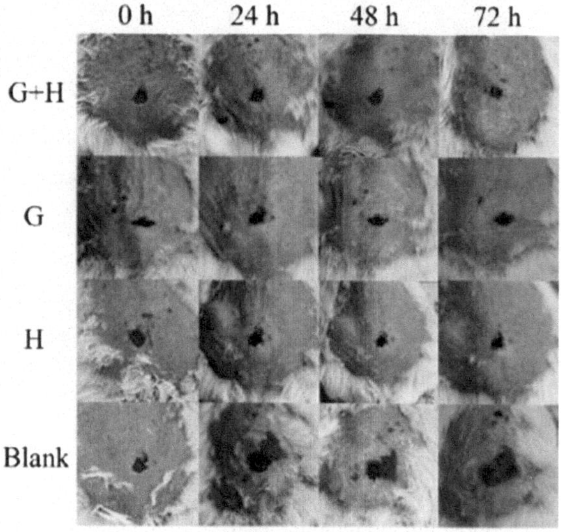

Fig. 15.24 Photographs of wound on the mice from the four groups at different times during the therapeutic process [78]. G + H: H_2O_2 + GQD-Band-Aid group; G: GQD-Band-Aid group; H: H_2O_2 + Blank-Band-Aid group; Blank: Saline + Blank-Band-Aid. Copyright © 2014 American Chemical Society

treated with H_2O_2 + GQD-Band-Aid did not show any erythema and edema during the experiment, and the wounds of mice formed scabs after 3 days of therapy in this group. While the other three groups appeared erythema and edema, and scabs were not found. It was noteworthy that the remaining bacteria of GQD-Band-Aid group was less than that of saline group, even though the GQDs did not display any antibacterial activity in vitro. The mechanism might be attributed to that the H_2O_2 released by injured cells with the presence of GQDs on the Band-Aid could suppress bacteria on the wounds.

The ultrathin graphitic carbon nitride (g-C_3N_4) has been reported to pose intrinsic peroxidase-like activity similar to that found in natural peroxidases. Zhenzhen Wang et al. fabricated a novel nanozyme based on g-C_3N_4@AuNPs (CNA) nanocomposites with dramatically enhanced peroxidase catalytic activity [51]. Kunming mice wound model was fabricated to evaluate the anti-infective property of CNA-based antibacterial system. As shown in Fig. 15.25, the CNA + H_2O_2 group could significantly accelerate wound healing process compared with control, g-C_3N_4, CNA, and H_2O_2 groups. Differing from the CNA + H_2O_2 group, different level of inflammatory reaction was found in other groups. Even worse, the wounds of mice that treated with control group and g-C_3N_4 group formed the biofilms due to the bacterial infection. However, the wound of the mice treated with both CNA and H_2O_2 was healed and started to grow hair around the wound.

The histological analyses results displayed that a complete and thickened epidermis was observed in CNA + H_2O_2 treated group compared with other groups. The results indicated that CNA + H_2O_2 system could significantly defeat bacterial infection and accelerate the healing rate of artificial wounds with H_2O_2 (10 mM), which could avoid side effects from high concentrations of H_2O_2.

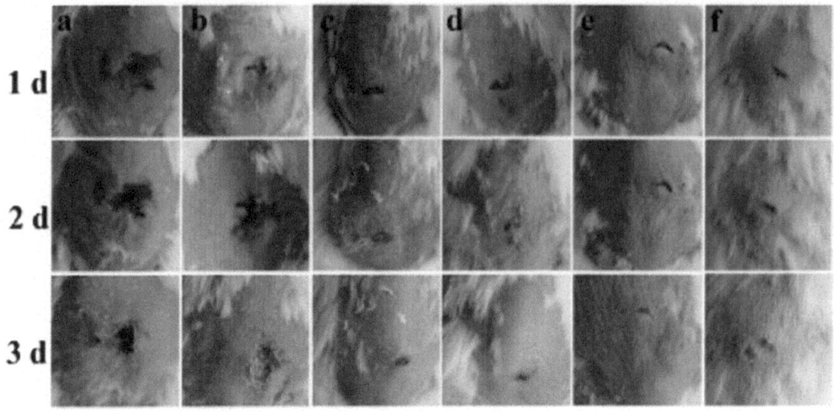

Fig. 15.25 Photographs of wound on the mice from the six groups at different times during the therapeutic process [51]. **a** control; **b** g-C_3N_4(20 mg/mL); **c** CNA (20 mg/mL); **d** H_2O_2 (100 mM); **e** g-C_3N_4(20 mg/mL) + H_2O_2 (100 mM); **f** CNA (20 mg/mL) + H_2O_2 (100 mM).

So far, MoS_2 nanomaterials have been used for killing bacteria and wound disinfection as their intrinsic peroxidase-like catalyst. Wenyan Yin et al. reported the development of polyethylene glycol functionalized MoS_2 nanoflowers (PEG-MoS_2 NFs) as wound healing material for bacterial infections [35]. A conception based on •OH produced by exogenous enhancing PTT to fight bacteria was carried out.

The authors evaluated the healing ability of PEG-MoS_2 in wound-infecting bacterial skin of mice (Fig. 15.26). Ampicillin-resistant *E. coli* infected wound of mice were treated with PBS, H_2O_2, MoS_2, MoS_2 + NIR, MoS_2 + H_2O_2, and MoS_2 + H_2O_2 + NIR respectively. The pictures of wound healing treatment groups and the histologic analyses were carried out at day 2 and day 5. Scars appeared in all treatment groups except the control group at the second day. The keratinocytes began to increase in the wound site for all treatment groups, and scars became smaller at the fifth day. Wound disappeared for synergistic MoS_2 + H_2O_2 + NIR group whereas be observed in other groups. More strikingly, a mostly entire epithelial tissue was found in the synergistic group. Overall, MoS_2 exhibited significant activity in fighting against drug-resistant bacterial infections of wounds.

Besides wound healing, nanozymes also demonstrated therapeutic effect for mouth ulcer. Yunhao Gu et al. reported that vitamin B_2 modified iron oxide nanozymes not only suppressed bacteria in oral, but also exerted anti-inflammatory effect in the loci of mouth ulcer [83]. In particular, the anti-inflammatory effect relied on the ROS scavenging by superoxide dismutase-like activity enhanced by vitamin B_2 modification. This work indicates that nanozymes may be used for accelerating healing of mouth ulcer.

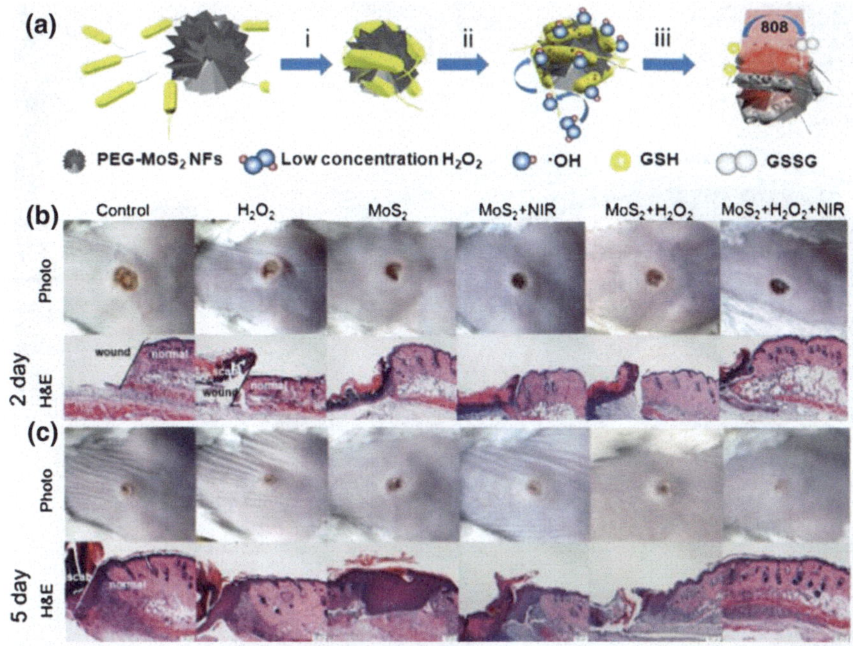

Fig. 15.26 **a** Schematic illustration of PEG-MoS_2 for peroxidase catalyst-photothermal synergistic eliminating of bacteria and wound healing [35]. Photographs of Ampicillin-resistant *E. coli* infected wound treated with PBS (control), H_2O_2, MoS_2, MoS_2 + NIR, MoS_2 + H_2O_2, and MoS_2 + H_2O_2 + NIR at **b** the second day and **c** the fifth day and their corresponding histologic analyses. Copyright © 2016 American Chemical Society

15.5 Nanozymes for In Vivo Infectious Treatment

Although successfully showing effects on wound healing, it is still unclear whether nanozymes perform antibacterial activity for the treatment of in vivo infection. Rourong Shi et al. recently reported that iron oxide nanozymes showed suppressive efficacy on *Salmonella enteritidis* under intracellular environment [84]. They confirmed that the antibacterial activity was through peroxidase-like activity which could be performed under the acidic environment of lysosome in the host cell due to the colocalization of iron oxide nanozymes and the bacteria. Furthermore, they demonstrated that iron oxide nanozymes can be used to suppress bacterial infection in the liver tissue of a chick by oral administration (Fig. 15.27). This work indicates that iron oxide nanozymes may be used as feed additive to control and prevent bacterial infection in veterinary field.

Besides *Salmonella enteritidis*, iron oxide naozymes were found to be able to disrupt biofilm formed by *Salmonella typhimurium* [85]. Yinyan Yin et al. reported that iron oxide nanozymes not only prevented the biofilm formation, but also destroyed the matured biofilm. And these antibacterial ability was also through peroxidase-like

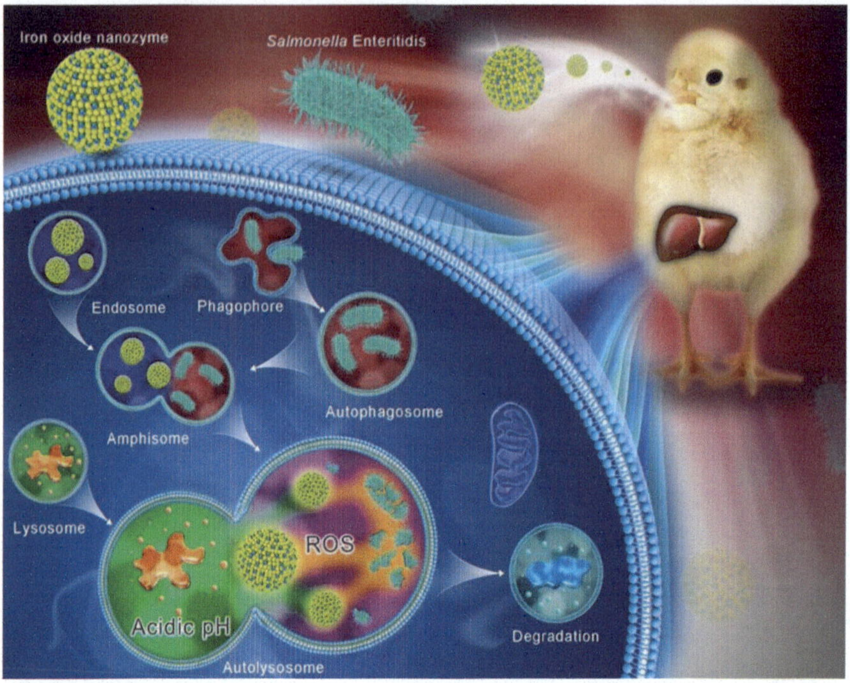

Fig. 15.27 The schematic of the process for iron oxide nanozymes suppressing intracellular *Salmonella enteritidis* in a chick model [84]. Copyright © Ivyspring International Publisher

activity to regulate intracellular ROS levels. This result indicates that iron oxide nanozymes can be used for preventing and treating relevant diseases caused by biofilm of *Salmonella typhimurium.*

15.6 Nanozymes for Antivirus

The antiviral activity of nanozymes has been investigated in the most recent work. Tao Qin et al. reported that iron oxide nanozymes were able to disrupt viral structure and deactivate influenza virus [86]. After incubation with iron oxide nanozymes, influenza virus lost infectivity quickly and the integrity of virus structure was significantly destroyed. Mechanism assay demonstrated that lipid peroxidation was induced by iron oxide nanozymes by mimicking lipoxidase-like activity (Fig. 15.28). The lipid envelope was required for this antiviral property. They proposed that the antiviral mechanism was that the induction of lipid peroxidation degraded the proximal proteins of virus including hemagglutinin, neuraminidase and matrix protein 1. The deactivated virus lost the ability to infect the host cell and did not cause flu symptom in mice model. Furthermore, this antiviral mechanism possesses a broad antiviral

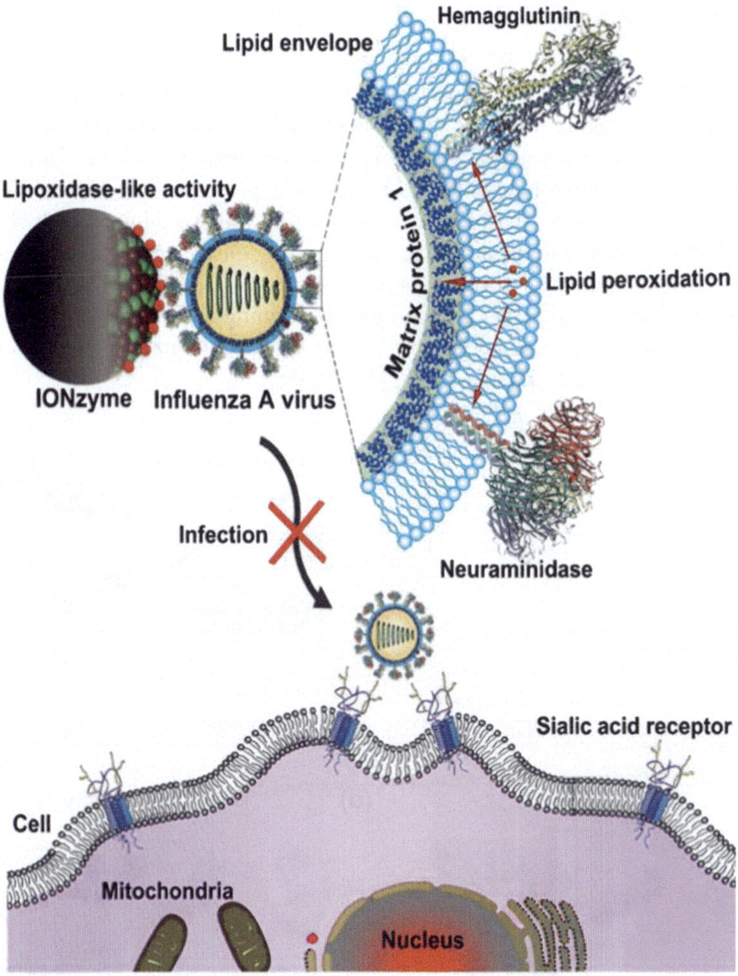

Fig. 15.28 The schematic of the antiviral mechanism of iron oxide nanozymes by inducing lipid peroxidation [86]. Copyright © Ivyspring International Publisher

spectrum to more than 11 subtypes of influenza A viruses. Based on these features, the authors designed a facemask interlaid with iron oxide nanozymes and demonstrated that the facemask showed the considerable antiviral efficacy by deactivating virus, showing an enhanced protection from infection.

15.7 Marine Anti-fouling

In 2012, Wolfgang et al. first reported an interesting application of nanozyme in the area of marine antibiofouling [25]. They found that Vanadium pentoxide (V_2O_5) nanowires have intrinsic catalytic haloperoxidase activity to prevent the formation of biofilm in a marine environment (Fig. 15.29). In the presence of H_2O_2 and Br^-, the nanowires catalyze the oxidation of Br^- into HOBr, which then react with H_2O_2 to form $^1O_2$, thus showing strong antibacterial activity against *S. aureus* and *Escherichia coli*, but without being toxic to marine biota. In situ experiment indicated that stainless steel plates covered with paints for boat hulls containing V_2O_5 nanowires prevented marine biofouling under seawater for up to 2 months, while untreated control plates without nanowires suffered from severe biofouling.

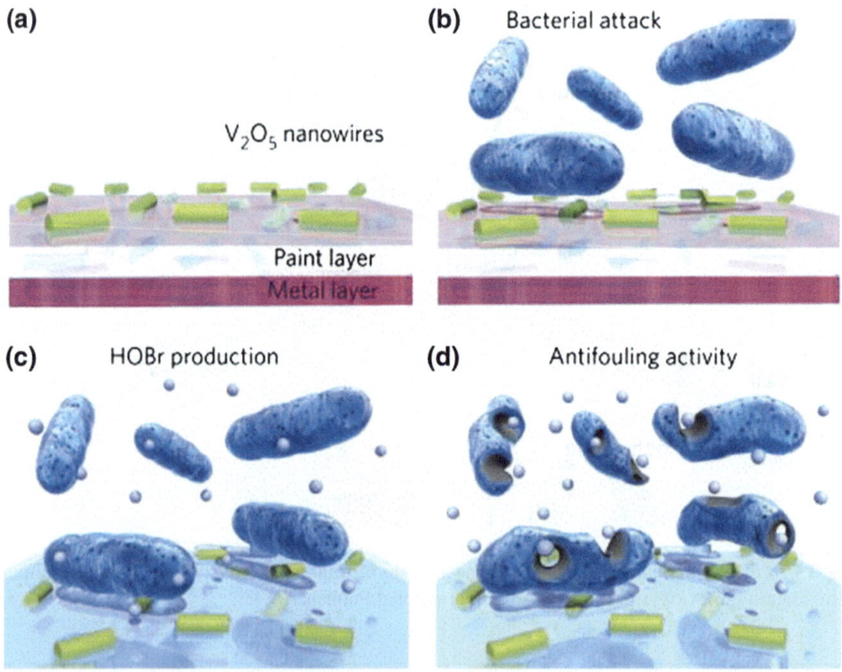

Fig. 15.29 Bactericidal properties of V_2O_5 nanowires. **a** Nanoparticles (yellow–green rods) are embedded in a matrix (paint) and applied onto a metal surface. **b** They can be attacked easily by bacteria. **c** The V_2O_5/paint nanocomposite displays an intrinsic biomimetic catalytic activity, as found in vanadium haloperoxidases (V-HPOs); that is, in the presence of substrates such as Br_2 and H_2O_2, small amounts of hypobromous acid (HOBr, small light blue spheres) are produced continuously. **d** The released HOBr interferes with the quorum sensing system of bacteria, preventing adhesion of the bacteria and biofilm formation. Copyright © 2018 Macmillan Publishers Limited, part of Springer Nature

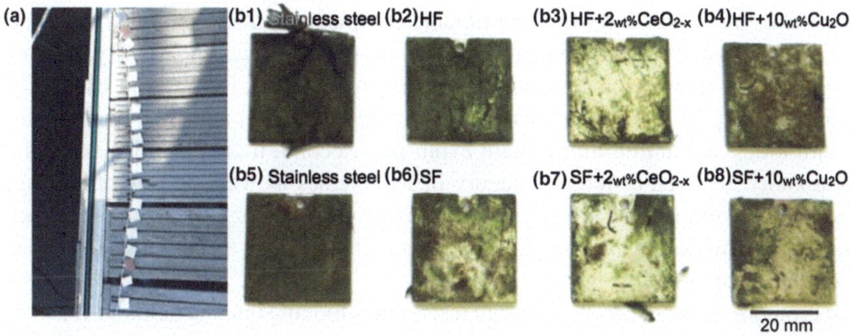

Fig. 15.30 **a** Offshore immersion site (Belgium, Ophoven-Kinrooi, Marina De Spaanjerd, Maas). Digital images of stainless steel plates (4 cm × 4 cm) covered with different paint formulations (HF, hard formulation and SF, soft formulation) for boat hulls with and without CeO_{2-x} nanorods. The plates were fixed statically to a boat bridge. **b** Selected samples after seven and half weeks of the static field immersion. **b1**, **b5** stainless steel plate; **b2**, **b6** HF and SF without biocide; **b3**, **b7** SF and HF containing 2 wt% of CeO_{2-x}; **b4**, **b8** SF and HF containing 10 wt% of Cu_2O [87]. Copyright © 1999-2018 John Wiley & Sons, Inc.

Recently, the same group revealed that CeO_{2-x} nanorods are also functional mimics of haloperoxidases with marine antibiofouling property [87]. The nanorods catalyze the oxidation of Br^- to HOBr under marine Br^- and H_2O_2 conditions, which related to the Ce^{4+}/Ce^{3+} redox reaction on the surface of nanorods (Fig. 15.30). The highly reactive HOBr intermediate can react to bromination of signaling compounds such as N-acyl homoserine lactones, thus impedes bacterial quorum sensing. Consequently, the paint containing CeO_{2-x} nanorods exhibited better antibiofouling performance than the commonly used biocidal pigment Cu_2O in laboratory and field tests, showing potential to be a useful alternative to conventional marine anti-fouling agents.

15.8 Conclusions and Perspective

Due to ever-increasing microbial resistance to the common disinfectants and antibiotics, numerous studies have been performed to improve antimicrobial strategies. Several valuable studies have been documented in the field of antibacterial nanoparticles in the recent years. The antimicrobial nanoparticles could be benefitted in the pharmaceutical and biomedical industries for sterilization of the medical devices [80, 88, 89].

As universal biocatalysts, natural enzymes play central roles in almost all the biological reactions among living systems. The emergence of nanozymes not only overturns the traditional concept of nanomaterials as inert substances, but also provides a new point of view for exploiting the biological effects of nanomaterials. While enriching the involved aspects of mimetic enzymes, it also extends the application of nanomaterials which cover from organic complexes to inorganic nanomaterials. Accompany with its remarkable mini size and outstanding biocompatibility, it cannot be denied that there are still several problems to be solved [2, 13, 42]. Firstly, the particular and profound mechanism of nanozymes applied in clinical treatment such as antibiotics and antibiofilm remain exploring. Second, there exist distinctions of cytotoxicity between different category of nanozymes as well as different dosage. What's more, unremitting efforts are needed to promote nanozymes from experimental study into marketing and practical application by simplify the therapy method and extend the adaptation disease.

It is our joint purpose to attract more researchers looking into the fields and broaden it by collecting and providing insightful knowledge of nanozymes. Combining unique characteristic of nanomaterials with catalytic function, the further researching direction is how to bring the double functional characteristics of nanozymes into full play [52]. Meanwhile create more nanozymes with magical functions and reveal its mechanism so as to apply to more fields including human health, environmental protection and bioenergy.

Acknowledgements This work was supported by the National Natural Science Foundation of China (Grant No. 81930050, 81671810), the Jiangsu Provincial Basic Research Program for the Natural Science Foundation (Grant No. BK20161333), the Foundation of the Thousand Talents Plan for Young Professionals and Jiangsu Specially-Appointed Professor.

References

1. Zheng K, Setyawati MI, Leong DT, Xie J (2017) Antimicrobial gold nanoclusters. ACS Nano 11(7):6904–6910
2. Wang LS, Gupta A, Rotello VM (2016) Nanomaterials for the treatment of bacterial biofilms. ACS Infect Dis 2(1):3–4
3. Koo H, Yamada KM (2016) Dynamic cell-matrix interactions modulate microbial biofilm and tissue 3D microenvironments. Curr Opin Cell Biol 42:102–112
4. He J, Zhu X, Qi Z, Wang C, Mao X, Zhu C, He Z, Li M, Tang Z (2015) Killing dental pathogens using antibacterial graphene oxide. ACS Appl Mater Interfaces 7(9):5605–5611
5. Hajipour MJ, Fromm KM, Ashkarran AA, Jimenez de Aberasturi D, de Larramendi IR, Rojo T, Serpooshan V, Parak WJ, Mahmoudi M (2012) Antibacterial properties of nanoparticles. Trends Biotechnol 30(10):499–511
6. Vu B, Chen M, Crawford RJ, Ivanova EP (2009) Bacterial extracellular polysaccharides involved in biofilm formation. Molecules 14(7):2535–2554
7. Hall-Stoodley L, Stoodley P (2009) Evolving concepts in biofilm infections. Cell Microbiol 11(7):1034–1043
8. Flemming HC, Wingender J (2010) The biofilm matrix. Nat Rev Microbiol 8(9):623–633
9. Flemming HC, Neu TR, Wozniak DJ (2007) The EPS matrix: the "house of biofilm cells". J Bacteriol 189(22):7945–7947
10. Ding Y, Zhou Y, Yao J, Szymanski C, Fredrickson J, Shi L, Cao B, Zhu Z, Yu XY (2016) In situ molecular imaging of the biofilm and its matrix. Anal Chem 88(22):11244–11252
11. Zhao Y, Chen Z, Chen Y, Xu J, Li J, Jiang X (2013) Synergy of non-antibiotic drugs and pyrimidinethiol on gold nanoparticles against superbugs. J Am Chem Soc 135(35):12940–12943

12. Giri K, Yepes LR, Duncan B, Parameswaran PK, Yan B, Jiang Y, Bilska M, Moyano DF, Thompson M, Rotello VM, Prakash YS (2015) Targeting bacterial biofilms via surface engineering of gold nanoparticles. RSC Adv 5(128):105551–105559
13. Beyth N, Houri-Haddad Y, Domb A, Khan W, Hazan R (2015) Alternative antimicrobial approach: nano-antimicrobial materials. Evid Based Complement Altern Med eCAM 2015:246012
14. Meeker DG, Jenkins SV, Miller EK, Beenken KE, Loughran AJ, Powless A, Muldoon TJ, Galanzha EI, Zharov VP, Smeltzer MS, Chen J (2016) Synergistic photothermal and antibiotic killing of biofilm-associated *Staphylococcus aureus* using targeted antibiotic-loaded gold nanoconstructs. ACS Infect Dis 2(4):241–250
15. Guo Q, Zhao Y, Dai X, Zhang T, Yu Y, Zhang X, Li C (2017) Functional silver nanocomposites as broad-spectrum antimicrobial and biofilm-disrupting agents. ACS Appl Mater Interfaces 9(20):16834–16847
16. Rojas-Andrade MD, Chata G, Rouholiman D, Liu J, Saltikov C, Chen S (2017) Antibacterial mechanisms of graphene-based composite nanomaterials. Nanoscale 9(3):994–1006
17. Jan T, Iqbal J, Ismail M, Badshah N, Mansoor Q, Arshad A, Ahkam QM (2014) Synthesis, physical properties and antibacterial activity of metal oxides nanostructures. Mater Sci Semicond Process 21:154–160
18. Qu K, Shi P, Ren J, Qu X (2014) Nanocomposite incorporating V_2O_5 nanowires and gold nanoparticles for mimicking an enzyme cascade reaction and its application in the detection of biomolecules. Chemistry 20(24):7501–7506
19. Maurer-Jones MA, Gunsolus IL, Meyer BM, Christenson CJ, Haynes CL (2013) Impact of TiO_2 nanoparticles on growth, biofilm formation, and flavin secretion in *Shewanella oneidensis*. Anal Chem 85(12):5810–5818
20. Jankauskait LV, Vitkauskien LA, Lazauskas A, Baltrusaitis J, ProsyLevas I, AndruleviLius M (2016) Bactericidal effect of graphene oxide/Cu/Ag nanoderivatives against *Escherichia coli*, *Pseudomonas aeruginosa*, *Klebsiella pneumoniae*, *Staphylococcus aureus* and Methicillin-resistant *Staphylococcus aureus*. Int J Pharm 511(1):90–97
21. Gao L, Liu Y, Kim D, Li Y, Hwang G, Naha PC, Cormode DP, Koo H (2016) Nanocatalysts promote *Streptococcus mutans* biofilm matrix degradation and enhance bacterial killing to suppress dental caries in vivo. Biomaterials 101:272–284
22. Gao L, Giglio KM, Nelson JL, Sondermann H, Travis AJ (2014) Ferromagnetic nanoparticles with peroxidase-like activity enhance the cleavage of biological macromolecules for biofilm elimination. Nanoscale 6(5):2588–2593
23. Masadeh MM, Karasneh GA, Al-Akhras MA, Albiss BA, Aljarah KM, Al-Azzam SI, Alzoubi KH (2015) Cerium oxide and iron oxide nanoparticles abolish the antibacterial activity of ciprofloxacin against gram positive and gram negative biofilm bacteria. Cytotechnology 67(3):427–435
24. Charbgoo F, Ahmad MB, Darroudi M (2017) Cerium oxide nanoparticles: green synthesis and biological applications. Int J Nanomed 12:1401–1413
25. Natalio F, Andre R, Hartog AF, Stoll B, Jochum KP, Wever R, Tremel W (2012) Vanadium pentoxide nanoparticles mimic vanadium haloperoxidases and thwart biofilm formation. Nat Nanotechnol 7(8):530–535
26. Li J, Zhou H, Wang J, Wang D, Shen R, Zhang X, Jin P, Liu X (2016) Oxidative stress-mediated selective antimicrobial ability of nano-VO_2 against Gram-positive bacteria for environmental and biomedical applications. Nanoscale 8(23):11907–11923
27. Diggikar RS, Patil RH, Kale SB, Thombre DK, Gade WN, Kulkarni MV, Kale BB (2013) Silver-decorated orthorhombic nanotubes of lithium vanadium oxide: an impeder of bacterial growth and biofilm. Appl Microbiol Biotechnol 97(18):8283–8290
28. Yousefi M, Dadashpour M, Hejazi M, Hasanzadeh M, Behnam B, de la Guardia M, Shadjou N, Mokhtarzadeh A (2017) Anti-bacterial activity of graphene oxide as a new weapon nanomaterial to combat multidrug-resistance bacteria. Mater Sci Eng C Mater Biol Appl 74:568–581

29. Dai X, Zhao Y, Yu Y, Chen X, Wei X, Zhang X, Li C (2017) Single continuous near-infrared laser-triggered photodynamic and photothermal ablation of antibiotic-resistant bacteria using effective targeted copper sulfide nanoclusters. ACS Appl Mater Interfaces 9(36):30470–30479
30. Xu W, Xie W, Huang X, Chen X, Huang N, Wang X, Liu J (2017) The graphene oxide and chitosan biopolymer loads TiO_2 for antibacterial and preservative research. Food Chem 221:267–277
31. Tian Y, Cao H, Qiao Y, Meng F, Liu X (2014) Antibacterial activity and cytocompatibility of titanium oxide coating modified by iron ion implantation. Acta Biomater 10(10):4505–4517
32. Kalita S, Kandimalla R, Sharma KK, Kataki AC, Deka M, Kotoky J (2016) Amoxicillin functionalized gold nanoparticles reverts MRSA resistance. Mater Sci Eng C Mater Biol Appl 61:720–727
33. Tang J, Chen Q, Xu L, Zhang S, Feng L, Cheng L, Xu H, Liu Z, Peng R (2013) Graphene oxide-silver nanocomposite as a highly effective antibacterial agent with species-specific mechanisms. ACS Appl Mater Interfaces 5(9):3867–3874
34. Huang F, Gao Y, Zhang Y, Cheng T, Ou H, Yang L, Liu J, Shi L, Liu J (2017) Silver-decorated polymeric micelles combined with curcumin for enhanced antibacterial activity. ACS Appl Mater Interfaces 9(20):16880–16889
35. Yin W, Yu J, Lv F, Yan L, Zheng LR, Gu Z, Zhao Y (2016) Functionalized nano-MoS_2 with peroxidase catalytic and near-infrared photothermal activities for safe and synergetic wound antibacterial applications. ACS Nano 10(12):11000–11011
36. Peng MY, Zheng DW, Wang SB, Cheng SX, Zhang XZ (2017) Multifunctional nanosystem for synergistic tumor therapy delivered by two-dimensional MoS_2. ACS Appl Mater Interfaces 9(16):13965–13975
37. Cao F, Ju E, Zhang Y, Wang Z, Liu C, Li W, Huang Y, Dong K, Ren J, Qu X (2017) An efficient and benign antimicrobial depot based on silver-infused MoS_2. ACS Nano 11(5):4651–4659
38. Maleki Dizaj S, Mennati A, Jafari S, Khezri K, Adibkia K (2015) Antimicrobial activity of carbon-based nanoparticles. Adv Pharm Bull 5(1):19–23
39. Maas M (2016) Carbon nanomaterials as antibacterial colloids. Materials 9(8):617
40. Zou X, Zhang L, Wang Z, Luo Y (2016) Mechanisms of the antimicrobial activities of graphene materials. J Am Chem Soc 138(7):2064–2077
41. Zhu J, Wang J, Hou J, Zhang Y, Liu J, Van der Bruggen B (2017) Graphene-based antimicrobial polymeric membranes: a review. J Mater Chem A 5(15):6776–6793
42. Szunerits S, Boukherroub R (2016) Antibacterial activity of graphene-based materials. J Mater Chem B 4(43):6892–6912
43. Park JY, Jeong HY, Kim MI, Park TJ (2015) Colorimetric detection system for *salmonella typhimurium* based on peroxidase-like activity of magnetic nanoparticles with DNA aptamers. J Nanomater 2015:1–9
44. Barzamini B, Moghbeli M, Soleimani NA (2014) *Vibrio cholerae* detection in water and wastewater by polymerase chain reaction assay. Int J Enteric Pathog 2(4)
45. Thiramanas R, Jangpatarapongsa K, Tangboriboonrat P, Polpanich D (2013) Detection of *Vibrio cholerae* using the intrinsic catalytic activity of a magnetic polymeric nanoparticle. Anal Chem 85(12):5996
46. Zhang L, Huang R, Liu W, Liu H, Zhou X, Xing D (2016) Rapid and visual detection of Listeria monocytogenes based on nanoparticle cluster catalyzed signal amplification. Biosens Bioelectron 86:1–7. https://doi.org/10.1016/j.bios.2016.05.100
47. Dennehy PH (2008) Rotavirus vaccines: an overview. Clin Microbiol Rev 21(1):198
48. Woo MA, Kim MI, Jung JH, Park KS, Seo TS, Park HG (2013) A novel colorimetric immunoassay utilizing the peroxidase mimicking activity of magnetic nanoparticles. Int J Mol Sci 14(5):9999–10014
49. Dye C (2015) Ebola virus disease in West Africa–the first 9 months. N Engl J Med 372(2):189
50. Duan D, Fan K, Zhang D, Tan S, Liang M, Liu Y, Zhang J, Zhang P, Liu W, Qiu X, Kobinger GP, Gao GF, Yan X (2015) Nanozyme-strip for rapid local diagnosis of Ebola. Biosens Bioelectron 74:134–141

51. Wang Z, Dong K, Liu Z, Zhang Y, Chen Z, Sun H, Ren J, Qu X (2017) Activation of biologically relevant levels of reactive oxygen species by Au/g-C_3N_4 hybrid nanozyme for bacteria killing and wound disinfection. Biomaterials 113:145–157
52. Gao L, Fan K, Yan X (2017) Iron oxide nanozyme: a multifunctional enzyme mimetic for biomedical applications. Theranostics 7(13):3207–3227
53. Chen Z, Ji H, Liu C, Bing W, Wang Z, Qu X (2016) A multinuclear metal complex based dnase-mimetic artificial enzyme: matrix cleavage for combating bacterial biofilms. Angew Chem 55(36):10732–10736
54. Yang X, Yang J, Wang L, Ran B, Jia Y, Zhang L, Yang G, Shao H, Jiang X (2017) Pharmaceutical intermediate-modified gold nanoparticles: against multidrug-resistant bacteria and wound-healing application via an electrospun scaffold. ACS Nano 11(6):5737–5745
55. Chen W-Y, Chang H-Y, Lu J-K, Huang Y-C, Harroun SG, Tseng Y-T, Li Y-J, Huang C-C, Chang H-T (2015) Self-Assembly of antimicrobial peptides on gold nanodots: against multidrug-resistant bacteria and wound-healing application. Adv Func Mater 25(46):7189–7199
56. Ayaz Ahmed KB, Subramanian S, Sivasubramanian A, Veerappan G, Veerappan A (2014) Preparation of gold nanoparticles using Salicornia brachiata plant extract and evaluation of catalytic and antibacterial activity. Spectrochim Acta Part A Mol Biomol Spectrosc 130:54–58
57. Comune M, Rai A, Chereddy KK, Pinto S, Aday S, Ferreira AF, Zonari A, Blersch J, Cunha R, Rodrigues R, Lerma J, Simoes PN, Preat V, Ferreira L (2017) Antimicrobial peptide-gold nanoscale therapeutic formulation with high skin regenerative potential. J Control Release Off J Control Release Soc 262:58–71
58. Tao Y, Ju E, Ren J, Qu X (2015) Bifunctionalized mesoporous silica-supported gold nanoparticles: intrinsic oxidase and peroxidase catalytic activities for antibacterial applications. Adv Mater 27(6):1097–1104
59. Mohamed MM, Fouad SA, Elshoky HA, Mohammed GM, Salaheldin TA (2017) Antibacterial effect of gold nanoparticles against *Corynebacterium pseudotuberculosis*. Int J Vet Sci Med 5(1):23–29
60. Zhao Y, Ye C, Liu W, Chen R, Jiang X (2014) Tuning the composition of AuPt bimetallic nanoparticles for antibacterial application. Angew Chem Int Ed 53(31):8127
61. Gao L, Zhuang J, Nie L, Zhang J, Zhang Y, Gu N, Wang T, Feng J, Yang D, Perrett S, Yan X (2007) Intrinsic peroxidase-like activity of ferromagnetic nanoparticles. Nat Nanotechnol 2(9):577–583
62. Fan K, Wang H, Xi J, Liu Q, Meng X, Duan D, Gao L, Yan X (2016) Optimization of Fe_3O_4 nanozyme activity via single amino acid modification mimicking an enzyme active site. Chem Commun 53(2):424–427
63. Mahmoudi M, Hofmann H, Rothen-Rutishauser B, Petri-Fink A (2012) Assessing the in vitro and in vivo toxicity of superparamagnetic iron oxide nanoparticles. Chem Rev 112(4):2323–2338
64. Gupta AK, Gupta M (2005) Synthesis and surface engineering of iron oxide nanoparticles for biomedical applications. Biomaterials 26(18):3995–4021
65. Durmus NG, Taylor EN, Kummer KM, Webster TJ (2013) Enhanced efficacy of superparamagnetic iron oxide nanoparticles against antibiotic-resistant biofilms in the presence of metabolites. Adv Mater 25(40):5706–5713
66. Gumy D, Morais C, Bowen P, Pulgarin C, Giraldo S, Hajdu R, Kiwi J (2006) Catalytic activity of commercial of TiO_2 powders for the abatement of the bacteria (*E. coli*) under solar simulated light: influence of the isoelectric point. Appl Catal B Environ 63(1–2):76–84
67. Ananpattarachai J, Boonto Y, Kajitvichyanukul P (2016) Visible light photocatalytic antibacterial activity of Ni-doped and N-doped TiO_2 on *Staphylococcus aureus* and *Escherichia coli* bacteria. Environ Sci Pollut Res Int 23(5):4111–4119
68. Vijayalakshmi K, Sivaraj D (2016) Synergistic antibacterial activity of barium doped TiO_2 nanoclusters synthesized by microwave processing. RSC Adv 6(12):9663–9671
69. Wang L, He H, Yu Y, Sun L, Liu S, Zhang C, He L (2014) Morphology-dependent bactericidal activities of Ag/CeO_2 catalysts against *Escherichia coli*. J Inorg Biochem 135:45–53

70. Yu J, Ma X, Yin W, Gu Z (2016) Synthesis of PVP-functionalized ultra-small MoS_2 nanoparticles with intrinsic peroxidase-like activity for H_2O_2 and glucose detection. RSC Adv 6(84):81174–81183
71. Pandit S, Karunakaran S, Boda SK, Basu B, De M (2016) High antibacterial activity of functionalized chemically exfoliated MoS_2. ACS Appl Mater Interfaces 8(46):31567–31573
72. Yang X, Li J, Liang T, Ma C, Zhang Y, Chen H, Hanagata N, Su H, Xu M (2014) Antibacterial activity of two-dimensional MoS_2 sheets. Nanoscale 6(17):10126–10133
73. Bowen WH, Koo H (2011) Biology of *Streptococcus mutans*-derived glucosyltransferases: role in extracellular matrix formation of cariogenic biofilms. Caries Res 45(1):69–86
74. Wu X, Tan S, Xing Y, Pu Q, Wu M, Zhao JX (2017) Graphene oxide as an efficient antimicrobial nanomaterial for eradicating multi-drug resistant bacteria in vitro and in vivo. Colloids Surf B 157:1–9
75. Chen S, Quan Y, Yu Y-L, Wang J-H (2017) Graphene quantum dot/silver nanoparticle hybrids with oxidase activities for antibacterial application. ACS Biomater Sci Eng 3(3):313–321
76. Liu S, Zeng TH, Hofmann M, Burcombe E, Wei J, Jiang R, Kong J, Chen Y (2011) Antibacterial activity of graphite, graphite oxide, graphene oxide, and reduced graphene oxide: membrane and oxidative stress. ACS Nano 5(9):6971–6980
77. Wang H, Li P, Yu D, Zhang Y, Wang Z, Liu C, Qiu H, Liu Z, Ren J, Qu X (2018) Unraveling the enzymatic activity of oxygenated carbon nanotubes and their application in the treatment of bacterial infections. Nano Lett 18(6):3344–3351
78. Sun HJ, Gao N, Dong K, Ren JS, Qu XG (2014) Graphene quantum dots-band-aids used for wound disinfection. ACS Nano 8(6):6202–6210
79. Yan Z, Bing W, Ding C, Dong K, Ren J, Qu X (2018) A H_2O_2-free depot for treating bacterial infection: localized cascade reactions to eradicate biofilm in vivo. Nanoscale 10:17656–17662
80. Zhao X, Wu H, Guo B, Dong R, Qiu Y, Ma PX (2017) Antibacterial anti-oxidant electroactive injectable hydrogel as self-healing wound dressing with hemostasis and adhesiveness for cutaneous wound healing. Biomaterials 122:34–47
81. Thet NT, Alves DR, Bean JE, Booth S, Nzakizwanayo J, Young AE, Jones BV, Jenkins AT (2016) Prototype development of the intelligent hydrogel wound dressing and its efficacy in the detection of model pathogenic wound biofilms. ACS Appl Mater Interfaces 8(24):14909–14919
82. Dai X, Guo Q, Zhao Y, Zhang P, Zhang T, Zhang X, Li C (2016) Functional silver nanoparticle as a benign antimicrobial agent that eradicates antibiotic-resistant bacteria and promotes wound healing. ACS Appl Mater Interfaces 8(39):25798–25807
83. Gu YH, Huang YX, Qiu ZY et al (2019) Vitamin B_2 functionalized iron oxide nanozymes for mouth ulcer healing. Sci China Life Sci. https://doi.org/10.1007/s11427-019-9590-6
84. Shi SR, Wu S, Shen YR et al (2018) Iron oxide nanozyme suppress intracellular *Salmonella Enteritidis* growth and alleviates infection in vivo. Thernostics 8(22):6149–6162
85. Yin YY, Wu MX, Qin T et al (2019) Effects of iron-based nanozymes on *S. typhimurium* biofilm. Progr Biochem Biophys 46(06):587–595
86. Qin T, Ma RN, Yin YY et al (2019) Catalytic inactivation of influenza virus by iron oxide nanozyme. Theranostics. https://doi.org/10.7150/thno.37380
87. Herget K, Hubach P, Pusch S et al (2017) Haloperoxidase mimicry by CeO_{2-x} nanorods combats biofouling. Adv Mater 29(4):1603823
88. Zhao Z, Yan R, Yi X et al (2017) Bacteria-activated theranostic nanoprobes against methicillin-resistant *Staphylococcus aureus* infection. ACS Nano 11(5):4428–4438
89. Zhao R, Lv M, Li Y, Sun M, Kong W, Wang L, Song S, Fan C, Jia L, Qiu S, Sun Y, Song H, Hao R (2017) Stable nanocomposite based on pegylated and silver nanoparticles loaded graphene oxide for long-term antibacterial activity. ACS Appl Mater Interfaces 9(18):15328–15341

Chapter 16
Nanozymes for Environmental Monitoring and Treatment

Jiuyang He and Minmin Liang

Abbreviations

ABTS	2,2′-azino-bis(3-ethylbenzo-thiazoline-6-sulfonic acid) diammonium salt
AChE	Acetylcholinesterase
Ag^+	Silver ions
CHO	Choline oxidase
CoOxH-GO	Cobalt hydroxide/oxide-modified graphene oxide
γ-FeOOH	Lepidocrocite
GNPs	Gold nanoparticles
GQDs/Fe_3O_4	Graphene quantum dot and Fe_3O_4
Hg^{2+}	Mercury ion
MNPs	Magnetic nanoparticles
NPs	Nanoparticles
OPs	Organophosphate compounds

16.1 Introduction

Environment pollution problems are becoming a global issue that affects not only our normal way of life but also may even threaten our future existence. Issues such as global warming, ozone layer destruction, acidic rain, toxic chemical pollution, etc., are so serious environmental problems that are urgent to be solved. Two current fundamental difficulties are achieving sensitive monitoring and high efficient removal

J. He · M. Liang (✉)
Experimental Center of Advanced Materials School of Materials Science & Engineering, School of Materials Science & Engineering, Beijing Institute of Technology, Beijing 100081, China
e-mail: mmliang@bit.edu.cn

X. Yan (ed.), *Nanozymology*, Nanostructure Science and Technology,
https://doi.org/10.1007/978-981-15-1490-6_16

of environmental pollutants and toxic compounds that have been accumulating over time. Enzymes are highly efficient natural catalysts existed in living organism that have already provided effective approaches for dealing with these environmental problems. However, natural enzymes have poor stability and are easy to inactivate under unphysiological conditions, that seriously affect their environment application.

Nanozymes are nanomaterials with intrinsic enzyme-like activity, which are distinct from that "nanozymes" with natural enzyme-immobilized on nanomaterials. Ever since the first evidence that ferromagnetic nanoparticles (NPs) possess intrinsic peroxidase-like activity was reported in 2007 [1], more than hundreds of nanozymes have been reported, and raised increasing attention owing to their high catalytic activity, low cost, high stability and extensive suitability for a broad spectrum of application from environment detection to pollutants treatment [2]. With the emergence of the new concept "nanozymology", nanozymes have now become an emerging new field connecting nanotechnology and biology that opens up a wider range of applications for nanomaterials [3].

In contrast to natural enzymes, nanozymes have the following distinct advantages: (1) High stability. Inorganic nanomaterials are less fragile than natural bio-enzymes. Therefore, nanozymes can be used under a wider range of pH (1–12) and temperature (4–90 °C) conditions. Natural enzymes are generally deactivated under such extreme conditions. (2) Low cost. The production process of natural enzymes is complex and companied by high cost, whereas inorganic nanomaterials can be produced easily with high efficiency, and at relatively low cost. (3) Reusability. Nanozymes are recyclable with no significant loss in catalytic activity in subsequent cycles. (4) Easy multi-functionalization. Nanozymes have a sufficiently large specific surface area for various ligand conjugation to achieve surface functionalization of nanozymes.

The above characterizations of nanozymes have led scientists and technologist to develop novel nanozymes-based technologies and approaches for environmental monitoring, pollutant removal and wastewater treatment, etc.

16.2 Environmental Pollutant Detection

Since the discovery of nanozymes, they have been widely used for protein, nucleic acid, metal ion, and environment small-molecule detection. In Yan's initial work, they first developed an iron oxide nanozyme-based immunoassay for the detection of hepatitis B virus [1], and followed by developing an iron oxide nanozyme-based strip for the detection of Ebola virus [4]. Compared with the traditional colloidal gold strip method, the developed nanozyme strip achieved 100-fold times higher sensitivity due to the high catalytic activity and the unique physicochemical properties of iron oxide nanozymes. By connecting the physicochemical properties and catalytic activity of nanozymes, many nanozymes-based platform technologies have been developed for the following environmental contaminants detection (Table 16.1).

Table 16.1 Nanozymes used for contaminants detection

Nanozymes	Contaminants	Linear range	Limit of detection (LOD)	References
Fe_3O_4 NPs	Glucose	50–1000 μM	30 μM	[5]
Fe_3O_4 NPs	Organophosphate	Not available	1 nM Sarin, 10 nM methyl-paraoxon, and 5 μM acephate	[6]
Au/Fe_3O_4/GO hybrid NPs	Hg^{2+}	1–50 nM	0.15 nM	[7]
Pt NPs	Hg^{2+}	0–120 nM	7.2 nM	[8]
Gold nanozyme-based paper chip	Hg^{2+}	0.02–2000 ng	0.06 ng (30 μg L^{-1})	[9]
Glutathione capped Pd NPs	Ag^+	10–100 nM	1.2 nM	[10]
His-Au NPs	Cu^{2+}	20 nM–2 μM	20 nM	[11]
Fe_3O_4 NPs	Zn^{2+}	60 nM–5 μM	60 nM	[12]
Au NPs	Kanamycin	1–100 nM	4.52 nM	[13]
CoO_xH/graphene oxide nanohybrid	Cyanide	100 nM–100 μM	22 nM	[14]
Pd@Au nanorods	Malathion	Not available	60 ng mL^{-1}	[15]
NiO hexagonal NPs	Parathion	0.1–30 μM	0.024 μM	[16]

16.2.1 Hydrogen Peroxide

Hydrogen peroxide is a partially reduced form of oxygen that is able to form superoxide and hydroxyl radicals. These generated oxidative molecules in turn have significant roles within the cell for lipid oxidation and DNA damage as well as inactivating natural enzymes. Therefore, the detection of H_2O_2 plays critical roles in biology, medicine, and environmental protection. So far, numerous nanozymes including Fe_3O_4, Co_3O_4, Cu_2O, FeS, CeO_2, $CoFe_2O_4$, $BiFeO_3$, $MnFe_2O_4$, CdS, FeSe, FeTe, $ZnFe_2O_4$ NPs, graphene oxide, fullerene, and carbon nanotubes have been applied for hydrogen peroxide detection [17].

Wei and Wang were the first to develop nanozymes-based colorimetric methods for H_2O_2 detection using Fe_3O_4 NPs (Fig. 16.1) [5]. In their work, Fe_3O_4 nanozymes catalyzed the oxidation of the substrate of 2,2′-azino-bis(3-ethylbenzo-thiazoline-6-sulfonic acid) diammonium salt (ABTS) by H_2O_2 to the oxidized colored product, providing a colorimetric detection of H_2O_2. In their work, as low as 3×10^{-6} mol L^{-1} H_2O_2 was detected with a linear range from 5×10^{-6} to 1×10^{-4} mol L^{-1} by using the nanozyme colorimetric method.

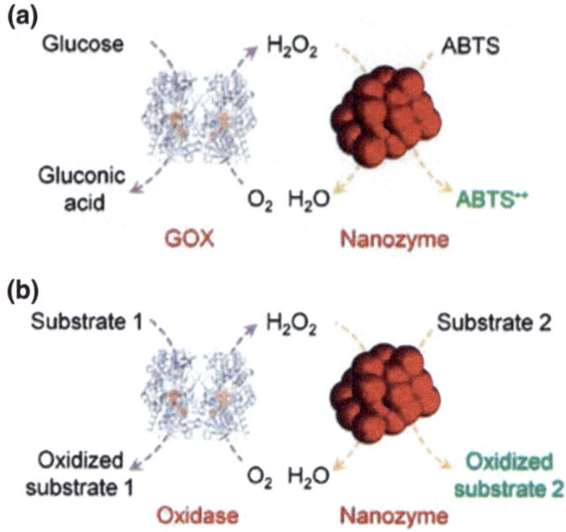

Fig. 16.1 **a** Nanozyme as peroxidase mimic for colorimetric sensing of H_2O_2 and glucose when combined with glucose oxidase. **b** The sensing format in **(a)** could be extended to other targets (substrate 1 here) when combined with an appropriate oxidase. Numerous transduction signals can be adopted for sensing (such as colorimetric, fluorometric and chemiluminescent signals when the corresponding substrates are used; and electrochemical signals when a nanozyme is immobilized on an electrode). Reprinted from Ref. [18]. Copyright (2013) Royal Society of Chemistry

16.2.2 *Organophosphorus Pesticide and Nerve Agents*

Rapid and sensitive screening for the extensively used organophosphorus pesticides and highly toxic nerve agents is in urgent demand. Liang et al. developed a Fe_3O_4 nanozyme-based colorimetric assay for the rapid detection of organophosphorus pesticide and nerve agent. The natural enzyme acetylcholinesterase (AChE) and choline oxidase (CHO) catalyze the formation of H_2O_2 in the presence of acetylcholine, which then activates Fe_3O_4 nanozymes to catalyze the oxidation of colorimetric substrates to produce a color reaction (Fig. 16.2) [6]. After incubation with organophos-

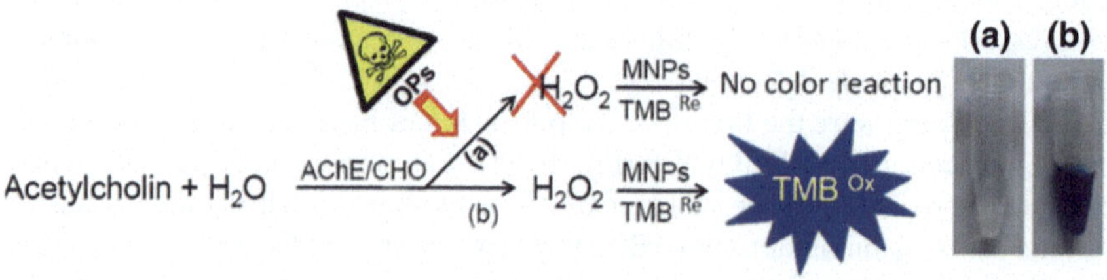

Fig. 16.2 Organophosphate compounds (OPs) detection on magnetic nanoparticles (MNPs) peroxidase mimetic-based colorimetric assay. Reprinted with permission from Ref. [6]. Copyright (2013) American Chemical Society

phate compounds (OPs), the enzyme AChE activity was inhibited and produced less H_2O_2, resulting in a decreased catalytic oxidation of colorimetric substrates over Fe_3O_4 nanozyme, accompanied by a drop in color intensity. In the work, as low as 1 nM Sarin, 10 nM methyl-paraoxon, and 5 μM acephate could be detected via the developed nanozyme-based assay.

16.2.3 Metal Ions

Detecting metal ions in the environment is highly significant, the accumulation and storage of heavy metals in living organisms is potentially able to transfer through the food chain into humans. Mercury ion (Hg^{2+}) is widely recognized as one of the most hazardous pollutant in the environment and its accumulation into humans can produce damage to central nervous system [19]. Mercury pollution comes from industrial and household waste, such as volcanic emissions, gold mining, chlorine alkali industry. Yang et al. developed an efficient Au/Fe_3O_4/graphene oxide hybrid nanomaterial-based nanozyme strategy for highly sensitive detection and rapid removal of Hg^{2+} in aqueous solutions [7]. The developed nanozyme strategy achieved the ultrasensitive detection of Hg^{2+} with a detection limit as low as 0.15 nM. Importantly, the Hg^{2+} can be quickly (within 30 min) and efficiently (>99% elimination efficiency) removed by the simple application of an external magnetic field, and furthermore the Au/Fe_3O_4/graphene oxide could be subsequently reused at least 15 times, with the elimination efficiency remaining high (>96%).

Fu et al. developed a BSA-based nucleation template to synthesize Pt-based peroxidase nanomimetics and achieved Hg^{2+} detection with a detection limit of 7.2 nM and a linear response range of 0–120 nM in aqueous solution without significant interference from other metal ions [8]. The developed Pt sensing system is potentially applicable for quantitative determination of Hg^{2+} in drinking water. Kwon et al. also developed a gold nanozyme-based paper chip for colorimetric detection of Hg^{2+} and achieved a detection limit of 0.06 ng with a linear response range of 0.02–2000 ng (Fig. 16.3) [8].

Silver ions (Ag^+) are mainly from industrial wastes with over 2500 tons of emissions annually. Ag^+ is able to coordinate with amine, imidazole, carboxylate and thiol groups of proteins, and results in serious neurotoxicity. The U.S. Environmental Protection Agency announced that Ag^+ could cause significant neurotoxicity in mammals and microorganisms under the concentration higher than 1.6 nM. Fu et al. developed a glutathione-stabilized palladium nanomimetic for colorimetric assay of silver ions [10]. In their work, Ag^+ selectively binds to Pd^0 species through metallophilic interactions and induces an apparent aggregation of Pd NPs. It is the first report that Ag^+ could significantly inhibit the peroxidase catalytic activity of Pd nanozymes. By using this mechanism, the developed Pd nanozyme was employed to explore Ag^+ detection and achieved a detection limit of 1.2 nM. The developed sensing system is potentially applicable for quantitative detection of Ag^+ in drinking water as well as Ag NPs in aqueous solution.

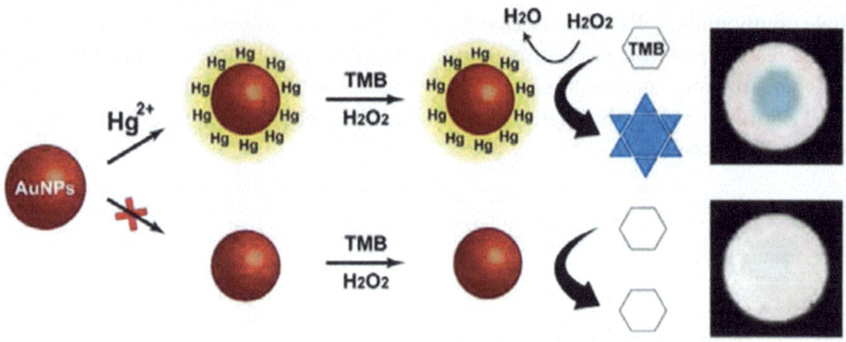

Fig. 16.3 Schematic illustration of the gold nanozyme-based paper chip sensing mechanism for Hg^{2+} detection. Reprinted with permission from Ref. [9]. Copyright (2013) Springer Nature

Copper ion is an essential element in organism that plays an important role in various biological systems. Not only does it participate in the synthesis of numerous enzymes it is also found in the metabolism process [20]. However, exceeding amount of copper ion can inhibit enzyme activity and interfere with the normal biological oxidation process. Liu et al. demonstrated a facile strategy to detect copper ions by using the peroxidase-like activity of gold nanocluster [11]. By adding Cu^{2+}, the enzyme-like activity of gold nanocluster was inhibited, allowing for the detection of Cu^{2+} with a detection limit of 0.1 nM and a linear range of 1–100 nM.

By using iron oxide NPs to mimic peroxidase, Liu et al. synthesized a fluorescent polydopamine that could be used to detect Zn^{2+} [12]. In their work, Zn^{2+} could enhance the fluorescence intensity of polydopamine at excitation wavelength of 360 nm, achieving a sensitive Zn^{2+} detection with a detection limit of 60 nM.

16.2.4 Toxin and Organic Pollutants

Kanamycin is an antibiotic, which is regularly used in humans as well as cell culture to treat bacterial infections. However, kanamycin exhibits serious side effects to humans, which seriously limits its clinical usefulness. Bansal et al. developed a surface plasmon resonance sensor for kanamycin detection by combining the intrinsic peroxidase-like activity of gold nanoparticles (GNPs) with the high affinity and specificity of an ssDNA aptamer (Fig. 16.4), where the aptamer molecules have a high affinity and specificity to kanamycin attached to GNPs [13]. When the target molecules are present, the aptamers are released from GNPs and adhere to kanamycin, and simultaneously free GNPs performing catalytic activity. The developed assay provided a visual readout for kanamycin detection within 3–8 min and achieved a limit of detection of 1.49 nM.

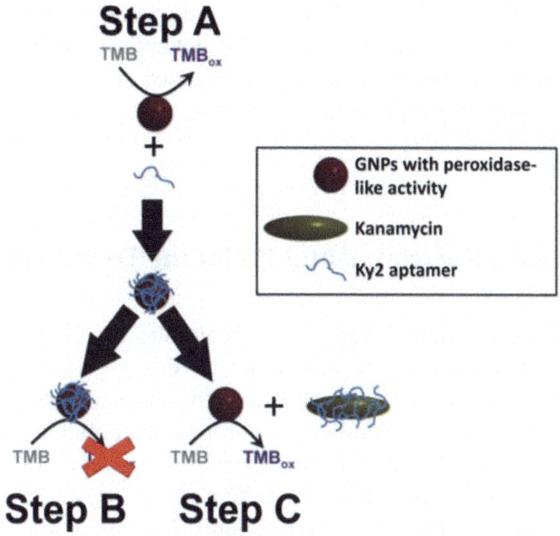

Fig. 16.4 Schematic representation of the "turn-off/turn-on" nanozyme activity of aptamer-functionalized GNPs for the detection of Kanamycin. Reprinted with permission from Ref. [13]. Copyright (2014) Royal of Chemistry

Cyanide is deadly toxic to human due to its strong binding to heme that will paralyze cellular respiration. Lien et al. achieved a visual detection of cyanide ions by the developed membrane-based nanozyme assay [14]. They developed a simple one-step synthesis of well-dispersed amorphous cobalt hydroxide/oxide-modified graphene oxide (CoOxH-GO), which possesses the peroxidase-like catalytic activity. Cyanide could significantly inhibit the catalytic activity of CoOxH-GO nanozymes, allowing for the detection of cyanide in water sample. This assay provided a limit of detection of 32 nM. Due to the high stability of this assay, it allows the cyanide detection even in high salt without interference.

Singh et al. developed a colorimetric sensor for the sensitive malathion detection by utilizing the palladium-gold bimetallic nanozyme [15]. Palladium-gold nanozymes show excellent peroxidase mimetic activity with the color substrates of o-phenylenediamine in the presence of hydrogen peroxide. Importantly, the developed palladium-gold nanozyme is stable over a broad temperature range (4–70 °C) and shows high peroxidase activity from 2 to 6 pH, and thus can achieve malathion detection under such extreme conditions. The assay's lowest detection limit was 60 ng mL^{-1} and showed no cross-reaction with other organophosphates or metal salts.

Khairy et al. synthesized a meso-/macro-porous NiO nanozyme and achieved parathion pesticide detection based on nanometer-sized nickel oxide-modified screen-printed electrodes [16]. This NiO nanozyme-based sensor could be used in a wide concentration range from 0.1 to 30 μM with a low detection limit of 24×10^{-9} mol L^{-1}. The developed NiO nanozyme sensor was utilized for detection of parathion in water, urine and vegetable samples because of the stability of the prepared NiO nanozyme.

When iron oxide nanozymes and oxidative enzymes were co-entrapped within mesoporous carbon, a highly efficient and robust electrochemical sensor was developed. The nanocomposite was then used to construct highly efficient electrodes for sensing several phenolic compounds, which showed great promise for environmental monitoring [21].

16.3 Environmental Pollutant Treatment

The development of highly efficient, robust and low-cost catalysts for environmental pollutant removal in wastewater is very important to our environment and health. The peroxidase-like catalytic activity of nanozymes has been extensively explored for catalytic degradation of environmental pollutants, such as phenol, rhodamine B, aniline, methylene blue and xylenol orange etc. (Table 16.2). Using nanozymes as catalysts for environmental pollutant treatment shows several advantages, such as low cost, easy preparation, high stability, low environmental impact, and recyclability.

16.3.1 Iron-Based Nanozymes for Pollutant Treatment

The most widely studied nanozymes are iron and iron oxide nanozymes, which demonstrate high degradation efficacy and has been extensively applied for environmental pollutant treatment. Thus far, the following pollutants removal and degradation have been achieved by using iron-nanozyme technology.

In 2008, for the first time nanozymes were used for pollutant treatment by using iron oxide NPs as peroxidase-like mimics [22]. In the study, phenols were removed from the wastewater by using the peroxidase activity of Fe_3O_4 nanozymes, which catalyzed the formation of hydroxyl radical and then catalytically degraded phenols. Under the condition of 16 °C, pH = 3, with the addition of hydrogen peroxide, Fe_3O_4 nanozymes could effectively remove 85% of the phenols, while the nanozymes can be recycled and reused several times.

Based on the above study, Cai et al. also used Fe_3O_4 NPs-based peroxidase activity to remove contaminants from wastewater in 2009 [24]. In addition to phenol, the nanozymes-based methods have also been used for aniline treatment. When the concentrations of the NPs and hydrogen peroxide reached 5 mg mL^{-1} and 1.2 mol L^{-1} respectively, the organic compounds could be completely cleared within 6 h, while the suitable pH conditions could be extended to acidic and neutral conditions which are greatly expanding their scope of future applications.

From 2009 to 2011, Tang et al. published three articles which successively achieved pollutant removal using the peroxidase-like catalytic activity of Fe_3O_4 NPs and studied the mechanism of catalysis of the prepared Fe_3O_4 nanozymes (Fig. 16.5) [25, 26, 48]. In their work, the authors explored a new method of preparing the ferromagnetic NPs with high catalytic activity by using ultrasonic radiation and advanced

Table 16.2 Nanozymes used for environmental pollutant decomposition

Nanozymes	Pollutant	Treatment efficiency (%)	References
Fe_3O_4 NPs	Phenol	85	[22, 23]
Fe_3O_4 NPs	Phenol	100	[24]
	Analine	100	
Fe_3O_4 NPs	Rhodamine B	90	[25]
Fe_3O_4 NPs	Sulfamonomethoxine	100	[26]
Fe_3O_4 NPs	Methylene blue	96	[27]
Fe_3O_4 NPs	Biphenol A	95	[28]
Fe_3O_4 NPs	2,4-DCP	73	[29]
Nanoparticulate zero-valent iron	4-chloro-3-methyl phenol	100	[30]
Fe_3O_4/CeO_2 hybrid NPs	4-chlorophenol	100	[31]
$Fe_2(MoO_4)_3$ NPs	Acid Orange II	94.1	[32]
Humic acid coated Fe_3O_4 NPs	Sulfathiazole	90	[33]
Fe_3O_4/MSU-F-C	Phenol/As(III)	Not available	[34]
Fe_3O_4 on CNTs	Acid Orange II	94.0	[35]
Carbon capsule encapsulated Fe3O4	Methylene blue	99	[36]
	Congo red	Not available	
	Phenol	Not available	
Fe_3O_4 porus nanospheres	Xylenol orange	100	[37]
$BiFeO_3$ NPs	Rhodamine B	95.2	[38]
Fe_3O_4/GQDs composites	Phenolic compound	80.3	[39]
CeO_2/γ-Fe_2O_3 composites	Organophosphate	100	[40]
Au/Fe_3O_4/MoS_2Cas aerogel	Mercury	95	[41]
Fe_3O_4 @ALG/Fe NPs	Norfloxacin	100	[42]
γ-FeOOH 2D nanosheets	Phenol	80	[43]
Cubic BN	Rhodamine B	86	[44]
CuO NPs	Phenol	100	[45]
Co_3O_4 nanorodes	Methylene blue	97	[46]
VO_x nanoflakes	Rhodamine B	95	[47]

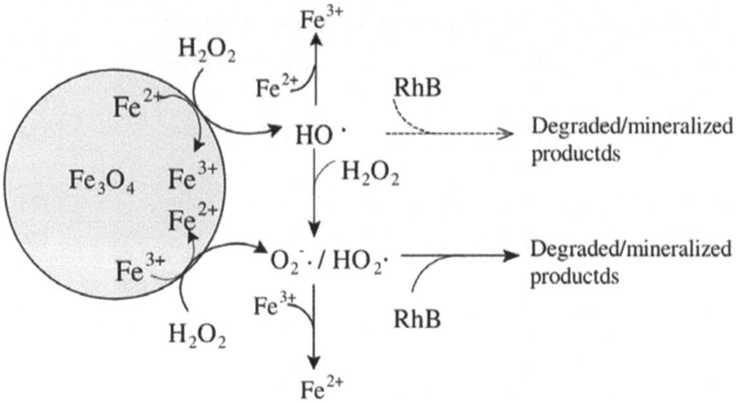

Fig. 16.5 Mechanism for the activation of H_2O_2 on the peroxidase-like Fe_3O_4 NPs for the degradation of organic pollutants. Reprinted with permission from Ref. [25]. Copyright (2014) Elsevier

reverse coprecipitation method. At pH 5.4 and temperature of 40 °C, the sonochemically synthesized ferromagnetic NPs were observed to be able to remove 90% of Rhodamine B at the concentration of 0.02 mmol L^{-1} within 1 h with an apparent rate constant of 0.034 min^{-1}, which is 12.6 folds of that (0.0027 min^{-1}) over the Fe_3O_4 NPs prepared using the conventional reverse coprecipitation method [25, 26, 48]. In the following work, the authors validated the optimal pH and temperature of the Fe_3O_4 nanozyme-based catalytic oxidation reaction, as well as the optimal reaction system [25, 26, 48]. The reaction could be carried out in the range of pH = 3–9, and the optimized nanozyme system can clear up to 95% of Rhodamine B within 15 min at the temperature of 55 °C and pH 5.0. In addition, Tang et al. further studied the catalytic mechanism of the Fe_3O_4 nanozymes to degrade sulfonamide-pyrimidine [25, 26, 48]. They found that Fe_3O_4 magnetic NPs could activate sulfate anion to produce free radicals and subsequently to degrade the contaminants. Zou et al. subsequently synthesized the magnetic NPs by using the reverse coprecipitation method reported by Tang et al., and used the prepared magnetic NPs to catalyze the production of hydroxyl radical degradation of methylene blue [27].

In 2012, Fang et al. studied the removal of Biphenol A by using the peroxidase-like catalytic activity of ferromagnetic nanozymes and achieved a clearance rate up to 95% [28]. In 2015, Cheng et al. reported the catalytic oxidation removal of 4-chlorophenol, which is a priority pollutant that widely exists in the environment but is recalcitrant towards chemical and biological degradation, using the catalytic activity of their synthesized Fe_3O_4 NPs [49]. The authors studied the interaction between Fe_3O_4 NPs and 4-chlorophenol and revealed that the synthesized Fe_3O_4 NPs show high catalytic activity even after being used several times. In addition, they demonstrated that the acidic conditions are favorable for the dechlorination of 4-chlorophenol. Furthermore, the results also showed that 4-chlorophenol could also be degraded under neutral and alkali conditions, which greatly expand its application in the complex environments.

16.3.2 Non-iron-Based Nanozymes for Pollutant Treatment

In addition to the use of iron-based nanomaterials for pollutant decontamination, there are other metal oxides, non-metallic nanozymes that have recently been utilized for environmental application.

In 2013, Feng et al. developed a cupric oxide nanoparticle with peroxidase activity to remove phenol contaminants. In the optimum condition, phenols can be completely cleared within 35 min [45]. In the work of Tang et al. [46], porous Co_3O_4 nanorod/graphene oxide hybrid nanomaterials were used for methylene blue removal by using the peroxidase-like activity of the hybrid nanomaterials. Compared to the individual Co_3O_4 NPs or graphene oxide, the hybrid nanomaterials exhibited the greatly increased pollutant adsorption capacity and enhanced peroxidase activity. In 2014, Jiang et al. developed a bismuth bromide nanozyme for the removal of methylene orange organic dye pollutant [50]. In 2015, Janoš et al. showed that the CeO_2 nanomaterials have phosphatase activity [40], which could be used for the rapid degradation of organophosphate pesticide parathion methyl and certain nerve agents. They reported that the composites retained the good magnetism and exhibited high efficient degradation properties with a reaction half-life of approximately ten minutes and the degree of conversion approaching 100%. In 2016, Mugesh found that the vacancy of cerium oxide NPs could simulate the catalytic activity of enzyme phosphotriesterase for the rapid degradation of neurotoxicity, such as paraoxon, parathion, and malathion, etc., to reduce or even eliminate their harm to human body [51]. They demonstrated that the hydrolytic effect of the nanozymes is due to the synergistic activity between both Ce^{3+} and Ce^{4+} ions located in the active site-like hotspots. Xu et al. also reported the peroxidase activity of vacant vanadium oxide nano-films, which could be used for the degradation of rhodamine pollutants [47]. Importantly, the authors suggested that the different oxygen deficiency in nanozymes can mimic the complexity and functionality of natural enzymes.

In addition to the metallic nanozymes, Qu et al. also developed an organic-inorganic hybrid nanoflower for organic pollutants removal [52]. Compared with the individual organic or inorganic nanozymes, the hybrid nanomaterials showed excellent durability, thermal stability, structural stability and the controllable structural composition.

In 2016, Yang et al. for the first time reported the peroxidase-like catalytic activity of cubic boron nitride for the decomposition of H_2O_2 and generating hydroxyl radicals [44]. The kinetics studies showed that the catalytic efficiency of the prepared cubic boron nitride is superior to the natural peroxidase counterparts. Furthermore, the prepared cubic boron nitride can be reused up to 5 times and retain its catalytic activity after incubation at extremes of pH and temperature. This work inspired the development of nitride-based nanozymes.

16.4 Optimization of Nanozymes

The application of nanozymes has shown a great advantage for removing environmental contaminants, which combine the high oxidative efficiency with the magnetic separation characteristics of magnetic NPs, and also exhibit the characteristics of high catalytic activity, mild application conditions and convenient recovery. However, nanozymes have their own technical difficulties when employed for actual environmental application. For example, the catalytic activity of nanozymes still need to be further improved to adapt to the demand for large quantity of pollutants treatment in the actual environment. Over the past decade, researchers have established various nanomaterials-based nanozymes by using various modifications to improve the catalytic activity of nanozymes, so as to be better applied for pollutants removal in the actual environment.

Zuo et al. designed a carbon nanotube/magnetic nanoparticle nanocomposite for the highly efficient catalytic oxidation of phenols and their removal from industrial wastewaters [23]. The synthesized nanocomplex retains the magnetic properties of individual magnetic NPs and thus can be effectively separated under an external magnetic field. In addition, the formation of the nanocomposite enhanced the intrinsic peroxidase-like activity of the magnetic NPs that achieved ~99% of phenol removal efficiency after a treatment time of 10 h. More importantly, the generated insoluble polyaromatic products can be readily separated from wastewater.

Wen et al. also used the Fe_3O_4 NPs dispersed carbon nanotubes to improve their peroxidase activity and applied their high catalytic activity to degrade orange II [35]. Fe_3O_4 NPs were grown in situ on carbon nanotubes by a solvothermal method and the formed Fe_3O_4 NPs uniformly deposited on nanotubes with an average diameter of approximately 7 nm, showing higher catalytic activity than the pure Fe_3O_4 NPs. In their work, 94.0% of Orange II at 0.25 mmol L^{-1}, pH = 3.5 was degraded on Fe_3O_4/carbon nanotubes nanocomplex in the presence of H_2O_2 at 30 °C. The authors demonstrated that the high degradation activity of Fe_3O_4/carbon nanotubes was attributed to the uniform Fe_3O_4 NPs growing on the side walls of the carbon nanotubes and the synergetic effect between Fe_3O_4 and carbon nanotubes. Importantly, the Fe_3O_4/carbon nanotubes maintained their activity at temperatures as high as 65 °C, and presented high reusability and stability even after eight uses, showing that the Fe_3O_4/carbon nanotubes-catalyzed degradation is a promising technique for wastewater treatment.

Environmental contaminant removal by using carbon nanomaterials combined with the ferromagnetic particles is a promising strategy to improve the catalytic activity of iron-based nanozymes. Li et al. prepared a mesoporous carbon capsule encapsulated with ferromagnetic NPs to degrade a variety of contaminants in the wastewater, such as methylene blue, congo red and phenol [36]. The mesoporous structure provided the specific surface of up to 1570 $m^2\ g^{-1}$, and the total pore volume of up to 3.02 $cm^3\ g^{-1}$ that contribute to high adsorption capacities for organic contaminants in aqueous media. In addition, the nanocomposites are superparamagnetic at room temperature with a saturation magnetization of 5.5 emu/g, which provides the prerequisite for the fast magnetic separation in wastewater treatment application. The water treatment results showed that the maximum removal capabilities of the prepared nanocomposites for methylene blue, congo red, and phenol reached 608.04, 1656.9 and 108.38 mg g^{-1}, respectively, showing more effective removal capacity than the commercial activated carbon.

The advantage of the mesoporous structure was also directly utilized for porous magnetite nanosphere preparation with high catalytic activity for the degradation of xylenol orange in aqueous solution with H_2O_2 as oxidant [37]. In contrast with traditional ferromagnetic NPs, the prepared porous magnetite nanosphere exhibited higher magnetic and peroxidase activity and showed a degradation rate constant at room temperature of 0.056 min^{-1}. In addition, the catalytic activity of the porous magnetite nanospheres decreases very slightly after seven cycles of the catalysis experiment. Therefore, the prepared porous nanospheres could serve as effective recyclable nanozymes for xylenol orange treatment.

Zhang et al. also used the composites of graphene quantum dot and Fe_3O_4 NPs (GQDs/Fe_3O_4) for phenolic contaminants removal [39]. The prepared GQDs/Fe_3O_4 composites showed superb peroxidase-like activity, much higher than individual GQDs, or individual Fe_3O_4 NPs. Importantly, the GQDs/Fe_3O_4 composites exhibited a higher stability and reusability than natural peroxidases. In the work, using GQDs/Fe_3O_4 composites for the catalytic removal of phenolic compounds from aqueous solutions was explored with nine phenolic compounds and the results showed better or comparable removal efficiencies for the phenolic compounds when compared to native horseradish peroxidase under the same conditions. The extraordinary catalytic performance and physical properties of the prepared GQDs/Fe_3O_4 composites render it practically useful for industrial wastewater treatment.

Doped metallic nanozymes, such as Au/Fe_3O_4/graphene, and Au/Fe_3O_4/MoS_2 hybrid nanomaterials exhibited higher catalytic activity, stability and separation capability than the individual metallic nanozymes while applied for the removal of Hg^{2+} in aqueous solutions [7, 41]. The hybrid nanostructure affords strong chemical attachments between the metallic NPs and graphene, allowing for this hybrid nanozyme to sensitively detect Hg^{2+} showing a detection limit as low as 0.15 nM. Furthermore, the deposition of Hg^{2+} on the surface of the nanozymes could be quickly (within 30 min) and efficiently (>99% elimination efficiency) removed by the simple application of an external magnetic field and then the nanozymes could be subsequently reused at least 15 times, with the elimination efficiency remaining high (>96%).

By modification of the ferromagnetic NPs with Al pillared bentonite, catalytic activity of iron-based nanozymes were significantly improved [53]. Compared with the plain oxide ferromagnetic particles, the modification provided more pollutant adsorption sites for ferromagnetic particles, and exhibited more efficient pollutant removal. There is also application of using graphene-templated formation of two-dimensional lepidocrocite (γ-FeOOH) nanostructures to improve the catalytic activity of iron nanozymes for highly efficient degradation of phenols in wastewater [43].

16.5 Conclusions and Perspective

It has been a decade since the intrinsic enzyme-like activity of nanomaterials was first discovered. By combining the unique physicochemical properties and catalytic characteristics, nanozymes showed great promise for environmental monitoring and treatment [18, 54]. So far, nanozymes have successfully applied for the removal of a variety of pollutants, including industrial wastewater containing reduced organic pollutants, pharmaceutical wastewater, ammunition wastewater treatment, etc. However, as an emerging new field, there are also a lot of challenges for nanozyme-based environment monitoring and pollutants removal. First of all, although the catalytic activity of some nanozymes is higher than natural enzymes, there is still a pressing need to improve their catalytic activity to treat large quantities of pollutants in the actual environment. Secondly, even though the detection limit for organic and metal ion pollutants is reduced to the low nanomolar concentration range, nanozyme-based sensors are susceptible to interference while detecting various wastewater samples due to the complex contaminants in wastewater. Thus, nanozyme-based sensors need to improve the detection stability. Thirdly, the peroxidase-like activity of nanozymes is mostly utilized for pollutants removal. However, the peroxidase-like nanozymes only exhibit high catalytic activity at low pH, indicating that the nanozyme-based treatment techniques can only be used in acidic wastewater. Therefore, although the area of nanozymes has been developed substantially for environmental application, future breakthroughs in nanozyme technology still need to be realized to be able to utilize nanozyme technology in large-scale application in actual environmental treatment.

Acknowledgements This work was supported by the National Natural Science Foundation of China (81722024 and 81571728), the National Key R&D Program of China (2017YFA0205501), the Key Research Program of Frontier Sciences (QYZDY-SSWSMC013), and the Youth Innovation Promotion Association (2014078).

References

1. Gao L, Zhuang J, Nie L, Zhang J, Zhang Y, Gu N, Wang T, Feng J, Yang D, Perrett S, Yan X (2007) Intrinsic peroxidase-like activity of ferromagnetic nanoparticles. Nat Nanotechnol 2(9):577–583
2. Shin HY, Park TJ, Kim MI (2015) Recent research trends and future prospects in nanozymes. J Nanomater 1:1–11
3. Ragg R, Tahir MN, Tremel W (2016) Solids of bio: inorganic nanoparticles as enzyme mimics. Eur J Inorg Chem 13–14:1906–1915
4. Duan D, Fan K, Zhang D, Tan S, Liang M, Liu Y, Zhang J, Zhang P, Liu W, Qiu X, Kobinger GP, Gao GF, Yan X (2015) Nanozyme-strip for rapid local diagnosis of Ebola. Biosens Bioelectron 74:134–141
5. Wei H, Wang E (2008) Fe_3O_4 magnetic nanoparticles as peroxidase mimetics and their applications in H_2O_2 and glucose detection. Anal Chem 80(6):2250–2254

6. Liang M, Fan K, Pan Y, Jiang H, Wang F, Yang D, Lu D, Feng J, Zhao J, Yang L, Yan X (2013) Fe_3O_4 magnetic nanoparticle peroxidase mimetic-based colorimetric assay for the rapid detection of organophosphorus pesticide and nerve agent. Anal Chem 85(1):308–812
7. Zhang S, Li H, Wang Z, Liu J, Zhang H, Wang B, Yang Z (2015) A strongly coupled Au/Fe_3O_4/GO hybrid material with enhanced nanozyme activity for highly sensitive colorimetric detection, and rapid and efficient removal of Hg^{2+} in aqueous solutions. Nanoscale 7(18):8495–8502
8. Li W, Chen B, Zhang H, Sun Y, Wang J, Zhang J, Fu Y (2015) BSA-stabilized Pt nanozyme for peroxidase mimetics and its application on colorimetric detection of mercury(II) ions. Biosens Bioelectron 66:251–258
9. Han KN, Choi JS, Kwon J (2017) Gold nanozyme-based paper chip for colorimetric detection of mercury ions. Sci Rep 7(1):2806
10. Fu Y, Zhang H, Dai S, Zhi X, Zhang J, Li W (2015) Glutathione-stabilized palladium nanozyme for colorimetric assay of silver (I) ions. Analyst 140(19):6676–6683
11. Liu Y, Ding D, Zhen Y, Guo R (2017) Amino acid-mediated 'turn-off/turn-on' nanozyme activity of gold nanoclusters for sensitive and selective detection of copper ions and histidine. Biosens Bioelectron 92:140–146
12. Liu B, Han X, Liu J (2016) Iron oxide nanozyme catalyzed synthesis of fluorescent polydopamine for light-up Zn^{2+} detection. Nanoscale 8(28):13620–13626
13. Sharma TK, Ramanathan R, Weerathunge P, Mohammadtaheri M, Daima HK, Shukla R, Bansal V (2014) Aptamer-mediated 'turn-off/turn-on' nanozyme activity of gold nanoparticles for kanamycin detection. Chem Commun (Camb) 50(100):15856–15859
14. Lien CW, Unnikrishnan B, Harroun SG, Wang CM, Chang JY, Chang HT, Huang CC (2018) Visual detection of cyanide ions by membrane-based nanozyme assay. Biosens Bioelectron 102:510–517
15. Singh S, Tripathi P, Kumar N, Nara S (2017) Colorimetric sensing of malathion using palladium-gold bimetallic nanozyme. Biosens Bioelectron 92:280–286
16. Khairy M, Ayoub HA, Banks CE (2018) Non-enzymatic electrochemical platform for parathion pesticide sensing based on nanometer-sized nickel oxide modified screen-printed electrodes. Food Chem 255:104–111
17. Weerathunge P, Sharma TK, Ramanathan R, Bansal V (2017) Nanozyme-based environmental monitoring. In: Advanced environmental analysis: applications of nanomaterials, vol 2. The Royal Society of Chemistry, pp 108–132
18. Wei H, Wang E (2013) Nanomaterials with enzyme-like characteristics (nanozymes): next-generation artificial enzymes. Chem Soc Rev 42(14):6060–6093
19. Ha E, Basu N, Bose-O'Reilly S, Dórea JG, McSorley E, Sakamoto M, Chan HM (2017) Current progress on understanding the impact of mercury on human health. Environ Res 152:419–433
20. Osredkar J, Sustar N (2011) Copper and zinc, biological role and significance of copper/zinc imbalance. J Clinic Toxicol S 3:001
21. Kim MI, Ye Y, Won BY, Lee J, Park HG (2011) A highly efficient electrochemical biosensing platform by employing conductive nanocomposite entrapping magnetic nanoparticles and oxidase in mesoporous carbon foam. Adv Funct Mater 21(15):2868–2875
22. Zhang J, Zhuang J, Gao L, Zhang Y, Gu N, Feng J, Yang D, Zhu J, Yan X (2008) Decomposing phenol by the hidden talent of ferromagnetic nanoparticles. Chemosphere 73(9):1524–1528
23. Zuo X, Peng C, Huang Q, Song S, Wang L, Li D, Fan C (2009) Design of a carbon nanotube/magnetic nanoparticle-based peroxidase-like nanocomplex and its application for highly efficient catalytic oxidation of phenols. Nano Res 2(8):617–623
24. Zhang S, Zhao X, Niu H, Shi Y, Cai Y, Jiang G (2009) Superparamagnetic Fe_3O_4 nanoparticles as catalysts for the catalytic oxidation of phenolic and aniline compounds. J Hazard Mater 167(1–3):560–566
25. Wang N, Zhu L, Wang D, Wang M, Lin Z, Tang H (2010) Sono-assisted preparation of highly-efficient peroxidase-like Fe_3O_4 magnetic nanoparticles for catalytic removal of organic pollutants with H_2O_2. Ultrason Sonochem 17(3):526–533

26. Yan J, Lei M, Zhu L, Anjum MN, Zou J, Tang H (2011) Degradation of sulfamonomethoxine with Fe_3O_4 magnetic nanoparticles as heterogeneous activator of persulfate. J Hazard Mater 186(2–3):1398–1404
27. Jiang J, Zou J, Zhu L, Huang L, Jiang H, Zhang Y (2011) Degradation of methylene blue with H_2O_2 activated by peroxidase-like Fe_3O_4 magnetic nanoparticles. J Nanosci Nanotechnol 11(6):4793–4799
28. Huang R, Fang Z, Yan X, Cheng W (2012) Heterogeneous sono-Fenton catalytic degradation of bisphenol A by Fe_3O_4 magnetic nanoparticles under neutral condition. Chem Eng J 197:242–249
29. Xu L, Wang J (2012) Fenton-like degradation of 2,4-dichlorophenol using Fe_3O_4 magnetic nanoparticles. Appl Catal B: Environ 123–124:117–126
30. Xu L, Wang J (2011) A heterogeneous Fenton-like system with nanoparticulate zero-valent iron for removal of 4-chloro-3-methyl phenol. J Hazard Mater 186(1):256–264
31. Xu L, Wang J (2012) Magnetic nanoscaled Fe_3O_4/CeO_2 composite as an efficient Fenton-like heterogeneous catalyst for degradation of 4-chlorophenol. Environ Sci Technol 46(18):10145–10153
32. Tian SH, Tu YT, Chen DS, Chen X, Xiong Y (2011) Degradation of Acid Orange II at neutral pH using $Fe_2(MoO_4)_3$ as a heterogeneous Fenton-like catalyst. Chem Eng J 169(1–3):31–37
33. Niu H, Zhang D, Zhang S, Zhang X, Meng Z, Cai Y (2011) Humic acid coated Fe_3O_4 magnetic nanoparticles as highly efficient Fenton-like catalyst for complete mineralization of sulfathiazole. J Hazard Mater 190(1–3):559–565
34. Chun J, Lee H, Lee SH, Hong SW, Lee J, Lee C, Lee J (2012) Magnetite/mesocellular carbon foam as a magnetically recoverable fenton catalyst for removal of phenol and arsenic. Chemosphere 89(10):1230–1237
35. Deng J, Wen X, Li J (2016) Fabrication highly dispersed Fe_3O_4 nanoparticles on carbon nanotubes and its application as a mimetic enzyme to degrade Orange II. Environ Technol 37(17):2214–2221
36. Zhang Y, Xu S, Luo Y, Pan S, Ding H, Li G (2011) Synthesis of mesoporous carbon capsules encapsulated with magnetite nanoparticles and their application in wastewater treatment. J Mater Chem 21(11):3664–3671
37. Zhu M, Diao G (2011) Synthesis of porous Fe_3O_4 nanospheres and its application for the catalytic degradation of xylenol orange. J Phys Chem C 115(39):18923–18934
38. Luo W, Zhu L, Wang N, Tang H, Cao M, She Y (2010) Efficient removal of organic pollutants with magnetic nanoscaled $BiFeO_3$ as a reusable heterogeneous Fenton-like catalyst. Environ Sci Technol 44(5):1786–1791
39. Wu X, Zhang Y, Han T, Wu H, Guo S, Zhnag J (2014) Composite of graphene quantum dots and Fe_3O_4 nanoparticles: peroxidase activity and application in phenolic compound removal. RSC Adv 4(7):3299–3305
40. Janoš P, Kuran P, Pilařová V, Trögl J, Šťastný M, Pelant O, Henych J, Bakardjieva S, Životský O, Kormunda M, Mazanec K, Skoumal M (2015) Magnetically separable reactive sorbent based on the CeO_2/γ-Fe_2O_3 composite and its utilization for rapid degradation of the organophosphate pesticide parathion methyl and certain nerve agents. Chem Eng J 262:747–755
41. Zhi L, Zou W, Chen F, Wang B (2016) 3D MoS_2 composition aerogels as chemosensors and adsorbents for colorimetric detection and high-capacity adsorption of Hg^{2+}. ACS Sustain Chem Eng 4(6):3398–3408
42. Niu H, Dizhang Meng Z, Cai Y (2012) Fast defluorination and removal of norfloxacin by alginate/Fe@Fe_3O_4 core/shell structured nanoparticles. J Hazard Mater 227–228:195–203
43. Peng C, Jiang BW, Liu Q, Guo Z, Xu ZJ, Huang Q, Xu HJ, Tai RZ, Fan CH (2011) Graphene-templated formation of two-dimensional lepidocrocite nanostructures for high-efficiency catalytic degradation of phenols. Energy Environ Sci 4(6):2035–2040
44. Chen TM, Xiao J, Yang GW (2016) Cubic boron nitride with an intrinsic peroxidase-like activity. RSC Adv 6(74):70124–70132
45. Feng YB, Hong L, Liu AL, Chen WD, Li GW, Chen W, Xia XH (2013) High-efficiency catalytic degradation of phenol based on the peroxidase-like activity of cupric oxide nanoparticles. Int J Environ Sci Te 12(2):653–660

46. Zhang Z, Hao J, Yang W, Lu B, Ke X, Zhang B, Tang J (2013) Porous Co_3O_4 nanorods-reduced graphene oxide with intrinsic peroxidase-like activity and catalysis in the degradation of methylene blue. ACS Appl Mater Interfaces 5(9):3809–3815
47. Zeb A, Xie X, Yousaf AB, Imran M, Wen T, Wang Z, Guo HL, Jiang YF, Qazi IA, Xu AW (2016) Highly efficient Fenton and enzyme-mimetic activities of mixed-phase VOx nanoflakes. ACS Appl Mater Interfaces 8(44):30126–30132
48. Wang N, Zhu L, Wang M, Wang D, Tang H (2009) Sono-enhanced degradation of dye pollutants with the use of H_2O_2 activated by Fe_3O_4 magnetic nanoparticles as peroxidase mimetic. Ultrason Sonochem 17(1):78–83
49. Cheng R, Li G, Chneg C, Shi L, Zheng X, Ma Z (2015) Catalytic oxidation of 4-chlorophenol with magnetic Fe_3O_4 nanoparticles: mechanisms and particle transformation. RSC Adv 5(82):66927–66933
50. Li L, Ai L, Zhang C, Jiang J (2014) Hierarchical {001}-faceted BiOBr microspheres as a novel biomimetic catalyst: dark catalysis towards colorimetric biosensing and pollutant degradation. Nanoscale 6(9):4627–4634
51. Vernekar AA, Das T, Mugesh G (2016) Vacancy-engineered nanoceria: enzyme mimetic hotspots for the degradation of nerve agents. Angew Chem Int Ed Engl 55(4):1412–1416
52. Huang Y, Ran X, Lin Y, Ren J, Qu X (2015) Self-assembly of an organic-inorganic hybrid nanoflower as an efficient biomimetic catalyst for self-activated tandem reactions. Chem Commun (Camb) 51(21):4386–4389
53. Wan D, Li WB, Wang GH, Chen K, Lu LL, Hu Q (2015) Adsorption and heterogeneous degradation of rhodamine B on the surface of magnetic bentonite material. Appl Surf Sci 349:988–996
54. He W, Wamer W, Xia Q, Yin JJ, Fu PP (2014) Enzyme-like activity of nanomaterials. J Environ Sci Health C Environ Carcinog Ecotoxicol Rev 32(2):186–211

Chapter 17
Beyond: Novel Applications of Nanozymes

Sheng Zhao, Sirong Li and Hui Wei

Abbreviations

BES	Bioresorbable electronic stent
BIS	Methylene-bis-acrylamide
CAT	Catalase
CuO:GNS	CuO:Graphene nanosphere
DAF	Diaminofluorescein
DMAA	N,N-dimethylacrylamide
FPD	Fluorescent polydopamine
GNS	Graphene nanosphere
GOx	Glucose oxidase
HNO	Nitroxyl
HRP	Horseradish peroxidase
HUVECs	Human umbilical vein endothelial cells
NIR	Near infrared reflection
NO	Nitric oxide
NPs	Nanoparticles
OX	Oxidase
PEI	Percutaneous coronary interventions
PLA	Polylactic acid
POX	Peroxidase
ROS	Reactive oxygen species

Sheng Zhao and Sirong Li are equally contributed.

S. Zhao · S. Li · H. Wei (✉)
Department of Biomedical Engineering, College of Engineering and Applied Sciences, Nanjing National Laboratory of Microstructures, Jiangsu Key Laboratory of Artificial Functional Materials, Nanjing University, Nanjing 210093, Jiangsu, China
e-mail: weihui@nju.edu.cn
URL: http://weilab.nju.edu.cn

X. Yan (ed.), *Nanozymology*, Nanostructure Science and Technology,
https://doi.org/10.1007/978-981-15-1490-6_17

SEM	Scanning electron microscope
4-AAP	4-amino-antipyrine

Since Yan's report of iron oxide nanoparticles (Fe_3O_4 NPs) as peroxidase mimics [1], nanozymes have been extensively explored for a variety of applications, ranging from in vitro detection and therapeutics to environment protection and antibiofouling [2–7]. These interesting applications have been covered in previous chapters. In this chapter, we will discuss the innovative applications of nanozymes in chemical synthesis, biomedical devices, and logic gates.

17.1 Chemical Synthesis

The essence of catalysts or enzymes is their ability to form active transition state and thus lower activation energy. Likewise, nanozymes, the functional nanomaterials with enzyme-like activities, are capable of inducing reactive species such as radicals or cationic intermediates [8, 9]. By utilizing the ability to generate reactive species, nanozymes-based chemical synthesis has been successfully conducted.

Fabrication of bioactive hydrogel. To avoid the toxicity brought by traditional initiators in polymerization/gelation, nanozymes have been applied into biomedical polymeric materials. Wang et al. employed CuO, a peroxidase mimicking nanozyme, to prepare poly(DMAA-cross-BIS) hydrogel [10]. As shown in Fig. 17.1,

$$\bullet\!-\!Cu(II) + H_2O_2 \rightleftharpoons \bullet\!-\!Cu(III) + OH^- + {}^{\bullet}OH \qquad (1)$$

$$RH + {}^{\bullet}OH \rightleftharpoons R^{\bullet} + H_2O \qquad (2)$$

$$R^{\cdot} + \bullet\!-\!Cu(III) \rightleftharpoons R^+ + \bullet\!-\!Cu(II) \qquad (3)$$

$$R^{\bullet} + O_2 \rightleftharpoons ROO^{\bullet} \qquad (4)$$

$$ROO^{\bullet} \rightleftharpoons R^+ + O_2^{-\bullet} \qquad (5)$$

$$R_1CH = CHR_2 + {}^{\bullet}OH \rightleftharpoons R_1COCH - {}^{\bullet}CHR_2 \qquad (6)$$

$$H_2O_2 \overset{\text{CuO NPs}}{\rightleftharpoons} O_2 + H_2O \qquad (7)$$

Fig. 17.1 Reactions in CuO NPs-initiated polymerization. Adapted with permission from [10]. Copyright (2017) Royal Society of Chemistry the peroxidase-like CuO produced $^{\bullet}OH$ in the presence of H_2O_2 following a Fenton-like reaction (Eq. 17.1) [10]. The $^{\bullet}OH$ then reacted with RH (a reductive substrate) to generate $R^{\bullet}$ (a carbon radical) (Eq. 17.2). The $R^{\bullet}$ subsequently reacted with the vinyl group of the DMAA monomer to produce a monomer radical, which then initiated the chain propagation (Eq. 17.6). However, the polymerization would be terminated since $R^{\bullet}$ could be eliminated by reacting with the CuO or O_2 (Eqs. 17.3–17.5). Moreover, the catalytic decomposition of H_2O_2 into O_2 with the CuO nanozymes would further inhibit the polymerization, summarized as quenching effect (Eq. 17.7).

To overcome the quenching effect of oxygen on radical polymerization, glucose oxidase (GOx) with glucose was applied to consume oxygen. The GOx catalyzed glucose oxidation not only produced H_2O_2 but also suppressed the oxygen inhibition effect (Fig. 17.2). It should be noted that a suitable amount of oxygen was needed for the initiation of the polymerization. Therefore, the fine-tuning of oxygen concentration was critical to polymerization induced by CuO nanozyme. Under the optimal conditions, the transparent hydrogen was obtained (Fig. 17.2c). Due to the generation of $^{\bullet}OH$, the nanozyme-hydrogel exhibited interesting antibacterial activities.

It should be noted that the current strategy still utilized natural GOx to consume oxygen. With the development of GOx-like nanozymes, an enzyme-free strategy would be expected in the near future.

Synthesis of fluorescent polydopamine (FPD). In addition to direct gelation initiated by CuO nanozymes, a fluorescent polydopamine was synthesized via Fe_3O_4

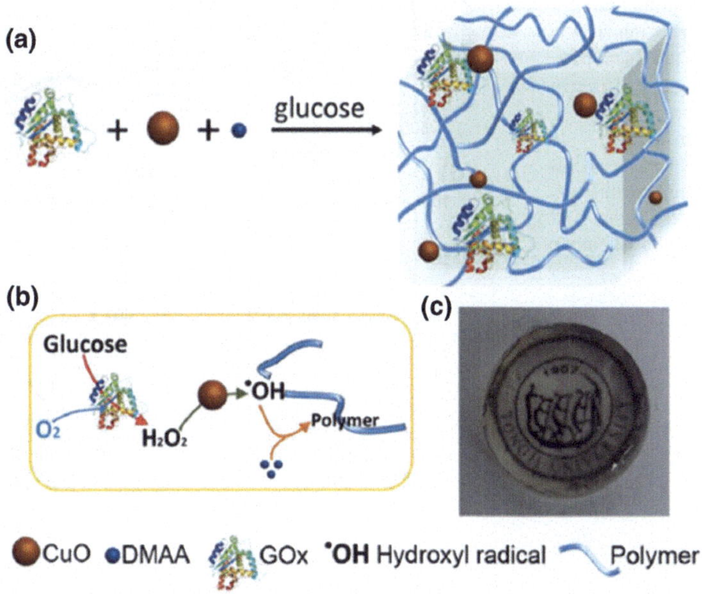

Fig. 17.2 Nanozyme–gel formation and initiation mechanism. **a** Formation of nanozyme–gel. **b** Initiation mechanism of the nanozyme–gel. **c** Nanozyme–gel on the top of a printed logo. Reprinted with permission from [10]. Copyright (2017) Royal Society of Chemistry

nanozyme catalysis. After screening a variety of peroxidase-like nanozymes (i.e., Fe_3O_4, Fe_2O_3, CeO_2, CoO, Co_3O_4, NiO, AuNPs, and graphene oxide) and non-catalytic TiO_2, Liu et al. found that only Fe_3O_4 could oxidize dopamine into polydopamine with robust fluorescent intensity in the presence of H_2O_2 [11].

The as-prepared FPD was then used to construct a light-up Zn^{2+} sensor as the fluorescent intensity of FPD excited at 360 nm was selectively enhanced by Zn^{2+} while no intensity changed under 470 nm excitation (Fig. 17.3). With the developed sensor, as low as 60 nM of Zn^{2+} could be detected. Moreover, Zn^{2+} in 1% diluted serum was also successfully determined, suggesting the potential practical use of the fabricated FPD sensor.

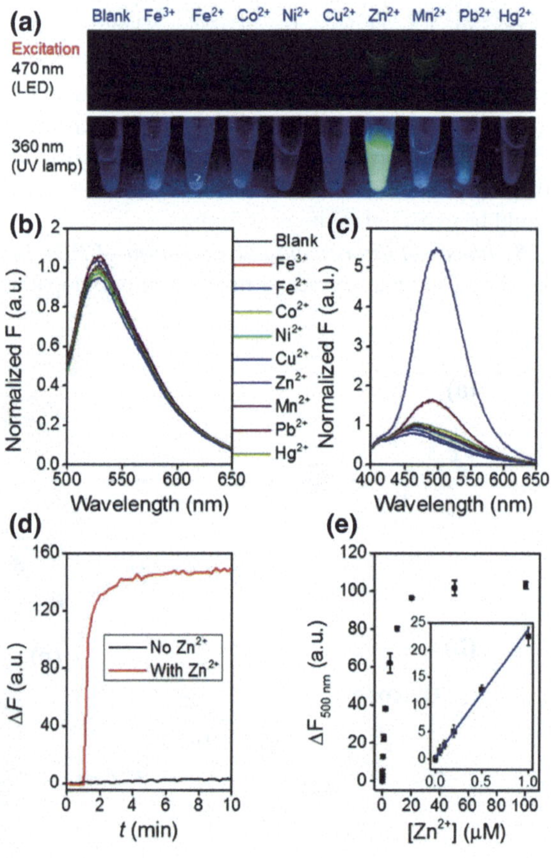

Fig. 17.3 Detection of Zn^{2+} using FPD. **a** Fluorescence images of FPD in the presence of various metal ions (10 μM) under two excitation wavelengths (470 nm and 360 nm). The corresponding fluorescence spectra under **b** 470 nm and **c** 360 nm excitation, respectively. **d** Kinetics of fluorescence variation upon adding Zn^{2+}. **e** Fluorescence intensity as a function of Zn^{2+} concentration. The inset shows the linear part with a fitting line. The error bars represent standard deviations from three independent measurements. Reprinted with permission from [11]. Copyright (2016) Royal Society of Chemistry

17.2 Biomedical Devices

Besides the biomedical detection and therapies, the nanozyme-based biomedical equipments and devices are another promising fields for potential translation. In this section, we will discuss some recent studies about nanozyme-based biomedical devices.

Anticoagulation devices. Huang, Duan, Meyerhoff, and co-workers demonstrated that the graphene–haemin–GOx conjugates could be incorporated within polyurethane to fabricate a long-lasting antithrombotic coating for blood-contacting biomedical devices [12]. As shown in Fig. 17.4, the covalently linked GOx would cat-

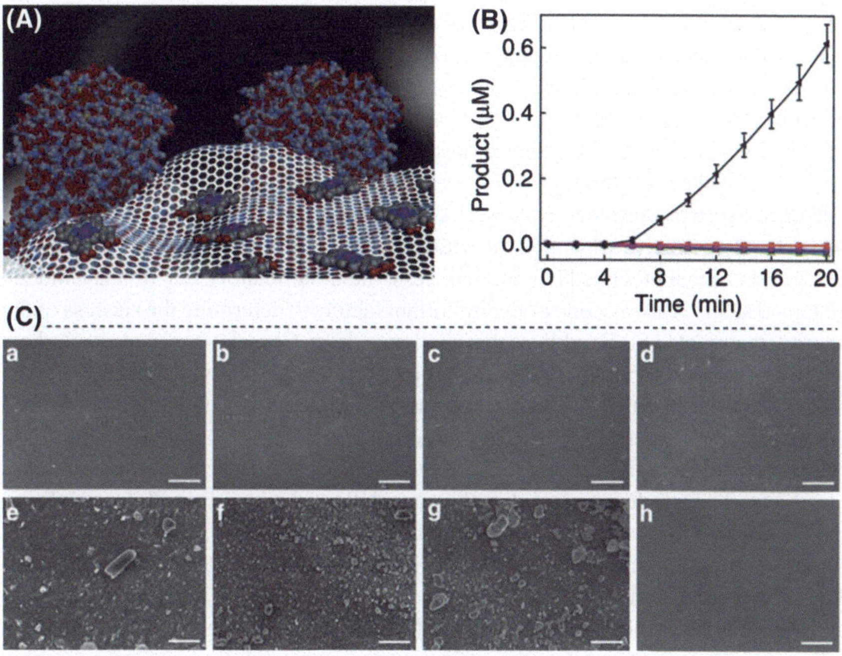

Fig. 17.4 **A** Schematic illustration of graphene–haemin–GOx conjugates. **B** HNO generation catalyzed by graphene–haemin–GOx conjugates. Graphene–haemin–GOx catalyzed HNO generation and control experiments. The production of HNO was quantified using a DAF assay. Black line, graphene–haemin–GOx in glucose and L-arginine; red line, graphene–haemin in glucose and L-arginine; blue line, graphene–GOx in glucose and L-arginine; green line, graphene–haemin–GOx in glucose; pink line, graphene–haemin–GOx in L-arginine. The size of the error bars represents the minimum to maximum values measured from at least three independent experiments. **C** Antithrombotic behavior of biocompatible films (polyurethane) containing graphene–haemin–GOx conjugates. Scanning electron microscopic (SEM) images of as-formed films containing **a** graphene, **b** graphene–haemin, **c** graphene–GOx and **d** graphene–haemin–GOx; and the respective films after immersing into platelet-rich rabbit blood plasma for 3 days: **e** graphene, **f** graphene–haemin, **g** graphene–GOx and **h** graphene–haemin–GOx. Scale bars, 10 mm. Adapted with permission from [12]. Copyright (2014) Nature Publishing Group

alyze the oxidation of glucose to locally generate H_2O_2. The conjugated monomeric haemin molecules would then catalyze the oxidation of L-arginine to nitroxyl (HNO) with H_2O_2. By this integrated design, the graphene–haemin–GOx conjugates enabled the cascade reactions to in situ produce HNO in the physiological conditions (Fig. 17.4B). Moreover, an antiplatelet film was fabricated by embedding the cascade catalysts into polyurethane (a biocompatible polymer) to illustrate the generation of HNO for sustained antithrombotic activity. Compared with control films of graphene, graphene–haemin or graphene–GOx, only films containing graphene–haemin–GOx showed effective antithrombotic activity after immersion into blood plasma (Fig. 17.4C).

In addition, Gu, Yan, Chen, and co-workers prepared a blood-contacting biomaterial that used polyethylenimine molecule as the linker to immobilize L-cysteine onto graphene nanosheets [13]. This L-cysteine-functionalized graphene film could catalytically decompose the exogenous or endogenous NO donors to generate NO, which would in turn inhibit the platelet activation/aggregation and reduce platelet adhesion. Those efforts indicate a new strategy to fabricate anticoagulant medical equipments and devices by integration of molecular catalysts and nanozymes on graphene (or other substrates).

Anti-inflammation stents. As is well-known, percutaneous coronary interventions is one of most effective ways to treat arterial obstructions and endothelial injuries in the current clinical practice [14]. Nonetheless, the inflammatory reaction around the implanted stents is always one of the important factors to determine the success of the stenting treatment [15]. To address this challenge, Kim, Choi, Hyeon, and co-workers recently developed a bioresorbable electronic stent integrated with therapeutic NPs for endovascular disease [16]. In their bioresorbable electronic stent, they introduced cerium oxide NPs (ceria NPs) as anti-inflammatory catalysts because of their high surface-to-volume ratio and reactive oxygen species (ROS) scavenging capability (Fig. 17.5). The results showed that the ceria NPs embedded in a PLA (polylactic acid) film exhibited an excellent ROS scavenging activity in vitro by incubating human umbilical vein endothelial cells (HUVECs) with the ceria NPs embedded PLA films under 50 μM H_2O_2 and in vivo by implanting the integrated stent in a canine common carotid artery. Because of self-regenerative redox cycles between Ce^{3+} and Ce^{4+} oxidation states, the ceria nanozyme ensured the sustaining catalytic scavenging of the endovascular ROS and thus eliminated ROS-induced inflammation. This study has shown the promising potential of ROS scavenging nanozymes for future biomedical engineering.

17.3 Logic Gates

Logic gates are concatenations involving reflections, after a set of judgements or treatments according to the initial inputs. With signals in the form of 0 or 1 (binary),

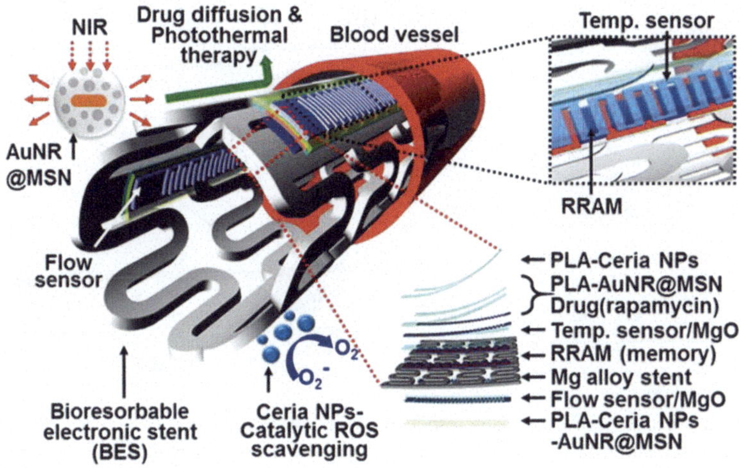

Fig. 17.5 Bioresorbable electronic stent (BES). Schematic illustration of the BES (left), its top view (top right), and the layer information (bottom right). The BES includes bioresorbable temperature/flow sensors, memory modules, and bioresorbable/bioinert therapeutic NPs. The therapeutic functions are either passive (ROS scavenging) or actively actuated (hyperthermia-based drug release) by near infrared reflection (NIR) exposure. Reprinted with permission from [16]. Copyright (2015) American Chemical Society

sophisticated devices can be fabricated obeying specific logic flow. Inspired by computer science, researchers have developed molecular logic gates by using different molecular substrates as inputs so that products serve as outputs, making smart chemistry achievable [17–19].

Benefited from the development of nanotechnology and computation, various nanomaterials-based logic gates have been established [19–21]. Despite the substantial progress, it remains challenging to realize reset in a system of logic gates using chemical reactions or interactions as inputs. Several studies have shown that nanozymes may have the potential to tackle this challenge.

Cerium oxide NPs based logic gates. By making use of the self-regenerative redox cycles between Ce^{3+} and Ce^{4+} oxidation states, Qu, Ren, and co-workers have developed resettable logic gates with CeO_2 NPs and natural enzymes (Fig. 17.6) [22]. When the Ce^{3+} in CeO_2 NPs was oxidized into Ce^{4+} (such as with H_2O_2), the color of CeO_2 NPs was changed from colorless to yellow (i.e., from "off" to "on" stage). Moreover, the Ce^{4+} could be reduced to Ce^{3+} by slightly heating (i.e., the logic gate could be reset thermally). With this interesting resettable property, three logic gates (i.e., AND, OR, and INHIBIT) were fabricated. For example, using GOx and β-galactosidase as the inputs, an AND logic gate was achieved. β-Galactosidase would catalytically convert lactose into galactose and glucose. The produced glucose would then be catalytically oxidized to produce H_2O_2 by GOx. The H_2O_2 then catalytically oxidized colorless CeO_2 NPs to generate yellow colored CeO_2 NPs, achieving the logic operation of AND (Fig. 17.6b).

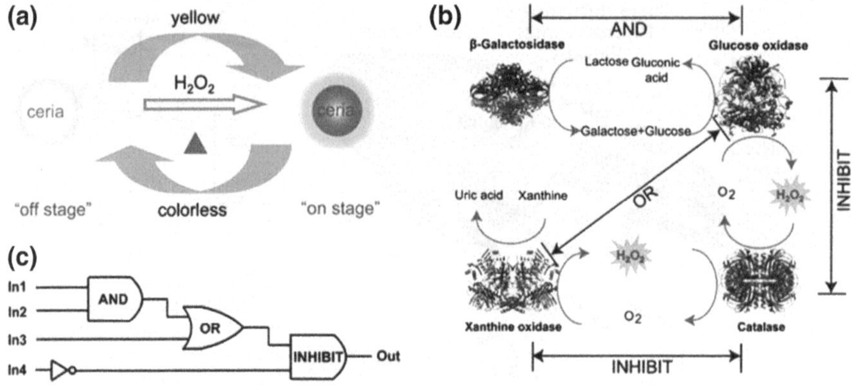

Fig. 17.6 **a** Illustration of a thermally responsive switch based on CeO_2 NPs. **b** The operation of logic gates based on biocatalytic reactions. **c** Logic circuitry for the integrated logic system. In1 = β-Gal, In2 = GOx, In3 = XO, In4 = CAT. Reprinted with permission from [22]. Copyright (2012) John Wiley & Sons

Au NPs based logic gates. Since then, various nanozyme-based logic gates have been investigated. In 2013, Huang et al. found that the multiple enzyme-like activities (i.e., peroxidase-, oxidase-, or catalase-like activities) of Au NPs could be modulated by various metal ions, such as Ag^+, Bi^{3+}, Pb^{4+}, and Hg^{2+}. By exploring the different effects of these metal ions on the enzyme-like activities of Au NPs, OR, AND, INHIBIT, and XOR logic gates were developed [20]. For example, an OR logic gate was constructed by modulating the catalase-like activity of Au NPs with Bi^{3+} and Hg^{2+} (Fig. 17.7).

CuO:Graphene based logic gates. Mobin et al. designed an AND logic gate for colorimetric cholesterol sensing (Fig. 17.8) [23]. They showed that the CuO:Graphene nanosphere (CuO:GNS) possessed good peroxidase-like activity. When coupled with cholesterol oxidase, the cholesterol oxidase would first catalyze the oxidation of cholesterol to produce H_2O_2. Then, the CuO:GNS nanozymes catalyzed the oxidation of 4-AAP or phenol with H_2O_2 to generate colorimetric products as outputs. With the fabricated logic gate, cholesterol could be selectively detected over several interfering substances (such as glucose, sucrose, and urea).

Summary and outlook. In summary, the innovative applications of nanozymes in chemical synthesis, biomedical devices, and logic gates have been discussed. While nanozymes have been successfully explored and researched for wide ranges of applications (including the ones discussed here), the "killer" applications do not arrive yet. Therefore, to fulfill the promise of nanozymes, it is necessary to make the best use of multi-disciplinary cross-cutting advantages (biology, chemistry, materials, physics, medicine, mathematics, computers, etc.) to search for suitable applications cross-disciplinary fields.

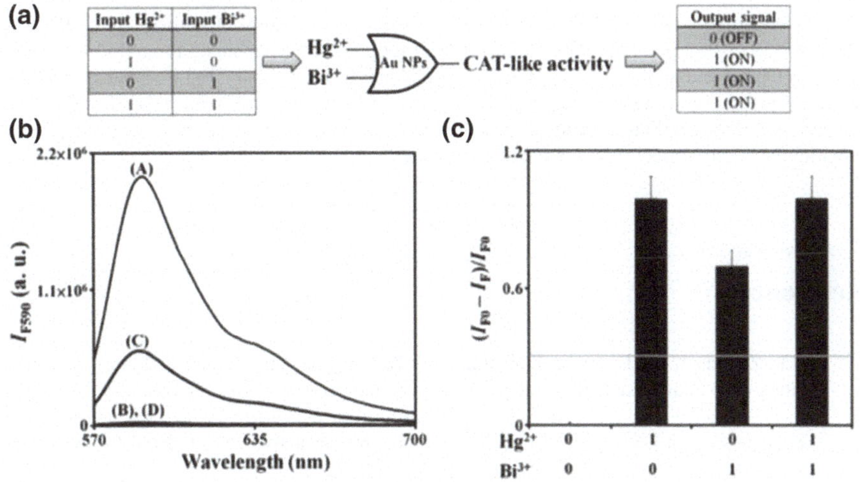

Fig. 17.7 OR logic gate constructed using Hg^{2+}/Bi^{3+} ions and the CAT-like activity of the Au NPs. **a** Schematic representation of the OR logic gate "Hg^{2+}/Bi^{3+}(OR)–Au NP_{CAT}". **b** Fluorescence spectra from the CAT-like active assay of 5.0 mM Tris–borate (pH 7.0) containing Au NPs (750 pM) in the (A) absence and (B–D) presence of (B) Hg^{2+} (10 μM), (C) Bi^{3+} (10 μM), and (D) Hg^{2+} (10 μM) and Bi^{3+} (10 μM). **c** Bar diagram of output signals $[(I_{F0} - I_F)/I_{F0}]$ in response to four different combinations of two inputs, Hg^{2+} (0, 10 μM) and Bi^{3+} (0, 10 μM), where I_{F0} and I_F represent the fluorescence intensities of the solution at 590 nm in the absence (0 μM) and presence (10 μM) of an input (Hg^{2+}, Bi^{3+}). The threshold line is set to the 0.3-fold signal output of true (on). Reprinted with permission from [20]. Copyright (2013) Royal Society of Chemistry

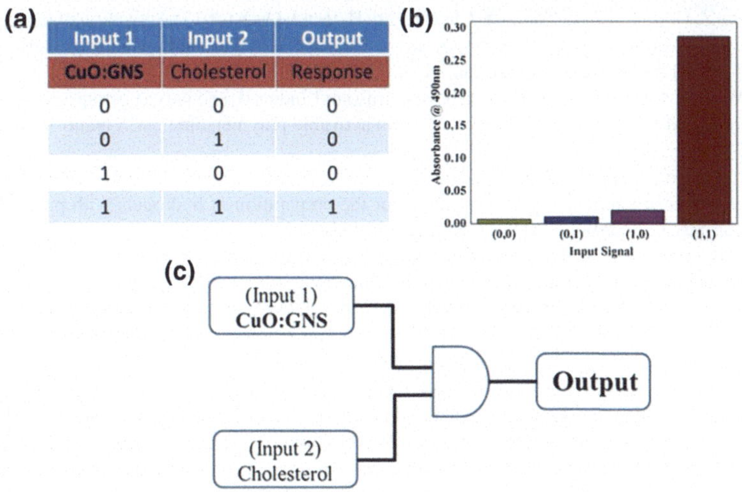

Input 1	Input 2	Output
CuO:GNS	Cholesterol	Response
0	0	0
0	1	0
1	0	0
1	1	1

Fig. 17.8 Logic gate implementation. **a** Truth table of AND logic. **b** Absorbance at 490 nm at different input signals. **c** Logic symbol of AND gate. Reprinted with permission from [23]. Copyright (2017) Elsevier

Acknowledgements We thank National Natural Science Foundation of China (21722503 and 21874067), 973 Program (2015CB659400), PAPD program, Shuangchuang Program of Jiangsu Province, Open Funds of the State Key Laboratory of Analytical Chemistry for Life Science (SKLA-CLS1704), Open Funds of the State Key Laboratory of Coordination Chemistry (SKLCC1819), Fundamental Research Funds for the Central Universities (021314380103), and Thousand Talents Program for Young Researchers for financial support.

References

1. Gao L, Zhuang J, Nie L, Zhang J, Zhang Y, Gu N, Wang T, Feng J, Yang D, Perrett S (2007) Intrinsic peroxidase-like activity of ferromagnetic nanoparticles. Nat Nanotechnol 2(9):577–583
2. Li S, Huang Y, Liu J, Wang E, Wei H (2018) Nanozymes in analytical chemistry: From *in vitro* detection to live bioassays. Prog Biochem Biophys 45(2):129–147
3. Wang X, Hu Y, Wei H (2016) Nanozymes in bionanotechnology: from sensing to therapeutics and beyond. Inorg Chem Front 3(1):41–60
4. Wei H, Wang E (2013) Nanomaterials with enzyme-like characteristics (nanozymes): next-generation artificial enzymes. Chem Soc Rev 42(14):6060–6093
5. Wu J, Li S, Wei H (2018) Integrated nanozymes: facile preparation and biomedical applications. Chem Commun 54:6520–6530
6. Wu J, Li S, Wei H (2018) Multifunctional nanozymes: enzyme-like catalytic activity combined with magnetism and surface plasmon resonance. Nanoscale Horiz. https://doi.org/10.1039/C8NH00070K
7. Wang X, Guo W, Hu Y, Wu J, Wei H (2016) Nanozymes: next wave of artificial enzymes. Springer, New York
8. Wang Z, Dong K, Liu Z, Zhang Y, Chen Z, Sun H, Ren J, Qu X (2017) Activation of biologically relevant levels of reactive oxygen species by Au/g–C_3N_4 hybrid nanozyme for bacteria killing and wound disinfection. Biomaterials 113:145–157
9. Li W, Su C, Chang Y, Lin Y, Yeh C (2016) Ultrasound-induced reactive oxygen species mediated therapy and imaging using a Fenton reaction activable polymersome. ACS Nano 10(2):2017–2027
10. Ye Y, Xiao L, He B, Zhang Q, Nie T, Yang X, Wu D, Cheng H, Li P, Wang Q (2017) Oxygen-tuned nanozyme polymerization for the preparation of hydrogels with printable and antibacterial properties. J Mater Chem B 5(7):1518–1524
11. Liu B, Han X, Liu J (2016) Iron oxide nanozyme catalyzed synthesis of fluorescent polydopamine for light-up Zn^{2+} detection. Nanoscale 8(28):13620–13626
12. Xue T, Peng B, Xue M, Zhong X, Chiu C-Y, Yang S, Qu Y, Ruan L, Jiang S, Dubin S (2014) Integration of molecular and enzymatic catalysts on graphene for biomimetic generation of antithrombotic species. Nat Commun 5:3200
13. Du Z, Dou R, Zu M, Liu X, Yin W, Zhao Y, Chen J, Yan L, Gu Z (2016) Nitric oxide-generating L-cysteine-grafted graphene film as a blood-contacting biomaterial. Biomater Sci 4(6):938–942
14. Silber S, Albertsson P, Avilés FF, Camici PG, Colombo A, Hamm C, Jørgensen E, Marco J, Nordrehaug J-E, Ruzyllo W (2005) Guidelines for percutaneous coronary interventions: the task force for percutaneous coronary interventions of the european society of cardiology. Eur Heart J 26(8):804–847
15. Jukema JW, Verschuren JJ, Ahmed TA, Quax PH (2012) Restenosis after PCI. Part 1: pathophysiology and risk factors. Nat Rev Cardiol 9(1):53
16. Son D, Lee J, Lee D, Ghaffari R, Yun S, Kim S, Lee J, Cho H, Yoon S, Yang S (2015) Bioresorbable electronic stent integrated with therapeutic nanoparticles for endovascular diseases. ACS Nano 9(6):5937–5946

17. De Silva AP, Dixon IM, Gunaratne HN, Gunnlaugsson T, Maxwell PR, Rice TE (1999) Integration of logic functions and sequential operation of gates at the molecular-scale. J Am Chem Soc 121(6):1393–1394
18. De Silva AP, Uchiyama S (2007) Molecular logic and computing. Nat Nanotechnol 2(7):399–410
19. Gupta T, Van Der Boom ME (2008) Redox-active monolayers as a versatile platform for integrating boolean logic gates. Angew Chem Int Ed 120(29):5402–5406
20. Lien C, Chen Y, Chang H, Huang C (2013) Logical regulation of the enzyme-like activity of gold nanoparticles by using heavy metal ions. Nanoscale 5(17):8227–8234
21. Liu D, Chen W, Sun K, Deng K, Zhang W, Wang Z, Jiang X (2011) Resettable, multi-readout logic gates based on controllably reversible aggregation of gold nanoparticles. Angew Chem Int Ed 50(18):4103–4107
22. Lin Y, Xu C, Ren J, Qu X (2012) Using thermally regenerable cerium oxide nanoparticles in biocomputing to perform label-free, resettable, and colorimetric logic operations. Angew Chem Int Ed 51(50):12579–12583
23. Sharma V, Mobin SM (2017) Cytocompatible peroxidase mimic CuO: graphene nanosphere composite as colorimetric dual sensor for hydrogen peroxide and cholesterol with its logic gate implementation. Sensor Actuat B-Chem 240:338–348

Chapter 18
Nanozymology: Perspective and Challenges

Lizeng Gao, Hui Wei, Xiyun Yan and Xiaogang Qu

As introduced in this book, the field of nanozymes has been evidenced a rapid developing since 2007 [1]. The new concept of nanozymes has been internationally recognized as a new generation of artificial enzymes/enzyme mimics. So far, over 300 different types of nanomaterials have been reported with enzymatic activity. Importantly, the mimicking activities are not limited in oxidative-reductive system and have been extended to DNase, protease, and phosphatase. These achievements make it possible for nanozymes to be versatile in practical applications by combing the enzyme-like activities and other nanoscale properties, such as electricity, magnetism, fluorescence, etc. Therefore, as a multifunctional combination, nanozymes will be more powerful than natural enzymes and other traditional enzyme mimics or artificial enzymes by showing their unreplaceable characterizations. We believe nanozymes will represent a new generation of artificial enzymes and provide a wide

---

Part of the content in this chapter has been reproduced from Ref. 2 with permission from The Royal Society of Chemistry

The original version of this chapter was revised: The information about citations are updated. The correction to this chapter is available at https://doi.org/10.1007/978-981-15-1490-6_19

L. Gao · X. Yan
CAS Engineering Laboratory for Nanozyme, Key Laboratory of Protein and Peptide Pharmaceutical, Institute of Biophysics, Chinese Academy of Sciences, 15 Datun Road, Beijing 100101, China

H. Wei
Department of Biomedical Engineering, College of Engineering and Applied Sciences, Nanjing National Laboratory of Microstructures, Jiangsu Key Laboratory of Artificial Functional Materials, Nanjing University, Nanjing 210093, Jiangsu, China

X. Qu (✉)
Laboratory of Chemical Biology and State Key Laboratory of Rare Earth Resource Utilization, Chinese Academy of Sciences, 5625 Renmin Street, Changchun 130022, Jilin, China
e-mail: xqu@ciac.ac.cn

X. Yan (ed.), *Nanozymology*, Nanostructure Science and Technology,
https://doi.org/10.1007/978-981-15-1490-6_18

variety of enzyme mimics by using well-established nanotechnology, this will certainly accelerate the development of nanozymology in basic research and practical applications.

Although developing fast, many critical challenges remain unsolved, including the nature of nanozyme to mimic natural enzyme, the possibility to imitate all activities of natural enzymes, comparable enzymatic selectivity and controllable enzymatic performance in vivo. However, these challenges represent the future trends in the research of nanozymology. Here we propose the following subjects as the perspective of nanozymes, and challenges in the field for the years to come.

18.1 Fundamental Principles and Mechanisms [2–4]

a. Nature of nanozymes as enzyme mimics. A key question is that why nanomaterials can mimic enzyme activities. Currently most nanozymes are made from inorganic nanomaterials, which are obviously different with protein enzymes in the composition. At nanoscale, both nanozymes and natural enzymes have complicated nanostructure which severs as the scaffold to support the catalysis. But nanozymes often do not have the same active sites found in natural enzymes. Although some metal oxide nanozymes contain similar transitional metals serving as active center in natural enzymes, carbon nanozymes lack these features and still perform high catalytic activity. Therefore, the challenge is to investigate if there are common active counterparts in nanozymes to perform enzyme-like activities. Besides, "*based on the Sabatier principle and the more quantitative scaling relationships, descriptors of catalytic activities can be identified and the corresponding activity (and selectivity) maps can be obtained. The Sabatier principle has been successfully used to search for catalysts for heterogeneous catalysis (such as in ammonia synthesis and electrocatalysis). Inspired by such a success, as well as encouraged by our recent results, here we propose that general design rules for high-performance nanozymes can be established from the electronic point of view by identifying suitable descriptors of activities. These descriptors can be the adsorption (binding-free) energy of a reactant (intermediate) on catalysts; the dissociation energy of a reactant (intermediate) on catalysts; d band for transition metal; e_g occupancy, Op band center, metal–oxygen hybridization (bond strength), and charge-transfer gap for transition metal oxides, etc.*" [2]. We are aware that current studies focus on developing new nanozymes or improving the activities of nanozymes. However, for natural enzyme, there is always activator or inhibitor which for regulation of the enzymatic activities in biological system. The activators (or co-factors) or inhibitors to nanozymes are rarely discovered, although some small chemicals, such as ATP, Hg^{2+}, or I^-, have been found to affect the activity in a positive or negative way. The understanding of these co-factors and inhibitors would help to regulate nanozyme activity for in vitro and in vivo applications.

b. Rational design of nanozymes. It has been found that the components, morphologies, crystal facets, surface modifications, can affect the catalytic activity of nanozymes. It is necessary to elucidate the relationship between these aspects and catalytic activities, which will make it possible to precisely predict and tune the activity of nanozymes. To this end, both experimental and computational studies should be combined together to construct a universal structure-activity model. Importantly, the investigations on single atom (active site) nanozymes and single nanozyme catalysis in real-time characterization should be introduced to the field. Therefore, a standard procedure for preparation and characterization for one specific nanozyme is highly desirable for practical applications. Furthermore, "*for a fair evaluation of different nanozymes, standards should be established to quantitatively determine their catalytic activities. Recently, Yan, Liang, and co-workers proposed standardized assays for peroxidase-like nanozymes, which would promote the development of peroxidase-like nanozymes based bio-detection. In the future study, we expect standards for other types of nanozymes to be set as well. Notably, specific surface area, a unique characteristic of nanomaterials, plays an important role in catalytic activities of nanozymes, especially for those with porous structures. Therefore, it is necessary to take specific surface area into account when setting the standards*" [2].

c. "***Expanding the types of nanozymes and beyond***. *Currently, the types of reactions catalyzed by nanozymes have been expanded from redox to hydrolysis and a few others. But they are still not wide enough to cover all the important enzymatic reactions. For further exploring the types of nanozymes, a fast and possible method is to create the analogues of active centers in natural enzymes, and then incorporate them into MOFs or other nanomaterials for mimicking the catalytic activities*" [2]. For an even wider view, does nanozyme corelate to molecular evolution, the origin of life, and even exist and play roles in early living system? "*The fabrication of proto-organelles and -cells should be considered for future studies. The study of the potential implications of nanozymes in prebiotic chemistry and origin of life would be another challenging but rewarding research area. In addition, by taking advantage of physiochemical properties of nanomaterials, the development of multifunctional nanozymes should be another interesting and challenging topic in the future. Besides catalysis, nanomaterials endow nanozymes with more functions including magnetic, optical, and thermal properties, allowing more potential applications for ultra-sensitive sensing, sustainable chemistry, and multi-modality therapy; the mimics of protein scaffold of an enzyme, which is important for the selectivity and efficiency of an enzymatic reaction, has largely not been studied yet. Moreover, some enzymes only act properly in their native environments (such as residence within a lipid membrane). The mimics of such environment were rarely reported*" [2]. Furthermore, since nanomaterials possess rich surface chemistry, this feature can be used for the design of stereo-selective nanozymes and isonanozymes. Chiral nanozyme may achieve enantioselective specificity, like or even better than natural enzymes; by modulation of morphology and surface functional groups, one can change nanozyme activity with pH, ionic strength, and different microenviroments, expressing a family of isonanozymes. Therefore, all these new strategies should be explored to address these challenges in the future study of nanozymes [2].

d. Another critical issue is how to improve substrate specificity of nanozymes, which is a fundamental property of natural enzymes. "*Although the combination* *of natural enzymes and nanozymes together could partially solve this problem, the stability and cost of the whole catalyst system were sacrificed because of the natural enzymes. Thus, by learning from nature, incorporating certain recognition mechanisms, such as building protein-like (or aptamer-like) binding pockets and molecular imprinting of selective substrate pockets, should be promising approaches to nanozymes with high specificity. Constructing asymmetric nanomaterials or using bioorthogonal nanozymes could also be considered as an optional choice in the future. Fine-modulation the interaction of substrate with nanozymes by engineering the nanomaterials could be an alternative way to fabricate specific nanozymes. On the other hand, some nanozymes have multienzymatic activities, which have been demonstrated to be helpful for therapeutics in previous chapters. However, in certain cases, one unavoidable type of catalytic properties would cause some potential side effects, which should be carefully investigated to obtain an effective window for therapy*" [2].

18.2 Applications of Nanozymes in Biomedical Treatments and Other Fields [3, 4]

a. Although nanozymes have been demonstrated well performed at bench, it is still unknown whether they can perform the same functions in the living system. All these should be carefully identified before any clinical consideration. Importantly, one nanozyme may possess several enzyme-like activities under different conditions. It needs to ensure that the identical performance of nanozymes can be achieved in in vivo. Therefore, the strategies to monitor nanozyme activity and regulate the activity under proper in vivo microenvironment are required for biomedical applications. One has to be aware that once nanozyme enters the cell, besides performing enzymatic activity, it may interfere in other biochemical or signal pathways. For instance, nanozyme may affect a metabolic pathway which is often composed by multiple enzymes to make cascade reactions. It may also induce or suppress cellular signal pathways. These influences may be demonstrated in the relevant cytotoxicity. The bioeffects on a cell or an in vivo system need to be carefully evaluated according to the activities of nanozymes and other nanoscale features just like nanomaterials. Another issue should be considered that nanozymes may react with biological components, such as serum proteins, lipids, polysaccharides, to form corona on the surface once they enter circulation system or tissue. The corona coating may affect the activity of nanozymes. Importantly, it is possible that biological reactants may interact with nanozyme surface and induce deactivation or degradation of nanozyme. Therefore, new methods are needed to characterize the corona and other interactions in order to ensure nanozymes performing desired activity in such complicated biosystem.

b. For* in vivo *therapeutic application, targetability and controllability are critical for nanozymes. It needs to ensure nanozymes precisely localized at the diseased sites and delivered into a proper microenvironment to allow them to perform the desired activity. We are aware that nanozymes are multifunctional nanomaterials possessing enzymatic activities and other special properties, such as magnetism, fluorescence, photothermal, photoacoustic, or photodynamic property. The latter properties may directly enhance or control the enzymatic activities as external stimulus. Importantly, it is possible to combine all these features to achieve theranostic goal with one nanozyme. To this end, the pharmacokinetics needs to be carefully investigated for clinical translation if nanozymes are used as a medicine. The administration of nanozyme agents by topical or injectable solution or as components of implants may be considered during designing and preparing nanozymes. Therefore, it may need careful evaluation according to the criteria in pharmacology. Another important issue has to be pointed, like other nanomaterials, nanozymes are distributed into different organs or tissues once entering an in vivo system. Degradation will occur simultaneously and exist for a long term. The biodistribution and biodegradation determine the therapeutic effect, biocompatibility and the long-term biosafety. The preclinical study will be beneficial for nanozymes translation into clinics.

c. Nanozymes used for antibacterial and environmental treatment. Recent studies have demonstrated [4] that nanozymes can perform antibacterial activity by utilizing their peroxidase activity to generate toxic ROS, such as iron oxide nanozymes, cerium nanozymes, vanadium oxide nanozymes, etc. Importantly, these nanozymes show promising efficacy in eliminating biofilm and biofouling. However, the ability to induce new resistance by nanozymes needs to be carefully evaluated. Furthermore, the influence on microbial ecology needs to be considered as nanozymes enter environment and may change the microecological balance in long term. Similarly, although nanozymes show promising applications in environmental treatment by generating free radicals to decompose organic and biological pollutants, the prerequisite is that nanozymes cannot be a new source of pollutant. Some nanozymes are made from noble metals or rare earths or containing these environment-unfriendly elements. Therefore, careful evaluation is needed for the environmental applications of nanozymes.

d. Ready for applied to plants, animals, or even for synthesis in which natural enzymes cannot accomplish by showing nanozyme irreplaceable? Most nanotechnologies or nanomaterials including nanozymes are used to tackle the problems in biomedicine related to human diseases. In contrast, the fields of agriculture and veterinary have not been benefited widely. Can nanozymes be used to improve food production or quality? Is it possible to use nanozymes as enzyme alternatives to work in extreme conditions (pH, temperature, organic phase) for drug synthesis to demonstrate their advantages? Can nanozymes also provide potential strategies to treat diseases in animals and plants? These will further extend their applications to improve human health and life quality in the global ecology.

In summary, the field of nanozymes is developing very fast. It combines multiple disciplines together, such as material science, chemistry, biochemistry, bioengineering, and biomedicine. There are a lot of opportunities and challenges in designing high quality and applicable nanozymes. We believe it will bring positive effect on scientific research and human life in three aspects: (i) a basic biomimetic concept for material design, (ii) a new point of view on life origin from inorganic to organic system, and (iii) applicable enzyme mimics to improve life quality from biomedicine to environments.

References

1. Gao LZ, Zhuang J, Nie L, Zhang JB, Zhang Y, Gu N, Wang TH, Feng J, Yang DL, Perrett S, Yan X (2007) Intrinsic peroxidase-like activity of ferromagnetic nanoparticles. Nat Nanotechnol 2(9):577–583
2. Wu J, Wang X, Wang Q, Lou Z, Li S, Zhu Y, Qin L, Wei H (2019) Nanomaterials with enzyme-like characteristics (nanozymes): next-generation artificial enzymes (II). Chem Soc Rev 48(4):1004–1076
3. Cormode DP, Gao L, Koo H (2018) Emerging biomedical applications of enzyme-like catalytic nanomaterials. Trends Biotechnol 36(1):15–29
4. Huang Y, Ren J, Qu X (2019) Nanozymes: classification, catalytic mechanisms, activity regulation, and applications. Chem Rev 119(6):4357–4412

Correction to: Nanozymology: Perspective and Challenges

Lizeng Gao, Hui Wei, Xiyun Yan and Xiaogang Qu

Correction to:
Chapter 18 in: X. Yan (ed.), *Nanozymology,* Nanostructure Science and Technology, https://doi.org/10.1007/978-981-15-1490-6_18

In the original version of this book, the following information about citation was missed. Now, the missing information was updated.

"Part of the content in this chapter has been reproduced from Ref. 2 with permission from The Royal Society of Chemistry."

The cited contents were in italic font with quotation marks and corresponding reference (Ref. 2). Ref. 2 was also updated.

The chapter and book have been updated with the changes.

The updated version of this chapter can be found at
https://doi.org/10.1007/978-981-15-1490-6_18

X. Yan (ed.), *Nanozymology*, Nanostructure Science and Technology,
https://doi.org/10.1007/978-981-15-1490-6_19

Index

A
Active site, 17, 22, 25–31, 33–35
Antibacterial, 491, 492, 496, 498–502, 505, 506, 508–512, 514, 515, 517, 520, 521
Antibiofilm, 510, 522
Antibodies, 398, 401, 402, 408
Anti-fouling, 520, 521
Anti-oxidation, 479
Artificial enzymes, 5–9, 13

B
Bacterial detection, 492
Biofilm, 204–206, 208, 209, 244–247, 250, 251, 253
Biofouling, 199, 241, 244, 245, 247, 249, 251, 252
Biological enzyme mimetics, 283–286, 288–290, 309
Biomedical applications, 106, 107, 132
Biomedical devices, 546, 549, 552
Biosensors, 396–400, 403, 408, 410, 411, 418–420

C
Carbon nanomaterials, 174, 176, 179, 181
Catalysis, 3, 6, 7, 9, 10, 13, 14
Catalysis mechanism, 281, 318
Catalytic mechanism, 21, 23, 26, 32, 35, 37
Characterization, 84, 85, 92, 95

D
Deoxyribonucleic Acid (DNA), 407–414, 418

E
Environment detection, 528
Enzymatic activities of nanozyme, 451
Enzyme-like activities, 106–108, 113, 120, 129–131
Enzyme mimetics, 48, 50, 53, 63
Enzyme modification, 368, 371
Enzymes, 396–403, 405, 406, 408, 409, 413, 415–420

F
Fluorescence, 400, 410, 413, 414
Functional nanomaterials, 80, 546

H
Haloperoxidase, 199
Haloperoxidase activity, 42
Hybrid nanozymes, 368, 369, 374, 375, 381, 382, 384

I
Iron oxide, 106, 107, 111–115, 117, 119–121, 123–125, 127, 129–132

K
Kinetics, 17, 21–26, 37, 38

L
Logic gates, 546, 550–553

X. Yan (ed.), *Nanozymology*, Nanostructure Science and Technology,
https://doi.org/10.1007/978-981-15-1490-6

M
Metal–Organic Frameworks (MOFs), 143, 144, 146–158, 164
Michaelis–Menten, 22–26

N
Nanoceria nanozyme, 280–283, 286–304, 306–311, 314–319
Nanocomposite, 347–354, 358
Nanomaterials, 3, 5, 6, 8, 9, 14, 15
Nanoparticles, 401, 405, 410, 414, 418, 419
Nanotechnology, 5, 8, 12, 15
Nanozymes, 3–15, 143, 147–150, 154, 155, 157–159, 161, 163, 164
Nanozymes support
Nanozymology, 13–15
Noble-metal, 332, 333, 341, 347, 349, 351, 354–356, 358, 359

O
Oxidase activity, 42, 46, 60
Oxidative stress, 460, 461, 463, 466, 471, 473, 474, 479, 480, 483

P
Peroxidase, 173–175, 179–181, 183, 184, 186
Peroxidase activity, 42, 47, 61, 67–69
Pollutant treatment, 534, 537
Preparation, 80, 81, 85, 87
Prussian Blue Nanoparticles (PBNPs), 159–163

R
Reactive Oxygen Species (ROS), 150, 158, 163
Recyclability, 534
Redox potential regulation
Robustness, 51, 87, 172, 198

S
Superoxide dismutase, 173–179, 186
Superoxide dismutase activity, 42

T
Tumor imaging, 439
Tumor *in vitro* diagnosis, 428
Tumor therapy, 440–443, 449–451

The manufacturer's authorised representative in the EU is Springer Nature Customer Service Centre GmbH, Europaplatz 3, 69115 Heidelberg, Germany. If you have any concerns regarding our products, please contact ProductSafety@springernature.com

Printed and bound by CPI Group (UK) Ltd, Croydon, CR0 4YY
25/09/2026
02222886-0009